20th Joint Workshop on Electron Cyclotron Emission and Electron Cyclotron Resonance Heating 2018 (EC-20)

EPJ Web of Conferences Volume 203 (2019)

Greifswald, Germany
14 - 17 May 2018

Editors:

Emanuele Poli
Heinrich Laqua
Johan Oosterbeek

ISBN: 978-1-5108-8308-6

Printed from e-media with permission by:

Curran Associates, Inc.
57 Morehouse Lane
Red Hook, NY 12571

Some format issues inherent in the e-media version may also appear in this print version.

For additional information, please contact EDP Sciences – Web of Conferences at the address below.

EDP Sciences – Web of Conferences
17, Avenue du Hoggar
Parc d'Activité de Courtabœuf
BP 112
F-91944 Les Ulis Cedex A
France

Phone: +33 (0) 1 69 18 75 75
Fax: +33 (0) 1 69 28 84 91

contact-edps@webofconferences.org

Additional copies of this publication are available from:

Curran Associates, Inc.
57 Morehouse Lane
Red Hook, NY 12571 USA
Phone: 845-758-0400
Fax: 845-758-2633
Email: curran@proceedings.com
Web: www.proceedings.com

TABLE OF CONTENTS

Preface

Emanuele Poli[1], *Heinrich* Laqua[2], and *Johan* Oosterbeek[2]

[1]Max-Planck-Institut für Plasmaphysik, Garching bei München, Germany
[2]Max-Planck-Institut für Plasmaphysik, Greifswald, Germany

The 20th Joint Workshop on Electron Cyclotron Emission and Electron Cyclotron Resonance Heating was held in the Alfried Krupp Wissenschaftskolleg, Greifswald, Germany, on May 14-17, 2018. The local organization was provided by the Max-Planck-Institut für Plasmaphysik Greifswald.

The programme was organized around the traditional four main topics represented by EC theory, EC experiments, EC diagnostics and EC-related technology. The workshop was attended by 84 participants from 15 countries. There were 11 invited and 28 oral presentations. The contributions on the four topics listed above were summarized at the end of the workshop by A. Ram, Y.-S. Bae, H. Oosterbeek and J. Jelonnek, respectively. The present proceedings collect a total of 57 papers.

The quality of the presentations, the liveliness of the discussions after the talks and during the poster sessions, and the presence of many young scientists testifies not only the intense research activity in the field of EC waves and related technologies, but also its attractiveness for new researchers. This is expected to be even more the case in the future, given the key role that EC heating and diagnostic systems are expected to play in a fusion reactor. Furthermore, the strong interaction between the participants, in conjunction with the numerous attendance, confirmed the importance of this workshop series as a forum allowing leading experts in the field to discuss present results and future plans at a level of detail hardly achievable in a different context.

The next meeting (EC21) is planned at the ITER premises in spring 2020. The sites of EC20 and EC21, Greifswald and Cadarache, emblematically represent the tight connection between the work of the EC-community and the current fusion research programme. Both W7-X, the latest major fusion experiment to go into operation, and ITER, the next crucial step on the way to controlled nuclear fusion, rely on EC waves for their very operation. We hope that these proceedings display a useful snapshot of the status of our progress in the field.

Many persons deserve a special thank for their support before, during and after the workshop. In particular, the local organizing committee provided a smooth and flawless management of the meeting and a very interesting social programme. The summary speakers had to sacrifice a considerable part of their time in Greifswald for the preparation of their talks and did an excellent job. Last but not least, we gratefully acknowledge the effort of the reviewers of the papers appearing in the present proceedings.

The members of the Local Organizing Committee were
Heinrich Laqua (chair, scientific programme)

Hans Oosterbeek (scientific coordinator)
Maria Radau (conference secretary, registration)
Claudia Schönian (conference secretary, registration)
Beate Kemnitz (public relations)
Karin Schier (finances)
Andrea Kleiber (website)

The members of the International Programme Committee were
Emanuele Poli (IPP Germany, chair)
Max Austin (IFS University of Texas at Austin, USA)
Young-Soon Bae (NFRI, Korea)
Dhiraj Bora (IPR, India)
Gregory Denisov (IAP Nizhny Novgorod, Russia)
Daniela Farina (CNR Milano, Italy)
Franco Gandini (ITER Organization)
Mei Huang (SWIP, China)
Shin Kubo (NIFS, Japan)
John Lohr (GA San Diego, USA)
Burkhard Plaum (IGVP University of Stuttgart, Germany)
Braj K. Shukla (IPR, India)
Victor Udintsev (ITER Organization)

Emanuele Poli, Heinrich Laqua, Hans Oosterbeek (editors)
Garching/Greifswald, January 2019

Electron Cyclotron collisionless interaction during EC-assisted tokamak start-up

Daniela Farina[1,*]

[1]Istituto di Fisica del Plasma, CNR, via Cozzi 53, 20125 Milano, Italy

Abstract.
In the initial phase of a plasma discharge with EC-assisted breakdown, the wave-particle interaction is nonlinear and wave trapping provides the mechanism for transition from a very low energy state to a much larger energy state. A Hamiltonian adiabatic approach provides the condition for the energy variation to occur in a rigorous way, together with quantitative estimate as a function of the wave frequency, harmonic number, polarisation and EC power and beam width, for the first, and second cyclotron harmonic. The interaction is highly localized in space close to the EC resonance. The spatial profile of the energy gain are reported together with the estimates of the width of the radial region where the interaction takes place in the case of a tokamak configuration. The analysis provides a physics based description of the process as well as indications for an optimized experimental setup.

1 Introduction

In ITER it is foreseen to startup the plasma injecting Electron Cyclotron (EC) waves to reduce the volt-second consumption of the main transformer during breakdown. This method is used routinely in stellarators as well as in other tokamaks.

Various physical mechanisms occur during r.f. assisted plasma startup, like wave-particle interaction, particle collisions with both neutrals and ionised plasma, particle confinement and plasma build-up, and the overall process is quite complex. The analysis presented here is focussed on the very first phase of the process, i.e., on the EC wave interaction with the seed electrons in the vacuum chamber, which takes place before collisions start to play a role. The electron dynamics under the action of a localised electromagnetic (e.m.) field is investigated theoretically aiming to determine conditions and parameters for which the electrons can achieve enough energy to start the ionisation process. A detailed investigation of this problem has been reported recently in [1, 2]. Here, the theoretical analysis is summarised for sake of completeness and some new results are presented which are of practical interest to design the experimental setup.

During the initial phase of the EC-assisted breakdown, electrons are assumed to behave as free particles at room temperature moving along the static magnetic field, under the action of a localised microwave beam. As it was first pointed out long ago in [3], such cold electrons can acquire a large energy via nonlinear wave trapping for beam power values like those injected in fusion machines (100 kW - 1 MW in single rf source), whenever the e.m. amplitude is large enough and/or the time of flight in the EC beam is long enough to exceed the nonlinear trapping period.

The wave particle nonlinear interaction process is conveniently investigated within a Hamiltonian adiabatic formalism, following the theoretical approach developed in [3–9] and adopted in [1, 2], which allows the derivation of the conditions at which the particles gain energy in single beam crossing in a rigorous way and the quantitative dependence of the energy variation on wave frequency, harmonic number, polarisation and EC power and beam width. The analysis applies to timescales satisfying the following conditions

$$\tau_{cycl} \ll \tau_{trap} \ll \tau_{flight} \ll \tau_{coll}, \qquad (1)$$

being the cyclotron period, the trapping time in the wave field, the time flight in the beam and the collision time, respectively. In ITER typical parameters are $\tau_{cycl} \simeq 6 \times 10^{-12}$ s, $\tau_{trap} \approx 10^{-8}$ s, $\tau_{flight} \approx 1 - 5 \times 10^{-6}$ s, $\tau_{coll} \approx 5 \times 10^{-4} - 5 \times 10^{-3}$ s, Equation (1) identifies a nonlinear adabiatic interaction regime.

2 Nonlinear interaction

In a tokamak, EC assisted startup occurs in the magnetic null region close to the EC cold resonance.

The electron motion can be modelled in a simple way as that of a free charged particle in a given static, uniform and constant magnetic field $\mathbf{B}_0 = B_0\mathbf{e}_z$ (representing the vacuum toroidal field), interacting with a localised e.m. beam with frequency ω close to a harmonic of the cyclotron frequency $\Omega = |q_e|B_0/m$. The motion equations read

$$\frac{d\mathbf{x}}{dt} = \mathbf{v}, \qquad \frac{d\mathbf{p}}{dt} = q_e\mathbf{E}(\mathbf{x}, t) + q_e\mathbf{v} \times [\mathbf{B}_0 + \mathbf{B}(\mathbf{x}, t)], \quad (2)$$

*e-mail: farina@ifp.cnr.it

where $\mathbf{p} = m\gamma\mathbf{v}$ is the kinetic momentum, $\mathbf{v}$ the electron velocity, $\gamma = \sqrt{1 + p^2/m^2c^2}$ the relativistic factor, $\mathbf{x}$ the coordinate vector, q_e the (negative) electron charge and $\mathbf{E}(\mathbf{x}, t)$ and $\mathbf{B}(\mathbf{x}, t)$ the electric and magnetic field of the wave, with $\mathbf{E}(\mathbf{x}, t) = E[e_x \sin\chi\,\hat{\mathbf{x}} - e_y \cos\chi\,\hat{\mathbf{y}} + e_z \sin\chi\,\hat{\mathbf{z}}]$, $\chi = k_\perp x + k_\parallel z - \omega t$ the wave phase, $\mathbf{e} = \mathbf{E}/|\mathbf{E}|$ the wave polarisation vector, and $k_{\perp,\parallel}$ the wave vector components perpendicular and parallel to the static magnetic field. The EC beam is assumed to have a Gaussian-like profile with wave electric field amplitude $E = E_M \exp(-r^2/w^2)$, being $E_M = 2[\eta_0 P/(\pi w^2)]^{1/2}$ its maximum value (in SI units), r the beam radius, w the local beam width, P the incident EC power and $\eta_0 \simeq 376.73$ ohm the free space impedance.

In the Coulomb gauge with zero scalar potential, the electric $\mathbf{E}$ and magnetic $\mathbf{B}$ fields are related to the vector potential $\mathbf{A}(\mathbf{x}, t) = \mathbf{A}_0(\mathbf{x}) + \mathbf{A}_1(\mathbf{x}, t)$ as $\mathbf{E} = -\partial\mathbf{A}/\partial t$, and $\mathbf{B} = \nabla \times \mathbf{A}$, and the electron motion is described by the Hamiltonian function

$$H(\mathbf{x}, \mathbf{P}) = mc^2\gamma = mc^2(1 + p^2/m^2c^2)^{1/2}, \qquad (3)$$

being $\mathbf{P} = \mathbf{p} + q_e\mathbf{A}$ the canonical momentum conjugate to $\mathbf{x}$. Dimensionless quantities and variables are used, with time, length, momentum, energy and Hamiltonian, and vector potential normalised over Ω^{-1}, $c\Omega^{-1}$, mc, mc^2, and $|q_e|/mc$, respectively.

Following the approach described e.g. in [8], the Hamiltonian transforms via canonical transformations to a time-independent Hamiltonian function H with two degrees of freedom, valid for ω close to a cyclotron harmonic $n\Omega$. At first order in the normalised wave amplitude $\epsilon(z) = |q_e|E(z)/mc\omega \ll 1$, it reads

$$H(z, \theta, \bar{P}_z, I) = \gamma - \nu_n I = H_0(\bar{P}_z, I) + \epsilon(z)H_1(z, \theta, \bar{P}_z, I) \quad (4)$$

where $\nu_n = \nu/n = \omega/n\Omega$, $H_0(\bar{P}_z, I) = \Gamma - \nu_n I$, $H_1(z, \theta, \bar{P}_z, I) = \Theta_n(\bar{P}_z, I)/\Gamma \cos n\theta$, with I and θ action-angle variables, $\bar{P}_z = P_z - N_\parallel \nu_n I$ the canonical momentum conjugate to z, and $\Gamma = (1 + 2I + P_z^2)^{1/2}$. For the unperturbed system, $\epsilon(z) = 0$, θ is the phase of the gyromotion, $I = p_\perp^2/2$, and $P_z = p_z$. The polarisation term Θ_n reads $\Theta_n = \sqrt{2I}[e_- J_{n-1}(b) + e_+ J_{n+1}(b)] + P_z e_z J_n(b)$ with $e_\pm = (e_x \pm e_y)/2$ and J_n the Bessel function of argument $b = \nu N_\perp \sqrt{2I}$. Hereafter, two normal modes in vacuum are considered with the same polarization as the Ordinary (OM) and eXtraordinary (XM) modes of cold plasmas in the limit of zero density.

Due to the constancy of the Hamiltonian, the following relationships hold between the initial and final electron states

$$\begin{aligned} \Delta\gamma &= \nu_n \Delta I \equiv \nu_n \Delta(p_\perp^2/2), \\ \Delta p_z^2 &= \Delta\gamma[\Delta\gamma - 2(1/\nu_n - \gamma_0)], \end{aligned} \qquad (5)$$

so that the computation of ΔI provides both the total and the parallel final energy. The first relation holds also at any time, $\delta\gamma = \nu_n \delta I$, and in addition one has $\delta P_z = \delta\bar{P}_z + N_\parallel \nu_n \delta I$. Outside the beam region, the parallel canonical momentum $\bar{P}_z$ is a constant of motion.

The main features of the interaction process can be studied solving the Hamiltonian equation of motion for thermal electrons at room temperature $T_e = 0.03$ eV with initial conditions outside the EC beam illumination region. In Figure 1 the behaviour of the kinetic energy $W = mc^2(\gamma - 1)$ versus the spatial coordinate z is shown for two electrons with same initial conditions except for the initial phase value θ, for second harmonic, XM polarised beam in oblique propagation. While the electrons cross the EC beam, they become non linearly trapped in the wave field, experience large energy variations with maximum value at $z = 0$ where the wave field amplitude is maximum. When the e.m. field amplitude decreases ($z > 0$) the electrons are detrapped for the same E value, and exit the beam region with final energy either quite large (Figure 1a) or equal to the initial one (Figure 1b).

The numerical investigation of the electron motion reported in [1] showed that a cold electron may gain a significant amount of energy when it flows along a magnetic field line under the action of a localised EC beam at frequency close to the first and second cyclotron harmonics. Under adiabatic conditions, the energy that a cold electron gains after beam crossing is quite large and independent of the beam power and width, provided that the electron is non linearly trapped in the wave field. At fixed beam parameters, the gained energy increases with the frequency mismatch from the resonance up to a critical value beyond which is zero. Trapping occurs if the maximum electric field amplitude exceeds a threshold value that depends on the harmonic number, wave frequency, injection angle and mode polarisation.

Whenever the flight time in the beam region is larger than the trapping time in the wave field and the e.m. amplitude variation is sufficiently slow, the Hamiltonian is characterised by a couple of fast action-angle variables (θ, I) and a couple of slow canonically conjugate variables $(z, \bar{P}_z)$, and the action integral $J(z, \bar{P}_z) \equiv n/2\pi \oint I\, d\theta$ is an adiabatic invariant of the motion, that breaks only at the separatrix crossing between trapped and untrapped orbits. Breaking of the invariant via trapping/untrapping allows to connect different regions in phase space and provides the mechanism for net energy variation to occur. This mechanism for EC nonlinear interaction is the core physical phenomenon [3–9].

A few critical conditions need to be fulfilled in order to give rise to adiabatic nonlinear energy gain: i) the initial status, namely the electron parallel momentum, the frequency ratio ν_n and the local $N_\parallel$, correspond to a "resonant case" for the unperturbed motion, such that two different energy states (low and high respectively) exist for the same value of the constant of motion H; ii) connection between these two energy states occurs via nonlinear trapping, which implies a threshold condition on the wave amplitude; iii) in addition, at trapping/detrapping the separatrix splits the phase space of the fast variables in three regions, corresponding to the trapped motion, and to the two free streaming motion at different energy levels.

The process can be sketched in the phase space of the fast variables (I, θ) (see e.g. Figure 1 in [4] and Figure 4 in [2]), whenever a resonant condition is satisfied and the Hamiltonian is double-valued. Outside the beam region the electron lies on a given branch of the Hamiltonian H_0.

When it flies in the illumination region along the magnetic field line, it experiences an increasing field amplitude ϵ and is eventually trapped in the wave field at some ϵ_s value if lower than the maximum value ϵ_M. The lower and upper branches of the Hamiltonian corresponding to same value merge together at the separatrix crossing. For decreasing ϵ values detrapping occurs at ϵ_s and the particle may leave the separatrix either on the upper or lower branch, so that after beam crossing both branches can be populated. A net energy variation occurs with a given probability only if after the last separatrix crossing the particle will lie on the other branch of the Hamiltonian.

This process is illustrated in Figure 2 where the results of the same simulations shown in Figure 1 are reported in the "phase" space W, θ (note that W is proportional to the action I conjugate to θ). Initially the particle follows a trajectory quite close to the horizontal axis and, after crossing of the separatrix, it oscillates around the elliptic point at $\theta = 0$. When the electric field amplitude decreases the particle is untrapped, and two different situations can occur depending on the Hamiltonian branch on which the particle lies. After exiting the illumination region, the particle describes a horizontal line in phase space either (a) at large energy or (b) at very low energy (note that the maximum energy is $\approx 7 \times 10^3$ its initial value so that the details of the motion at low energy are barely seen when linear scale is used). The structure of the phase space (I, θ) for the first three harmonics is not discussed in detail here and can be found in [1] for the first three harmonics.

An example of motion for which nonlinear wave trapping does not occur is shown in Figure 3, which refers to the same parameters as in Figures 1, and 2, but at lower power. A similar behavior is observed increasing the frequency mismatch from the resonance at fixed power. After crossing the EC beam the electron recovers exactly the same energy as initially.

The above mechanism can be described formally within the Hamiltonian formalism, making use of the constancy of the Hamiltonian and of the adiabatic invariant in each stage of the motion, i.e., $J(\epsilon, \bar{P}_z) = I_0, I_f$ in the initial and final phases respectively, in which electrons are not trapped (for $\epsilon \le \epsilon_s$), and $J(\epsilon, \bar{P}_z) = J_s = A_s/2\pi$ in the trapped phase (for $\epsilon_s \le \epsilon \le \epsilon_M$), where A_s is the area of the phase space region inside the separatrix at its crossing. If trapping does not occur, $J(\epsilon, \bar{P}_z) = I_0$ at any ϵ. The total variation of the action variable I is equal to the jump of the adiabatic integral at the separatrix

$$\Delta\gamma = \nu_n(I_f - I_0) = \nu_n(|J_f(\epsilon_s, \bar{P}_{zs})| - |J_0(\epsilon_s, \bar{P}_{zs})|) \quad (6)$$

where $\bar{P}_{zs}$ is the $\bar{P}_z$ value at the separatrix, and $J_{0,f}(\epsilon, \bar{P}_z)$ is the action integral before and after the separatrix crossings. The ϵ and $\bar{P}_z$ values at the separatrix are related to the initial (or final) momentum values by means of the following system of equations

$$\begin{aligned} I_{0,f} &= |J_{0,f}(\epsilon_s, \bar{P}_{zs})|\,, \\ H_0(I_{0,f}, \bar{P}_{z0,f}) &= H_s(\epsilon_s, \bar{P}_{zs})\,, \end{aligned} \quad (7)$$

where H_s is the Hamiltonian function at the separatrix.

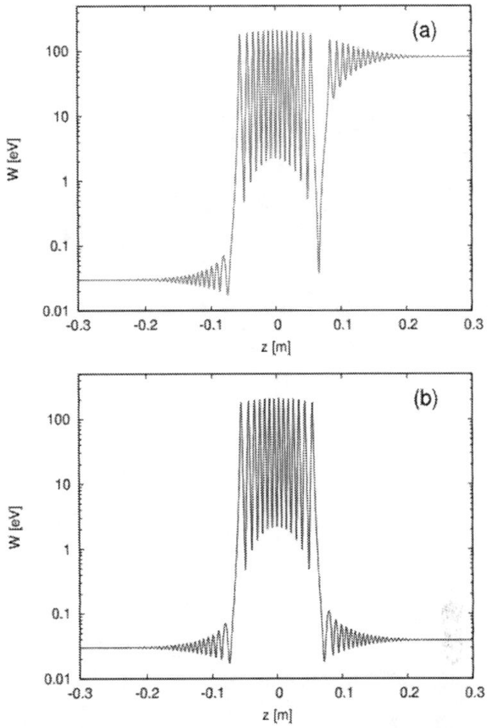

Figure 1. Electron energy $W = mc^2(\gamma - 1)$ versus spatial coordinate z, obtained solving the Hamiltonian equations of motion for two different initial phase values χ, (a) and (b), respectively). The chosen parameters are: EC power $P = 1$ MW, beam width $w = 10$ cm, frequency ratio $2\Omega/\omega = 1$, for second harmonic XM polarisation in oblique propagation with $N_\parallel = 0.3$.

Although the physical process has been reduced to the solution of system of two equations only (7), it is still not straightforward to solve in general. However, in the case under investigation analytical solutions can be found [1] making simplifying approximations that are well justified for the case of very cold electrons: i) weakly relativistic approximation, ii) negligible initial energy, iii) canonical momentum $\bar{P}_z$ almost constant for most cases.

As demonstrated in [1], the nonlinear interaction is effective only for the first and the second cyclotron harmonics, this fact being related to the specific topology of the phase space. In fact the phase space of the fast variables θ, I evolves for increasing perturbation parameter ϵ from a pendulum-like structure to a more complicate structure when finally the elliptic and hyperbolic points merge together at a critical value ϵ_{cn} beyond which there is no separatrix (see Figures 9 and 10 in [1]). It can be shown that trapping of low energy electrons occurs for ϵ values slightly lower that the critical one, $\epsilon \lesssim \epsilon_{cn}$. For $\epsilon = \epsilon_{cn}$ the adiabatic invariant takes the very simple expression $J_s = 2I_{rs}$ where I_{rs} is the resonant action value, solution of the EC resonant condition $\partial H_0/\partial I = 0$ that reads $\gamma - n\Omega/\omega - N_\parallel p_z/mc = 0$ outside the beam region. Under the above conditions the nonlinear energy variation expe-

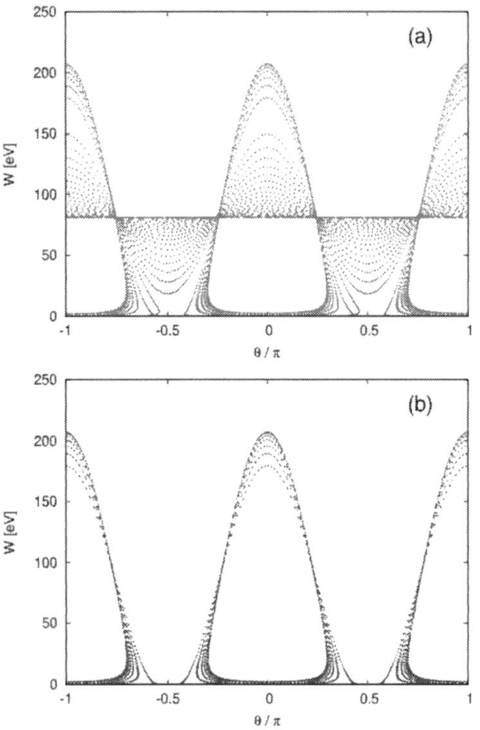

Figure 2. Same simulations as in Figure 1 in the "phase" space (W, θ).

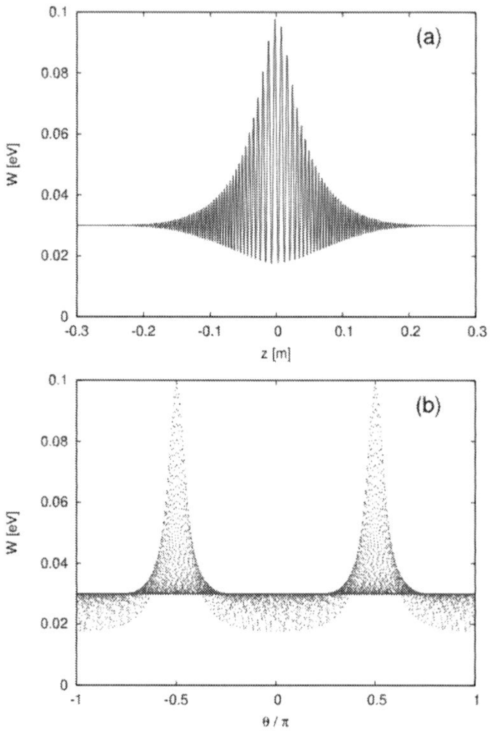

Figure 3. Example of motion for a non trapped particle in the (W, z) plane ((a)), and (W, θ) plane ((b)). Same parameters as in Figure 1, at lower power $P_{EC} = 300$ kW.

rienced by cold electrons reads

$$\Delta\gamma = 2\nu_n(I_{rs} - I_0) \approx 2I_{rs}, \tag{8}$$

The energy gain is proportional to the distance from the resonant action in momentum space (at $\bar{P}_z \approx constant$) and is maximum for particle with the lowest possible perpendicular energy I_0. This result is sketched in Figure 4 for the case of extraordinary mode in perpendicular propagation (see in [8] for a more detailed analysis). The lowest blue curve denotes the initial condition at low energy, the dashed black curve the EC resonant condition, and the red upper curve the final energy state after nonlinear interaction.

As it is clear from Eq.(8), the final energy state is independent of the electric field amplitude (i.e. of the beam power and width) for given ratio of the wave frequency to the cyclotron frequency and propagation angle. It is also independent of the detailed motion in the illumination region, provided trapping occurs. The overall process occurs with a given probability q, with $0 \leq q \leq 1$. The transition probability has been estimated numerically assuming that the initial particle distribution in parallel and perpendicular momentum is a Maxwellian function with temperature $T_e = 0.03$ eV, with initial phases θ distributed uniformly. The simulations have been performed for a large number of particles and the probability q has been estimated as the fraction of particles undergoing a net variation for given $p_\parallel, p_\perp$. In Figure 5a the result of a simulation is reported

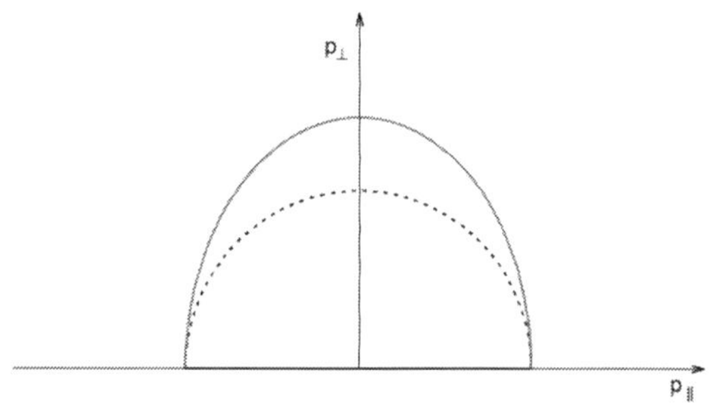

Figure 4. The energy variation process is sketched in the $(p_\parallel, p_\perp)$ plane for XM in perpendicular propagation. Initially the cold electrons lie on the blue line close to the axis origin. The dashed black curve is the resonant condition $\gamma = n\Omega/\omega$, and the red curve is the final energy given by Eq. (8).

in the momentum space for second harmonic extraordinary mode in perpendicular propagation and 48000 particles (i.e. $40 \times 40 \times 30$ for $p_\perp$, $p_\parallel$, and θ). Note that the momentum is normalised to the initial thermal momentum. The resonant condition and the final energy are not plotted in the same scale, the resonant momentum being $\approx 58p_{th}$ and the final energy $\sim 2.5 \times 10^3\,T_e$. Probability q is quite close to 1 for most initial conditions satisfying trapping condition (i.e., the dynamics of the majority of

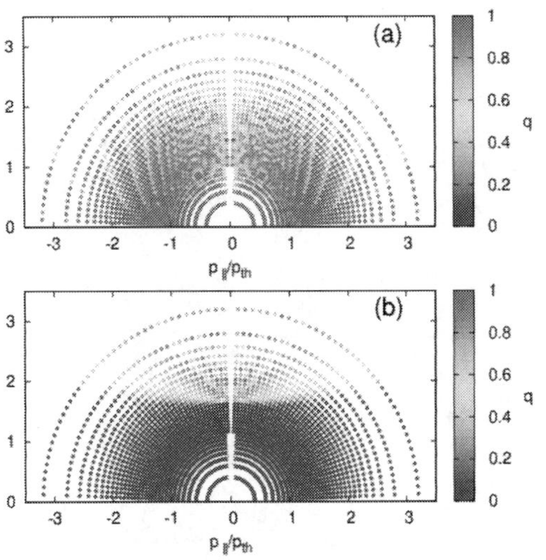

Figure 5. Probability transition q to a large energy state reported in the momentum plane of initial conditions $p_{\parallel 0}, p_{\perp 0}$. For each couple of momentum values, the evolution of $N_\theta = 30$ particles with uniformly distributed phases θ is followed through the EC illumination region. Probability q is defined as the number of particles that experience the finite energy variation divided by N_θ. XM2 perpendicular injection, with $P_{EC} = 1$ MW and $1-\omega/2\Omega = 10^{-4}$ (a), and $1-\omega/2\Omega = 1.22\times10^{-4}$ (b).

the particles is that illustrated in Figures 1a and 2a). If the frequency distance from the resonance is increased keeping the field amplitude fixed, the coldest particles are no more trapped and q drops to zero in a large region of the momentum space as it is shown in Figure 5b. It should be noted that the understanding of the observed numerical results for the probability transition q require an additional careful theoretical investigation which is beyond the scope of this paper.

3 Energy gain

The energy W gained by an electron interacting with a localized EC beam is given by

$$W = mc^2\Delta\gamma \cong 2mc^2 I_r(\bar{P}_{zs}) \tag{9}$$

where the resonant action reads in the weakly relativistic approximation

$$I_r(\bar{P}_z) \cong \frac{1-\nu_n + N_\parallel \bar{P}_z - \bar{P}_z^2/2}{1-N_\parallel^2} \tag{10}$$

The gained energy W is a function of the frequency ratio ν_n, of the parallel refractive index $N_\parallel$ and of the parallel momentum at the separatrix crossing $\bar{P}_{zs}$. Note that the explicit expressions in Eqs.(9,10) are valid in the limit $1-\nu_n \ll 1$ which is well verified for the pre-ionization parameters. As shown in [1], the variation of the canonical momentum $\bar{P}_z$ is negligible except for the first harmonic, oblique propagation, where it is finite although not large. Thus, one can safely use the initial parallel canonical momentum $\bar{P}_{z0}$ in Eq.(10), i.e. $I_{rs} \approx I_r(\bar{P}_{z0})$, except in the case $n = 1, N_\parallel \neq 0$, where I_{rs} can be estimated as

$$I_{rs} \simeq \frac{1-\nu}{1-N_\parallel^2 \mp N_\parallel[(1-N_\parallel^2)/3]^{1/2}} \tag{11}$$

in the limit $|1-\nu| \gg |\bar{P}_{z0}|$. The two solutions correspond to particles passing and reflected by the EC nonlinear ponderomotive potential.

The energy W increases linearly with the frequency ratio up to a critical value ϵ_{cn}, where it achieves its maximum value beyond which it drops to zero. The critical frequency value corresponds to electrons crossing the separatrix for critical field value equal to the peak value $\epsilon_M = |q_e|E_M/m\omega$

$$\epsilon_M = \epsilon_{cn} \equiv \beta_{cn} I_{rs}^{2-n/2}(1-N_\parallel^2)/T_n \tag{12}$$

where $\beta_{c1} = (2/3)^{3/2}$, $\beta_{c2} = 1/2$ and the polarization term $T_n = N_\perp^{n-1}(e_x - e_y + e_z N_\perp \bar{P}_z)/2$. The maximum energy gained at the critical frequency ratio $\omega/n\Omega$ is implicitly defined by Eq.(12) and reads

$$W_{max} = 2mc^2\left(\frac{\epsilon_M T_n}{\beta_{cn}(1-N_\parallel^2)}\right)^{2/(4-n)} \tag{13}$$

For oblique propagation with finite $N_\parallel$ the ordinary mode yields results similar to extraordinary mode, while energy values lower by the factor $\beta_{th}^{2/(4-n)}$ are found for perpendicular propagation.

4 Spatial localisation

4.1 Radial profile

As shown in Eqs.(9) and (10), at dominant order the gained energy W depends linearly on the ratio between the wave frequency and the local cyclotron frequency $\nu_n \equiv \omega/n\Omega$. In the null region of a tokamak, the ratio ν_n is equal simply to the ratio between the local major radius R and the radius of the cold cyclotron resonance R_{ECn}, defined by $\omega = n\Omega(R_{ECn})$, so that the dependence of W on the major radius can be obtained straighthforward.

Examples of the radial profile in ITER of the energy W are shown in Figure 6 for XM2 and $N_\parallel = 0$ for two values of the input power and in Figure 7 for XM2 at fixed power for two $N_\parallel$ values. The analytical estimate derived in Sec. 3 is compared with the numerical result obtained solving the Hamiltonian equations of motion for a uniformly distributed set of initial phase values θ and same thermal initial action $I_0 = \beta_{th}^2/2$ and $P_{z0} = \beta_{th}$. The agreement between the numerical simulations and the analytical estimates is extremely good, note that no free parameters are used here.

Making use of the relationship

$$\epsilon_M \simeq 2.043\times10^{-3}(P/w^2)^{1/2}/f, \tag{14}$$

the maximum electron energy (13) can be expressed in terms of power P in MW, frequency f in GHz and beam

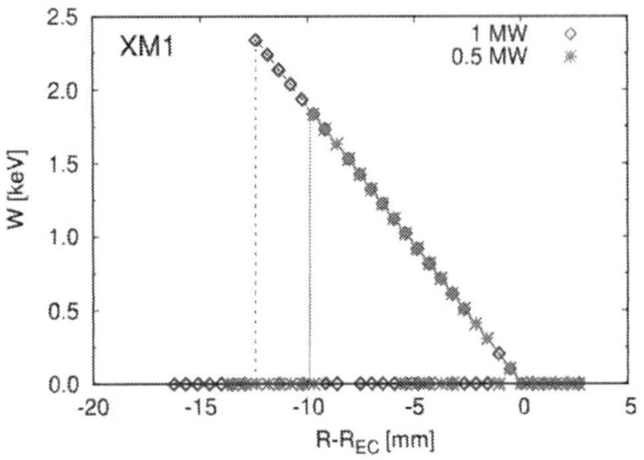

Figure 6. Radial energy profile for XM1 perpendicular propagation and P_{EC} = 0.5, 1 MW in ITER, with $R_{EC} \approx 5.41$ m at full field $B_0 = 5.3$ T.

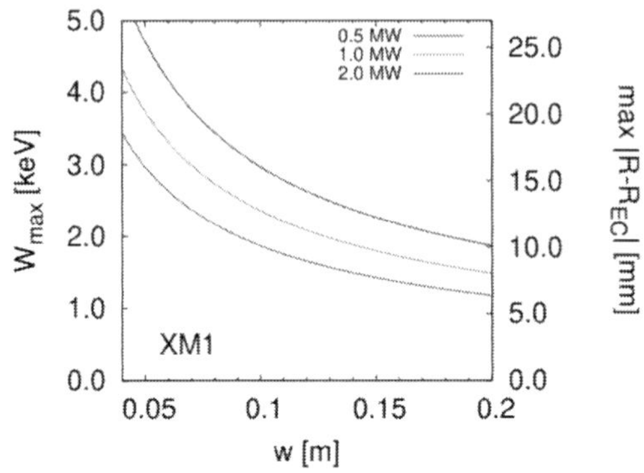

Figure 8. Maximum gained energy (left) and radial width of the interaction region (right) versus local beam width for 170 GHz, XM1 perpendicular propagation and various input power values.

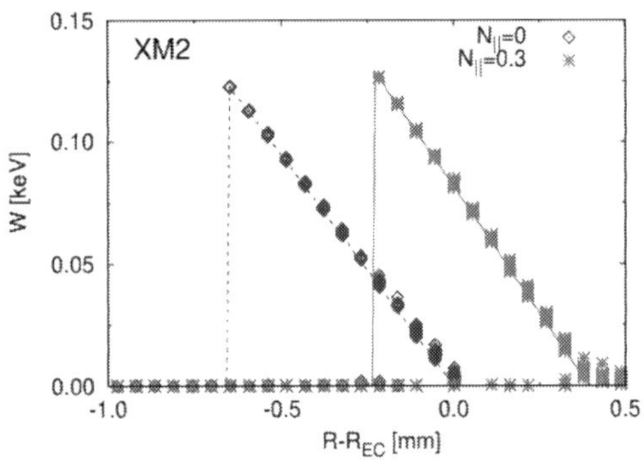

Figure 7. Radial energy profile for XM2 for $N_{\|}$ = 0, 0.3 and P_{EC} = 1 MW in ITER ($R_{EC2} = R_{EC}$ at half field $B_0 = 2.65$ T.)

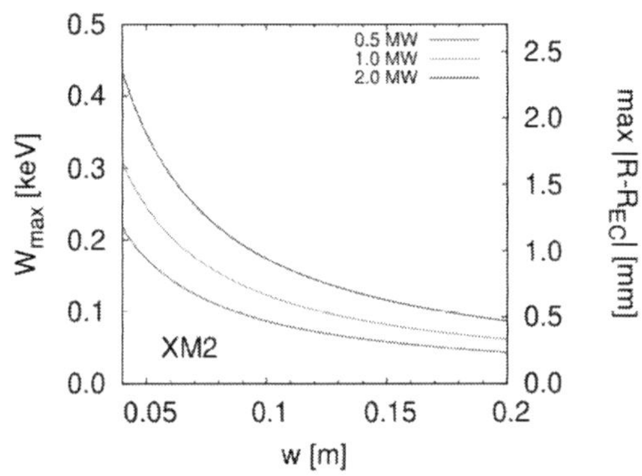

Figure 9. Same as in Figure 8 for XM2 in perpendicular propagation.

spot sizes in m. In case of XM in perpendicular propagation one obtains

$$W_{max,n=1} \cong 15.55\, P^{1/3}/(fw)^{2/3} \quad [\text{keV}]$$
$$W_{max,n=2} \cong 2.09\, P^{1/2}/(fw) \quad [\text{keV}]. \qquad (15)$$

The width of the radial region in which the energy gain process occurs is given by

$$0 < 1 - R/R_{ECn} < W_{max,n}/2mc^2. \qquad (16)$$

The maximum electron energy Eq.(15) and the radial width of the interaction region Eq.(16) are shown in Figures 8 and 9 versus the local beam size for various input power in the ITER case for XM1 and XM2 in perpendicular propagation and injected frequency 170 GHz. Note that both the final energy and the radial width are much larger for first harmonic than for second harmonic (approximately by one order of magnitude for the ITER case).

4.2 2D Gaussian beam

Let us consider a more realistic although simple injection configuration and a two dimensional EC beam, introducing a rectified tokamak geometry with cartesian coordinate system (x, y, z) related to the Tokamak system (R, ϕ, Z) as follows: $x = R - R_{EC}$, $y = Z$ and z along ϕ. For simplicity, hereafter we present results for the case of propagation perpendicular to the magnetic field $\mathbf{B}_0$, in the horizontal plane, i.e., $\mathbf{k}$ along x axis.

The EC Gaussian beam is assumed to be astigmatic in the general case, with local beam width both in the y and z directions, and electric field amplitude given by

$$E(y, z) = E_M\, g_y(y)\, g_z(z)$$

with $g_y(y) = \exp(-y^2/w_y^2)$ and $g_z(z) = \exp(-z^2/w_z^2)$. On a given magnetic field line, i.e. at fixed x and y in this simplified geometry, an electron will experience a normalised e. m. amplitude variation $\epsilon = \epsilon_M g_y(y) g_z(z)$. The results derived in the previous sections are still valid applying the transformations $\epsilon_M \rightarrow \epsilon_M g_y(y)$, and $w^2 \rightarrow w_y w_z$ [in Eq.(14)].

Both the maximum energy gain and the width of the radial interaction region exhibit a Gaussian dependence on the coordinate y

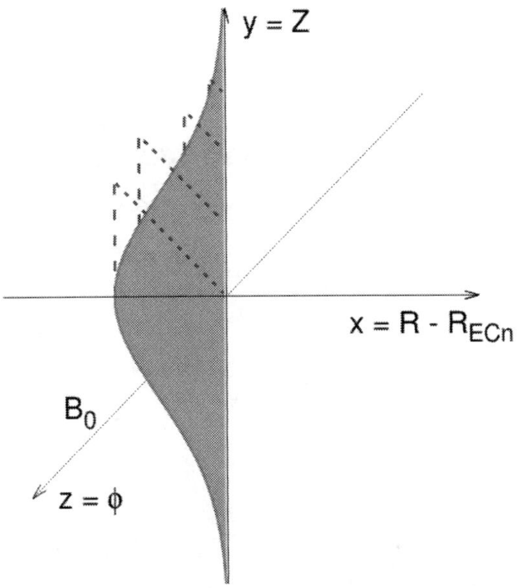

Figure 10. Region $\mathcal{R}_{xy}$ in (x, y) plane where nonlinear EC interaction takes place (red filled region). Note that the picture is expanded in x direction. The blue dashed lines are the energy profiles for different values of the y coordinate.

$$
\begin{aligned}
W_{max,1}(y) &= 3mc^2[\epsilon_M g_y(y)/2]^{2/3} \\
W_{max,2}(y) &= 2mc^2 \epsilon_M g_y(y) \,.
\end{aligned} \tag{17}
$$

and

$$\Delta R(y) \equiv |R_{ECn} - R| < R_{ECn}\, W_{max,n}(y)/2mc^2 \tag{18}$$

A sketch of the region $\mathcal{R}_{xy}$ in the $x - y$ "poloidal" plane where condition (18) is verified is shown in Figure 10 (filled area). The dashed curves represent the energy spatial variation, with Gaussian shape $g_y(y)$. The picture is not realistic since it has been stretched in the x direction, the x scale being usually much smaller than the y scale in experimental conditions, as it is clear from Eq. (18) and Figures 6, and 7.

An average energy value $\overline{W}_n$ can be introduced integrating W over the region $\mathcal{R}_{xy}$ in the $x - y$ plane where condition (16) is satisfied (see Figure 10)

$$\overline{W}_n = \frac{\int\int dxdyW}{\int\int dxdy} \tag{19}$$

where the denominator represents the area A_n of the interaction $\mathcal{R}_{xy}$. For a Gaussian amplitude profile it is found that

$$\overline{W}_n = W_{max,n}(y=0)/2^{3/2}. \tag{20}$$

which reads as a function of power P, frequency and beam spot sizes

$$
\begin{aligned}
\overline{W}_{n=1} &\simeq 5.50\,(P/w_y w_z)^{1/3}/f^{2/3}\ [\text{keV}] \\
\overline{W}_{n=2} &\simeq 0.74\,(P/w_y w_z)^{1/2}/f\ [\text{keV}]
\end{aligned} \tag{21}
$$

The area A_n of the region $\mathcal{R}_{xy}$ (measured in m^2) scales as

$$
\begin{aligned}
A_{n=1} &\simeq 3.3 \times 10^{-2}\, R_{EC}(P\, w_y^2/w_z)^{1/3}/f^{2/3} \\
A_{n=2} &\simeq 3.6 \times 10^{-3}\, R_{EC}(P\, w_y/w_z)^{1/2}/f\,.
\end{aligned} \tag{22}
$$

Note that both the average and the maximum energy depends on the beam spot sizes via their product.

For a uniform and constant density of seed electrons, the number of electrons N acquiring a finite energy can be assumed to be proportional to the area A_{xy} and to scale with beam local size as

$$
\begin{aligned}
N &\propto (w_y^2/w_z)^{1/3} \quad \text{for} \quad n = 1 \\
N &\propto (w_y/w_z)^{1/2} \quad \text{for} \quad n = 2
\end{aligned} \tag{23}
$$

5 Conclusive remarks

The analysis presented here provides analytical estimates which can give hints for the design of the experimental configuration of EC-assisted plasma breakdown. It is shown that electrons at room temperature can achieve finite energy via non linear collisionless wave-particle interaction. After being heated these electrons will collide with neutrals and ionise the gas and the plasma whenever their energy exceeds the ionization energy. Thus it is required that the gained energy is large enough to give rise to ionization (say ~ 30 eV) and that a sufficient number of electrons are heated. This analysis is a first step towards a physics based approach to EC pre-ionization, and estimates (21) and (23) may give a practical contribution to its optimization.

This work was partially carried out within the framework of the ECHUL consortium, supported by the F4E grant F4E-GRT-615. The views and opinions expressed herein do not necessarily reflect those of the European Commission or the ITER Organisation.

References

[1] D. Farina, *Nucl. Fusion*, **58** 066012 (2018)

[2] D. Farina, *EPJ Web Conf.*, **157** 03012 (2017)

[3] E. V. Suvorov, M. D. Tokman, *Sov. J. Plasma Phys.* **14** 557 (1988)

[4] W. M. Nevins, T. D. Rognlien and B I Cohen *Phys. Rev. Lett.* **59** 60 (1987)

[5] I. A. Kotelnikov and G. V. Stupakov *Phys. Fluids B* **2** 881 (1990)

[6] I. A. Kotelnikov and G. V. Stupakov *J. Plasma Phys.* **45** 19 (1991)

[7] D. Farina, R. Pozzoli and D. Ronzio *Plasma Phys. Control. Fusion* **33** 485 (1991)

[8] D. Farina and R. Pozzoli *Phys. Fluids B* **3** 1570 (1991)

[9] D. Farina, R. Pozzoli and M. Romé *Phys. Fluids B* **3** 3065 (1991)

Anomalous absorption in ECRH experiments due to the parametric low-threshold excitation of localized UH waves

Alexei Yu. Popov, Evgeniy Z. Gusakov

Ioffe Institute, St. Petersburg 194021, Russia

Abstract. We describe the theoretical model which interprets the anomalous phenomena, *i.e.* the generation of backscattering signal observed in the ECRH experiments at TEXTOR, TCV, TJ-II, ASDEX-UG, LHD and FTU, as a consequence of the excitation of the parametric decay instability (PDI) leading to anomalous damping of the pump wave. The PDI power-threshold is shown to be extremely low due to the localization of both or one daughter upper hybrid (UH) waves in presence of a nonmonotonic (hollow) density profile, which is often observed in the ECRH experiments due to the magnetic island or the density pump-out effect. In the case of the extraordinary wave pump the model predicts substantial (up to 25%) anomalous absorption in the electron channel and explains the anomalous ion acceleration by the generation of secondary low frequency waves which directly transfer the pump power to the ion component. The possibility of anomalous absorption of the O-mode pump in the ECRH experiment due to the parametric excitation of trapped UH wave is also discussed and the anomalous absorption rate at the 10% level is predicted.

1 Introduction

Electron cyclotron resonance heating (ECRH) and current drive is widely used in toroidal plasmas and is considered for application in ITER for heating and neoclassical tearing mode control. The parametric decay instabilities (PDIs), which can accompany the ECRH experiments, are believed to be deeply suppressed by the huge energy loss of daughter waves from the decay region, according to the predictions of theory developed in 80th [1]. However, during the last decade numerous ECRH experiments performed at TEXTOR, TCV, TJ-II, ASDEX-UG, LHD, T-10 and FTU [2-13] have demonstrated excitation of the anomalous phenomena. The clearest evidence of the nonlinear phenomena excitation was obtained at TEXTOR [2, 3] where the strong backscattering signals down–shifted in frequency and amplitude modulated at the frequency of magnetic island were observed. A convincing demonstration of the anomalous ion acceleration during the ECRH pulse under conditions when the energy exchange between the ion and electron components is negligible was obtained at TCV [4,5].

In the present paper we describe the theoretical model taking into account, as distinct from the conventional theory [1], the presence of a non-monotonic density profile, which always exists on the discharge axis or may be present due to the magnetic island or the density pump-out effect. We interpret the generation of backscattering signal and the anomalous ion heating, as a result of secondary nonlinear processes that accompany a primary low – threshold two–upper-hybrid (UH) – plasmon PDI of the pump X2-mode wave. The threshold of primary PDI is shown [14, 15] to be smaller than the one predicted in [1] due to the trapping of at least one UH wave in presence of the non-monotonic density profile. The primary PDI appears to be absolute due to the finite-size of a pump beam. Its growth enhancing the UH wave fluctuations from the thermal noise level is saturated in our theory due to both the secondary decays of the daughter UH wave that leads to excitation of the secondary UH and ion Bernstein (IB) waves [16] and the pump wave depletion. The threshold of this spontaneous parametric frequency down-conversion can be easily overcome for the radially trapped secondary UH waves. The couplings of different daughter UH waves are responsible in the theory for the generation of backscattering signals [17]. This mechanism appears capable of reproducing the fine details of the frequency spectrum of anomalously backscattered X wave and the absolute value of the observed backscattering signal in TEXTOR experiment. It also predicts substantial (up to 25%) anomalous absorption in the electron channel and explains the anomalous ion heating at TCV by the generation of secondary IB waves, which directly transfer the pump power to the ion component.

The results of a model experiment, demonstrating strong anomalous absorption of the X-mode pump in a plasma filament due to the two-plasmon decay [18] and thus confirming the theory, are discussed.

The possibility of anomalous absorption of the O-mode pump in the ECRH experiment due to parametric excitation of the trapped UH wave [19,20] is discussed as well.

* Corresponding author: a.popov@mail.ioffe.ru

2 Model

We assume that the three-wave nonlinear interaction leading to the low-threshold PDI excitation occurs in a small volume in the vicinity of a local maximum of plasma density profile illuminated by the pump beam where the daughter UH waves/wave are localized. Therefore, it seems to be reasonable to utilize throughout the paper, describing different PDIs, the Cartesian coordinate system (x,y,z) with its origin located at the local maximum of the UH frequency profile, the x-axis directed along the flux coordinate and (y,z) imitating accordingly the coordinates perpendicular to and aligned with the magnetic field line on the magnetic surface.

2.1. Two-UH-plasmon parametric decay of an extraordinary wave

We start with consideration of the low-power threshold two-UH-plasmon decay of a beam of the extraordinary (X) waves

$$\mathbf{E} = \mathbf{e}_0 \frac{A_0}{2} \exp\left(i \int_{-\infty}^{x} k_x(\omega_0, x')dx' - i\omega_0 t \right) + c.c. \quad (1)$$

propagating perpendicular to the magnetic surface along the inhomogeneity direction and its saturation owing to the pump depletion and the cascade of secondary decays. In (1) $\mathbf{e}_0 = i\mathbf{e}_x g(\omega_0)/\varepsilon(\omega_0) + \mathbf{e}_y$ is the polarization vector, $k_x = \omega_0 / c\sqrt{\varepsilon - g^2/\varepsilon}$ is the local wavenumber, A_0 is the slowly varying amplitude, ε, g stand for the components of the cold-plasma dielectric tensor. A set of equations describing the two-step cascade of decays reads

$$\begin{aligned}
&\left(\frac{\partial^2}{\partial x^2} + k_x^2(\omega_0)\right) A_0 = \frac{\omega_0^2}{c^2}\kappa_p^2 \frac{\varphi_1^* \varphi_2}{H} \\
&\hat{D}(\varphi_1) = \kappa_p^2 \frac{A_0^*}{H}\varphi_2 - \kappa_s^2 \frac{|e|}{T_e}\varphi_1'^* \varphi_2' \\
&\hat{D}(\varphi_2) = \kappa_p^2 \frac{A_0}{H}\varphi_1 \\
&\hat{D}(\varphi_1') = \kappa_s^2 \frac{|e|}{T_e}\varphi_1^* \varphi_2' \\
&\hat{D}(\varphi_2') = \kappa_s^2 \frac{|e|}{T_e}\varphi_1 \varphi_1'
\end{aligned} \quad (2)$$

where $\varphi_{1,2}$ are the potentials of primary UH waves, $\varphi_{1,2}'$ are the potentials of secondary UH and ion Bernstein (IB) waves and $\kappa_{p,s}^2$ are the coupling coefficients for both decays presented explicitly in [21]. In weakly inhomogeneous plasma the integral operators $\hat{D}$ are taking the following form:

$$\begin{aligned}
&\hat{D}(\varphi) = \int_{-\infty}^{t}\int_{-\infty}^{\infty} dt' d\mathbf{r}' D(t-t', \mathbf{r}, \mathbf{r}')\varphi(\mathbf{r}', t'), \\
&D(t-t', \mathbf{r}, \mathbf{r}') = \int_{-\infty}^{\infty} \frac{d\mathbf{q}d\omega}{(2\pi)^4} D(\omega, \mathbf{q}, \mathbf{r}+\mathbf{r}') e^{i\mathbf{q}(\mathbf{r}-\mathbf{r}') - i\omega(t-t')}
\end{aligned} \quad (3)$$

with their kernel $D(\omega, \mathbf{q}) = q^2 + \chi_e + \chi_i$ being the local dispersion function of the corresponding electrostatic waves [22], obtained in the homogeneous plasma theory and χ_e, χ_i standing for the familiar expressions of the electron and ion susceptibilities

$$\chi_j = \frac{2\omega_{pj}^2}{\upsilon_{tj}^2}\left(1 - \frac{\omega}{q_z \upsilon_{tj}} \sum_{m=-\infty}^{\infty} Z\left(\frac{\omega - m\omega_{cj}}{q_z \upsilon_{tj}}\right) \exp\left(-\frac{q_\perp^2 \upsilon_{tj}^2}{2\omega_{cj}^2}\right) I_m\left(\frac{q_\perp^2 \upsilon_{tj}^2}{2\omega_{cj}^2}\right)\right) \quad (4)$$

In (4) $j=e,i$, $q_{\perp,z}$ are the wavenumber components perpendicular to and along the magnetic field, $\upsilon_{te,i}$ are the electron and ion thermal velocities, $\omega_{ce,i}$ and $\omega_{pe,i}$ are the electron and ion cyclotron and Langmuir frequencies, I_m is the modified Bessel function of the first kind and $Z(\xi) = 2\exp(-\xi^2)\int_0^{\xi}\exp(t^2)dt - i\sqrt{\pi}\exp(-\xi^2)$. In (2) we have assumed that only one primary UH wave experience frequency down-conversion. Then, we assume the nonlinear coupling is weak. This allows utilizing the perturbation theory approach [23], in the first step of which we drop all the RHS of equations (2) and seek the solutions of the homogeneous wave-equations. In particular, the trapped UH waves potentials are derived by means of the WKB approximation as

$$\begin{aligned}
&\varphi_{1,2}(\mathbf{r},t) = \phi_{n,m}(\mathbf{r})\exp(iq_y y \pm i\omega_{n,m}t)/2 + c.c., \\
&\varphi_2'(\mathbf{r},t) = \phi_p(\mathbf{r})\exp(-i\omega_p t)/2 + c.c.
\end{aligned} \quad (5)$$

where $\omega_m = \omega_0 - \omega_n$, $\omega_1' = \omega_n - \omega_p$, $c.c.$ stands for the term derived from the first one by the complex conjugation,

$$\phi_n = \frac{C_n \exp\left(i\int_{x_{nl}^*}^{x} q_{nx}^+ d\xi - i\frac{\pi}{4}\right)}{\sqrt{L_n^+(x)}} + \frac{C_n \exp\left(i\int_{x_{nl}^*}^{x} q_{nx}^- d\xi + i\frac{\pi}{4}\right)}{\sqrt{L_n^-(x)}} \quad (6)$$

is the eigenfunction, $C_n = const$, $q_{nx}^{\pm}$ is a root of the local UH wave dispersion relation, $L_n^{\pm}$ stands for the optical length of the wave path between two turning points (x_{nl}^*, x_{nr}^*). The frequencies ω_n, ω_m and the poloidal wave-number $q_y^{n,m}$ obey simultaneously the quantization conditions. The frequency of secondary UH wave ω_p obeys the quantization condition as well

$$\sum_{s=\pm}\int_{x_{pl}^*}^{x_{pr}^*} q_{fx}^s(\omega_p, \xi)d\xi = \pi(2p+1) \quad (7)$$

Equation (6) describes the UH wave localized in the x-direction at the maximum of the UH frequency profile.

In the second step of the perturbation procedure we take into consideration the nonlinear interactions and the pump energy loss due to the UH plasmons excitation. We utilize the envelope amplitude method and keep in mind the boundary condition for the pump wave

$$A_0(\mathbf{r})\Big|_{x\to-\infty} = \sqrt{\frac{8P_0}{w^2\upsilon_{gx}}}\exp\left(-\frac{y^2+z^2}{2w^2}\right) \quad (8)$$

with $\upsilon_{gx}(x)$, P_0, w being the wave group velocity, pump power and beam width. Finally we get

$$\begin{aligned}
&\left(\frac{\partial}{\partial t} - u_n\frac{\partial}{\partial y} + i\Lambda_n\frac{\partial^{=2}}{\partial z^2}\right)a_n = v_0^*\exp\left(-\frac{y^2+z^2}{2w^2}\right)a_m \\
&-\left(v_d^*\frac{|a_m|^2}{a_T^2} + v_s^*\frac{|a_p|^2}{a_T^2}\right)a_n \\
&\frac{\partial a_m}{\partial t} + u_m\frac{\partial a_m}{\partial y} - i\Lambda_m\frac{\partial^2 a_m}{\partial z^2} = v_0\exp\left(-\frac{y^2+z^2}{2w^2}\right)a_n - v_d\frac{|a_n|^2}{a_T^2}a_m
\end{aligned}$$

$$\frac{\partial a_p}{\partial t} + u_p \frac{\partial a_p}{\partial y} - i\Lambda_p \frac{\partial^2 a_p}{\partial z^2} = \nu_s \frac{|a_n|^2}{a_T^2} a_p \qquad (9)$$

where

$$a_{n,m,p} = |C_{n,m,p}| \left(\omega_{n,m,p} < D_{n,m,p\omega} > / 16\pi \right)^{1/2}, \qquad (10)$$

$< D_{n,m,p\omega} > = < \partial D / \partial \omega|_{\omega_{n,m,p}} >$, $<\ldots>$ stands for averaging over the localization region of the corresponding UH plasmons

$$<\ldots> = \int_{-\infty}^{\infty} dx \ldots |\phi_{n,m,p}(x)|^2 . \qquad (11)$$

$|a_{n,m,p}|^2$ yield their energy, $u_{n,m,p} = < \partial D / \partial q_y|_{\omega_{n,m,p}} >$ are the convective velocities in the poloidal direction, $\Lambda_{n,m,p} = < \partial^2 D / \partial (2q_z^2)|_{\omega_{n,m,p}} >$ are the diffraction coefficients along the magnetic field,

$$\nu_0 = i\sqrt{\frac{2P_0}{w^2 H^2 v_{gx}}} \int_{-\infty}^{\infty} dx \frac{\kappa^2(x)\phi_n^*(x)\phi_m(x)}{\sqrt{< D_{1\omega} >< D_{2\omega} >}} \exp\left(-i\int^x k_x(x'')dx''\right),$$

$$\nu_d = \frac{4\pi a_T^2}{H^2 c} \times$$

$$\int_{-\infty}^{\infty} dx \int_{-\infty}^{x} dx' \frac{\kappa^2(x)\kappa^2(x')}{< D_{1\omega} >< D_{2\omega} >} \phi_n(x)\phi_m^*(x)\phi_n^*(x')\phi_m(x') \exp\left(i\int_{x'}^{x} k_x(x'')dx''\right),$$

$$\nu_s = \frac{4\pi a_T^2}{H^2} \left[\frac{c\omega_{pe}^2}{(\omega_1^2 - \omega_{ce}^2)(\omega_1'^2 - \omega_{ce}^2)} \frac{\omega_1 \omega_1'}{|\omega_{ce}|} \right]_{x_{uh}}^2 \times$$

$$\int_{-\infty}^{\infty} dx \frac{q_{1x}(x) q_{1x}'(x) q_{Ix}^2(x)}{\sqrt{< D_{1\omega}(x) >< D_{1\omega}'(x) >}} \frac{\phi_n^*(x)\phi_p'^*(x)}{\sqrt{D_{Iq}(x)}} \times$$

$$\int_{x}^{\infty} dx' \frac{q_{1x}(x') q_{1x}'(x') q_{Ix}^2(x')}{\sqrt{< D_{1\omega}(x') >< D_{1\omega}'(x') >}} \frac{\phi_n(x')\phi_p'(x')}{\sqrt{D_{Iq}(x')}} \exp\left(i\int_{x'}^{x} q_{Ix}(x'')dx''\right)$$

are the coupling coefficients [21] describing the primary (ν_0) and secondary (ν_s) decays, the pump wave depletion (ν_d) and

$$a_T = \sqrt{T_e / (\pi w^2)} \qquad (12)$$

Further, we analyze (9) analytically and then solve numerically. We start with consideration of the growth-rates of primary and secondary decays and get an estimation of the daughter wave saturation levels based on the energy-balance formulation. For the conditions under consideration the convective energy losses of the primary and secondary UH plasmons from the decay layer in the poloidal direction are much smaller than their energy losses in a form of diffraction in the toroidal direction, *i.e.* $u_{n,m,p} w / \Lambda_{n,m,p} << 1$ and $\Lambda_{n,m,p} \approx \Lambda$. In this case the growth-rate of the primary two-UH-plasmon decay can be evaluated on the basis of analysis carried out in [14] as

$$\gamma = \sqrt{|\nu_0|^2 - \left(\pi(2\zeta+1)/2w\right)^4 \Lambda^2} \qquad (13)$$

with ζ being the unstable eigenmode number. As soon as the energy of primary plasmons growing exponentially reaches a level which exceeds the threshold of secondary decay instability, the energy of secondary plasmons has to start growing as well. Eventually, the nonlinear system relaxes to the stationary regime in which the saturation level of primary plasmons number $|a_n^s|^2$ can be evaluated from a condition indicating saturation of the secondary decay described by the last equation in (9). Namely, the growth rate $\nu_s |a_1|^2$ of this decay should be compensated by the secondary UH plasmon diffractive loss rate from the pump beam spot, characterized by a typical time scale $\tau = \pi w^2 / \Lambda$. The corresponding level is given by

$$|a_n^s|^2 \approx a_T^2 / (\tau \nu_s) \qquad (14)$$

The number of the primary plasmons $|a_m^s|$ in the saturation regime could be estimated using the second equation of (9) as

$$|a_m^s|^2 \approx \frac{|\nu_0|^2}{\tau^{-2} + |\nu_d|^2 |a_n^s|^4 / a_T^4} |a_n^s|^2 \qquad (15)$$

The energy of secondary plasmons $|a_p^s|^2$ saturates at a level balancing the nonlinear pumping rate $\nu_0 |a_m|$ and the loss rate due to the secondary decay of first (a_n) primary UH wave $\nu_s |a_p|^2 |a_n|$. This saturation level can be estimated as

$$|a_p^s|^2 \approx \frac{a_T^2}{|\nu_s|} \frac{|\nu_0|^2}{\sqrt{\tau^{-2} + |\nu_d|^2 |a_n^s|^4 / a_T^4}} . \qquad (16)$$

Then, we solve (9) numerically in a 2D box $2y_B \times 2z_B$ assuming the periodic boundary conditions and an initial thermal-noise-level of the UH waves. The imposed boundary conditions imitate the toroidal geometry and describe the multiple transitions of UH plasmons through the 2D box. However, it is worth noting that equations (9) were derived ignoring the magnetic field poloidal dependence. Thereby, the meaningful solution of (9) is an intermediate asymptotic dependence valid at the time much less than the transition time of the daughter waves through the 2D box. We assume the initial thermal noise level (12) of all plasmons and use the parameters $w = 1$ cm , $T_e = 600$ eV , $P_0 = 600$ kW which correspond to the TEXTOR experimental conditions. The results of numerical solution of equations (9) are shown in Fig. 1 where the temporal evolution of the average energy of different daughter plasmons in the spot $\langle |a_n(t)|^2 \rangle_p = \int \frac{dydz}{2\pi} |a_n(y,z,t)|^2 \exp\left(-\frac{y^2+z^2}{w^2}\right)$ of a pump wave is presented. In the early stage of PDI one can see an exponential growth of the energy of primary plasmons with the growth rate being adequately described by the analytical equation (13) (short dotted line). When the energy of primary plasmons is large enough so that the threshold of the secondary decay instability appears to be overcome, the energy of secondary plasmons starts to grow as well. Eventually, the energies of primary and secondary plasmons achieve their saturation levels. It is seen that the saturation levels of the daughter plasmons derived analytically (equation (14) – lower thin dashed horizontal line, equations (15) and (16) – upper thin dashed horizontal lines which are merged) are in a reasonable agreement with the results of numerical modeling.

The power transferred to the UH plasmons at the pump power well above the threshold ($P_0 >> P_0^{th}$) can be estimated based on expressions for the number of plasmons at the PDI saturation (14)-(16) and assuming their diffractive losses from the decay region. The corresponding estimation takes a form $\Delta P \propto (|a_n^s|^2 + |a_m^s|^2 + |a_p^s|^2)/\tau \propto |v_0|^2$. Therefore, we can conclude that the anomalously absorbed power should be proportional to the pump power and thus the anomalous absorption rate $\Delta P / P_0$ at large threshold excess should be constant. This conclusion is confirmed by the results of numerical solution which predicts at the pump power much larger than the PDI power threshold, $P_0^{th} = 5\ \text{kW}$, the anomalous absorption of X - mode wave at a level of 18%, independently on the pump power. Thus, for the parameters under consideration up to 110 kW are transferred to the daughter UH waves. Subsequently, this power fraction is deposited into the electron component far from the pump wave second harmonic electron cyclotron resonance layer. This can make the power deposition profile much wider than it is predicted by the linear theory and can create an illusion of the fast electron energy redistribution. In other words, this nonlinear absorption mechanism can be responsible, at least partly, for the so-called nonlocal electron transport in the on-axis X2-mode ECRH experiments at T-10 and LHD [12,13]. We can also estimate the pump power fraction absorbed by the daughter IB waves as $\Delta P_1 = 6.3\ \text{kW}$. The anomalous absorption leads to the generation of a population of the suprathermal ions. Thus, the proposed mechanism can explain the experimental observations of the ion acceleration in the on-axis X2-mode ECRH experiments at TCV and TJ-II [7-10].

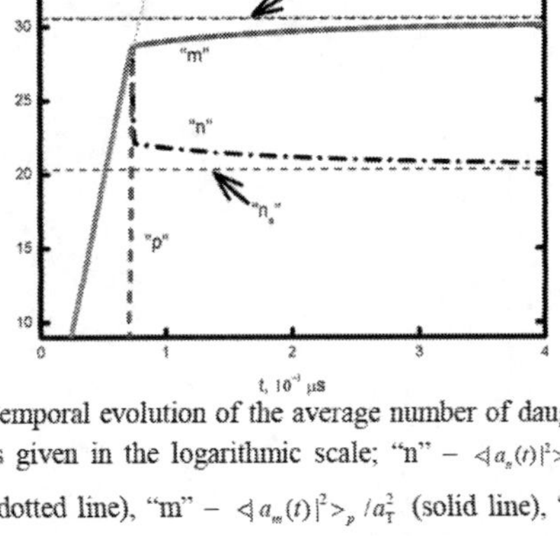

Fig. 1. Temporal evolution of the average number of daughter plasmons given in the logarithmic scale; "n" – $\langle|a_n(t)|^2\rangle_p / a_T^2$ (dashed dotted line), "m" – $\langle|a_m(t)|^2\rangle_p / a_T^2$ (solid line), "p" - $\langle|a_p(t)|^2\rangle_p / a_T^2$ (dashed line); "n_s" – equation (14) (lower thin dashed horizontal line), "m_s" and "p_s" - equation (15) and equation (16) (merged upper thin dashed horizontal lines). The gain coefficient $2\gamma t$ (short dotted line) of the primary decay waves (13).

The nonlinear coupling of different UH waves can lead to the generation of microwave signals with the frequencies smaller than the pump frequency. In this way the proposed nonlinear mechanism explains in detail the anomalous backscattering effect observed at TEXTOR [2]. In particular, the frequency spectrum structure and the spectral power density of the backscattering signal predicted by this theoretical model appear to be in a reasonable agreement with those measured by the ECE and CTS techniques [2,3].

In order to examine the mechanism of anomalous absorption described in this section the model experiment on the linear device "Granite" in Minsk was proposed and then performed [18]. The huge anomalous absorption of microwave power was observed in the plasma column when the density was close to the UHR value for half the pump frequency which validates the main theoretical predictions concerning the possibility of the low-power-threshold two-UH-plasmon PDI of extraordinary wave and its role in the energy balance.

Summarizing, we would like to emphasize that there are two key points of the model developed. The first point is the trapping of the UH waves at the local maximum of a nonmonotonic density profile. The second one of them is the possibility of an additional localization of the UH waves propagating in the opposite directions on the magnetic surface within the spot of a finite-width pump beam. In the forthcoming section we consider the possibility of low-power-threshold ordinary wave decay into a low hybrid wave (LH) and a radially trapped UH wave. We demonstrate that even one radially trapped UH daughter wave can be nonlinearly localized on the magnetic surface due to the finite-width pump, that leads to excitation of the most dangerous absolute PDI.

2.2. Low-threshold parametric decay of an ordinary wave

In this section we analyse the low-threshold PDI of an obliquely propagating ordinary wave. The parametric decay leads to the excitation of the obliquely propagating LH wave and the UH wave. The latter appears to be 3D localized due to the nonmonotonic density profile and the finite-width beam. We analyze also the transition of primary PDI to the saturation regime due to the secondary decay instability of primary UH wave. We focus on the case of the one-step down-conversion leading to the low-threshold excitation of the IB and radially trapped UH waves. We derive a set of equations describing the cascade of decays and then solve it numerically under the conditions similar to the on-axis O1-mode ECRH experiments at FTU [6] in which the nonmonotonic density profile arises due to the density pump-out effect from the ECR layer.

The pump ordinary wave can be represented by means of the WKB approximation as follows

$$\mathbf{E} = \mathbf{e}(x) A_0 \exp\left(i\int^x k_x(x',k_z)dx' + ik_z z - i\omega_0 t\right)/2 + c.c. \quad (17)$$

where $\mathbf{e}$ is a polarization vector with its components $e_x / e_y = -i\omega_{ce}/\omega_0$ and $e_x / e_z \approx \sqrt{\eta}\, k_z c/\omega_0$, η is the parallel component of the cold-plasma dielectric tensor, A_0 is the amplitude defined already in (8) and $k_x \approx \omega_0\sqrt{\eta}\left(1 - k_z^2c^2/2\omega_0^2\right)/c >> k_z$. A set of equations describing the primary decay of the pump wave (17) with the fixed amplitude A_0 and subsequent secondary decay of the trapped UH wave reads

$$\hat{D}(\varphi_1) = \kappa_p^{2*} \frac{A_0^*}{2H} \varphi_2 - \kappa_s^{2*} \frac{|e|}{2T_e} \varphi_1'^* \varphi_2'$$
$$\hat{D}(\varphi_2) = \kappa_p^{2} \frac{A_0}{2H} \varphi_1 \tag{18}$$
$$\hat{D}(\varphi_1') = \kappa_s^{2*} \frac{|e|}{2T_e} \varphi_1^* \varphi_2'$$
$$\hat{D}(\varphi_2') = \kappa_s^{2} \frac{|e|}{2T_e} \varphi_1' \varphi_1$$

where $\varphi_1 \propto \exp(i\omega_1 t)$ and $\varphi_2 \propto \exp(-i\omega_2 t)$ are the potentials of primary UH and LH waves, $\varphi_1' \propto \exp(-i\omega_1' t)$ and $\varphi_2' \propto \exp(i\omega_2' t)$ are the potentials of secondary UH and IB waves. In equations (18) the amplitude of pump wave A_0 is assumed to be fixed, the integral operators on their left-hand-sides are determined in the same way as in (3), κ_p^2 and κ_s^2 are the coupling coefficients of the waves involved in the primary and secondary decays presented explicitly in [20] and [24], correspondingly. Assuming the nonlinear coupling is weak we treat the set of equations (18) using the perturbation procedure [23]. In its first step we neglect the nonlinear pump. The solutions of homogeneous equations describing the trapped UH waves can be presented in the following form

$$\varphi_1 = C_n \phi_n(\mathbf{r}) \exp(i\omega_n t) + c.c.,$$
$$\varphi_1' = C_m \phi_m(\mathbf{r}) \exp(i q_{1z}' z - i\omega_m t) + c.c. \tag{19}$$

where $C_{n,m} = \text{const}$, the eigenfunctions $\phi_{n,m}(x)$ and eigenfrequency $\omega_{n,m}$ are derived by means of the WKB approximation in the same way as in equations (6) and (7). The choice $q_{1y,z} = 0$, $q_{1y}' = 0$ allows us to minimize the energy losses of UH plasmons from the decay region. Then, we treat equations in (18) describing the daughter LH and IB waves. Utilizing the perturbation procedure we neglect their right-hand-sides and seek the solution of homogeneous equations by means of the WKB approximation

$$\varphi_2 = \frac{C_\text{I}}{|D_{2q}(x)|^{1/2}} \exp\left(i\int^x q_{2x}(\omega_2, q_{2z}, x') dx' + i q_{2z} z - i\omega_2 t\right)$$
$$\varphi_2' = \frac{C_\text{I}'}{|D_{2q}'(x)|^{1/2}} \exp\left(i\int^x q_{2x}(\omega_2', q_{2z}', x') dx' + i q_{2z}' z + i\omega_2' t\right) \tag{20}$$

where the amplitudes C_I, C_I' are constant, $q_{2x}(\omega_2, q_{2z})$, $q_{2x}' = q_{2x}(\omega_2', q_{2z}')$ are the solutions of the local dispersion relations $D(\omega_2) = 0$ and $D(\omega_2') = 0$, $q_{2z} = k_z$, $q_{2z}' = q_{1z}'$. The choice $q_{2y}, q_{2y}' = 0$ minimizes the low frequency waves energy losses perpendicular to the magnetic field along the axis $\mathbf{e}_y$. The pre-exponential factors $\propto 1/|D_{2q}(x)|^{1/2}$, $\propto 1/|D_{2q}'(x)|^{1/2}$ with D_{2q} and D_{2q}' being defined as follows $D_{2q} = \partial D / \partial q_x |_{q_{2x}(\omega_2, x), \omega_2}$, $D_{2q}' = \partial D / \partial q_x |_{q_{2x}(\omega_2', x), \omega_2'}$ ensure the conservation of the wave energy fluxes along the direction of inhomogeneity.

In the second step of the perturbation procedure we take into account the nonlinear wave coupling. The nonlinear three-wave interactions lead to the variation of amplitudes $C_{n,m}$ and C_I, C_I' in such a way that $C_\text{I}, C_\text{I}' \to C_\text{I}(\mathbf{r}), C_\text{I}'(\mathbf{r})$ and $C_{n,m} \to C_{n,m}(t, y, z)$. Using the envelope amplitude method we can get a set of the coupled equations

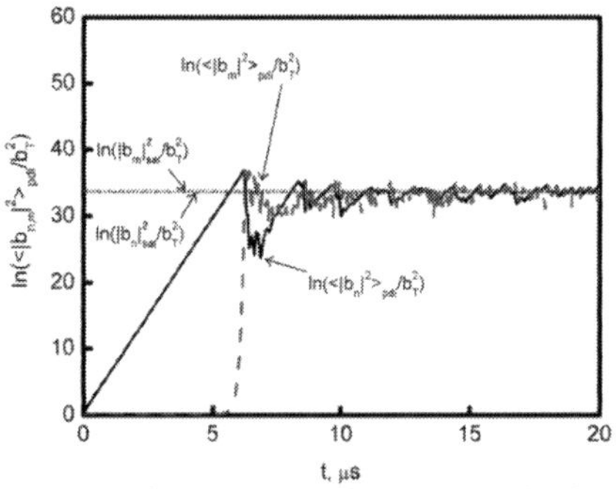

Fig. 2. Temporal evolution of the averaged density of the primary (solid curve) and secondary (dashed curve) UH plasmon energies given in the logarithmic scale. The dash dotted line is $2\nu_{0,0}t$ with $\nu_{0,0}$ determined in (31), the dash-dot-dotted curve is the dependence (35). Two horizontal lines (visually merged) are the saturation levels (36) and (37).

$$\left(\upsilon_{\text{I}x} \frac{\partial}{\partial x} + i\Lambda_{\text{I}y} \frac{\partial^2}{\partial y^2} + \upsilon_{\text{I}z} \frac{\partial}{\partial z}\right) C_\text{I}(\mathbf{r}) = \tag{21}$$
$$-\kappa_p^2 \frac{|D_{2q}|^{1/2}}{D_{2\omega}} \frac{A_0}{2H} C_n \phi_n(x) \exp\left(i\int^x (k_x - q_{2x}) dx'\right),$$

$$\left(\frac{\partial}{\partial t} + i\Lambda_{ny} \frac{\partial^2}{\partial y^2} + i\Lambda_{nz} \frac{\partial^2}{\partial z^2}\right) C_n =$$
$$\frac{\kappa_p^{2*}}{D_{n\omega}} \frac{A_0^*}{2H} \int_{-\infty}^{\infty} \frac{dx}{|D_{2q}|^{1/2}} \phi_n^*(x) C_\text{I}(\mathbf{r}) \exp\left(i\int^x (q_{2x} - k_x) dx'\right) - \tag{22}$$
$$i \frac{|e|}{2T_e} \frac{\kappa_s^{2*}}{D_{n\omega}} C_m^* \int_{-\infty}^{\infty} \frac{dx}{|D_{2q}'|^{1/2}} \phi_n^*(x) \phi_m^*(x) C_\text{I}'(\mathbf{r}) \exp\left(i\int^x q_{2x}' dx'\right),$$

$$\left(\upsilon_{\text{I}x}' \frac{\partial}{\partial x} + i\Lambda_{\text{I}y}' \frac{\partial^2}{\partial y^2} + \upsilon_{\text{I}z}' \frac{\partial}{\partial z}\right) C_\text{I}'(\mathbf{r}) = \tag{23}$$
$$i\kappa_s^2 \frac{|e|}{2T_e} \frac{|D_{2q}'|^{1/2}}{D_{2\omega}'} C_n C_m \phi_n(x) \phi_m(x) \exp\left(-i\int^x q_{2x}' dx'\right),$$

$$\left(\frac{\partial}{\partial t} + i\Lambda_{my} \frac{\partial^2}{\partial y^2} - \upsilon_{mz} \frac{\partial}{\partial z}\right) C_m =$$
$$\frac{i|e|}{2T_e} \frac{\kappa_s^{2*}}{D_{m\omega}} C_n^* \int_{-\infty}^{\infty} \frac{dx}{|D_{2q}'|^{1/2}} \phi_m^*(x) \phi_n^*(x) C_\text{I}'(\mathbf{r}) \exp\left(i\int^x q_{2x}' dx'\right) \tag{24}$$

where $\upsilon_{\text{I}x,z}$, $\upsilon_{\text{I}x,z}'$, $\Lambda_{\text{I}y}$ and $\Lambda_{\text{I}y}'$ are the local group velocity components and the diffraction coefficient of the daughter LH and IB waves, $\Lambda_{ny,z}$ are the coefficients describing the primary UH waves diffraction in the corresponding directions averaged over the localization region, Λ_{my}, υ_{mz} are the averaged diffraction coefficient and group velocity of the secondary UH wave, $D_{n,m\omega} = <\partial D/\partial\omega>_{\omega_{n,m}}$, $D_{1\omega} = \left|\partial D/\partial\omega\right|_{\omega_2}$, $D'_{1\omega} = \left|\partial D/\partial\omega\right|_{\omega'_2}$. The averaging procedure of an arbitrary function is determined as in equation (11). For the parameters under consideration $(w/l_d)(\upsilon_{1x}/\upsilon_{\Sigma}) \approx 2.1$, $(w/l'_d)(\upsilon'_{1x}/\upsilon'_{\Sigma}) \approx 5.3$ where $l_d = \left|d\Delta K/dx\right|_{x_d}^{-1/2}$ and $l'_d = \left|d\Delta K'/dx\right|_{x'_d}^{-1/2}$ are the lengths of the three-wave resonance layers and x_d, x'_d are the points where the decay conditions hold. Therefore the convective losses of low frequency waves from decay regions across the magnetic field are dominant and we can neglect all terms on the left-hand-side of equations (15), (17) apart from those proportional to the x - derivatives. Then, we integrate (21) and (23) with the boundary conditions $C_1\big|_{x\to-\infty} = 0$ and $C'_1\big|_{x\to\infty} = 0$ that yields

$$C_1 = C_n \frac{\kappa_p^2 A_0}{2H} \int_{-\infty}^{x} \frac{d\xi}{\left|D_{2q}\right|^{1/2}} \phi_n(\xi) \exp\left(i\int^{\xi}(k_x - q_{2x})dx'\right)$$
$$C'_1 = -i\kappa_z^2 \frac{|e| C_n C_m}{2T_e} \int_{x}^{\infty} \frac{d\xi}{\left|D'_{2q}\right|^{1/2}} \phi_n(\xi)\phi_m(\xi) \exp\left(-i\int^{\xi} q'_{2x} dx'\right) \quad (25)$$

Substituting (25) into the right-hand-sides of (22), (24) and introducing the dimensionless functions

$$b_{n,m} = \left|C_{n,m}\right| \left(\frac{\omega_{n,m}}{16\pi} D_{n,m\omega}(\omega_{n,m})\right)^{1/2} \quad (26)$$

we arrive at a system of equations for UH waves

$$\left(\frac{\partial}{\partial t} + i\Lambda_{ny}\frac{\partial^2}{\partial y^2} + i\Lambda_{nz}\frac{\partial^2}{\partial z^2}\right) b_n = \left(\gamma_0 \exp\left(-\frac{y^2}{w^2} - \frac{z^2}{w^2}\right) - \gamma'_0(b_m)\right) b_n$$
$$\left(\frac{\partial}{\partial t} + i\Lambda_{my}\frac{\partial^2}{\partial y^2} - \upsilon_{mz}\frac{\partial}{\partial z}\right) b_m = \gamma'_0(b_n) b_m \quad (27)$$

where

$$\gamma_0 = \frac{\left|\kappa_p^2\right|^2 \left|A_0\right|^2}{4H^2 D_{n\omega}\left|D_{1q}\right|} \int_{-\infty}^{\infty} dx \int_{-\infty}^{x} d\xi \varphi_n(x)^* \varphi_n(\xi) \exp\left(i\int_{\xi}^{x}(q_{2x} - k_x)dx'\right),$$

$$\gamma'_0(b_{n,m}) = \frac{\left|\kappa_z^2\right|^2}{n_e T_e} \frac{\left|b_{n,m}\right|^2}{\omega_{\Sigma}^{n,m} D_{n\omega} D_{m\omega} \left|D_{1q}(x)\right| r_{de}^2} \times$$
$$\int_{-\infty}^{\infty} dx \varphi_n^*(x) \varphi_m^*(x) \int_{x}^{\infty} d\xi \varphi_n(\xi) \varphi_m(\xi) \exp\left(i\int_{\xi}^{x} q'_{2x} dx'\right)$$

and r_{de} is the Debye length. The parametric decay instability starts with the stage of exponential growth in which the amplitude of primary wave is still small and the secondary decay is yet not excited. In this case the system of equations (27) reduces to the following equation

$$\left(\frac{\partial}{\partial t} + i\Lambda_{ny}\frac{\partial^2}{\partial y^2} + i\Lambda_{nz}\frac{\partial^2}{\partial z^2}\right) b_n = \gamma_0 \exp\left(-\frac{y^2}{w^2} - \frac{z^2}{w^2}\right) b_n \quad (28)$$

Solving it we readily get the exponentially growing solution which is presented as

$$b_n(t,y,z) = b_T \exp\left(\nu_{p,s} t + i\delta\omega_{p,s} t\right) \psi_p(y) \psi_s(z), \quad (29)$$

$$\psi_p(\xi) = \exp\left(-\frac{\xi^2}{2\delta_{\xi}^2}\right) H_p\left(\frac{\xi}{\delta_{\xi}}\right), \ \xi = y, z \quad (30)$$

are the eigenfunctions, H_p is the Hermitian polynomials and $\delta_{\xi} = \Lambda_{\Sigma\xi}^{1/4} w^{1/2} \exp(-i\pi/8 - i\arg\gamma_0/4)/\sqrt[4]{\gamma_0}$ and

$$\begin{pmatrix} \nu_{p,s} \\ \delta\omega_{p,s} \end{pmatrix} = \begin{pmatrix} \mathrm{Re}(\gamma_0) \\ \mathrm{Im}(\gamma_0) \end{pmatrix} -$$
$$\begin{pmatrix} \cos\left(\arg(\gamma_0/2) - \pi/4\right) \\ \sin\left(\arg(\gamma_0/2) - \pi/4\right) \end{pmatrix} \sqrt{\left|\gamma_0\right|} \left((2p+1)\sqrt{\frac{\Lambda_{ny}}{w^2}} + (2s+1)\sqrt{\frac{\Lambda_{nz}}{w^2}}\right) \quad (31)$$

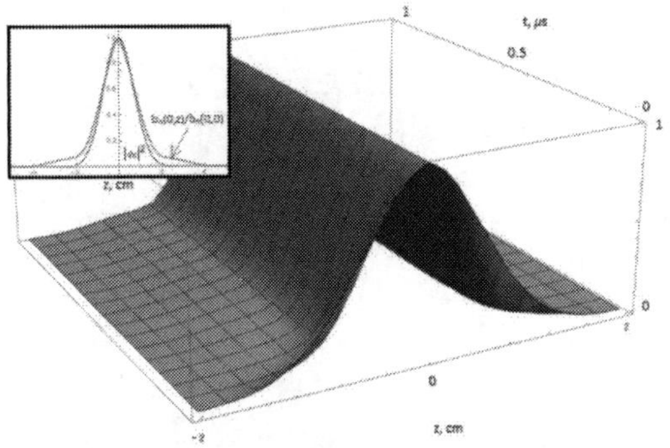

Fig. 3. The spatial distribution of UH wave $|b_n(0,z)|^2/|b_n(0,0)|^2$. Zoom gives the comparison of the function derived numerically with the analytical dependence (30).

The PDI power-threshold P_0^{th} corresponding to the imaginary part of the eigen frequency being equal to zero is determined by the balance between the diffractive loss of UH waves ($\Lambda_{ny,z}/w^2$) and the effectiveness of their nonlinear generation ($\mathrm{Re}(\gamma_0)$). It can be found out solving the following equation

$$\mathrm{Re}\left(\gamma_0\left(P_0^{th}\right)\right) = \cos\left(\arg\left(\gamma_0\left(P_0^{th}\right)/2\right) - \pi/4\right) \times$$
$$\sqrt{\left|\gamma_0\left(P_0^{th}\right)\right|} \left((2p+1)\sqrt{\frac{\Lambda_{ny}}{w^2}} + (2s+1)\sqrt{\frac{\Lambda_{nz}}{w^2}}\right) \quad (32)$$

Thus, the primary decay of the pump ordinary wave (17) leads to the nonlinear transfer of pump power to the localized UH wave (29). It enhances the UH fluctuations up from the thermal-noise-level

$$b_T = \sqrt{\frac{T_e}{\pi w^2}} \quad (33)$$

to such a level $|b_n|$ that the inequality

$$\gamma'_0(|b_n|) \geq \max\left(\frac{\Lambda_{my}}{\pi w^2}, \frac{\upsilon'_{mz}}{w}\right) \quad (34)$$

holds and the power threshold of secondary PDI appears to be overcome. Then, the UH wave b_n starts not only to grow due to the primary instability but at the same time to transfer a part of pump power to the secondary wave. When the secondary PDI power threshold is exceeded significantly, the temporal gain coefficient can be estimated as follows

$$\ln\left(<\left|b_m(t)\right|^2>_{\mathrm{pdi}}/b_T^2\right) \approx \gamma'_0\left(<\left|b_n(t)\right|^2>_{\mathrm{pdi}}\right)/\gamma_0 \quad (35)$$

where the averaging procedure over the spot in which the decay occurs is defined as in equation (11). If the PDI saturation results in a stationary regime, the saturation levels of daughter waves can be estimated assuming the nonlinear pumping in the primary decay is balanced by the energy losses due to the secondary decay $\gamma_0 = \gamma'_0\left(\left|b_m\right|_{\mathrm{sat}}\right)$ and the pumping in the secondary decay $\gamma_0'\left(|b_n|_{\text{sat}}\right)$ is compensated by the effect of the secondary wave energy loss from the spot of a pump beam. Finally, we get

$$|b_m|^2_{\text{sat}} = \frac{\gamma_0}{\omega_n}\varepsilon_{\text{T}}\,, \tag{36}$$

$$|b_n|^2_{\text{sat}} = \max\left(\frac{\Lambda_{my}}{\pi w^2\omega_m}, \frac{\upsilon_{mz}}{w\omega_m}\right)\varepsilon_{\text{T}}\,, \tag{37}$$

$$\varepsilon_{\text{T}} = \frac{n_e T_e}{\left|\kappa_s^2\right|^2}\frac{\omega_n D_{n\omega}\omega_m D_{m\omega}\left|D'_{2q}\right| r_{de}^2}{\int_{-\infty}^{\infty} dx \int_x^{\infty} d\xi \varphi_n^*(x)\varphi_m^*(x)\varphi_n(\xi)\varphi_m(\xi)\exp\left(i\int_\xi^x q_{2x}(\omega_2', x')dx'\right)}.$$

Then, we can determine the fraction of anomalous power absorbed by the daughter decay waves as follows

$$\Delta P_{\text{E}} \approx \frac{\Lambda_{nz}}{\pi w^2}|b_n|^2_{\text{sat}} + \frac{\upsilon_{mz}}{\sqrt{\pi}w}|b_m|^2_{\text{sat}} \tag{38}$$

In (38) we have assumed that at the dominant losses of primary and secondary waves from the pump spot are along the magnetic field. It is worth noting that the pump power absorbed by secondary plasmons (the second term in (38)) should be proportional to the pump power, whereas the power fraction absorbed by primary plasmons is not dependent on it. Therefore well above the PDI threshold $P_0 >> P_0^{th}$ the second term in (38) dominates and the fraction absorbed anomalously is proportional to the pump power $\Delta P_{\text{E}} \propto P_0$.

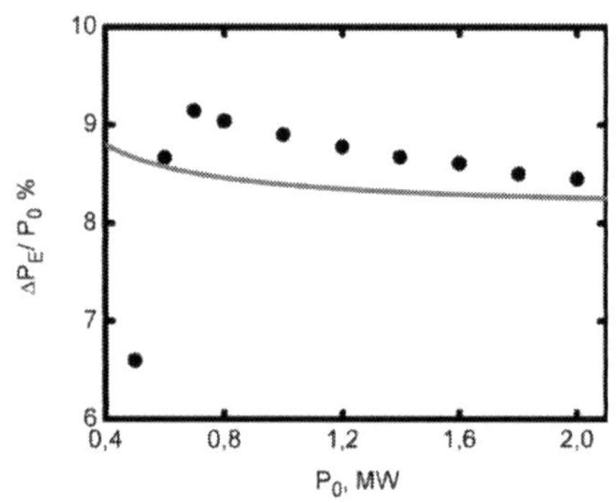

Fig. 4. The anomalous absorption rate derived numerically (scattered symbols) and predicted by the asymptotic equation (38) valid at $P_0 >> P_0^{th}$ are shown; $P_0^{\text{th}} = 125$ kW.

Then, we solve equations (27) numerically in a 2D box $2y_B \times 2z_B$. We impose the periodic boundary conditions improving stability of the numerical scheme. The initial conditions for the UH wave amplitudes are assumed to be determined by the thermal-noise level $|b_{n,m}(0)| = b_{\text{T}}$ (see (33)), whereas their spatial distribution is supposed to be homogeneous. Solving for the incident pump power $P_0 = 600$ kW and $w = 1$ cm beam we get the dependences which are shown in figure 2. In it one can see the temporal evolution of the averaged density of the primary (solid curve) and secondary (dashed curve) UH plasmon energies given in the logarithmic scale. In the early stage of primary decay going on approximately up to 6 μs the growth of the primary UH wave energy is perfectly described by the analytical dependence $2\nu_{0,0}t$ shown by the dash-dotted line with $\nu_{0,0}$ having been determined in (31). For the parameters under consideration the growth-rate is $2\nu_{0,0} = 5.9$ MHz and the power threshold of primary decay is $P_0^{\text{th}} = 125$ kW. In this regime the primary daughter wave retains its spatial distribution invariable (see figure 3). As it is shown in a zoom in figure 3 it is well described by the analytical dependence (30). When the secondary PDI power threshold is exceeded the energy of secondary wave (dashed curve) starts to grow. The growth is reasonably described by the analytical dependence (35). Then, we can see the temporal relaxation of the strong nonlinear phenomenon and the transition of the instability into the saturation regime. The saturation levels of the daughter waves are quite well described by the analytical estimations (36) and (37), at least in the logarithmic scale. Then, we can calculate the daughter waves total energy

$$W_d = \int_{-y_B}^{y_B} dy \int_{-z_B}^{z_B} dz\left(|b_n(y,z)|^2 + |b_m(y,z)|^2\right) \tag{39}$$

Finding the temporal derivative of the dependence (39) in the saturation regime, which appears to be linear, allows us to estimate the power fraction gained by the daughter UH waves at the level $\Delta P_{\text{E}} \approx 52$kW. Subsequently, this power fraction is deposited into the electron component far from the pump wave absorption region predicted by linear theory. We can also estimate the pump power fraction absorbed by the LH and IB waves $\Delta P_{\text{I}} = 0.9$ kW. This anomalous absorption can lead to the generation of a population of the suprathermal ions. In figure 4 the anomalous absorption rate derived numerically (scattered symbols) and predicted by the asymptotic equation (38) valid at $P_0 >> P_0^{th}$, $P_0^{\text{th}} = 125$kW are depicted. As it is seen, the model developed in this paper predicts at the high pump power anomalous absorption rate of 8.4 %. This value is lower than in the case of the X-mode decay producing UH plasmons analysed in [21] due to the specific direction of the pump electric field (almost parallel to the magnetic field). The low value of the anomalous absorption rate provides an argument in favour of omitting the pump depletion in our analysis.

3 Conclusions

We have described the theoretical model, which interprets the anomalous phenomena experimentally observed in the X2-mode and O1-mode ECRH experiments, such as the generation of backscattering signals and the fast ion generation, as a consequence of the low-threshold PDI of a pump wave leading to generation of localized UH daughter waves in presence of a nonmonotonic (hollow) density profile. The natural consequence of this model is also the absolute instability excitation in presence of a finite-width pump beam and its saturation at a substantially high level due to the secondary decay phenomena and the pump depletion. The developed model also results in the anomalous pump wave absorption at a level of 10% or higher.

We have investigated the X2-mode wave decay into two 3D localized UH waves and analysed the transition of primary instability into the saturation regime due to both the pump wave depletion and the secondary decays of daughter UH wave, leading to the excitation of secondary IB/LH and UH waves. The threshold of this spontaneous parametric frequency down-conversion is shown to be easily overcome for the secondary UH waves trapped at the local maximum of a density profile. The model predicts substantial RF power transfer to the UH waves (up to 25%). Subsequently, the UH waves power is deposited into the electron component far from the pump wave electron cyclotron resonance layer. This makes the power deposition profile much wider than it is predicted for the pump wave itself. The anomalous absorption of the pump power by the IB or LH waves leads to the generation of a population of the suprathermal ions, which can explain the experimental observations of the ion acceleration in the X2-mode ECRH experiments at TCV and TJ-II [7-10].

Nonlinear coupling of different UH waves can lead to the generation of microwave signals at frequencies smaller than the pump frequency. Thus, the nonlinear mechanism allows explaining the anomalous backscattering effect observed at TEXTOR [2] in detail. In particular, the frequency spectrum structure and the spectral power density of the backscattering signal predicted by this theoretical model appear to be in a reasonable agreement with those measured by the ECE and CTS techniques [2,3].

We have also investigated the pump O-mode wave anomalous absorption due to the parametric excitation of trapped UH waves. The instability power threshold ($P_0^{th} = 125\text{ kW}$) can be exceeded by the MW microwave beams utilized usually for auxiliary heating in the O1-mode ECRH experiments. However, its growth-rate is much smaller than the growth-rate of X-mode PDI. This is due to the weak coupling of the ordinary wave to the electrostatic waves. Nevertheless, the derived growth-rate of O-mode PDI is large enough allowing this nonlinear phenomenon to play an important role in the ECRH power balance. In order to estimate the fraction of pump power absorbed anomalously we have analyzed the saturation regime of primary instabilty. We assumed the UH wave secondary decay into the localized UH waves and IB waves to be a presumably most important nonlinear mechanism responsible for the transition to saturation regime. We have estimated the power fraction gained by the daughter UH waves in the O1-mode ECRH experiments at FTU, in which the 600 kW pump beams are utilized. Numerical solution gives the power fraction $\Delta P_E \approx 52\text{kW}$ absorbed via the UH daughter waves. This power fraction is deposited into the electron component far from the pump wave absorption region predicted by linear theory.

It is worth noting that the powerful ECRH utilizing ordinary waves is planned for application in ITER. It should allow the neoclassical tearing mode control leading to reduction of its influence on the confinement in this thermonuclear reactor. According to the analyses performed in this paper, if the nonmonotonic density profile will arise in ITER at the edge of the discharge, the low-threshold parametric decay of the pump wave to the 3D localized UH wave could be possible playing a role in the ECRH power balance. This stimulates further analysis of the O1-mode parametric decay under the conditions which will be typical for ITER.

The financial support of the Ioffe Institute is acknowledged.

1. M. Porkolab, B. I. Cohen, Nucl. Fusion **28**, 239 (1988)
2. E. Westerhof, S. Nielsen, J. W. Oosterbeek, *et al.*, Phys. Rev. Lett. **103**, 125001 (2009)
3. S. K. Nielsen, M. Salewski, E. Westerhof, et al., Plasma Phys. Control.Fusion **55**, 115003 (2013)
4. S. Kubo, M. Nishiura, K. Tanaka et al., Rev. Sci. Instrum. **81**, 10D535 (2010)
5. W. A. Bongers, W. Kasparek, N. Doelman et al., EPJ Web Conf. **32**, 03006 (2012)
6. A. Bruschi, E. Alessi, W. Bin et al., Nucl. Fusion **57**, 076004 (2017)
7. A. N. Karpushov, S. Coda, and B. P. Duval, Proc.of 33rd EPS Conference **301**, P–1.152 (2006)
8. S. Coda for the TCV Team, Nucl. Fusion **55**, 104004 (2015)
9. D. Rapisarda, B. Zurro, V. Tribaldos et al., Plasma Phys. Controlled Fusion **49**, 309 (2007)
10. B. Zurro, A. Baciero, V. Tribaldos et al., Nucl. Fusion **53**, 083017 (2013)
11. M. Martínez, B. Zurro, A. Baciero et al., Plasma Phys. Control. Fusion **60**, 025024 (2018)
12. V. F. Andreev, A. A. Borschegovskij et al., Plasma Phys. Control. Fusion **58**, 055008 (2016)
13. N. Tamura, S. Inagaki, K. Ida et al., Proc. 3rd EFDA Transport Topical Group Meeting O5-03 (2010)
14. A. Yu. Popov, E. Z. Gusakov, Plasma Phys. Control. Fusion **57**, 025022 (2015)
15. A. Yu. Popov, E. Z. Gusakov, European Phys. Lett. **116**, 45002 (2016)
16. E. Z. Gusakov, A. Yu. Popov, Plasma Phys. Control. Fusion **59**, 025005 (2017)
17. E. Z. Gusakov, A. Yu. Popov, Physics of Plasmas **23**, 082503 (2016)
18. L. V. Simonchik, A. B. Altukhov, *et al.*, EPJ Web of Conferences **149**, 03013 (2017)
19. E. Z. Gusakov, A. Yu. Popov *et al.*, Plasma Phys. and Control. Fusion **59**, 075002 (2017)
20. E. Z. Gusakov, A. Yu. Popov, Physics of Plasmas **25**, 012101 (2018)
21. E. Z. Gusakov, A. Yu. Popov, Plasma Phys. Control. Fusion **60**, 025001 (2018)
22. D. G. Swanson, Plasma Waves Academic Press, Boston (1989)
23. E. Z. Gusakov, V. I. Fedorov, Sov. J. Plasma Phys. **5**, 463 (1979)
24. E. Z. Gusakov, A. Yu. Popov, Journal of Experimental and Theoretical Physics **127**, 155–166 (2018)

3D Full-Wave modelling and EC mode conversion in realistic plasmas

Pavel Aleynikov[1,*] and *Nikolai B.* Marushchenko[1]

[1]Max-Planck-Institut für Plasmaphysik, Greifswald, Germany

Abstract. The wave physics of O-X conversion in overdense W7-X plasma is discussed. For this study, a new 3D, cold plasma full-wave code has been developed. The code takes advantage of massive parallel computations with Graphics Processing Units (GPU), which allows for up to 100 times faster calculations than on a single-CPU. A 3D calculation of the O-X conversion is demonstrated. We discuss limitations of the mode conversion scenario within the capabilities of the existing ECRH system in W7-X, and demonstrate an optimised conversion scenario in which the launching antenna location is altered. The conversion efficiency of the optimised scenario is predicted to be >85%.

1 Introduction

Electron Cyclotron Resonance Heating (ECRH) is the main plasma heating mechanism in Wendelstein 7-X (W7-X) Stellarator. It is provided by 10 gyrotrons at 140 GHz (corresponding to the second harmonic cyclotron resonance at 2.5 T) with the power of 1 MW each. X- and the O-modes were successfully used in a wide range of operation scenarios: X-mode for low and moderate densities (up to the cutoff at $1.2 \cdot 10^{20}\mathrm{m}^{-3}$), and O2-mode for higher densities (up to $2 \cdot 10^{20}\mathrm{m}^{-3}$). Possible operation at yet higher densities would involve double mode-conversion from O- to slow X- and to Bernstein-mode, i.e. an OXB-scenario. The physics of O-X conversion is outside of applicability of the routinely used geometrical optics approximation (WKB-theory) and should be considered within a full-wave approach.

This work reports on the development of a new 3D cold plasma full-wave code, named CUWA. The code utilizes the Finite Difference Time Domain (FDTD) technique [1, 2] and has an interface with the ray-tracing code TRAVIS [3]. The computation domain is "minimized" around the WKB-trajectory obtained from the ray-tracing code by means of the Convolutional Perfectly Matched Layer (CPML) technique [4]; the background magnetic field is recovered from pre-computed 3D equilibrium data. The code takes advantage of massive parallel computations with Graphics Processing Units (GPUs), which allows for up to ×100 acceleration over a single-CPU.

2 Numerical model

FDTD (also known as Yee's method) is a standard technique for full-wave simulations in various branches of physics [1, 2]. In fusion plasmas, it has been successfully applied to study wave propagation, O-X conversion [5], reflectometry [6] and the scattering of EC waves on plasma turbulence [7]. In this work we mostly follow the FDTD method as given in Chapter 11.3 of [2], with two important differences: first, we modify the current discretisation scheme and, second, we amend the differential operators of Maxwells equations in the boundary layers in accordance with the CPML technique [4] (see below).

The system of equations been solved is Maxwell's equations with addition of a "cold plasma" response current equation (**J**) given by the electron law of motion:

$$\begin{cases} \frac{\partial}{\partial t}\mathbf{B} = -\nabla \times \mathbf{E}, \\ \frac{\partial}{\partial t}\mathbf{E} = c^2\nabla \times \mathbf{B} - \mathbf{J}/\varepsilon_0, \\ \frac{d}{dt}\mathbf{J} + \nu\mathbf{J} = \varepsilon_0\omega_p^2\mathbf{E} - \mathbf{J} \times \omega_c, \end{cases} \quad (1)$$

where $\omega_p^2 = n_e e^2/\varepsilon_0 m_e$ is the plasma frequency, ν is electron collisional frequency and $\omega_c = |e|\mathbf{B}_0/m_e$ is the cyclotron frequency corresponding to the background magnetic field, $\mathbf{B}_0$.

In the FDTD method the field components are discretised on staggered grids in space and time. Note that is it common to discretize the plasma-response curren, **J**, in such a way that its J_x, J_y and J_z components are co-located in space [2, 5]. This facilitates the current update calculation. However, we find the stability of this scheme to be unsatisfactory, in particular when applied to the CMPL region. Therefore we have implemented a slightly more computationally demanding but more stable scheme where **J** is discretized in the same way as the wave electric field, **E**. This implies that the the current update equation will involve interpolation of its components.

3 Perfectly matched layer

Minimization of the computation domain plays a critical role in efficient 3D computations. A physical absorber layer with a non-zero collisionality ($\nu > 0$) is often used in plasma physics computations to truncate the computation

*e-mail: pavel.aleynikov@ipp.mpg.de

domain [5, 7]. In such a case one would need to amend the computation domain with layers of "many-wavelengths" thickness at each boundary ensuring smallness of reflected signals. This may become problematic in 3D calculations when the domain of interest itself has the size of a few wavelengths. Amending such a domain with six boundary layers may significantly reduce the overall efficiency.

A so-called CPML boundary [4] is instead implemented in our full-wave code to truncate the computation domain. An efficient CPML requires only a few extra Yee cells in each direction - i.e. a fraction of a single "wavelength". CPML also requires two auxiliary (virtual) 3D fields (ψ) for each component of E and B, i.e. 12 auxiliary fields in total. However, these auxiliary fields are defined only within the narrow CPML layers, resulting in a negligible overhead in the practical setup.

The conceptual idea behind CPML [4] can be summarised in the following way (further details and a rather complete overview of various PMLs can be found in the book [8]). CPML is an efficient numerical implementation of Complex Frequency Shifted PML [9], in which the differential operators in Eqs. (1) are replaced with the convolution:

$$\frac{\partial}{\partial u} \rightarrow \mathcal{L}^{-1}\left\{\frac{1}{\kappa_u + \frac{\sigma_u}{\alpha_u + i\omega\varepsilon_0}}\right\} * \frac{\partial}{\partial u}, (u = x, y, z), \quad (2)$$

where $\mathcal{L}^{-1}\{F(\omega)\}$ denotes inverse Laplace transform and $(\kappa_u, \alpha_u, \sigma_u)$ are free parameters accounting for coordinate stretching in real and complex plains and an "effective conductivity". PML given by Eq. (2) introduces no reflections because it can be viewed as a vacuum with complex-stretched coordinates. In other words, with Eq. (2) the solution of Eq. (1) is *analytically* continued into the complex plain with a subsequent mapping of the complex coordinates back to real space but with a "complex material".

In order to demonstrate CPML performance for the FDTD realisation described above we design a benchmark which represents a general anisotropic scenario. In this scenario, a divergent Gaussian beam propagating through the plasma escapes from the computation domain through a strongly nonuniform boundary. The entire computation domain is surrounded by the narrow CPML layer. This is shown in Fig. 1, the wave is launched vertically from the origin $(0, 0)$. The plasma density ranges from 0 to approximately 80% of the O-mode cut-off with a gradient directed diagonally (bottom right to top left). It can be seen that at the top PML layer the wave-front incident angle varies from almost normal to quite oblique, ensuring extended benchmarking. The resulting reflection error is estimated as a maximum value of the electric field amplitude recorded in the bottom-right quarter of the domain. Note that the maximum relative error reported in the original work [4] is $\approx 5 \cdot 10^{-4}$. Similar reflection is observed in our benchmark case (a wave with the cyan contours $[10^{-4}; 10^{-3}]$). Note the presence of another wave going out through the top right corner (contours $[10^{-5}; 10^{-4}]$). This spurious wave is associated with an inaccuracy of the Gaussian beam setup at the bottom boundary. Frequently, CPML outperforms the initial condition accuracy (for instance in a typical case where PML lies in vacuum) then

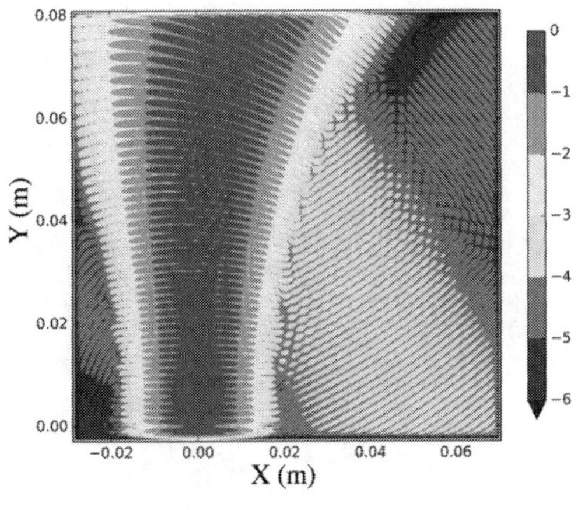

Figure 1. A snapshot of the decimal logarithm of the wave E-field magnitude for a CPML benchmark case.

the latter becomes the dominant noise source (i.e. introduces $\sim 10^{-4}$ error in field amplitude). Note that the relative errors are squared when the intensity or mode power is of interest. The above example is calculated with the resolution of 12 Yee cells per vacuum wavelength and a CPML size of 12 cells.

4 Mode conversion examples

One of the applications of our code is the calculation of the O- to X-mode conversion efficiency in 3D plasmas.

Figure 2 shows an example of a full-wave 2D calculation of the O-X conversion process in a realistic W7-X plasma. The O-mode, launched from vacuum (bottom right corner of the left plot), propagates into the plasma and is partly reflected off the corresponding cut-off layer, where a part of the O-mode is converted into a slow X-mode, which propagates toward the upper-hybrid resonance. Eventually, this X-mode will be converted to a Bernstein wave. The reflected part of the O-mode has a "hole" (due to the non-uniform O-X conversion). The perpendicular component of the refractive index is computed with the use of a windowed Fourier transform and shown in Fig. 2 (right). The incoming O-mode, the reflected O-mode, and the slow-X mode (which settles near the Upper Hybrid resonance layer) can be clearly distinguished in this diagram, which appears to be in a good agreement with expectations from WKB theory (i.e. dispersion curves), validating FDTD solution. Note that the slow-X mode is eventually dissipated numerically without leaving the computation domain. This dissipation is of purely numerical nature. It happens when the wavelength of the mode becomes shorter than 10 computational cells.

Figure 3 shows an example of a 3D calculation similar to the above scenario (the wave is launched from the bottom of the domain in y direction, the slow X- mode is

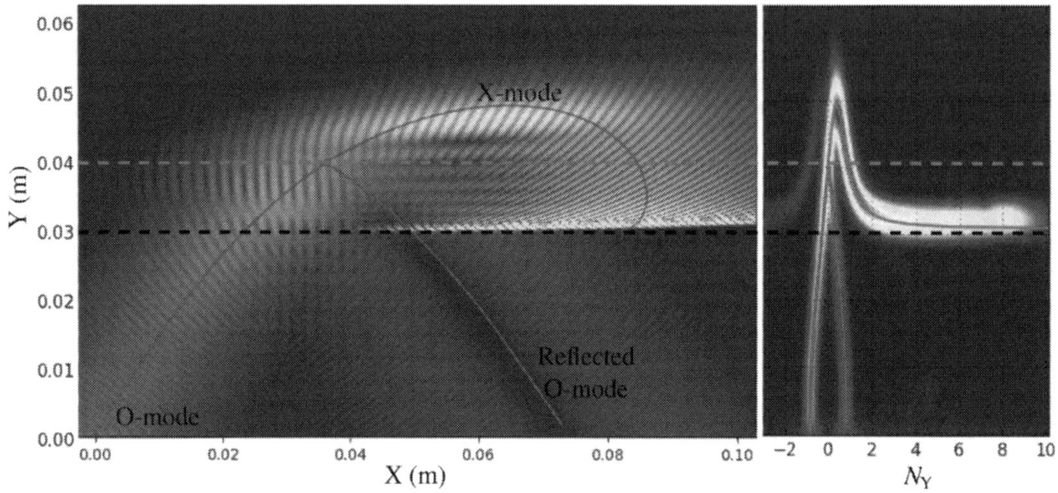

Figure 2. An example of an O-X mode conversion calculation. Left plot: snapshot of the wave electric field (color-coded), ray tracing (solid red). The antenna is located in the bottom left corner. The O-mode cutoff and Upper Hybrid resonance layers are indicated with red and the black dashed lines, respectively. Right plot: windowed Fourier transform of the wave field, i.e wave refractive index (color-coded) and the refractive index predicted by ray-tracing (solid).

contained within the computation domain, the reflected O-mode power distribution is shown on the plane $x = const$). The decimal logarithm of the time-averaged E^2 is colour-coded. The sequence of dark-colored points represents the corresponding WKB trajectory for the reference ray. It is evident from the peculiar cross-section of the reflected wave that the 3D geometry needs to be appropriately accounted for.

To estimate the power conversion efficiency, we compute surface integrals of the time-averaged Poynting vector (i.e. the intensity) over every boundary of the computation domain. In the above case, the OX conversion efficiency can be calculated as the difference between the intensities of the input (S_{in}) and reflected (S_{ref}) waves. An estimation of the OX conversion efficiency is made as follows for W7-X-relevant ECRH beam parameters and plasma geometry. A Gaussian beam approximation is used as an initial condition. A beam of 2 cm radius and a phase front curvature of 1 m approximates the ECRH beam at the plasma entrance. The density profile is chosen arbitrarily (with a normalized gradient of $\approx$ 20m^{-1} near the cutoff). The antenna launcher location was fixed, corresponding to the upper port "A, C1" in W7-X nomenclature. Then, the launcher azimuth and altitude angles were varied to find direction for highest OX conversion. We find that for the given conditions (equilibrium, density profile, launcher position and beam focusing) the maximum conversion $(S_{\text{in}} - S_{\text{ref}})/S_{\text{in}}$ reaches 48% (the middle tile in Fig. 3 (right)).

5 Optimised conversion scenario in W7-X

It appears that ~50% is a typical maximimum conversion fraction for realistic W7-X beam parameters (focusing and width were optimized) and launcher positions. Refs. [10–12] suggest that conversion can be improved by a careful design of the beam wavefront. In this report we pursue an alternative optimisation strategy.

We take advantage of the fact that stellarator equilibrium "offers" a wide variety of magnetic field curvatures on a cutoff surface (flux surface), and we search for a location which would minimise the thickness of the evanescent layer, weighted over the beam radial profile. The thickness of the evanescent layer at the O-mode cutoff surface is estimated using the results of analytical theory [13], where the thickness is taken to be the distance (in wave-vector space) between the WKB solutions of the two modes. This estimate is known to agree very well with calculations in a flat 2D case [5].

The result of this calculation is demonstrated in Fig. 4 (left). The function representing the Gaussian-weighted gradient of the thickness of the evanescent layer is plotted on the cutoff surface, which is parametrized by the poloidal and the toroidal angles. A few minima of this function are marked with red dots. A different optimal value of parallel refractive index $N_{\parallel}^{\text{opt}}$ corresponds to each minimum due to the variation of the magnetic field strength on the surface.

It is easy to understand the "meaning" of these special points. We recall that the efficient conversion occurs when the "turning" points for the wave-packets of the two modes (O- and slow X-) are close to each other. Formally, the conversion condition is satisfied at the intersection of two surfaces: the O-mode cutoff surface and the X-mode turning surface parameterised by the plasma frequency (ω_p), the cyclotron frequency (ω_c), and the wave parallel refractive index $N_{\parallel}$. For a fixed equilibrium the in-

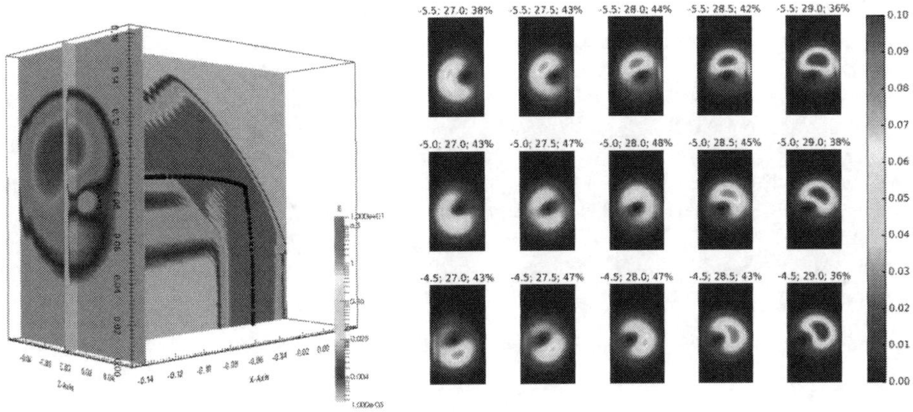

Figure 3. 3D calculation of the scenario similar to the one described in Fig. 2 (left). Reflected O-mode shape for various aiming angles of the launched beam (right).

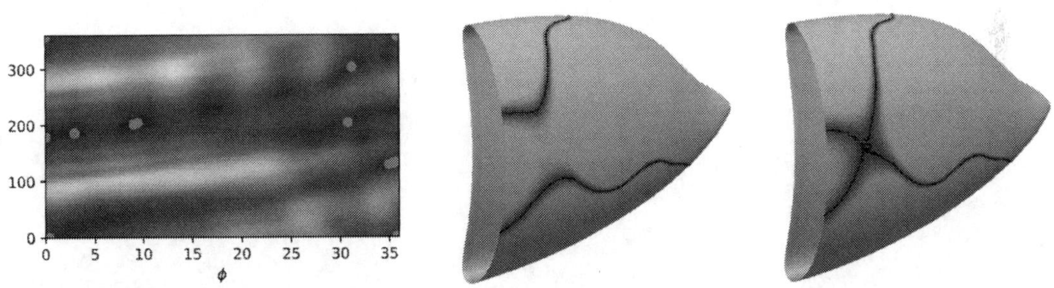

Figure 4. Left: Location of the optimal conversion regions (red dots) on the cutoff surface (ϕ and θ are toroidal and poloidal angles). The function representing the weighted gradient of the thickness of the evanescent layer is colorcoded. Middle: Conversion region for $N_{\parallel}^{opt} \approx 0.564$. Right: Optimised efficient conversion region for $N_{\parallel}^{opt} \approx 0.5632$ ($\phi \approx 3.5$ and $\theta \approx 190$ on the left plot).

tersection curve (3D-shaped, in general) will vary with $N_{\parallel}$. Fig. 4 (middle) shows this intersection for $N_{\parallel}^{opt} \approx 0.564$, where the surfaces cross each other obliquely and thus the evanescent layer is narrow only close to the intersection (if the beam has a constant $N_{\parallel}$ structure). When $N_{\parallel}$ corresponds to one of the optimised locations, the intersection has a more complex structure (a saddle point in an example of Fig. 4 (right)), ensuring minimisation of the evanescent layer weighted over a finite Gaussian beam.

The location of the conversion region in plasmas (and thus $N_{\parallel}^{opt}$) uniquely defines the O-mode ray which arrives to this point within the WKB approximation. Therefore the beam launching location and angles can be computed with a backward ray-tracing calculation (staring from the conversion region). When the initial conditions for the beam outside of plasma are known, the full-wave calculations are done to predict the conversion efficiency. The maximised conversion efficiency predicted with our full-wave code for a case of Fig. 4 (right) reaches a remarkable 85%.

Note that such an algorithm for an identification of an optimised conversion scenario implies that the wave launcher location can be chosen freely. This is generally not the case in W7-X. Therefore our future plans include a study in which various W7-X equilibria are analysed in order to identify those where the optimal conversion regions are accessible within the capabilities of the existing ECRH system.

References

[1] A. Taflove, S.C. Hagness, *Computational Electrodynamics: The Finite-Difference Time-Domain Method* (ARTECH HOUSE INC, 2005), ISBN 1580538320,

[2] U.S. Inan, R. A.Marshall, *Numerical Electromagnetics: The FDTD Method* (Cambridge University Press; 1 edition, 2011), ISBN 052119069X,

[3] N. Marushchenko, Y. Turkin, H. Maassberg, Computer Physics Communications **185**, 165 (2014)

[4] J.A. Roden, S.D. Gedney, Microwave and optical technology letters **27**, 334 (2000)

[5] A. Köhn, Á. Cappa, E. Holzhauer, F. Castejón, Á. Fernández, U. Stroth, Plasma Physics and Controlled Fusion **50**, 085018 (2008)

[6] J.H. Irby, S. Horne, I.H. Hutchinson, P.C. Stek, Plasma Physics and Controlled Fusion **35**, 601 (1993)

[7] T.R.N. Williams, A. Köhn, M.R.O. Brien, R.G.L. Vann, Plasma Physics and Controlled Fusion **56**, 075010 (2014)

[8] J.P. Berenger, *Perfectly Matched Layer (PML) for Computational Electromagnetics* (Morgan and Claypool Publishers, 2007), ISBN 1598290827,

[9] M. Kuzuoglu, R. Mittra, IEEE Microwave and Guided wave letters **6**, 447 (1996)

[10] A.G. Shalashov, E.D. Gospodchikov, Plasma Physics and Controlled Fusion **52**, 115001 (2010)

[11] E.D. Gospodchikov, T.A. Khusainov, A.G. Shalashov, Plasma Physics and Controlled Fusion **54**, 045009 (2012)

[12] H. Weitzner, Physics of Plasmas **24**, 022502 (2017)

[13] E. Mjølhus, Journal of Plasma Physics **31**, 7 (1984)

 https://doi.org/10.1051/epjconf/201920301004

Top Launch for Higher Off-axis Electron Cyclotron Current Drive Efficiency

Xi Chen[1,*], *Ron* Prater[1], *Craig* Petty[1], *John* Lohr[1], *David* Su[1], *Lang* Lao[1], and *Vincent* Chan[2]

[1]General Atomics, P.O. Box 85608, San Diego, CA 92186-5608, USA
[2]University of Science and Technology of China, Hefei, Anhui 230026, China

Abstract. Efficient off-axis current drive is crucial for economic, steady-state tokamak fusion power plants. "Top Launch" electron cyclotron current drive (ECCD) is one promising method for driving strong off-axis current to achieve the desired broad current profile for the Advanced Tokamak regime. New simulations show that by launching the electron cyclotron (EC) waves downwards (or upwards) nearly parallel to the resonance plane with a large toroidal steering, higher current drive efficiency can be obtained at large radii owing to the large Doppler shift, wave-particle interactions on HFS of the plasma, and the long absorption path. Implementations on CFETR and DIII-D are simulated identifying clear benefits and optimal configurations. The design of a prototype test being installed on DIII-D is presented, which will experimentally validate whether top launch ECCD can be an improved efficiency current drive technique for future fusion reactors.

1 Introduction

Studies of many tokamak reactors have shown that off-axis current drive is a requirement for a steady-state device in the Advanced Tokamak (AT) regime [1, 2, 3, 4]. The off-axis current drive needs to be efficient because, if the efficiency is too low, 1) the required current drive to sustain the plasma configuration, i.e., the elevated q_{min} in the 'high-q_{min}' scenario might not be attainable due to limited source power; or 2) it will need too much power, preventing a net electricity goal or increasing cost of electricity since fusion gain is linearly proportional to current drive efficiency $Q=P_{fus}/P_{aux}\sim\zeta_{CD}$.

A suitable method of off-axis current drive with the efficiency and location required for economical AT reactors has yet to be demonstrated. While neutral beam current drive (NBCD) can be an efficient non-inductive current drive scheme, neutral beams are large objects that are challenging to make mechanically consistent with other systems in the tokamak reactor environment (e.g., the vacuum vessel, breeding blanket, neutron shielding, etc.). The large tangency and larger port size needed to achieve off-axis current drive increases the challenges. In addition, NBCD is efficient only at high voltage (i.e., 1MeV) and long pulse high voltage beam development is very challenging [5, 6]. Radio frequency (RF) wave injection is another common non-inductive current drive technique. Among these techniques, lower hybrid current drive usually has the highest efficiency. However, in fusion plasma conditions, the absorption of lower hybrid waves is so strong that it usually drives current only in the pedestal region (ρ>0.85). Fast wave has moderate efficiency but is core localized, plus there are concerns about the presence and survival of the antenna in the reactor environment. Current drive using electron cyclotron waves [7] does not require the antenna structures be placed near the plasma since the waves propagate in vacuum. ECCD launched from the low field side (LFS) can drive current at $\rho\sim0.5$-0.7 in reactor-like plasmas but it has low efficiency due to electron trapping in the magnetic well. Increasing the applied wave frequency can improve the current drive efficiency by moving the primary resonance layer to the high field side (HFS) where the impact of trapped particles is less, but parasitic second (or third) harmonic absorption can become problematic.

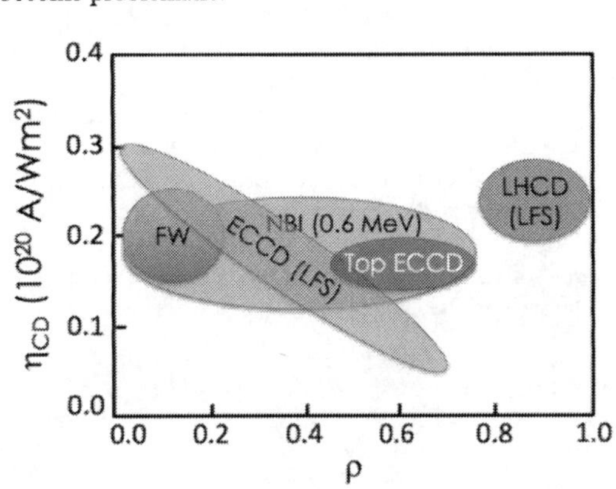

Figure 1: Cartoon illustration of current drive efficiency of various current drive methods: NBI at 0.6MeV (0.6MeV used here as an average of beam voltages used in various future device studies; LHCD (conventional LFS launch); Fast wave; ECCD (conventional LFS launch) and Top launch ECCD

* Corresponding author: chenxi@fusion.gat.com

Several new off-axis current drive techniques have been proposed, such as 'Helicon' or 'Whistler' waves (ultra-high-harmonic fast wave), which is predicted to have excellent penetration for high density and high temperature plasmas, as well as high field side (HFS)-launch lower Hybrid (slow wave) current drive that is predicted to dramatically reduce plasma-wall (+antenna) interaction issues, improve wave coupling and penetration, and result in a current profile better suited for steady-state scenarios than conventional outside-launch lower hybrid. More details on these two techniques are given in Ref. [8].

The development and qualification of physics and technologies that deliver off-axis current drive (CD) with the profile and high efficiency under reactor-like conditions needed to achieve high stability and high overall energy gain will greatly enhance steady-state fusion power plants. This work presents the modeling investigation and planned experimental validation of a non-standard ECCD technique – top launch ECCD. Comparisons of top launch and conventional outside launch ECCD modelled for DEMO [9] and FNSF-AT [10] suggest that the top launch approach can substantially increase the interactions of EC waves with higher energy electrons and thus the ECCD efficiency at mid-radius. We consistently find 35% or more improvement in ECCD efficiency for CFETR [11] baseline scenario and even higher increases for some cases on the DIII-D tokamak. The concept of top launch ECCD scheme is described in section 2.1. Its application for reactors is illustrated in section 2.2 using CFETR as an example and its experimental test planned on DIII-D tokamak is presented in section 2.3 followed with summary in Section 3.

2 Top launch ECCD

Electron cyclotron current drive (ECCD) is based on production of anisotropic electron resistivity resulting from selective heating of electrons traveling in one toroidal direction [12]. These accelerated electrons undergo less collisions and contribute more to the toroidal current than their unheated counterparts traveling in the opposite direction.

Top launch ECCD uses electron cyclotron waves injected from top (or bottom) of the vessel at a major radius less than the plasma axis, in a nearly vertical plane with a large toroidal steering, to achieve higher current drive efficiency. The launch geometry allows more selective heating so that the wave energy is preferentially damped on high energy electrons.

2.1 Why does top launch give higher ECCD efficiency than conventional outside launch?

To maximize the ECCD efficiency, one would want to 1) have the EC waves interact with only high energy electrons since they are less collisional; 2) ensure enough interactions such that the wave energy is fully absorbed since there are fewer high energy electrons. The top launch ECCD approach achieves this through the combination of the launch location and steering geometry which allows the EC waves to stay close, but not too close, to the cold resonance layer, and stay there for longer time. The wave energy is almost all damped on these high energy electrons.

In the top launch concept, the EC wave is launched from the top or bottom of the tokamak on the high field side (HFS) of the magnetic field axis and nearly parallel to the primary cold resonance layer, whereas in the conventional ECCD concept the wave is injected from the low field side (LFS), on or off the outer midplane and towards the cold resonance. Another unique feature of the top launch concept is that the EC wave is steered with a much larger toroidal angle than for LFS launch. Some features of the top launch ECCD scheme that differ from the conventional ECCD scheme are shown in Fig. 2, where the wave trajectories of these two schemes are compared in the poloidal and toroidal planes. In both examples, the ECCD deposition profile peaks at ρ~0.55 and the primary resonance is fundamental O-mode.

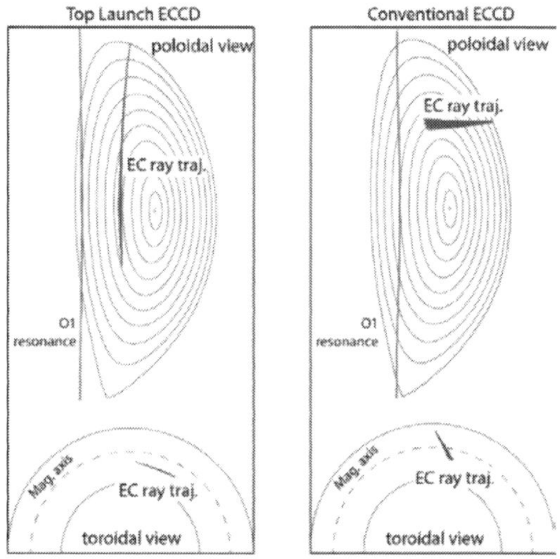

Figure 2: Typical ray trajectory comparison between top launch ECCD and conventional ECCD in poloidal and toroidal views. (from CFETR study described in Section 2.2)

The physics features that lead to this increase of ECCD efficiency using top launch ECCD concept are as follows:

- Stronger toroidal steering forces a larger Doppler shift, which benefits current drive.

Electron cyclotron waves can be absorbed by electrons with cyclotron frequency satisfying:

$$\omega = \frac{l\Omega_e}{\gamma} + k_{\parallel} v_{\parallel} \quad (1)$$

where ω is the frequency of applied EC wave, Ω_e is the electron cyclotron frequency, l is the harmonics, γ is the relativistic factor, $k_{\parallel}$ and $v_{\parallel}$ are $\vec{k}$ and velocity parallel to $\vec{B}$, $k_{\parallel}v_{\parallel}$ gives the Doppler shift. The wave absorption occurs when the difference between ω and the local Ω_e is nulled by the Doppler shift. If the frequency difference remains large because the wave trajectory does not reach the cold resonance, then the wave energy is damped on faster, less collisional electrons that give a higher CD efficiency.

Fig. 3 compares the power absorption and the frequency difference between ω and the local Ω_e along the ray trajectory for the two cases shown in Fig. 2. It shows that the frequency difference in the conventional LFS ECCD is rapidly decreasing and relatively small for the deposition region, but it is much larger and nearly constant along the whole ray trajectory for top launch ECCD which means the wave/particle interactions stay in high energy electrons (larger $v_{\parallel}$).

- The wave trajectory is nearly parallel to the resonance layer, approaching the resonance much more slowly than the more perpendicular injection of the conventional LFS launch, which results in a long absorption path. As is illustrated in Fig. 3, the absorption path for top launch is ~20 times that of LFS launch. Although the density of high-energy electrons selected by the large Doppler shift is small, the long path enables sufficient wave interaction with these more energetic electrons to damp all of the EC wave power.

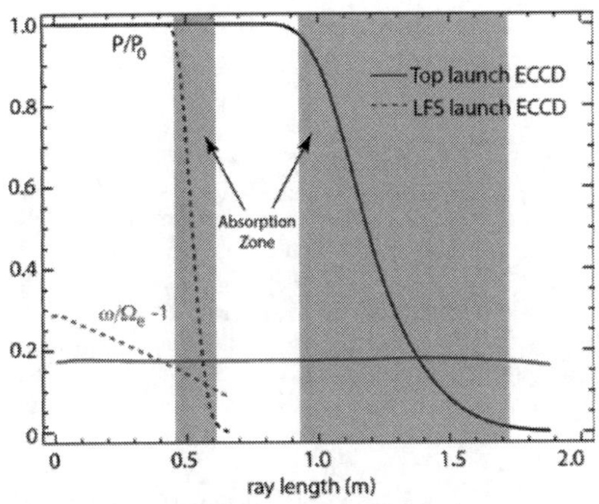

Figure 3: Power absorption fraction (P0 is initial power) and the difference frequency (blue) between applied EC wave frequency (ω) and local electron cyclotron frequency (Ωe) for top launch (solid) and LFS launch (dashed) ECCD

- The wave-plasma interactions take place on the HFS of the flux surface where the effect of mirror trapping of current-carrying electrons is greatly reduced [13, 14].

Although top launch ECCD approach is also limited to cut off density as the conventional outside launch, it has better accessibility than outside launch at high density benefit from the long absorption path.

2.2 Top launch ECCD scheme for reactors

As an example of how the top launch ECCD scheme can work in reactors, this section models the ECCD using the equilibrium and kinetic profiles of the China Fusion Engineering Test Reactor (CFETR) [11] baseline scenario. CFETR is designed to operate with a large bootstrap fraction, approaching 80%, but non-inductive support of the remaining current is still required. The parameters of the machine are major and minor radius of the machine R_0=5.7m and a=1.6m, plasma current Ip = +10 MA, toroidal magnetic field B_T = +5T (Ip and B_T are in the same direction), central density ne(0) = 7.49×10^{19}/m3, and central temperature Te(0) = 23.33 keV (Fig. 4).

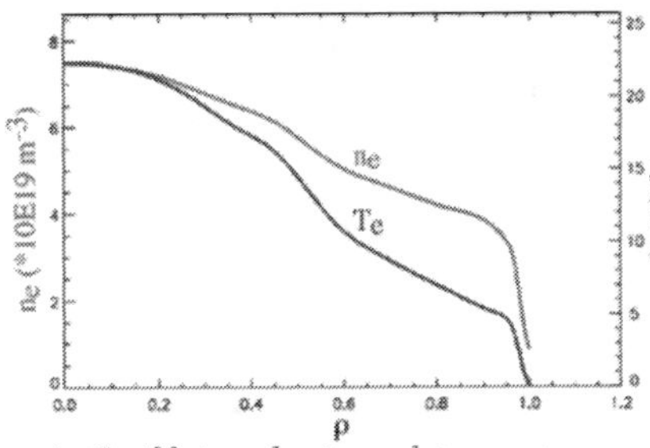

Figure 4: Equilibrium density and temperature profiles of CFETR baseline scenario

A top launch ECCD scheme can be a valuable approach to increase the off-axis current drive in CFETR. Using the applied EC frequency and launch location as free parameters, a systematic study is carried out to identify the conditions for optimum off-axis ECCD profile with high dimensionless current drive efficiency [15] at specified locations (needed to reduce power consumption). Several launch locations and frequencies are studied as illustrated in Fig. 5. For a given launch location and frequency, the launching polar and azimuthal angles are scanned.

The optimal injection angle is determined by which the I_ECCD is highest at the target deposition location. The polar and azimuthal angles are the Euler angles for EC wave launch: the polar angle is that between the EC ray and the vertical z-axis, roughly representing the poloidal steering; the azimuthal angle is the angle between the plane containing z and the EC ray away from a reference plane, roughly representing the toroidal steering. Figure 6 shows such a scan from the LFS outside launch and the optimal injection is determined as (Polar, Azimuthal) = (93.9, 208.8) degree. A sample of the results calculated using the TORAY ray tracing code [16, 17] is listed in Table 1 where the optimal aiming (polar and azimuthal angles) (not shown) is different for different launch locations and/or frequencies. The CD efficiency ζ_{ec}□s derived from normalizing the standard theoretical efficiency (Jec/Pec) to $n_e v_{th}/nkT v_e$ [15].

$$\zeta_{ec} = \frac{e^3}{\varepsilon_0^2}\left(\frac{I_{ec}Rn_e}{P_{ec}kT_e}\right) = 3.27\frac{I_{ec}(A)R(m)n_{19}(m^{-3})}{P_{ec}(W)T_e(keV)} \quad (2)$$

Using the dimensionless current drive efficiency allows comparison with other machines. Table 1 shows that a top launch ECCD scenario with ~190GHz launching from the top of the vessel (R=5.45 m, z=4.0 m) can drive current >30% more efficiently than conventional LFS launch at mid-radius.

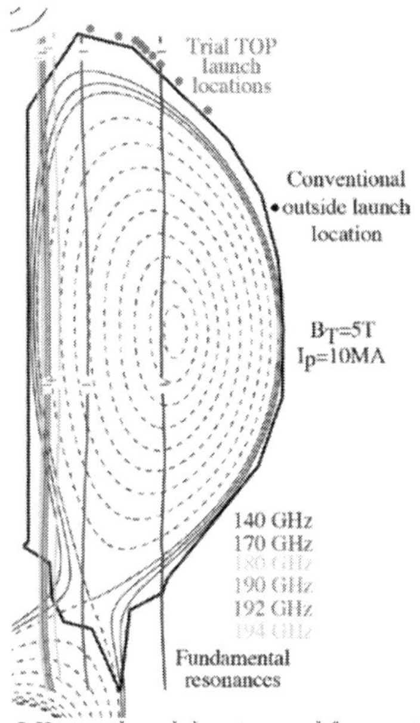

Figure 5: Various launch locations and frequencies studied for high off-axis ECCD efficiency on CFETR.

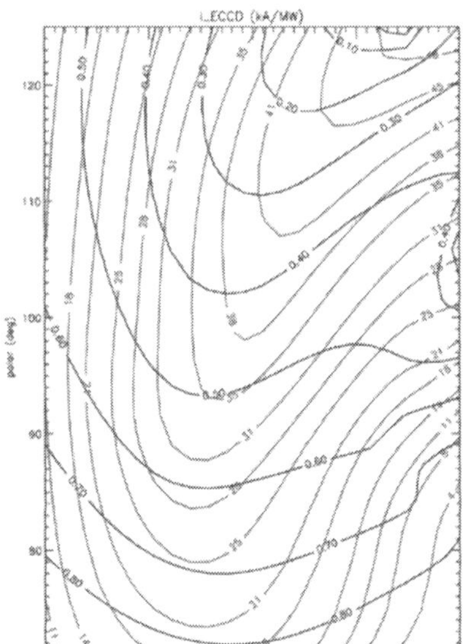

Figure 6: Contour plot of the EC driven current from a scan of launching polar and azimuthal angles, the level of I_ECCD (red) at each radial location (blue). EC wave launched from the LFS outside launch at 170GHz in CFETR baseline scenario.

*Table 1: Sample of peak ECCD efficiency from LFS vs Top launch ECCD for CFETR**

Launch location (R, Z) (meter)	Applied EC frequency (GHz)	ρ ~ 0.5	
		ECCD (kA/MW)	CD efficiency
LFS Launch	170	35.4	0.275
(7.20, 2.04)	140	30.0	0.235
	190	36.0	0.284
	190	46.8	0.370
Top Launch	194	47.9	0.378
(5.45, 4.0)	192	47.0	0.372

* off-axis NBI current drive is about 100kA/MW with 0.6MeV NB for this scenario

2.3 Top launch ECCD for DIII-D tokamak

2.3.1 Systematic physics feasibility study

In preparation for a physics test of top launch ECCD on DIII-D, a physics scoping study similar to that for CFETR has been done. The ECCD efficiencies at mid-radius from various launch locations are evaluated at the two applied frequencies (110GHz and 117.5GHz, 2nd harmonic X-mode) that are available on DIII-D. The scoping study is performed using the equilibrium and kinetic profiles from six DIII-D discharges and one modelled DIII-D steady state discharge with the majority being high q_{min} AT scenarios in double null plasma shape (investigated launch locations and the plasma shapes are illustrated in Fig. 7). Compared to the existing LFS launch, the TORAY ray tracing code found 50% to 100% higher off-axis ECCD efficiency from top launch in some of these plasmas (examples shown in Table 2). For optimized launch angles, the ECCD efficiency generally differs less than 10% between 110 GHz or 117.5 GHz.

*Table 2: Sample of peak ECCD efficiency from LFS vs Top launch ECCD for DIII-D**

Shot	out 110	out 117	v1 110	v1 117	port1 110	port1 117	port2 110	port2 117	port6 110	port6 117	port7 110	port7 117
	ξ	ξ	ξ	ξ	ξ	ξ	ξ	ξ	ξ	ξ	ξ	ξ
147634	0.16	0.16	0.26	0.3	0.26	0.28	0.25	0.27	0.25	0.28	0.26	0.29
146821	0.10	0.15	-	-	-	0.27	-	0.26	-	0.31		
133103	0.14	0.17	-	-	0.14	0.27	-	0.23	0.15	0.26		
122976	0.16	-	-	0.25	0.22	0.29	0.25	0.29				
Model	0.18	0.05	-	-	0.24	-	0.28	-				

* - means inaccessible; blank box means not studied

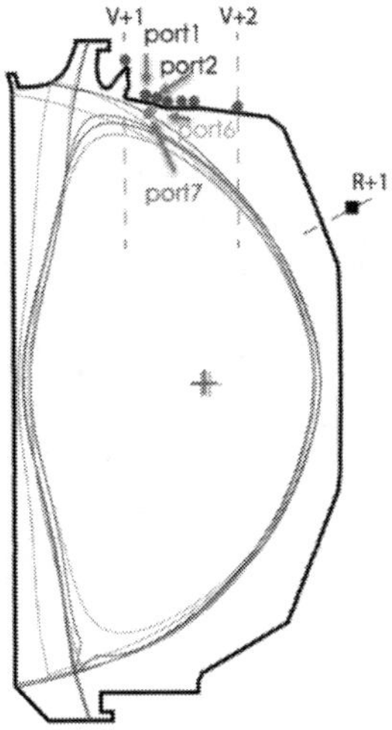

Figure 7:The launch locations and plasma shapes used in the physics feasibility study of Top launch ECCD on DIII-D

These improvements in off-axis ECCD efficiency accrued from top launch on DIII-D have also been confirmed by the quasi-linear Fokker-Planck CQL3D code [18]. Figure 8 shows an example where the current drive efficiency nearly doubles using top launch over the LFS launch. CQL3D results also reveal a modest (~10%) quasilinear increases in ECCD efficiency for top launch as the power is scanned from 0.01MW to 3MW in the top the launch ECCD scheme while no quasilinear increase is observed for outside launch scheme.

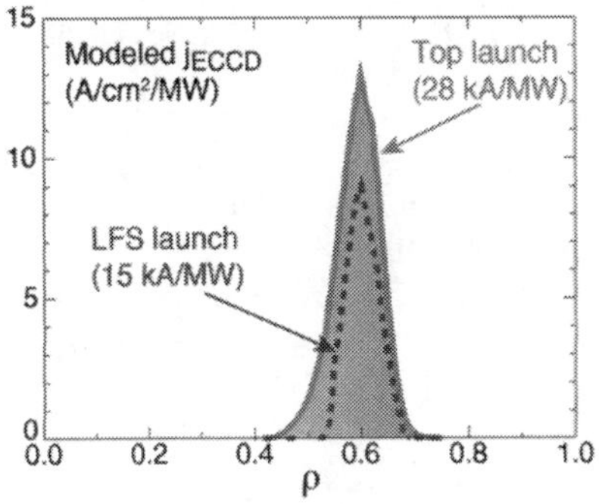

Figure 8:Driven current density profile for top launch and LFS launch for a DIII-D discharge modeled using CQL3D

2.3.2 Prototype fixed-injection top launch ECCD system on DIII-D

A prototype top launch ECCD system is being planned on the DIII-D tokamak to test the top launch concept, establish physics basis for future burning plasma devices and advance the AT program of DIII-D at lower cost. Due to the limitations in the existing port structure, the prototype top launch launcher (Fig. 9) will have a fixed injection angle from a given port location (port 7 shown in Fig. 7). A waveguide switch will be used to connect an existing gyrotron to either the LFS or the top launcher.

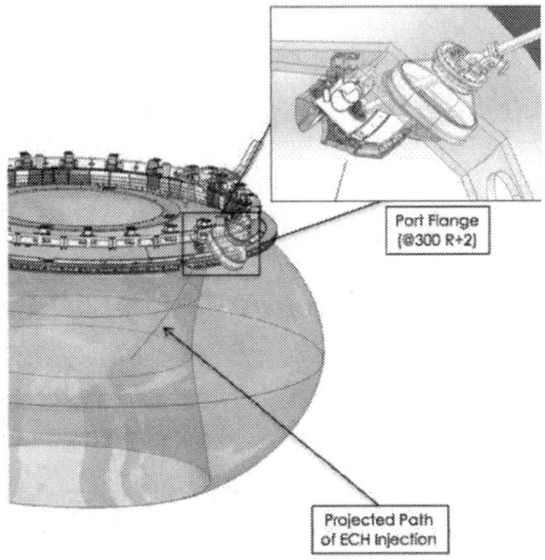

Figure 9: Preliminary design of the prototype fix-injection top launch launcher

Top launch ECCD scheme is sensitive to launching angle. Deviation of 5 degrees from the optimal aiming in some direction could lead to large reduction in the current drive efficiency or inaccessible to the desired location. Therefore, the accuracy of aiming of the fixed-injection launcher is set to 1 degree which limits any reduction in current drive efficiency and ECCD to be within 15% and 20% respectively and the movement of the deposition location to be within $\Delta\rho = 0.03$.

2.3.3 Additional parameters to tune the ECCD and efficiency with fixed-injection

With fixed-injection, magnetic field strength can be a useful parameter to move the deposition location since the resonance layer location moves with respect to the launch position. For example, if the need arises to change between a 117.5GHz and 110GHz gyrotron, then B_T can be decreased to keep the deposition location fixed with only a small reduction in ECCD efficiency. Plasma location and shape are additional parameters that can be adjusted. For example, the current drive efficiency in a "D" shape plasma is higher than in an elliptical plasma because the wave interacts with electrons on the same flux surface for a longer time which results in a greater distortion of the Maxwellian distribution, thus more driven current.

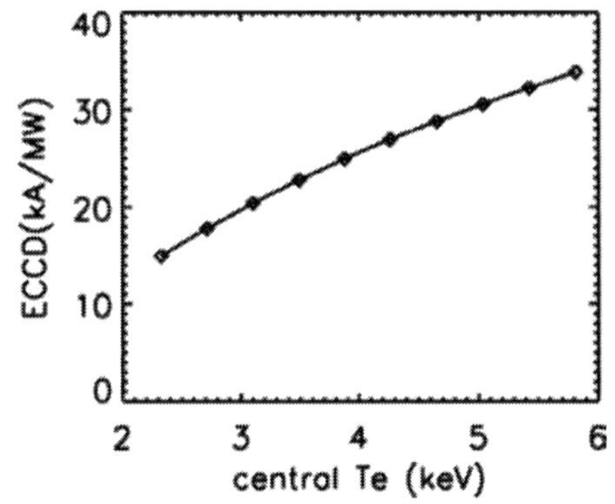

Figure 10: ECCD driven from top launch increases with the electron temperature

ECCD from the conventional outside launch has been shown to increase with electron temperature, and a similar feature is also observed for top launch ECCD as illustrated in Figure 10. Since the larger effect of the electron temperature is removed by the normalization in Eq (2), it has less effects on the dimensionless current drive efficiency which mainly comes in through the quasilinear effect. Thus, increasing the electron temperature at fixed-injection angle increases the driven ECCD which can improve the signal to noise ratio for the measurements.

3 Summary

Steady-state operation in a cost effective fusion power plant requires efficient off-axis current drive but present approaches have limitations. Simulations show that significantly higher current can be driven at mid-radius using the top launch ECCD approach compared to conventional low field side launch. To verify the physics basis and develop the engineering of this technique, a prototype top launch ECCD system is being designed and planned to operate on the DIII-D tokamak in 2019 which, if successful, will help to advance the AT program of DIII-D at lower cost since more off-axis current will be driven with the same gyrotron power.

The top launch ECCD scheme is sensitive to changes in operating conditions, i.e., magnetic field and EC wave injection angle. As the proof-of-principle prototype top launch ECCD system on DIII-D will have fixed injection, the deposition location will be controlled mainly through changing the magnetic field strength, plasma shape and electron temperature. The top launch ECCD test can be done with either a 110GHz or 117.5GHz gyrotron, and use of a waveguide switch between the outside LFS and top launchers means that a dedicated gyrotron is not needed. This prototype top launch system will also allow us to test the hot plasma effect which is not included in the simulation.

The Work supported in part by DE-FC02-04ER54698 and GA-USTC contract AHCC15B11459A, DIII-D data shown here can be obtained in digital format at https://fusion.gat.com/global/D3D_DMP. This report was prepared as an account of work sponsored by an agency of the United States Government. Neither the United States Government nor any agency thereof, nor any of their employees, makes any warranty, express or implied, or assumes any legal liability or responsibility for the accuracy, completeness, or usefulness of any information, apparatus, product, or process disclosed, or represents that its use would not infringe privately owned rights. Reference herein to any specific commercial product, process, or service by trade name, trademark, manufacturer, or otherwise does not necessarily constitute or imply its endorsement, recommendation, or favoring by the United States Government or any agency thereof. The views and opinions of authors expressed herein do not necessarily state or reflect those of the United States Government or any agency thereof.

References

1. T.S. Taylor, et al., Plasma Phys. Control. Fusion **36**, B229 (1994)
2. F. Najmabadi, et al., FED **38**, 3 (1997)
3. A.C.C. Sips, et al., Phys. Plasma **22**, 021804 (2015)
4. V.S. Chan, et al., Nucl. Fusion **46**, 451(2006)
5. R.S. Hemsworth, et al., New J. Phys. **19**, 025005 (2017)
6. P. Sonato, et al., New J. Phys. **18**, 125002 (2016)
7. R. Prater, Phys. Plasma **11**, 2349 (2004)
8. R.I. Pinsker, *et al.*, this conference
9. E. Poli, et al., Nucl. Fusion **53**, 013011 (2013)
10. A. M. Garofalo, et al., Nucl. Fusion **54**, 073015 (2014)
11. Y.X. Wan, et al., Overview of the present progresses and activities on the Chinese fusion engineering test reactor (preprint : IAEA Fusion Energy Conf., Kyoto, Japan, 2016)
12. N.J. Fisch, A.H. Boozer, Phys. Rev. Lett. **45**, 720 (1980).
13. G. Schmidt, Physics of High Temperature Plasmas, p 22 (Academic, New York, 1966)
14. C.C. Petty, et al., Nucl. Fusion **43**, 700 (2003)
15. T.C. Luce, et al., Phys. Rev. Lett. **83**, 4550 (1999)
16. K. Matsuda, IEEE Trans. Plasma Science **17**, 6 (1989)
17. Y.R. Lin-Liu, V.S. Chan and R. Prater, Phys. Plasma **10**, 4064 (2003)
18. R.W. Harvey and M.G. McCoy, Proc. IAEA Technical Committee Meeting, p 498 (Montreal, Canada, 1992)

The deteriorating effect of plasma density fluctuations on microwave beam quality

Alf Köhn[1,2,*], *Max E.* Austin[3], *Michael W.* Brookman[3], *Kenneth W.* Gentle[3], *Lorenzo* Guidi[2], *Eberhard* Holzhauer[1], *Rob J.* La Haye[4], *Jarrod B.* Leddy[5], *Omar* Maj[2], *Craig C.* Petty[4], *Emanuele* Poli[2], *Antti* Snicker[2,6], *Matthew B.* Thomas[5], *Roddy G. L.* Vann[5], and *Hannes* Weber[2]

[1]Institute of Interfacial Process Engineering and Plasma Technology, University of Stuttgart, Stuttgart, Germany
[2]Max Planck Institute for Plasma Physics, Garching, Germany
[3]Institute for Fusion Studies, University of Texas at Austin, TX, USA
[4]General Atomics, PO Box 85608, San Diego, CA, USA
[5]York Plasma Insitute, Department of Physics, University of York, York, U.K.
[6]Department of Applied Physics, Aalto University, Aalto, Finland

Abstract. Turbulent plasma edge density fluctuations can broaden a traversing microwave beam degrading its quality. This can be a problem for scenarios relying on a high spatial localization of the deposition of injected microwave power, like controlling MHD instabilities. Here we present numerical estimations of the scattering of a microwave beam by density fluctuations over a large parameter range, including extrapolations to ITER. Two codes are used, the full-wave code IPF-FDMC and the wave kinetic equation solver WKBeam. A successful comparison between beam broadening obtained from DIII-D experiments and corresponding full-wave simulations is shown.

1 Introduction

Electromagnetic waves in the microwave frequency range have become an indispensable tool for fusion experiments based on the magnetic confinement concepts of tokamak and stellarator. *Electron cyclotron resonance heating* (ECRH) and *current drive* (ECCD), see e.g. Refs. [1, 2], allow to transfer power in the MW-regime [3] to the plasma. On the other hand, active and passive microwave diagnostics occupying only very little space on the inner wall are routinely used in today's experiment [4] and will become even more important in future reactors [5].

Heating and diagnostics suffer, however, both from plasma density fluctuations existing at the plasma boundary. These fluctuations can reach levels of up to 100 % [6] and potentially spoil heating efficiencies and result in ambiguous diagnostics results. This is in particular a problem for the stabilization of so-called *neo-classical tearing modes* (NTMs), a magneto-hydrodynamic instability arising from small perturbations in the plasma current profile [7, 8]. If not taken care of, they can result in disruptions which are to be avoided at all costs in large-scale tokamaks like ITER. One way to stabilize the NTMs consists in localized current drive in order to restore the original current profile. ECCD has been successfully used to provide this current [9]. Using numerical tools to evaluate the consequences of edge density turbulence on the quality of a microwave beam injected for NTM stabilization is the topic of this paper.

*e-mail: koehn@igvp.uni-stuttgart.de

The interaction of electromagnetic waves with plasma density fluctuations is an important topic since the very beginning of plasma physics: in the 1930, radio waves emitted from ground were found to be strongly disturbed after reflection at the ionosphere [10]. Likewise, radio waves emitted by distant rotating neutron stars experience phase disturbances when passing the terrestrial ionosphere which makes detection on Earth complicated. In the 1950s, this problem, which is still an issue for today's satellite communication [11], was tackled by describing the ionosphere as a layer of thin phase screens [12]. This approach relies on small perturbation levels, and locally strong density irregularities provide still a challenge in modeling wave propagation across the ionosphere [13].

In the 1980s high-power microwave sources became available and the interaction of injected microwaves and plasma edge density fluctuations in fusion-relevant scenarios was studied with geometrical-optics tools [14, 15]. Due to the potentially negative consequences for NTM stabilization in ITER, this topic has gained significant traction since 2009 [16–21].

2 Numerical tools

2.1 The full-wave code IPF-FDMC

The full-wave code IPF-FDMC is based on the finite-difference time-domain scheme. It solves Maxwell's equation and the fluid equation of motion of the electrons on a 2D Cartesian grid. The evolution equations for the wave magnetic field **B**, the wave electric field **E**, and the current density **J** of the wave read:

$$\frac{\partial}{\partial t}\mathbf{B} = -\nabla \times \mathbf{E} \quad (1)$$

$$\frac{\partial}{\partial t}\mathbf{E} = c^2\nabla \times \mathbf{B} - \mathbf{J}/\epsilon_0 \quad (2)$$

$$\frac{\partial}{\partial t}\mathbf{J} = \epsilon_0\omega_{pe}^2\mathbf{E} - \omega_{ce}\mathbf{J} \times \hat{\mathbf{B}}_0 - \nu_e\mathbf{J}, \quad (3)$$

with c the speed of light, ϵ_0 the vacuum permittivity, $\omega_{pe} = \sqrt{n_e e^2/(\epsilon_0 m_e)}$ the electron plasma frequency, $\omega_{ce} = |e|B_0/m_e$ the electron cyclotron frequency, $\hat{\mathbf{B}}_0$ the unit vector into the direction of the background magnetic field, and ν_e an electron collision frequency (as a dissipation mechanism). More details about the code can be found in Refs. [22, 23].

2.2 The WKBeam code

The WKBeam code solves the wave kinetic equation in the presence of random density fluctuations embedded in a slowly varying background plasma density. The fluctuations are included by applying a scattering operator whose derivation is based on the *Born approximation* [25]. The wave kinetic equation to be solved can be cast into a form such that the scattering operator depends only on the *correlation function* of the random density field and thus remains valid even for short-scale fluctuations [26]. Details about the WKBeam code and the derivation of the underlying equations can be found in Refs. [20, 23, 27]

Since the derivation of the scattering operator relies on the Born approximation, it is expected to become invalid for large fluctuation levels. More precisely,

$$\left[\langle(\delta n_e/n_{e,0})^2\rangle\right]^{1/2} \omega_{pe}^2/\omega_0^2 \ll 1 \quad (4)$$

needs to be fulfilled for the Born approximation to hold. Therefore the combination of fluctuation level and background density needs to be taken into account. The validation of the Born approximation in WKBeam is investigated in detail in Ref. [23].

2.3 The simulation set-up

To benchmark WKBeam with the full-wave code IPF-FDMC, a 2D computational domain is chosen resembling part of a poloidal cross section in a toroidal magnetic confinement device. The background plasma density profile is linearly increasing as

$$n_{e,0}(x) = \frac{n_{e,\max}}{x_1 - x_2}(x_1 - x), \quad (5)$$

with $n_{e,\max} = 0.65 \cdot n_{e,\mathrm{cut}}$ (where $n_{e,\mathrm{cut}}$ is the O-mode cut-off density of the injected microwave), x the radial coordinate, $x_1 = 2.45$ m the position where the density profile starts to rise until a position of $x_2 = 2.30$ m. A layer of fluctuations is added to the background profile, where the envelope of the fluctuation amplitude $F(x)$ is described by a Gaussian. The full electron plasma density profiles then reads

$$n_e(x,z) = n_{e,0}(x)\left(1 + F(x)\delta n_e(x,z)\right), \quad (6)$$

with z the vertical coordinate and $\delta n_e(x,z)$ the random field which is generated by a truncated sum of Fourier-like modes:

$$\delta n_e(x,z) = \sum_i^{M_i}\sum_j^{M_j} A_{i,j}\cos\left[k_{x,i}x + k_{z,j}z + \varphi_{i,j}\right], \quad (7)$$

with $A_{i,j}$ the amplitudes of the modes and $\varphi_{i,j}$ independent random phases. There is no dependence on time as the turbulent density fluctuations appear to be frozen in the time frame of the microwave.

Figure 1 shows a contour plot of the resulting plasma density. Note that this is an actual sample used in the full-wave simulations as input. To properly simulate the effect of fluctuations in the full-wave code, an ensemble-average is required with the ensemble consisting of a series of density profiles each being a sample of the same random density field. We use synthetic turbulence to ensure that the statistics of the random field are the same as those assumed in WKBeam.

A Gaussian beam in O-mode polarization is injected with its focal point located inside of the grid (as can be seen in Fig. 1). The simulation parameters are listed in Table 1. They correspond to ASDEX Upgrade (AUG) parameters scaled-down by approximately a factor of 3 (the reduction in frequency and hence increase in wavelength allows for coarser numerical grids to be used, strongly reducing the required computational resources). Only the ratio λ/w_0 is roughly an order of magnitude larger in the benchmark case as for the AUG parameters. According to Ref. [28], the results from the benchmark case serve therefore as an upper boundary for the AUG case although the dependence of beam broadening on the beam size was found to be small.

Table 1. Simulation parameters for the benchmark together with typical ASDEX-Upgrade values

	benchmark	AUG
f	50 GHz	140 GHz
$\lvert B_0\rvert$	1 T	2.5 − 3 T
$L_\perp$	5 mm	2 − 10 mm
w_0	1 cm	3 cm

2.4 Data analysis

The efficiency of the NTM stabilization depends on the good localization of the current driven by the injected microwaves, as explained in the introduction. If the power deposition is too broad, the resulting current filament will be too broad to restore the original plasma current profile. Assuming as a first order approximation that the power deposition width corresponds to the beam width of the microwave at the location of absorption (for a detailed discussion about the correlation between microwave beam width and deposition profile width, see Ref. [31]), the additional broadening of the beam, as compared to the case without turbulence, is chosen as the quantity of interest to be compared between the two codes.

As a first step, both codes were compared for beams propagating in vacuum, which yielded excellent agreement. Next, the plasma profiles as described by Eq. (5) were considered, i.e. without turbulent density fluctuations (again, excellent agreement is found [23]). The resulting beam widths are used as reference positions to which the beam widths obtained from scenarios with fluctuations included are compared. In detail, Gaussians are fitted to the ensemble-averaged beam cross sections with the beam width being one fit parameter. The obtained values are then normalized to the cases without fluctuations.

3 Simulation results

3.1 Illustration of beam broadening

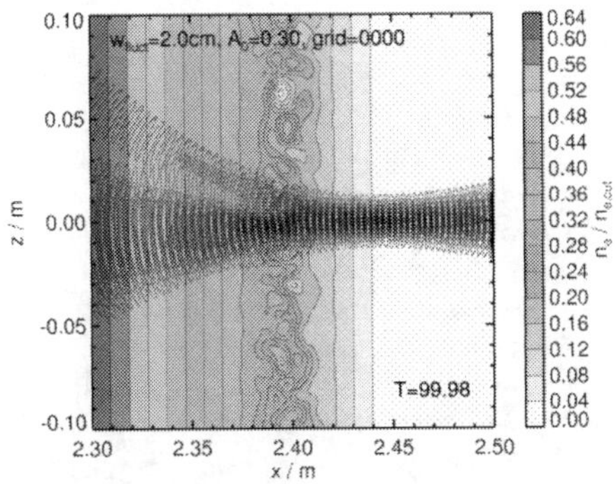

Figure 1. Full-wave simulation grid with colored contour levels indicating the electron plasma density and contour lines indicating the absolute wave electric field. Shown is a sample from the ensemble with the average fluctuation parameters given in the plot. The snapshot is taken from a video published at [24].

The effect of a layer of turbulent plasma density fluctuations on a microwave beam can be nicely illustrated with full-wave simulations, as shown in Fig. 1. The beam is injected from the right hand side and has its focal point approximately where the density profile starts to rise. A splitting of the injected beam is seen in this example, strongly perturbing the quality of the beam. The interested reader is referred to Ref. [24] where the video belonging to the snapshot shown in Fig. 1 can be seen.

For a quantitative data analysis, the average over the full ensemble of the full-wave simulations needs to be taken. This work uses an ensemble size of $N = 3000$ which is large enough to ensure that the statistical error is small compared to the averages. Figure 2 shows the ensemble-averaged beam cross section from full-wave simulations together with the corresponding result from WKBeam calculations. The turbulence parameters are the same as for the single sample shown in Fig. 1. A first thing to notice is that the average beam does not change its direction of propagation. Hardly any difference of WKBeam to the full-wave solution can be seen in this representation. Both signals experience a small broadening (together with a reduction of the peak amplitude) as compared to the case without turbulence.

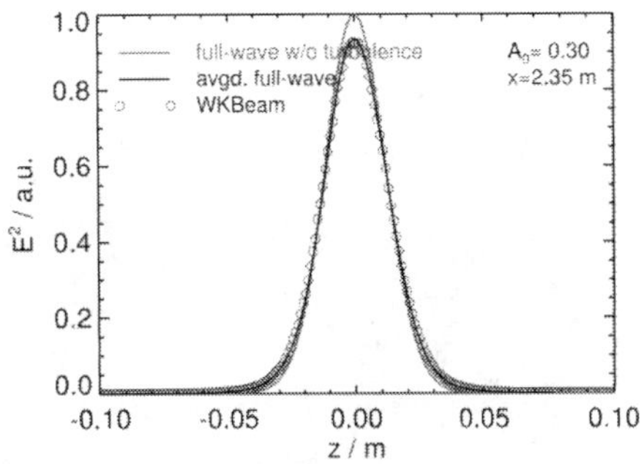

Figure 2. Beam cross section at a radial position of $x = 2.35$ m from full-wave simulations (ensemble-averaged) and WKBeam calculations.

3.2 Scanning the fluctuation parameters

In both codes, IPF-FDMC and WKBeam, the parameters defining the turbulent plasma density fluctuations were varied over a wide parameter range. This was done to explore the ability of the codes to analyze current and future devices, as well as to explore the validity of the Born approximation, expressed in Eq. (4), and thus of WKBeam. The fluctuation strength, the background density, and the width of the fluctuation layer were varied in a series of parameter scans. Due to the limited space in this paper for the proceedings, we will only present a selection of the results here and like to refer the interested reader to the full paper [23].

Figure 3 (left) shows the beam broadening deduced from full-wave simulations as a function of fluctuation level and background density at the location of the fluctuations. It can be clearly seen that the broadening is a function of both parameters, as expected. The contour plot on the right hand side shows the deviation of the corresponding WKBeam calculations with respect to the full-wave simulations. Note that WKBeam consistently overestimates the broadening. Even for large fluctuation levels of 50 %, the disagreement to the full-wave solution is below 10 %. Recalling the criterion to be fulfilled for the Born approximation to hold, Eq. (4), it is possible to draw a rough estimation: with the y-axis corresponding to the first factor and the x-axis to the second factor, the resulting product of the two quantities should be below 0.15 to keep the overestimation of WKBeam below 10 %.

Note that those results were obtained for a width of the fluctuation layer of approximately $w_0 = 2\,\mathrm{cm} \approx 3.3\,\lambda_0$, with λ_0 the vacuum wavelength of the injected microwave. According to the results discussed in Ref. [28], the beam broadening scales linearly with the width of the fluctuation layer if the fluctuation level is not too large. The observed scaling of the beam broadening may serve as estimations for current experiments or predictions for future experiments.

An important result of the benchmark study is the applicability of WKBeam for modern tokamaks. For the ASDEX Upgrade tokamak, for example, edge density values reach at maximum $0.2 \cdot n_{e,\mathrm{cut}}$ [29] and maximum fluctuation

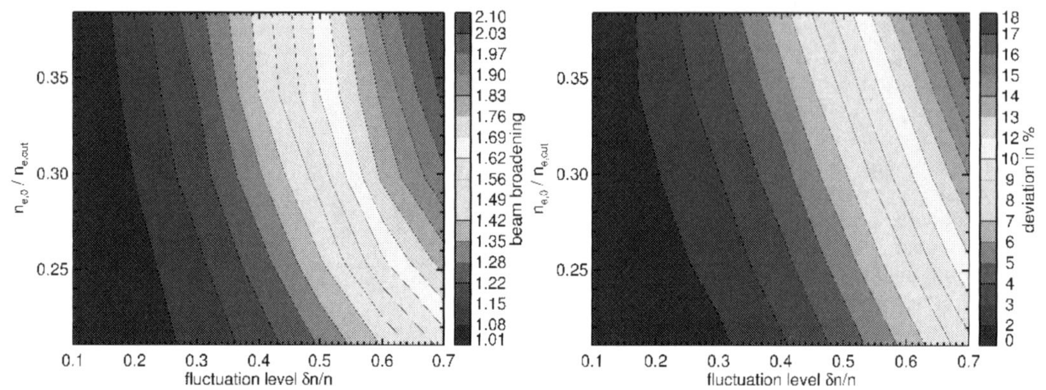

Figure 3. (Left) Beam broadening as obtained from full-wave simulations as a function of the background density at the position of the fluctuation layer and as a function of the fluctuation level. (Right) Overestimation of the corresponding WKBeam results to the full-wave solution.

levels of 20 % [30]. With the resulting control parameter given by Eq. (4) being below the stated value of 0.15 (and by checking Fig. 3 (right)) WKBeam can be reliably applied for this scenario. Considering ITER parameters [20], the overestimation of WKBeam calculations is on the order of 1 % due to the correspondingly low (normalized) density. Note that the deviation of the WKBeam results does not depend on the width of the fluctuation layer [23]. Due to the long propagation length of the injected microwave beam in ITER, this is an important information strengthening the reliability of WKBeam.

3.3 Extrapolations towards ITER

Mitigation and suppression of NTMs is of vital importance for future tokamak experiments as explained in the introduction. In ITER, the EC upper launcher system is designed for this purpose [32]. Recently, the importance of quantifying the influence of plasma density fluctuations on the efficiency of NTM stabilization in ITER has been highlighted [31]. Due to the short spatial scales of the density fluctuations, their effect cannot be accounted for by tools based on the geometrical-optics approximations (which are otherwise commonly used to design and describe microwave heating scenarios in fusion experiments). The WKBeam code was deliberately developed to handle this. Estimating the beam broadening due to plasma density fluctuations in ITER using WKBeam is the topic of a paper published earlier this year [20]. The main result of that paper will only be briefly discussed here.

As shown in Section 3.2, the strength of the scattering of the microwave beam, precisely the beam broadening, depends on a number of parameters of the turbulent density fluctuations. Choosing a "correct" set of parameters is therefore important to get reliable results. The term "correct" refers to parameters describing the flat top phase of the ITER baseline scenario [33]. The spatial profile of the fluctuations' amplitude and the perpendicular correlation length of the density structures used in the modeling are based on experimental surveys [34]. The fluctuations amplitude profile we are using has a constant strength of 2 % in the core and 20 % in the scrape-off layer, corresponding to an H-mode discharge.

Comparing first the power deposition profiles deduced from WKBeam and from the beam tracing code TORBEAM for the case without fluctuations, they reveal excellent agreement. Including fluctuations, which cannot be done with codes like TORBEAM [35], a significant broadening of the deposition profile is found. Taking the effect of ballooning into account, i.e. a poloidal variation of the density fluctuation amplitude, an overall broadening of approximately a factor of 2 is obtained. Considering the installed microwave power in ITER, it can be concluded that the most dangerous NTMs expected to occur in the 15 MA $Q = 10$ standard H-mode scenario can still be stabilized based on the current model of the density fluctuations. As mentioned in the beginning of this section, a detailed discussion of the results is found in Ref. [20].

Besides the EC upper-launcher system, ITER is also equipped with an equatorial EC launcher [36] aiming to influence the sawtooth cycle which is thought to affect the onset of NTMs. It is thus of interest to estimate the deteriorating effect of density fluctuations on this scenario as well. The WKBeam code predicts a less severe beam broadening (referring to the upper-launcher calculations) of a factor of 1.2 – 2.5, as discussed in detail in Ref. [37]. Note that the broadening is also of less importance here as the requirement on the localization of the power deposition is reduced for influencing the sawtooth cycle.

3.4 Cross-polarization scattering

An interesting effect occurs at small plasma densities, where the dispersion surfaces of O- and X-mode are not well separated: mode coupling can occur. Triggered by fluctuations at low background plasma density values, this *cross polarization scattering* can be problematic: mode scattered from one polarization to the other not only reduces the heating or current drive efficiency due to the original polarization but the other polarization can itself deposit power at non-optimal positions or locally damage wall components if being reflected at a cut-off. Thus, the scattering operator in WKBeam has been extended to also included the possibility of mode scattering [37, 38]. As the underlying assumptions for deriving the scattering operator are, however, only marginally fulfilled for the parameters where mode scattering can actually be relevant, a benchmark against full-wave simulations is required. Here, we present preliminary results of cross-polarization scattering obtained from full-wave simulations in the course of this (ongoing) benchmark study.

The configuration of the simulations differs from the geometry described in Section 3.1: it is optimized to result in a recognizable amount of mode-scattering. The wave frequency is increased to 110 GHz and the background magnetic field has a component pointing into the 2D simulation domain, $B_{tor} = -1$ T, and a vertical component of $B_z = 1$ T. The correlation length of the density structures is $L_\perp = 5$ mm and the layer of fluctuations is shifted towards lower densities as shown in Fig. 4. Due to the oblique orientation of the background magnetic field, the injected O-mode is now elliptically polarized. As also shown in Fig. 4, the simulation domain includes the right-hand cut-off, which is a cut-off for the X-mode, in order to serve as a filter for that part of the microwave beam which is scattered from the injected O-mode to the X-mode: the X-mode is reflected at the cut-off and can thus be easily detected. To visually separate the reflected X-mode from the injected O-mode, an appropriate injection angle is chosen such that the two modes do not overlap.

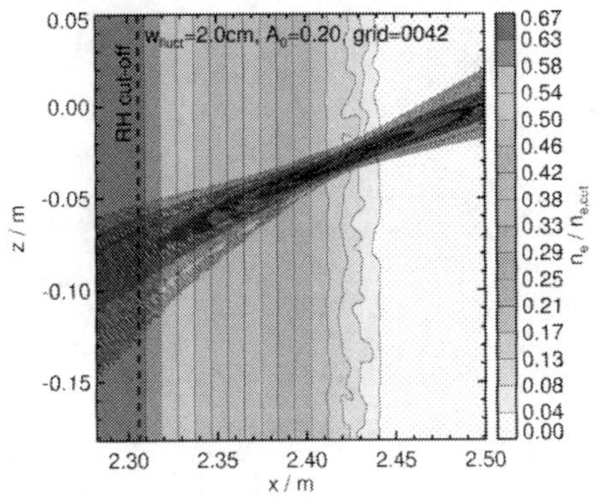

Figure 4. Contour plot of the electron plasma density (colored) and of the absolute value of the wave electric field as obtained from full-wave simulations intending to study cross-polarization scattering due to density fluctuations at low plasma densities.

Although no reflection at the right-hand cut-off is seen in Fig. 4 at first glance, a small amount of cross-polarization scattering does actually occur in this case. It is approximately 0.40 % which is too small to be visible in this representation. As expected, even for the case without density fluctuations, there is a small amount of coupling to the other mode. From the full-wave simulations, a value of approximately 0.36 % is deduced for this geometry (which can be reduced by injecting wider beams with a narrower angular spectrum). The fluctuations thus result only in a small absolute increase of the wrong mode content.

WKBeam calculations yield an X-mode content larger by approximately a factor 2–3. Due to the assumptions being applied when deriving the scattering operator, this deviation is expected to change with the background density (into the direction that lower densities, when the modes are "more" degenerated, result in less reliable results). A systematic comparison between full-wave simulations and WKBeam is currently underway and will be published in a separate paper.

As stated in Refs. [37, 38], the cross-polarization scattering values obtained from WKBeam thus serve as an upper bound. Together with the preliminary results from full-wave simulations, cross-polarization due to density fluctuations seems to be on the same order in ITER as the polarization uncertainty due to the launcher. It is therefore not expected to result in severe problems.

4 Simulation-experiment comparison

The microwave power deposition profiles can be experimentally determined using an electron cyclotron emission diagnostic in power-modulated discharges. An accompanying transport analysis separates the effect of profile broadening due to diffusion and due to other effects like a turbulent edge layer. Broadening of the injected EC beam of up to a factor of 2.5 could be determined in DIII-D this way, as presented on the last EC-conference [39]. Inspired by these experiments, we elaborated the possibility of performing simulations resembling the experiment as closely as possible. An ensemble of plasma density profiles including turbulence was created with the BOUT++ code [40] using the Hermes model [41]. Equilibrium profiles obtained from DIII-D discharges were used as input parameters. The fluctuation levels obtained from the BOUT++ simulations were confirmed to correspond to those found in the experiment.

The ensemble of density profiles including plasma density fluctuations was then used as input for 3D full-wave simulations using the EMIT-3D code [42]. The simulation domain does not include the full propagation path from the emitting antenna to the place of absorption in the plasma center. It is instead restricted to the area where the strongest fluctuation levels are observed as this region is thought to be responsible for the observed broadening. Making the assumption that the perturbed (ensemble-averaged) beam can be approximately described by a Gaussian (and that the simulation domain is sufficiently far away from the beam waist), its width can be linearly extrapolated to the location of absorption. Comparison with the experimentally deduced beam broadening is possible this way, yielding excellent quantitative agreement over a wide range of fluctuation levels (realized by different experimental scenarios). As an example, for the case of an L-mode shot, an experimental broadening of 2.7 ± 0.3 compared to the case without fluctuations was determined. The value obtained from full-wave simulations, 2.5 ± 0.3, illustrates the excellent agreement. The results are presented and discussed in detail in a paper submitted recently [43].

5 Summary

We have shown that turbulent edge density fluctuations can lead to significant broadening of an injected microwave beam. The stabilization of NTMs requires good spatial localization of the beam and the presented parameter scans of the fluctuation parameters may serve to quantitatively estimate the reduction in efficiency. Full-wave simulations have been used to benchmark WKBeam calculations, where very good agreement was found of a large parameter range. In particular for the expected density fluctuations in ITER, the deviations of the WKBeam are negligible. This supports the results from the corresponding calculations for ITER, performed with WKBeam, predicting that NTM stabilization is still possible in ITER taking into account the effect of fluctuations. A first estimation of the effect of cross-polarization scattering has been performed, which seems to not be an issue in ITER. Finally, a comparison between beam broadening obtained from a series of DIII-D experiments and from full-wave simulations of these experiments was presented yielding excellent agreement.

Acknowledgments

This work has been carried out within the framework of the EUROfusion Consortium and has received funding from the Euratom research and training programme 2014–2018 under grant agreement No. 633053. The views and opinions expressed herein do not necessarily reflect those of the European Commission.

The authors are indebted to the efforts of the open-source software community.

References

[1] M. Bornatici *et al.*, Nucl. Fusion **23**, 1153 (1983).
[2] V. Erckmann *et al.*, Plasma Phys. Control. Fusion **36**, 1869 (1994).
[3] M. Thumm, Plasma Phys. Control. Fusion **45**, A143 (2003).
[4] H.-J. Hartfuss, *Fusion Plasma Diagnostics with mm-Waves* (Wiley-VCH, Weinheim, 2013)
[5] F. A. Volpe, JINST **12** C01094 (2017).
[6] S J Zweben *et al.*, Plasma Phys. Control. Fusion **49** S1 (2007).
[7] R. J. La Haye, Phys. Plasmas **13**, 055501 (2006).
[8] H. Zohm *et al.*, Nucl. Fusion **47**, 228 (2007).
[9] W. Kasparek *et al.*, Nucl. Fusion **56**, 126001 (2016).
[10] H. G. Booker *et al.*, Phil. Trans. Royal Society A **242** 579 (1950).
[11] L. C. Andrews *et al.*, Applied Optics **34** 7742 (1995).
[12] A. Hewish, Proc. Royal Society A **209** 81 (1951).
[13] H. J. Strangeways *et al.*, Radio Science **49** 899 (2015).
[14] E. Ott *et al.*, Phys. Fluids **23** 1031 (1980).
[15] F. R. Hansen *et al.*, Nucl. Fusion **28**, 769 (1988).
[16] C. Tsironis, Phys. Plasmas **16**, 112510 (2009).
[17] Y. Peysson *et al.*, Plasma Phys. Control. Fusion **53** 124028 (2011).
[18] A. Ram *et al.*, Phys. Plasmas **23**, 022504 (2016).
[19] Z. C. Ioannidis *et al.*, Phys. Plasmas **24**, 102115 (2017).
[20] A. Snicker *et al.*, Nucl. Fusion **58** 016002 (2018).
[21] O. Chellaï *et al.*, Phys. Rev. Lett. **120** 105001 (2018).
[22] A. Köhn, *Investigation of microwave heating scenarios in the magnetically confined low-temperature plasma of the stellarator TJ-K* (Ph.D. thesis).
[23] A. Köhn *et al.*, Plasma Phys. Control. Fusion **60** 075006 (2018).
[24] A. Köhn *et al.*, TIB AV-Portal (2018). doi:10.5446/36093
[25] M. Born and E. Wolf, *Principles of Optics: Electromagnetic Theory of Propagation, Interference and Diffraction of Light (7th ed.)* (Cambridge University Press, Cambridge, 1999) p. 695
[26] S. W. McDonald, Phys. Rev. A **43**, 4484 (1991).
[27] H. Weber *et al.*, EPJ Web of Conf. **87** 01002 (2015).
[28] A. Köhn *et al.*, Plasma Phys. Control. Fusion **58** 105008 (2016).
[29] M. Willensdorfer *et al.*, Plasma Phys. Control. Fusion **52** 114026 (2012).
[30] A. Medvedeva *et al.*, Plasma Phys. Control. Fusion **59** 125014 (2017).
[31] E. Poli *et al.*, Nucl. Fusion **55** 013023 (2015).
[32] M. Henderson *et al.*, Nucl. Fusion **48** 054013 (2008).
[33] V. Parail *et al.*, Nucl. Fusion **53** 113002 (2013).
[34] G. Conway, Plasma Phys. Control. Fusion **50** 124026 (2008).
[35] E. Poli *et al.*, Comp. Phys. Comm. **136** 90 (2001).
[36] M. Henderson *et al.*, Phys. Plasmas **22** 021808 (2015).
[37] A. Snicker *et al.*, Plasma Phys. Control. Fusion **60** 014020 (2018).
[38] L. Guidi *et al.*, J. Phys.: Conf. Series **775** 012005 (2016).
[39] M. Brookman *et al.*, EPJ Web of Conf. **147** 03001 (2017).
[40] B. Dudson *et al.*, Comp. Phys. Comm. **180** 1467 (2009).
[41] B. Dudson *et al.*, Plasma Phys. Control. Fusion **59** 054010 (2017).
[42] T. R. N. Williams, Plasma Phys. Control. Fusion **56** 075010 (2014)
[43] M. Brookman, Phys. Rev. Lett. *submitted*

Reduced field Scenario with X3 heating in W7-X

Nikolai B. Marushchenko[1,*], *Pavel* Aleynikov[1], *Craig D.* Beidler[1], *Andreas* Dinklage[1], *Joachim* Geiger[1], *Per* Helander[1], *Heinrich P.* Laqua[1], *Henning* Maassberg[1], *Yuri* Turkin[1], and the W7-X Team

[1]Max Planck Institute of Plasma Physics, EURATOM Association, Wendelsteinstr. 1, Greifswald, DE17491, Germany

Abstract. In the present work, an ECRH scenario with reduced magnetic field 1.75 T is considered. For 140 GHz, this field corresponds to X3 heating. The high mirror-ratio magnetic configuration, $B_{01}/B_{00} \simeq 0.24$, was considered as one from most attractive for long-pulse operation with low bootstrap current. Since X3 wave mode can be effectively absorbed only in sufficiently hot plasmas, a preheating stage is necessary, and the requirements for target plasmas suitable for starting X3 have been studied. Different ways to establish target plasmas are also discussed, in particular, augmenting X3 heating with X2 beams at 105 GHz.

1 Introduction

At the near-term focus of the W7-X scientific program are experiments to help assess stellarator optimization in view of economic operation of a stellarator fusion power plant [1]. The high-level scientific goals of W7-X are the demonstration of improved neoclassical confinement as well as improved confinement of fast ions, plasma stability up to a volume-averaged β of 5%, and a stiff magnetic equilibrium to facilitate the island divertor concept [2] while achieving steady-state operation.

W7-X is equipped with 10 gyrotrons, which generate RF beams of 140 GHz with power up to 1 MW or (optionally) of 105 GHz with 0.5 MW each, respectively [3, 4]. The quasioptical system in W7-X has been well tested for the main frequency, 140 GHz, and operation with different frequencies requires an additional adaptation. Scenarios with 105 GHz however have been included in the list of possible options.

The main ECRH scenarios are designed for operation at $B_0 \simeq 2.5$ T and 140 GHz with the X2-mode for moderate density, $n_e < 1.2 \times 10^{20}$ m^{-3}, and O2- mode for high density, $n_e < 2 \times 10^{20}$ m^{-3}. The launch-ports are close to the "bean-shaped" poloidal plane of the W7-X magnetic configuration, where the maximum of B is located and the radial dependence of B is tokamak-like. Additionally, two remote steering launchers (RSL) are installed near the triangular plane, i.e. near the minimum of B.

Already during the initial experimental campaigns, the main ECRH/ECCD scenarios have been well tested [5]. In particular, the possibility of heating with a pure multi-pass O2-mode scenario was demonstrated. This scenario was performed in three steps: start-up by X2, then adding the O2 beams and, finally, ramping-up the density and switching off the X2 beams. High power ECRH heating with multipass O2 beams has been performed due to a set of Tungsten-coated holographic tiles installed on the wall opposite to the ECRH launchers for refracting back about 90% of power [6]. For the same purpose, a non-holographic polished steel liner installed on the outboard side.

Several magnetic configurations were tested, including the high-mirror configuration with a mirror ratio of about 10%. At the same time, configurations with mirror ratios between 15% and 24%, which have specific relevance as candidates for long-pulse operation due to their small bootstrap currents [7], exceed present technical limits at nominal field, 2.5 T [8], and were not cleared for operation in the nearest experimental campaigns. However, for scenarios with reduced magnetic field these configurations can be applied. The reduction of the field also opens an additional option for experiments at high-β plasmas addressing fundamental MHD predictions and ECRH with the X3-mode has to be considered as the primary option.

Applications of the X3-mode in W7-X was considered for different goals already; see, for example, the references [3] and [9] (in the last, X3 at 140 GHz launched near the minimum of B in high-mirror configuration was investigated with respect to selective heating of trapped/passing electrons). Similarly to O2 scenarios, X3 heating cannot be used for start-up. It is therefore anticipated that the scenario with X3-mode first requires the establishment of a target plasma with density and temperature required for stable X3 operation. An important task is to find the general requirements for such target plasmas.

At the moment, three possible methods are under consideration for the role of creating the target-plasma: NBI, ICRH and ECRH with X2 at the frequency 105 GHz. The assessment of plasma start-up with pure NBI in W7-X was investigated already in detail in Ref. [15]. It was found that the time for target-plasma establishment is too long (about 4 s) and even exceeds the time of NBI operation. Investigation of ICRH start-up and creating the target-plasma needs to be done.

*e-mail: nikolai.marushchenko@ipp.mpg.de

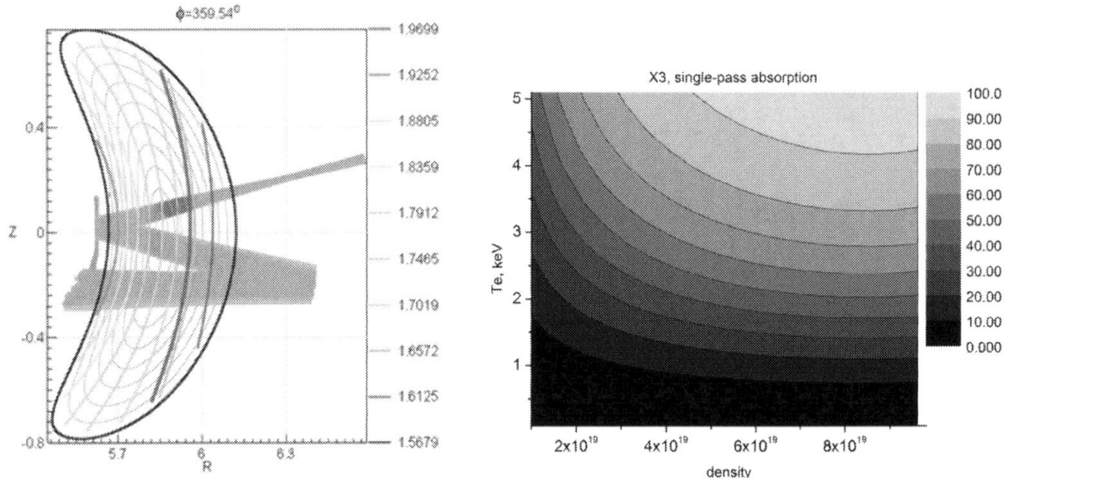

Figure 1. *Left*: $R - Z$ projection of the RF beam that performs X3 (140 GHz) heating. Three passes are shown, with reflections from the mirror and vacuum chamber. *Right*: For the beam at the left, a single-pass absorption of the X3- mode is shown as a function of density and electron temperature.

In this paper we analyze the conditions necessary for stable operation of X3 heating in the magnetic configurations with very high mirror ratios and consider the mixed X2 (105 GHz) and X3 (140 GHz) ECRH scenarios. All simulations are performed with the ray-tracing code TRAVIS [10] coupled to the transport code NTSS [11].

2 X3 heating and requirements for target plasmas in W7-X

In order to obtain initial information about the effectiveness of X3 ECRH at 140 GHz, the cyclotron absorption must be scanned over plasma parameters. For this purpose, the magnetic configuration with highest mirror ratio, $B_{01}/B_{00} = 24\%$, designated UEM, was applied, and the magnetic field magnitude, $B_0(0) = 1.75$ T, was adjusted to maximize the deposition near the axis. Typically, plasmas in W7-X are optically gray for the X3-mode and several passes are required for complete absorption of the RF beam. Fig. 1 (left) shows the geometry of the X3-beam launched through the port near the "bean-shaped" plane.

In order to control the deposition profile and to avoid any damage for the reflecting mirror from the 1 MW beam, the single-pass absorption for X3-mode in steady-state must exceed 70%. Fig. 1 (right) shows the fraction of absorbed heating power from which one can see that the low-density range, $(1–5)\times10^{19}$ m^{-3}, seems to be less preferable for X3-mode since quite high temperature, $T_e \gtrsim 3$ keV, appears to be required for sufficient microwave absorption. However, since the plasma heating leads to better absorption when the temperature increases, a more refined analysis has been conducted taking into account the dynamics of the electron temperature into account starting from very small values of it.

To this end, we perform transport simulations (NTSS + TRAVIS) for scenarios at a plasma density range $(1 - 5) \times 10^{19}$ m^{-3} (n_e-profile fixed) and 5 MW X3 heating from six RF-beams with launch geometry similar to that shown in Fig. 1 (left). It is assumed that neoclassical transport is dominant within the bulk plasma, while the periphery is characterized by anomalous transport proportional to $1/n_e$. The neoclassical transport coefficients were calculated by the DKES code [12]. Simulations were performed as follows: it was assumed that the target plasma had been created in some manner and only the initial density and temperatures, $n_e(0)$ and $T_{e,i}(0)$, were changed from run to run. For every RF beam, three passes were taken into account and for rather low T_e only a small fraction of power was absorbed. Simulations yielding very low absorption are not of interest. Actually, we select only the cases with a rapid electron temperature growth and omit other cases. If the growth of T_e appears, absorption of the X3-mode will increase as it is proportional to $n_e T_e^2$. A plasma build-up time is defined as the characteristic time interval $\Delta t_{1/2}$ for which $P_{abs} \geqslant P_{inj}/2$ is reached. Further evolution of the plasma parameters in the phase $t > \Delta t_{1/2}$ is usually very rapid as a physical process, but sometimes requires rather time consuming calculations.

In Fig. 2 (left), $\Delta t_{1/2}$ is plotted as a function of the initial $n_e(0)$ and $T_e(0)$ values. As a margin for acceptability of the target plasmas parameters, which comes from safety requirements for the reflecting mirrors, $\Delta t_{1/2} = 0.1$ s (shown by red) is chosen. The domain with $\Delta t_{1/2} > 0.1$ s which lies in Fig. 2 (left) below this line, is qualified as not acceptable. It is found that the minimum start-up criterion, $\Delta t_{1/2} < 0.1$ s, is practically equivalent to the requirement $T_e > 0.7$ keV for target plasmas.

It is necessary to mention here that if establishment of a hot plasma occurs, the final steady state ($t \gg \Delta t_{1/2}$ and $P_{abs} \simeq P_{inj}$) found by simulations for the given density is practically independent of the initial temperatures. In Fig. 2 (right), an example of the steady-state is shown

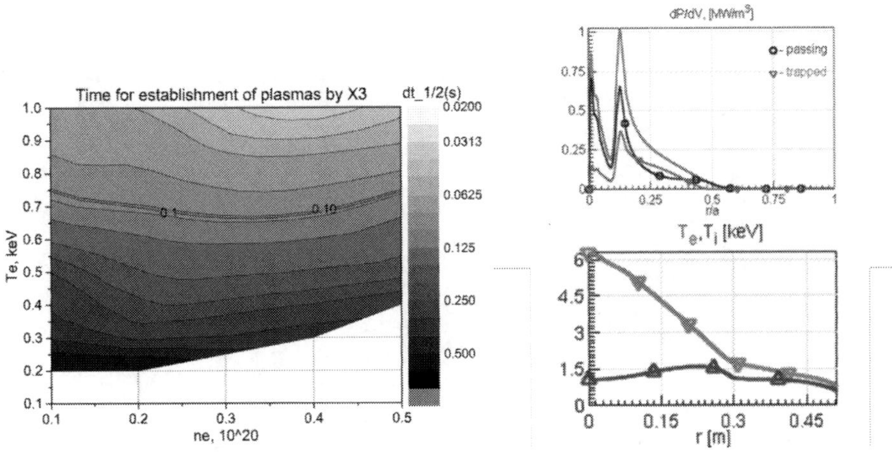

Figure 2. *Left*: The plot of $\Delta t_{1/2}$, i.e. the time interval required for evolution from the initial condition to the point where the half of launched power is absorbed; final steady state for plasmas with 5 MW heating by X3-mode. The red line corresponds to the critical value $\Delta t_{1/2}$ = 100 ms. *Right*: For fixed density, $n_e(0) = 3 \times 10^{19}$ m^{-3}, the deposition profile, dP_{abs}/dV (red for total, blue and green for passing and trapped electrons, respectively), and temperature profiles, $T_{e,i}$, are shown for the final steady state, $t \gg \Delta t_{1/2}$. The fractions of power absorbed by trapped and passing electrons are also shown.

which is reached for $n_e(0) = 3 \times 10^{19}$ m^{-3} with 5 MW of pure X3 heating. The solution obtained corresponds to the CERC (core electron root confinement) regime [13] which is characterized by large positive radial electric field, E_r, peaked T_e and low T_i due to a collisional decoupling of electrons and ions. One can find also that the integral absorbed power is about equally distributed between the trapped and passing electrons, which, potentially, might play a role for generation of non-local effects in the transport. Indeed, since the trapped electrons in stellarators, being locked within the toroidal mirrors, do not spread the absorbed power toroidally over the magnetic surface, the local absorbed power density is significantly increased (by a factor of five, at least), and some specific kinetic effects may be generated (additional convective losses [14], for example). Nevertheless, there is no clarity to date about the consequences of this effect.

The main conclusion from this consideration is the following: for the high-mirror magnetic configuration (UEM) with reduced magnetic field, B_0 = 1.75 T, target plasmas suited for X3 heating (a total power of 5 MW was checked) must be preheated up to temperatures higher than 0.7 keV. This value has almost no dependence on the initial density. This requirement must be taken into account for any method used to create a target plasma for ECRH with the X3-mode.

3 Target plasma with X2 (105 GHz) start-up

In this section we consider possibilities to create the target plasmas with help of the X2-mode at 105 GHz. Again, while the gyrotrons can be used for generation of 105 GHz without any problems (only the efficiency is reduced), the quasi-optical system needs to be tested and might require an additional optimization for this frequency. Below, we suppose that this option is available.

For simulations, we apply the same magnetic configuration and the same magnetic field, i.e. UEM with B_0 = 1.75 T. For X2 heating, four RF beams of 105 GHz with the total power 2 MW are considered. Since the cyclotron resonance is strongly shifted to the high-field-side (HFS) and quasi-perpendicular launch seems to be rather ineffective (only plasma periphery can be heated), a moderate obliqueness for the beams needs to be used, $N_\parallel \simeq 0.25 - 0.3$. This case corresponds to off-axis heating at $r/a \simeq 0.5 \pm 0.2$ (see Fig.3). As expected for X2, the plasma is optically thick and the single-pass absorption is 100%. Since the launch is oblique and located close to the maximum of B, only passing electrons absorb the power.

Transport simulations show that the temperature profiles established in steady-state are flat and $T_e \simeq T_i \simeq$ 1.4 keV, which is completely satisfactory for X3 operation. This scenario, as common with off-axis heating, is characterized by central "ion-root" confinement with negative radial electric field and with collisionally well coupled electrons and ions.

In addition to the steady-state simulations, one also needs to check the breakdown with RF heating beams. In W7-X, the typical time-delay between switch-on the power and plasma breakdown for quasi-perpendicular RF beams is about 10 ms [5]. However, since the energy of the seed electrons required for cyclotron interaction increases with obliqueness ($E_0 \simeq m_e c^2 N_\parallel^2/2$, as follows from the resonance condition), a similar tendency is expected for the time-delay, and, at first glance, oblique launch is not favorable for breakdown. At the same time, there are exper-

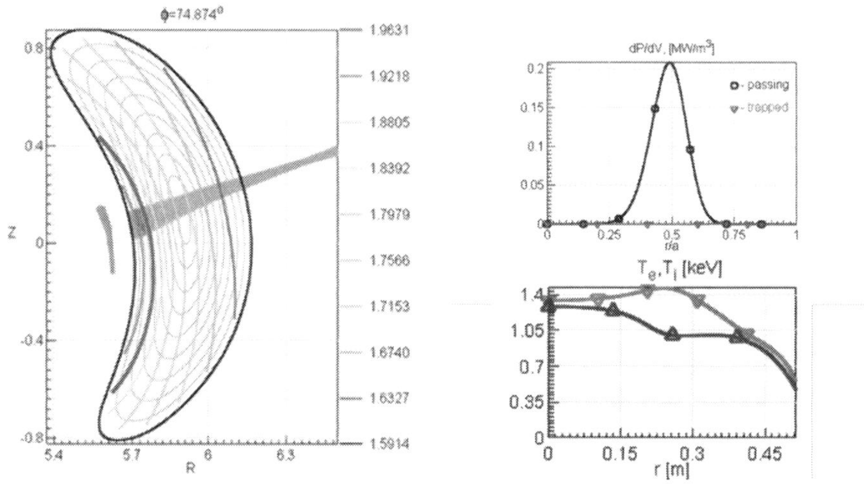

Figure 3. *Left*: $R-Z$ projection of the RF beam applied for X2 (105 GHz) heating. *Right*: Deposition profile, dP_{abs}/dV, and temperature profiles, $T_{e,i}$, calculated for the same density as above, $n_e(0) = 3 \times 10^{19}$ m^{-3}. The results correspond to the final steady state.

imental investigations of the breakdown for both stellarators [16] and tokamaks [17] which confirm that its dependence on obliqueness is rather weak. Here, we assume that the time-delay for X2 breakdown is acceptable and focus only on steady-state parameters which must be compared with requirements for target plasmas.

Apart from the scenario considered for ECRH start-up, it seems realistic also to simplify the establishment of the initial plasma by using for the X2 breakdown (105 GHz) the field $B_0 = 1.87$ T. In this case, the location of the resonance is much more favorable and practically on-axis deposition can be obtained. As a 2nd step, a transition to parameters necessary for operation with the X3-mode needs to be done, i.e. ramping down the magnetic field while simultaneously increasing the obliqueness of the RF beams. Since the rate of change of the magnetic field is limited to $|dB_0/dt| < 0.24$ T/min, the time required for the transition from 1.87 T to 1.75 T can be estimated as $\Delta t > 30$ s.

As a preliminary result from these considerations, one can conclude that establishment and preheating of target plasmas suited for stable X3-operation at reduced magnetic field, X2 ECRH start-up using beams with a frequency of 105 GHz seems quite realistic.

4 Summary

For the magnetic configuration UEM characterized by very high mirror-ratio, $B_{01}/B_{00} \simeq 0.24$, the scenario with reduced magnetic field, $B_0 \simeq 1.75$ T, was considered. Pure X3 heating at 140 GHz was numerically investigated with help of the coupled ray-tracing code TRAVIS [10] and transport code NTSS [11]. General requirements for target-plasmas necessary for X3 operation were studied.

Despite incomplete single-pass absorption multipass ($\leqslant$ 3 passes) heating allows deposition profiles well localized within the plasma core. The expected neoclassical transport regime is CERC. At the same time, the X3-mode alone appears to be insufficient for ECRH start-up and other methods need to be applied for creation of the target plasma with parameters necessary for X3 operation.

A scan over the initial electron density and temperature was performed for 6 RF beams with 5 MW total power. It was found that stable operation with the X3-mode is possible if the target plasma is preheated up to electron temperatures exceeding 0.7 keV and this value has only a weak dependence on the initial density. If this condition is satisfied, the time interval required for evolution to the final steady-state with X3 heating does not take more than 100 ms. This value meets with technical requirements to the safety of the mirrors used for beam reflection.

As an example of possible target-plasma creation for X3 operation, an X2 scenario at the frequency 105 GHz was considered. Transport and ray-tracing simulations demonstrate that the temperatures in steady-state reach $T_e \simeq T_i \simeq 1.4$ keV, be matching with the requirements found. This scenario is based on the assumption that the high obliqueness of the RF beams does not significantly reduce the breakdown efficiency. As an alternative, if breakdown far from the axis by oblique RF beams is problematic, the magnetic field $B_0 \simeq 1.87$ T for X2 breakdown at 105 GHz can be applied since the cyclotron resonance is located close to the axis and no obliqueness required. In order to make a transition to X3 operation, the magnetic field needs to be ramped down (the required time is about of 30 s).

As a preliminary result from this consideration, we can conclude that the establishment and preheating of target plasmas suitable for stable X3-operation at reduced magnetic field by X2 ECRH start-up using the RF beams at the optional frequency of 105 GHz is quite realistic. Further options for target-plasma generation appear to be difficult (NBI) or need to be assessed (ICRH).

References

[1] F. Warmer et al., Plasma Phys. Control. Fusion **58**, 074006 (2016)

[2] J. Nührenberg and E. Strumberger, Contrib. Plasma Phys. **32**, No. 3/4, 204 (1992)

[3] V. Erckmann et al., Fusion Sci. Technol. **52**, 291 (2007)

[4] G. Dammertz et al., IEEE Trans. Plasma Sci. **34**, 173 (2006) (https://ieeexplore.ieee.org/document/1621285)

[5] R. C. Wolf et al., Nucl. Fusion **57**, 10202 (2017)

[6] K.J. Brunner et al., EPJ Web of Conferences **149**, 03011 (2017) (https://doi.org/10.1051/epjconf/201714903011)

[7] A. Dinklage et al. Nature Physics **14**, 855 (2018) (https://doi.org/10.1038/s41567-018-0141-9)

[8] Th. Rummel et al., IEEE Trans. Plasma Sci. **46**, No. 5, 1517 (2018) (https://ieeexplore.ieee.org/document/8327848)

[9] N. B. Marushchenko et al., EPJ Web of Conferences **87**, 01007 (2015) (http://dx.doi.org/10.1051/epjconf/20158701007)

[10] N. B. Marushchenko, Y. Turkin, H. Maaßberg, Comp. Phys. Commun. **185**, 165 (2014)

[11] Y. Turkin et al., Phys. Plasmas **18**, 022505 (2011)

[12] S. P. Hirshman et al., Phys. Fluids **29**, 2951 (1986)

[13] M. Yokoyama et al., Nucl. Fusion **47**, 121 (2007)

[14] S. Murakami et al., *J. Plasma Fusion Res. Ser.* **1**, 122 (1998).

[15] D. Gradic et al., Nucl. Fusion **55**, 033002 (2015)

[16] K. Nagasaki et.al., Nucl. Fusion **45**, 13 (2005)

[17] J. Stober et al., Nucl. Fusion **51**, 083031 (2011)

EC absorption efficiency in ITER at one-third nominal magnetic field strength

Daniele Micheletti*, *Daniela* Farina, and *Lorenzo* Figini

Istituto di Fisica del Plasma - CNR, 20125 Milano, Italy

Abstract. During the first phase of the ITER Pre-Fusion Power Operations (PFPO-1), the Electron Cyclotron Radio Frequency system will be the main heating and current drive source [1]. Evaluations of the L-H power threshold [2] have shown that H-mode is unlikely to be achieved in PFPO-1 scenarios at half-field (B_0=2.65T). For this reason, the use of one-third field scenarios in an early phase might help to access H-mode. This means that the EC Heating (ECH) system, with gyrotrons operating at 170 GHz, would then rely on n=3 harmonic resonance. Using a lower frequency, namely 104 GHz, could allow the exploitation of n=2 harmonic resonance, improving heating efficiency and facilitating EC-assisted breakdown which has not been demonstrated yet for n=3 harmonic. This work presents a preliminary evaluation of the performances of an EC system using the two frequencies in a B_0=1.8T scenario (obtained by adequately rescaling a reference half-field scenario [3]) via a parametric scan of density/temperature profiles, to cover the range of possible values expected to occur during the operations and provide an estimation of the parameter space region where the use of each frequency allows a good absorption efficiency.

1 Introduction

The initial phase of the ITER research plan, after the First Plasma and before the D and DT operations, consists of non-activated H/He plasmas in which the L-to-H-mode transition power thresholds will be investigated [1]. In the first part of this phase, named Pre-Fusion Power Operation 1 (PFPO-1), the Heating and Current Drive system (H&CD) will consist of a 20 MW Electron Cyclotron Radio Frequency (ECRF) system, and a possible 10 MW Ion Cyclotron RF system. During the PFPO-1 the magnetic field will start at half the nominal field strength, that is B_0 = 2.65 T, and then proceed towards full-field (B_0 = 5.3 T). It has been shown [2] that H-mode access at half-field in this scenario is unlikely due to the high power threshold. For this reason, a one-third field scenario (B_0 = 1.8 T) has been proposed, since it could help to overcome the L-to-H threshold and access H-mode. The ITER gyrotrons will operate at f = 170 GHz, taking advantage of n=1 harmonic resonance at the nominal field and n=2 harmonic resonance at half-field. Using a magnetic field at one-third strength means that the heating system would rely on the lower efficiency of n=3 harmonic resonance absorption. Using dual-frequency gyrotrons working both at 170 GHz and at a lower frequency, namely 104 GHz, would allow to exploit the n=2 harmonic resonance absorption at 1.8 T and its higher absorption efficiency at the beginning of the plasma discharge, so to increase the temperature till the point at which the absorption of 170 GHz EC waves by the n=3 harmonic reaches a safe level. Besides, n=3 harmonic EC-assisted breakdown has not been achieved up to now.

The present work aims at presenting a preliminary evaluation of the performances of an EC system operating at the two frequencies (170/104 GHz) in a one-third magnetic field scenario. The ITER ECRF system, the reference scenario and the modelling framework used for the analysis are briefly described in Section 2. The performances of the EC launchers for f = 170 GHz are evaluated in Section 3 by calculating absorption efficiency and access to inner plasma region for a wide range of electron density (n_e) and temperature (T_e) values. The analysis is repeated for the lower frequency f = 104 GHz in Section 4. The results of the analysis are summarised in Section 5.

2 Modelling framework

2.1 ITER ECRF system

The ITER ECRF system consists of 5 launchers: 1 Equatorial Launcher and 4 Upper Launchers. The Equatorial Launcher (EL) includes 3 mirrors (TOP, MID, LOW), while each Upper Launchers (UL) includes 2 mirrors, Upper and Lower Steering Mirror (USM, LSM). The launchers are connected to 24 gyrotrons working at f = 170 GHz, providing 1 MW each (0.83 MW effectively injected into the plasma due to losses along the transmission line). During PFPO-1 all 24 gyrotrons will be available, for a total of 20 MW injectable into the plasma. The gyrotrons output can be

* Corresponding author: micheletti@ifp.cnr.it

split among the launchers: each of the EL mirrors can inject 8 beams into the plasma, while the UL mirrors uses 4 beams each.

2.2 Plasma reference scenario

The analysis presented in the following sections is based on the ITER half-field scenario (B_0 = 2.65 T, I_p = 7.5 MA, He plasma, H-mode) [3] rescaled to one-third field (B_0 = 1.8 T, I_p = 5 MA), keeping constant safety factor q and Greenwald density fraction. Fig. 1 shows T_e and n_e profiles of the resulting scenario (the same scaling factor was used for both profiles). In the following sections, the scaled profiles will be referred to using their central plasma values, $T_{e,0} = T_e(\rho=0)$ and $n_{e,0} = n_e(\rho=0)$.

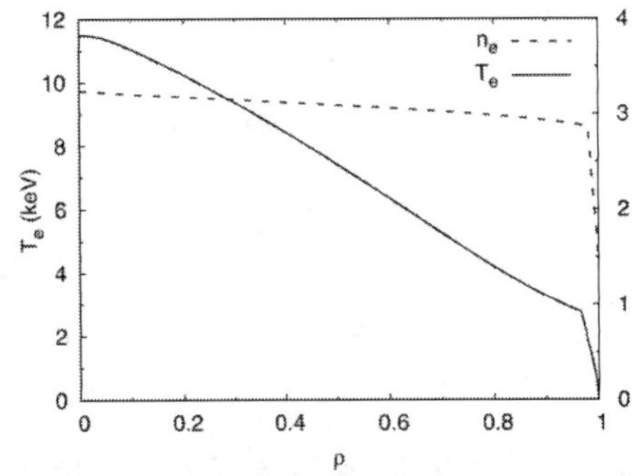

Fig. 1. T_e and n_e profiles for B_0=1.8T scenario.

2.3 Simulations and optimal injection angle

The ECH&CD calculations have been carried out using the beam-tracing code GRAY [4], using a representative 1 MW X-mode polarized beam for each mirror. Tables 1a,b list the mirror positions, the injection angles and the beam parameters at launch (width and radii of curvature along directions perpendicular to propagation) used.

Table 1a. Beam starting position and injection angles.

	R (m)	z (m)	α (°)	β (°)
TOP	9.394	1.192	-20÷15	-20
MID	9.394	0.620	0÷35	25
LOW	9.394	-0.004	-10÷25	25
USM	6.999	4.414	40÷65	20
LSM	7.054	4.178	30÷55	20

Table 1b. Starting beam waist and radii of curvature.

	w_x (cm)	w_y (cm)	R_x (cm)	R_y (cm)
TOP	2.65	2.76	-63.53	-210.6
MID	2.65	2.76	-63.53	-210.6
LOW	2.65	2.76	-63.53	-210.6
USM	5.05	5.05	-318.6	-318.6
LSM	4.81	4.81	-200.1	-200.1

The poloidal (α) and toroidal (β) injection angles, defined from the cylindrical components of the wavevector at launch $\boldsymbol{k} = (k_r, k_\Phi, k_z)$ as

$$\tan\alpha = k_z / k_r \qquad (1)$$

$$\sin\beta = k_\Phi / |\boldsymbol{k}| \qquad (2)$$

impact on the absorption efficiency. The values of β correspond to the most recent design of the EC system [1], while α cover the full steering range available for each mirror. Fig. 3 shows an example of beam paths for f = 170 GHz. It can be noticed that both n=3 and n=4 harmonic are in the plasma, on HFS and LFS.

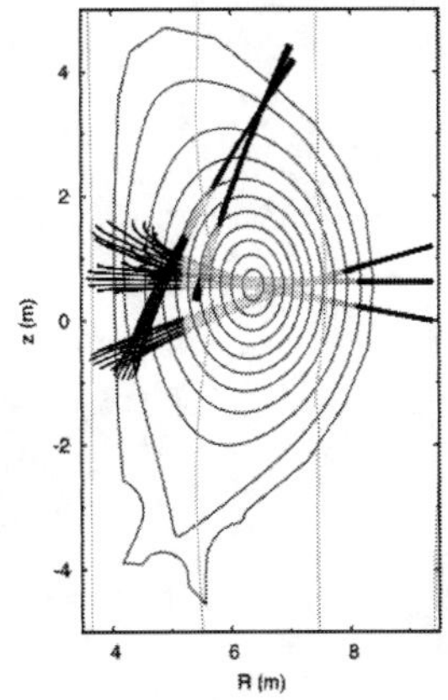

Fig. 2. Propagating 170 GHz beams for UL and EL, with harmonic resonance locations (n=3,4 inside the plasma).

To cover the range of possible values expected to occur during the discharge process in PFPO-1, n_e and T_e profiles have been varied by means of an independent parametric scaling, keeping constant B and I_p, as

$$n'_e(\rho) = f_n \cdot n_e(\rho) \qquad (3)$$

$$T'_e(\rho) = f_T \cdot T_e(\rho) \qquad (4)$$

with the scaling factors $0.05 \leq f_n, f_T \leq 1$.

The optimal poloidal steering angle α_{opt}, defined as the angle that maximizes the absorbed power fraction, has been determined for each mirror via a scan on the steering range for different T_e, n_e scaling combinations, and has then been used throughout the subsequent analysis. As shown in Fig. 3, α_{opt} is located at the extreme of the steering range corresponding to the beam pointing toward the centre of the plasma. This is true for every mirror and profile combination.

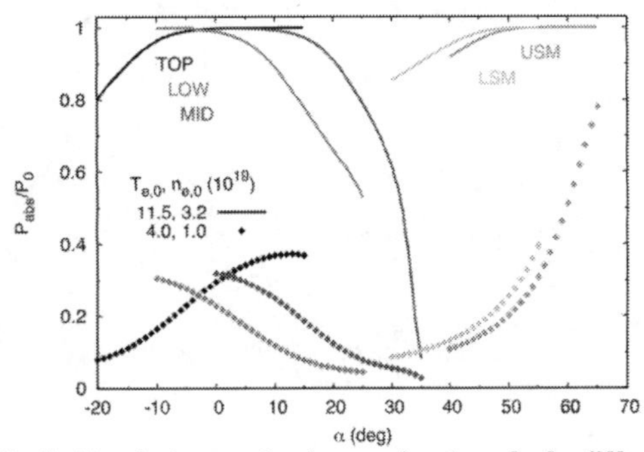

Fig. 3. Absorbed power fraction as a function of α for different T_e, n_e scaling.

3 Performances at 170 GHz

3.1 Power absorption efficiency

The analysis, as written in the previous section, has been performed for a wide range of temperature and density scaling factors. Fig. 4a,b show the dependence of absorption efficiency on electron temperature and density for two mirrors, while in Fig. 5 corresponds to slices at constant $n_{e,0}$ (for all mirrors). As expected, the efficiency increases more rapidly with T_e than n_e (optical thickness for n=3 harmonic proportional to $n_eT_e^2$). It can be seen that for MID (similar results obtained for TOP, LOW and LSM) a large fraction of the injected power is not absorbed at typical temperature values of an ohmic plasma.

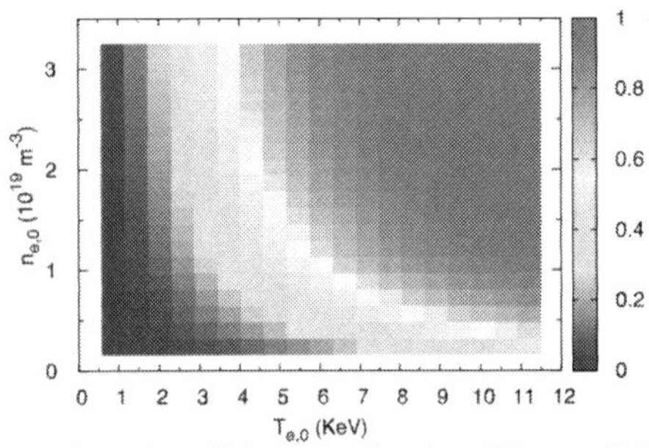

Fig. 4a. Absorption efficiency as a function of $T_{e,0}$, $n_{e,0}$ (MID).

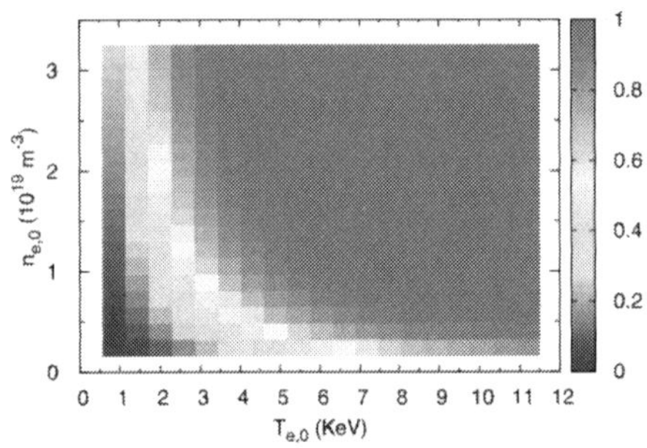

Fig. 4b. Absorbed efficiency as a function of $T_{e,0}$, $n_{e,0}$ (USM).

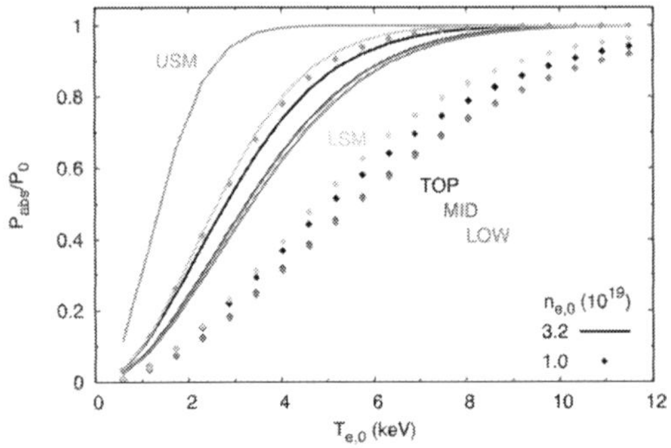

Fig. 5. Absorbed efficiency as a function of $T_{e,0}$ for different $n_{e,0}$ scaling; lines correspond to slices at constant $n_{e,0}$ in Fig.4a,b and equivalent maps for other mirrors.

As reported in Table 2, a good absorption efficiency (90% of input power absorbed) can be reached only for $T_{e,0} \geq 5$ keV for EL mirrors and around this value for LSM. Only USM allows to reach a good absorption already at $T_{e,0} \geq 3$ keV for $n_{e,0} \geq 2 \cdot 10^{19}$ m^{-3}, but the threshold temperature increases rapidly with decreasing density. This mirror alone, however, would allow the use of only 16 of the 24 available beams (13.3 MW out of 20).

Table 2. $T_{e,0}$ for 90% absorption for different density scaling.

	$n_{e,0}$ (m^{-3})	
	3.2·10^19	**1.0·10^19**
TOP	5.5 keV	10.2 keV
MID	6.2 keV	10.9 keV
LOW	6.4 keV	10.9 keV
USM	2.7 keV	5.1 keV
LSM	4.9 keV	9.2 keV

The difference in efficiency between MID/LOW and TOP, is due to the different toroidal injection angles. Beams from UL mirrors are more focused than those from EL and are nearly tangent to the magnetic surfaces, being able to travel for a longer path in the resonance region. Moreover, USM is able to reach deeper into the plasma thanks to its higher poloidal angle.

The low absorption efficiency for low temperature values means that only a fraction of the injected power is available for heating, thus leaving a relevant amount of residual power free to reach the wall on the HFS of the machine, where it is reflected and partially converted to O-mode polarization. This fraction, around 20% of incident power, is then able to pass through the plasma undergoing minimal absorption and, for beam originating from EL mirrors, can possibly reach the LFS wall and ports.

3.2 Power deposition profiles and CD

The presence of the n=4 harmonic inside the plasma can impact on the accessibility to the plasma inner regions and also on the Current Drive (CD) efficiency, although if the goal is pure heating a low CD is not an issue. The power deposition profiles, as shown in Fig. 6, indicate that the deposition region for the UL mirrors are significantly narrower ($\Delta\rho_{tor}$<0.1) than for those of EL. This is due to the fact that the beams propagating form UL mirrors enter the plasma already behind the n=4 harmonic resonance and do not interact with it.

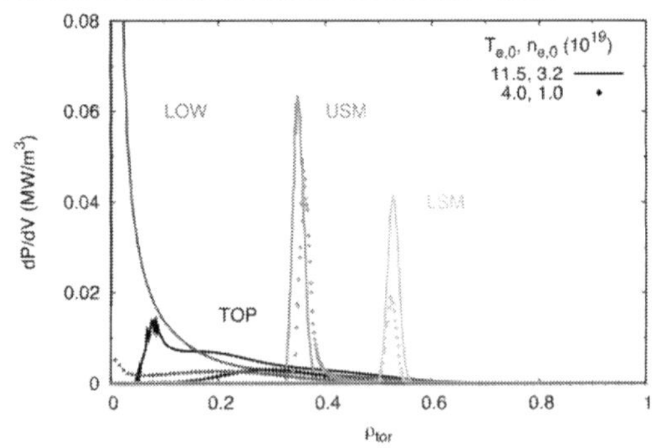

Fig. 6. Power deposition profile for different T_e, n_e scaling ($T_{e,0}$ in keV, $n_{e,0}$ in m^{-3}).

The fact that the beams are nearly tangent to the magnetic surfaces in the region of n=3 harmonic resonance means that the spreading of the power deposition region depends more on the local beam width than density and temperature scaling, which impact on it mainly through their effect on local optical thickness. The beams launched form EL mirrors, conversely, cross both n=4 and n=3 harmonic resonance regions and are nearly perpendicular to the magnetic surfaces, presenting broader profiles. EL mirrors however are able to reach regions of plasma at lower radii than UL, that is restricted to ρ_{tor} > 0.3-0.5. The interaction of n=4 harmonic with the EL mirrors becomes evident when evaluating the fraction of power it absorbs compared to the total absorbed power, as shown in Fig. 7a,b.

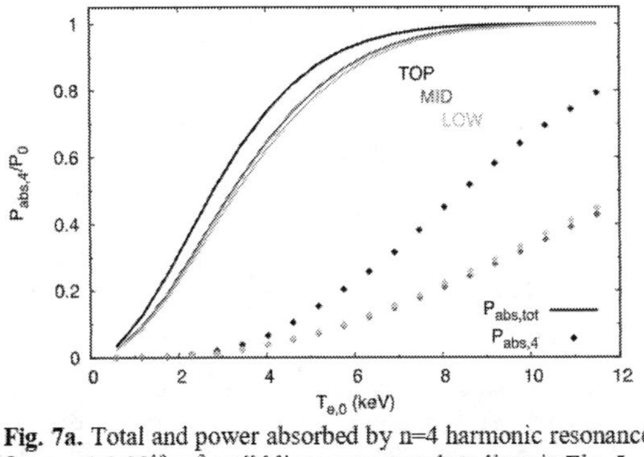

Fig. 7a. Total and power absorbed by n=4 harmonic resonance for $n_{e,0}$=3.2·10[19] m[-3]; solid lines corresponds to lines in Fig. 5.

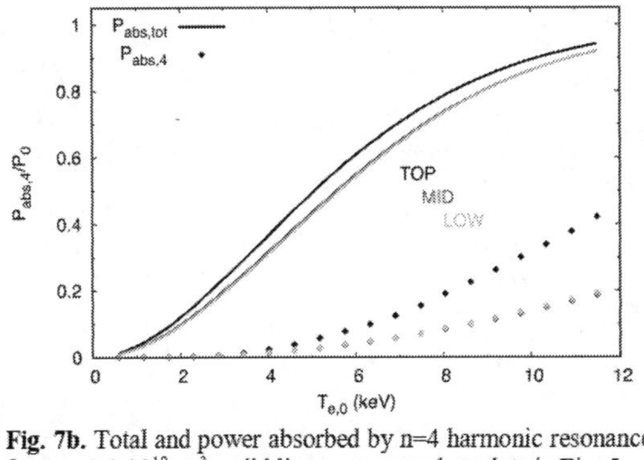

Fig. 7b. Total and power absorbed by n=4 harmonic resonance for $n_{e,0}$=1.0·10[19] m[-3]; solid lines corresponds to dots in Fig. 5.

The absorption for EL mirrors can become significant with increasing T_e. The most affected is the TOP mirror, for which $P_{abs,4}/P_0 \geq 50\%$ for $T_{e,0} > 8.5$ keV.

This might be an issue for Current Drive applications. The n=4 harmonic absorption, in fact, has two detrimental effects on the driven current (I_{CD}): it reduces the power that can reach the higher efficient n=3 harmonic resonance; it generates a current flowing in the opposite direction of the I_{CD} generated by n=3. Its CD efficiency, however, is very low compared to n=3 harmonic and thus the opposite I_{CD} is very low or negligible.

The reduction of power reaching the lower harmonic can have a greater effect: both n=3 harmonic CD efficiency and n=4 harmonic power absorption efficiency increase with T_e. At high T_e, if the n=3 CD efficiency is not able to compensate the loss of power due to n=4 absorption, the I_{CD} it generates can decrease, as shown in Fig. 8 for the LOW and TOP mirrors (similar results for MID) starting from $T_{e,0}$ ~ 8-9 keV.

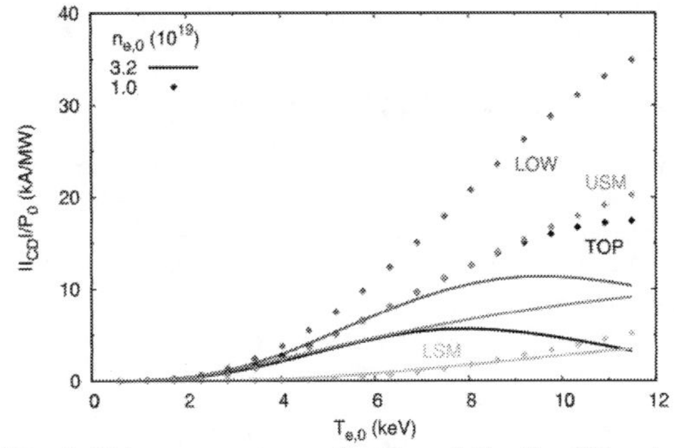

Fig. 8. Driven current as a function of $T_{e,0}$ for different n_e scaling ($n_{e,0}$ in m[-3]).

4 Performances at 104

4.1 Power absorption efficiency

The same analysis described in the previous section has been performed for the three EL mirrors simulating the propagation of X-mode polarized beams with f = 104 GHz (the use of dual-frequency gyrotrons is foreseen, if realized, only for the three EL mirrors). The beam parameters are those used in the previous section, since a design of an EC system operating at f = 104 GHz is not available at the moment.

The use of the lower frequency, as said before, would allow to take advantage of the n=2 harmonic resonance, which for B_0 = 1.8 T and f = 104 GHz is located on the HFS near the centre of the plasma. For this part of the analysis the temperature scaling has been extended down to 0.005 times the unscaled profile ($T_{e,0}$ ~ 50 eV), in order to identify the range of T_e, n_e values for which the EC waves are not completely absorbed on the first pass in the plasma. In fact, the simulations have shown a complete absorption of the injected power for all but the lowest density and temperature values in the previously used scaling range, as can be guessed from Fig. 9.

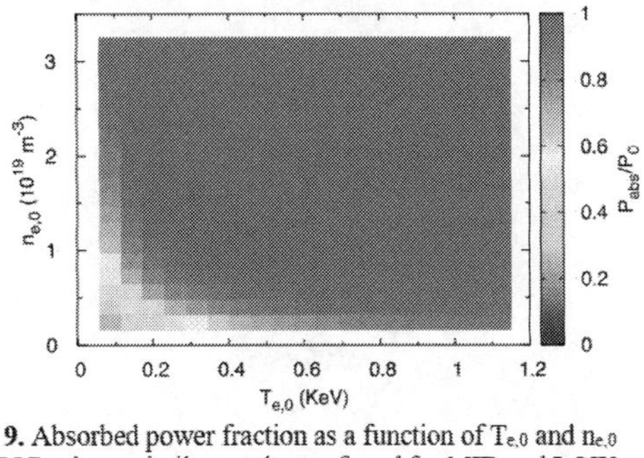

Fig. 9. Absorbed power fraction as a function of $T_{e,0}$ and $n_{e,0}$ for TOP mirror; similar results are found for MID and LOW.

The α_{opt} values determined for f = 170 GHz have been used also for the lower frequency. Using 104 GHz EC waves would thus allow full absorption even for temperature values found in an ohmic plasma, with no risk of residual power propagating around the vessel. At high density there can be a decrease in absorption due to refraction effects, especially for beams not pointing towards the plasma centre ($n_{e,0}/n_{cut-off} \sim 0.5$ for the highest density value considered in this analysis).

4.2 Power deposition profiles and CD

The n=3 harmonic is located outside the plasma, and shouldn't therefore have a significant interaction with the injected beams. The analysis shows that there can be a small fraction of injected power absorbed by n=3 harmonic resonance, due to high energy electrons, but it always remains under 5% of total absorbed power, even for the highest temperature values considered. Fig. 10 shows the power density absorption profile for the different EL mirrors, similarly as Fig. 6. As for f = 170GHz, the profiles are still wider than UL, albeit narrower than for the higher frequency ($\Delta\rho_{tor}$~0.1-0.15) due to the fact that only one harmonic resonance is involved. As in Fig. 6, it is clear that density and temperature scaling does not impact significantly on the power deposition width while does influence the radial deposition location, and that the EC waves are still able to reach the inner regions of the plasma.

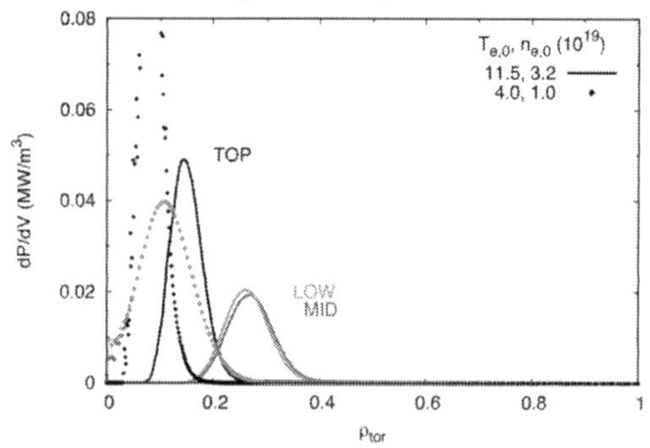

Fig. 10. Power deposition profile for different $T_{e,0}$, n_e scaling ($T_{e,0}$ in keV, $n_{e,0}$ in m^{-3}).

The n=2 harmonic resonance higher Current Drive efficiency in conjunction with the higher absorption result in a higher driven current than f = 170 GHz case, as shown in Fig. 11.

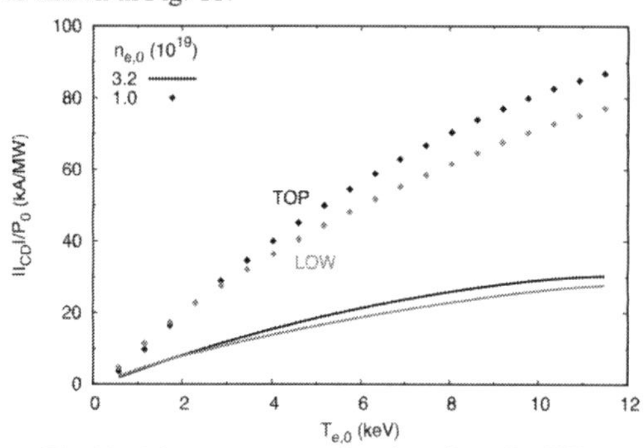

Fig. 11. Driven current as a function of $T_{e,0}$ for different n_e scaling ($n_{e,0}$ in m^{-3}).

For f = 104 GHz only one harmonic resonance (n=2) contribute significantly to absorption. This avoids the effects causing a decrease in I_{CD} at high temperatures, unlike the higher frequency.

5 Conclusions

This study provided a preliminary estimation of the performances of an ECRF system using the two frequencies in a one-third magnetic field scenario.

The analysis performed has shown that the 170 GHz ECRF system of ITER can ensure good absorption efficiency also for one-third field operations, provided the electron temperature is higher than approximately 5 keV for typical density values of the order of $3 \cdot 10^{19}$ m^{-3}. At lower temperature, the unabsorbed fraction of the injected power shines through the plasma hitting the HFS wall and can potentially reach the LFS wall and ports. The only exception is the USM, which already shows good absorption at lower temperature ($T_{e,0} \geq 3$ keV for $n_{e,0} \geq 2 \cdot 10^{19}$ m^{-3}), easily reachable with ohmic heating. This means that at least 13.3 MW out of the 20 MW available to the EC system can be used to help raising the temperature to the point where the other launchers are effective.

The use of dual-frequency gyrotrons operating also at 104 GHz would allow to obtain a complete power absorption starting from much lower temperature and density values, allowing to bridge the gap between the ohmic plasma conditions at the beginning of the discharge and the range of efficiency of the 170 GHz system.

The use of f = 104 GHz would also allow a higher Current Drive efficiency with respect to f = 170 GHz, which has limited CD efficiency, especially for EL mirrors at high temperature ($T_{e,0} \geq$ 8-9 keV for $n_{e,0} \geq 3 \cdot 10^{19}$ m^{-3}), due to the presence of n=4 harmonic in the LFS of the plasma

A complete assessment of the performance would need a more in-depth analysis, accounting for mutual influence between plasma profiles evolution and EC power absorption.

References

1. M. Henderson, et al, Phys. Plasmas **22**, 021808 (2015)
2. Y.R. Martin, et al, Journal of Physics : conference Series **123**, 012033 (2008)
3. V. Parail, et al, Nucl. Fusion **53**, 113002 (2013)
4. D. Farina, Fusion Sci. Technol. **52**, 154 (2007)

Fast evaluation of the current drive efficiency by electron cyclotron waves for reactor studies

Emanuele Poli[1,*], *Maximilian* Müller[1], *Hartmut* Zohm[1], and *Michael* Kovari[2]

[1]Max-Planck-Institut für Plasmaphysik, Garching bei München, Germany
[2]CCFE, Culham Science Centre, Abingdon, UK

Abstract. The determination of the current driven by electron cyclotron waves is usually performed employing ray/beam tracing codes, which require as an input the magnetic equilibrium, the electron density and the electron temperature profiles on one side and the beam injection parameters on the other. In the frame of systems-code applications, however, a different approach is needed, as some of the required input quantities are not available. Here, a procedure to evaluate the achievable ECCD efficiency for given global reactor parameters is proposed. It relies on a single numerical evaluation of the current drive efficiency (based on the adjoint method and including momentum-conserving corrections) for suitably chosen input values. The results are shown to be in good agreement with the full numerical optimization of the ECCD efficiency for a number of reactor-relevant scenarios. As described in this paper, this approach does not include the effect of parasitic absorption from higher cyclotron harmonics, which becomes important starting from electron temperatures of the order of 30 keV.

1 Introduction

The calculation of the current driven by auxiliary heating systems is just a small part of the operations performed by systems codes like PROCESS [1, 2] in order to assess the performance of a fusion reactor for a given set of global parameters. While in most applications electron-cyclotron current drive (ECCD) is calculated by ray/beam tracing codes [3] employing as an input some given values for the magnetic equilibrium (two-dimensional in tokamaks), the density and the electron temperature profiles (one-dimensional), and the EC beam parameters (launch position, angles and frequency), in a systems code some of these data are not available. In particular, the antenna parameters which would yield the optimum current drive for a given set of global parameters are not known. This makes the usual strategy based on tracing the beam path through the plasma and exploring the relevant parameter space [4] not applicable, even if the numerical burden were manageable. In this contribution, a procedure is proposed to evaluate the current driven by EC waves through a single numerical determination of the ECCD efficiency according to the adjoint method [5] for a suitably selected set of input values deduced from the global machine parameters. The related predictions are compared to extensive optimization loops performed with the code TORBEAM [6, 7] for a number of reactor-relevant scenarios. The computational effort of the proposed approach, which amounts basically to the numerical quadrature of two one-dimensional integrals (one integrand being itself known in integral form), is assumed here to be affordable within systems code applications.

*e-mail: Emanuele.Poli@ipp.mpg.de

2 Theoretical background

In order to establish the notation and introduce the discussion to follow, the basic features of the interaction between EC waves and electrons around the cyclotron resonance (or one of its harmonics) are recalled in this section. The energy carried by the wave is transferred to the particles in the plasma satisfying the relativistic resonance condition [8])

$$\omega - \frac{n\Omega}{\gamma} - k_\parallel v_\parallel = 0, \qquad (1)$$

where ω is the frequency of the injected wave, n is the harmonic number, $\Omega = eB/m$ is the cyclotron frequency of the electrons, $\gamma = \sqrt{1 + u_\parallel^2 + u_\perp^2}$ is the relativistic Lorentz factor (with $\mathbf{u} = \mathbf{p}/mc$ the normalized momentum) and $k_\parallel$ is the component of the wave vector parallel to the magnetic field. Introducing $N_\parallel = ck_\parallel/\omega$ and $\overline{\Omega} = \Omega/\omega$, the resonance condition becomes

$$\gamma - N_\parallel u_\parallel - n\overline{\Omega} = 0. \qquad (2)$$

In momentum space, the resonance curve for given values of $\overline{\Omega}$ and $N_\parallel$ is hence a half-ellipse (only values $u_\perp > 0$ are considered)

$$(1 - N_\parallel^2)\left(u_\parallel - \frac{N_\parallel}{1 - N_\parallel^2} n\overline{\Omega}\right)^2 + u_\perp^2 = \frac{n^2\overline{\Omega}^2}{1 - N_\parallel^2} - 1. \qquad (3)$$

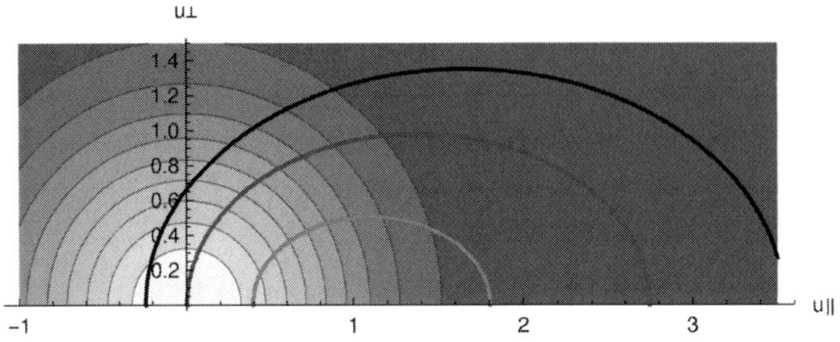

Figure 1. Relativistic resonance curves for $N_\parallel = 0.7$ and different values of $\overline{\Omega}$ (red: $\overline{\Omega} = 0.8$; blue: $\overline{\Omega} = 1$; black: $\overline{\Omega} = 1.2$). Maxwellian isocontours centred at $(u_\parallel, u_\perp) = (0, 0)$ are also shown.

which intersects the $u_\parallel$-axis at

$$u_{\parallel\mp} = \frac{nN_\parallel\overline{\Omega} \mp \sqrt{n^2\overline{\Omega}^2 - (1 - N_\parallel^2)}}{1 - N_\parallel^2}. \tag{4}$$

The resonance condition for a given harmonic n can be satisfied if

$$n\overline{\Omega} = \sqrt{1 - N_\parallel^2} \tag{5}$$

or for larger values of $n\overline{\Omega}$. For low-field side (LFS) injection, Eq.(5) defines the so-called pinch point [9] as the first point in resonance

$$u_{\parallel pp} = \frac{nN_\parallel\overline{\Omega}}{1 - N_\parallel^2} = \frac{N_\parallel}{\sqrt{1 - N_\parallel^2}} \tag{6}$$

which can be seen as a degenerate case of the ellipse given by Eq.(3).

Fig. 1 shows the resonance ellipses for three different values of $\overline{\Omega}$. Also shown are the contour levels of a Maxwell distribution function centred at the origin. The strongest wave-particle interaction can be assumed to take place around $u_{\parallel-}$, the lower intercept of the resonance curve with the $u_\parallel$-axis, as this is the point on the resonance curve where most electrons are available (for scenarios with negative $N_\parallel$, the role of $u_{\parallel-}$ and $u_{\parallel+}$ is swapped and the strongest interaction is localized around $u_{\parallel+}$). For a given harmonic, the position of $u_{\parallel-}$ decreases (for LFS injection schemes) from the pinch point to $u_\parallel = 0$ (if the cold resonance $\overline{\Omega} = 1$ is reached) and then crosses the origin. Since the collisionality decreases with increasing particle energy $\mathcal{E} = mc^2(\gamma - 1)$, optimum ECCD conditions are reached when sufficient absorption is provided on high-velocity electrons, as discussed in the following.

3 ECCD Optimization

Since the energy range of the resonant electrons is crucial for this study, here the electron energy for a standard ECCD application (injection onto the $q = 3/2$-surface with frequency 170 GHz as foreseen in ITER for neoclassical-tearing-mode stabilization [10]) is compared with that of a maximum-ECCD case, see Fig. 2. In the latter case, the optimum ECCD conditions have been found by scanning the injection angles for different positions of the antenna and different wave frequencies [4]. For the stabilization of neoclassical tearing modes, where a good localization of the deposited power is beneficial [11], the wave energy is deposited on slower electrons ($\mathcal{E}/T_e \lesssim 2$), the absorption is strong and the radial extent of the absorption region is small (left panel). On the contrary, under conditions of high current drive, the wave-particle resonance is satisfied on electrons whose energy is several times larger than the electron temperature (right panel). As the number of electrons with such a high energy is low, the absorption is distributed across a much wider radial range. It is noticed that the optimum-ECCD case requires a wave frequency of 210 GHz, confirming previous studies [9]. The reason for this high frequency can be understood by recalling Eq.(6), which shows that increasing $N_\parallel$ shifts the resonance to higher $u_\parallel$ and hence increases the energy of the current-carrying electrons. Since parallel refractive index and wave frequency are connected at the pinch point through Eq.(5), a high value of $N_\parallel$ must be accompanied by an increase in the wave frequency in order to ensure that the resonance does not move into the LFS.

It is interesting to note that the energy of the resonant electrons at the position of maximum absorption, defined as the position along the beam axis at which the transfer of energy from the wave to the plasma is highest and marked by a black cross in Fig 2, is found for a value of $\mathcal{E}/T_e \equiv f_T \simeq 4$ under conditions of optimum ECCD (here $\mathcal{E}$ is defined as $mc^2(\sqrt{1 + u_{\parallel-}^2} - 1)$, according to the discussion in the previous section, and T_e is the electron temperature), as shown by the black cross in the right panel of Fig. 2. An analysis of other reactor-relevant plasma scenarios (see parameters in Table 1) leads to similar values of $\mathcal{E}/T_e$ for resonant electrons around the location of maximum absorption. This can be explained by the fact that optimum current drive is achieved by moving the resonance to the highest electron energies still compatible with sufficient absorption, as discussed above. As shown in [12],

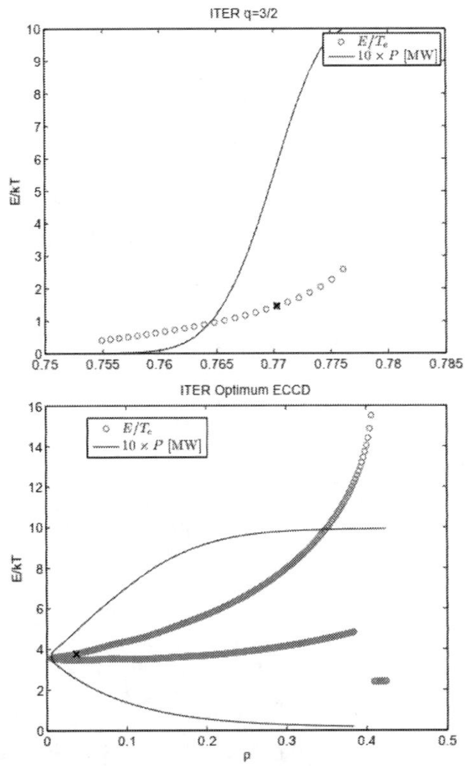

Figure 2. Comparison of the energy of the resonant electrons (calculated as $\mathcal{E} = mc^2(\gamma - 1)$ with $\gamma = \sqrt{1 + u_{\parallel -}^2}$, see text), normalized to the electron temperature, for a standard ITER situation (deposition on the $q = 3/2$-surface, injected frequency 170 GHz, launch position $(R, Z) = (705.4, 417.8)$ cm, toroidal injection angle $\beta = 20°$, left panel) and an ITER high-CD scenario (central deposition, injected frequency 210 GHz, launch position $(R, Z) = (810, 320)$ cm, toroidal injection angle $\beta = 37°$, right panel). The black crosses denote the position of maximum absorption. Also shown is the power carried by the beam, multiplied by 10 for the ease of comparison (black curves).

estimating analytically the absorption coefficient for the O-mode at the fundamental harmonic and imposing that the absorption takes place within a given fraction $\Delta\rho$ of the minor radius of the machine one can determine the maximum energy of the resonant electrons as a function of the plasma parameters. Imposing $\Delta\rho \simeq 0.2$ leads to a ratio $\mathcal{E}/T_e \simeq 4$ for values of temperature, density and magnetic field typically envisaged for a reactor. This condition is used in the approach described in the next section to fix the electron energy in the resonance condition and thus impose a constraint on ω and $N_\parallel$.

4 Evaluation of the optimum ECCD efficiency from global parameters

As explained in Sec. 1, an extensive search for the maximum achievable ECCD efficiency is not feasible within the optimization loop typical of reactor systems codes, because of the related numerical burden and because these codes do not provide, as a rule, the input needed by ray/beam tracing codes. The procedure adopted here with the aim of evaluating the efficiency achievable by ECCD for given global plasma parameters relies on a single numerical evaluation of the current drive efficiency, as implemented in the routine written by Lin-Liu for the TORAY-GA code [13], modified to include the momentum-conserving scheme developed by Marushchenko [14]. The parameters for the call are determined around the point of maximum absorption, since the procedure described at the end of the previous section provides a constraint on the energy of the resonant electrons at that position. Now we discuss how to determine the values of ω and $N_\parallel$. As it turns out, most of the input quantities needed by the current-drive routine can be inferred on the basis of the global parameters available from systems codes or calculated from the dispersion relation once ω and $N_\parallel$ have been fixed.

Recalling the discussion in Sec. 2, it is assumed that the wave-particle interaction occurs around the low-energy end of the resonance curve, i.e. around $u_{\parallel -}$. This allows also the determination of the relativistic factor in the resonance condition as $\gamma = \sqrt{1 + u_{\parallel -}^2}$. From Eq. (2), the parallel refractive index can then be expressed as a function of the wave frequency as

$$N_\parallel = \frac{1}{u_{\parallel -}} \left(\sqrt{1 + u_{\parallel -}^2} - \frac{n\Omega}{\omega} \right). \tag{7}$$

In order to determine ω, one can invoke the requirement on the localization of the absorption employed to fix the electron energy. If R_a is the major-radius coordinate of the position of maximum absorption, the extension along R of the absorption region can be determined as $\Delta R = a\Delta\rho \cos\chi$, where a is the minor radius and χ is the poloidal angle (measured from the midplane) of the absorption position. Since propagation from an elevated position is beneficial for ECCD in a reactor [4], the value of χ has been fixed here to $\chi = 60°$. Approximating now ΔR with the distance $R_{pp} - R_a$, where R_{pp} is the major radius of the pinch point, one can determine the ratio R_{pp}/R_a as

$$\frac{R_{pp}}{R_a} = 1 + \frac{a\Delta\rho}{R_a} \cos\chi. \tag{8}$$

Assuming a scaling $B \propto 1/R$, the position R_{pp} fulfils by definition the relation

$$\frac{R_{pp}}{R_a} = \frac{n\Omega(R_a)}{\omega \sqrt{1 - N_\parallel^2}}, \tag{9}$$

see Eq. (5). Assuming the absorption to be sufficiently localized that $N_\parallel$ can be taken as constant in the region of interest (as supported by beam tracing simulations), one can substitute Eq. (7), with $\Omega = \Omega(R_a)$, into the previous equation, to obtain a quadratic equation for the frequency shift $\omega/n\Omega(R_a)$. The relevant solution is

$$\frac{\omega}{n\Omega(R_a)} = \sqrt{1 + u_{\parallel -}^2} + |u_{\parallel -}| \sqrt{1 - \frac{R_a}{R_{pp}}}. \tag{10}$$

Table 1. Global tokamak parameters employed in the analysis of the ECCD efficiency. Additionally, a half-temperature DEMO1 case has been considered.

	DEMO1	DEMO2 nflat	DEMO2 npeak	ITER
R_0 [cm]	907	749.9	749.9	620
a [cm]	292	288.5	288.5	201
B_0 [T]	5.66	5.627	5.627	5.3
n_{e0} [10^{19} m^{-3}]	10.50	9.94	17.99	10.56
T_{e0} [keV]	33.25	31.17	23.50	24.49
$Z_{\rm eff}$	2	4.1823	4.13	1.7617

This determines in turn the value of $N_\parallel$ through Eq. (7). Since $\gamma = f_T T_e/mc^2 + 1$, from the previous equations it results that the frequency shift increases nearly linearly with T_e, while $N_\parallel$ scales with the square root of T_e. This result would suggest that higher and higher electron temperatures could be exploited to push the energy of the resonant electrons to higher and higher values (by correspondingly increasing $N_\parallel$ and hence $u_{\parallel pp}$) and thus improve the ECCD efficiency further and further. Unfortunately, the Doppler shift cannot be increased indefinitely because of the competing absorption of the next harmonic (for the scenarios considered here, the second O-mode harmonic), which limits the ECCD improvement with T_e to a range where the (local) temperature remains approximately below 30 keV [12]. This effect is still not included in the model described above.

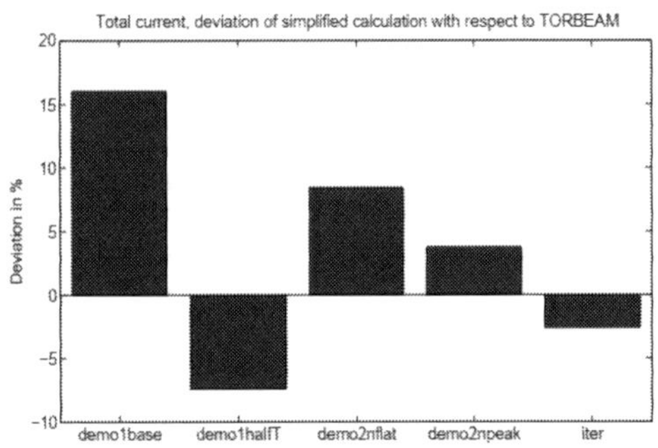

Figure 3. Deviation (in percent) between the estimated value of the current drive (calculated from a single evaluation of the ECCD efficiency, using Eqs. (7) and (10) to determine the required input) with respect to the current found in explicit beam tracing ECCD optimizations for the scenarios described in Table 1. The label demo1base (baseline) refers to the DEMO1 parameters reported in the table, while demo1halfT refers to the same case with the temperature divided by two.

Eqs. (7) and (10) have been implemented in a module which computes the input parameters needed for the numerical evaluation of the current drive efficiency. The results are summarized in Fig. 3. For each of the scenarios considered here, it can be seen that the procedure described above matches closely the results of a complete TORBEAM optimization, the difference between the computed and the estimated current being below 15%. The highest deviation is found for the case with the highest temperature, as expected, and is originated by the lack of a model for the saturation of the ECCD efficiency at high temperatures, as discussed above. This limitation has been removed in a work performed after the preparation of the present paper [15]. It should be stressed that given the reduction of the computational burden (a single evaluation of the current drive efficiency as compared to an optimization over a multi-dimensional parameter space via complete beam-tracing calculations), an error of the order of 10% can be considered as satisfactory. Other parameters, in particular the wave frequency corresponding to maximum ECCD, are found to be estimated quite accurately by the present model as well.

References

[1] M. Kovari *et al.*, Fus. Eng. Des. **89**, 3054 (2014).
[2] M. Kovari *et al.*, Fus. Eng. Des. **104**, 9 (2016).
[3] R. Prater *et al.*, Nucl. Fusion **48**, 2349 (2008).
[4] E. Poli *et al.*, Nucl. Fusion **53**, 013011 (2013).
[5] T. M. Antonsen, K. R. Chu, Phys. Fluids **25**, 1295 (1982).
[6] E. Poli *et al.*, Comp. Phys. Comm. **136**, 90 (2001).
[7] E. Poli *et al.*, Comp. Phys. Comm. **225**, 36 (2018).
[8] M. Bornatici *et al.*, Nucl. Fusion **23**, 1153 (1983).
[9] R. W. Harvey *et al.*, , Nucl. Fusion **37**, 69 (1997).
[10] G. Ramponi *et al.*, Nucl. Fusion **48**, 054012 (2008).
[11] H. Zohm *et al.*, Plasma Phys. Control. Fusion **49**, B341 (2007).
[12] G. R. Smith *et al.*, Phys. Fluids **30**, 3633 (1987).
[13] Y. R. Lin-Liu *et al.*, Phys. Plasmas **10**, 4064 (2003).
[14] N. B. Marushchenko *et al.*, Nucl. Fusion **48**, 054002 (2008).
[15] E. Poli *et al.*, Phys. Plasmas **25**, 122501 (2018).

Current drive by high intensity, pulsed, electron cyclotron wave packets

Abhay K. Ram[1,*], *Kyriakos* Hizanidis[2,**], and *Richard J.* Temkin[1,***]

[1]Plasma Science and Fusion Center, Massachusetts Institute of Technology, Cambridge, MA 02139. USA.
[2]National Technical University of Athens, 157 73 Zographou, Greece.

Abstract. The nonlinear interaction of electrons with a high intensity, spatially localized, Gaussian, electromagnetic wave packet, or beam, in the electron cyclotron range of frequencies is described by the relativistic Lorentz equation. There are two distinct sets of electrons that result from wave-particle interactions. One set of electrons is reflected by the ponderomotive force due to the spatial variation of the wave packet. The second set of electrons are energetic enough to traverse across the wave packet. Both sets of electrons can exchange energy and momentum with the wave packet. The trapping of electrons in plane waves, which are constituents of the Gaussian beam, leads to dynamics that is distinctly different from quasilinear modeling of wave-particle interactions. This paper illustrates the changes that occur in the electron motion as a result of the nonlinear interaction. The dynamical differences between electrons interacting with a wave packet composed of ordinary electromagnetic waves and electrons interacting with a wave packet composed of extraordinary waves are exemplified.

1 Introduction

In the early 1990s, high power, pulsed, microwaves, in the electron cyclotron (EC) frequency range, were used in the Microwave Tokamak Experiment (MTX) [1] for heating and for current generation in high density plasmas. Bursts of pulses were generated by a free-electron laser at a frequency of 140 GHz with powers up to 2 GW, and pulse durations of 20-30 ns. It was noted in [1] that "extrapolation to reactor heating and current drive predicts high efficiency." In the intervening years, significant advances have occurred in the technology of high power millimeter wave sources. Gyrotrons now routinely achieve megawatt power levels in continuous wave operations – 140 GHz for W7-X and 170 GHz for ITER. Relativistic gyrotrons can operate in a pulsed mode, in the millimeter wavelength range, at power levels ranging from 10 to 100 MW [2]. These developments have led to our studies on the interaction of electrons with high peak power EC fields, while, simultaneously, aiming to optimize current drive in fusion plasmas.

The nonlinear interaction of electrons with high intensity wave packets in the EC frequency range is described by the relativistic Lorentz equation. The wave packet is spatially localized perpendicular to its direction of propagation. We find that the interaction of electrons with an extraordinary EC wave is very different from the interaction with an ordinary wave. Consequently, the polarization of the EC beam has an important effect in the wave-particle interaction. The components of the electron momentum, along and across the magnetic field, are a function of the wave power and the direction of wave propagation. The interaction of electrons with a wave packet is significantly different from that with a large amplitude plane wave [3]. The interaction with a spatially confined wave packet leads to a wider variety of nonlinear phenomena including spatial modifications to the temperature profile as well as affecting MHD activity. Recently, a study of the nonlinear interaction of electrons with a Gaussian beam, in a low temperature plasma, has concluded that EC waves could be used in the pre-ionization phase for start-up of tokamak operations [4]. The theoretical and computational studies show that the nonlinear interaction with the beam provides sufficient energy to electrons so as to ionize the plasma.

In the usual quasilinear theory, the electron motion is given by linear theory which ignores any nonlinear effects, for example, due to trapping or the ponderomotive force. The limitations of quasilinear theory and its applicability to even the present-day EC heated plasmas has been questioned [5, 6]. The reason being that trapping of electrons in waves cannot be neglected or ignored. Our studies show that electrons interacting with EC beams do not follow linear orbits – a bedrock of quasilinear theory. In addition, our calculations lend credence to the claims made in [1], that high powered EC waves will lead to efficient heating and current drive in fusion plasmas through nonlinear effects. The theoretical studies of wave-particle interactions discussed in this paper, along with the advances in the technology of high peak power sources, may offer a breakthrough in achieving steady-state operations in reactor scale plasmas.

*e-mail: abhay@mit.edu
**e-mail: kyriakos@central.ntua.gr
***e-mail: temkin@psfc.mit.edu

2 Description of the Gaussian beam

The EC Gaussian wave packet is assumed to be a superposition of plane waves having the same angular frequency ω_0. Each plane wave satisfies the local plasma dispersion relation and its polarization is obtained by using the cold plasma permittivity in Maxwell's equations. Among the set of plane waves, consider the one with the maximum amplitude of the Poynting flux. Let $\mathbf{k}_0$ is the wave vector of this particular plane wave and $\hat{\mathbf{s}}_0$ be the unit vector along the direction of the group velocity; $\hat{\mathbf{s}}_0$ essentially determines the axis of the beam. The time-averaged Poynting flux of the Gaussian beam is given by,

$$\langle \mathbf{S} \rangle = \frac{P_0}{\pi w^2} \exp\left(-\frac{|\mathbf{r}_{0\perp}|^2}{w^2} \right) \hat{\mathbf{s}}_0. \qquad (1)$$

Here P_0 is the total power flowing along the beam, $w/\sqrt{2}$ is the beam radius normal to $\hat{\mathbf{s}}_0$, and $\mathbf{r}_{0\perp}$ is the position vector transverse to the beam – the beam axis being at $\mathbf{r}_{0\perp} = 0$.

The electric field associated with the Gaussian beam is derived from the time-averaged Poynting flux of each associated planar wave. After some algebraic manipulations, we obtain the following expression for the electric field at position $\mathbf{r}$ and time t,

$$\mathbf{E}_{RF}(\mathbf{r}, t) = E_0 \operatorname{Re}\left[e^{-i\omega_0 t} \frac{1}{N} \sum_{\mathbf{n}} E_n \tilde{\mathbf{e}}_n \exp(i\mathbf{k}.\mathbf{r}) \right], \qquad (2)$$

where Re indicates the real part of the expression inclosed within the brackets, and N is the total number of plane waves included in the representation of the Gaussian beam. The summation is over the normalized wave vectors $\mathbf{n} = c\mathbf{k}/\omega_0$ where c is the speed of light. In the Cartesian coordinate system $\mathbf{n} = (n_x, n_y, n_z)$ with the z direction being along the magnetic field and x being the direction of the plasma inhomogeneity. It is important to realize that the summation is actually a double sum – the components of $\mathbf{n}$ are not all independent since each plane wave satisfies the cold plasma dispersion relation. In Eq. 2, $\tilde{\mathbf{e}}_n$ is the electric field polarization, normalized to a unit norm, of the n-th plane wave, and

$$E_n = e^{\left(-w^2 |\mathbf{k} - \mathbf{k}_0|^2 / 2\right)}. \qquad (3)$$

where $\mathbf{n}_0 = c\mathbf{k}_0/\omega_0$. The amplitude E_0 is,

$$E_0 = \sqrt{\frac{2P_0}{\pi w^2}} \left(\frac{\mu_0}{\epsilon_0} \right)^{1/4} S \qquad (4)$$

where ϵ_0 and μ_0 are the free space permittivity and permeability, respectively, and

$$S = \left(\frac{1}{\pi} \frac{w^2 \Delta k_x^S \Delta k_y^S}{\operatorname{Re}\{\hat{\mathbf{s}}_0.\mathbf{n}_0 - (\hat{\mathbf{s}}_0.\tilde{\mathbf{e}}_n)(\tilde{\mathbf{e}}_n^*.\mathbf{n}_0)\}\, C_N} \right)^{1/2}, \qquad (5)$$

and,

$$C_N = \frac{1}{N} \sum_n \exp\left[\frac{w^2}{2} \left\{ (\mathbf{k} - \mathbf{k}_0).\mathbf{k}_\perp^S - |\mathbf{k} - \mathbf{k}_0|^2 \right\} \right]. \qquad (6)$$

In Eqs. 5 and 6, k_x^s, k_y^s, and $\mathbf{k}_\perp^s$ are defined with respect to $\hat{\mathbf{s}}_0$. If we define the unit vectors in the Cartesian coordinate system with respect to the beam as $(\hat{\mathbf{x}}^S, \hat{\mathbf{y}}^S, \hat{\mathbf{s}}_0)$, then k_x^s and k_y^s are components of $\mathbf{k}$ along $\hat{\mathbf{x}}^S$ and $\hat{\mathbf{y}}^S$, respectively. Correspondingly, for the Gaussian beam, Δk_x^S and Δk_y^S are the spreads in the wave vectors perpendicular to $\hat{\mathbf{s}}_0$. Note that $\hat{\mathbf{y}}^S$ is equivalent to $\hat{\mathbf{y}}$ in the laboratory frame.

3 Interaction of electrons with the Gaussian beam

The interaction of electrons with the spatially localized Gaussian beam is given by the Lorentz equation,

$$\frac{d\mathbf{p}}{dt} = -e \left\{ \mathbf{E}_{RF} + \frac{\mathbf{p}}{m_e \gamma} \times (\mathbf{B}_0 + \mathbf{B}_{RF}) \right\} \qquad (7)$$

where $\mathbf{p}$ is the electron momentum, e is the magnitude of the electron charge, $\mathbf{B}_0$ is the uniform background magnetic field,

$$\gamma = \sqrt{1 + \frac{\mathbf{p}.\mathbf{p}}{(m_e c)^2}} \qquad (8)$$

and, m_e is the rest mass of the electron. The RF wave magnetic field $\mathbf{B}_{RF}$ is obtained from (2) by replacing $\tilde{\mathbf{e}}_n$ in the numerator by $\mathbf{n} \times \tilde{\mathbf{e}}_n/c$ – this follows from Faraday's law.

4 Numerical results and discussion

In order to illustrate the effect of a Gaussian beam on the dynamics of electrons interacting with it, we assume the following parameters. The wave frequency and the plasma density are assumed to be 170 GHz and 2×10^{20} m^{-3}, respectively. The local magnetic field is chosen to be such that the electron cyclotron frequency is slightly less than the wave frequency, i.e, $B_0 \approx 6.07$ T. The width of the Gaussian beam is set to $w = 12$ cm with $n_{0\parallel} = 0.2$ being the wave number parallel to the direction of the magnetic field. The perpendicular wave number $n_{0\perp}$ is determined from the local cold plasma dispersion relation. All these quantities are needed to define the wave vectors and the polarizations of the planar waves that make up the Gaussian beam. The ambient magnetic field is along the $\hat{\mathbf{z}}$ direction and the beam $\mathbf{n}_0$ lies in the $x - z$ plane. The normalization of the electron velocity is with respect to the thermal velocity – the temperature is taken to be 15 keV. The relation between the magnitudes of the velocity and the momentum is $u = p/(m_e c)$.

In the numerical runs that are displayed below, the initial x, y, and z components of the electron momentum are $p_x/p_T = p_y/p_T = 2.5$, and $p_z/p_T \equiv p_\parallel/p_T = 4$, respectively. Here p_T is the magnitude of the electron thermal momentum. The center of the coordinate system $x = y = z = 0$ lies on the axis of the Gaussian beam. Figures 1 and 2 show the normalized parallel and perpendicular speeds of an electron as it passes through a 1 MW beam with an ordinary mode polarization. The electron is initially located outside the beam at $z = -25$ cm and is moving in the positive z direction. As the electron traverses the beam the effect of the planar waves that comprise the Gaussian beam is quite apparent. The electron oscillates in the potential of each plane wave as it moves along the z-direction. The parallel and perpendicular momenta change rapidly, but the electron exits the beam with the same momenta as when it entered the beam. However, this changes when the beam power is increased. Figures 3 and 4 illustrate the corresponding results when the beam power is increased to 10 MW. Inside the beam, the motion of the electron is not as oscillatory and coherent as it was for 1 MW of beam power. There is a lack of symmetry around the center of the beam that is evident in Figs. 1 and 2. The net result is that the electron has gained parallel and perpendicular speed by the time it exits the beam – the combined effect implying an increase in the plasma current.

The dynamics of electrons changes significantly if the beam has the polarization of the extraordinary EC wave. In Figs. 5 and 6, the normalized parallel and perpendicular components of the velocity are plotted as function of time for the extraordinary wave polarization. The other parameters are the same as in Figs. 1 and 2. The motion of the electron inside the beam is random and different from the oscillatory motion in Figs. 1 and 2. The increase in the two components of the velocity after the electron exits the beam is large, and we would expect the corresponding change in the current carried by the electron to be substantial. The changes are even more sizable when the beam power is 10 MW. Figures 7 and 8 illustrate these changes. Again, the electron behaves differently inside the beam than when the wave polarization corresponds to the ordinary EC wave.

Another interesting aspect of the nonlinear interaction of electrons with a Gaussian beam is the difference in the dynamics when the electrons enter the beam from different directions. This effect manifests itself as the beam is propagating obliquely with respect to the magnetic field. In Figs. 9 and 10, we plot the normalized components of the velocity when the initial location of the electron is at $z = 25$ cm and, initially, $p_{\|}/p_T = -4$. In other words, the electron enters the beam from the positive z side with its parallel speed directed along the $-\hat{\mathbf{z}}$ direction. Figure 9 shows the evolution $u_{\|}$ as a function of time. The electron enters the beam from the right – defined with respect to $z = 0$. Inside the beam, the parallel speed changes sign, i.e., the electron gets reflected, and exits the beam from the same side as it enters. This reflection is due to the nonlinear ponderomotive force caused by the spatial variation in the beam profile. The interaction of the electron with the beam is spatially restricted as it does not traverse through the entire width of the beam. This also limits the increase in the magnitude of the perpendicular velocity, as is evident when comparing Fig. 10 with Fig. 8.

The results displayed in Figs. 7 – 10 elucidate the importance of nonlinear effects in the interaction of electrons with a Gaussian beam. Let us define left and right with respect to the center of the beam at $z = 0$. Electron 1 entering the beam from the left transits through the beam and exits on the right with an increase in its velocity along the magnetic field line as well as gaining energy from the beam (Figs. 7 and 8). Electron 2 has the same initial energy and momentum as electron 1 except that its starting momentum parallel to the magnetic field is in a direction opposite to the starting parallel momentum of electron 1. Electron 2 is initially located to the right of the beam. During its transit through the beam, the parallel velocity reverses sign and the electron exits the beam from the same right side that it had entered the beam. The energy and the magnitude of the parallel velocity have increased relative to their initial values (Figs. 9 and 10). Electrons 1 and 2, which were initially traveling in opposite directions, are now moving in the same direction parallel to the magnetic

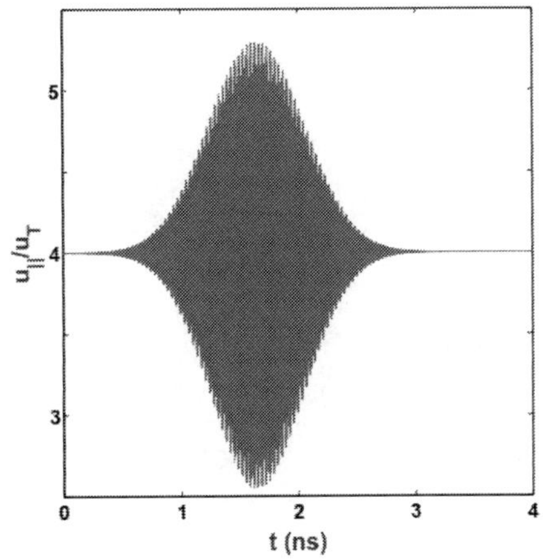

Figure 1. The evolution of $u_{\|}/u_T$ as a function of time. The initial position of the electron is outside the beam and it is moving along the positive z direction. The Gaussian beam has the polarization of the ordinary EC wave and power of 1MW. The other parameters are as indicated in the text.

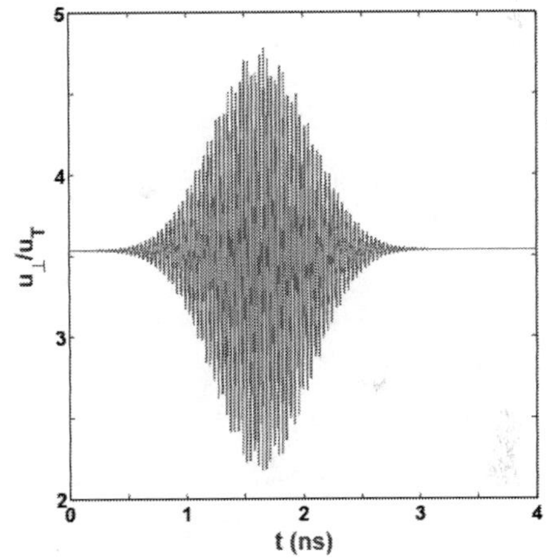

Figure 2. The evolution of $u_{\perp}/u_T$ as a function of time for the same parameters as in Fig. 1; $u_{\perp}/u_T = \sqrt{p_x^2 + p_y^2}/p_T$

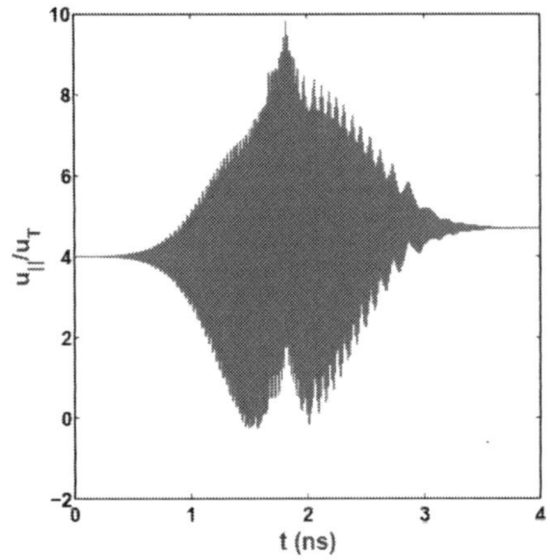

Figure 3. The parameters are the same as in Fig. 1 except that the beam power is 10 MW.

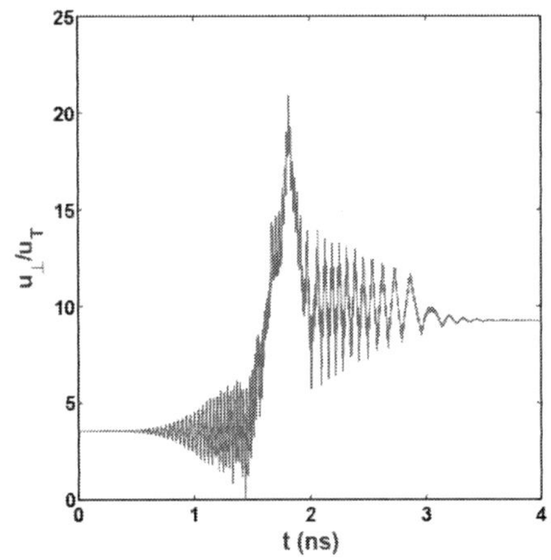

Figure 4. Same as Fig. 2 except that the beam power is 10 MW.

field. The practical consequence of this is an enhancement in the net plasma current.

5 Acknowledgement

AKR was supported by US DoE Grants DE-FG02-91ER54109 and DE-SC0018090.

References

[1] E. B. Hooper, et al., "MTX Final Report," [https://www.osti.gov/scitech/servlets/purl/10194124] (1994).

[2] M. Thumm, "State-of-the-Art of High Power Gyro-Devices and Free Electron Lasers," KIT Scientific Report 7735 (2016).

[3] W. M. Nevins, T. D. Rognlien, and B. I. Cohen, Physical Review Letters **59**, 60 (1987).

[4] D. Farina, Nuclear Fusion **58**, 066012 (2018).

[5] R. Kamendje, S. V. Kasilov, W. Kernbichler, and M. F. Heyn, Physics of Plasmas **10**, 75 (2003).

[6] R. Kamendje, S. V. Kasilov, W. Kernbichler, I. V. Pavlenko, E. Poli, and M. F. Heyn, Physics of Plasmas **12**, 012502 (2005).

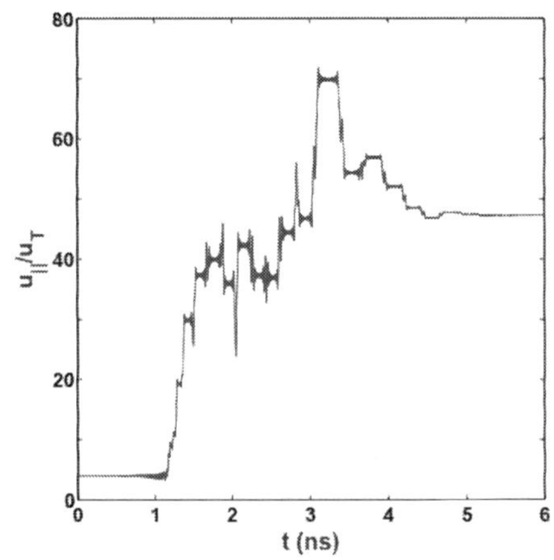

Figure 5. Same as Fig. 1 except that the beam polarization is of the extraordinary EC wave.

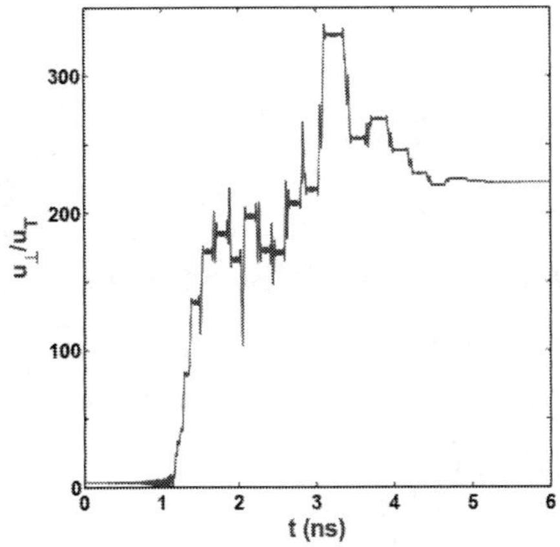

Figure 6. Same as Fig. 2 except that the beam polarization is of the extraordinary EC wave.

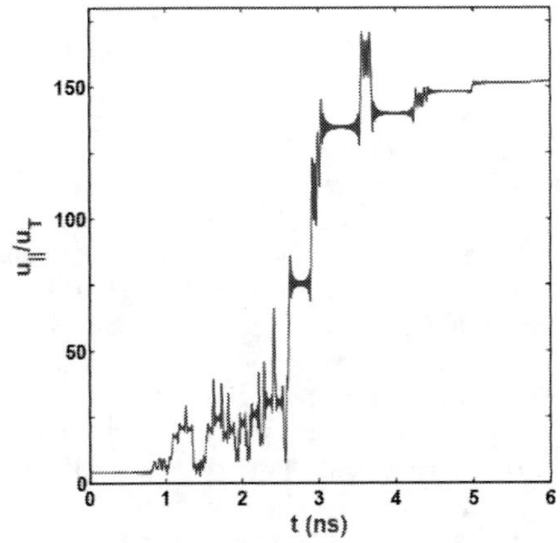

Figure 7. Same as Fig. 5 except that the beam power is 10 MW

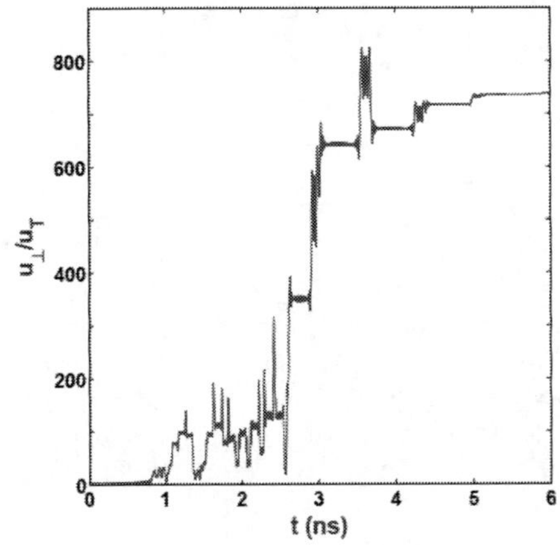

Figure 8. Same as Fig. 6 except that the beam power is 10 MW

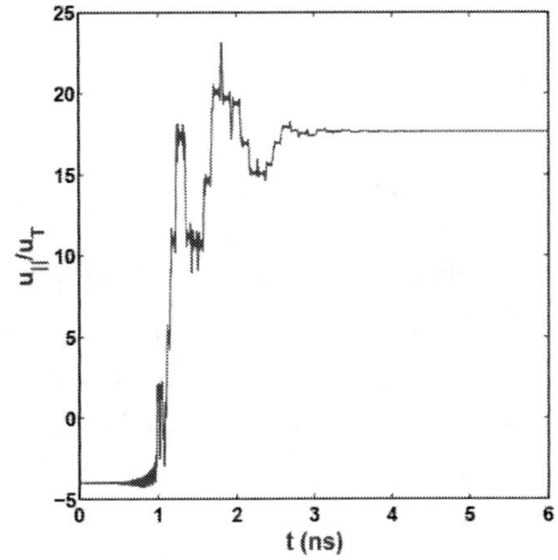

Figure 9. The beam polarization corresponds to the extraordinary EC wave and the beam power is 10 MW. The initial location and magnitude of the parallel velocity of the electron are z = 25 cm and $u_\parallel/u_T = -4$, respectively. All the other parameters are as described in the text.

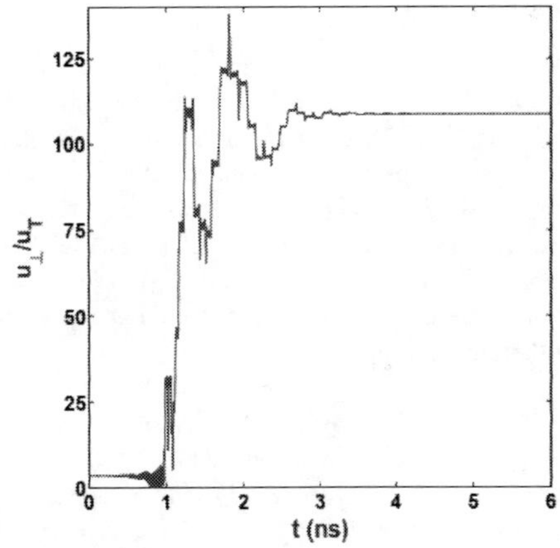

Figure 10. A plot of $u_\perp/u_T$ as a function of time for the same parameters as for Fig. 9.

Particle-in-cell simulations of parametric decay instabilities near the upper hybrid layer

M G Senstius[1,*], *S K* Nielsen[1], and *R G L* Vann[2]

[1]Department of Physics, Technical University of Denmark, Fysikvej, DK-2800 Kgs. Lyngby, Denmark
[2]York Plasma Institute, Department of Physics, University of York, York YO10 5DD, United Kingdom

Abstract. The particle-in-cell (PIC) code EPOCH is used to simulate parametric decay instabilities (PDIs) converting a 105 GHz microwave X-mode pump wave into electrostatic daughter waves at the upper hybrid (UH) layer of a fusion plasma in 1D. The resulting fields are analyzed, identifying modes in f- and k-space, and estimating their spectral power as a function of pump wave intensity. Both linearly and nonlinearly converted modes are identified and their characteristics agree with literature. A dipole approximation employed in literature appears to be unjust.

1 Introduction

Observations of strong scattering shifted in frequency in several recent collective Thomson scattering (CTS) experiments[1] has led to investigations to quantify the scattering and uncover the mechanisms that cause it. If not anticipated, the scattering can be strong enough to damage microwave equipment in current reactors such as ASDEX Upgrade. CTS diagnostics are expected to be installed in ITER and a better understanding of the anomalous scattering is therefore necessary.

The observed scattering occurs when the CTS gyrotron beam reaches the UH layer and it has characteristics associated with PDIs. An investigation of the daughter waves of such a PDI process may lead to efficient heating schemes and there is experimental evidence to suggest that it is possible to suppress the PDI by angling the CTS microwave beam in certain ways. Better knowledge of this PDI is expected to also benefit investigations of other PDIs such as two plasmon decay during 2nd harmonic ECRH, which may have a significant impact on its heating efficiency.

Using a numerical code to simulate a warm fusion plasma where the anomalous scattering is expected to occur is a convenient tool to characterize the scattered modes on a number of parameters, not only in frequency. The PDIs of interest are expected to also rely on both wave and particle dynamics so a PIC code should be able recreate the observed scattering and test theoretical models describing the scattering as the product of a PDI.

2 Parametric decay theory

The PDIs of interest are described theoretically in [2]. They are three wave interactions where a large amplitude electromagnetic X-mode pump wave, of f_0 and $\mathbf{k}_0$, decays into a low frequency warm lower hybrid (LH) daughter wave, f_1 and $\mathbf{k}_1$, and a slightly downshifted electron Bernstein wave (EBW), f_2 and $\mathbf{k}_2$. Subsequent recombination of the pump wave and the low frequency daughter produces an upshifted daughter wave, f_3 and $\mathbf{k}_3$. These processes must conserve energy and momentum which translates to conservation of frequency and wave vector, i.e. $f_0 = f_1 + f_2$, $\mathbf{k}_0 = \mathbf{k}_1 + \mathbf{k}_2$ and $f_3 = f_0 + f_1$, $\mathbf{k}_3 = \mathbf{k}_0 + \mathbf{k}_1$ for the decay and combination process respectively. Both f_2 and f_3 are then shifted by f_1 relative to f_0. Using a dipole approximation, the electromagnetic X-mode pump wave is shown to decay into two electrostatic daughter waves which for perpendicular propagation to the magnetic field in 1D are described by the dispersion relations

$$\omega_1^2 \approx \omega_{LH}^2\left[1 + 3\left(\frac{\omega_{pe}^2}{4\omega_{UH}^2} + \frac{T_i}{Z_i T_e}\frac{\omega_{UH}^2}{\omega_{pe}^2}\right)k_1^2 r_{Le}^2\right], \quad (1)$$

$$\omega_2^2 \approx \omega_{UH}^2 - \omega_{pe}^2\frac{k_2^2 r_{Le}^2}{1 - \omega_{pe}^2/(3\omega_{ce}^2)}. \quad (2)$$

Here $\omega_{ce} = eB/m_e$, $\omega_{pe} = \sqrt{n_e e^2/(\epsilon_0 m_e)}$, $\omega_{pi} = \sqrt{n_i Z_i^2 e^2/(\epsilon_0 m_i)}$, $\omega_{UH} = \sqrt{\omega_{ce}^2 + \omega_{pe}^2}$ and $\omega_{LH} = |\omega_{ce}\omega_{pi}/\omega_{UH}|$ are the electron cyclotron, electron plasma, ion plasma, upper hybrid (UH) and LH angular frequency respectively, T_i, T_e are ion and electron temperature in electron volts, Z_i is the ion atomic number, $r_{Le} = v_{te}/(\sqrt{2}|\omega_{ce}|)$ is the electron Larmor radius, and $v_{te} = \sqrt{2T_e/m_e}$ is the thermal velocity. These waves are referred to as warm LH and UH waves as those are their frequencies in the cold limit. Above a pump wave amplitude threshold, the processes become unstable and increase nonlinearly with pump wave amplitude. The wave enhancement of an X-mode wave incident on the UH layer in a magnetically confined fusion plasma means that the

*e-mail: mgse@fysik.dtu.dk

threshold may be exceeded at CTS diagnostic power levels.
This conversion process of the pump wave, however, is not the only one that is expected to occur near the UH layer. As the X-mode pump wave moves closer to the UH layer it gets increasingly electrostatic. Eventually coinciding with an EBW, the pump wave is linearly converted from X-mode to EBW[4] and is reflected.

3 Numerical Setup

The simulations are run using the PIC code EPOCH[3] version 4.9.0 and plasma parameters are chosen to resemble the region around the UH layer of a warm deuterium plasma. The numerical domain is a 1D domain in the x-direction, split into three regions. The left and right regions are completely homogeneous for easier mode identification and the middle region bridges the parameters of the other two regions. The background magnetic field and temperature are homogeneous and given by $B_0 = B_{0y} = 3.35$ T and $T_e = T_i = 0.3$ keV. The plasma is a deuterium plasma with a density profile, $n(x) = n_e(x) = n_i(x)$, given by

$$n(x) = \begin{cases} n_l, & x < x_l \\ n_l + (n_r - n_l)\dfrac{x - x_l}{x_r - x_l}, & x_l \leq x < x_r \\ n_r, & x_r \leq x \end{cases} \tag{3}$$

where $n_l = 3.4 \times 10^{19}$ m^{-3}, $n_r = 2.7 \times 10^{19}$ m^{-3}, $x_l = 4.00$ cm, $x_r = 4.35$ cm. An $f_0 = 105$ GHz beam is propagated from the left boundary located at $x = 0$ using a pseudo-current at the boundary. Particles leaving at the boundaries are replaced by new thermally distributed ones. It is currently only possible to excite a beam with transverse polarization in EPOCH. An X-mode beam is desired so the beam is polarized in the z-direction which then gives rise to an x-component also. Note that the X-mode beam is injected from the high density side as it is unable to propagate in the region between the UH layer and the R-cutoff. This region is found on the right side of the UH layer with the used parameters and launching an X-mode beam from this side would therefore not work. It seems, however, that exciting the X-mode in the described manner in EPOCH also generates what appears to be an unwanted EBW at the same frequency in addition to the X-mode wave at the boundary. The EBW has a much greater wave number and propagates significantly slower. This wave is likely excited to compensate for the missing x-component when exciting the X-mode. In that case, the unwanted EBW can be decreased to some degree by starting at an even higher density at the expense of a larger computational domain. To avoid unwanted effects arising from significant discontinuities in the fields, the beam power is ramped up from zero using an arctan-function with a characteristic time scale of $\tau_{ramp} = 0.20$ ns. The amplitude of the pump beam varies between simulations to investigate the nonlinearity of the PDI. The beam is expected to reach the UH layer in the middle region at $x_{UH} = 4.27$ cm where $n_e(x_{UH}) = 2.87 \times 10^{19}$ m^{-3}. The entire domain is 5.35 cm, and contains $n_x = 6600$ grid points and $n_{part} = 60000 n_x$ pseudoparticles, of which half are electrons and the other half are deuterons. The time step is 0.95 of what is allowed by the CFL criterion, the output time step is $\delta t = (10.3 f_0)^{-1} = 9.25 \times 10^{-13}$ s and the total simulation time is almost 46.5 ns of which the first 10 ns is disregarded in most of the analysis as this time is spent propagating the pump from the boundary to the UH layer.

4 Identification of modes

4.1 k-space

Figure 1a shows a continuous wavelet transform (CWT) of E_x into k-space, covering the last 1 cm of the first homogeneous region as well as the entire inhomogeneous region which includes the UH layer. As expected, the pump wave is seen to propagate to the UH layer where it is amplified and waves of different k are scattered back into the homogeneous region. Some waves also seem to be present beyond the UH layer. In the homogeneous region, k_0 is seen to be much smaller than that of the back scattered waves, however, near the UH layer, the wave number of the pump wave increases while the wave number of the back scattered waves decrease and become comparable. This suggests that the dipole approximation used in [2] might not be valid. With the dipole approximation, the wave number of the daughter waves are expected to differ only by sign, i.e. $k_2 \approx -k_1$. Without the dipole approximation, the selection rules change and the frequency shift might not agree with theory then.

4.2 Frequency space

An FFT into f-space near the pump wave is shown in figure 1b. A large peak at the pump frequency $f_0 = 105$ GHz is clearly visible and around this peak, $f_1 = 0.78 \pm 0.03$ GHz from it, are two smaller peaks. Since they are found symmetrically around the pump frequency and shifted approximately by the lower hybrid frequency, $f_1 \sim f_{LH} = 0.71$ GHz, the smaller peaks are suspected to be the daughter waves at $f_2 \approx 104.22 \pm 0.03$ GHz and $f_3 \approx 105.78 \pm 0.03$ GHz.

4.3 2D FFT into frequency and k-space

Plots showing selected parts of a 2D FFT of E_x into f- and k-space are shown in figure 2 (a)-(b). Figure 2 (a) shows the squared spectrum near the pump frequency. Symmetrically around $k = 0$, lines are observed resembling the dispersion relations of warm X-mode connecting to EBWs at the UH layer. A number of peaks can be spotted, one coinciding with the X-mode pump wave and two other at the same frequency but on the EBW branches. The left EBW wave has a positive group velocity and is the unintended one created at the boundary whereas the right wave has negative group velocity and is the one originating from around the UH layer. Around the peak and along the EBW line, a number of smaller peaks separated by a similar distance in f and k can be seen. The closest two are likely to

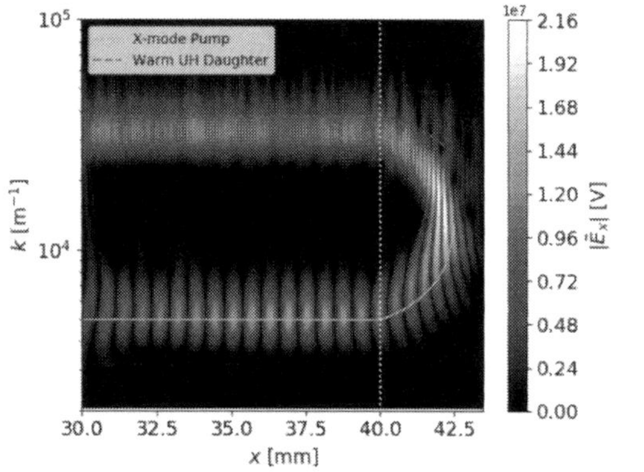

(a) CWT of E_x leading up to the UH resonance at t = 30 ns, using the Morlet wavelet as base function. The dotted white line indicates $x = x_l$. The solid yellow line is the solution to warm X-mode propagation[5], and the dashed green line is the solution to the back scattered warm UH mode from equation (2).

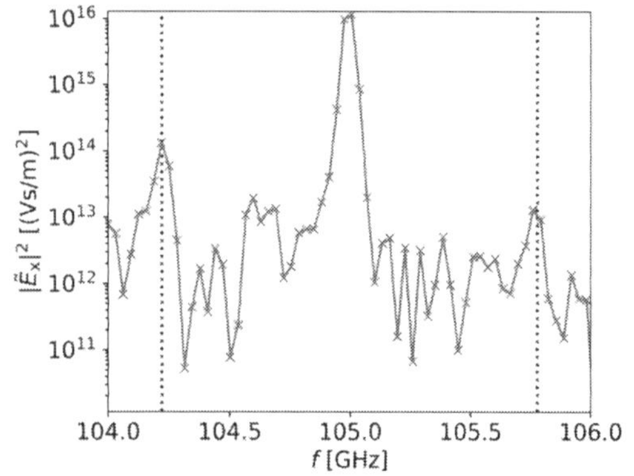

(b) Squared FFT of a 31.75 ns long E_x signal for an input intensity of I_{pump} = 16.7 MW/m^2. Only the spectrum near the pump wave frequency and the expected UH daughter wave is shown. The plot is centered around the pump at f_0 = 105.00 GHz and the two vertical dotted lines placed symmetrically around the pump wave mark the two PDI products f_2 and f_3.

Figure 1: Plots showing transforms of the longitudinal electric field component, E_x, into k-space (a) and f-space (b).

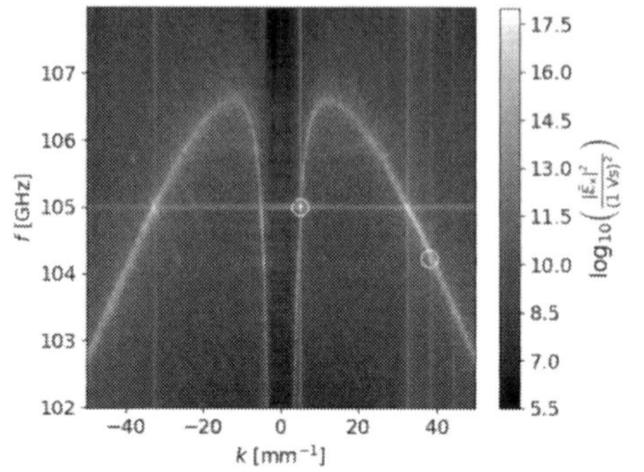

(a) Spectrum near the pump frequency.

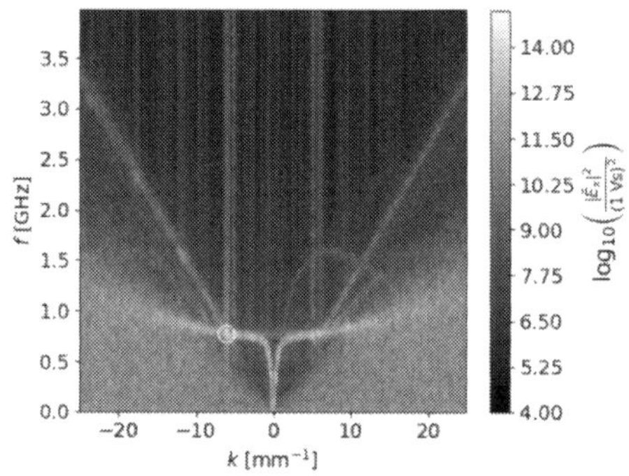

(b) Spectrum near the lower hybrid frequency.

Figure 2: 2D fast Fourier transform of E_x in $t \in [10.0, 46.7]$ ns and $x \in [2, 4]$ cm into f- and k-space. The white circles are the expected location for the pump beam (a), the warm UH daughter wave (a) and the warm LH daughter wave (b).

be the daughter waves also observed in figure 1b. Using the observed frequency shift to calculate k_2 from equation (2) places it on the EBW branch as expected. Figure 2 (b) shows the squared spectrum near the warm LH frequency. Using the same frequency shift to calculate k_1 from equation (1) places it close to albeit not quite at the observed peak. Judging from the peaks and the dispersion curves observed in both plots and comparing their wave numbers with figure 1a, the backscattered waves at much greater wave number than k_0 seen in figure 1a are comprised mainly of a linearly converted EBW and the warm UH daughter wave from the nonlinear PDI.

4.4 Group velocity

By considering $v_g = \partial\omega/\partial k$ along the dispersion curves in figure 2 (a), the linearly converted EBW and the warm UH daughter wave are expected to propagate at a similar speed and tracking the position of the backscattered wavefront in time is expected to give a reasonable estimate of the group velocity for the warm UH daughter wave. The wavefront is tracked in the first homogeneous region by fitting a shifted arctan-function to $\langle|E_x|\rangle$ averaged over 60 gridpoints. Fitting a first order polynomial to the wavefront position in time gives the group velocity $v_g = -(7.51 \pm 0.01) \times 10^5$ m/s. The expected group velocity from equation (36) in [2] is $v_{g,theo} = 10.66 \times 10^5$ m/s. A similar approach gives $v_{g,pump} = (5.962 \pm 0.006) \times 10^6$ m/s for the pump wave. Note that the very small errors are calculated as standard deviations which do not take systematic errors into account. The group velocity of the back scattered waves is

https://doi.org/10.1051/epjconf/201920301010

much closer to the theoretical group velocity of the UH daughter than to the incoming X-mode pump beam.

4.5 Pump intensity dependence

As a measure of the power contained in a particular mode, the $|\tilde{E}_x|^2$ spectra shown in figures 2 (a)-(b) are numerically integrated around selected modes in boxes of dimension 940 m^{-1} along the k-axis and 0.22 GHz along the f-axis. The intensity of the pump wave is calculated as $I = \hat{\mathbf{e}}^* \cdot \partial \mathbf{M}/\partial\omega \cdot \hat{\mathbf{e}}\,|E|^2 v_{g,pump}$, where $\hat{\mathbf{e}}$ is the unit vector pointing in the direction of the electric field and $\mathbf{M}$ is the Maxwell operator as defined in equations (5.138)-(5.139) in [6]. Values are taken from the first homogeneous region of the domain and $\mathbf{M}$ is assumed to be cold as it is not right at a resonance. This is done for all simulations and the integrals are normalized to their value at the lowest pump intensity simulation to give a relative power. The found values are shown in figure 3. The normalized power of the EBW mode at the pump frequency appears to follow that of the pump wave and grows linearly with pump intensity, implying that it is indeed linearly converted at the UH layer. The suspected PDI products all grow faster than linearly and whilst it is difficult to determine the shape of the curves with only three points, they are perhaps getting close to a saturation level.

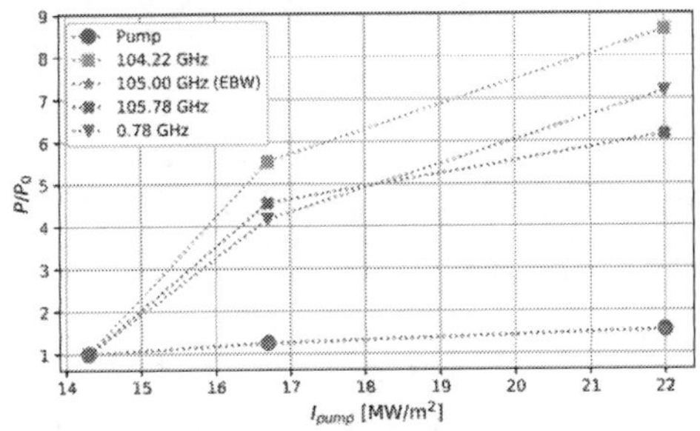

Figure 3: Normalized integrated spectral density (see figure 2) for modes indicated in the legend, plotted against pump wave intensity. The fraction P/P_0 means that the values are normalized to that of the lowest intensity simulation. Different series are not comparable in magnitude.

5 Discussion

The PIC code EPOCH successfully recreated scattering peaks shifted in frequency relative to a microwave pump, resembling peaks observed experimentally. It was found that the scattering increases with a nonlinear dependence on the pump wave intensity as a PDI would. Whilst theoretical predictions about the scattering characteristics seemed to agree to the order or better, implying that it is indeed caused by a PDI at the UH layer, the dipole approximation employed in [2] to derive equations (1)-(2) does not seem to hold as seen in figure 1a. Tracking the backscattered wavefront to get the group velocity of the warm UH wave yielded a result on the same order as the theoretical prediction under the assumption that it propagates at a similar speed as the linearly converted EBW since the amplitude of the warm UH wave is likely insignificant in comparison to the linearly converted EBW. This might be better to do at greater intensities. The box integration method for obtaining a measure of power in the modes is not ideal as it cuts off the tails of the peaks and is affected in particular by neighboring modes, dispersion lines etc. Assuming, however, that the peak is the most significant contribution to the integral and that nothing else changes in its vicinity as the pump intensity changes, the method does the job and will provide a measure in which growth of the modes can be observed. More simulations are needed to establish a threshold pump intensity and to explore the possibility of a saturation level of the daughter waves.

6 Acknowledgments

The work was supported by a grant from VILLUM FONDEN and by EUROfusion through cpu time on the Marconi HPC cluster.

References

[1] S K Nielsen *et al* Plasmas Phys. Control. Fusion **55** 115003 (2013)

[2] S K Hansen, *et al*, Plasma Phys. Control. Fusion **59** 105006 (2017)

[3] T D Arber *et al*, Plasma Phys. Control. Fusion **57** 113001 (2015)

[4] H P Laqua, Plasma Phys. Control. Fusion **49** R1 (2007)

[5] Richard Fitzpatrick, *Plasma Physics: An Introduction*, CRC Press, Taylor & Francis Group (2015)

[6] Donald Gary Swanson, *Plasma Waves, 2nd Edition*, CRC Press (2003)

Numerical description of mode coupled waves in inhomogeneous magnetized plasmas

Kota Yanagihara[1,*], *Shin* Kubo[1,2,**], and *Toru* Tsujimura[2]

[1]Nagoya University, Nagoya, Aichi, Japan
[2]National Institute for Fusion Science, Toki, Gifu, Japan

Abstract. Geometrical optics (GO) ray tracing has been widely used for a description of electron cyclotron resonance waves in inhomogeneous magnetized fusion plasmas. However, this reduced approach is not correct in sufficient low density plasmas with a sheared magnetic field, where mode coupling between two electromagnetic-like cold plasma modes can occur. Here, we extend a ray tracing method based on the first-order theory of extended geometrical optics (XGO), which captures mode coupled complex amplitude between O and X mode along the ray trajectory. In our approach, reference ray is calculated with ray equation to satisfy the lowest-order part of XGO theory and an evolution of complex amplitude profile along the reference ray is calculated with partial differential equation derived from first-order terms. Calculation results performed by extended ray tracing are in good agreement with 1D full wave analysis. By introducing second-order terms into our numerical approach, diffraction will be treated.

1 Introduction

Geometrical optics (GO) ray tracing has been used for a description of coherent wave propagation in inhomogeneous media, and therefore has been incorporated into a number of numerical codes for electron cyclotron resonance heating (ECRH) simulations in fusion plasmas. These codes treat two eigen plasma modes independently and can not manage the coupling between them. Such treatment is valid in sufficiently high density plasmas, where dispersion relations of two cold plasma modes are well separated. However, dispersion relations of two cold plasma modes are not sufficiently separated in low density plasmas, and therefore these modes will be easily mixed by sheared external magnetic field. Unfortunately, because this is not an unique condition in the peripheral region of fusion plasmas, a more general technique is needed for a more realistic EC wave description. Here, we developed a new ray tracing code based on the extended geometrical optics (XGO) theory [1–4]. Our extended code can calculate the O and X mode coupled vector complex amplitude along the ray. In this paper, a brief outline of a calculation scheme with 1st order XGO theory (Sec.2) and comparison between results performed by our approach and one-dimensional full wave (1DFW) analysis [5, 6] (Sec.3) are presented. Summary is given in Sec.4.

*e-mail: yanagihara.kohta@nifs.ac.jp
**e-mail: kubo.shin@nifs.ac.jp

2 Modeling

2.1 XGO equation

XGO equation in the first-order limit is,

$$\left[\mathcal{E} + i\boldsymbol{\Gamma} - \boldsymbol{U} - i\mathbf{v}\cdot\nabla - \frac{i}{2}(\nabla\cdot\mathbf{v})\right]\boldsymbol{a} = 0 \qquad (1)$$

where

$$\begin{aligned}
\mathcal{E} &= \boldsymbol{\Xi}^\dagger \boldsymbol{D}_H \boldsymbol{\Xi}\\
\boldsymbol{\Gamma} &= \boldsymbol{\Xi}^\dagger \boldsymbol{D}_A \boldsymbol{\Xi}\\
\boldsymbol{U} &= -\left[\boldsymbol{\Xi}^\dagger (\partial_{k_\alpha}\boldsymbol{D})(\partial_{x^\alpha}\boldsymbol{\Xi})\right]_A\\
\mathbf{v} &= \boldsymbol{u} + \tilde{\mathbf{v}} = \partial_{\boldsymbol{k}}\mathcal{E}_0 + \tilde{\mathbf{v}}\\
\tilde{\mathbf{v}} &= i\boldsymbol{\Xi}^\dagger(\partial_{\boldsymbol{k}}\boldsymbol{D}_A)\boldsymbol{\Xi} - \left(\partial_{\boldsymbol{k}}\boldsymbol{\Xi}^\dagger\right)\boldsymbol{D}_H\boldsymbol{\Xi}\\
&\quad -\boldsymbol{\Xi}^\dagger\boldsymbol{D}_H(\partial_{\boldsymbol{k}}\boldsymbol{\Xi})\\
\nabla\cdot\mathbf{v} &\approx \nabla\cdot\boldsymbol{u} = \partial_{x^\alpha}\partial_{k_\alpha}\mathcal{E}_0 + \partial_{k_\alpha}\partial_{k_\beta}\mathcal{E}_0\cdot\partial_{x^\alpha}k_\beta.
\end{aligned}$$

This basic vector equation is approximated wave equation up to first-order of GO parameter $\epsilon = \lambda/L \ll 1$, where λ is a wave length and L is a scale length of the inhomogeneity. The derivation and details of this theory are presented in [1]. Here, we present just an essence of our calculation. $\boldsymbol{D}(\boldsymbol{x},\boldsymbol{k})$, $\boldsymbol{\Xi}$, $\mathcal{E}$, and $\boldsymbol{a}$ are dispersion matrix, arrayed eigenvectors matrix of $\boldsymbol{D}_H(\boldsymbol{x},\boldsymbol{k})$ for active modes, diagonal matrix of eigenvalues of $\boldsymbol{D}_H$ for active modes, and vector complex amplitude for each active mode, correspondingly. Averaged eigenvalue $\mathcal{E}_0$ is lowest-order term and is qualified as Ray-Hamiltonian. This is a very important technique for our approach. Therefore, details are mentioned below. Other terms are first-orders. $\boldsymbol{\Gamma}$ is dissipation but becomes zero in cold plasmas, $\mathbf{v}\cdot\nabla + (\nabla\cdot\mathbf{v})/2$

represents Poynting vector, and $\boldsymbol{U}$ gives polarizational effects including mode coupling. The index H and A denote the Hermitian and anti-Hermitian parts.

In the derivation of XGO, whole wave field $\boldsymbol{\Psi}$ is defined as

$$\boldsymbol{\Psi} = \sum_s \boldsymbol{\xi}_s a^s e^{i\theta(\boldsymbol{x})} = \boldsymbol{\Xi a} e^{i\theta(\boldsymbol{x})} \quad (2)$$

where s is a number of eigenmodes, $\boldsymbol{\xi}_s$ is eigenvector for eigenmode s, and $\theta(\boldsymbol{x})$ is rapidly oscillating phase function. Wave vector $\boldsymbol{k}$ is usually defined as normal vector of phase front. We also define wave vector as $\boldsymbol{k} \equiv \nabla\theta$.

Then, we introduced an unfamiliar definition, that is, "active." Roughly speaking, when appropriate wave vector $\boldsymbol{k}$ is given, the modes which can be excited in the medium are active modes. If branches of dispersion relations for each eigenmode are well separated, only one mode should be excited for given $\boldsymbol{k}$. However, if branches are very close, two modes can be excited simultaneously. We can model the former case with XGO equation in one active limit, where complex amplitude $\boldsymbol{a}$ is treated as a scalar amplitude for only one active mode a. However, modeling of two active modes, which treats complex amplitude as a two-dimensional amplitude for convertible two active modes, is suitable for the latter case. More detailed accounts of modeling of two active modes are presented in [1, 3].

2.2 Formulation for two active modes

Let us consider the O and X mode mixed waves, which can be modeled by our theory with the assumption that two eigenmodes are active. Such complicated waves can be excited in sufficiently low density plasmas where two eigenvalues of dispersion matrix $\boldsymbol{D}_H$ are almost degenerated. (In vacuum, they are degenerated completely.) Also, interaction between these two cold plasma modes can be easily caused by a shear of a magnetic field, because dispersion relations of both modes are close to each other. The evolution of wave field in such a case is calculated as follows.

First, we use conventional ray equation for a conservation of the lowest-order part of XGO as,

$$\frac{\mathrm{d}\boldsymbol{r}}{\mathrm{d}\tau} = \frac{\partial \mathcal{E}_0}{\partial \boldsymbol{k}} \quad , \quad \frac{\mathrm{d}\boldsymbol{k}}{\mathrm{d}\tau} = -\frac{\partial \mathcal{E}_0}{\partial \boldsymbol{r}} \quad (3)$$

where $\boldsymbol{r}$, $\boldsymbol{k}$, $\mathcal{E}_0$, and τ are cartesian coordinate, wave vector, averaged active eigenvalue of $\boldsymbol{D}_H$, and small step along a ray trajectory, respectively. $\mathcal{E}_0 = O(1)$ is decomposed value from diagonal eigenvalue matrix $\boldsymbol{\mathcal{E}}$ with deviation $\tilde{\boldsymbol{\mathcal{E}}} = O(\epsilon)$ by the following calculation.

$$\begin{aligned}\boldsymbol{\mathcal{E}} &= \begin{pmatrix} \mathcal{E}_1 & 0 \\ 0 & \mathcal{E}_2 \end{pmatrix} \\ &= \begin{pmatrix} \frac{\mathcal{E}_1+\mathcal{E}_2}{2} & 0 \\ 0 & \frac{\mathcal{E}_2+\mathcal{E}_1}{2} \end{pmatrix} + \begin{pmatrix} \frac{\mathcal{E}_1-\mathcal{E}_2}{2} & 0 \\ 0 & \frac{\mathcal{E}_2-\mathcal{E}_1}{2} \end{pmatrix} \\ &= \mathcal{E}_0 \boldsymbol{I} + \tilde{\boldsymbol{\mathcal{E}}} \end{aligned} \quad (4)$$

The way to choose initial $\mathcal{E}_0$ is as follows, that provisional wave vectors $\boldsymbol{k}_1$ which satisfies $\mathcal{E}_1 = 0$ and $\boldsymbol{k}_2$ for $\mathcal{E}_2 = 0$ are calculated by using well known and corresponding dispersion relations [7]. Then, averaged wavevector $\boldsymbol{k} = (\boldsymbol{k}_1 + \boldsymbol{k}_2)/2$ is calculated and by substituting $\boldsymbol{k}$ into Eq. [4], $\mathcal{E}_0$ and $\tilde{\boldsymbol{\mathcal{E}}}$ are given. Magnitudes of $\mathcal{E}_1$ and $\mathcal{E}_2$ can be very different and $\tilde{\boldsymbol{\mathcal{E}}}$, which is deviation from the averaged part of $\boldsymbol{\mathcal{E}}$, can approach the lowest-order magnitude in high density plasmas. In such cases, the assumption $\tilde{\boldsymbol{\mathcal{E}}} = O(\epsilon)$ is violated. Thus, this model works well in sufficiently low density plasmas, where $\mathcal{E}_1$ and $\mathcal{E}_2$ are almost degenerated. However, when $\mathcal{E}_1$ and $\mathcal{E}_2$ are very different, its branches of dispersion relation are also very different. It means that two active modes cannot excited simultaneously for a given $\boldsymbol{k}$. Therefore, in high density plasmas, it seems difficult to emerge mode coupling, and you should model such cases with XGO in one active limit. The lowest-order system gives a reference ray trajectory. Also, first-order terms of XGO equation including $\tilde{\boldsymbol{\mathcal{E}}}$, which expresses the deviation from a reference ray, give mode coupled wave field structure along a ray trajectory.

Here, it is convenient to introduce the curvilinear coordinates $(\zeta, \boldsymbol{\varsigma})$ for a calculation of complex amplitudes around the reference ray, where ζ is the longitudinal and $\boldsymbol{\varsigma}$ is the transverse coordinate as,

$$\begin{pmatrix} \zeta & \varsigma^1 & \varsigma^2 \end{pmatrix} = \begin{pmatrix} \frac{\dot{\boldsymbol{r}}}{|\dot{\boldsymbol{r}}|} & \frac{\dot{\boldsymbol{r}} \times (\ddot{\boldsymbol{r}} \times \dot{\boldsymbol{r}})}{|\dot{\boldsymbol{r}} \times (\ddot{\boldsymbol{r}} \times \dot{\boldsymbol{r}})|} & \frac{\dot{\boldsymbol{r}} \times \ddot{\boldsymbol{r}}}{|\dot{\boldsymbol{r}} \times \ddot{\boldsymbol{r}}|} \end{pmatrix}. \quad (5)$$

Then, by mapping into ray based curvilinear coordinates and a few simplification, $\mathbf{v} \cdot \nabla$ in original coordinates is rewritten as

$$\mathbf{v} \cdot \nabla \approx \pm\partial_\tau + \left(\tilde{\mathbf{v}}^\alpha \frac{\partial \varsigma^\mu}{\partial x^\alpha} \right) \partial_{\varsigma^\mu} = \pm\partial_\tau + \boldsymbol{\delta V}^\mu \partial_{\varsigma^\mu}. \quad (6)$$

Also, since we consider waves in cold magnetized plasma where $\boldsymbol{D}_A$ is vanishing, dissipation term $\boldsymbol{\Gamma}$ becomes zero. Eventually, partial differential equation for mode coupled vector complex amplitudes is obtained from higher-order terms of Eq. [1] as,

$$\pm\frac{\partial \boldsymbol{a}}{\partial \tau} = \left(i\tilde{\boldsymbol{\mathcal{E}}} - i\boldsymbol{U} + \frac{1}{2}(\nabla \cdot \mathbf{v}) \right) \boldsymbol{a} + \delta \boldsymbol{V}^\mu \frac{\partial \boldsymbol{a}}{\partial \varsigma^\mu}. \quad (7)$$

The lowest-order balance with Eq. [3] gives a reference ray, and first-order balance with Eq. [7] gives an envelope. This model is valid when the magnitude of $\tilde{\boldsymbol{\mathcal{E}}}$ stays in higher-order. In other words, this model works well in sufficiently low density plasmas. (e.g., a peripheral region of fusion plasmas.) We presented equations for a numerical approach up to the first order theory of XGO, which does not include diffraction. (Analytical approach is presented in [1].) However, we can introduce diffraction by using second-order theory. Actually, our numerical approach facilitates to include the diffraction. A more general method which captures mode coupling and diffraction simultaneously will be discussed elsewhere.

3 Numerical result

In this section, some numerical results simulated by a newly developed ray tracing code based on a model that treats two active modes, which is described in Sec. 2, and by 1DFW analysis code are presented.

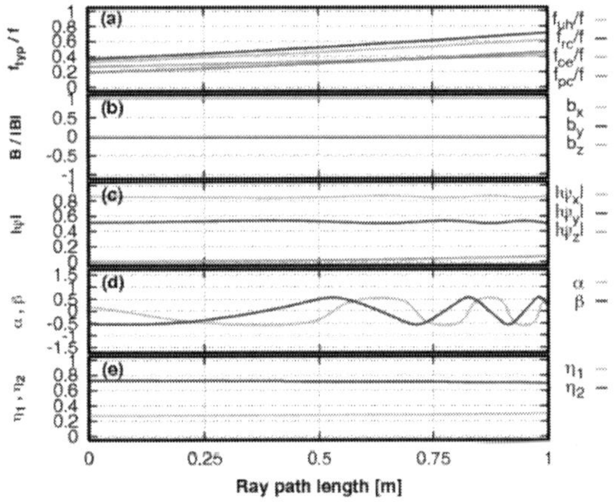

Figure 1. Calculation results of extended ray tracing in cold plasma slab with a magnetic field shear period $\tau_s = 1000$. Upper hybrid resonance, right hand cutoff, cyclotron resonance, and plasma cutoff normalized frequencies (a), x, y, and z components of magnetic field unit vector (b), x, y, and z components of amplitude intensity (c), polarization state angle α and β (d), and power fraction of two modes (e) are shown as a function of ray path length.

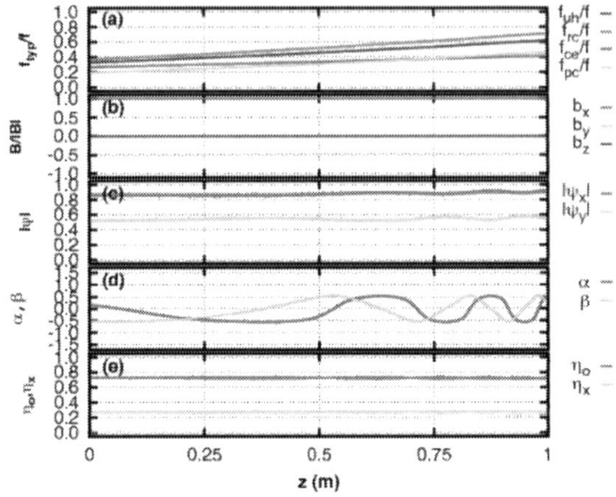

Figure 2. Calculation results of 1DFW in cold plasma slab with a magnetic field shear period $\tau_s = 1000$. (a) - (e) have the same meaning with Fig. 1

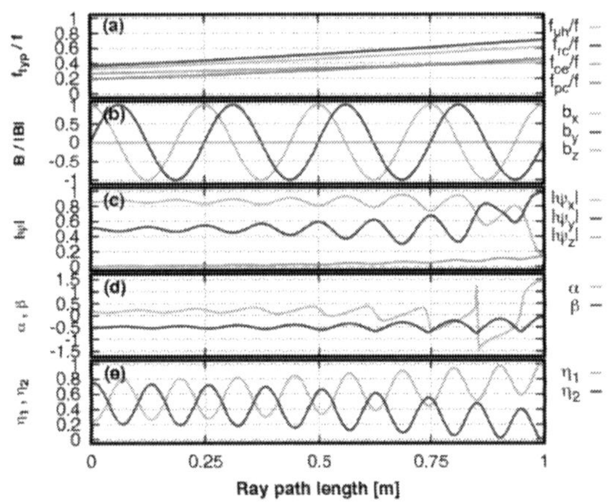

Figure 3. Calculation results of extended ray tracing in cold plasma slab with a magnetic field shear period $\tau_s = 0.25$. (a) - (e) have the same meaning with Fig. 1

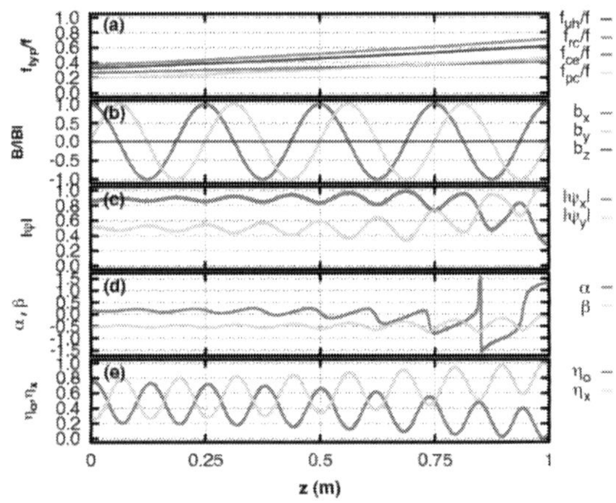

Figure 4. Calculation results of 1DFW in cold plasma slab with a magnetic field shear period $\tau_s = 0.25$. (a) - (e) have the same meaning with Fig. 1

Test wave is excited at $z = 0.0$ [m] and propagates into slab plasma along z axis. Here, wave frequency is 77.0 [GHz] and polarization state of vector complex amplitude is defined initially as follows,

$$\boldsymbol{\psi} = \begin{pmatrix} \cos\alpha \cdot \cos\beta - i \cdot \sin\alpha \cdot \sin\beta \\ \sin\alpha \cdot \cos\beta + i \cdot \cos\alpha \cdot \sin\beta \\ 0 \end{pmatrix} \tag{8}$$

where $(\alpha, \beta) = (10.0, -30.0)$ (deg) in these cases. In ray tracing code, excited vector wave is mapped into eigenmode representation with $\boldsymbol{a} = \Xi^{\dagger}\boldsymbol{\psi}$ and integrated by paraxial equation along a ray. 1DFW code calculates an evolution of vector wave field $\boldsymbol{\psi}$ directly. Profiles of electron density and external magnetic field, depending only on z direction, are

$$n_e(z) = n_{e0} \cdot \exp\left[-(z - z_c)^2/\lambda_n{}^2\right] \tag{9}$$

$$\boldsymbol{B}(z) = B_0 \cdot \exp\left[-(z - z_c)^2/\lambda_b{}^2\right] \times \begin{pmatrix} \cos(2\pi(z - z_c)/\tau_s) \\ \sin(2\pi(z - z_c)/\tau_s) \\ 0 \end{pmatrix} \tag{10}$$

where $n_{e0} = 1.5 \times 10^{20}$ $[m^{-3}]$, $B_0 = 2.0$ [T], density scale length $\lambda_n = 2.0$, magnetic field strength scale length $\lambda_b = 4.0$, and $z_c = 4.0$ are the same for Fig. 1 - 4.

Fig. 1 - 4 show the evolutions of wave parameters (amplitude, polarization states, and mode contents) of an electromagnetic-like O and X mode coupled wave along its trajectory in a cold slab plasma. Fig. 1 and 3 are results of extended ray tracing code, and Fig. 2 and 4 are results of 1DFW code [5, 6]. Fig. 1 and 2 are calculated for the weakly sheared magnetic field (magnetic field shear period $\tau_s = 1.0 \times 10^3$), and Fig. 3 and 4 are calculated for the strong one ($\tau_s = 2.5 \times 10^{-1}$).

Upper hybrid resonance, right hand cutoff, cyclotron resonance, and plasma cutoff frequencies normalized by test wave frequency (a), each component of the external magnetic field unit vector (b), each component of the wave field intensity (c), local α and β of mode coupled wave (d) and power fraction of two modes (e) as a function of ray path length from excited point are showen in Fig. 1 - 4, respectively. (a) shows that waves do not cross any resonance or cutoff layer. Local α and β of the combined modes are plotted in (d), which are given by

$$\alpha = \frac{1}{2}\tan^{-1}(\tan 2\gamma \cos\delta) \quad (11)$$

$$\beta = \frac{1}{2}\sin^{-1}(\sin 2\gamma \sin\delta) \quad (12)$$

where $\gamma = \tan^{-1}(|\psi_y|/|\psi_x|)$ and $\delta = \arg\psi_y - \arg\psi_x$. Power fraction in (e) is given by

$$\eta_q = \cos^2(\alpha - \alpha_q)\sin^2(\beta - \beta_q) + \sin^2(\alpha - \alpha_q)\cos^2(\beta + \beta_q) \quad (13)$$

where q denotes the each cold plasma mode and (α_q, β_q), which is polarization state angle for each mode, can be obtained by applying Eq. [11,12] for each mode field ψ_q. Here, η_1 and η_2 in Fig. 1 and 3 correspond to η_x and η_o in Fig. 2 and 4, respectively. Initial mode fraction is $\eta_{1,x} : \eta_{2,o} \sim 0.2 : 0.8$ in Fig. 1 - 4. However, The fraction of two modes are reversed in Fig. 3 and 4. Comparison between the calculation for a weakly sheared magnetic field (Fig. 1 and 2) and for a strong one (Fig. 3 and 4) demonstrates that mode coupling is caused by a sheared magnetic field, and the coupling depends on the shear. Then, good agreement between extended ray tracing (Fig. 1 and 3) and 1DFW analysis (Fig. 2 and 4) demonstrates that extended ray tracing captures mode coupling very well. Here, our new approach does not reduce refraction compared to 1DFW, and conventional GO ray tracing cannot calculate mode coupled vector waves in anisotropic media. Therefore, extended ray tracing has a great advantage for simulations of waves in peripheral regions of fusion plasmas.

4 Summary

Extended ray tracing based on first-order XGO theory is a prospective technique for describing the electromagnetic-like vector waves in sufficient low density plasmas with a sheared magnetic field. Such a situation is realized in a peripheral region of fusion plasmas. Mode coupling estimation performed by extended ray tracing shows good agreement with the result of 1DFW analysis. Diffraction will be treated, by introducing second-order terms of XGO into our numerical approach. A more general technique which captures refraction, mode coupling, and diffraction simultaneously will be discussed in forthcoming papers [8].

Acknowledgments

The authors deeply thank I. Y. Dodin of Princeton Plasma Physics Laboratory for his invaluable discussions.

References

[1] I. Y. Dodin et. al., Physics of Plasmas **24**, 122116 (2017).

[2] D. E. Ruiz and I. Y. Dodin, Physics of Plasmas **24**, 055704 (2017).

[3] D. E. Ruiz and I. Y. Dodin, Physics Letters A **379**, 2337 (2015).

[4] D. E. Ruiz and I. Y. Dodin, Physical Review A **92**, 043805 (2015).

[5] S. Kubo et. al., AIP Conference Proceedings, 1689, 090006-090009 (2015).

[6] T. I. Tsujimura et. al., Nuclear Fusion **55**, 123019-123030 (2015).

[7] T. H. Stix, Waves in Plasmas, AIP, New York, (1992).

[8] I. Y. Dodin and K. Yanagihara, to be submitted Physics of Plasmas.

Recent ECRH/ECCD experiments aiming for higher density and temperature operations in the LHD

Hiroe Igami[1*], *Shin* Kubo[1,2], *Takashi* Shimozuma[1], *Yasuo* Yoshimura[1], *Toru* Ii Tsujimura[1,3], *Sakuji* Kobayashi[1], *Yoshinori* Mizuno[1] *Hidenori* Takubo[1], *Kenji* Tanaka[1], *Masayuki* Yokoyama[1,3], *Ryosuke* Seki[1], *Ichihiro* Yamada[1], *Ryo* Yasuhara[1], *Hayato* Tsuchiya[1], *Katsumi* Ida[1,3], *Mikiro* Yoshinuma[1,3], *Tatsuya* Kobayashi[1,3], *Satoshi* Ohdachi[1,3], *Masaki* Osakabe[1,3], *Tomohiro* Morisaki[1,3] and the LHD Experiment Group

[1]National Institute for Fusion Science, National Institutes of Natural Sciences, Toki 509-5292, Japan
[2]Nagoya University, Nagoya 464-8603, Japan
[3]SOKENDAI (Graduate University for Advanced Studies), Toki 509-5292, Japan

Abstract. In LHD, real-time control of the incident EC wave polarization and quick response microwave bolometer for monitoring the stray radiations have been developed for efficient and safe operation of the high power and long pulse ECRH/ECCD. As a high power ECRH/ECCD application aiming for high density, ECRH has been demonstrated up to 85% of the cutoff density by the fundamental X-mode excitation in HFS with use of a horizontal port antenna located in LFS. As another application aiming for high temperature, the effect of the control of the rotational transform with use of the ECCD on the sustainment of the e-ITB is investigated. It has been suggested that higher local electron energy confinement time is obtained inside the e-ITB with placing the m/n = 2/1 magnetic island near the edge of the e-ITB compared to the case when the 2/1 island is vanished in the plasma.

1 Introduction

The electron cyclotron resonance heating (ECRH) and current drive (ECCD) have various important roles in fusion oriented magnetically confined plasmas. Development of the gyrotrons aiming for high power and long pulse operation has expanded the possibility of applications of the ECRH/ECCD. With increasing of the available gyrotron power, to reduce the heat load caused by non-absorbed stray radiation by improvement of the heating efficiency with monitoring the stray radiation has become more important from the viewpoint of safety. Moreover, to save the required total power oscillated by gyrotrons contributes to save the money to build and run the plasma devices including future fusion reactors. In addition to the improvement of the transmission efficiency, the improvement of the heating efficiency is also important.

In the Large Helical Device (LHD), the operation region toward both higher temperature and higher density has been expanded with increasing the available heating power, optimizations of the operation conditions, and progress in studying the underlying physics [1]. After the start of the deuterium operation, the highest record of the ion temperature, $T_i = 10$ keV with the formation of the ion internal transport barrier (i-ITB), was achieved through several optimizations of the operation conditions, for example, wall condition, magnetic configuration, and timing of the impurity pellet injection [2]. As one of the optimizations of the operation conditions, the ECRH discharge cleaning with Helium gas puffing contributed to form the peaked density profile that is preferable for realizing high ion temperature by reducing the wall recycling. Through the experiments, the isotope effect on the confinement characteristics has been indicated in both the i-ITB plasmas [2] and the electron internal transport barrier (e-ITB) plasmas [3]. When the highest $T_i = 10$ keV was obtained with formation of the i-ITB, the electron temperature T_e was only 3.5 keV. Since both T_i and T_e are high in the fusion reactor, studies on the confinement and the transport characteristics of ions and electrons in such a plasma are important. In LHD, $T_i \sim 6$ keV and $T_e \sim 8$ keV were already achieved simultaneously with simultaneous formations of the i-ITB and the e-ITB with neutral beam injection (NBI) of 26 MW and the centralized ECRH of 5.1 MW [4]. To expand the operation region of the simultaneous pursuit of high T_i and T_e, we must have a better understanding of the role of ECRH/ECCD on the formation and the sustainment of the internal transport barriers (ITB)s.

Expanding the operation region toward high density is also important. In a heliotron type fusion reactor named FFHR, the central magnetic field strength B(0)=6 T, density $n(0) = 6 \times 10^{20}$ m^{-3} and temperature T(0) =8.5 keV were proposed as a set of final operating parameters [5]. Since the central density exceeds the cutoff density of 170 GHz that is adopted as the ITER gyrotron frequency, establishment of the effective fundamental ECRH in the core and the peripheral region from the low density regime at the plasma initiation to the final over-dense

* Corresponding author: igami.hiroe@nifs.ac.jp

regime is desired because it is considered difficult to produce a high power and long pulse gyrotron of the second harmonic frequency (> 300 GHz) at this time.

In this paper, the experimental apparatuses are explained first in section 2. In section 3, recent progress of control and measurement systems developed for efficient and safe operation of the ECRH are introduced. In section 4, applications of the ECRH/ECCD aiming for high density and high temperature operations are reported. The summary is in section 5.

2 Experimental Apparatuses

2.1 The Large Helical Device (LHD)

LHD is the second largest superconducting helical device with poloidal/toroidal period numbers of 2/10. The plasma major radius is 3.42-4.1 m and the average minor radius is approximately 0.6 m [6]. The magnetic field strength at the magnetic axis R_{ax} = 3.6 m can be raised up to 2.85 T by super-cooling of the helical coils. Three negative ion-based neutral beams (NBs) of 180 keV with total heating power of ~16 MW (hydrogen) / ~20 MW (deuterium) are injected tangentially to the magnetic axis to generate and heat the plasma. Two positive ion-based neutral beams of 60 keV and 80 keV with total heating power of ~12 MW (hydrogen) / ~20 MW (deuterium) are also injected perpendicularly to the magnetic axis. The electron cyclotron resonance heating (ECRH) with the total heating power of 5.4MW is also used to generate and heat the plasma. Further information about the ECRH system is described below.

2.2 Present ECRH system

The present ECRH system consists of one 82.7 GHz (0.45 MW/ 2 s, GYCOM) diode gyrotron, three 77 GHz (1-1.5 MW/2 s, 0.3 MW/CW, TETD) triode CPD gyrotrons, two 154 GHz triode CPD gyrotrons (1 MW/2 s, 0.5 MW/CW, TETD), six evacuated transmission lines using corrugated waveguides of 88.9 mm diameter, and six quasi optical mirror antennas. Two of the antennas are installed in a top port named 5.5-U and the other four antennas are installed in a horizontal port named 2-O. As mentioned above, the maximum simultaneous input power is 5.4 MW for short pulse (< 2 seconds) operations. During the last experimental campaign, two of the gyrotrons failed. One gyrtron failed because of the water leakage from a piping joint at the exit of the gyrotoron cavity in the oil tank. However, this gyrotron is already ready for operation after the component replacement. Another one gyrotron failed because its relief window broke. This gyrotron will be operated in the next experimental campaign after repair in the factory.

3 Development of control and measurement systems for efficient and safe operation of the ECRH

3.1 Real time control of the EC wave polarization

In the peripheral region of the LHD plasma, the ratio of the O mode and the X mode of the EC wave significantly changes during the propagation because of the existence of the strong magnetic shear and the electron density that is widely distributed even outside of the last closed flux surface (LCFS). For efficient ECRH, optimization of the polarization of the launched EC wave with taking into account this effect is necessary to excite pure O/X mode at the electron cyclotron resonance (ECR) layer. In the transmission line of the 77 GHz wave launched from one of the top port antennas, the real-time polarization control under real-time acquisition of the density profile is performed with real-time field programmable gate array (FPGA) processing with using CompactRIO (cRIO) [cRIO-9035] and its input/output modules made by National Instruments and its software through LabView FPGA programming. This optimization method and experimental results are reported precisely in Ref. [7]. The deviation of the rotation angle of each λ/4 and λ/8 elliptically polarizer from the optimum angle can be corrected with the time rate of change of the polarization rotator at about 40°/sec.

3.2 Stray radiation measurement with use of a quick response microwave bolometer

In LHD, sniffer probes were used for the stray radiation measurement [8]. However, they were removed before the start of deuterium operation because the neutrons generated as fusion products in the deuterium operation degrade the sensitivity of the microwave detectors made from semiconductors. The microwave bolometer [9] has been used to estimate the accumulated energy of the stray radiation [10]. Different from the sniffer probe, the existing microwave bolometers which consist of a microwave absorber and a thermocouple cannot respond to the fast change of the stray radiation intensity. To estimate the stray radiation intensity in several ten milliseconds we have developed a quick response microwave bolometer as an alternative to the sniffer probes.

3.1.1 Design of a quick response microwave bolometer

The microwave absorber of our quick response microwave bolometer consists of Al_2O_3-TiO_2 (87%/13%) coated copper cylinder whose diameter and height are 10 mm. The copper cylinder is divided into two blocks to

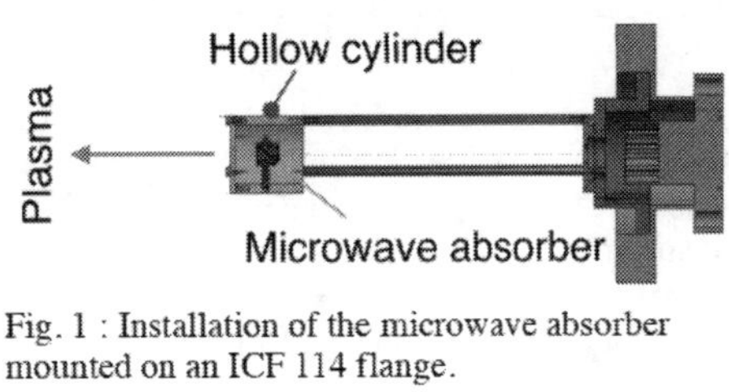

Fig. 1 : Installation of the microwave absorber mounted on an ICF 114 flange.

squeeze the super-fine sheathed thermocouple whose diameter is 0.25 mm tightly with screws. As shown in Fig. 1, this copper cylinder is mounted in a hollow cylinder made from aluminium alloy with three-point support by set screws to avoid the metallization and to shade the radiations directly come from the plasma. The output of the sheathed thermocouple is transmitted via vacuum feedthroughs mounted on an ICF 114 flange and a compensating lead wired to a fast thermocouple signal convertor [MS3701, MTT Corporation].

3.1.2 Experimental results

For the first step, a quick response microwave bolometer was installed in a horizontal port where four of the ECRH antennas are installed. Fig.2 shows discharge waveforms when the ECRH of 2.5 MW in total was performed from 3.30 s to 5.30 s. The plasma was initiated just after the simultaneous start of neutral beam injection (NBI) and ECRH with fast increases of the electron temperature and density. The temperature of the microwave bolometer increases within 15 ms after the start of ECRH and does not increase any more after the stop of ECRH although the plasma radiation remains the same level. Therefore, this system has almost no sensitivity of plasma radiations. The averaged increasing rate of the temperature dT/dt during the ECRH is 3.11 Celsius degrees/sec. Since dT/dt =1.43 degrees/sec. for the power flux of 10 kW/m^2 is expected with assuming that the absorption rate is 0.82 and the emissivity is 0.4 at the surface, the local power flux at the bolometer is estimated to be 22 kW/m^2. The local power flux per unit power is 8.8 kW/m^2/MW. The total ECRH efficiency estimated by the ray-tracing code "LHDGauss" [11] with using post-processing three-dimensional (3D) equilibrium mapping obtained from experiments is about 90% without the start-up phase of the plasma.

For the case of 10 ms 154 GHz, 0.8 MW EC wave launching to the plasma where no ECR layer exists the

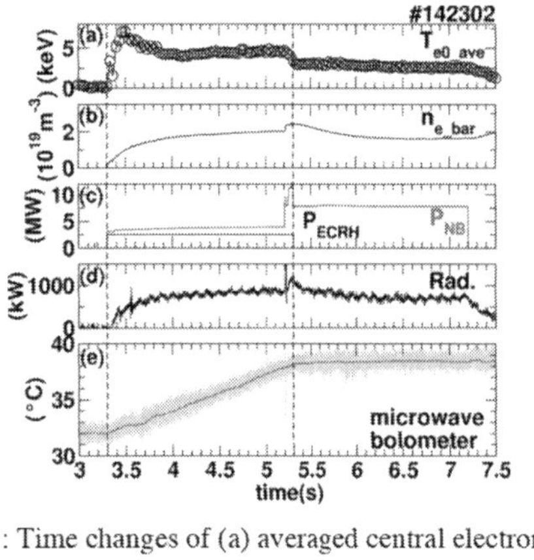

Fig.2 : Time changes of (a) averaged central electron temperature, (b) line averaged electron density, (c) total input power of ECRH and total input power of NBs, (d) radiation power from the plasma, and (e) temperature of the microwave bolometer, raw (grey) and low-pass filtered (f_{cutoff}=40Hz, red) output of the thermocouple signal convertor.

microwave bolometer could respond even within 5 ms with dT/dt =10 degrees/sec. The estimated power flux per unit power is 87 kW/m^2/MW, which is approximately ten times larger than the above case of 90 % ECRH efficiency. Although it has been pointed out that the stray radiation intensity measured close to the ECRH antenna does not directly reflect the ECRH efficiency because the stray radiation is not isotropic, these experimental results may suggest the order of the stray radiation. Although the microwave bolometer can respond within 15 ms, irregular voltage fluctuations come into the signal line of this system frequently. Considering the typical time width of the voltage fluctuation, the increase of temperature caused by the strong stray radiation larger than 28 kW/m^2, or the stray radiation accumulated for more than 50 ms can be discriminated from the voltage fluctuation.

4 ECRH experiments to aim high density and temperature operations

4.1 ECRH by the fundamental X-mode and/or the EBW excited by direct oblique launching from the LFS

4.1.1 Launching method to realize fundamental X-mode propagation at the HFS of the ECR layer

For usual ECCD in LHD, the fundamental ordinary (O) mode is launched obliquely to the external magnetic field from the horizontal port antenna to access the ECR layer from the low magnetic field side (LFS) as shown in Fig. 3. While a ray tracing calculation with taking into account the density profile even outside the last closed flux surface (LCFS) (see Fig.4) suggests that the

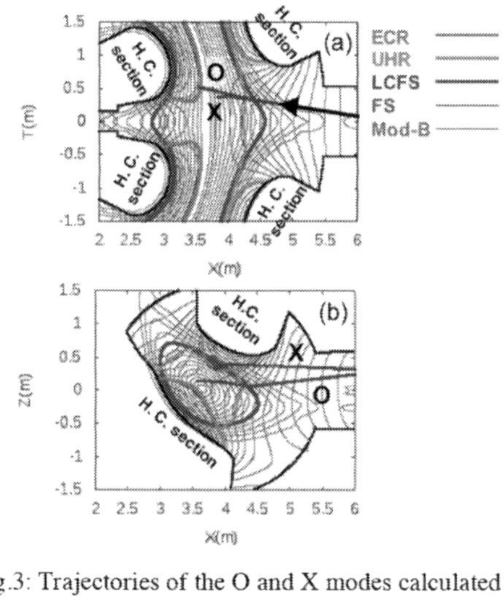

Fig.3: Trajectories of the O and X modes calculated by ray-tracing with use of the fitted profile shown in Fig.4. (a): projection on the equatorial plane. (b): projection on a plane vertically sliced along the launched beam indicated by a black arrow drawn in (a). Contours of the ECR layer, UHR layer, flux surfaces, LCFS and Mod-B surfaces are also plotted.

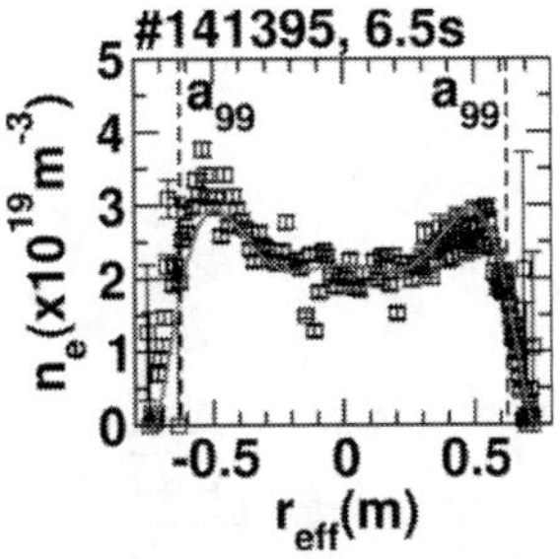

Fig. 4: The electron density profile obtained by Thomson scattering measurement (square) and its polynomial fitted profiles (solid line) plotted versus the effective minor radius "r_{eff}". The dashed lines represent the a_{99} surface inside which 99% of the electron kinetic energy is included and almost coincides with the LCFS.

extraordinary (X) mode launched from the same antenna can access the high field side (HFS) of the ECR layer with aiming for the upper space located slightly below the neighboring helical coil. This fast X (FX) mode propagates at HFS with a finite parallel refractive index ($N_{//}$) that changes along the propagation because of the inhomogeneity of the external magnetic field. If the FX mode passes the ECR layer with a finite $N_{//}$, then strong power absorption is expected. On the other hand, if $N_{//}$ becomes close to zero, the power absorption rate is reduced. The non-absorbed power of the FX mode reaches the upper hybrid resonance (UHR) layer and excites the electron Bernstein wave (EBW) via the linear mode conversion process. ECRH by the EBW is expected in that case. In the parameter space of $\Omega_e/\omega > 1$, where Ω_e is the electron cyclotron angular frequency and ω is the wave angular frequency, the FX-mode can propagate until the electron density reaches twice the plasma cutoff density in maximum. Therefore, ECRH from the under-dense to the over-dense regime is expected without changing of the launching direction.

4.1.2 Experimental results

In the experiment, we started to launch the 77 GHz EC wave in the manner shown in Fig. 3 in parallel with NB 2,3, (tangential) and 4 (perpendicular) just after the stop of NB 1 (tangential) and 5 (perpendicular) as shown in Fig. 5 (a), (b). The magnetic configuration was (R_{ax}, B_t) = (3.6 m, 2.75 T), where R_{ax} is the position of the magnetic axis, where the magnetic field strength is B_t. The electron density is maintained at about 35% of the cutoff density. The right handed circularly polarization is selected as the incident wave polarization to couple with the X-mode. As shown in Fig.5 (b)-(d), the electron cyclotron emission (ECE) and the stored energy were modulated with following the 23 Hz/100 % ECRH power modulation. Fig. 6 shows the result of the FFT analysis of ECE signals of each channel with taking the differences from their time trends and normalizing by the time averaged values. It is suggested that the power absorption occurs in the peripheral region where the peak of the modulation amplitude (Fig.6-(a)) and the bottom of the phase delay

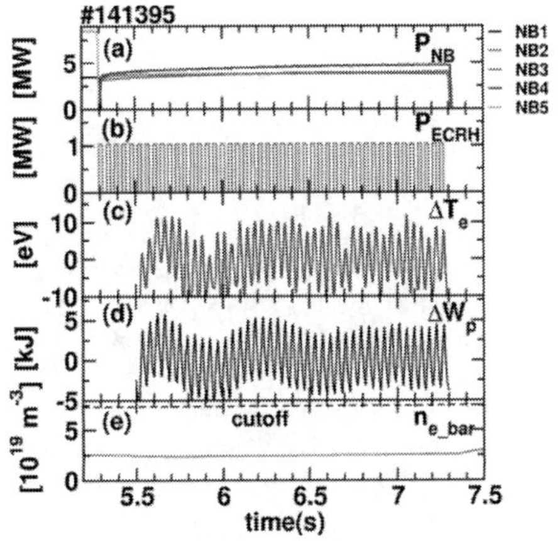

Fig.5 : Time changes of (a) input powers of the NBs, (b) input power of the modulated ECRH, (c) difference of the ECE (measured at R=4.216 m) from the time trend between 5.5 and 7.3 s, (c) difference of the stored energy from the time trend between 5.5 and 7.3 s, and (e) line averaged electron density.

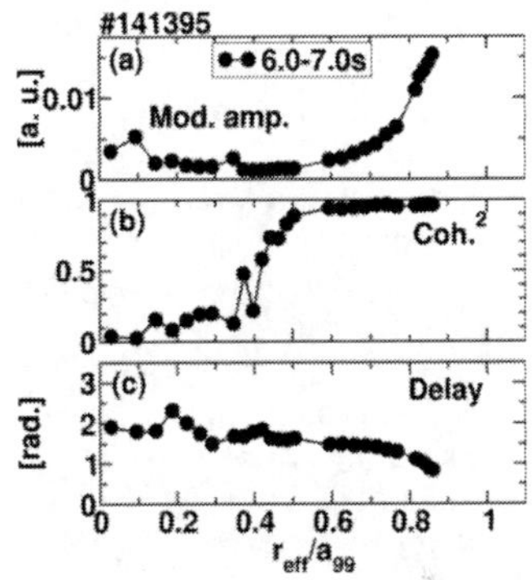

Fig.6 : (a) modulation amplitude normalized by the time average, (b) coherence function, and (c) phase delay are plotted versus the normalized minor radius for the 23 Hz component obtained by FFT analysis of each ECE channel from 6.0 s. to 7.0 s.

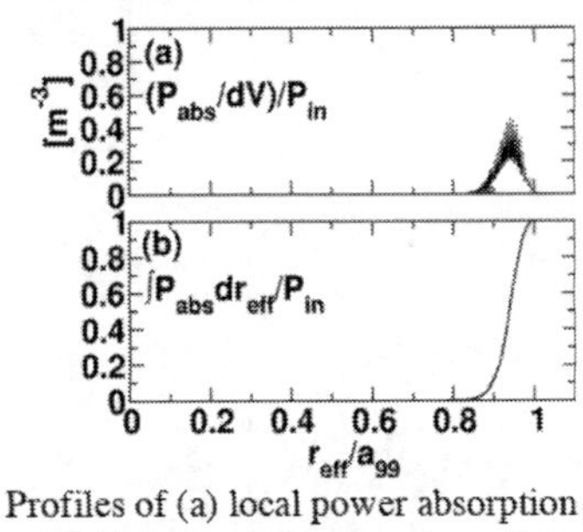

Fig.7 : Profiles of (a) local power absorption per unit volume and (b) integrated power absorption, plotted versus the normalized minor radius. These profiles are obtained by the ray tracing calculation with use of the fitted electron density profile shown in Fig. 4

(Fig.6- (b)) are located. The ray-tracing calculation with using the equilibrium mapping data associated with the electron temperature and density profiles at 6.0 s in this discharge suggests that the power absorption of the fundamental X-mode occurs in the peripheral region (Fig.

7). The absorbed power of the ECRH (P_{abs}) can be estimated from the difference of the time derivative of the stored energy between before and after power input for the ECRH as follows.

$$P_{abs} = \left.\frac{dW_p}{dt}\right|_{t_{on}+t_{delay}}^{t_{on}+t_{delay}+\Delta t} - \left.\frac{dW_p}{dt}\right|_{t_{on}-\Delta t}^{t\_on} \quad (1)$$

In the above equation, W_p is the stored energy obtained by the diamagnetic flux loop signal, *t_on* is the time when the modulated ECRH is turned on, and Δt (= 1 ms) is the time for linear fitting to calculate the time derivative dW_p/dt. To take into account the response time of the diamagnetic flux loop signal, dW_p/dt is calculated *t_delay* (= 5 ms) later from *t_on*. After taking the conditional average of W_p during the power modulation, the ECRH efficiency P_{abs}/P_{in} = 70 % is obtained, where P_{in} is the input power. However, the ray-tracing calculation at the beam center indicates 100 % power absorption. There are some possible reasons for reducing the heating efficiency. One of the possibilities may be caused from the broadening of the launched beam. A fraction of the launched beam at the wall side may encounter the vacuum vessel wall and cannot penetrate into the plasma. Another fraction in the opposite side of the wall encounters the right handed cutoff (RC). Since the reflection occurs at the RC to some extent, the power fraction that couples with the FX mode at HFS is reduced. However, as it was pointed in the previous studies [12, 13], the rest fraction can transmit through the thin evanescent region between the RC and the UHR layer and couple with the FX mode. The ECRH efficiency obtained in this experiment is improved compared to the previous studies with use of the similar launching method with use of the bottom port antennas. The power fraction that can reach the HFS might be improved with use of the horizontal port antenna. In addition, the deviation of the incident wave polarization from the optimum polarization for exciting the pure X-mode might be also possible.

In another discharge, the electron density was ramped up during the ECRH as shown in Fig. 8-(e). The 77 GHz

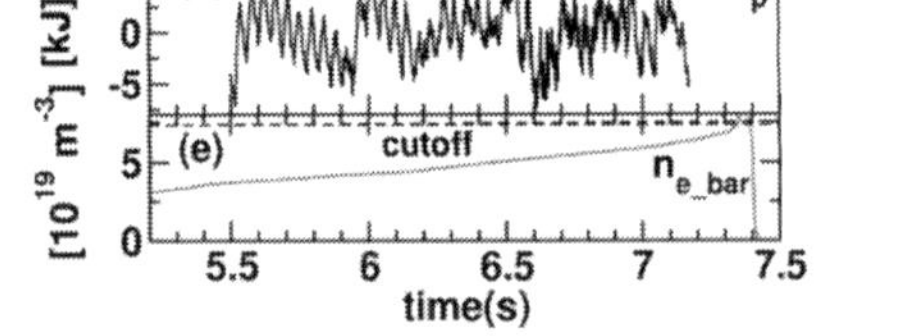

Fig.8 : Plots similar to Fig. 5. The electron density was ramped up. Differences of the ECE and the stored energy from their time trends were taken between 5.5 s to 7.17 s. when NB2 stopped.

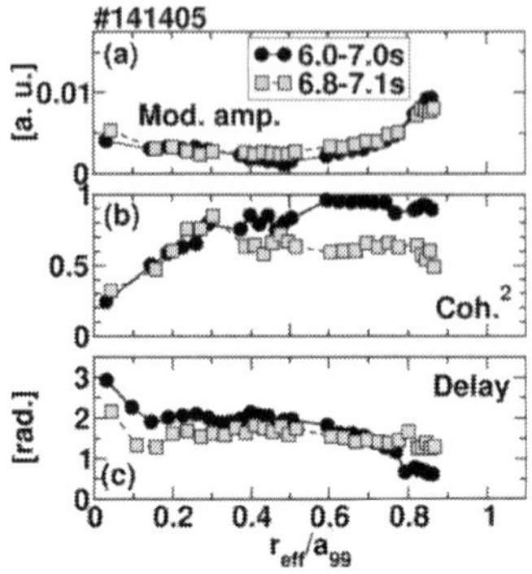

Fig.9 : Profiles similar to Fig. 6 obtained by FFT analysis of each ECE channel from 5.5 s. to 5.8 s. (circles) and from 6.8 s. to 7.1 s. (squares).

EC wave was launched in the manner shown in Fig. 3 with 23 Hz/100 % power modulation. It was observed that the ECE and the stored energy were modulated with following the modulated ECRH until 7.17 s. when the line averaged electron density reaches 85 % of the cutoff density. Since the MHD instability appears and grows with the increase of the density in this magnetic field configuration of (R_{ax}, B_t) = (3.6 m, 2.75 T), it is difficult to increase the electron density more than the cutoff density of 77GHz. In Fig.9, similar to the case above, the peak of the normalized amplitude and the bottom of the phase delay are located in the peripheral region locally in the time range of 5.5-5.8 s. Meanwhile, it is suggested that the power absorption region spreads across a wide area in the time range of 6.8-7.1 s. because the bottom of the phase cannot be found clearly. To demonstrate the ECRH in the over-dense plasma with this launching method effectively, the magnetic axis outside 3.6 m should be selected to suppress the MHD instabilities and to place the ECR layer closer to the magnetic axis with increasing the magnetic field strength by super-cooling operation of the helical coils.

4.2 Control of the e-ITB performance with controlling the rotational transform by ECCD

The formation of the ITB is necessary to realize a high temperature core plasma in helical systems. It is known that the electron ITB (e-ITB) is formed when the applied power of the centralized ECRH is above the threshold value in helical systems. Although the threshold power tends to increase with the increase of the electron density of the target plasma, it has been pointed out that the threshold power can be different even for the similar density if the profile of the rotational transform, $\iota/2\pi$, is different [14, 15]. It has also been pointed out that the required ECRH power to sustain the similar central electron temperature in the e-ITB is different [15]. It is important to investigate what type of $\iota/2\pi$ profile is better for sustaining the e-ITB inside which higher electron temperature and higher density is realized over a wider region with less power input. From this point of view the performance of the e-ITB is characterized by the local energy confinement time defined inside $r_{eff} \leq 0.25$ m and investigated for various $\iota/2\pi$ profiles controlled by ECCD and neutral beam current drive (NBCD).

4.2.1 Control of $\iota/2\pi$ by ECCD and NBCD

In the experiment where (R_{ax}, B_t) = (3.6 m, 2.75 T), one of the tangential negative ion-based counter NBs was injected to measure $\iota/2\pi$ profile experimentally by the motional Stark effect (MSE) spectroscopy [16,17]. The inductive current that compensates the toroidal current driven by this counter NBI increases $\iota/2\pi$ in the core region. This inductive current can be cancelled by the opposite tangential negative co NBI of similar input power. We call this combination of co and counter tangential NBIs as the "balanced NBI" here. The co ECCD increase $\iota/2\pi$ and the counter (cntr.) ECCD decrease $\iota/2\pi$ in the core region. With the counter ECCD superimposed on the balanced NBI, $\iota/2\pi$ in the core decreased close to 1/3 (Fig.10 (a)). With the co ECCD superimposed on the balanced NBI, $\iota/2\pi$ in the core increased near but at slightly lower than 1/2 (Fig.10 (b)). With the co ECCD superimposed on the counter NBI $\iota/2\pi$ was changed from1/2 (Fig.10 (c)-1) to far beyond 1/2 (Fig.10 (c)-2) during a discharge.

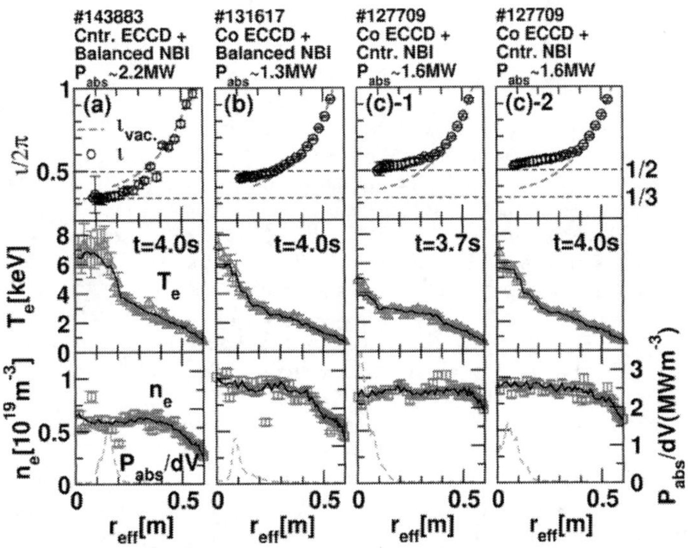

Fig.10 : Profiles of (Top) measured and vacuum $\iota/2\pi$, (Middle) electron temperature, (Bottom) electron density, and power density absorbed by electrons plotted versus effective minor radius.

4.2.2 Evaluation of the e-ITB performance with a partial-coverage energy confinement time in the core region

In the fusion reactor, higher temperature and density should be realized with less input power in wider region. In this paper, we would like to introduce the partial-coverage electron energy confinement time as a measure of inclusive e-ITB performance in the core region where $r_{eff} \le 0.25$ m ($\tau_{e\_reff\le0.25m}$) as follows.

$$\tau_{e\_r_{eff}\le0.25m} \equiv \int_0^{0.25} \left\{ \frac{3}{2} n_e T_e \left(\frac{dV}{dr_{eff}} \right) / P_{abs} \right\} dr_{eff} \quad (2)$$

P_{abs} in eq. (2) is the sum of the ECRH and NBI powers absorbed by electrons. The former is calculated by the "LHDGauss" and the latter is calculated by the "FIT3D" [18] codes. Both the "LHDGauss" and the "FIT3D" codes use the post-processing three-dimensional (3D) equilibrium mapping obtained from the experiment.

4.2.3 Co ECCD superimposed on the counter NBI

Fig. 11 shows the time change of $\tau_{e\_reff\le0.25m}$ when the co ECCD was superimposed on the counter NBI. Both $\tau_{e\_reff\le0.25m}$ and ECE signals shown in Fig.12 near the core show sudden drops and quick recoveries before and after

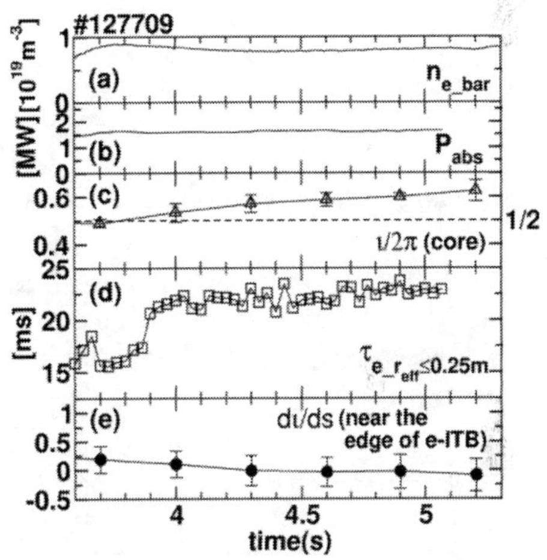

Fig.11 : Time changes of (a) line averaged density, (b) the total absorption power in $r_{eff} \le 0.25$ m, (c) $\iota/2\pi$ in the core region (averaged in 0.10 m $\le r_{eff} \le$ 0.20 m), (d) $\tau_{e\_reff\le0.25m}$, and (e) magnetic shear near the edge of the e-ITB (averaged in 0.25 m $\le r_{eff} \le$ 0.30 m). The co ECCD was superimposed on the counter NBI.

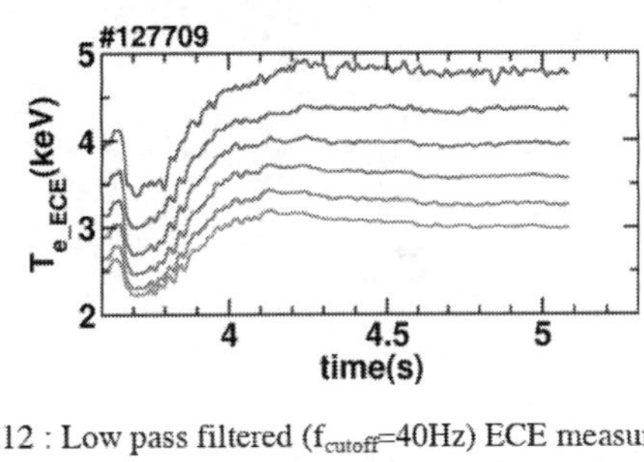

Fig.12 : Low pass filtered (f_{cutoff}=40Hz) ECE measured at 0.144 m $\le r_{eff} \le$ 0.25 m in the same discharge of Fig.11.

$\iota/2\pi$ near the core crosses 1/2. At the bottom of the temperature drop (3.7 s), $\iota/2\pi=1/2$ is realized with gentle gradient (weak magnetic shear) at the core region and the flattening of the electron temperature around the e-ITB is observed as shown in Fig.10 (c)-1. It was reported that the stochastization can occur in the core region when the weak magnetic shear is realized near $\iota/2\pi=1/2$ and causes the flattening of the density and temperature profiles [19-22]. It was also reported that the required ECRH power to form and sustain the e-ITB is higher when the stochastic region exists in the core region of the target plasma [15]. In this experiment, it is suggested that the reduction of $\tau_{e\_reff\leq 0.25m}$ is caused by the partial stochastization around the e-ITB. From 3.5 s to 4.0 s, harmonic magnetic fluctuations of ~2.5 kHz, ~5kHz, ~7.5 KHz, and 10 KHz are observed. On the contrary, after 4.0 s, only magnetic fluctuation of ~2.5 kHz remained. This indicates that the MHD instabilities which restrict the increase of the electron temperature are well suppressed after the quick recovery from the partial stochastization. Moreover, $\tau_{e\_reff \leq 0.25m}$ increases slightly after the magnetic shear near the edge of the e-ITB changes from positive to negative. Here, the magnetic shear is defined as $d\iota/ds \equiv \rho/\iota(d\iota/d\rho)$ and $\rho \equiv r_{eff}/a_{99}$, where a_{99} is the flux surface inside which 99% of the electron kinetic energy is included. It is speculated that the reduction of the magnetic shear near the edge of the e-ITB may affect the radial heat transport to some extent in this case as the previous studies (e.g., [23]) pointed out.

4.2.4 Counter ECCD superimposed on the balanced NBI

When the counter ECCD was superimposed on the balanced NBI, $\iota/2\pi=1/3$ was realized near the core with gentle gradient as shown in Fig 10-(a). $\iota/2\pi$ remains to be 1/3 and $\tau_{e\_reff\leq 0.25m}$ decreases as time passes. Wideband magnetic fluctuations up to 6 kHz observed during the sustainment of the e-ITB suggest the existence of MHD instabilities which may restrict the increase of the electron temperature and resultant decrease of $\tau_{e\_reff \leq 0.25m}$. Actually, rapid rises and drops of low pass filtered (< 40 Hz) ECE signals shown in Fig. 13 indicate the local collapse of the electron temperature profile in the core region. Note that these rises and drops of the low pass filtered ECE signals are observed independently from the perpendicular NBI (every 0.4 s from 3.6 s) for 20 ms superimposed for the ion temperature measurement by charge exchange spectroscopy (CXS).

4.2.5 Co ECCD superimposed on the balanced NBI

In the case of co ECCD superimposed on the balanced NBI, the highest $\tau_{e\_reff\leq 0.25m}$ ~30 ms is obtained as shown in Fig. 15. As shown in Fig. 10-(b) $\iota/2\pi = 1/2$ is located near the edge of the e-ITB at the early phase of the discharge. Then $\iota/2\pi = 1/2$ moves to the central region. It was reported that the existence of $\iota/2\pi = 1/2$ near the edge of the e-ITB can realize higher central electron temperature [14]. And it is pointed out that the strong shear of the radial electric field at the boundary of the magnetic island in the helical plasma can contribute to the reduction of transport through turbulence suppression by ***E* x *B*** shear [24]. This mechanism can explain that higher $\tau_{e\_reff\leq 0.25m}$ is obtained with the existence of $\iota/2\pi = 1/2$ near the edge of the e-ITB compared to the case when $\iota/2\pi = 1/2$ is vanished by co ECCD superimposed on the counter NBI, as described above. However, $\tau_{e\_reff\leq 0.25m}$ decreases and increases repeatedly as shown in Fig. 13. A sudden drop (4.2 s) and gradual decrease (4.7 s) with fluctuations are also observed in low pass filtered (< 40

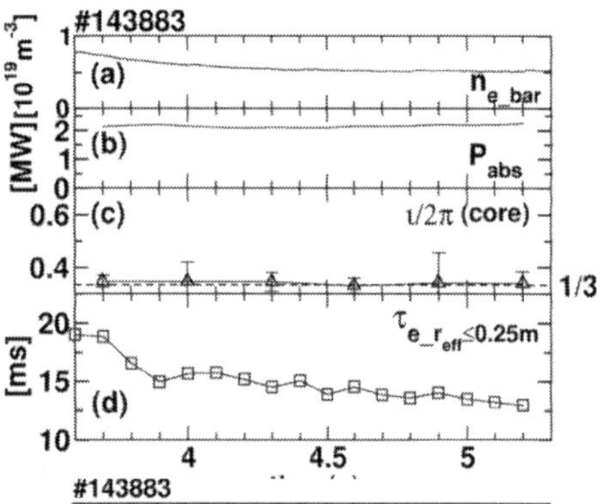

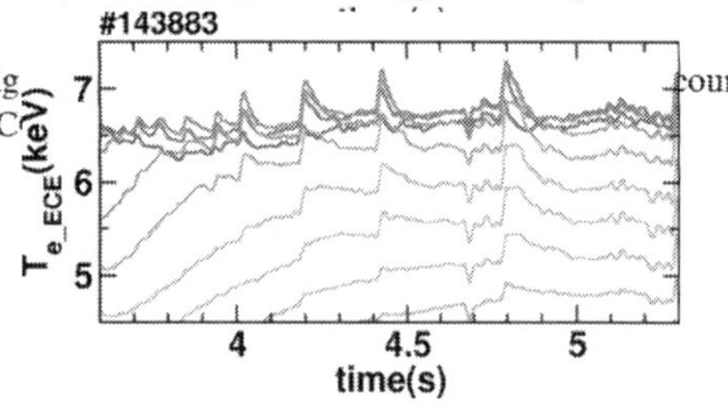

Fig.14 : Low pass filtered (f_{cutoff}=40Hz) ECE measured at 0.07 m ≤ r_{eff} ≤ 0.24 m in the same discharge of Fig.13.

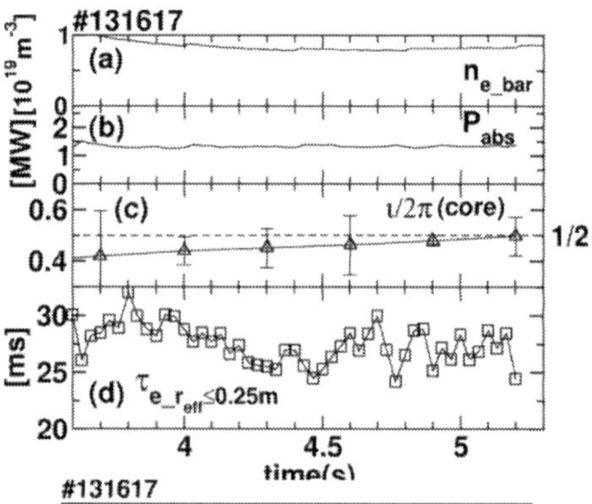

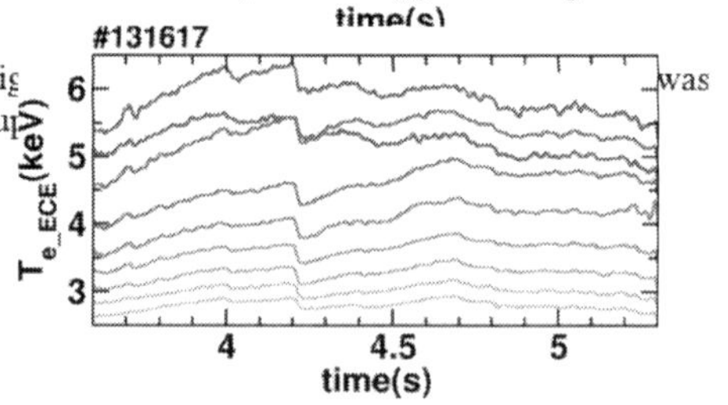

Fig.16 : Low pass filtered (f_{cutoff}=40Hz) ECE measured at 0.02 m ≤ r_{eff} ≤ 0.23 m in the same discharge of Fig.15.

https://doi.org/10.1051/epjconf/201920302001

Hz) ECE signals independently from the perpendicular NBI (every 0.4 s from 3.6 s) for 20 ms. Since several harmonic magnetic fluctuations are observed from 3.5 s to 4.5 s and wideband (10 kHz -17 kHz) fluctuations with frequency chirping down are observed from 4.5 s to 5.0 s, the existence of MHD instabilities that prohibit the increase of the electron temperature is indicated also in this case.

4.2.6 Discussion

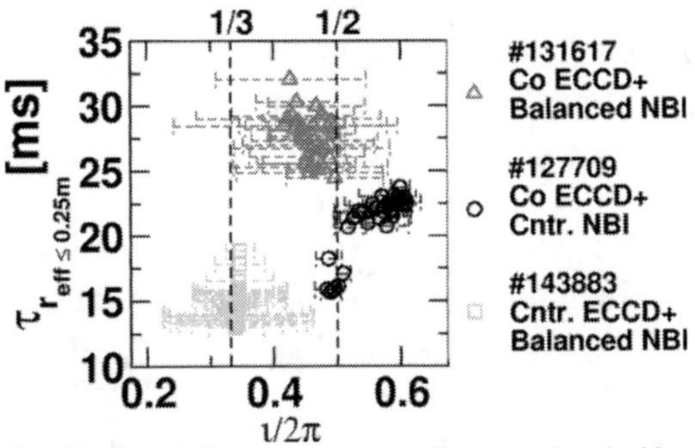

Fig. 17 : Local electron energy confinement time inside r_{eff} =0.25 m ($\tau_{e\_reff \le 0.25m}$) plotted versus $\iota/2\pi$ near the core (averaged in in 0.10 m $\le r_{eff} \le$ 0.20 m).

In Fig. 17, $\tau_{e\_reff \le 0.25m}$ obtained by the above three types of current drive is plotted versus $\iota/2\pi$ which is averaged near the core (0.1 m $\le r_{eff} \le$ 0.2 m). When $\iota/2\pi$ near the core is close to the 1/2 or 1/3 rational surface where the magnetic shear is weak, $\tau_{e\_reff \le 0.25m}$ decreases to the lowest level although the e-ITB is formed in such cases as shown in Fig. 10. MHD instabilities which occur near the rational surfaces prohibit the increase of the electron temperature. To realize high electron temperature with high density, such $\iota/2\pi$ profiles should be avoided.

When the Co ECCD is superimposed to the balanced NBI, the highest $\tau_{e\_reff \le 0.25m}$ is obtained. The improvement of $\tau_{e\_reff \le 0.25m}$ by the existence of the m/n=2/1 magnetic island near the edge of the e-ITB is clearly indicated compared to the case of the co ECCD superimposed to the counter NBI in which $\iota/2\pi$=1/2 is vanished in the plasma. However, $\tau_{e\_reff \le 0.25m}$ cannot be kept stable with the existence of $\iota/2\pi$=1/2 which causes MHD instabilities. In contrast, $\tau_{e\_reff \le 0.25m}$ can be kept stable without MHD instabilities by vanishing $\iota/2\pi$ = 1/2 and increases slightly after the magnetic shear near the edge of the e-ITB decreases from the positive to negative. To validate the effect of the magnetic shear on the e-ITB performance, such an $\iota/2\pi$ profile is favorable.

5 Summary

With the development of high power and long pulse gyrotron, application of the ECRH/ECCD has been expanded toward higher temperature and density. Improvement of the ECRH efficiency and reliable monitoring of the stray radiation has become more important from both economic and safety viewpoints. In LHD, real-time controlling of the incident wave polarization has been successfully demonstrated and a quick response microwave bolometer developed as a substitute for the sniffer probe has been able to respond to the change of the stray radiation intensity within 20 ms. As an application aiming for high density operation, ECRH up to 85 % of the cutoff density has been demonstrated by X-mode excitation in HFS with use of an antenna located in LFS. As an application aiming for high temperature operation, the effect of the control of the rotational transform on the sustainment of the e-ITB has been investigated with ECCD for seconds. It has been clearly derived that the existence of the m/n=2/1 island improves the confinement inside e-ITB compared to the case when the m/n=2/1 island is vanished in the plasma.

This work has been supported by the NIFS LHD project under ULRR701, ULRR804, ULRR025, and JSPS KAKENHI Grant Number 18K03590.

References

1. Y. Takeiri et al., *Nucl. Fusion* **57**, 102023 (2017)
2. H. Takahashi et al., *Nucl. Fusion* **58**, 106028 (2018)
3. F. Warmer et al., *Nucl. Fusion* **58**, 106025 (2018)
4. H. Takahashi et al, *Nucl. Fusion* **57**, 086029 (2017)
5. O. Mitarai et al., *Plasma Fusion Res.* **2** 021 (2007)
6. O. Motojima et al., *Physics of Plasmas* **6** 1843 (1999)
7. T. Ii Tsujimura et al., *Fusion Eng. Des.* **131** 130 (2018)
8. H. Igami et al., *Rev. Sci.c Instrum.* **77** 10E931 (2006)
9. M. Hirsh et. al., *AIP Conference Proceedings* **1612**, 39 (2014)
10. S. Marsen et al., *Nucl. Fusion* **57**, 086014 (2017)
11. T. Ii Tsujimura, et al., *Nucl. Fusion* **55**, 123019 (2015)
12. H. Igami, et al., *Nucl. Fusion* **49**, 115005 (2009)
13. H. Igami, et al, *Plasma Fusion Res.* **7** 2402110 (2012)
14. T. Shimozuma et al., *Nucl. Fusion* **45** 1396 (2005)
15. H. Igami, et al., *EPJ Web of Conferences* **32**, 02006 (2012)
16. K. Ida, et al., *Rev. Sci. Instrum.* **76** (2005) 05350
17. T. J. Dobbins et al., *Rev. Sci. Instrum.* **88**, 093518 (2017)
18. S. Murakami et al., Trans. Fusion Technol. **27**, 256 (1995)
19. K. Ida et al., *Phys. Rev. Lett.* **100** (2008) 045003
20. K. Ida et al., *Plasma Phys, Control. Fusion* **57** (2015) 014036
21. K. Ida et al., *Nature Communications* **6**, 5816 (2015)
22. K. Ida et al., *New Journal of Physics* **15** 013061 (2013)
23. Y. Kishimoto et. al, *Plasma Phys, Control. Fusion* **41** A663, (1999)
24. K. C. Shaing et al., *Nucl. Fusion* **55** (2015) 125001

Overview of W7-X ECRH Results

H.P. Laqua[1], J. Baldzuhn[1], H. Braune[1], S. Bozhenkov[1], K.J. Brunner[1], Ye.O. Kazakov[2], S.Marsen[1], D. Moseev[1], T.Stange[1], R.C. Wolf[1], M. Zanini[1] and Wendelstein7-X Team

[1] *Max-Planck-Institute for Plasma Physics, Greifswald, Germany*
[2] *Laboratory for Plasma Physics, LPP-ERM/KMS, Brussels, Belgium*

Abstract In its second operation phase (OP1.2a) W7-X was equipped with full 3d island divertor and an upgraded ECRH-system of 10 gyrotrons with a total port through power of 7 MW. The combination of pellet fueling and high density ECRH enabled to explore density above 1 10^{20} m^{-3}. In particular with the O2-polarisation combined with a multi-pass reflector tile system a density of up to 1.4 10^{20} m^{-3} was achieved. At slightly lower densities high core beta values and record values of the fusion tripple product of 0.66 10^{20} m^{-3} keVs for stellarators were reached. In addition routine plasma start-up and ECRH wall conditioning were performed. The island divertor enables to demonstrate the intrinsic steady state capabilities of W7-X, where stationary discharges of up to 30s were demonstrated being only limited by the heat capacity of the uncooled divertor. With the flexible ECRH launch system current density profile variations were used for MHD stability investigations. Here by fine-tuning of the ECCD profile different MHD activity could be triggered.

Introduction

The main objective of the optimized stellarator Wendelstein 7-X (W7-X) is the demonstration of steady-state plasma operation at fusion relevant plasma parameters thereby verifying that the stellarator is a viable fusion power plant concept. The W7-X design is the result on an elaborate optimization procedure which different optimization criteria like neoclassical transport losses, high beta stability, low internal currents and a sufficiently good fast particle confinement at high beta values [1]. W7-X is a low shear stellarator, a concept that avoids the iota profile crossing major rational values. They are positioned at the plasma edge, where the 4/5, 5/5 or 6/5 island structures are selectively used for the island divertor operation [2].
W7-X was heated by ECRH exclusively in the operation phase OP1.2a. The ECRH system consists of 10 gyrotrons, with an output power of 0.6-1.0 MW each at a frequency of 140 GHz, a quasi optical transmission line and flexible microwave launchers inside the plasma vessel [3]. The over-all transmission efficiency was estimated to be ~ 94%. The ECRH system was commissioned for 1800s operation at full power. It already demonstrated all requirements, which are necessary for a high-performance steady state operation at reactor relevant parameters at W7-X. The 140 GHz ECRH used the second harmonic resonance at 2.5T with an absorbed power of up to 7 MW. There are in total 12 front steering launchers Inside the plasma vessel, which allow a flexible ECRH launch in both toroidal and poloidal direction for EC current drive (ECCD) and off axis power deposition. In addition W7-X is equipped with two remote steering launchers (see Fig.1), a reactor relevant concept without movable elements in the plasma chamber and an extremely high through-put power density of 100MW/m^2. The high reliability is guaranteed by a FPGA based fast control system, which controls the timing and ratio of the body and cathode voltage at the gyrotrons. In particular it is also able to recognize and recover a gyrotron

Corresponding author: H.P.Laqua *laqua @ ipp.mpg.de*

mode loss, which allows a reliably gyrotron operation at maximum power.

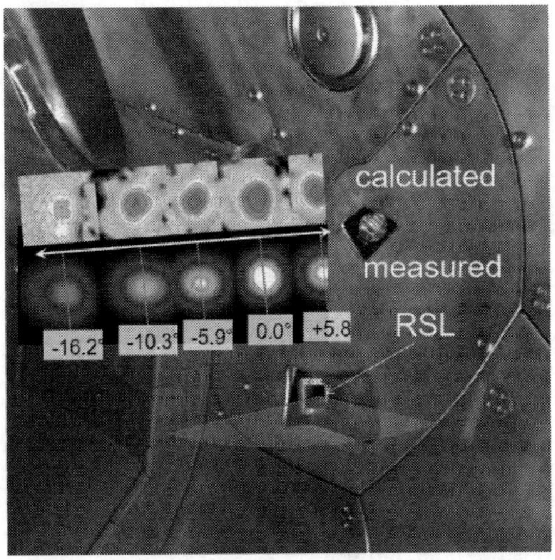

Fig1: Remote steering antenna mouth in the plasma vessel. Comparison of measured and calculated beam pattern for different launch angles.

The W7 -x fue lling system consists of a feedback-controlled gas fueling with valves in each of the 5 W7 -X modules enabling to fuel Hydrogen, Helium and other gases if required. Furthermore W7-X is equipped with a pellet injector, which injected up to 60 hydrogen ice pellets trough both a high -field-side and low - field-side injection position.

Plasma start-up and EC-wall conditioning.

The plasma start -up was exclusively performed by ECRH. The vacuum chamber was prefilled by a gas puff about 100 ms before the ECRH start. The break-down in the hydrogen or helium gas was achieved within less than 10 ms with typically 2 gyrotrons beams with 1.7 M W power in total and focused at the plasma center. Here the ECRH-stray radiation was decaying exponentially indicating a well absorbing plasma on the axis, where the ECRH -beams are crossing the second harmonic EC-resonance at 2.5T. This plasma was expanding until it reaches the separatrix. W ith an additional gas puff the density was then increased to typically 1.5 $10^{19}m^{-2}$ within 100 ms and density feedback system takes over for the regular experimental program. The plasma start -up was successfully achieved from day one with nearly 100% relia bility for the X2 -heated plasma. In the case of O2 -heated plasmas the central magnetic field had to be slightly increased up to 2.65 T, which necessitated more power for the off-axis start-up. Here the plasma start -up have been achieved within less than 50ms with 3 gyrotrons and 2.5 MW ECRH power . The reliable and efficient operation was guaranteed by a set of ECRH protective diagnostics, which were specially established for the ECRH requirements. In particular the ECRH stray radiation measurement turned out to be a reliable interlock signal to prohibit ECRH operation with insufficient plasma absorption. Furthermore the position and launch direction of each beam was measured by infrared cameras observing its thermal footprint at the heat shield tiles opposite of the antennas (see Fig.2).

The W7-X plasma chamber was baked-out at 160°c for two weeks before the experimental operation. Afterwards glow discharge cleaning with hydrogen and helium was conducted. During the experimental campaign daily helium glow disc harge cleaning was only possible in the morning before the magnetic field was r amped-up. For wall conditioning during the experimental day with a permanent magnetic field of 2.5T helium cleaning discharges with ECRH were conducted to remove the otherwise accumulating hydrogen from the plasm a-wetted surfaces. Several cleaning scenarios were compared. The "pulse train" method turned out as the most effective. Here a sequence of ten 1.5s long ECRH -plasmas in Helium for every 30s was used the recovery of density control.

ECCD and long pulse operation.

W7-X operation does not necessitate toroidal plasma current for operation. Therefore it is not equipped with any tool for inductive plasma current generation. The toroidal plasma current origins only from the bootstrap effect or is non-inductively driven by the ECRH (ECCD). The temporal development of the plasma current is defined by the current diffusion in the plasma (skin time), which is of the order of 1 s, and the L/R time of the whole plasma ring, which is of the order of 30 s. Therefore bootstrap current and ECCD studies requires long pulse operation of at least 10 s, but discharge length of up to 30s were achieved. The maximum pulse length was only limited by the heat capacity of the uncooled test divertor tiles. The minimisation of the bootstrap current was one of the main W7-X optimisation criteria. The comparison of the bootstrap evolution in different magnetic configuration verified experimentally the optimisation predictions [4]. Here any ECCD current has to be suppressed, therefore the ECRH launch was precisely to the perpendicular direction at the plasma axis, where the power is absorbed. For the ECCD studies the ECRH beam were steered in oblique direction with an angle of up to 12 ° with respect to the perpendicular direction. The polarisation was adjusted adequately to insure pure X-mode injection. The ECCD current was driven in co-and counter direction. A maximum current of 14 kA was achieved after 17s . This current did not reach its stationary valued yet but was limited by a strong MHD activity with core plasma confinement collapse. We suppose the very localized current density profile deformed the rational transform (iota) profile such that main rational resonance iota=1 was crossed and the confinement was destabilized. Detailed theoretical investigations are ongoing. The MHD activity expressed as saw-tooth like collapse. The collapse frequency and inversion radius was controlled by localized ECCD. In particular in the so-called high iota configuration, where the iota profile is above but very close to the iota=1 value at the core, even small ECCD current generated strong MHD-activity.

High density O2-ECRH operation

W7-X expressed a rather low density of 0.4 10^{19} m^{-3} limit for hydrogen, which is mainly determined by the power balance of the heating power and the radiation at the plasma edge. Even though it is expected that with improved wall conditioning and boronisation less radiation and thus higher densities will be achievable, a separation of core and edge fuelling is beneficial, since both core confinement and divertor performance can be optimized separately.

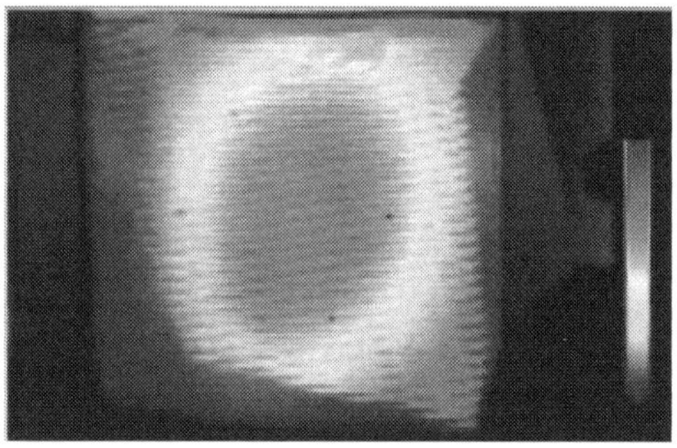

Fig. 2: Infrared footprint of a ECRH beam at the O2-reflector tile with holographic grating at its front.

Therefore W7-X was equipped with an pulsed pellet injector in order to investigate pellet fuelling of ECRH plasmas. With pellet injection the gas fuelling limit could be easily overcome and record densities of up to 1.4 10^{20} m^{-3} could be achieved. This density is of course above the cut-off density (1.2 10^{20} m^{-3}) of the X2 mode for the 140 GHz ECRH. Therefore the second harmonic O-mode (O2) with a cut-off density of 2.4 10^{20} m^{-3} was be used. Here the single pass absorption was incomplete and no plasma start-up was possible with O2. Therefore a special O2 scenario was developed. First the plasma vessel was equipped with tungsten covered molybdenum tiles with holographic gratings on its front side as shown in Fig 2. These reflector tiles enabled at least three controlled passes through the plasma core increasing the total absorption to above 90%.

In order to deposit the ECRH power in the hot core and to compensate the plasma beta and Shafranov shift the central magnetic field must be slightly increase up to 2.62 T, which made the plasma start-up even more difficult since now the ECRH power is deposited off-axis at low beta. Therefore as shown in Fig.3 the plasma start-up and the first density ramp was performed by 4 ECRH beam in X2-polarisation. During the density ramp-up phase of 2s the polarisation of these beam had been changed to O2-mode and the density was increased by pellet injection to above the X2 cut-off. At the same time the other O2-beam have been switched-on and sustained the high density plasma during the pellet injection over 2 s. In addition to the density measured by Thomson scattering and the laser interferometer, the achievement of the high density was also clearly approved by the vanishing of different ECE-signals due to their cut-off. In Fig. 3 the X2 ECE of the 141 GHz channel is strongly reduced in the high density phase, while the electron temperature measured by Thomson scattering remains high. It has to be mentioned that this operation is already close to the envisaged future steady state high-performance plasma scenario.

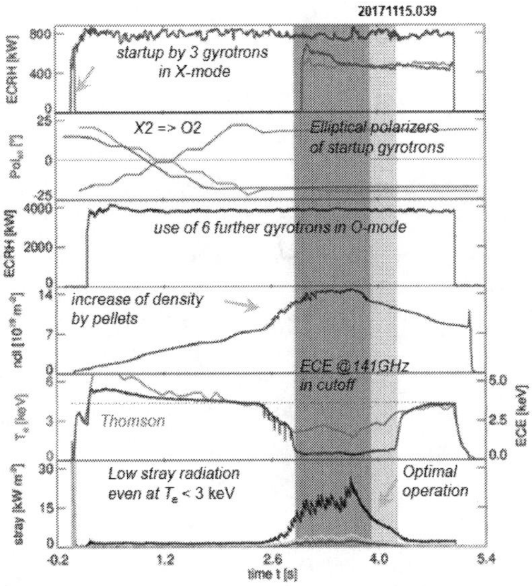

Fig.3: High density plasma scenario with O2-ECRH and pellet injection. The second row shows the chance of polarizer position for the 3 "start-up" gyrotron beams, when their polarisation was switched from X2 to O2.

High performance discharge.

At densities close below the X2 cut-off pellet injection was combined also with X2-ECRH (see Fig. 4). Here the discharge scenario was optimized in respect to the highest ion temperature, beta value and tripple product. Therefore just after glow discharge cleaning an X2 heated hydrogen gas fuelled plasma was established at a density of 0.3 10^{19} m^{-3}. Subsequently, a set of pellets increased the density up to 1 10^{20} m^{-3} and the heating power was increased up to 5 MW. After the pellet injection in the so-called reheat mode the highest plasma performance was achieved. The electron and ion temperature equilibrated already during the pellet phase and the ion temperature increases up to 3.8 keV with ECRH only. The confinement time increased up to 220 ms in maximum. The diamagnetic plasma energy reached a record value of 1.1 MJ with a resulting triple product was 0.66 10^{20} m^{-3} keV.

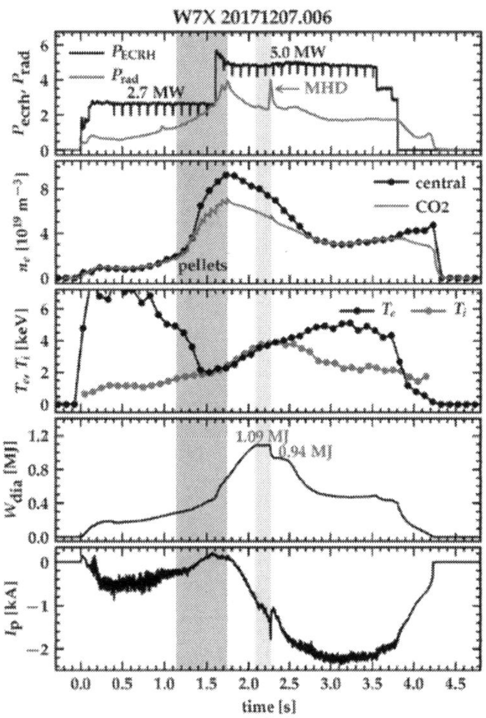

Fig 4: High performance discharge with record beta and triple product. At the top there is the ECRH power and the radiated power. In the next boxes the central Thomson density and the line-averaged density followed by the electron temperature form the Thomson scattering and the ion temperature measured by the x-ray spectroscopy are shown. At the fourth row there is the diamagnetic energy and the toroidal plasma current at the bottom.

Collective Thomson Scattering

The 140 GHz ECRH beams were also used for the collective Thomson scattering (CTS) [5]. Two scattering geometries were available. One in the beam shape section, which uses the movable ECRH front steering launcher, and one in the triangular shape section, which uses the combination of the remote steering launcher for the high power beam and the a quasi-optical remote steering antenna, which was designed by the Stuttgart university, for the receiving beam. The signal transmission towards the detector was performed with an unused ECRH Transmission line in reversed direction. This so-called F1-line is a spare line for further ECRH upgrades. The receiver system, which was built by the Danish technical university (DTU), was placed into the unused F1 gyrotron box. An un-cooled version of the F1 transmission line was built-up and used to measure the scattering signal. It is remarkable that the μW level CTS signal was partly transmitted with the same multi-beam mirrors, that are simultaneously used for the ECRH beams in the MW power level. Nevertheless, no perturbation by the heating beams was detected. The system was successfully commissioned and first ion temperatures were measured. A CTS spectrum is shown in Fig.5.

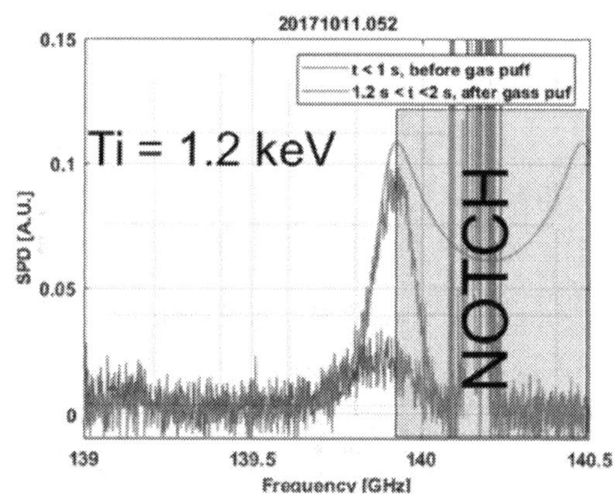

Fig5: CTS spectrum at low density (bleu) and medium density (red) in a Helium plasma.

Summary and conclusion

The W7-X ECRH system has successfully demonstrated its versatility in the operation phase OP1.2a. It provided the plasma start-up and the wall conditioning routinely. It sustained plasma discharges of up to 30s, which also enabled to demonstrate its ECCD capability. Most remarkable are the results of the combination of ECRH and pellet injection. Here record plasma densities were reached with the O2 multi-pass scheme. With X2-ECRH slightly below the X2-cutoff density record values of beta, ion temperature and triple product for ECRH plasmas were achieved. Finally the ECRH-beams were also successfully used for the CTS-diagnostic.

References:

[1] Grieger G. et al *Physics optimization of stellarators* 1992 Phys. Fluids B4 2081
[2] A. Peacock, et al., Fusion Eng. Des. 84, 1475 (2009).
Progress in the design and development of a test divertor (TDU) for the start of W7-X operation
[3] V. Erckmann et al.(2007) *Electron Cyclotron Heating for W7-X: Physics and Technology*, Fusion Science and Technology, 52:2, 291-312, DOI: 10.13182/FST07-A1508
[4].A. Dinklage et al. *Magnetic configuration effects on the Wendelstein 7-X stellarator*, Nature Physics
https://doi.org/10.1038/s41567-018-0141-9
[5] D. Moseev et al. *Collective Thomson Scattering Diagnostic at Wendelstein 7-X* submitted to review of scientific instruments

Study of off-axis current drive using EC top-launch in KSTAR

Young-soon Bae[1,*]

[1]National Fusion Research Institute, 169-148 Gwahak-ro, Yuseong-gu, Daejeon 34133, Korea

Abstract. The off-axis current drive using electron cyclotron (EC) wave is typically very low due to the low density and temperature and electron trapping effect when it is launched from outside midplane. However, the heating and current drive by EC wave is being regarded as a essential element as an off-axis current drive source together with lower hybrid current drive (LHCD) for advanced tokamak operation research in KSTAR in future. Therefore, the reliable and high efficient ECCD using top launch has been studied for two different launch schemes of down-shift and up-shift resonance with various EC frequencies which will be available in the future KSTAR ECRH system. The ray tracing (GENRAY) simulation studies show that a broad ECCD profile peaked off axis is obtained with a high current drive efficiency for both schemes. It is almost twice as high as that of the outside midplane launch for the fundamental O-mode EC wave.

1 INTRODUCTION

KSTAR aims to develop advanced tokamak operation mode with enhanced broad off axis current drive using off-axis NBI and RF heating devices [1]. In the KSTAR ECRH system upgrade plan, we will install four 105/140 GHz gyrotrons and two 170 GHz gyrotrons with 1 MW output power each. In the pre-liminary study [2], vertical top launch of electron cyclotron (EC) wave shows improvement in EC current drive (ECCD) efficiency with broad ECCD profile peaked off-axis. It is understood that the vertical top launch of EC wave enables nearly vertical trajectories on the low field side from the cold resonance, called "up-shift" resonance. EC waves approaches the resonance slowly with high Doppler shift and waves interact with higher energy electrons during long absorption path, which are less collisional and provide larger ECCD. Also, the long absorption path allows the broad current profile. The other approach of the vertical top launch is the "down-shift" resonance when the EC wave is launched from the higher field side to the cold resonance. It is seen that there is advantage and disadvantage between the down-shift and up-shift vertical launch in terms of launch angle tolerance and current drive efficiency. It will be discussed in the following sections.

Eq. (1) is the Doppler-shifted relativistic EC resonance condition at l'th harmonic.

$$\omega = \frac{l\Omega}{\gamma} + k_{||}v_{||} \quad (1)$$

$$\frac{v_\perp^2}{c^2} = \left(1 - \frac{\omega^2}{l^2\Omega^2}\right) + 2n_{||}\frac{\omega^2}{l^2\Omega^2}\frac{v_{||}}{c} - \left(1 + n_{||}^2\frac{\omega^2}{l^2\Omega^2}\right)\frac{v_{||}^2}{c^2} \quad (2)$$

Where, γ is the relativistic factor, ω is EC wave frequency, Ω is the cyclotron resonance frequency, $k_{||}$ ($k_\perp$) is the parallel (perpendicular) component of the wave number of EC wave to the magnetic field, $n = kc/\omega = c/v$ is the refractive index, and v is the velocity of electron. Fig. 1 shows ellipses in velocity space using the Doppler-shifted relativistic resonance condition at the fundamental harmonic of 105 GHz with the toroidal magnetic field of 3.0 T at the major radius, R_0 = 1.8 m. Note that the corresponding the cold resonant position is 1.44 m.

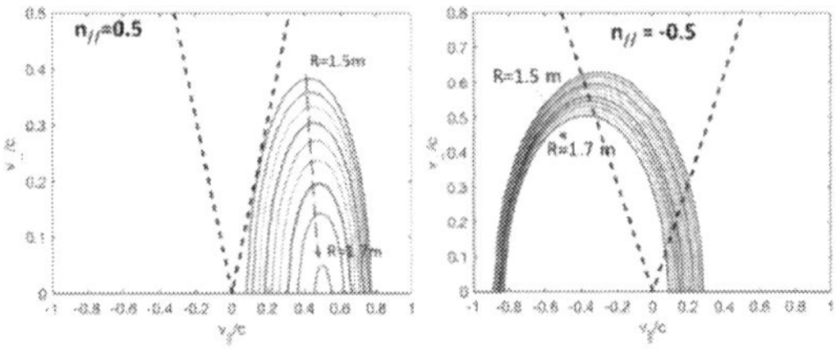

Fig. 1. Two different Doppler-shifted relativistic EC resonance curves using Eq. (2): (left) up-shift resonance and (right) down-shift resonance. The dot line is the trapping boundary. The arrow indicates different radial position of the vertical trajectory from R = 1.5 m to 1.7 m.

The up-shift resonance condition is obtained with the parallel wave vector $k_{||} > 0$ for the co-current drive in the negative direction (clock-wise, $v_{||} < 0$), and the down-shift resonance condition is obtained with the $k_{||} < 0$ and keeping the same co-current drive ($v_{||} > 0$). In the down-shift case, there are more relativistic resonant electrons with higher energy moving in the counter-current direction than the co-current moving electrons.

Upgrade plan of various EC frequencies (105/140/170 GHz) in KSTAR enables the vertical top launch of second harmonic 170 GHz EC wave using up-shift resonance with low toroidal magnetic field and using down-shift resonance with high toroidal magnetic field. In this paper,

* Corresponding author: ysbae@nfri.re.kr

 https://doi.org/10.1051/epjconf/201920302003

we discuss the ray tracing results for the case of the fundamental harmonic up-shift, and for the fundamental and second harmonic down-shift vertical top launches.

2 Ray Tracing Study

The EC wave propagation and power absorption is calculated using the ray tracing code, GENRAY [3] and Fokker-Planck code, CQL3D [4]. In the ray tracing simulation, the toray module in the genray code is used with the finite beam divergence at the launch position and 48 Gaussian zone. The equilibrium is basically double-null shape with elongation, κ = 2 with the plasma current I_p = 1.0 MA. The toroidal magnetic field is just scaled for the different approach of up-shift and down-shift vertical top launch. The plasma density and temperature profile is prescribed with typical H-mode profile obtained in KSTAR experiment as shown in Fig. 2.

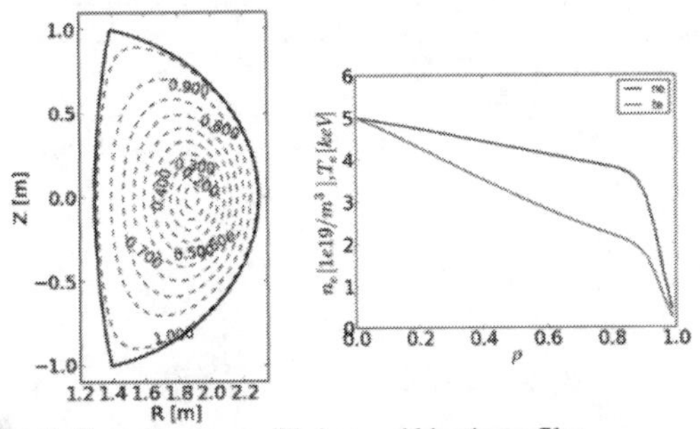

Fig. 2. The reference equilibrium and kinetic profiles.

2.1 Up-shift top launch of 105 GHz O1 wave

In previous calculation [2], the vertical top launch of 140 GHz X2 wave with B_0 = 2 T at the magnetic axis shows strong sensitivity of wave trajectory on both azimuthal and polar angles of injection. In this paper, the fundamental O-mode of 105 GHz wave is used for the new vertical top launch study with the toroidal magnetic field of 3 T at the magnetic axis (R_0 = 1.8 m). Fig. 3 shows the wave trajectory and driven current profile obtained from GENRAY simulation for a specific case of injection angles. In this case, the driven current has maximum, ~50 kA/MW at the given density of $n_e(0)$ = 7 x 10^{19} m-3 and temperature of $T_e(0)$ = 7 keV.

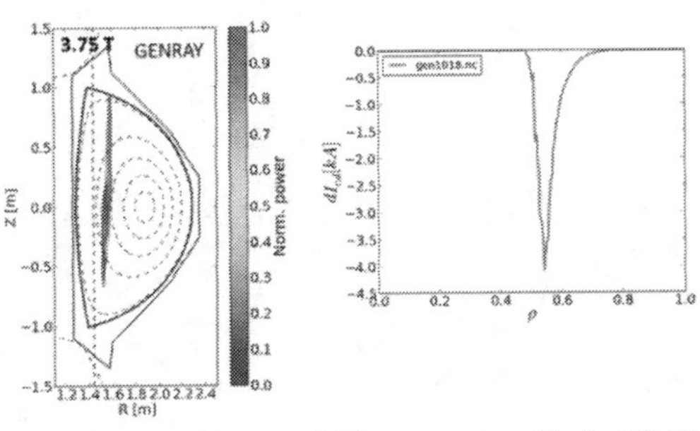

Fig. 3. The ray trajectory and driven current profile for 105 GHz O1 wave obtained from GENRAY code with toray module and 48 Gaussian zone.

The dimensionless current drive efficiency η = $I_{CD}R<n_e>/P_{abs}$ = 50 kA x 1.8 m x 5.65 x 10^{19} /1 MW = 0.51 x 10^{19} $Am^{-2}W^{-1}$ for this case. In fig. 2, the quantity plotted for current drive is dI_{cd} = $j_{tor}dA$ in each GENRAY's 200 radial zone. The divergence of the injected wave is defined with half width at half-power, α = 1.5 deg. Fig. 4 shows the power absorption and current drive as the polar angle is scanned for each azimuthal angles and for the launcher mirror pivot position of R = 1.6 m and Z = 1.2 m. The launcher mirror pivot position is chosen to have the current profile at the mid-radius. The polar angle (θ) is scanned by step of 0.5 deg for each azimuthal angle (ϕ) from 100 deg to 107 deg by step of 1 deg. The scanned range shown in fig. 4 is determined to have the power absorption ratio larger than 95%. For 100 deg toroidal injection angle, the power absorption and current drive decreases by 3% and 20%, respectively when the deviation of -5 deg in the polar injection angle from the angle where we have the maximum current drive and full power absorption.

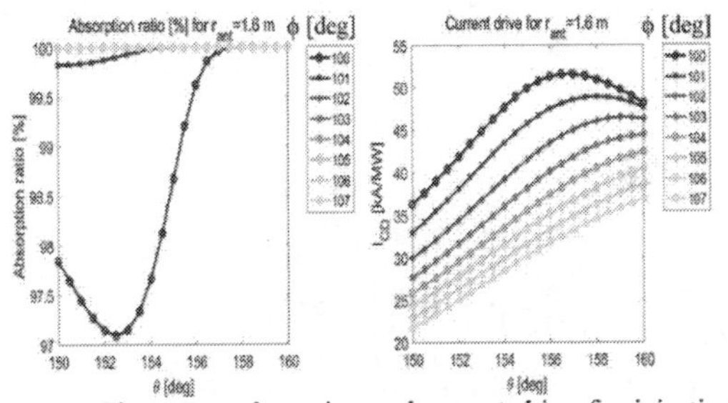

Fig. 4. The power absorption and current drive for injection angles.

When the toroidal injection angle increases, the power absorption larger than 99% is obtained but the current drive decreases because waves are quickly absorbed at the far off-axis as seen in Fig. 5.

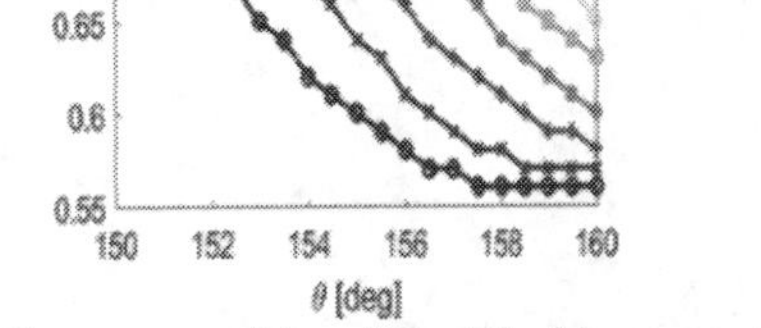

Fig. 5. The average radial position of the driven current profile.

Fig. 6 shows the driven current profile and ray trajectory of 105 GHz O1 wave as the plasma density is scanned from 3 x 10^{19} m^{-3} to 10 x 10^{19} m^{-3} with fixed central electron temperature of 7 keV. As expected, the wave is quickly damped at far off-axis for the higher density. The current drive is obtained with 40-105 kA/MW for $n_e(0)$ = 3-10 x 10^{19} m^{-3}. The corresponding dimensionless current drive efficiency, η = 0.45 – 0.58 x 10^{19} $Am^{-2}W^{-1}$. The dimensionless current drive efficiency averaged in the range of the density scan is almost same as the one in Fig. 3.

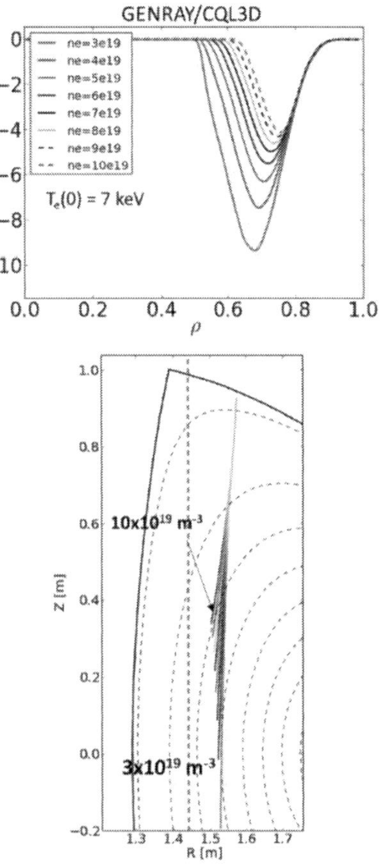

Fig. 6. The EC-driven current profiles and ray trajectories obtained GENRAY and CQL3D code as the plasma density increases. The vertical green dot line indicates the cold resonance line.

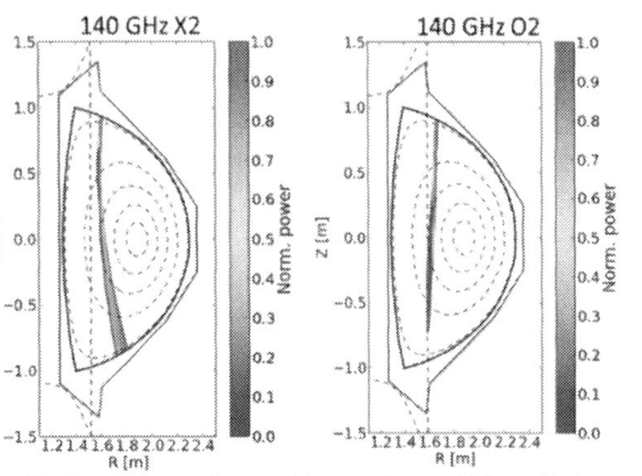

Fig. 7. Comparison of second harmonic up-shift vertical top launch of 140 GHz X2 and O2 waves for $n_e(0)$ = 7 x 10^{19} m^{-3} and $T_e(0)$ = 7 keV.

Fig. 7 shows the comparison study between two different polarizations of second harmonic up-shift vertical top launch of 140 GHz EC wave. Both cases have similar current drive efficiency of ~ 25 kA/MW. But X2 wave is refracted outward from the cold resonance and its power is not fully absorbed. Note that the launching pivot location is adjusted to get the maximum current drive for each X2 and O2 wave in this simulation.

2.2 Down-shift top launch

The ray tracing simulation of the vertical top launch at the high field side from the cold resonance location shows the ray trajectory is less sensitive to the wave injection angles. The ray always goes toward the resonance layer for the moderate range of injection angles, then its power is fully absorbed. The less sensitivity of the ray trajectory on the injection angles and full power absorption are considered as the big advantage in the down-shift top launch. Fig. 8 shows the high-field side top-launch for the fundamental harmonic X-mode of 105 GHz (X1) and second harmonic X-mode of 170 GHz (X2) with the same launch position. The toroidal magnetic field at the magnetic axis (R_0 = 1.8 m) is 3.75 T and 3.0 T for 105 GHz X1 and 170 GHz X2 wave.

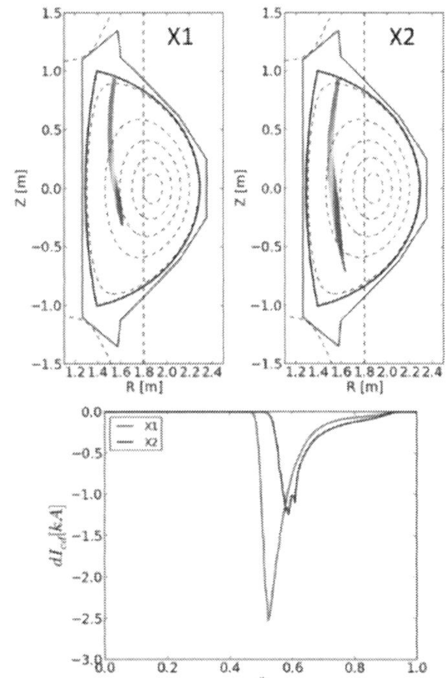

Fig. 8. The ray trajectories of 105 GHz fundamental harmonic X-wave and 170 GHz second harmonic X-wave, and the current drive profile of each case.

The X1 wave penetrates more deeply to the plasma center and damped with narrow width without X-mode cutoff. Note that the KSTAR tokamak is built with an enough superconductor engineering margin to be operated at the higher toroidal magnetic field than 3.5 T [5]. Each current drive efficiency is 44 kA/MW and 25 kA/MW. This value is comparable to the up-shift fundamental 105 GHz O-wave and 140 GHz second harmonic X- and O-waves.

Fig. 9 shows the RF-driven quasi-linear diffusion coefficient in the velocity space with the parallel velocity and perpendicular velocity normalized the thermal velocity for up-shift and down-shift vertical top launch cases investigated in the ray tracing simulation. As expected, the population of the resonant electrons in the velocity space is subject to the harmonic numbers and up-shift or down-shift resonances.

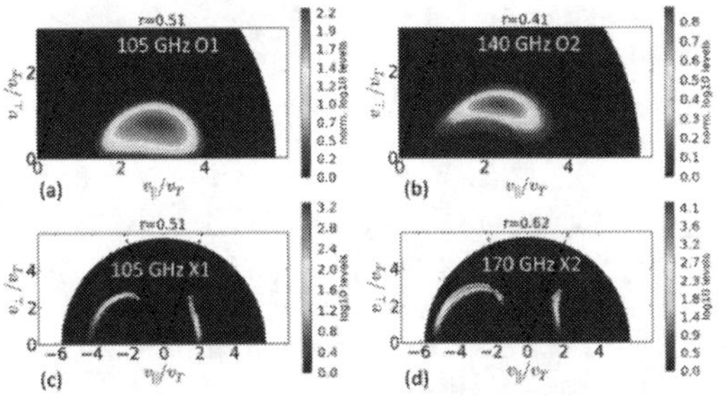

Fig. 9. The quasi-linear diffusion coefficient (D_{QL}) in velocity space obtained from CQL3D simulation for (a) the up-shift 105 GHz fundamental harmonic O-wave, (b) the up-shift 140 GHz second harmonic O-wave, (c) the down-shift 105 GHz fundamental harmonic X-wave, and (d) the down-shift 170 GHz second harmonic X-wave.

The current drive efficiency between the vertical top launch and outside midplane launch is compared using the wave of the same frequency and same harmonic resonance, i.e., the 105 GHz O1-wave. But, we used different toroidal magnetic field at the magnetic axis (R_0 = 1.8 m) in order to have the same off-axis peak location of the driven current profile: 3.0 T for the vertical top launch and 3.75 T for outside midplane launch.

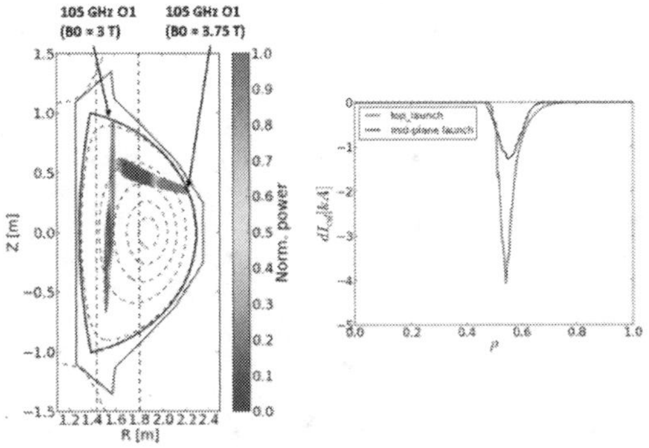

Fig. 10. The comparison of vertical top launch and outside midplane launch for the same harmonic EC frequency with the same density and temperature profiles with $n_e(0) = 7 \times 10^{19}$ m^{-3}, $T_e(0)$ = 7 keV.

Fig. 10 shows the comparison of ray tracing and current drive profile of two launches with same density and temperature profiles. The current drive efficiencies are ~ 50 kA/MW and ~ 24 kA/MW for top launch and outside launch, respectively. The current drive efficiency of the top launch is twice as high as that of the outside launch. Not like the X-mode wave, the O-mode wave power is slowly damped and the driven current profile is broad for the outside midplane launch as seen in fig. 10 because the O-mode absorption rate is much smaller than X-mode.

3. Design consideration of top launcher

At present, the divertor upgrade in the KSTAR tokamak is in progress in the design. The divertor upgrade, as well as new concept of the off-axis current drive, is required for the reactor relevant research activity in the next operation phase in the KSTAR.

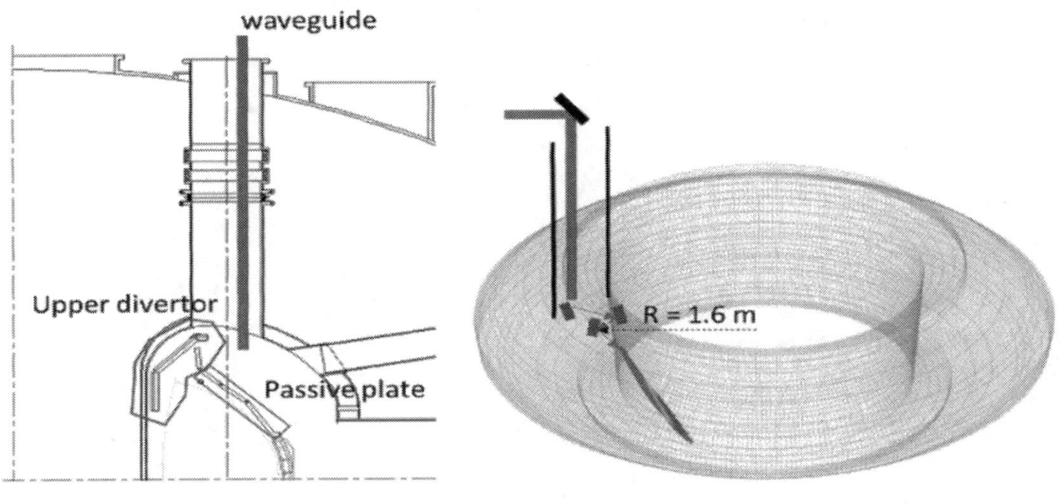

Fig. 11. The interface of the vertical top launche with upper divertor and passive plate and the conceptual drawing of the waveguide insertion and three mirrors.

For the new divertor, the single-null configuration is likely to be decided to easily accommodate the top ECRH launcher. The only concern of the vertical top launcher is how to protect the launcher from the 'vertical displacement event (VDE)' and plasma disruption. The other issue is the interface with the upper passive stabilizer. Fig. 11 shows the interface of the divertor and the upper passive plate with the conceptual configuration of the top launcher. The waveguide is inserted through the vertical port because the upper slant port is not avaliable any more. Therefore, at least three mirrors are needed to provide the scanning of both toroidal and poloidal injection angles. The all mirrors are protected by the upper passive plate and an ad-hoc protection plate with wave penetration hole.

4. Summary and discussion

This paper presents the further ray tracing simulation study on the vertical top launch in KSTAR following the previous ray tracing simulations focused on the second harmonic X-wave. This study is focused on the reliable and high efficient ECCD using two different launch schemes of up-shift and down-shift electron cyclotron resonance with various EC frequencies which will be available in the future KSTAR ECRH system. The up-shift electron cyclotron resonance is defined that EC wave is launched at the low field side from the cold electron cyclotron resonance location. The down-shift one is the high-field side launch from the cold resonance location. It shows a possibility of the fundamental O-mode wave for the up-shift scheme using 105 GHz in KSTAR with the toroidal magnetic field of 3 T at the magnetic axis. The dimensionless current drive efficiency for the fundamental harmonic down-shift scheme is 0.5 x 10^{19} $Am^{-2}W^{-1}$, twice as high as the outside midplane launch case. When we compare the second harmonic X and O mode for the up-shift case, the second harmonic O mode is more reliable in terms of the power absorption ratio because the ray path moves toward the resonance location. For the down-shift scheme, the 105 GHz fundamental O-mode will be also possible if the toroidal magnetic field can be raised to 3.75 T at the magnetic axis in KSTAR. It has an advantage of less sensitivity of ray trajectory on the wave injection angles and then ensuring the full power absorption with similar current drive efficiency as that of the up-shift scheme. The 140 GHz and 170 GHz second harmonic X-wave is also possible choice for the down-shift scheme with the toroidal magnetic field ranged from 2.5 T to 3.0 T which is favourable magnetic field range in the nominal KSTAR high β experiments. The CQL3D Fokker-Planck calculation for each top launch case shows that the current drive increases by about 20% with broader profile. It results from the non-Maxwellian effect and hence a bit of longer ray path. Considering the synergy effect for the outside midplane launch, the 105 GHz fundamental O-mode EC wave for the top launch will be favourable because the second harmonic resonance exists at the magnetic axis (B_0 = 3 T) for 170 GHz EC wave launched from the outside midplane. In conclusion, the top-launch approach of ECCD will be an efficient current drive source for the off-axis current drive experiments for the future KSTAR research aiming to the development of the advanced tokamak operation scenario. But for the DEMO device, the top ECCD launch should be carefully assessed through the further modeling studies in the aspect of the compatibility with DEMO operation scenario with very high temperature plasma.

References

1. Y.S. Bae et al., "NUBEAM simulation for off-axis NBI design in KSTAR," American Physical Society (Nov. 16-20, 2015, Savannah, Georgia).
2. Y.S. Bae et al., "Ray tracing study on top ECCD launch in KSTAR," EPJ Web of Conferences **157**, 03003 (2017)(DOI: 10.1051/epjconf/201715703003).
3. Smirnov A.P. and Harvey R.W. Report CompX-2000–01 Ver 2, (www.compxco.com/genray.html).
4. R.W. Harvey and M.G. McCoy, The CQL3D Code, Proc. IAEA TCM on Advances in Sim. and Modeling of Thermonuclear Plasmas, pp. 489-526, Montreal, (1992), available through USDOC/NTIS No. DE93002962; see also, http://www.compxco.com/cql3d.html CQL3D Manual.
5. K. Kim et al., "Status of the KSTAR superconducting magnet system development," Nucl. Fusion **45**, 783 (2005).

* Corresponding author: ysbae@nfri.re.kr

Optimization of HF- injection at the 2nd harmonic of ECRH on T-10 tokamak in order to obtain high energy content in plasma

A. Borschegovskiy[1,*], *S.Neudatchin*[1], *I. Pimenov*[1], *V. Trukhin*[1], *M. Dremin*[1], *A. Kislov*[1], *Yu. Pavlov*[1]

[1]*NRC "Kurchatov Institute", pl. AkademikaKurchatova 1, Moscow, 123182, Russia*

Abstract. Non-traditional scheme of HF-injection has been used in T-10 tokamak for ECRH at the 2nd harmonic of ECR in X-mode. Two HF-launcher systems, with focusing beams, injected power in opposite directions under toroidal angles +/-20°. Input power from each launcher was about the same (P_{hf} =0.8 ÷ 0.85MWt). Absorbed energy was deposed in central area of plasma column.
The new phenomenon has been found in some shots with W-limiter and Li-coating in the regime mentioned above. The spontaneous rise of the electron density nearly in the all plasma column occurs simultaneously with the rise of Te in the wide region (0<r/a<0.8) at the steady-state of ECRH. The value of the energy confinement time abruptly rises by ~15%. The absolute value of the electron heat and density fluxes reduces abruptly in the whole plasma column similarly to that of at non-local (or global) L-H transitions found earlier in JET and JT-60U. The accumulation of tungsten and light impurities is absent.
In a series of shots, without L-H transitions, measurements of X-ray spectra (PHA data) showed a significant difference in comparison with the ECCD co-injection by two gyrotrons.

1 Introduction

The main part of experiments on the ECR plasma heating are concentrated on local energy deposition of HF – power in a rather narrow zone (ECRH – perpendicular launching) and generation of non-inductive current (ECCD – oblique launching at toroidal angle ~ - 30°).
Non-inductive ECR current (ECCD) is ultimately based on an anisotropic velocity distribution function of electrons in the longitudinal direction of motion. In this case, microwave power is transmitted to the transverse component of the energy of electrons moving in the direction of wave propagation, which obviously leads to a distortion of the electron energy distribution function. Electrons, moving in the opposite direction, do not participate in the ECR heating and their energy distribution function remains Maxwellian due to Coulomb collisions.
Note that the mean value of the kinetic energy of particles in the distribution function, deformed towards higher energies, turns out to be greater than the same value in the Maxwellian distribution (see Fig.1).

$$\int_0^\infty f(v)\frac{mv^2}{2}dv > \int_0^\infty f_{max}(v)\frac{mv^2}{2}dv \quad (1)$$

Therefore in order to increase the electron energy in the ECR area, HF power was injected by two gyrotrons simultaneously in two opposite toroidal directions (co+contr injection). Thus, electrons moving in both directions had resonance with the wave (see Fig.2).
The new scheme of ECRH at the 2nd harmonic in X-mode with simultaneous co-ECCD (toroidal angle+20^0) and contr-ECCD (toroidal angle-20^0) and focused beams was used in the series of shots (called later as co+contr). The new phenomenon has been found in some shots with W-limiter and Li-coating in the above mentioned regime.

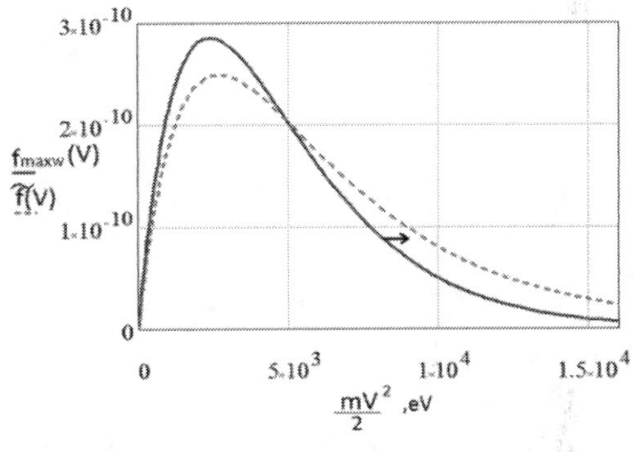

Fig. 1. Velocity distribution functions in terms of kinetic energy.

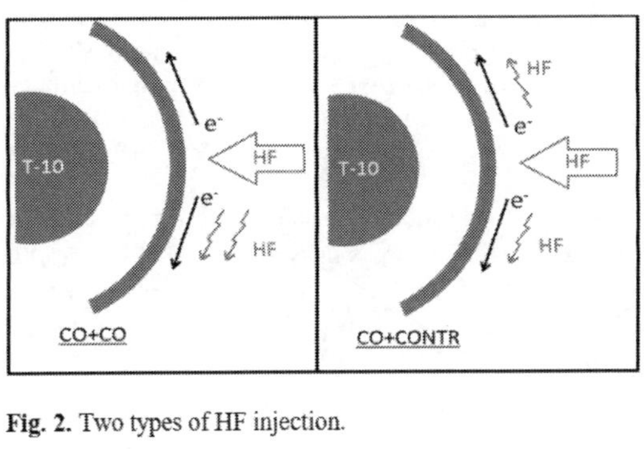

Fig. 2. Two types of HF injection.

*borschegovsky@yandex.ru

The spontaneous rise of the electron density nearly in the all plasma column occurs simultaneously with the rise of Te in the wide region (0<r/a<0.8) at the steady-state phase of ECRH. The value of the energy confinement time abruptly rises by ~15% remaining global L-H transition. Initially, non-local (global) confinement bifurcations were observed at 'fast' L–H–L transitions in some normal magnetic shear regimes of JET [1-4] and JT-60U [5]. The abrupt in time (ms timescale) and wide in space profiles of the electron and ion heat diffusivities variation $\delta\chi e,i$ (up to 90% of plasma volume or at 0.3<r/a<1) were found. The profile of $\delta\chi e$ hardly penetrates into the RS region for the strong ITB. The $\delta\chi e$ profile follows the position of q_{min} (minimum of q) and penetrates the RS region deeper for the weak ITB than for the strong one, suggesting 'global' edge–core connection penetrating through the position of q_{min} (up to shear value equal to ≈ -1.5) on a ms timescale.

2 T-10 tokamak and Gyrotron set-up

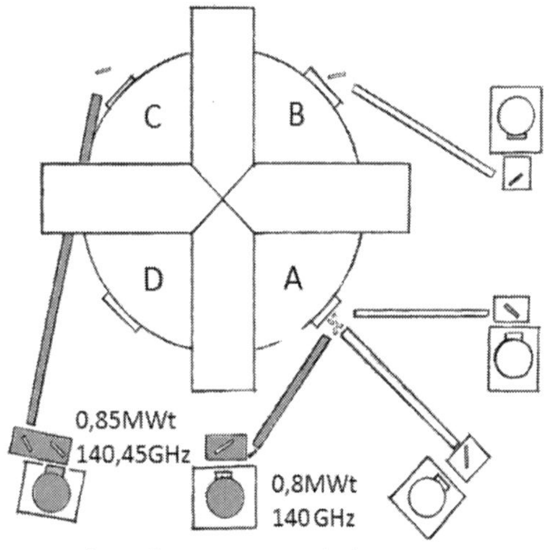

Fig. 3. Major radius R=1,5m, W-limiter a=0,3m. "A" and "C" ports are supplied with scanning launchers with focusing beam. Beam waist ρ=0,8cm.

Microwave radiation was injected at the stationary stage of the plasma discharge (650ms-850ms) by two gyrotrons simultaneously at an angle of 20 degrees in opposite directions on the second harmonic of ECR in X- mode. The zone of ECRH is located between the radius of the relativistic cut-off and the position of the cold resonance, which is about 7cm for T-10 gyrotron set-up.

With the toroidal magnetic field Bt(0)=2.42÷2.44T, the energy deposition has located in the central area of plasma column. Plasma current Ip=220kA, <n> = $2{\cdot}10^{13}\,cm^{-3}$.

3 Non-local L-H transition at co+contr ECCD

The timetraces of Te at the various radial positions are shown on the Fig. 4. Two gyrotrons with P_{ECRH} =1.4 MW are switched-on at t=700 ms. The arrow shows the start of Te rise. The Fig.5 shows the behavior of the central line averaged density measured by interferometer. The electron density starts to rise in all plasma volume. Fig.6 represents evolution of the ne profiles before and 20 ms after transition (only 6 channels of the laser interferometer were in operation). In our opinion, the spontaneous rise of Te and n_e represents L-H transition. The rise of (6+Li/2) appears at the same time and $\delta(\partial(6+Li/2)/\partial t)$ = 0.25 MW (the value of the energy content W was not measured in this shot). The value of the energy confinement time τE abruptly rises by ~15%. We restore the value of W (shot 68348) with the value W measured in similar shots without clear transition (e.g. 68355) by comparison of Te and ne profiles. Even after L-H transition the value of τE is equal to the predictions of ITER L-mode scaling (21 ms). The rise of Te is not symmetric at LFS and HFS due to the increase of Shafranov shift.

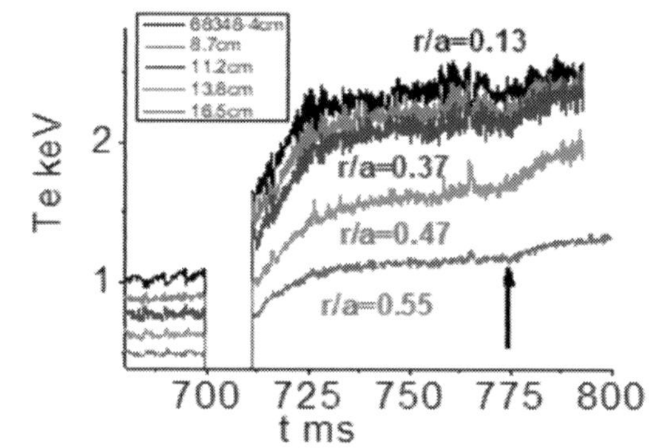

Fig. 4. Timetraces of Te in shot 68348, the start of Te rise is shown by arrow.

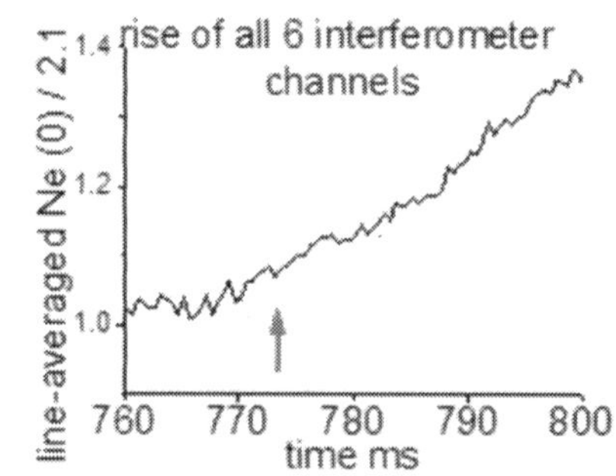

Fig.5. Timetraces of Ne (0) / 2.1 line-averaged in shot 68348, the start of Ne rise is shown by arrow.

The Figure 7(a) shows the profiles of Te before and 20 ms after L-H transition. Besides the obvious asymmetry, internal transport barrier (ITB) forms at r = -15cm and r=+18 cm. The Figure 7(b) represents the radial profile of $\delta(\partial Te/\partial t)$ at the time of the transition multiplied by 10 ms. Asymmetry at LFS and HFS is observed also. Since the gradient of Te nearly not varies at the time of L-H transition, the simple expression (2) allows us to calculate the jump of the electron heat flux δΓe from the simple equation (S is the enclosed surface, <> means averaging in a few ms time interval after transition) [1-6].

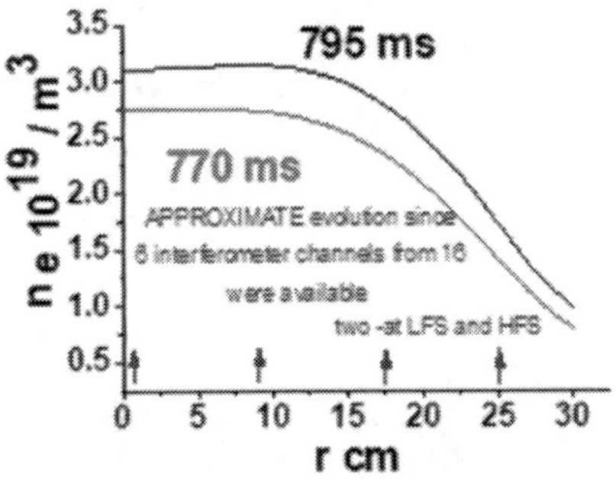

Fig. 6. Density profiles before and 20 ms after L-H transition.

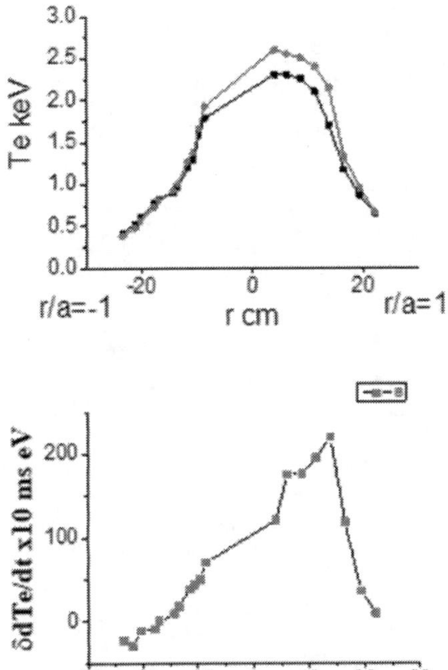

Fig.7. (a) Profiles of Te before and 20 ms after L-H transition (b) The radial profile of $\delta(\partial Te/\partial t)$ x 10ms at the L-H transition.

$$<\delta\Gamma e(r)> S(r) = -ne <\delta\chi e> \nabla(Te)S(r) = -3/2\int_0^{V(r)} \delta <(\partial(neTe)/\partial t)> dV \qquad (2)$$

The profile of $\delta\Gamma_e(r)$ is shown on the Figure 8. The rise of ne is important in this particular case since $\delta(\partial Te/\partial t)/Te < \delta(\partial ne/\partial t)/ne$ in the outer half of plasma column.

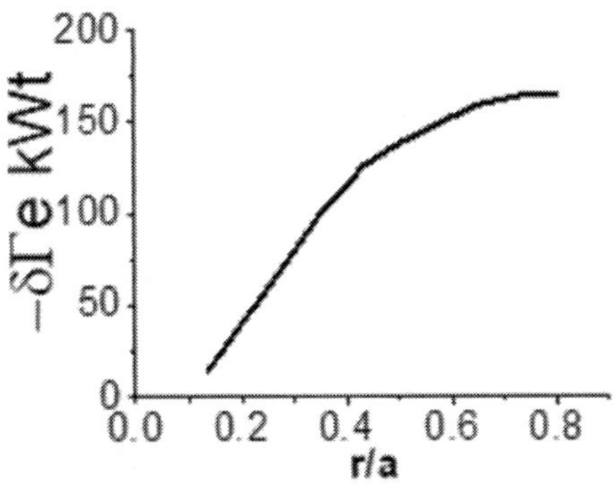

Fig 8 Radial profile of $\delta\Gamma_e(r)$ at the L-H transition

The profile of $\delta\Gamma_e(r)$ shown on the Figure 8 is very wide in space reminding that of at non local L-H transitions in JET and JT-60U [1-6], see detail in the Introduction above. The ion energy rises also since Ti (not well measured in shot 64348) at least did not drop after transition and n_e raised. The 15% rise of τ_E at the time of L-H transition shown above does not contradicts with the joint reduction of electron and ion heat fluxes. The electron density flux reduces in all plasma volume since the density rises in all plasma column. The level of the plasma radiation measured at the various radial positions with AUX bolometers (sensitive to the radiation of the tungsten [7]) drops also. The density of the tungsten ions reduces. The accumulation of the light impurities is absent since SXR data shows proper rise (linear to small values of δTe and square to the rise of ne).

4 Co+co and co+contr injection

Shots in which L-H transition was not observed were selected for a detailed comparison of the plasma parameters for co+co and co+contr HF injection. Experiments, in particular, showed that the increment of the average electron concentration $n_e(0)$ is higher in the case of co+contr injection in comparison with co+co, with the same "pump-out" effect, arising after gyrotrons start. Density feedback control was operated identically. The average value of electron density continued to grow during of co+contr injection and reaches a value greater than the initial value (see Fig.9). Also, the change of the shape and period of sawtooth oscillations was observed.

For the purity of the experiment, a series of pulses with an additional gas-puffing was carried out in the case of co + co injection in order to obtain the evolution of a density identical to co+contr case (see Fig.10).

As a result, with the same time evolution of electron density in the central region of the plasma, which varied from $n_e(0)$ =3.0•10^{13}/cm^3 to $n_e(0)$= 3.8•10^{13}/cm^3 during the time interval with the linear rise of the density (see Fig. 10), the central electron temperature measured by ECE was Te(r/a=0.15) =2.4 keV at the quasi-stationary stage for both HF injection methods (see Fig.11).Thus, the observed value of $n_e \times T_e$ in the central region of the plasma column was the same.

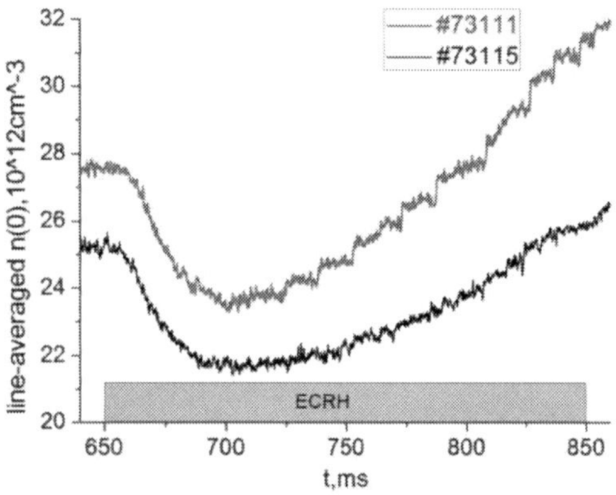

Fig. 9. Line-averaged n_e (0) time evolution, shot #73111-co+contr ECCD, shot #73115- co+co ECCD.

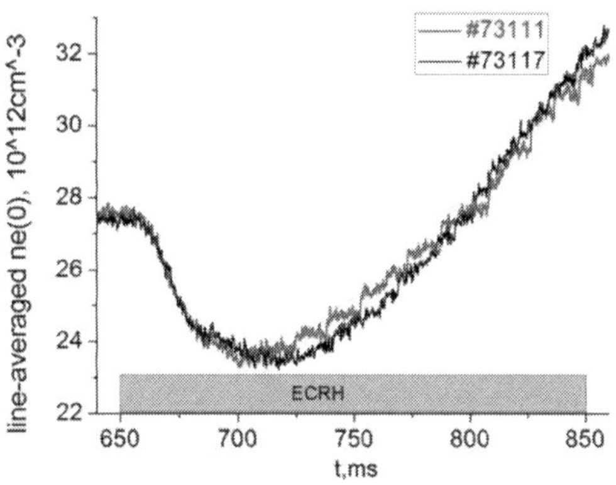

Fig.10. Line-averaged n_e (0) time evolution. (#73111-co+contr),(#73117- co+co with additional gas-puffing).

Spectrum analysis was carried out in the range of 8-14 keV because of the poor statistics at energies above 14-16keV.

However, the measurements of X-ray spectra (PHA data) showed a significant difference. (see Fig.12).

The approximated "temperatures", which were made after averaging of the PHA statistics, have values of 3.1keV for co+co injection and 3.56keV for co+contr injection (see Fig.13).

If we take into account that the measurements on the second harmonic ECE for collision plasma are very weakly dependent on the longitudinal energy of electrons, and X-ray spectrum is practically isotropic up to 100keV, we can assume that a significant part of the electron energy is contained in the longitudinal component. Moreover, in the case of co+contr injection longitudinal energy of the electrons is greater than for co+co injection. If we suppose, for example, that the mean value of kinetic energy of the longitudinal component is close to the transverse value, the energy content in the resonance zone increases by ≈30%.

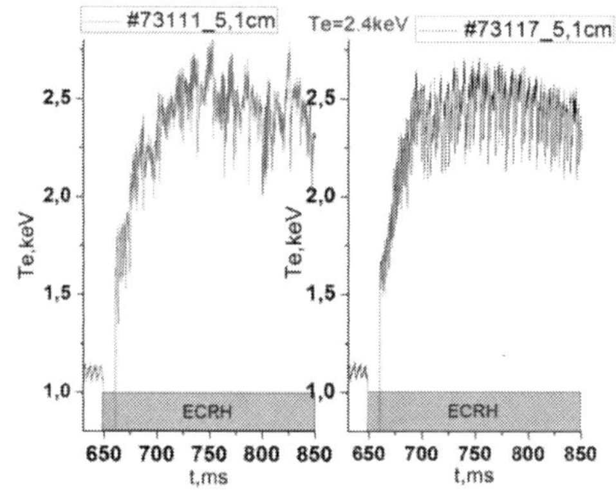

Fig.11. Time evolution of Te ECE (5cm) in comparable shots #73111 (co+contr) and #73117 (co+co).

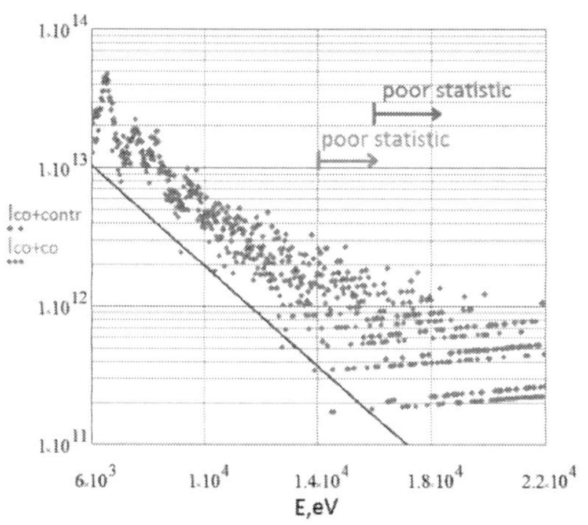

Fig. 12. •X-ray spectrum (#73111- co+contr, red circles), •X-ray spectrum (#73117- co+co, blue circles).

Conclusion

The new phenomenon has been found in some shots with W-limiter (and Li-coating) in the plasmas with two EC-beams with power injected in opposite directions under approximately equals toroidal angles and EC-power 1.5 MW. The spontaneous rise of the electron density in the all plasma column occurs simultaneously with the rise of Te in the wide region (0<r/a<0.8).We treat it as "non-local or global L-H transition" found earlier at JET [1-4] and in various regimes of JT-60U [5-6]. The value of D-beta falls, the positive value of dW/dt appears and the value of the energy confinement time τE abruptly rises by ~15%. The absolute value of the electron heat and density fluxes reduces abruptly in the whole plasma column. The accumulation of tungsten and light impurities is absent. By our up-to-date knowledge, the non-local L-H transitions never have been described in limiter tokamaks with circular cross-section.

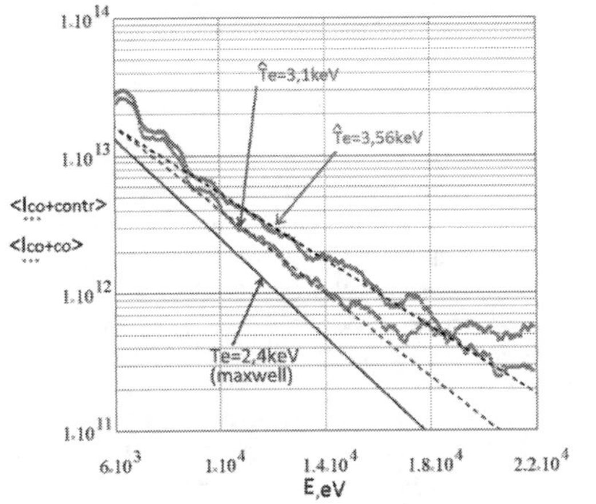

Fig.13. Approximated values of "electron temperature". Brown line corresponds to the temperature for Maxwellian distribution.

Preliminary analysis of X-ray spectra (PHA) shows, that in the case of co+contr injection the central area of plasma, where HF is deposited, becomes significantly non-Maxwellian.
We assume, that this phenomenon related to higher mean value of longitudinal energy of electrons in the resonance zone.
Co+contr injection provide more energy content in the central area of plasma column in comparison with co+co injection.
The energy value can be higher in relation to the energy in the Maxwell distribution up to 20% -30 %.
Resonance zone can be expanded by application of additional HF injections with slightly changed toroidal and poloidal angles and also frequency.
This heating method, together with the possible L-H transition can yield even greater energy content of the plasma.

We indebted to the members of T-10 team participated in the experiments described in the present paper.

References

1. S.V. Neudatchin, J.G. Cordey and D.J.Muir 20th EPS Conf. on Control. Fus. and Plasma Phys. (Lisboa,) vol. **I** (Geneva : EPS), p 83 (1993)
2. S.V. Neudatchin, J.G. Cordey and D.J.Muir The time behaviour of the electron conductivity during L-H and H-L transitions in JET JET Rep. JET-P (93)-58 (Jet Joint Undertaking Abingdon Oxforshire UK) (1993)
3. J.G.Cordey, D.J. Muir, S.V. Neudatchin et al Plasma Phys. Control. Fusion **36** A267 (1994)
4. J.G. Cordey, D.J. Muir, S.V. Neudatchin et al, Nucl. Fusion **35** 505 (1995)
5. S.V. Neudatchin, T. Takizuka, H. Shirai et al., Japan J. Appl. Phys. **35** 3595 (1996)
6. S. V. Neudatchin, T. Takizuka, H. Shirai et al., Plasma Phys. Control. Fusion **44** A383-389 (2002)

7. V.A. Vershkov et al Nucl Fusion 2017 347

Side emissions during EC injection for PDI studies in FTU tokamak

Alessandro Bruschi[1,*], *Edoardo* Alessi[1], *Benedetta* Baiocchi[1], *William* Bin[1], *Ocleto* D'Arcangelo[2], *Francesco* Fanale[1], *Lorenzo* Figini[1], *Cristian* Galperti[3], *Saul* Garavaglia[1], *Giuseppe* Gittini[1], *Gustavo* Granucci[1], *Giovanni* Grosso[1], *Lev* Lubyako[4], *Cristina* Mazzotta[2], *Vittoria* Mellera[1], *Alessandro* Moro[1], *Francesco* Orsitto[5], *Fabrizio* Pallotta[1], *Giuliano* Rocchi[2], *Alessandro* Simonetto[1], *Umberto* Tartari[1], *Onofrio* Tudisco[2]

[1]Istituto di Fisica del Plasma, Consiglio Nazionale delle Ricerche, via R. Cozzi 53, 20125 Milano, Italy
[2]ENEA Fusion and Nuclear Safety Department, C.R.Frascati, via E. Fermi 45, 00044 Frascati (Roma), Italy
[3] Swiss Plasma Center, Ecole Polytéchnique Fédérale de Lausanne, Lausanne, Switzerland
[4]Institute of Applied Physics, 46 Ulyanov st., Nizhny Novgorod, 603950, Russia
[5]Consorzio CREATE, Università degli Studi di Napoli, Napoli, Italy

Abstract The evidence of Parametric Decay Instabilities (PDI) excited by the ECH power injected in O-Mode has been explored in FTU Tokamak, using the Collective Thomson Scattering (CTS) diagnostic. The experiments show evidences to support the hypothesis of low-threshold excitation of waves generated by PDI mechanisms, formerly proposed in the case of 2nd harmonic X-mode injection in TEXTOR and ASDEX-U. Theoretical analysis predicts low-threshold parametric decay also for O-mode pump-wave injection, which can be injected in FTU at frequencies close to the first Harmonic EC resonance. Experiments were made at different magnetic fields, injecting the 140 GHz probe and observing the emission from the second antenna of the EC launcher in poloidally symmetric and asymmetric configurations, in presence of MHD islands. The signal is detected by the CTS radiometers, with a fast digitizer allowing the spectral reconstruction at very fine time and frequency scales. Different types of emissions are studied in detail, comparing them with the magnetic island rotation frequency in different plasma conditions. In order to locate the plasma volume originating the emissions, a new antenna and receiving line has been installed.

1 Introduction

The presence of emissions due to parametric decay instabilities (PDI) during the injection of the ECH power at the 1[st]Harmonic O-Mode has been investigated in the FTU Tokamak, using the renewed Collective Thomson Scattering (CTS) diagnostic [1-3].
The aim was to add experimental evidences to confirm the hypotheses of low-threshold excitation of waves generated by PDI mechanisms based on the observation of back-scattered power in the case of 2nd harmonic X-mode injection in TEXTOR [4 and reference therein], and on other observations in ASDEX-U [5]. Theoretical analysis [6,7] predicts the possibility of low-threshold parametric decay also for O-mode pump-wave injection, which can be explored at frequencies close to the first Harmonic resonance in FTU.
Experiments were made at magnetic fields of 4.7 T and 3.6 T, injecting the 140 GHz probe and observing the emission from the second antenna of the EC launcher in symmetric and asymmetric configurations (see section 1.1), in presence of MHD islands mainly stimulated by Ne injection. The signal is split in two orthogonal polarizations and detected with two channels of the CTS radiometer, with a fast digitizer connected to the IF of the two front-ends, recording the time traces of the signal at rates 3 to $6*10^9$ samples/s. The spectral reconstruction can be done offline by finely tuning the time windows of the Fourier Transform of samples, obtaining the possibility of very high time and frequency resolution.
Different types of emissions have been reported [8], some of which are studied in further detail, to exclude gyrotron spurious emission and to verify the synchronism with the magnetic island rotation. The plasma conditions in which the various types of emissions are found have been extensively studied. In order to locate the plasma volume originating the emissions, which would help to identify the causing phenomena, a new receiving antenna with a transmission line independent from the ECH lines and with a higher decoupling with respect to the probe has been installed.

1.1 Plasma and experimental conditions

The work has been performed in two main plasma conditions, the first at 4.7 T magnetic field in the vessel centre, with the resonant layer for the probe frequency shifted a few cm toward the plasma high-field side (see

* Corresponding author: bruschi@ifp.cnr.it

Fig.1, left), and the second at 3.6 T in the vessel centre, with the first-harmonic resonant layer for the probe frequency shifted at the plasma edge towards the high-field side (Fig.1, right). The plasma total current was respectively 500 kA and 350 kA in the two cases, in order to keep a similar q-profile, with similar location of the m:n=2:1 tearing mode. The MHD island at q=2 is often already present in the plasma, but for our purposes it is necessary to stimulate the increase in dimension by injecting a defined quantity of Ne impurities, that stiffen the profiles and stimulate the island size increase.

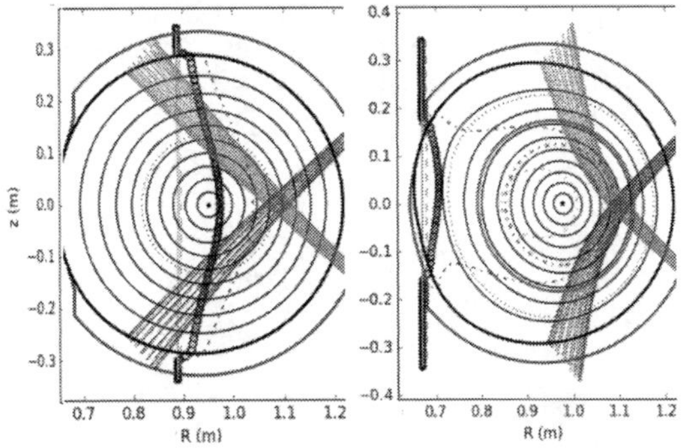

Fig. 1. Left: Plasma equilibrium and main EC resonances when operating at 4.7 field in the vessel centre. Probe beam (in red) is launched from the bottom right side, while the beam is received at the top right side (in blue) intersecting the probe on the equatorial plane. Right: the same with magnetic field at 3.6 T in the vessel center. The thin vertical (solid and dashed) lines correspond to the EC cold and warm resonances, while the Upper Hybrid (UH) resonance is marked by dots in colour.

In fact, the theoretical predictions of possible presence of low-threshold PDI rely on the possible inversion of the local density gradient profile, as in a well-grown island. The scope of the experiments is thus to verify the existence of the phenomenon and possibly quantify the power or plasma density threshold for its occurrence.

1.2 Probe launch and receiving conditions

The 140 GHz, 350 kW gyrotron beam used as the probe (in red in Fig.1) is launched from the lower mirror of the EC launcher located in FTU port 8, while the radiation is collected from the upper mirror of the same launcher. The launcher is set to an angle such that the received beam (in blue in Fig.1) intersects the probe on the equatorial plane (symmetric configuration) or, for some shots at 4.7 T, where the probe beam (launched horizontally) intersects the EC resonance (asymmetric configuration). The probe is launched with a toroidal angle of 5° in co-current direction, and usually the probe is swept toroidally back and forth from 1° to 9° in 200 ms, to put in evidence variations in the received signal with different amounts of beam overlapping, that could be taken as indication of emissions located in the superposition or "scattering" volume.

1.3 Receiver setup

In a first stage of experiments only one radiometer front-end was available, and it was set to receive the O-polarized wave; later a second front-end was connected to the line adding a wire-grid to split the polarization, and it was usually set to receive the X polarized wave. The radiation sent to the first radiometer was modulated at around 10 kHz to allow (constant) background noise subtraction in the subsequent conditioning by the filter bank section. The second radiometer was not modulated, in order to have a clean IF signal, to allow analysis of the Fast Digitizer signals free from modulation effects.

1.4 Data treatment and calibration

The calibration is done acquiring spectra from Ohmic shots (without EC power injection) in which the receiver line of sight intersects the EC resonance in an optically thick plasma region of known electron temperature, and, in the case of the digitizer raw data, also subtracting an offset due to noise, measured by receiver when blinded; details are given in [9]. A comprehensive database of observation has been realized, including all the performed shots, their main plasma parameters and for some of them, as temperature and density, the global (averaged) and local values (in the plasma centre and in the scattering volume). Also the main MHD island present in the plasma, m:n = 2:1, 3:2 or 3:1 modes has been recorded, together with the amplitude of the reference Mirnov Coil signal. This allowed the analysis of the statistical occurrence of the observations with respect to one or two chosen parameters.

2 Emissions detected by CTS

Several emissions of different type were detected in nearly 200 shots performed from 2014 to 2017. Some of them were found to be gyrotron side emission lines, detected also in the stray radiation picked-up in dedicated shots [10]. Other ones were found in 7.2 T shots performed in 2014, and reported in [8]; they may be of the same origin as those reported here, even though this has still to be confirmed by new experiments to be performed at 7.2 T. Here below we describe the emissions found in the latest experimental campaigns, and the studies of correlation of these emissions with the MHD activity (in particular, with the islands rotation frequency) and with the evolution of the other plasma parameters. In order to facilitate the description, the emissions have been divided in a few types, numbered as in Table 1, where a synthetic description is given. In the next sections we will describe a limited set of emissions, which have characteristics closer to the ones expected by PDIs and with enough statistical occurrence. In this respect, we concentrate on line emissions of types 2, and 7, since the description of emission of type 3 have not enough observations, and emissions of type 1, 4 and 6 correspond to the emissions already described in [8], respectively in sections 4.2, 4.1 and 4.3.

Table 1. Types of observed emission. Emissions of Type 2 and Type 7 are the ones studied in more detail in this paper.

n.	Type of spectrum	Min. duration (μs)	Max persistence (ms)	Δfrequency or range (GHz)
1	bursts of continuum	<1	variable	> (±) 1
2	line	<20	10-20	(-) (0.2-0.4)
3	broad line	>50?	~5	± (0.2-0.3)
4	line of variable frequency	>1	>20	- (0.5-1.2)
5	continuum	<50	<10	- (0.05-0.2)
6	multiple lines	<50	>15	± (0.05-0.4)
7	line	<100	<100	(±) (0.6-0.7)

2.1. Emissions from the gyrotron

As already reported in [10], in the spectra there are often side emissions that are originated by the gyrotron. These are put in evidence in the spectra taken when the receiving line was disconnected from the antenna and picked up the gyrotron stray radiation outside the launcher. This was presently done offline, with dedicated shots. A line with variable frequency, one GHz above the gyrotron frequency, is often found. Other emissions of the same appearance have been presently excluded from the analysis, waiting to confirm if they are originated (or not) from the gyrotron. With a deeper observation of the spectra of the stray radiation, in fact, also the emissions reported in section 4.4 in [8] as "periodic bursts at around 15 MHz from the probe" were observed in the gyrotron spectrum, thus revealing the weakness of the assumption of having a clean gyrotron spectrum, unless it is measured with the same high-sensitivity diagnostic apparatus.

With the future installation of a new receiving line, the aim is to receive simultaneously in two different channels the emission from the plasma and the emission from the gyrotron, to exclude any possible mis-interpretation of the origin of the emissions, that may also be due to gyrotron spectrum disturbances, induced by reflections or scattering of power into the plasma, that cannot be excluded in the present analysis.

2.2 Emissions of Type 2

These emissions (Fig.2) appear in shots with m:n=2:1 MHD rotating mode in different field configurations. The frequency indicated in the spectrogram is the frequency distance (in absolute value) from the local oscillator (LO) frequency, usually tuned on the gyrotron one. In this case we cannot distinguish the emissions higher or lower than the gyrotron frequency, since they appear superimposed. Occasionally the LO frequency is significantly shifted from the gyrotron frequency (as in the shot in Fig.12) to distinguish the upper and the lower sidebands from the probe.

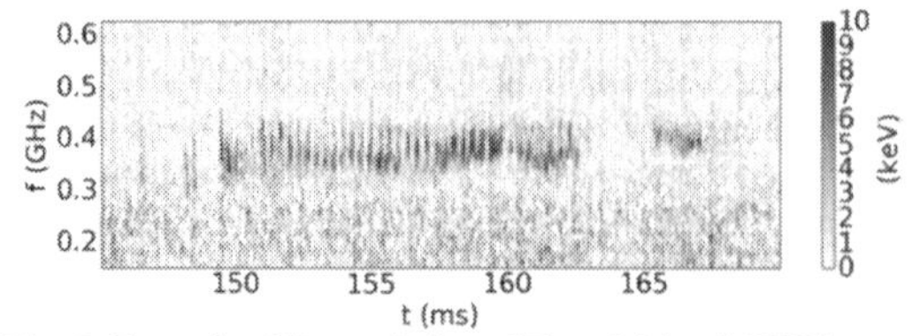

Fig. 2. Example of line-emission of Type 2 (shot #40753).

Type-2 emissions are more frequent (see Fig.3) at higher densities and when a higher dB/dt of the rotating island is measured by the reference Mirnov coil, i.e. when the magnetic island is larger, and thus the density gradient can be more easily found inverted, condition to facilitate the PDI occurrence as predicted by theory [6,7].

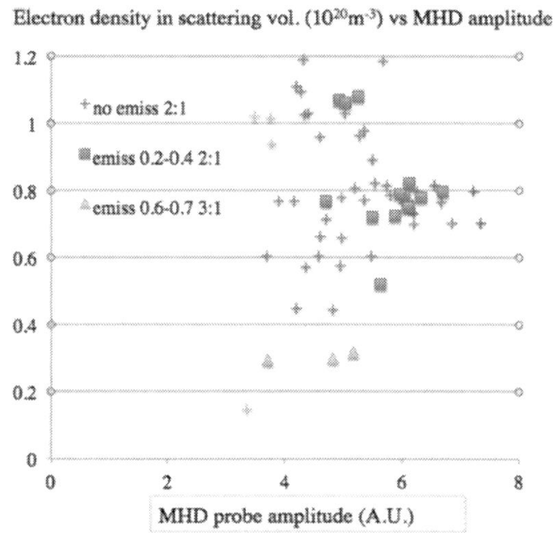

Fig. 3. Distribution in density and MHD amplitude of shots with emission of Type 2 (squares) and Type 7 (triangles) with respect to those without emission (crosses).

In any case emissions appear as spiky, with a few microseconds time-scale of temporal evolution (Fig.4).

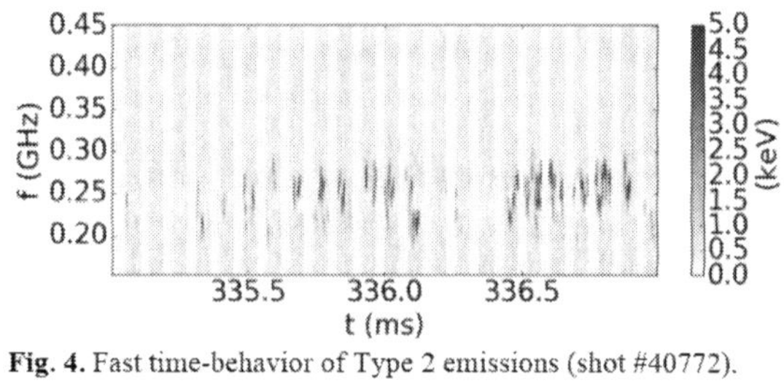

Fig. 4. Fast time-behavior of Type 2 emissions (shot #40772).

Emissions seem linked with the presence of a m:n=2:1 MHD island and their characteristic frequency shift may vary during the plasma evolution (from 0.25 GHz to 0.3 GHz and back to 0.2 GHz, as in the case of shot #40772 , in Fig.5, Top). In this particular case, the change may be correlated to the probe in the poloidal direction by a 3 degrees angle in 30 ms.

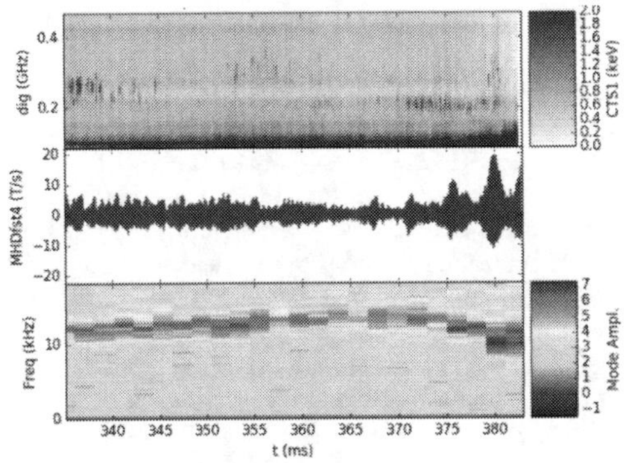

Fig. 5. Top: emissions for shot #40772, showing a variable frequency. Center: Mirnov coil signal, whose frequency spectrum (Bottom) indicates the presence of a m:n=2:1 mode rotating at a frequency above 10 kHz.

We studied the correlation of the emission amplitude with the Mirnov coil signal, to understand if the time evolution can be due to the rotation of the island. For this aim, their statistical appearance (average of power in the frequency band) was evaluated as a function of time and compared with the rotating MHD island phase (Fig.6). In two cases (#40771 and #40772) of asymmetric launch no correlation was found with the rotation frequency, while in another case, for a shot with symmetric launch at a relatively low density for FTU ($0.3*10^{20}$ m^{-3}, #40753) during plasma current ramp-up, the variation of the statistical appearance of the emissions seems to be correlated with the island rotation (Fig.6).

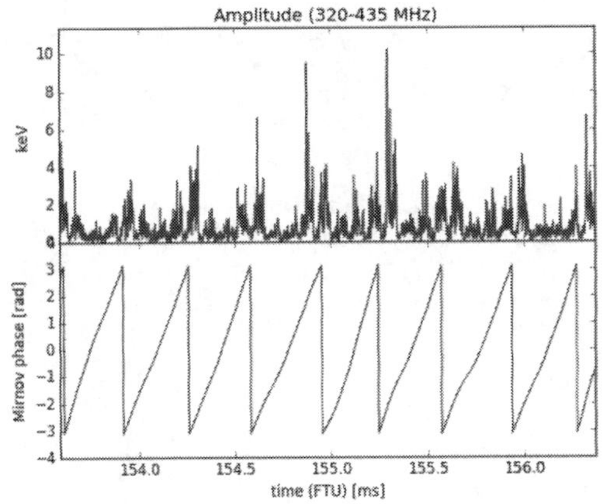

Fig. 6. Amplitude of emission between 0.32-0.435 MHz from the probe (Top), compared to the phase of the Mirnov coil signal (Bottom) for shot #40753: for this shot a correlation is found, while in other ones (#40771 and #40772) it is not.

Moreover, for shot #40753 the time spectrum of the amplitude of the line emission shows a trend similar to the MHD rotation evolution, i.e. slows down during the island rotation slowing down (Fig.7). In other terms it seems that the emissions are more frequent (or stronger) every time the island is in the same particular position during its rotation, even when changing rotation frequency.

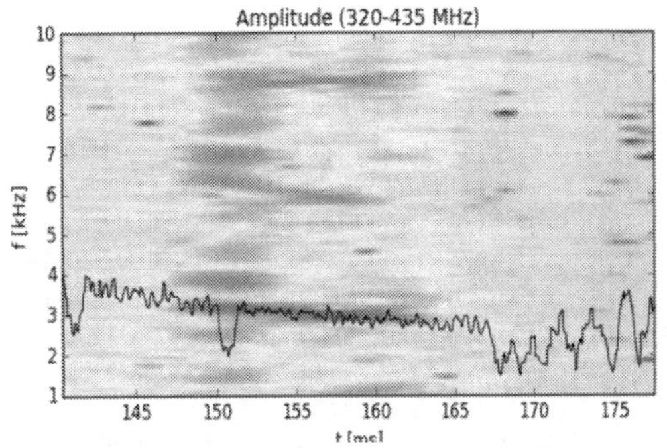

Fig. 7. Frequency spectrum of the emission amplitude in time, with the frequency of the Mirnov coil (black line) superimposed, showing a correlation of the variation of the averaged intensity of the emission with the rotation frequency of the island, during the island slowing down, for shot #40753

The configurations of the shots where a correlation is observed is different from the one in which it is not found. The launch configuration for shots #40771 and #40772, was asymmetric: the probe beam intersects the m:n=2:1 magnetic island in 2 points: the first in the low field side, close to the UH resonance position and then again at the EC resonance (Fig.8).

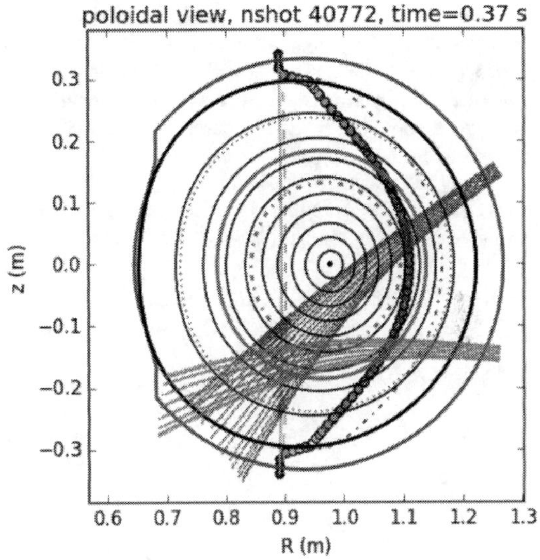

Fig. 8. Plasma equilibrium and resonances for shot #40772 with asymmetric launch: the probe beam intersects the island close to the UH resonance and then again the q=2 surface (solid circle in green) close to the EC resonance. The thin vertical (solid and dashed) lines correspond to the EC cold and warm resonances, while the UH resonance is marked by dots in colour. The q=1.5 and q=3 magnetic surfaces are also shown as dashed-dotted lines.

Moreover for these shots the temperature profile is rather flat. In the case of shot #40753 instead, the launching and plasma configuration is symmetric (Fig.9). The receiving line of sight crosses the launched beam close to the position of the q=2 magnetic surface. The position of the UH resonance is also close to the crossing.

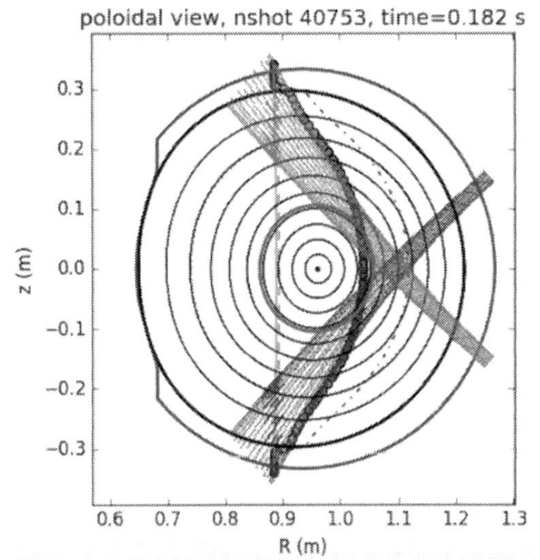

Fig. 9. Plasma equilibrium and main EC resonances for shot #40753 with symmetric launch: the probe beam (in red) and received beam intersect themselves close to the estimated position of the q=2 resonant surface on the equatorial plane, and not far from the UH resonance.

This is not the only difference from the shots #40771 and #40772: Also the temperature profile in the case of shot #40753 is much more peaked (see Fig.10). This suggests a possibly different mechanism or a different location of the Type 2 emissions in the two cases. The superposition of resonances and magnetic rational surfaces (the q=1.5 and 3 are also shown in Fig.8 and Fig.9 as dashed-dotted lines) increases the complication of the analysis aimed to find an explanation for the observed phenomena.

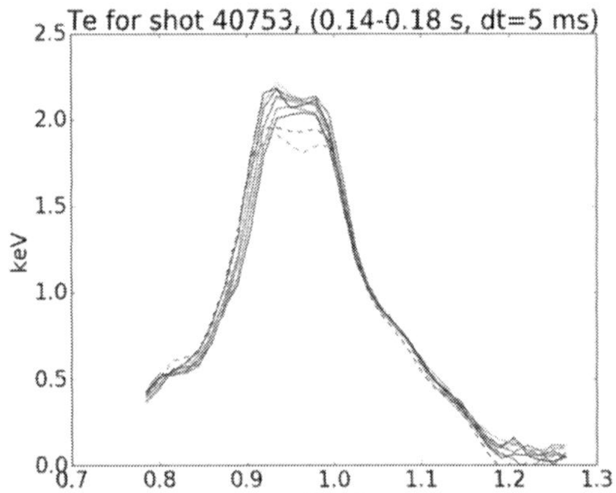

Fig. 10. Electron temperature profiles at different times during the Type 2 emissions of shot #40753, as reconstructed by the Michelson Interferometer.

In all cases of Type 2 emissions at 4.7 T, the frequency shift with respect to the probe is lower than the Lower Hybrid (LH) frequency in the plasma core. Emissions of Type 2 are found also in 3.6 T shots, with a different launch configuration and at higher densities. Shot #40282 is an example: for this shot (as for #40771 and #40772) no strict correlation of the emissions with the island rotation has been found, despite the presence of a m:n=2:1 MHD mode located at the beam crossing (see Fig.11). The UH resonance is not crossed by the beams but after a first bounce at the vessel. Also the LH cutoff is close to the crossing and the island position, and this may influence the emissions or the clean reception of the signal, since the (small and unwanted) part of the beam injected in the X polarization may be reflected in the direction of the received beam, even if a toroidal angle is given to the probe to prevent direct reflection.

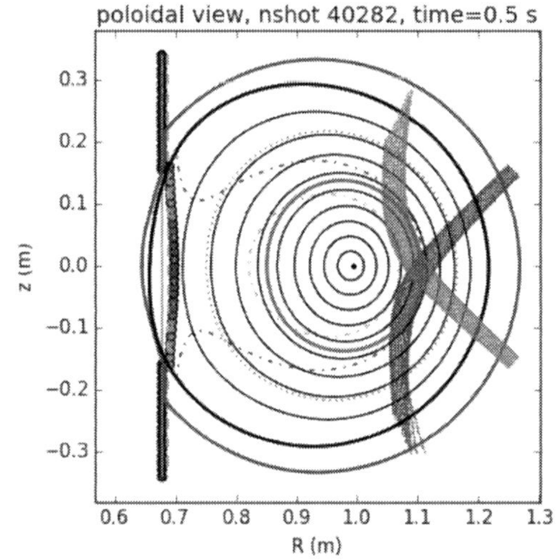

Fig. 11. Plasma equilibrium and main EC resonances for shot #40282 with symmetric launch: the probe beam (in red) and received beam intersect themselves close to the q=2 resonant surface on the equatorial plane and are refracted toward the vessel by the high-density plasma.

In this case, due to the shifted LO frequency, the emission is clearly seen only below the probe frequency (line at 0.8 GHz) at around 0.5 GHz from the LO frequency, as seen in Fig. 12.

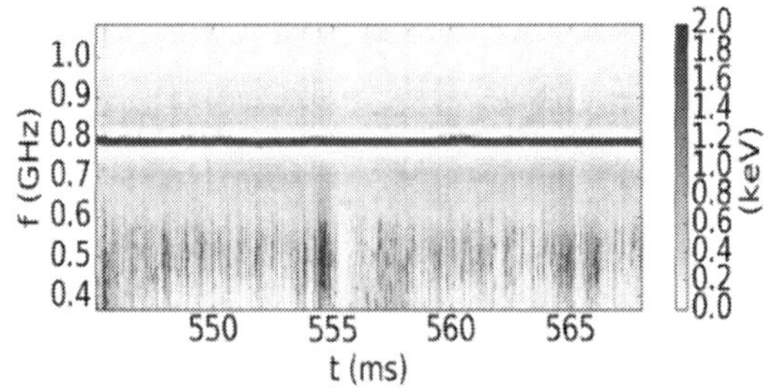

Fig. 12. Emissions below the probe frequency (shot #40282).

2.3 Emissions of Type 7

Another line emission type that is candidate as due to PDIs is Type 7. It has an intermittent character (as seen in Fig.13, where a 10 kHz chopping on the incoming radiation is also modulating the entire spectrum).

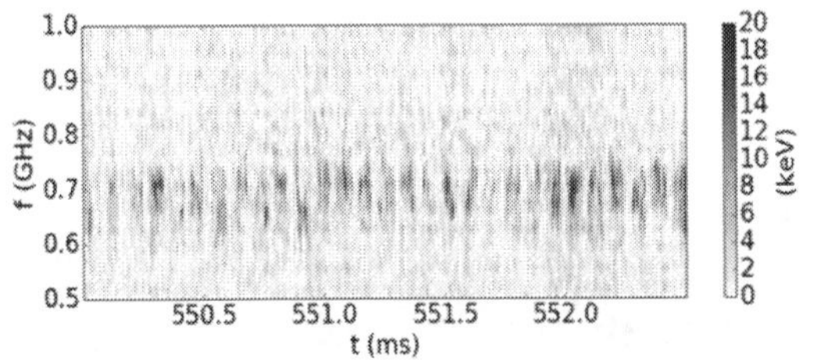

Fig. 13. Example of line-emission of Type 7 (shot #40735).

This emission is found in a few shots at lower density (triangles in Fig.3), characterised by the presence of a m:n=3:1 MHD island and by hollow temperature profiles (Fig.14).

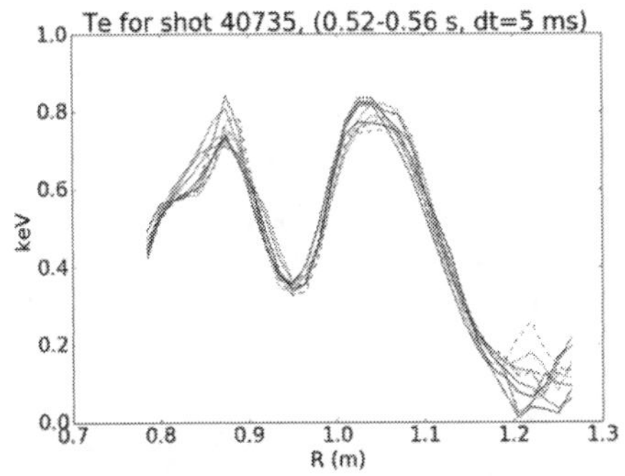

Fig. 14. Electron temperature profiles at different times during the Type 7 emissions of shot #40735, as reconstructed by the Michelson Interferometer. The regions at major radius R=0.83 m and R=1.08 m are perturbed (flattened) by the presence of a m:n=3:1 MHD magnetic island.

The presence of a m:n=3:1 MHD rotating island is confirmed by the analysis of the phases of Mirnov coils signals collected in different toroidal positions. The reconstructed launch and plasma equilibrium configurations are shown in Fig.15. In this case the receiving beam crosses the probe beam close to the q=3 rational surface where the MHD island is located. Before reaching the EC resonance (where they are presumably damped) the beams intersect again the q=3 surface, in the proximity of the UH resonance. For these shots the frequency shift of the detected emission is compatible with the LH frequency in the plasma core. The reconstruction of the density profiles for these shot is more complex, and will require a dedicated work.

3 Conclusions

The interpretation of the observations in terms of effects of secondary waves generated by Parametric Decay of the probe is still uncertain, given also the many way that the radiation can reach the resonances, for example after bouncing on the tokamak vessel. Certainly a comprehensive set of evidences point to phenomena occurring during probe injection (and thus during EC beam launch for heating and Current Drive) that can be studied with a diagnostic able to detect side emissions at a level of a the plasma thermal emission with a good spatial and a very high temporal/frequency resolution, as the improved CTS diagnostics recently operating on different devices. To locate and discriminate them from the gyrotron side emissions, independent "reference" receiving lines from plasma lines of sight and from the gyrotron stray radiation are necessary. A new receiving line, with an independent access on the equatorial plane and (alternatively) the capability to collect the gyrotron stray radiation before injection into the plasma [11] is presently installed on the FTU and connected to the CTS diagnostics, for future deeper investigations.

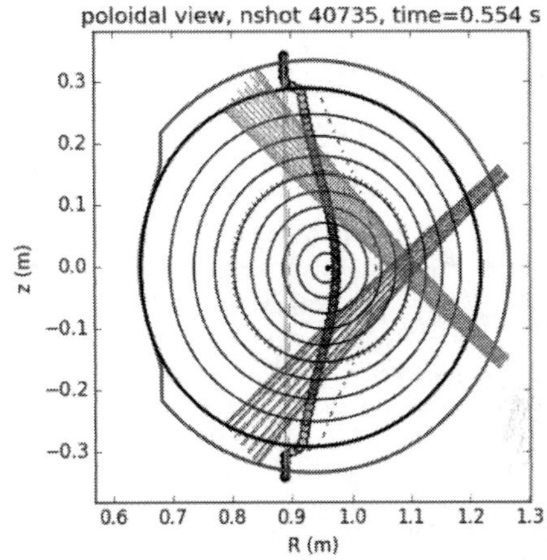

Fig. 15. Plasma reconstructed equilibrium and resonances for shot #40735 with symmetric launch: receiving beam crosses the probe beam close to the q=3 rational surface (dotted circle in green) where a MHD island exists. Before reaching the EC resonance (thin vertical line) the beams intersect again the q=3 surface, in the proximity of the UH resonance (dots in colour).

This work has been carried out within the framework of the EUROfusion Consortium and has received funding from the Euratom research and training programme 2014–2018 under grant agreement No 633053. The views and opinions expressed herein do not necessarily reflect those of the European Commission.

References

1. W. Bin et al., *JINST* **10**, P10007 (2015)
2. W. Bin et al., Rev. Sci. Instr. **87**, 11E507 (2016)
3. W. Bin et al., Fus. Eng. Des. **96–97**, 733–737 (2015)
4. E. Z. Gusakov and A Y Popov, Plasma Phys. Control. Fusion **60**, 025001 (2018)
5. S. K. Hansen, et al., Plasma Phys. Control. Fusion **59**, 105006 (2017)
6. E. Z. Gusakov, et al., Plasma Phys. Control. Fusion **59**, 075002 (2017)
7. E. Z. Gusakov and A. Y. Popov, Phys. Plasmas **25**, 012101 (2018)
8. A. Bruschi, et al., Nucl. Fusion **57**, 076004 (2017)
9. B. Baiocchi et al, ECPD 2017, (submitted to *JINST*).
10. A. Bruschi, et al., EPJ Web of Conferences **149**, 03017 (2017)
11. O. D'Arcangelo *et al, JINST* **13**, C01012 (2018)

Protection of the DIII-D Vessel, Diagnostics and Windows from Microwaves Injected by the ECH/ECCD System

Mirela Cengher[1], *John* Lohr[1], *Yuri* Gorelov[1], *Antonio* Torrezan[1], *Dan* Ponce[1], *Xi* Chen[1]

[1]General Atomics, 3550 General Atomics Ct. San Diego, California, USA

Abstract. The ECH/ECCD system on DIII-D benefits from protective measures set in place in order to reduce the likelihood of refracted rf power potentially reaching sensitive areas of the vacuum vessel or diagnostics. In addition to installation of shutters and valves, a density interlock prevents the rf injection into overdense plasma. Visible light cameras, light detectors, Langmuir probes and reflected power sniffers complete the array of diagnostics used for EC protection. Besides the 110 GHz gyrotron, a new higher frequency gyrotron at 117.5 GHz was added to the system during the 2018 experimental campaign. The new frequency allows for operation at higher plasma density and with better current drive efficiency at higher magnetic field than for 110 GHz. Injection in O-mode was used in some of the high density plasmas where the X-mode is refracted.

1 Introduction

Recent years have seen an increase in the large variety of experiments involving injection of 110 GHz EC power into the DIII-D tokamak plasma, with operational regimes approaching or exceeding the density limits for X-mode second harmonic resonance absorption. In these high-density plasmas, the EC beam can be refracted, possibly causing damage to EC launchers or diagnostic systems located behind high-transmission windows. DIII-D has a large array of protections to reduce the possibility of damage. In addition to operational limits for the line averaged plasma density, an array of diagnostics is used for EC protection, including visible light cameras which monitor the light in the launchers, light detectors, and reflected power sniffers, and Langmuir probes. Most recently, a new gyrotron operating at a higher frequency of 117.5 GHz has been added alongside the other operational 110 GHz, 1.0 MW class gyrotrons in the existing electron cyclotron heating and current drive (ECH/ECCD) complex on the DIII-D tokamak [1], all manufactured at Communications and Power Industries (CPI) [2] has added. The new frequency has a wider operating space in terms of plasma density and is more suitable for high density and higher field plasma, and with higher efficiency of current drive. O-mode second harmonic resonance electron cyclotron heating scenario was used successfully in high density plasmas.

2 EC beam refraction detection and protection

Protective measures are in place on DIII-D to reduce the likelihood of refracted rf power potentially reaching sensitive areas of the vacuum vessel or diagnostics. Ports at risk from an EC beam hit are fitted with shutters, isolation valves, filters, and low loss tangent windows are installed where necessary. Shutters and valves can pre-emptively isolate these sensitive ports during shots when the EC is injected in a plasma close to the overdense conditions that could result in unintentional rf beam refraction that could cause arcing and damage. In addition, rf filters reject the 110 GHz frequency from the EC system, so the millimeter waves at this frequency can not reach the sensitive electronics dedicated to other spectral ranges [3].

The transmission lines and launchers have the following detection systems installed: rf reflected power monitors acting as sniffers, CCD cameras, visible light monitors, waveguide temperature monitors and Langmuir probes. The Langmuir probes and temperature monitors detect plasma, arcs, or high EC power absorption in the launcher and can remove the EC permit and stop the EC injection if the launcher is at risk. The Langmuir probes are not always reliable since they have also responded to edge localized modes, disruptions, pellet injection, or instabilities. The CCD cameras are monitoring the launcher mirror area without an interlock, providing real-time information about areas of high luminosity or afterglow. The rf injection and absorption for the shots

Corresponding author: cengher@fusion.gat.com

with potential hot areas identified in the video images are carefully analyzed to eliminate the possibility of power refraction to the EC launchers. The visible light detectors also monitor the general launcher area and have a similar role as the video cameras, the difference being that their signal can be used as an interlock if necessary, without having any details about the areas where the light signal originates. The launcher waveguide temperature is also monitored with two types of detectors for anomalous temperature spikes that could indicate damage to the metal.

Reflected power monitors used as sniffers are installed in each transmission line at the last miter bend before the tokamak. These 4-port rf power monitors are sensitive to the transmission direction, wave polarization, and rf mode content [4]. A mirror with vacuum on both sides picks off the forward and reflected signals and is followed by a window. An orthomode transducer is then used to measure polarization or mode purity of the injected rf power. The sniffers respond only to the beam being refracted to the launcher port area of the vessel wall, including the case when the rf beam is launched from another port.

3 Interlocks and recovery

To further increase rf protection, the ECH group has implemented a complex set of diagnostics and interlocks. A combination of interlocks and a fault recovery system ensures the rf generation is stopped for the duration of low EC power absorption into the plasma and resumed when absorption conditions are optimal.

Interlock signals are sent by the Plasma Control System (PCS) using feedback from a number of diagnostics. The ECH injection into the tokamak is terminated under command from the PCS when the plasma parameters do not satisfy the pre-set interlock range.

3.1 Plasma density interlock

A real-time density interlock is implemented to stop the ECH if the density rises above a threshold calculated based on the plasma equilibrium reconstruction and electron temperature and density profile measurements. The interlock value is based on ray-tracing model calculations [5] and takes into account the amount of rf power expected to be absorbed into the plasma and the strike point of the refracted beam in case of low absorption.

When plasma becomes overdense for the injected EC beam, the density interlock stops rf generation, preventing beam refraction that could potentially cause damage to diagnostics or to the ECH launchers. The PCS uses the line-averaged electron density calculated in real-time and compares it to the threshold value, then sends a command to the ECH control system when the line average density reaches a value above the pre-set value. Above this plasma density the X-mode ECH beams would encounter the right hand cutoff before the resonance and would be refracted. The density interlock threshold is adjusted as needed based on ray-tracing of rf beam propagation at 110 GHz and 117.5 GHz.

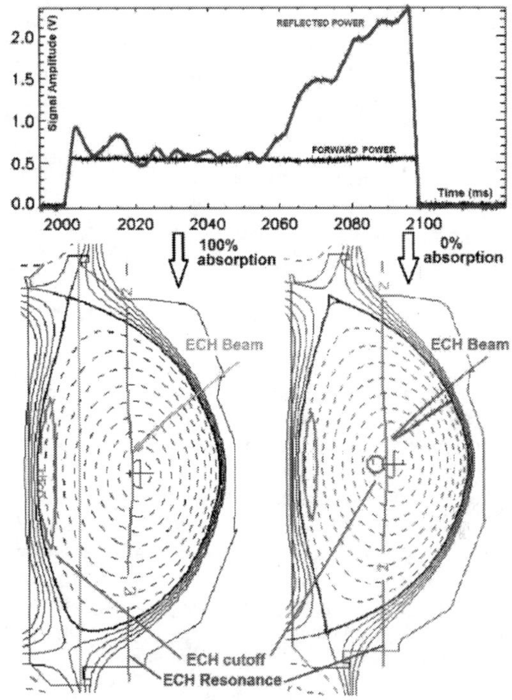

Fig.1. Ray tracing code location of the beam strike point is confirmed by the sniffer response. The sniffer signal (upper plot) shows increasing response as the ECH beam trajectory reaches back to the launcher in the lower right equilibrium reconstruction plot.

An example of the ray tracing of the ECH beam trajectories for such case is shown in Fig.1. The reflected power monitor response in the upper plot shows increasing response as the ECH beam starts to be refracted. The reconstructed plasma equilibrium and the beam trajectory in the lower right shows the refracted rf beam near the launch point monitored by the sniffer.

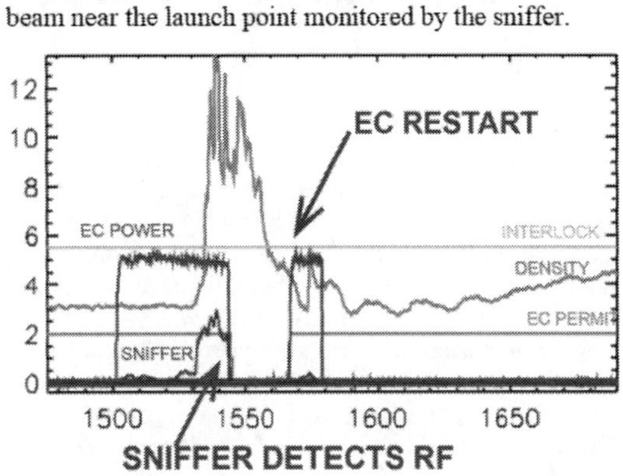

Fig. 2. Time evolution of a gyrotron shot with recovery during ECH injection showing the relevant signals.

Fig. 2 shows a case where the plasma density increased above the threshold and the interlock was engaged, stopping the rf generation. At the same time with the beginning of the ramp up of the reflected power, the line-average density exceeded the interlock limit and the EC control system stopped rf generation. The EC permit was lost with an 8 ms delay during which the sniffer detected the power refracted from the plasma. The rf generation was resumed once the density interlock is satisfied again.

A number of factors can influence the quality of beam trajectory tracing and the error in the density threshold calculation: the error bars in the density profile measurement and density profile fits, the precision of the aim of the ECH launchers, possible obstruction of the beam by the port boundary, the injected beam polarization, or the fact that strong refraction on density spikes is not considered in the model. The present 8 ms delay in response to interlock allows rf power to be refracted back to the launcher as seen in Fig. 2 and could end up causing damage even for these short duration periods. Work is under way to reduce this delay in the interlock response to less than 1 ms.

The ECH power is injected into the tokamak from the low field side using four dual launchers [6] and the RF beams can be steered ±20° in real time during one plasma shot. For the cases when mirrors are moving, the full range of motion is taken into consideration to determine the density interlock value, as the absorption condition can vary widely for a poloidal scan.

3.2 EC shot restart

After the interlock is engaged, the rf generation can be resumed later during the same shot if absorption conditions are satisfied, due to a newly added "restart" gyrotron control system feature, so the plasma shot can get full benefit of the ECH/ECCD injection. The new Real-Time Interruptible High-Resolution Control Waveform Generator [7] has made possible the recovery and restart of a gyrotron after a fault, as shown in Fig. 2. Previously, exceeding pre-determined operational limits, or exhibiting off-normal operation, would result in terminating the operation of the faulted gyrotron during that whole plasma shot.

3.3 Other interlocks

Other interlocks were implemented to stop EC injection when absorption is expected to be low or absent: plasma current, toroidal magnetic field, plasma early termination due to disruptions. These interlocks offer protection by stopping millimeter wave generation for scenarios such as when no plasma is present, or when the 2nd harmonic resonance layer is not in the plasma and the absorption at the 3rd harmonic resonance is low. The EC injection into the empty tokamak vessel is allowed for no longer than 250 ms to protect centerpost tiles and to prevent damage from the beam being reflected on the centerpost and then reaching a high risk port. If the plasma current in the first 250 ms is lower than a preset value one of the permits generated by the PCS is lost and interlock stops rf generation. The plasma current interlock as "no plasma" indicator is used to stop rf generation if the plasma shot is shorter than the requested EC pulse.

The real-time toroidal magnetic field B_t interlock stops rf generation if B_t drops below the value where the 2nd resonance is no longer in the plasma and can not be reached by the EC beam.

The launcher control system has interlocks that signal that the beam aiming request would be obstructed by the port boundary and the mirror movement is stopped when these limits are reached.

4 Expanded range of plasma regimes accessible for ECH/ECCD

For some launch configurations and high density plasmas the 110 GHz X-mode EC beam has low or no absorption. For some of these cases the risk was mitigated by launching an O-mode polarized wave instead of the X-mode, as shown in Fig. 3 A on the left. The new 117.5 GHz frequency allows the X-mode propagation and absorption for cases where a 110 GHz beam would be refracted, as shown in Fig. 3 B on the right.

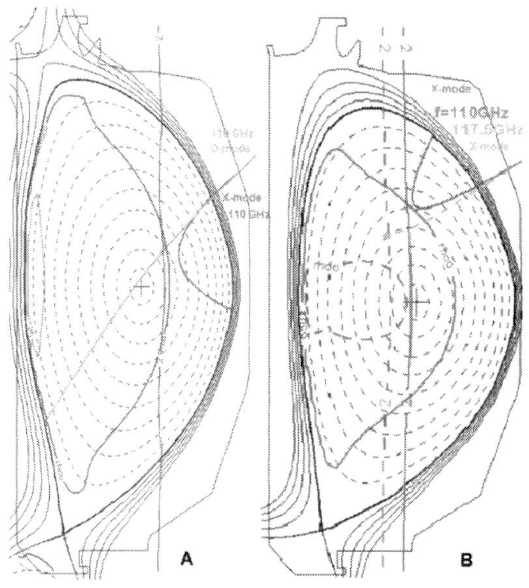

Fig. 3. Equilibrium reconstruction and ray tracing for: A. X-mode and O-mode (yellow) at 110 GHz (green), and B. X mode for 110 GHz (green) and 117.5 GHz (dashed yellow). The right-hand cutoff layer for the X-mode is shown in red. The resonance position is plotted in blue. Solid contours are used for 110 GHz and dashed contours for 117.5 GHz.

4.1 O-mode launch

O-mode second harmonic resonance plasma heating was used successfully for some of the cases where X-mode cutoff would have no power deposition in the plasma.

Two grooved mirror polarizers in the miter bends are used to control the elliptical polarization [8]. Polarizers can adjust polarization between shots to change between second harmonic O-mode and X-mode launch.

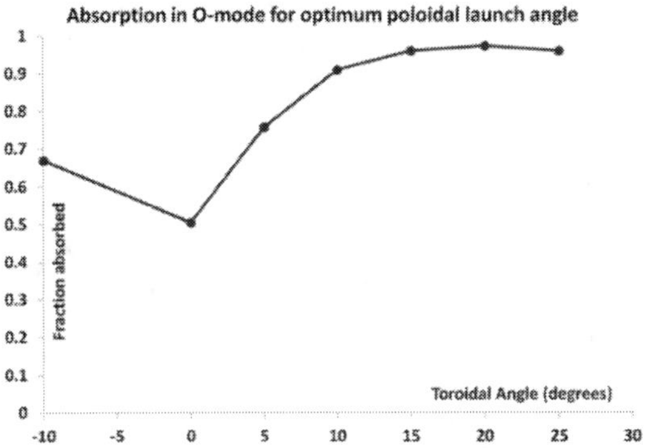

Fig. 4. Fraction of the O-mode power absorbed in the plasma versus the oblique launch angle for the optimum perpendicular launch angle.

O-mode launch has a number of drawbacks: incomplete absorption in the first pass, must be launched at an oblique angle for optimal heating power deposition (as shown in Fig. 4), cannot be used for current drive, and has a wide deposition profile. The advantage is the much higher density cutoff compared to the X-mode. Aiming was set to ensure absorption in the first pass, and also in the second pass after reflection on centerpost if possible and to avoiding hitting diagnostics on the vessel wall. O-mode can be used in limited situations because of limited range of plasmas where absorption in this scenario is high enough, above 80%, and because it cannot be used for current drive due to very low efficiency.

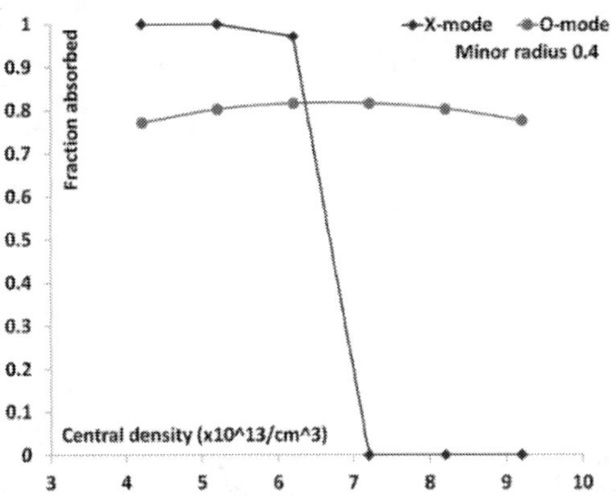

Fig. 5. Fraction of the O-mode and X-mode power absorbed in a single pass in the plasma versus the central electron plasma density for peak power deposition at normalized minor radius ρ=0.4.

The fraction of rf power absorbed in X-mode and O-mode for EC heating was investigated using TORAY ray-tracing. Fig. 5 shows the fraction of power absorbed at ρ=0.4 versus the central electron density. The full power launched in X-mode is absorbed only at low densities since the rf has access to the 2nd harmonic resonance. At electron densities above 6.5×10^{13} cm^{-3} the right-hand cutoff layer is encountered before the resonance, thus leading to an abrupt drop in the fraction of power absorbed in the plasma. The O-mode absorption is around 80 % in the first pass for the whole range investigated, so it can still be used (less efficiently) in some cases where the interlock would stop the X-mode launching into the plasma. A second pass could lead to extra absorption for the O-mode if the beam reflected from the centerpost is in the right mode, and is directed so it encounters the resonance.

4.2 Higher frequency X-mode launch

A new design depressed collector gyrotron operating at 117.5 GHz was used in the 2018 experimental campaign. The new frequency allows for rf injection into higher line-averaged plasma density, while providing a comparable efficiency for current drive, and better access to the resonance at higher magnetic field than the other frequency used in the system, 110 GHz.

Fig. 3 B shows a case where launched 117.5 GHz is absorbed in the plasma while 110 GHz beams are refracted due to different positions of the right-hand cutoff layers for the two frequencies. The 110 GHz power launched at larger steering angles (less central) would still be absorbed.

5 Summary

The DIII-D ECH system features a flexible versatile protection system that includes shutters, valves, windows, filters, cameras, light detectors, Langmuir probes, and rf monitors. The interlock protection has been extensively used and, as a result, rf injection for high risk conditions was avoided. The range of plasma regimes that can operate with EC power injection was extended by using second harmonic resonance O-mode injection and a new slightly higher frequency for the X-mode case. The EC shot recovery feature is a recent upgrade that increased reliability for shots with interlock interruptions.

Acknowledgments

This work was supported by the U.S. Department of Energy under DE-FC02-04ER54698.

DIII-D data shown in this paper can be obtained in digital format by following the links at https://fusion.gat.com/global/D3D_DMP.

process disclosed, or represents that its use would not infringe privately owned rights. Reference herein to any specific commercial product, process, or service by trade name, trademark, manufacturer, or otherwise does not necessarily constitute or imply its endorsement, recommendation, or favoring by the United States Government or any agency thereof. The views and opinions of authors expressed herein do not necessarily state or reflect those of the United States Government or any agency thereof.

References

[1] M. Cengher, *et al.*, IEEE Transactions on Plasma Science, vol. PP, **99** (2016), pp.1-6

[2] K.Felch, M.Blank, P.Borchard and S. Cauffman, EPJ Web of Conferences, **87**, 04006 (2015)

[3] J. Lohr, *et al.*, EPJ Web of Conferences **147**, 03003 (2017), pp1-6

[4] M.Cengher, *et al*, IEEE Transactions on Plasma Science, **42**, 7 (2014), pp.1964-1970

[5] K.Matsuda,IEEE Transactions on Plasma Science, **17** 1 (1989),p.6-11

[6] R.Ellis, *et al.*, Proc.14thTopical Conf. on RF Power in Plasma, AIP Conf. Proc., **595** (2011), pp. 318-321.

[7] D. Ponce *et al.*, Fusion Science snd Technology, **73** (2018), pp 1.

[8] J. Doane , Int.J. Infrared Millimeter Waves **13** (1992), p.1727

Observation and Modelling of the Onset of Parametric Decay Instabilities during Gyrotron Operation at ASDEX Upgrade

S.K. Hansen[1,2,*], *S.K.* Nielsen[2], *J.* Stober[1], *J.* Rasmussen[2], *M.* Stejner[2], and the ASDEX Upgrade Team[3]

[1]Max-Planck-Institut für Plasmaphysik, Boltzmannstraße, D-85748 Garching b. München, Germany
[2]Department of Physics, Technical University of Denmark, Fysikvej, DK-2800 Kgs. Lyngby, Denmark
[3]See Appendix in A. Kallenbach for the ASDEX Upgrade Team and the EUROfusion MST1 Team, Nucl. Fusion **57**, 102015 (2017)

Abstract. We investigate parametric decay instabilities (PDIs) occurring for gyrotron radiation near the upper hybrid resonance at the ASDEX Upgrade tokamak. The PDIs are observed through anomalous millimeter-wave scattering which is recorded using the high-resolution, fast acquisition collective Thomson scattering system installed at ASDEX Upgrade, and an experiment in which such observations are made during a scan of the toroidal magnetic field is performed. A previously published theoretical model is used to calculate the gyrotron power necessary to excite PDIs in the experiment; the theoretical model is capable of predicting whether or not PDIs will be observed at a given toroidal magnetic field with a high degree of accuracy.

1 Introduction

When high-power gyrotron beams are injected into magnetically confined fusion plasmas, the wave amplitude may in some cases become so large that the linear approximation breaks down and a parametric decay instability (PDI), which couples the incoming electromagnetic wave to two (electrostatic) plasma waves, is excited. Energy and momentum conservation in the three-wave process require the selection rules,

$$\omega_0 = \omega_1 + \omega_2, \quad \mathbf{k}_0 = \mathbf{k}_1 + \mathbf{k}_2, \tag{1}$$

to be satisfied; here, ω is the angular frequency, $\mathbf{k}$ is the wave vector, subscript 0 refers to the gyrotron wave, and subscripts 1,2 refer to the plasma waves excited by the PDI. In the cases of interest to us, the wave referred to by subscript 1 is a low-frequency wave whose frequency lies near the lower hybrid (LH) frequency, ω_{LH}, while the wave referred to by subscript 2 is a high-frequency wave whose frequency is close to that of the gyrotron. Because of the frequency selection rule, the high-frequency wave will be down-shifted by approximately ω_{LH} ($\sim 2\pi\times 1$ GHz in the main plasma of ASDEX Upgrade) from the gyrotron radiation, generally placing it outside the notch filters protecting the detectors used for millimeter-wave diagnostics such as ECE and collective Thomson scattering (CTS), causing it to hamper or potentially even damage these diagnostics. It is thus critical to understand PDIs in order to mitigate their potentially deleterious effects and, ideally, exploit the diagnostic tools that may be provided by them.

In this paper, we investigate PDIs occurring for X-mode radiation near the upper hybrid resonance (UHR), where $\omega_0 = \omega_{UH}$, with ω_{UH} being the UH frequency. PDIs are likely to occur for X-mode radiation near the UHR due to field enhancement effects, as noted by [1], who particularly considered the case of decay of the X-mode wave to a warm LH wave and an electron Bernstein wave, in connection with X1 ECRH of an optically thin resonance. The occurrence of PDIs in connection with X1 ECRH was observed in a number of small toroidal devices with optically thin resonances, e.g. Versator II [2], FT-1 [3], and TCA [4]. Such PDIs were also observed in connection with O1 ECRH at W VII-A [5] and TCA [4], which was attributed to conversion of a significant amount of power to X-mode upon reflection from the vessel walls; these observations indicate that the PDIs may also be of relevance during the start-up phase in ITER. The latter mechanism has also been used to explain PDIs observed during CTS experiments at ASDEX Upgrade [6–8], where the ECR is optically thin due to its location far on the high-field side in order to minimize the ECE background, and may also explain similar observations from LHD [9]. Besides the above cases, PDIs near the UHR have been shown to occur in connection with O-X-B heating of overdense plasmas, e.g. at W-7AS [10, 11], MAST [12], and TJ-K [13]; such PDIs may also be expected in connection with O-X-B heating at W-7X.

In the present work, we seek to validate the theoretical model from [8], which gives the gyrotron power threshold, P_0^{th}, that must be exceeded in order to excite PDIs in the CTS experiments at ASDEX Upgrade, by calculating the fraction of power converted to X-mode upon the reflection from the wall, the optical thickness of the ECR, the field enhancement near the UHR, and assuming decay of the X-mode wave to a warm LH wave and an electron Bernstein wave. To do this, O-mode pulses with a maximum power, $\max(P_0)$, of approximately 300 kW

*e-mail: Soeren.Kjer.Hansen@ipp.mpg.de

and $\omega_0 = 2\pi \times 105\,\text{GHz}$ are injected by a gyrotron from to the system described in [14], and the electromagnetic emission/scattering in O-mode at frequencies close to ω_0 is measured during the pulses using the fast acquisition CTS system presented in [15]. From the above observations, the occurrence of PDI features in a given pulse can be established and compared with whether or not $\max(P_0)$ exceeds P_0^{th} from [8], thus allowing the predictive value of the theoretical model to be assessed.

The paper is arranged as follows: in Section 2 the experimental setup and observations at ASDEX Upgrade are described; in Section 3 the results of the theoretical modelling are presented and compared with the observations; finally, in Section 4 we draw our conclusions.

2 Observations

The observations presented in this paper stem from ASDEX Upgrade discharge #34575, which is an H-mode deuterium discharge, during which the toroidal magnetic field, B_t, is scanned from $-2.66\,\text{T}$ to $-2.43\,\text{T}$; for reference, the central electron density and temperature in the discharge are approximately $5 \times 10^{19}\,\text{m}^{-3}$ and 4 keV, respectively. The most important effect of the B_t-scan for our purposes is the displacement of the ECR from inside the main plasma to the high-field side scrape-off layer which increases the fraction of power transmitted through the ECR by an order of magnitude, as will be shown in Section 3. The basic setup in ASDEX Upgrade discharge #34575 is seen in Figure 1, and the procedures used for obtaining the experimental results follow what was outlined in Section 1: during the B_t-scan 200 pulses (each of 2 ms duration, which is very short compared with the 3.5 s duration of the scan) with $\max(P_0) \approx 300\,\text{kW}$ are injected in O-mode from the low-field side, using a gyrotron with $\omega_0 = 2\pi \times 105\,\text{GHz}$ and fixed launcher settings. During each pulse, the electromagnetic scattering/emission signal in O-mode at frequencies close to ω_0 is recorded using the fast acquisition CTS system, which also has a fixed receiver located on the low-field side, and power spectra are calculated by performing fast Fourier transforms on 655 ns windows of the recorded signal.

The mean power spectrum obtained during a pulse at $B_t = -2.531\,\text{T}$, i.e. near the middle of the B_t-scan, is seen in Figure 2. In the figure, we clearly identify the gyrotron peak located inside a notch filter slightly below 105 GHz ($\omega_0 = 2\pi \times 104.93\,\text{GHz}$), as well as the down-shifted high-frequency peaks excited by PDIs around 104 GHz. In addition to the down-shifted peaks, peaks up-shifted by a similar amount are seen; these peaks originate from scattering of the gyrotron wave by the low-frequency plasma waves excited by PDIs, and satisfy the selection rules

$$\omega_3 = \omega_0 + \omega_1, \quad \mathbf{k}_3 = \mathbf{k}_0 + \mathbf{k}_1. \tag{2}$$

Apart from the peaks, we note the frequency-dependent ECE background which indicates the expected decrease of radiation temperature with increasing frequency (decreasing major radius of the ECR location). For reference, we plot $(\omega_0 \pm \omega_1)/(2\pi)$ and $(\omega_0 \pm \omega_{LH})/(2\pi)$ at the point where the PDI model of [8] leads to maximal amplification. Evidently, the simple estimate of the LH frequency underestimates the frequency shift of the PDI peaks relative to that of the gyrotron somewhat, while the warm plasma theory of [8] overestimates it somewhat; both these findings are robust throughout the B_t-scan, and can be attributed to finite plasma temperature and use of the dipole approximation, $\mathbf{k}_0 \approx \mathbf{0}$, respectively [8]. The mean spectrum in Figure 2 is representative of those close to the onset of PDIs; for pulses later in the B_t-scan more peaks occur, but in this paper we shall only be concerned with whether or not PDI peaks occur at all. Making only this distinction, the B_t-scan separates into three regions in terms of the types of spectra observed: for $B_t < -2.557\,\text{T}$ no spectra with PDI peaks are observed, for $B_t \in [-2.557\,\text{T}, -2.535\,\text{T}]$ spectra with and without PDI peaks are observed intermittently, and for $B_t > -2.535\,\text{T}$ all observed spectra contain PDI peaks; these observations are used for the comparison with the theory of [8] in Section 3.

3 Modelling

As mentioned in Section 2, the most important effect of the B_t-scan for PDI-purposes is to displace the ECR from inside the main plasma to the high-field side scrape-off layer, which should lead to a significant reduction of its optical thickness. This is indeed confirmed by calculations, based on [8], of the fraction of reflected power coupled to the plasma in X-mode, the fraction of power transmitted through the ECR from the gyrotron to the UHR (obtained by computing the optical thickness along the central beam rays, using a relativistic model described in [16, 17]), and the total fraction of gyrotron power reaching the UHR (obtained by multiplying the two previous fractions), shown in Figure 3: the X-mode fraction remains constant close to 0.1 until the end of the B_t-scan where it drops somewhat, while the fraction of power transmitted through the ECR increases from 0.1 at the beginning of the scan to 1 towards the end of the scan; the total fraction of gyrotron power reaching the UHR essentially follows the development of the transmitted fraction. We note a rather large uncertainty in the transmitted fraction due to uncertainties in the reconstructed magnetic equilibria from CLISTE [18], which determine the location of the last closed flux surface, as well as the electron temperature and density profiles from IDA [19], which vary rather quickly near the last closed flux surface and the values of which are primarily determined from measurements on the low-field side. We also note that Figure 3 only includes data points for pulses in which the central ray of the reflected X-mode beam reaches the UHR, which is necessary in order to apply the theory of [8].

To test the theory of [8] we compute P_0^{th} (the gyrotron power necessary to excite PDIs) and plot the values obtained in Figure 4. P_0^{th} shows a development which is approximately the inverse of that of the total fraction from Figure 3 during the B_t-scan; this is as expected, since the conditions at the UHR are roughly similar throughout the B_t-scan (it remains well inside the main plasma) and the main variation is the amount of power transmitted from the

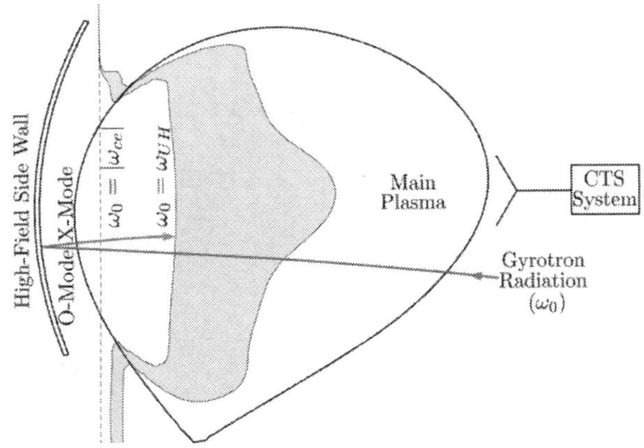

Figure 1. Experimental setup in ASDEX Upgrade discharge #34575. Gyrotron radiation, with $\omega_0 = 2\pi \times 105\,\text{GHz}$, is injected in O-mode from the low-field side, partially transmitted by the (optically thin) ECR ($\omega_0 = |\omega_{ce}|$) and reflected by the high-field side wall. After the reflection, part of the radiation re-enters the plasma in X-mode; if the ECR is not optically thick for the X-mode radiation, a significant amount of power may reach the UHR ($\omega_0 = \omega_{UH}$) where PDIs can be triggered due to field enhancement effects. The electromagnetic emission/scattering in O-mode at frequencies close to ω_0 is measured using the fast acquisition CTS system. No propagating X-mode radiation exists in the shaded region.

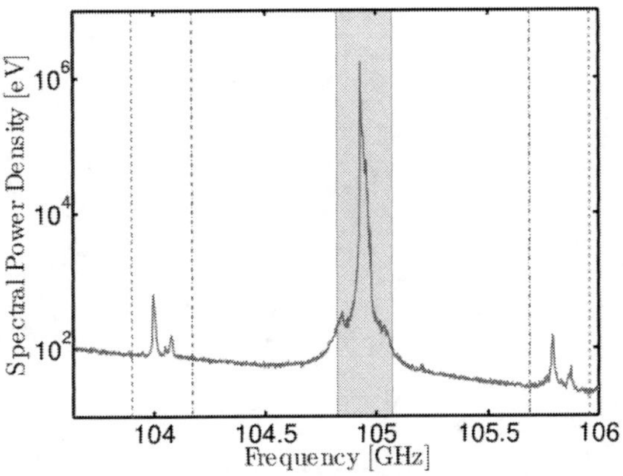

Figure 2. Mean spectrum recorded using the fast acquisition CTS system during a pulse at $B_t = -2.531$ T in ASDEX Upgrade discharge #34575. The gyrotron peak is clearly visible slightly below 105 GHz ($\omega_0/(2\pi) = 104.93$ GHz), and is located inside a notch filter stop band, marked by the shaded area, where the calibration giving the spectral power density in eV is rather uncertain. The peaks due to the PDI-generated plasma waves are shifted by 800 – 900 MHz, roughly $\omega_{LH}/(2\pi)$, relative to the gyrotron peak; both down-shifted and up-shifted peaks are visible in the spectrum. For reference, $(\omega_0 \pm \omega_1)/(2\pi)$ and $(\omega_0 \pm \omega_{LH})/(2\pi)$ from the theory of [8] are plotted as dashed and dashed-dotted lines, respectively; the peaks are located between the lines, as expected [8].

gyrotron to the UHR. In Figure 4 the pulses for which PDI peaks (and no PDI peaks) are observed are marked, along with $\max(P_0) = 308\,\text{kW}$. From [8], we expect that PDI peaks will be observed for $\max(P_0) > P_0^{\text{th}}$, and this condition is indeed seen to provide a good indicator for whether or not PDI peaks will be observed in Figure 4: for the 88 pulses with $\max(P_0) < P_0^{\text{th}}$, PDI peaks are only observed in 3 cases, while for the 61 cases with $\max(P_0) > P_0^{\text{th}}$, PDI peaks are only not observed in 5 cases. The theory of [8] thus seems to provide a quite reliable indication of whether or not PDI peaks will be observed for given gyrotron settings and plasma conditions, in spite of the large uncertainty on the fraction of power transmitted through the ECR.

4 Conclusions

We have investigated PDIs occurring for X-mode radiation near the UHR at ASDEX Upgrade. The PDIs were excited by injection of pulses of 105 GHz gyrotron radiation in O-mode from the low-field side, which was transmitted by the optically thin ECR, reflected by the high-field side wall, and partially converted to X-mode. Part of the reflected X-mode power was transmitted through the ECR, whose optical thickness could be varied from a significant

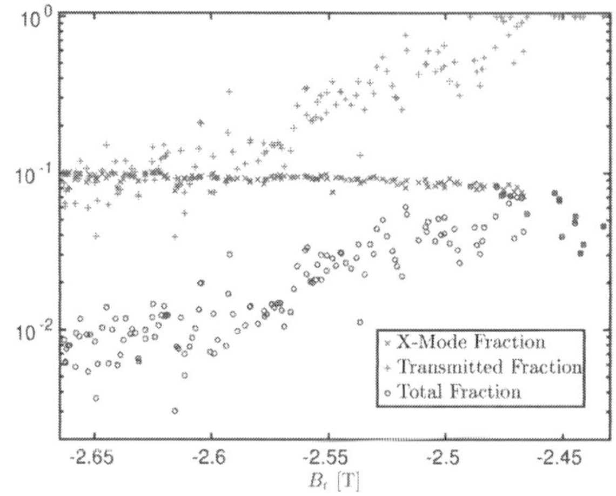

Figure 3. The fraction of reflected power coupled to the plasma in X-mode (×), the fraction of power transmitted by the ECR from the gyrotron to the UHR (+), and the total fraction of gyrotron power reaching the UHR (∘), during the B_t-scan in ASDEX Upgrade discharge #34575, modelled using the theory of [8]. The X-mode fraction remains relatively constant with values around 0.1 throughout most of the scan. The transmitted fraction increases from approximately 0.1 to 1 during the B_t-scan; the relative change in the total fraction is similar to that of the transmitted fraction.

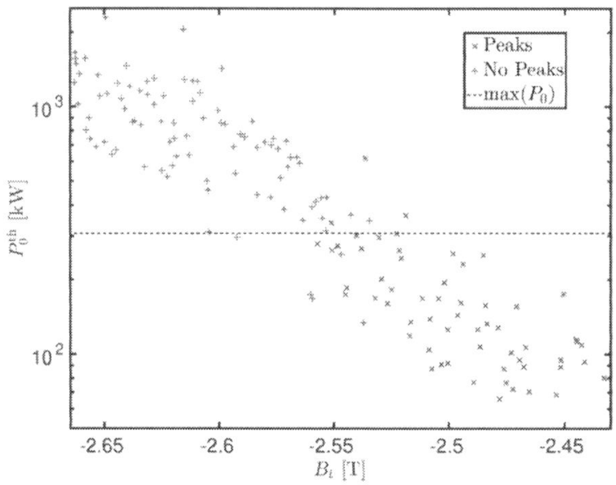

Figure 4. The development of P_0^{th} during the B_t-scan in ASDEX Upgrade discharge #34575, calculated using the theory of [8]. The pulses for which peaks are observed in the PDI frequency range (cf. Figure 2) are marked by ×, while the pulses for which no such peaks are observed are marked by +; $\max(P_0) = 308$ kW is marked by the dashed line. Evidently, the occurrence of PDI peaks is closely correlated with fulfilment of the condition $P_0^{th} < \max(P_0)$ provided by the theory of [8].

value to essentially zero by scanning B_t, and reached the UHR where it could excite PDIs due to field enhancement effects. The excitation of PDIs was observed by the occurrence of peaks shifted by roughly the LH frequency relative to the gyrotron peak in the spectra recorded using the fast acquisition CTS system. The occurrence of the PDI peaks was found to be well predicted by the theory of [8], which accounts for the fraction of power converted to X-mode at the wall reflection, the optical thickness of the ECR, field enhancement near the UHR, and models the PDI as decay of the X-mode waves into warm LH waves and electron Bernstein waves. Our findings indicate that the theory of [8] is capable of describing the onset of PDIs near the UHR in a magnetically confined fusion plasma, although it is only capable of providing limits of the frequency regions in which the PDI peaks occur due to the limitations of the dipole approximation on which it is based. It would thus seem reasonable to apply the theory of [8] to start-up scenarios for ITER in order to assess whether or not PDIs near UHR can be expected there; for the overdense plasmas expected in connection with O-X-B heating, the theory of [20] is presumably more appropriate.

Acknowledgements

This work was supported by a research grant (15483) from VILLUM FONDEN. This work has been carried out within the framework of the EUROfusion Consortium and has received funding from the Euratom research and training programme 2014–2018 under grant agreement No 633053. The views and opinions expressed herein do not necessarily reflect those of the European Commission.

References

[1] M. Porkolab, *Parametric Decay Instabilities in ECR Heated Plasmas*, in *Proc. 2nd Workshop Hot Electron Ring Physics (San Diego)*, edited by N.A. Uckan (National Technical Information Service, Alexandria, 1982), Vol. 1, p. 237

[2] F.S. McDermott, G. Bekefi, K.E. Hackett, J.S. Levine, M. Porkolab, Phys. Fluids **25**, 1488 (1982)

[3] D.G. Bulyginsky, V.K. Gusev, V.V. Djachenko, M.A. Irzak, M.Yu. Kantor, M.M. Larionov, L.S. Levin, G.A. Serebreny, N.V. Shustova, *ECR-Heating of Plasma in FT-1 Tokamak and Its Influence on the Ion Component*, in *Proc. 11th European Conf. Controlled Fusion and Plasma Physics (Aachen)* (European Physical Society, Mulhouse, 1984), Vol. 1, p. 457

[4] Z.A. Pietrzyk, A. Pochelon, R. Behn, A. Bondeson, M. Dutch, T.P. Goodman, M.Q. Tran, D.R. Whaley, Nucl. Fusion **33**, 197 (1993)

[5] R. Wilhelm, V. Erckmann, G. Janzen, W. Kasparek, G. Müller, E. Räuchle, P.G. Schüller, K. Schwörer, M. Thumm, the W VII-A team, Plasma Phys. Control. Fusion **26**, 1433 (1984)

[6] S.K. Nielsen, P.K. Michelsen, S.K. Hansen, S.B. Korsholm, F. Leipold, J. Rasmussen, M. Salewski, M. Schubert, M. Stejner, J. Stober et al., Phys. Scr. **92**, 024001 (2017)

[7] S.K. Hansen, S.K. Nielsen, M. Salewski, M. Stejner, J. Stober, the ASDEX Upgrade team, EPJ Web Conf. **149**, 03020 (2017)

[8] S.K. Hansen, S.K. Nielsen, M. Salewski, M. Stejner, J. Stober, the ASDEX Upgrade team, Plasma Phys. Control. Fusion **59**, 105006 (2017)

[9] S. Kubo, M. Nishiura, K. Tanaka, D. Moseev, S. Ogasawara, T. Shimozuma, Y. Yoshimura, H. Igami, H. Takahashi, T.I. Tsujimura et al., J. Instrum. **11**, C06005 (2016)

[10] H.P. Laqua, V. Erckmann, H.J. Hartfuß, H. Laqua, W7-AS Team, ECRH Group, Phys. Rev. Lett. **78**, 3467 (1997)

[11] H.P. Laqua, Plasma Phys. Control. Fusion **49**, R1 (2007)

[12] V. Shevchenko, G. Cunningham, A. Gurchenko, E. Gusakov, B. Lloyd, M. O'Brien, A. Saveliev, A. Surkov, F. Volpe, M. Walsh, Fusion Sci. Technol. **52**, 202 (2007)

[13] A. Köhn, G. Birkenmeier, A. Chusov, P. Diez, A. Feuer, U. Höfel, H. Höhnle, E. Holzhauer, W. Kasparek, S. Merli et al., Plasma Phys. Control. Fusion **55**, 014010 (2013)

[14] D. Wagner, G. Grünwald, F. Leuterer, A. Manini, F. Monaco, M. Münich, H. Schütz, J. Stober, H. Zohm, T. Franke et al., Nucl. Fusion **48**, 054006 (2008)

[15] M. Stejner, S. Nielsen, A.S. Jacobsen, S.B. Korsholm, F. Leipold, F. Meo, P.K. Michelsen, D. Moseev, J. Rasmussen, M. Salewski et al., Rev. Sci. Instrum. **85**, 093504 (2014)

[16] H. Bindslev, Plasma Phys. Control. Fusion **33**, 1775 (1991)

[17] H. Bindslev, Plasma Phys. Control. Fusion **34**, 1601 (1992)

[18] P.J. McCarthy, P. Martin, W. Schneider, Tech. Rep. IPP 5/85, Max-Planck-Institut für Plasmaphysik, Garching b. München (1999)

[19] R. Fischer, C.J. Fuchs, B. Kurzan, W. Suttrop, E. Wolfrum, the ASDEX Upgrade team, Fusion Sci. Technol. **58**, 675 (2010)

[20] E.Z. Gusakov, A.V. Surkov, Plasma Phys. Control. Fusion **49**, 631 (2007)

Tests of advanced RF off-axis current drive techniques on DIII-D

R.I. Pinsker[1,*], *X.* Chen[1], *J.M.* Lohr[1], *C.P.* Moeller[1], *M.* Porkolab[2], *M.W.* Brookman[1], *C.C.* Petty[1], *S.J.* Wukitch[2], *G.M.* Wallace[2] and *R.J.* Buttery[1]

[1]General Atomics, San Diego, California, USA
[2]Massachusetts Institute of Technology, Cambridge, Massachusetts, USA

Abstract. The establishment of reactor-relevant radiofrequency heating and current drive techniques is a focus of work on DIII-D in the next five-year period. This paper gives an overview of the planned experimental work in the areas of (1) nearly vertically launched ECCD, (2) 'helicon' (whistlers or fast waves in the lower hybrid range of frequencies) current drive, and (3) high-field-side-launch (HFS) lower hybrid (slow wave) current drive. Each of these techniques addresses the need for efficient off-axis current drive for a steady-state tokamak reactor to supplement the bootstrap current and to provide current profile control, and each will be experimentally assessed at a coupled power level of ~1 MW on DIII-D in the next few years.

1 Introduction

A steady-state tokamak fusion reactor must have most of the plasma current self-driven by the bootstrap effect to be economically practical. Reactor design studies since the 1990s have consistently shown that along with some 'seed current' near the magnetic axis, most of the remaining non-inductively-driven current is needed in the mid-radius region. For example, the ARIES-RS study proposed [1] a combination of fast waves in the lower hybrid range of frequencies (called 'High Frequency Fast Waves' in that work) and lower hybrid slow waves to drive the non-inductive, non-bootstrap current in the outer half of the minor radius. The efficiency of the current drive, both in terms of the driven current per Watt of power absorbed in the plasma core and of the efficiency of conversion of electrical power to current drive power and of coupling the power to the plasma (rf in the cases discussed here) is critical in determining the economics of the reactor. For this reason, the DIII-D Advanced Tokamak program is exploring multiple techniques in an effort to establish efficient means of off-axis non-inductive current drive.

In this paper, three such techniques that will be explored on DIII-D in the next few years are described: (1) nearly-vertically-launched electron cyclotron current drive at 110 and/or 117.5 GHz, (2) 'helicon' (also known as 'fast waves in the lower hybrid range of frequencies', 'high frequency fast waves', or 'whistlers') current drive at 0.5 GHz, and (3) high-field-side-launched slow lower hybrid wave current drive (HFS LHCD) at 4.6 GHz. The three cases are illustrated in Fig. 1. Each technique will be tested on DIII-D as a proof-of-principle with ~1 MW of coupled power. Such a power level is predicted to drive sufficient non-inductive current to quantitatively compare with theoretical models and thus qualify these techniques for next-step systems. In each case, such follow-on experiments would be possible in the future on DIII-D at power levels high enough that those systems would become viable tools for the further development of the Advanced Tokamak, along with the existing off-axis neutral beam systems on DIII-D.

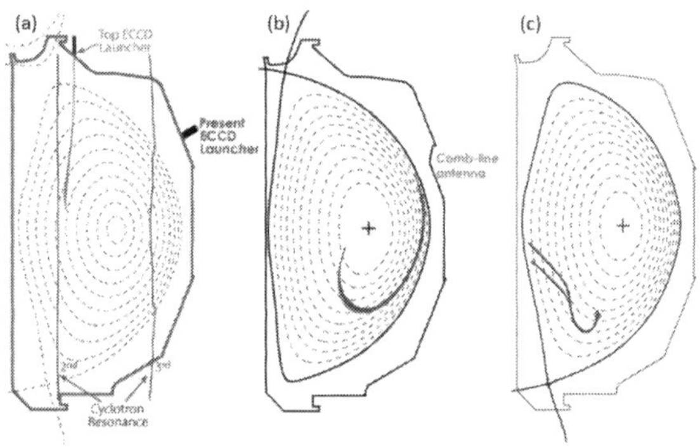

Fig. 1. Ray tracing illustrating the three current-drive techniques to be tested on DIII-D. (a) Nearly-vertically-launched ECCD, (b) Helicon current drive, and (c) high-field-side launch slow lower hybrid wave current drive.

2 Nearly-vertically-launched ECCD

Conventional low-field-side-launch electron cyclotron current drive (ECCD) such as is used on DIII-D (110 GHz, second harmonic X-mode) is an established technology with many applications in the DIII-D program [2]. Electron cyclotron current drive near the magnetic axis, arising from the Fisch-Boozer mechanism, on DIII-D has a predicted and observed

* Corresponding author: pinsker@fusion.gat.com

dimensionless current drive efficiency [3] of about 0.4, where the dimensionless current drive efficiency has been normalized for the power/particle and collisionality effects and is defined as:

$$\zeta_{EC} \equiv \frac{e^3}{\varepsilon_0^2}\frac{I_{EC} R\, n_e}{P_{EC} k T_e} = 3.27 \frac{I_{EC}(\mathrm{A}) R(\mathrm{m}) n_e(10^{19}\,\mathrm{m}^{-3})}{P_{EC}(\mathrm{W}) T_e(\mathrm{keV})}. \quad (1)$$

Reference [3] showed that the dimensionless current drive efficiency decreased as the deposition is moved to larger minor radius with the absorption location maintained to be in the midplane, due to the competing Ohkawa mechanism with opposite sign, and that this decrease is in agreement with theory. This tendency poses a problem for application of ECCD for off-axis current drive. Several authors have found that the off-axis ECCD efficiency could be significantly enhanced by changing the launch position and aiming so that the ray travels nearly vertically through the plasma at a major radius somewhat larger than that of the cold-plasma electron cyclotron absorption layer [4,5]. Use of a relatively large toroidal component to the k-vector, enabled by this geometry, allows damping on energetic electrons via the Doppler downshift and therefore more efficient current drive. Studies for DIII-D have shown a predicted improvement in dimensionless efficiency of up to a factor of two relative to outside launch for current drive at normalized minor radius of ρ=0.5-0.6, nearly recovering the efficiency of on-axis ECCD. To verify the predicted enhancement of current drive efficiency, a proof-of-principle experiment with a fixed launcher at the top of the DIII-D vacuum vessel powered by one or two existing gyrotrons will be carried out in the near future at the ~1 MW coupled power level. If the predicted improvement in current drive efficiency is experimentally demonstrated, a more flexible steerable launcher would be installed a few years later. Much more detail on this topic is given in the paper in these proceedings by X. Chen, *et al.*

3 Helicon current drive

Since the late 1970s, non-inductive current drive with a toroidally directive wave spectrum has been extensively studied in the lower hybrid range of frequencies (LHRF) $f_{ci} << f << f_{ce}$. For typical tokamak core plasma parameters, two wave branches can propagate in this frequency range: the quasi-electrostatic slow branch, usually simply referred to as the 'lower hybrid wave', and the quasi-electromagnetic 'fast wave', also known as the 'helicon' or 'whistler' wave. To date, the most efficient non-inductive current drive technique that has been successfully demonstrated is the use of the slow wave in the LHRF [6]. It has been recognized for a long time, though, that the Fisch mechanism of high current drive efficiency (scaling as $n_{\parallel}^{-2}$, with $n_{\parallel}$ the wave index of refraction along the static magnetic field) at electron temperatures of a few keV in present devices requires accessibility of waves with high parallel phase velocities (as high as ~0.7c corresponding to a low $n_{\parallel}\approx 1.5$) to the plasma core, either due to low density [7] or high magnetic fields [8]. A further requirement is that a fraction of the wave spectrum shift to larger $n_{\parallel}$ so that it could be absorbed by electron Landau damping (the condition $n_{\parallel} > 7/\sqrt{T_e[\mathrm{keV}]}$, or $n_{\parallel} > 2$ in today's plasmas). Upshift of a portion of the wave spectrum is necessary to fill the "spectral gap" in $n_{\parallel}$ and maintain an energetic electron tail pulled out from the bulk electron distribution, which is a process that follows from multi-pass toroidal ray propagation or possibly other mechanisms). We recognize, however, that this is likely not relevant for reactor-grade plasmas where the slow waves in the LHRF (henceforth simply referred to as "LH waves" here) would damp at the very edge of a large, hot, reactor-grade plasma [9] unless the waves could be launched on the high field side of the tokamak where, at typical aspect ratio, the magnetic fields are typically 70% higher than on the low field side (see Section 4 below).

For this reason, work in the 1980s began on attempts to use the other propagating wave in the LHRF for current drive, i.e. the fast LH wave [10] also known as the 'whistler' wave, or, more recently, as the 'helicon' [11]. Due to the quasi-electromagnetic polarization of this mode, with weaker rf electric field in the direction of the static magnetic field, the electron Landau damping is weaker, and the wave can reach the core of a reactor plasma with high electron beta before being strongly damped. However, in these early experiments [10] the extremely weak single-pass damping in low electron beta plasmas proved inadequate to lead to measurable current drive at high density. Recently, renewed interest in this idea led to new modeling studies [11,12]; it was found that DIII-D can now produce H-mode target plasmas with high electron beta for predicted strong single-pass damping of the helicon, and that an antenna concept proposed in 1992 [13] and developed in the mid-1990s [14] was well-suited to a proposed experimental program on DIII-D. The 'comb-line' traveling wave antenna can couple > 75% of the applied power to a highly directive spectrum peaked at $n_{\parallel}$(launch)~3, depending on the density profile in the far SOL. Since the details of the far SOL, in the neighborhood where the local electron density is a few times 10^{18} m^{-3}, are not well measured in DIII-D, the first step in an experimental evaluation of helicon current drive was to construct and install a low-power comb-line antenna and determine the wave coupling efficiency (set by the *far SOL* plasma conditions) in a regime in which the *core* parameters are consistent with strong first-pass absorption.

The preliminary experiment was performed in 2015-2016 with the result [15] that sufficiently strong coupling to yield 75% efficiency for a toroidally longer comb-line antenna were observed in a number of regimes, and most importantly, in a regime with full first-pass absorption predicted at mid-radius. Time histories of such a discharge are shown in Fig. 2. The lowest box shows the calculated first-pass absorption for 0.476 GHz helicon waves launched at $n_{\parallel}$=3 from a poloidal location 45 degrees above the outboard midplane (open red circles at discrete times) as a function of time; the Ohmic and L-mode portions before 800 msec in the discharge have negligible absorption (as in the 1980s experiments), while after the L/H transition signaled by the sudden drop in the D-alpha light and the consequent rise in electron beta the calculated first-pass absorption rises to 100% by 1000 msec. The quantity $\Delta\delta_a$ (defined in [15]) plotted in the uppermost box is proportional to the measured plasma loading per element in the low-power comb-line antenna; where this quantity is larger than the dashed red line, the coupling efficiency projected to a 30-element comb-line array exceeds 75%. This criterion is easily satisfied throughout the Ohmic and L-mode portions of the discharge, and is still satisfied between ELMs through most of the ELMy portion of the shot. Only for a brief period around 2300 msec, during which an instability momentarily reduces the far SOL density (correlated with visible changes in ELM character), does the antenna loading drop below the required level; for the quasi-stationary portion of the discharge from 2500-3500 msec, the criterion is satisfied at all times at an antenna/separatrix gap of about 9 cm.

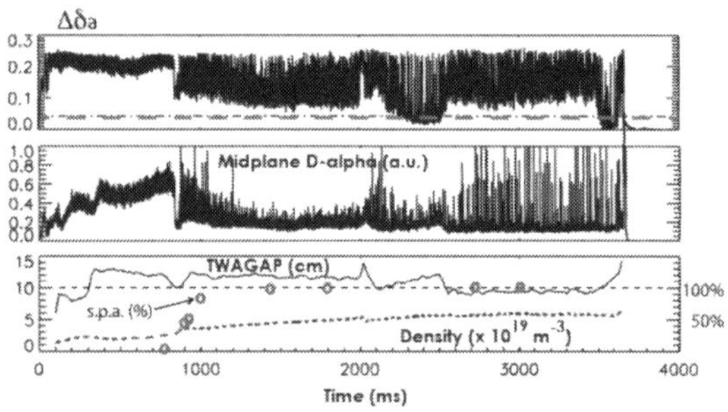

Fig. 2. DIII-D discharge 165908, showing a quantity proportional to the comb-line antenna plasma loading per element in the top box, the D-alpha light at the outboard midplane in the center box, and the distance between the separatrix and the antenna surface in cm ("TWAGAP" - solid line), the calculated single-pass absorption (s.p.a.) for 476 MHz helicons (red open circles), and the line-averaged electron density (dashed line) in the lower box.

A possible issue for excitation of the helicon waves is inadvertent direct excitation of the slow lower hybrid wave. Despite the difference in polarization between the two modes at the edge of the plasma, the fact that the cut-off density for the slow wave is so much lower than that of the helicon wave would suggest the possibility of parasitic slow wave excitation. Specifically, while the 0.476 GHz helicon wave begins to propagate at a density of about 1.6×10^{18} m^{-3} for DIII-D-like parameters, the slow wave cutoff is only 2.8×10^{15} m^{-3}, a density so low that it might be present immediately adjacent to the antenna surface during a discharge. Hence if a non-negligible component of rf electric field along the static magnetic field is present in the antenna near-field region, produced for example by a misalignment between the Faraday screen elements and the local static magnetic field, the amplitude of the fast wave fields at the cutoff might not be so much larger than the unintended slow wave fields, which would not have been reduced by evanescence. In addition to the measurements reported in [15], which suggested that for alignment between the magnetic field and the Faraday screen elements of better than about 10 deg the slow wave excitation appeared to be small, another piece of evidence supporting this conclusion is shown in Fig. 3. Here, direct measurements of the antenna loading deduced from fits to rf current probes in each of the twelve modules of the comb-line are shown from three times in a discharge similar to the one in Fig. 2, except the antenna/separatrix gap TWAGAP is 17 cm at around 3000 msec, compared with the 9-10 cm in Fig. 2. The slope on the semi-logarithmic plot of the probe amplitudes is inversely proportional to the antenna loading. The first measurement is a fiducial taken before the discharge starts to determine the loading due to Ohmic losses alone (no radiation), in blue. The data in green squares is in the ELMy H-mode portion of the discharge at large antenna/plasma gap, taken between two ELMs, just prior to an ELM. The slope is the same as without plasma, within the experimental error. This suggests that despite the fact that the density in the antenna near-fields probably exceeds the slow wave cutoff, the polarization must be sufficiently well defined that slow wave excitation is negligible. The ELM that occurs 1.4 msec later momentarily raises the density in the antenna near-field region above the FW cutoff and the coupling is strong, typical of what is observed in L-mode at a value of TWAGAP of ~10-12 cm. (Note that $\Delta\delta_A$=0.967-0.744=0.223, comparable to the value early in 165908 before the L/H transition, shown in Fig. 2.)

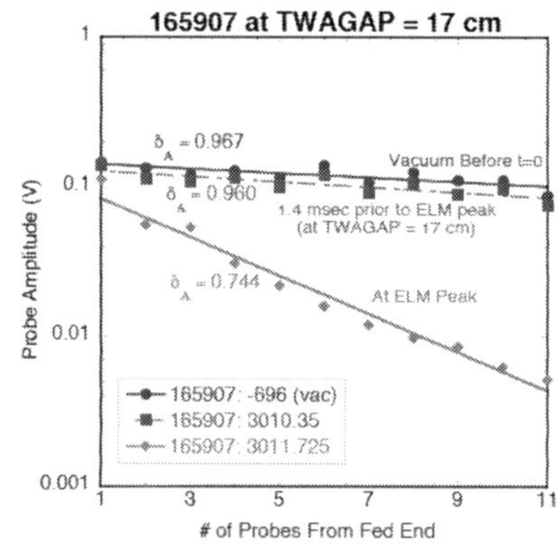

Fig. 3. Data from RF current probes built into the low-power comb-line modules from DIII-D discharge 165907, showing that the no-plasma loading (Ohmic dissipation only) is the same as in an ELMy H-mode between ELMs with a large antenna/plasma distance.

Two types of nonlinear effects on antenna coupling are (i) anomalously large loading at low power levels resulting from rectified sheath dissipation in the antenna near-fields and (ii) reduction of loading at high power due to ponderomotive density depletion in the antenna near-field. An effort was made to search for evidence of sheath dissipation in the low-power experiment by scanning the power from the maximum available power of ~0.2 kW down to zero and back up in a V-shaped notch 0.1 sec wide once during each plasma shot. No power dependence of the antenna loading was ever observed in these power notches. If sheath dissipation were a significant portion of the total loading at low power, the fractional contribution of the sheath dissipation would change in the region where the rf sheath voltage and the local electron temperature in volts are comparable [16], a region which should be accessible even at these low power levels. Hence we conclude that the observed antenna loading in the low-power experiments is predominantly the linear wave loading. The ponderomotive effects are expected to play a role only above power levels on the order of 0.1 MW, so any such effect is a subject that can be addressed only in high-power (MW level) experiments. Since ponderomotive effects would act to reduce the loading below the linear levels, we regard the loading observed in the low-power experiments as an upper limit.

Such MW-level experiments are planned for the near future at DIII-D. At present, a high-power version of the comb-line antenna is under construction, with the principal changes from the low-power prototype being: (1) changes in the way rf power is introduced into the fed element at the 'upstream' end of the comb-line to allow excitation at up to 1 MW, (2) increasing the number of modules from 12 to 30 (this already being taken into account when the level of the dashed line indicating 75% efficiency was set in Fig. 2), and (3) reduction of the Ohmic losses (skin effect) in the modules by replacing bolted connections in the low-power version with brazed ones in the high-power version (also taken into account in the 75% efficiency line in Fig. 2). The several hundreds of Watts of rf power in the low-power experiments was transmitted from the feedthrough on the tokamak to the powered end of the comb-line with small-diameter sealed dielectric-insulated coaxial semi-rigid cables; these will be replaced by striplines without any dielectric materials in the high-power antenna. The stripline center conductor will be supported by quarter-wave-long shorted stubs in the vacuum vessel.

A research and development program to support the design of the high-power antenna components is ongoing at DIII-D. A test stand has been constructed with a 10 kW solid-state UHF band transmitter, a vacuum chamber and a 0.1 T solenoid in which the vacuum vessel is placed (Fig. 4). Portions of a single module of the comb-line have been tested to investigate the effect of multipactor discharges and antenna conditioning procedures and to establish upper limits on the rf voltage that can be sustained within the module without high-voltage breakdown. Such testing is performed by enclosing the module under test within a conducting box to eliminate radiation and thereby maximize the resonant 'Q' factor, which enables reaching the rf electric fields expected in 1 MW operation of the full comb-line with only 10 kW of power. In another test performed in this facility, a section of stripline was configured as a half-wavelength-long resonator and tested with several kW of rf power, for characterization of multipactor and conditioning procedures. Preliminary tests of coatings intended to reduce the secondary electron emission coefficient and thus reduce or eliminate multipactor have been performed, without finding a clear advantage over bare copper alloy surfaces to date.

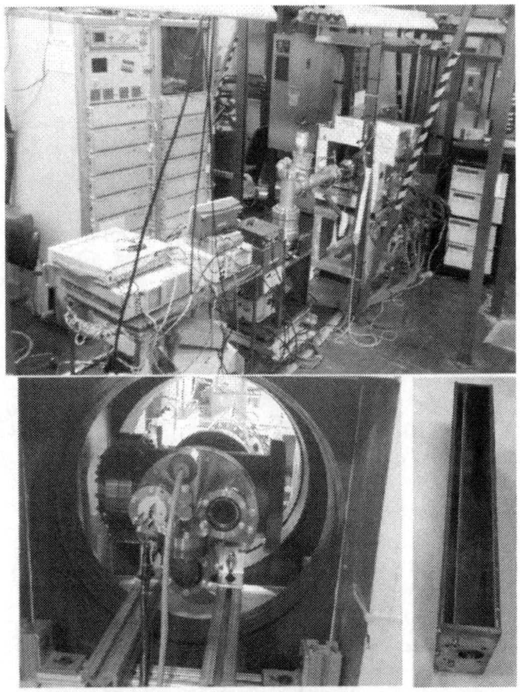

Fig. 4. Test stand for high-power helicon antenna components. Top: overall view of facility, with 10 kW rf source at left rear, test chamber mounted in solenoid at right-center, four magnet power supplies at right. Bottom left: view of end of tee of test chamber, with mounting flange for device under test at right, IR-transmissive window mounted in flange at left. Bottom right: half-wavelength section of test stripline with one sideplate removed. Power is coupled to the resonator with a single-turn loop inserted in circular hole in endplate.

The present schedule calls for installation of the 30-module high-power comb-line into the DIII-D vacuum vessel during the present Long Torus Opening, ending in spring 2019. In parallel, a single 1.2 MW klystron obtained from the Stanford Linear Accelerator Center (SLAC) along with its power supply will be installed and tested at DIII-D. Rather than installing extensive (and expensive) runs of rectangular waveguide, existing 230 mm-diameter 50 Ohm coaxial transmission line (formerly used for ICRF experiments at up to 120 MHz) will be used to convey power from the klystron to the tokamak, a distance on the order of 100 m. Measurements of a 55-m long section of this coax in reflection from a short at one end showed that the expected loss in the entire run of 230-mm coax at 476 MHz will be approximately 0.25 dB, or 5.6% in power. A waveguide-to-coax transition is required to utilize this existing transmission line. The high-power helicon current drive experiments are planned to commence in autumn 2019.

The goals of these experiments are to verify helicon current drive efficiency that is predicted by ray-tracing [12] and full-wave [17] models, (2) attempt to evaluate the importance of nonlinear effects such as parametric decay instability [18], and scattering of helicon waves from turbulent density fluctuations ('stimulated mode conversion' [19]), in addition to the non-linear effects on coupling discussed earlier. If possible deleterious nonlinear effects are ignored, the expected dimensionless current drive efficiency defined in Eq. 1 for the helicon in a discharge of the kind shown in Fig. 2 is $\zeta_{Helicon} \approx 0.55$ for current drive at $\rho \sim 0.5$-0.6, almost 50% more efficient than conventional ECCD at the magnetic axis in DIII-D, and hence much more efficient than conventional ECCD at mid-radius. In addition, the technological aspects of the comb-line antenna are non-trivial and themselves are an important motivation for the experiment: the 30-module comb-line will be the first MW-level traveling wave antenna tested in a magnetic confinement device, and as such is of interest for application in other frequency ranges [20].

4 High-field-side-launched LHCD

In recent years, some reactor studies have identified strong advantages to the placement of rf wave launchers on the high-field side (HFS, inboard midplane region) of the torus over the usual location in the outboard midplane region [21]. For a moderate aspect ratio torus, significant advantages of HFS launch over LFS launch for slow lower hybrid waves are particularly clear from a wave physics point of view. The highest magnetic field available in the torus is on the HFS edge, so that the minimum value of launched $|n_{||}|$ that can access the core is significantly lower than from the LFS. Hence a lower value of $|n_{||}|$ can be launched, which makes coupling easier for a given SOL density profile (reduced evanescence of the antenna near-fields) and the wave can reach a higher electron temperature before being strongly Landau damped. Also the lower value of $|n_{||}|$ yields a higher efficiency current drive efficiency as explained earlier. Another potential advantage of HFS launch is reduced local plasma turbulence and thus weaker plasma-materials interaction. This may prove important for launcher survivability in a steady-state reactor environment. The technical difficulty of introducing a launcher on the HFS into an already existing experiment is what has to date prevented an experimental assessment of the advantages of LHCD with a launcher so located. MIT participants in the DIII-D program are leading the effort to design, install and operate the world's first HFS LHCD experiment on the DIII-D tokamak [22]. In this project, a novel compact wave launcher has been designed for installation just below the midplane on the center-post of the DIII-D device, for launching a spectrum peaked at $n_{||} = 2.7$ at 4.6 GHz at power levels over 1 MW. The launcher system obtains the poloidal/vertical split between rows of waveguides with slotted waveguides, and the toroidal split is obtained with a version of the multijunction concept. These experiments will use a subset of the 4.6 GHz system that had been used on the C-mod tokamak prior to 2016; two groups of four 0.25 MW klystrons will be installed at DIII-D, for a total source power of 2 MW. This margin will allow for finite efficiency of the transmission lines between the klystrons and the torus, and for finite directivity of the launched spectrum. The computed directivity of the launcher is around 60%, depending somewhat on the assumed edge plasma profiles. In certain DIII-D AT regimes, the current drive efficiency predicted for this system with the combination of the GENRAY ray-tracing code and the CQL3D Fokker-Planck solver is relatively high. For example, Ref. [22] shows that for DIII-D shot 147634, an AT discharge with $\beta_N > 3.5$ and $q_{min} > 1.4$ (a broad current profile), the dimensionless efficiency predicted for the HFS LHCD system is $\zeta_{LH} \approx 1.8$, which results from 0.15 A/W driven at $\rho \sim 0.6$, where the electron temperature is $Te(\rho=0.6) = 2$ keV and the electron density is about 4.5×10^{19} m^{-3}. By comparison, the helicon system would be predicted to drive 0.07 A/W with a flat profile within $\rho \leq 0.6$ in this discharge.

More generally, the ranges of tokamak parameter space in which the helicon and the HFS LHCD systems are most useful are complementary, in that the AT regimes in which the LH system is most useful are those with lower density and higher toroidal field, while the helicon system drives current most off-axis in the cases with low toroidal field and higher density, while high electron temperature is maintained. An example of this complementarity is exhibited in Fig. 5, where the current driven by both systems was calculated at different times in a projection along the so-called "Super-H-mode path" to high density and high plasma pressure in DIII-D. At early times and lower density, the helicon drives current at the magnetic axis, and the LH drives current at $\rho \sim 0.6$, in each case at about 0.1 A/W, while at late times when the density is twice as high, the LH waves cannot access the core even from the HFS, while the helicon current drive is at $\rho \sim 0.4$, at an efficiency of about 0.03 A/W. (Note that for the purposes of this comparison, the

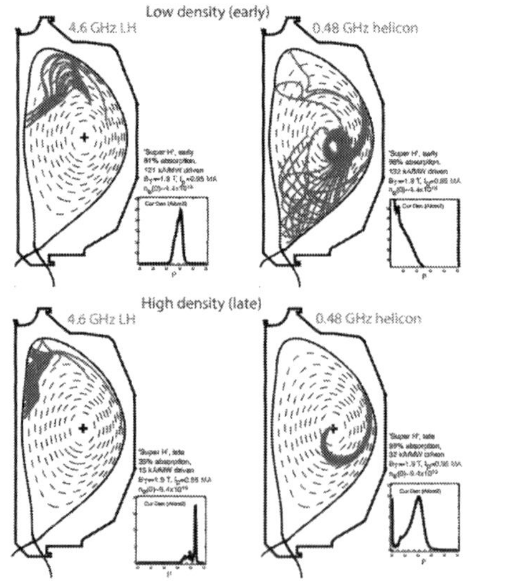

Fig. 5. Comparison of applicability of HFS LHCD and helicon current drive in the "Super-H-mode path" in DIII-D. Upper cases are at low density; lower cases are at twice the density.

location of the HFS LHCD launcher has been mirror-reflected with respect to the midplane from where it will actually be located.)

A point of practical significance for the implementation of the HFS LHCD on DIII-D is that in order to protect the feeding waveguides on the centerpost from the plasma, the thickness of the graphite tiles on the centerpost must be increased by 25 mm. Carrying out such a tile modification would mean that the wall/separatrix gap (the 'inside gap') for a geometrically identical plasma before and after the modification would differ by this amount. To investigate the effect of variation of the inside gap on confinement in AT discharges, an experiment was performed on DIII-D in which that gap was varied both dynamically and from shot to shot, in AT regimes of interest. Essentially no effect was observed, showing that installation of the modified tiles would not make comparisons of DIII-D discharges before and after the modification invalid.

Another practical point regarding the installation of the launcher was the fact that DIII-D discharges always start up as limited against the centerpost so that there were concerns that the presence of a metallic structure in the region where the plasma makes contact with the centerpost could lead to unacceptable metallic impurity influx. This concern was evaluated by fabrication and installation of a mockup (unpowered) launcher built of the same materials at the same poloidal location on the centerpost as the proposed launcher. Observations of the mockup during several weeks of DIII-D operations showed no significance metallic influx from the launcher under any condition studied. Recessing the metallic face of the launcher by ~3 mm behind the carbon protective tiles appears to have been sufficient to prevent any problem with high-Z impurity influx. Of course, such a test cannot provide information on rf-specific impurities that might be generated with a powered antenna.

The present schedule for the HFS LHCD project calls for completion of the in-vessel components (launcher, modified vessel protective tiles, feed waveguides) by autumn 2019, ready for installation in the following year, with the goal being achieving high-power operation of the system in late 2020.

5 Summary and conclusion

An exciting program to investigate the physics and technology of efficient off-axis rf current drive techniques is well underway at DIII-D. The three techniques under study are: nearly-vertically-launched electron cyclotron current drive, 'helicon' current drive (fast waves in the lower hybrid range of frequencies) at 0.5 GHz, and high-field-side-launched slow lower hybrid wave current drive at 4.6 GHz. Each technique will be tested at ~1 MW coupled power levels in the next few years. The most promising technique or techniques may be expanded in subsequent years from the proof-of-principle level to higher power levels to become useful tools for the Advanced Tokamak program on DIII-D.

This work supported by the United States Department of Energy under Award DE-FC02-04ER54698 and by US DoE Contract No. DE-FC02-01ER54648 under a Scientific Discovery through Advanced Computing Initiative. DIII-D data shown in this paper can be obtained in digital format by following the links at https://fusion.gat.com/global/D3D_DMP.

References

1. S.C. Jardin, *et al.*, Fusion Eng. Des. **38**, 27 (1997)
2. R. Prater and C.C. Petty, Fus. Sci. Tech. **48**, 1141 (2005)
3. C.C. Petty, *et al.*, Nucl. Fusion **43**, 700 (2003)
4. R. Prater, V. Chan, and A. Garofalo, Bull. Am. Phys. Soc. **57**, 188 (2012)
5. E. Poli, *et al.*, Nucl. Fusion **53**, 013011 (2013)
6. N.J. Fisch, Phys. Rev. Lett. **41**, 873 (1978)
7. J.E. Stevens, *et al.*, Nucl. Fusion **28**, 217 (1988).
8. M. Porkolab, *et al.*, Phys. Rev. Lett. **53**, 450 (1984)
9. V.S. Chan, *et al.*, Fus. Sci. Tech. **57**, 66 (2010)
10. R.I. Pinsker, in *Radio Frequency Power in Plasmas (Proc. 10th Topical Conf., Boston, MA, 1993)* (AIP, Melville, NY) p. 179
11. V.L. Vdovin, Plasma Phys. Rep. **39**, 95 (2013)
12. R. Prater, *et al.*, Nucl. Fusion **54**, 083024 (2014)
13. C.P. Moeller, *et al.*, in *Radio Frequency Power in Plasmas (Proc. 10th Top. Conf., Boston, MA, 1993)* (AIP, Melville, NY, 1994) p. 323
14. R.I. Pinsker, *et al.*, in *Fusion Technology 1996 (Proc. 19th Symp., Lisbon, Portugal, 1996)* (Elsevier, Amsterdam, 1997) p. 629
15. R.I. Pinsker, *et al.*, Nucl. Fusion **58**, 106007 (2018)
16. D.A. D'Ippolito and J.R. Myra, Phys. Plasmas **3**, 420 (1996)
17. C. Lau, *et al.*, Nucl. Fusion **58**, 066004 (2018)
18. M. Porkolab and R.I. Pinsker, EPJ Web of Conferences **157**, 03042 (2017)
19. P.L. Andrews, Phys. Rev. Lett. **54**, 2022 (1985)
20. R. Ragona and A. Messiaen, Nucl. Fusion **56**, 076009 (2016)
21. G.M. Wallace, *et al.* AIP Conf. Proceedings **1689**, 030017 (2015); doi: 10.1063/1.4936482
22. S.J. Wukitch, *et al.*, EPJ Web of Conferences **157**, 02012 (2017)

Beam tracing study for design and operation of two-pass electron cyclotron heating at ASDEX Upgrade

Martin Schubert[1,*], *Burkhard* Plaum[2], *Jörg* Stober[1], *Albrecht* Herrmann[1], *Walter* Kasparek[2], *Carsten* Lechte[2], *Fritz* Leuterer[1], *Francesco* Monaco[1], *Bernhard* Petzold[1], *Emanuele* Poli[1], *Stefan* Vorbrugg[1], and *Dietmar* Wagner[1]

[1] Max Planck Institute for Plasma Physics, Boltzmannstr.2, 85748 Garching, Germany
[2] Institut für Grenzflächenverfahrenstechnik und Plasmatechnologie, Universität Stuttgart, Pfaffenwaldring 31, 70569 Stuttgart, Germany

Abstract. The electron cyclotron resonance heating system at ASDEX Upgrade (AUG) is currently being extended to eight similar Gyrotrons in total. Each Gyrotron operates at 105 and 140 GHz and is designed for up to 1 MW millimetre wave output power. A substantial part of the AUG program will focus on experimental conditions, where the plasma density may be above the X-2 cut-off density at 140 GHz. In order to cope with the high density, the heating system will operate in the O-2 mode scheme with potentially incomplete absorption in the first pass. Reflecting gratings installed into the heat shield on AUG's inner column allow for a controlled second pass of the beam's unabsorbed fraction. Thermocouple measurements serve to control the beam position on the grating. The beam geometry is being finalized for the launchers #1-4. Beam propagation is simulated with the TORBEAM code and previous high density experiments are used as a database. The geometry is optimized using three criteria: central deposition, high absorption and robustness of the beam dump after the second pass. The experimental conditions, and the plasma electron density in particular, may vary such that the Gaussian beam parameters of the incoming beam on the grating deviate from the design values. It is proposed to model the effect of the grating with an equivalent ellipsoidal mirror. Laboratory measurements are shown, which support this model.

1 Introduction

Plasma electron densities up to $1 \cdot 10^{20}$ m^{-3} are regularly achieved in ASDEX Upgrade (AUG) experiments. Examples are high heat flux experiments [1] and fuelling with pellets [2]. In the latter case the pellet ablation temporary and locally increases the plasma electron density even further. With respect to the electron cyclotron resonance heating (ECRH) system, operating at 140 GHz, the density may exceed the X-2 cut-off density of $1.2 \cdot 10^{20}$ m^{-3}. A robust countermeasure is to change the launched polarisation at the given second harmonic to O-Mode (O-2), which doubles the cut-off density. Due to the finite absorption [3], there are several important constraints on the O-2 scheme: In a situation typical for an AUG experiment, the maximum optical thickness τ is around 1 (figure 1). This corresponds to approximately one third of the launched beam power leaving the plasma, the so-called 'shine-through', after the first pass. Moreover, the maximum τ is only obtained with a beam injection angle oblique to the magnetic field vector. Deviations between 10° and 30° from the perpendicular direction are necessary (figure 1). The optical thickness τ scales as T_e^2 [3]. The electron temperature T_e = 4 keV was used in figure 1. If T_e is significantly smaller, then O-2 mode heating might not be applicable due to lack of efficiency.

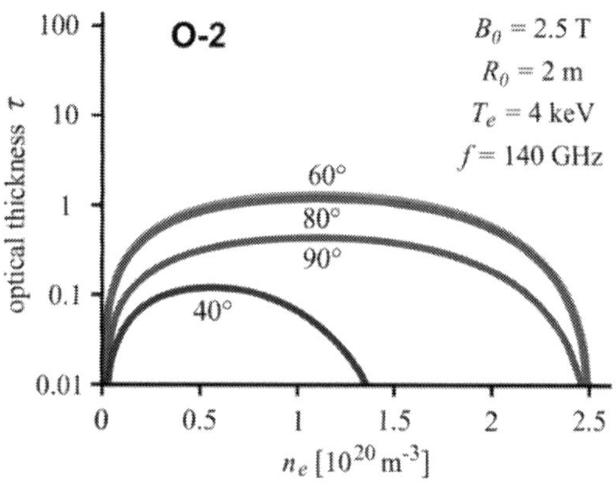

Fig. 1. Optical thickness τ of a model plasma for the second harmonic O-mode, function of the plasma electron density n_e according to [3]. This was evaluated for the launch angles 40°, 60°, 80° and 90° between **k** and **B**, and using parameters B_0, R_0, T_e, f of a typical ASDEX Upgrade experiment.

Even at nominal conditions at AUG, if one assumes a typical launched millimetre wave power of up to 900 kW with a compact beam diameter of a few centimetres and with the described plasma absorption properties, then the shine-through on the high field side corresponds to a power density on plasma facing components, which may

* Corresponding author: Martin. Schubert @ ipp. mpg. de

cause damage [4]. It was proposed [5, 6] to install a reflector at the position of the shine-through, which allows the unabsorbed fraction to make a controlled second pass through the plasma. In the design it should be taken into account that the second pass ends on a robust vessel component. The remaining unabsorbed beam power typically does not exceed 10% of the launched power and further beam propagation is usually not taken into account.

The two-pass heating scheme has exemplarily been demonstrated before [6]. With the persisting demand for ECRH in AUG experiments and with the on-going modernization of major ECRH components [7], it was decided to equip all launchers with the O-2 mode capability. This paper introduces the two-pass heating design for the launchers #1-4 in section 2. Beam propagation and absorption were simulated using the TORBEAM code [8].

The mirror gratings [5], which are used as reflectors, are designed and optimized for a given operating point. If the plasma parameters and the plasma electron density in particular are modified, then beam path and beam shape are changing. In section 3 it is proposed to model the reflection on the mirror grating in such case with an equivalent ellipsoidal mirror. This is tested in a laboratory mock-up of the reflection grating with a 140 GHz source and network analyser.

2 Electron cyclotron heating at AUG

With the modernization of AUG's first ECRH system, it is planned to have a final stage with 8 dual frequency (105/140 GHz) Gyrotrons [7] with up to 1000 kW millimetre wave output power per Gyrotron. During the last 2017 experimental campaign, 6 Gyrotrons were operational. Inside AUG, the millimetre wave power is launched at distinct positions (figure 2) and the beam direction can be controlled with movable mirrors [7]. Unabsorbed millimetre wave power is monitored with a set of detectors [9], including the AUG ECE system, cf. figure 2. All detectors are part of a fast interlock system, which can shut down Gyrotron operation in the case of a stray radiation excess.

The particularity of the O-2 mode shine-through is exemplarily visualized in figure 3 for AUG launcher #5. It is compared with the frequently used X-2 scheme in the same figure using launcher #8. The X-2 scheme has a nearly perfect absorption for a broad range of plasma parameters and launching geometries [3]. It is, however, only applicable for densities up to $1.2 \cdot 10^{20}$ m^{-3} at perpendicular incidence. The cut-off density can be significantly smaller, if the beam injection deviates from the perpendicular direction [3].

2.1 First pass and reflector position

The first step of the design is to determine the reflector position for a given launcher. Different positions are tested by simulating the beam path and the first pass absorption with the TORBEAM code [8] in the same model plasma. Depending on the position, the launch angles have to be adapted in order to make sure that the shine through beam after the first pass ends centrally on the reflector. Maximum possible absorption (section 1) depends on the angle between **k** and **B**, which is mainly determined by the toroidal launch angle. In our model plasmas, high first pass absorption correlates well with central power deposition. This is due to the generally peaked temperature profiles, the strong T_e dependence of the O-2 absorption, and the alignment of the absorption zone with the magnetic axis.

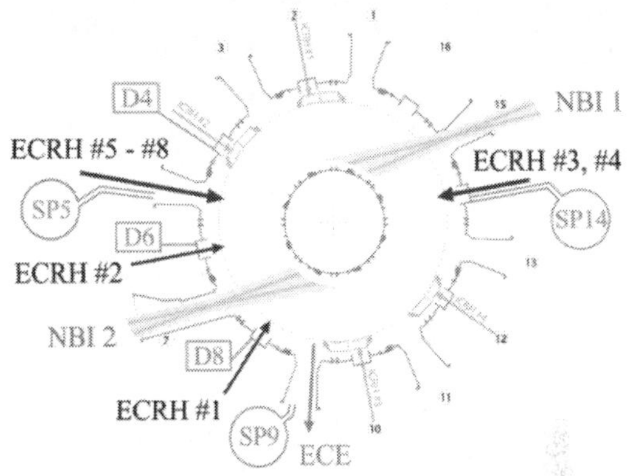

Fig. 2. Top view of the equatorial cross section through AUG with launching positions of different heating systems. ECRH launchers #5 - #8 are operational, incl. O-2. ECRH #1 and #2 are operational, launchers #3 and #4 are currently being upgraded. O-2 is being designed for launchers #1 - #4. Relevant detectors for unabsorbed radiation are shown in light colour. 'SP' Sniffer sphere, 'D' D-band pick-up waveguide, mounted on quartz vacuum window [4, 10].

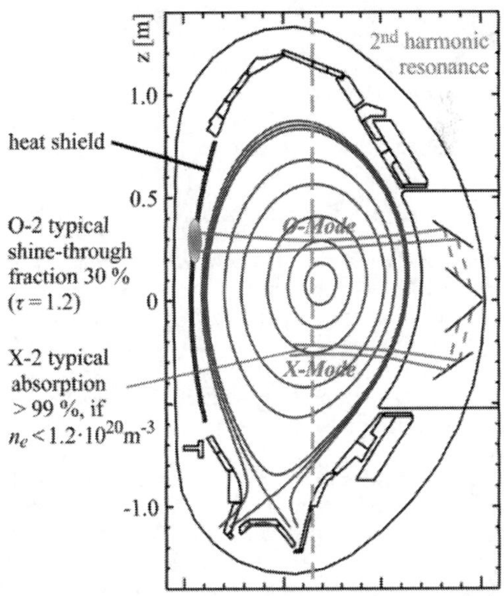

Fig. 3. Vertical cross section at the ECRH #5 - #8 launching position, visualising the 140 GHz beam path in a typical AUG experiment (B_0 = 2.5 T, T_e = 4 keV) with different polarisation. Upper launcher: ECRH #5 with O-mode, lower launcher: ECRH #8 with X-mode polarisation.

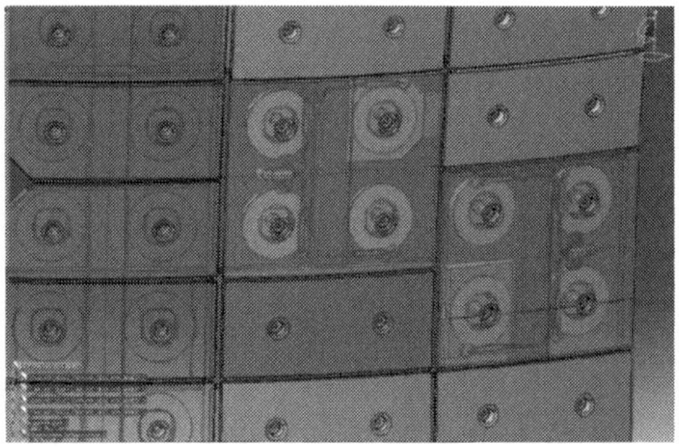

Fig. 4. Tiles of the AUG heat shield with double tile models for the reflectors in a CAD drawing. The left column is transparent in order to visualise the underlying support structure. Both reflector double tiles (middle and right column) are transparent to show the built-in thermocouple measurement.

TORBEAM input data were taken from AUG #30505, t= 3.3 sec, considered a model high power discharge with partial detachment. Characteristic data: 19.5 MW total heating power, I_p = 1.2 MA, B_t = -2.48 T, n_e = 1.15 · 10^{20} m^{-3}, and central T_e = 3.6 keV. The beam path may vary with different model plasmas. Given the alignment of the absorption zone with the magnetic axis is similar, then this variation, however, is generally small when compared to the variation that corresponds to different reflector positions.

Only those reflector positions are allowed, which fit into the heat shield on the central column inside AUG (figures 3 and 4). The heat shield consists of tiles, each with the dimensions of approximately 210 mm (width) and 90 mm (height). This is the 'grid' for possible reflector positions. Two adjacent tiles (vertically) are merged, such that the reflector has the approximate dimensions 210 mm x 180mm, which is considered sufficient for a good coverage of the width of a 140 GHz ECRH beam inside AUG. The resulting 'double tiles' are exemplarily shown in figure 4.

2.2 Second pass and beam dump

If the reflector has a high efficiency (section 3) and assuming similar values of the optical thickness for the first and second pass, then the unabsorbed power after the second pass will be of the order 7 – 8.5 % of the initially launched power (figure 5). This depends on T_e. The unabsorbed power may leave the plasma as a compact beam on the low field side. Hence one should take care that the in-vessel component, which faces this beam is robust enough. Typical values are 80 kW for the beam power and a beam radius of 5 cm corresponding to an area of the order of 80 cm^2 and 1 kW cm^{-2} on average. If the component has a metallic surface, then a large part of the beam will be reflected again. Usually the phase front is defocussing and with the diverging beam the power density goes down rapidly. Further propagation is, therefore, not taken into account.

In the vicinity of ECRH launchers #1-4 there are auxiliary low field side limiters (e.g. figure 6) which are able to withstand the given power density. The end point of the second path for each launcher was chosen such that, according to the model beam tracing, each beam leaving the plasma is horizontally centred on the auxiliary limiter next to the launcher.

The toroidal launch angle for the second pass is determined by the position of the auxiliary limiter. For different vertical launch angles, high second pass absorption correlates with the beam passing through the plasma centre. This is similar to the first pass. According to figure 5, controlled central deposition of typically 90% of the launched power is possible.

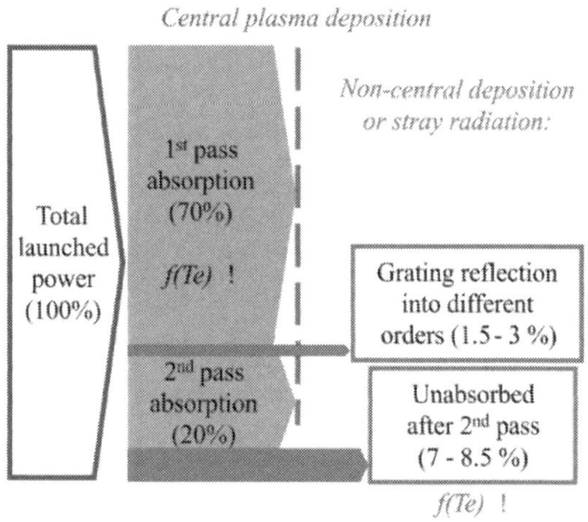

Fig. 5. Schematic power flow for the two-pass heating. These values are obtained on average with the different launching positions and the model absorption in AUG #30505 t= 3.3 s. Both 1st and 2nd pass absorption depend on T_e. The given numbers for the unabsorbed power are upper limits for the stray radiation, since it can be partly absorbed by the plasma.

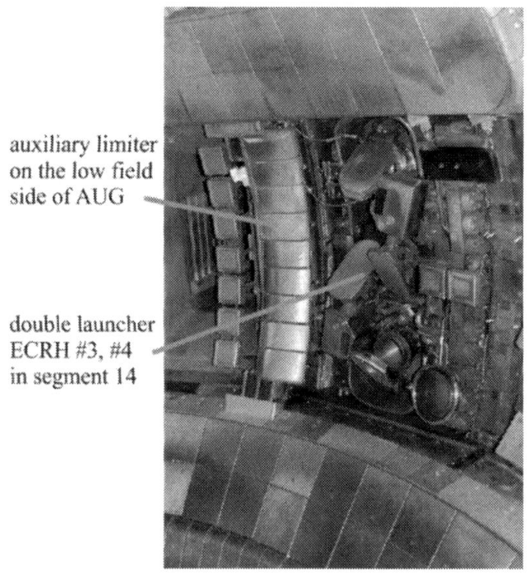

Fig. 6. Photo of the double launcher ECRH #3, #4 together with the auxiliary low field side limiter, both located in segment 14. The auxiliary limiter will act as 'beam dump' for the ECRH power which is unabsorbed after the 2nd pass.

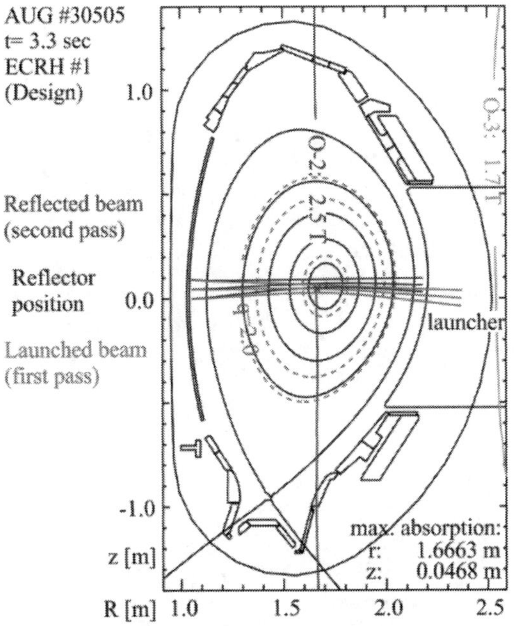

Fig. 7. AUG vertical cross section at the position of ECRH #1 to visualize the TORBEAM [8] results for the two-pass heating design. In this paraxial approximation, each beam is characterized by three lines: one for the centre of the beam and at each side one for the actual 1/*e* width of the Gaussian.

In order to avoid depolarization between the first and the second pass, the difference of the toroidal launch angles should be small. Normally this is achieved, if the toroidal distance between launcher and beam dump is small. The reflecting grating is optimized in order to preserve the polarisation state (section 3).

Beam tracing results for the two-pass heating design are exemplarily shown in a vertical cross section for ECRH #1 in figure 7. All launchers ECRH #1-4 are located in the midplane (z=0), the vertical cross sections of the corresponding beam tracing are, therefore, similar. The reflecting gratings (section 3) offer the choice of the refocussing properties. This is a free parameter in the two-pass design. Presently, we are aiming for similar focussing properties of the first and second pass beams.

An overview of the two-pass design for ECRH #1-4 is shown in figure 8 in a top view of the equatorial cross section through the AUG vacuum vessel. All toroidal launch angles for the first pass deviate between 12 and 15 deg from the perpendicular direction, maximizing the first pass plasma absorption. We note that in segment 14 both ECRH #3 and #4 have to share the same auxiliary limiter as the end point after the second pass. While for ECRH #1 and #2 the decision on the geometry is final, the decision on ECRH #4 is not yet taken. This may also have an impact on the geometry of ECRH #3.

Altogether and with respect to the toroidal launch angles and the optical depths, the design of the two-pass heating scheme for ECRH #1-4 follows closely the scheme for ECRH #5-8 [6, 10]. The latter has already been proved to be reliable in AUG high density experiments. The scheme for ECRH #1-4 features now horizontal beam paths. We would like to point out that the present two-pass design has a fixed geometry, corresponding to a fixed operating point. Large movements of the launcher mirrors are not allowed in this scheme unless additional reflectors are installed.

The intended thermocouple measurement positions on the reflecting gratings are shown (e.g. figures 4, 10c). These measurements are valuable for beam position verification and control [10].

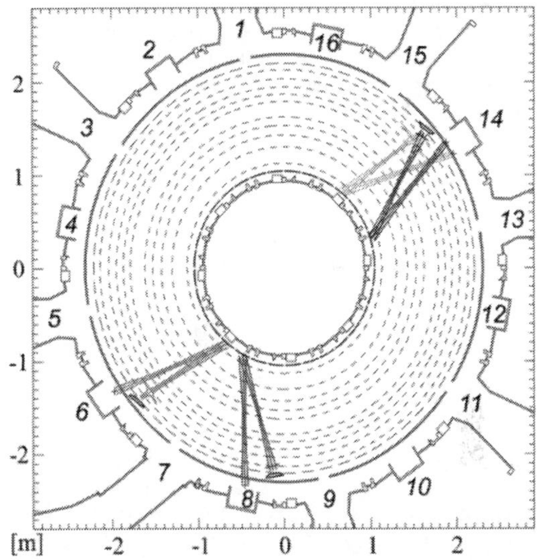

Fig. 8. Top view of the equatorial cross section through the AUG vacuum vessel, which consists of 16 segments. The outermost thin dashed line represents the plasma boundary on the low field side. The innermost thick dashes with the narrow gaps represent the tiles of the heat shield on the high field side. ECRH #1-4 launching positions are marked with a '+' sign and to the right of each launcher the midplane cross sections of the nearest auxiliary limiters are shown. The beam tracing according to the two-pass design is overplotted using colours. Segment 8: ECRH #1, blue. Segment 6: #2, red. Segment 14: #3 in black and #4 in green.

3 Reflecting gratings

The development of highly efficient reflecting gratings was driven by the need to reflect the beam off the tiles of the AUG heat shield (section 2.1), which results in an oblique angle between tile surface and beam direction. Both the incoming and the outgoing beam properties are design parameters and depend on the relative positions of launcher, reflector and beam dump, as well as on the beam refraction in the plasma. The polarisation and the Gaussian beam shape should be preserved. Machining issues have to be taken into account. In order to fulfil these pre-conditions, pioneering work has been carried out at the University of Stuttgart [5, 6].

3.1 Design properties of the gratings

The required boundary conditions could be fulfilled with a -3rd order grating reflection and an efficiency > 90% is possible on average [5]. Fine structures in the grating profile serve to minimize the difference of efficiencies between the TE and the TM polarisation, e.g. figure 9.

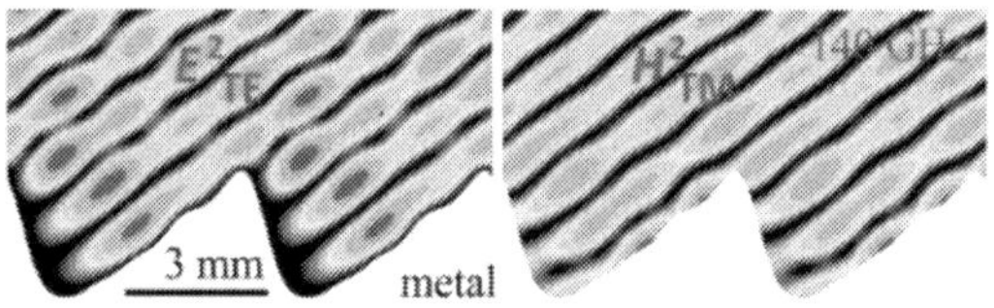

Fig. 9. Full wave simulation of the 140 GHz mm-wave reflection on an optimized Fourier grating [5] with θ_{IN}= 60°, θ_{OUT}= 50° and negative third order. Left: TE polarisation and calculated 89% efficiency for the reflection. Right: TM polarisation and 96.9% efficiency.

Each grating profile is optimized in the plane wave approximation. This corresponds to a local condition on the reflector. With a Gaussian beam shape, different positions on the reflector correspond to different optimized gratings. The surface structure of the full reflector is, therefore, interpolated from single optimized grating profiles and the phase front curvature of the Gaussian beam is taken into account by a curvature of the grooves of the grating [6]. The full reflector model is exemplarily shown in figure 10a and can be compared with the machined tile in figure 10b. The shown 'R7' reflector (for ECRH #7) is now newly installed for the upcoming 2018 experimental campaign at AUG and its beam tracing design has been described elsewhere [10].

In the final design of each grating, an average beam direction and shape is used, which is the result of averaging the beam tracing in a set of different model plasmas. For the 'R7' reflector in figure 10, the average beam properties were characterised as follows:

$$\mathbf{k}_{in} = [\ -0.51244,\ +0.30642,\ -0.80219\ ]$$

$$\mathbf{k}_{out} = [\ +0.49083,\ -0.48715,\ +0.72234\ ] \qquad (1)$$

The angle between $\mathbf{k}_{in}$ and $\mathbf{k}_{out}$ is γ = 11.4° and the coordinate system is like in figure 10a with x-axis horizontal, y-axis in vertical direction towards the upper edge of the reflector, and z-axis perpendicular to the reflector surface in the center. The Gaussian beam radii for both $\mathbf{k}_{in}$ and $\mathbf{k}_{out}$ are w_x = 5.9 cm, w_y = 7.6 cm, while the average data for the phase front curvature for the incoming beam are R_x = 150 cm and R_y = 128.5 cm, both defocussing. For the reflected beam, the phase front curvature was designed with R = -115 cm (both x and y) and we note that the negative sign represents a focussing phase front. For each refocussing in the phase front curvature, a characteristic focal length F can be introduced as in the following equation:

$$R_{out}^{-1} = R_{in}^{-1} - F^{-1} \qquad (2)$$

This means that two characteristic focal lengths can be attributed to the grating 'R7': F_x= 65.1 cm, F_y= 60.7 cm.

3.2 Laboratory measurements

The reflecting grating 'R7' has been designed and optimized with fixed parameters for the incoming and reflected Gaussian beam. When the two-pass scheme operates with various plasma conditions, the actual values of the incoming Gaussian beam may deviate from the design values.

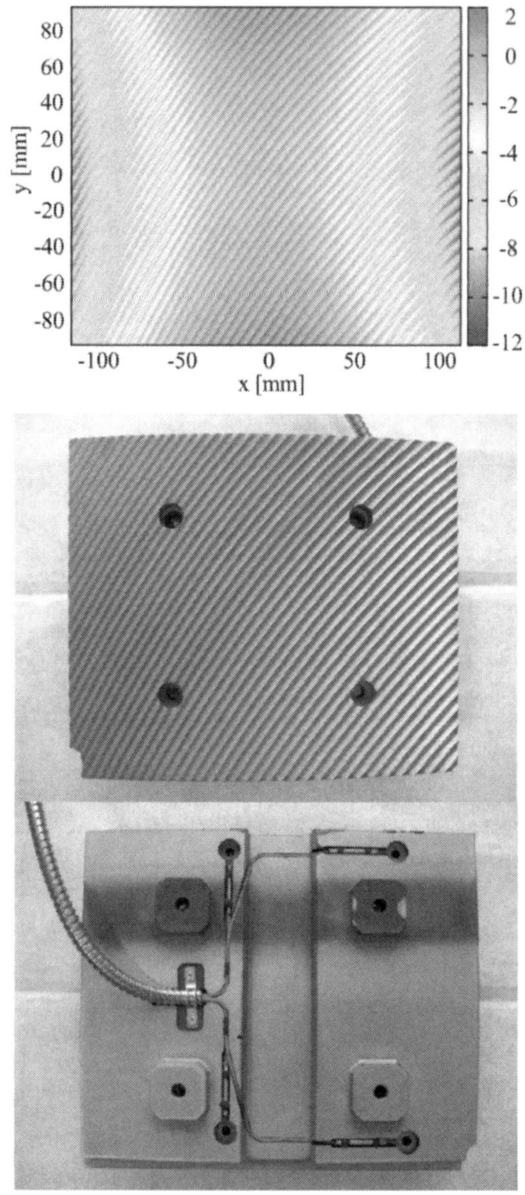

Fig. 10. Reflecting grating 'R7' with the design (a) on top and the machined and tungsten coated tile (b) in the middle. The grating was machined at the workshop of the University of Stuttgart. The photo (c) at the bottom shows the rear part of the reflector with the thermocouple instrumentation, which has been done at IPP Garching. The darker area is due to the mounting of the tile during the W coating procedure.

It is therefore to be tested, if the characteristic focal lengths, as introduced in equation (2) can be used to model the effect of the grating in such case. For this purpose we made laboratory measurements of the outgoing phase front curvature, function of incoming phase front curvature, as shown in figure 11. Since in the 'R7' case the angle γ between $\mathbf{k}_{in}$ and $\mathbf{k}_{out}$ is aligned nearly vertically, and since the reflection geometry on the laboratory table is horizontal, it should be noted that the original x-axis (figure 10a) approximately corresponds to the vertical direction in the laboratory, and the original y-axis to the horizontal one.

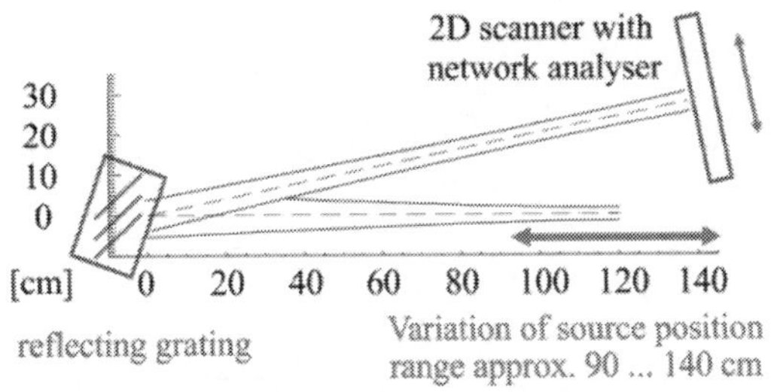

Fig. 11. Top view of the laboratory set-up for measuring the properties of the reflecting grating 'R7'. The source position was varied in order to change the incoming phase front curvature on the grating. For each source position, the reflected beam was scanned as indicated, which allowed to analyse the reflected beam. With these data the phase front curvature at the grating after the reflection could be determined.

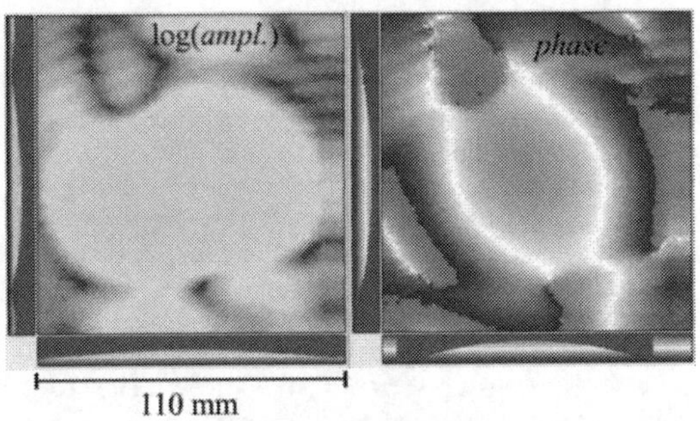

Fig. 12. Visualisation of amplitude (left, green: max, blue:min) and phase (right, colour coding $0\ldots2\pi$, blue via white to red) of the reflected beam in the scanner plane, as described in figure 11. The small insets on the lower and left edges of each contour plot are cross sections through the center of the 2D plot. According to the numerical analysis, these data correspond to a Gaussian beam content of 96.6% with beam waists w_0 = 1.41 cm (horizontal) and 1.56 cm (vertical). At the scanner position the beam is defocussing. The analysis yields 43.1 cm distance to the horizontal waist and 19.5 cm to the vertical beam waist. At the position of the grating this corresponds to phase front curvatures -107.3 cm (horizontal) and -132.7 cm (vertical) both focussing. The input phase front curvature was 139.2 cm (spherical, defocussing) at the grating.

The source (140 GHz) position was varied between 90 cm and 140 cm (figure 11). Both polarisations were tested. The incoming phase front curvature varied approximately by the same amount, since the grating is already in the far field of the source. This variation is also typical for different experimental situations at AUG. For each measurement of the output beam parameters (e.g. figure 12) the equivalent focal lengths were deduced. With 9 different source positions, the deduced horizontal focal length (laboratory, corresponding to F_y) was in the range 60.1 … 61.3 cm and vertical in the range 66.9 … 69.2 cm. Due to the specific geometry and as shown in figure 12, the principal axes of the beam cross section are not perfectly aligned with the axes of the laboratory system. This contributes to the error in the data analysis. Altogether we consider these results reasonably consistent with the design (section 3.1). In the scanned area the measured Gaussian mode content of the reflected beam was always of the order 95%.

The consistency of the set-up and the determination of the phase front curvature was verified with a spherical mirror with known focal length F = 80 cm. The scan of the output beam showed a circular Gaussian beam with 98.6% purity, and a change of the phase front curvature that corresponded to F values of 77.4 cm (horizontal curvature) and 79.8 cm (vertical curvature), respectively.

Future measurements could take into account a shift of the beam with respect to the reflector center. Direct measurements of the total efficiency are highly desirable, but very demanding.

4 Conclusions and Outlook

With the development of ECRH O-2 schemes for high density, the necessary simulation tools and laboratory testing equipment is being developed in parallel. The design for AUG ECRH #1-4 is being finalized and will make use of reflecting gratings. According to the presented measurement results, the characteristic focal lengths, attributed to each grating, seem to be adequate for modelling the reflection in the two-pass heating scheme. Focal lengths are free parameters and offer the possibility for future optimization.

The photos in figure 10 are a courtesy of Harald Schütz, IPP Garching. His careful assembly is gratefully acknowledged.

This work has been carried out within the framework of the EUROfusion Consortium and has received funding from the Euratom research and training programme 2014-2018 under grant agreement No 633053. The views and opinions expressed herein do not necessarily reflect those of the European Commission.

References

1. A. Kallenbach *et al*, Nucl. Fusion **55**, 053026 (2015)
2. P. Lang *et al*, Nucl. Fusion **58**, 036001 (2018)
3. V. Erckmann, U. Gasparino, Plasma Phys. Control. Fusion **36**, 1869-1962 (1994)
4. M. Schubert *et al*, EPJ Web Conf **87**, 02010 (2015)
5. O. Mangold, PhD thesis (Stuttgart University, 2009)
6. H. Höhnle *et al*, Nucl. Fusion **51**, 083013 (2011)
7. D. Wagner *et al*, J Inf Milli Terahz **37**, 45-54 (2016)
8. E. Poli *et al*, Comput. Phys Commun **136**, 90 (2001)
9. M. Schubert *et al*, EPJ Web Conf **32**, 02013 (2012)
10. M. Schubert *et al*, *43rd Conference Plasma Phys*, ECA **40A**, P1.026 (2016)

Construction of an EBW Heating System for the MST RFP

Andrew Seltzman[1,*], *Jay* Anderson[2], and *Cary* Forest[2]
[1]MIT Plasma Sciences and Fusion Center, 190 Albany St, Cambridge, MA 02139 USA
[2]University of Wisconsin, Dept. Of Physics, 1150 University Ave, Madison, WI 53706 USA

Abstract. The first observation of rf heating in a reversed field pinch (RFP) using the electron Bernstein wave (EBW) has been demonstrated on Madison Symmetric Torus (MST). The design of the RF heating system used in this experiment is described. Klystron and antenna limitations are examined during launch into the plasma.

1 EBW Heating Overview

The Electron Bernstein wave (EBW) presents an alternative for heating and current drive in overdense plasmas where conventional extraordinary (X-mode) and ordinary (O-mode) electromagnetic waves do not propagate past the periphery. The EBW is a short wavelength electrostatic wave excited by mode conversion of externally launched electromagnetic (O or X) modes [1,2]. There are three distinct conversion schemes that have successfully driven EBWs in toroidal plasma configurations. In the first case, a launched O-mode couples to the X mode at the O wave cutoff layer and then the X mode converts to the Bernstein mode with near 100% efficiency at the upper hybrid layer. This OXB scheme has been used to heat [3] and drive current [4] in stellarator plasmas, and heat tokamak [5] plasmas. Second, a high-field side X mode launch converted to Bernstein mode has also been shown to heat [6] and drive current [7] in the conventional tokamak. A third scheme has been utilized to accommodate the high beta plasma of the spherical tokamak (ST) by low field side launch of the X mode. In this case, the X mode must tunnel through a narrow evanescent region before conversion to the Bernstein mode with efficiency near 100% [8] for optimal edge density gradient scale length, $L_n=n_e(dn_e/dx)^{-1}$, where n_e is the electron density; heating of the ST plasma [9] has been observed.

Studies of EBW physics in the reversed field pinch (RFP) show efficient coupling, both through reciprocity in a blackbody emission measurement [10] and directly with optimization of a waveguide grill launching structure [11]. Ray tracing studies [12] predict accessibility of EBW heating and current drive over the outer half of the minor radius in Madison Symmetric Torus (MST) [13]. Full wave calculations of OXB mode conversion identify a feasible heating scenario in the RFX-Mod device considering realistic edge density fluctuations [14].

Recent studies [15] demonstrated RF heating in the RFP configuration using the EBW on Madison Symmetric Torus. A challenging heating environment exists in the RFP, where electromagnetic wave cutoffs occur within ~1cm of the edge and enclose the plasma volume due to an overdense plasma with no high field side. Outboard-launched X mode at 5.5 GHz is efficiently mode converted to the EBW in the naturally steep edge density gradient of the RFP, leading to absorption inside the last closed flux surface (LCFS) at a Doppler-shifted resonance ($\omega = n^*\omega_{ce} - k_\|^* v_\|$) for a broad range (n=1-7) of harmonics. Novel techniques were required to measure the suprathermal electron tail generated by EBW heating in the presence of intense ohmic heating; no net increase in plasma temperature is expected or observed.

2 MST EBW Heating System Design

An RF heating system designed for the EBW experiment included a radar klystron re-conditioned for long pulse operation, associated klystron control and protection systems, a waveguide antenna, and x-ray diagnostics. Operation at reduced power allowed long pulse operation of the klystron after conditioning. Design and testing of an antenna system revealed several inherent limitations in maximum launched power and coupling efficiency. Antenna designs evolved into a configuration that minimizes required porthole diameter, thereby reducing the spatial extent of the porthole induced field error. Protection systems allowed crowbarring of klystron voltage in the event or detection of an imminent arc, while limiting energy dissipated into the cathode. A feedback control system stabilized output power and reduced power noise induced by ripple in the input voltage provided to the klystron.

2.1 Klystron conditioning

A surplus CPI VKC-7762B radar klystron conditioned for long pulse operation provides RF power to the EBW heating experiment. The nominal operating specifications of the klystron are 5.5GHz, 18μs pulses at 1.3MW peak power with a 428pps rep rate; electrical requirements are ~75-78kV at ~38A. With proper conditioning, a klystron designed for short pulse operation is operated stably at significantly longer pulse length if peak output power is reduced and thermal limits are not exceeded. Klystron conditioning into a waveguide dummy load involved incremental increases in output power and pulse width, shown in **Fig 1**, while monitoring for klystron arcs and internal gas evolution monitored by the integrated ion pump. During the conditioning process, the klystron is pulsed at ~1 pulse

* Corresponding author: seltzman@mit.edu

per minute with ion pump current remaining below ~2-3μA; pulsing is delayed to allow pump down of evolved gasses if required.

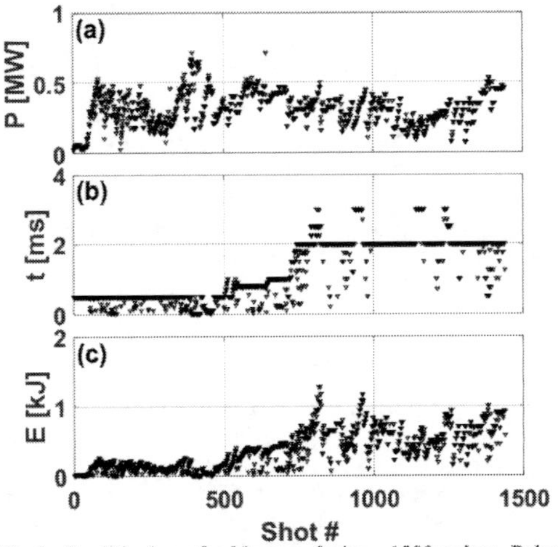

Fig 1. Conditioning of a klystron during ~1500 pulses. Pulse power is plotted in (a), pulse width is plotted in (b), and pulse energy is plotted in (c).

Internal arcing during the conditioning process is observed as a sharp spike in cathode current, typically concurrent with a burst of evolved gas within the klystron. Klystron output power decreases substantially preceding an arc and is used to trigger a crowbar spark gap. Following an internal arc during the conditioning process, output power is reduced to the last stable value and evolved gas is pumped down. In the event of continued arcs, both power and pulse width are further reduced until reliable operation is resumed. Klystron conditioning required ~1500 shots to achieve reliable operation for 2ms pulses at ~500kW. A longer pulse length of 3ms (limited by power supply capacitor bank) at lower power was used during the EBW experimental campaign due to power handling limitations of the waveguide antenna.

Reconditioning of a klystron cathode damaged due to arcing or age is required to recover klystron gain and stable operation. Prior to the final klystron power supply design where a crowbar spark gap and inductive-resistive (LR) snubber limits energy dissipation during an internal arc, energy dissipated from power modulator's 0.1μf filter capacitor into the klystron cathode could be as high as 280J. During initial testing, the klystron cathode sustained damage due to internal arcing. The dispenser type cathode utilizes an emissive material (BaO or equivalent) is impregnated into a porous tungsten matrix. Surface ablation of the emissive coating reduces emissivity, leading to unstable operation, and decrease in electron beam current and klystron output power.

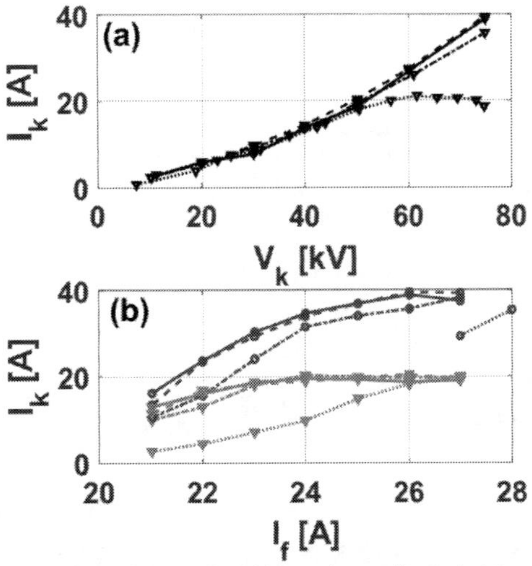

Fig 2. Klystron current for initial arc damaged cathode (..), and for conditioning times of 45min(-.), 105min(--), and 225min(-). Cathode current I_k vs cathode voltage V_k at 26A filament current is plotted in (a). Cathode current I_k vs filament current I_f at 50kV(red, ▼) and 75kV(blue, O) cathode voltage is plotted in (b).

Diffusing emissive material to the cathode surface by short-term operation of the filament above rated current successfully reconditioned a damaged dispenser cathode. Filament current was increased from the rated value of 26A AC to 30.5A AC for 225 minutes. Sweeps of klystron current at varying filament current and cathode voltage, shown in **Fig 2**, checked conditioning process at several intermediate intervals. Damaged cathodes exhibit an emission current roll off at a maximum less than the rated cathode current I_k, as shown in **Fig 2** (a) at rated filament current I_f, and a slope in I_k with respect to I_f at rated I_f as shown in **Fig 2** (b). A conditioned cathode exhibits I_k proportional to cathode voltage V_k at the rated I_f,V_k point, as well as a roll off in I_k as I_f reaches the rated value. Stable klystron operation with full recovery of rated beam current was observed after 105 minutes with no further improvement with subsequent high filament current conditioning.

2.2 Waveguide antenna

X-mode is launched radially from a 44.5mm ID cylindrical molybdenum waveguide antenna carrying the TE11 mode. The molybdenum waveguide is resistant to plasma bombardment and the low secondary electron emission coefficient further decreases arcing probability. A tapered waveguide section acts as a mode converter between the WR187 TE01 rectangular waveguide and the TE11 cylindrical waveguide. Power handling limits of the antenna in the presence of plasma was found to be ~150kW corresponding to 96MWm^{-2}, consistent with power limits in other RF heating experiments [16]. Operation at powers as high as 300kW has been observed with a higher probability of arcing in the vacuum section of the antenna body and pillbox window.

A pillbox type window, shown in **Fig 3** top, forms a vacuum break between the pressurized waveguide and vacuum section of the molybdenum antenna. The fused silica window is 4.76mm thick, and 111.1mm diameter. Pressure testing of two windows contaminated with non-removable sputtering caused window failure by fracturing at 60psi and 80psi differential pressure. During experimental campaigns, the pressurized section of the antenna operated reliably at 300kW without arcing when pressurized with 6 psi sulphur hexafluoride (SF_6) or 10 psi of air.

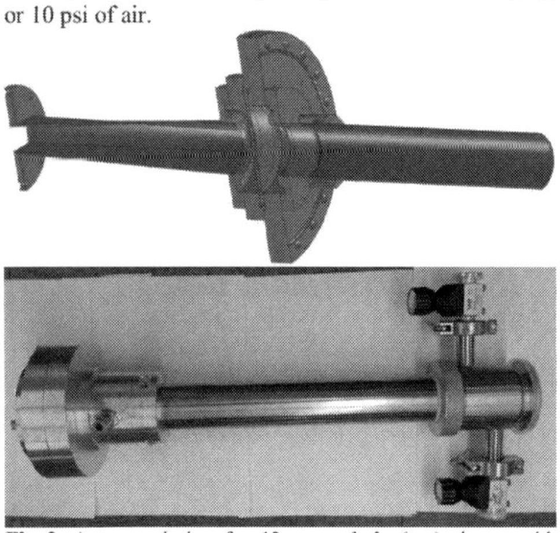

Fig 3. Antenna design for 12cm porthole (top) shown with bellows assembly and linear actuator hidden. Antenna for 5cm porthole (bottom) uses differentially pumped o-ring seal.

Magnetic field error introduced by the interruption of the current carrying shell by the antenna porthole leads to a substantial reduction of field (up to 50%) at the wall[17], with the perturbation decreasing to ~zero one porthole diameter radially inward. The effects of porthole field error include the reduction in radial accessibility due to the introduction of a higher harmonic resonance at the boundary. The spatial extent of the field is proportional to the porthole diameter, as shown in **Fig 4**. A first generation antenna using a 12cm porthole, **Fig 3** top, failed to heat on 1st and 2nd harmonic resonances, and frequently arced during 1st harmonic heating attempts. Reduction of magnetic field allowed successful EBW heating on 4th and 5th harmonics implying arcing during 1st harmonic attempts is due to introduction of an optically thick 2nd harmonic x-mode resonance within the antenna structure which induced breakdown. A second generation antenna, **Fig 3** bottom, used a differentially pumped sliding seal that allowed connection to a 5cm diameter porthole. The reduction in the spatial extent of the field error allowed for successful EBW heating on 1st and 2nd harmonics.

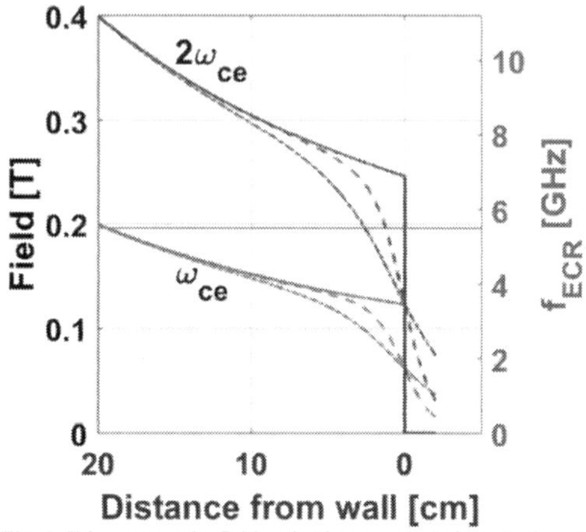

Fig 4. Edge magnetic field reduction on porthole axis due to porthole field error for a 320kA discharge. Magnetic fields are plotted in absence of a porthole (solid line), for 5cm (-- line) and 12cm (-. line) portholes corresponding to first harmonic (red) and second harmonic (blue) resonances (excluding Doppler shift). Launch frequency at 5.5GHz is marked with a horizontal red line.

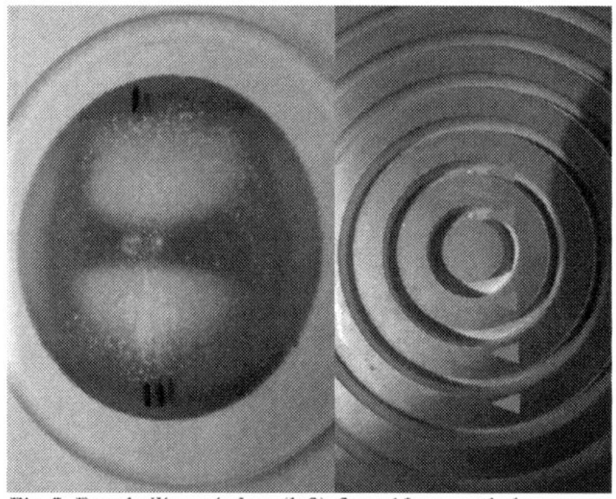

Fig 5. Fused silica window (left) from 12cm porthole antenna with sputtering from arcing in the vacuum side of the pillbox window. Grooved boron nitride limiter cover (right) used on 12cm porthole antenna. Cover shows signs of arcing (red triangle) parallel to the RF electric field near the center. RF electric field is oriented vertically in this figure.

Power handling capabilities of the antenna were improved with the removal of boron nitride ceramic covers, shown in **Fig 5** right, which inhibited the pump out of gas evolved during antenna conditioning and arcing, and acted as a breakdown point. Interruption of sputtered conductive coating from the plasma with a grooved surface improves cover lifetime, however addition of a cover was found to adversely affect power transmission and power handling capability. Electrical breakdown along the surface of the cover occurred near the center (marked in red triangles) in a direction parallel to the RF electric field.

Arcing within the vacuum side of the antenna sputters a conductive coating from the pillbox window

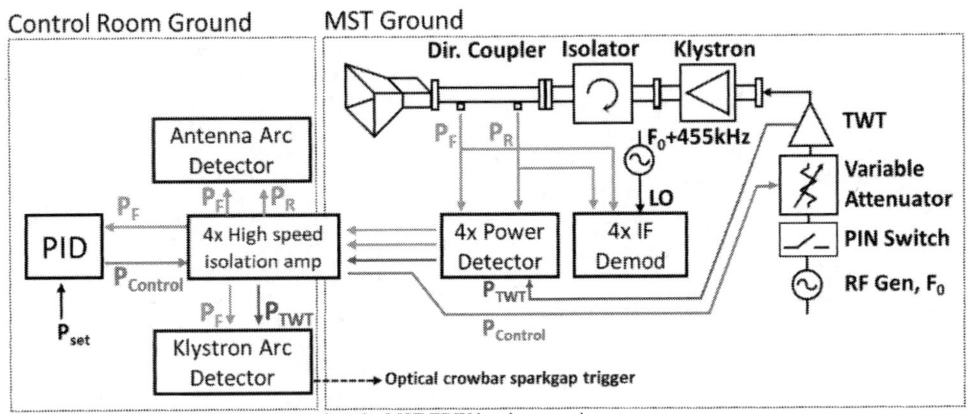

Fig 6. Layout of the RF control system used on the MST EBW heating experiment.

holder onto the window surface. Antennas with both stainless steel and aluminum pillboxes were tested resulting in sputtering of both types on metal onto the window, shown in **Fig 5** left. Cleaning with a phosphoric acid based cleaner (Oakite 33) successfully removed sputtered aluminum from the surface allowing the window to be returned to service; windows with stainless sputtering were not recoverable with acid based cleaning methods.

2.3 RF power stabilization and control

Adverse effects in klystron power stability due to voltage fluctuations from the HV power supply is corrected with active feedback methods. The voltage dependent gain of a klystron transposes voltage ripple and droop onto the output power. Voltage droop occurs over the 3ms pulse, while ripple exists at ~120kHz, the 6th harmonic of the power supply switching frequency injected by 3 phase full wave rectification. Noise reduction and improved power stability over the RF pulse duration has been achieved by implementing an automatic gain control (AGC) circuit with a proportional-integral-derivative (PID) control loop.

Layout of the RF system for the EBW experiment is shown in **Fig 6**. An HP-8673B generator supplies a stable 5.5GHz signal to drive the RF system. Generator output is switched with 10ns rise/fall times at signal level by an HP-11720A PIN diode switch. The windowed RF is fed through a Mini-Circuits HVA-73+ voltage variable attenuator (VVA) for fast power level control as part of the AGC system. The leveled RF signal is boosted up to 20W by a traveling wave tube (TWT) amplifier to drive the klystron; as part of an arc detection system, TWT output power (P_{TWT}) is monitored and compared to forward power out of the klystron. The klystron amplifies the RF power up to several hundred kW; power from the klystron passes through a CML engineering 25dB circulator functioning as an isolator to protect the klystron from power reflected from the plasma. A directional coupler between the circulator and waveguide antenna measures forward (P_F) and reflected (P_R) power which is split into power and phase detection systems.

A high speed diode detector based on an HP-33330b diode provides high bandwidth power measurement to allow fast arc detection and reduction of high frequency ripple. Attenuation scales the input power to the diode to 10mW/MW. Detector diodes have ground loop isolation on both shield and center coaxial connections and are high-pass filtered at 800MHz to reduce noise pickup. Detector voltage is buffered and amplified by an isolated AD828 OP amp; diodes are terminated with a 100ohm resistor in parallel with the amplifier input to improve rise time. Diode detector system achieves a 19ns delay in response to an RF pulse. Diodes are operated in non-linear mode to improve signal to noise ratio, greatly reducing noise pickup. Calibration of output voltage with a quadratic polynomial fit (R^2=0.9998) determines power from measured voltage. AGC stability and accuracy are not adversely affected by the small non-linearity in power detection since integral gain in the PID control minimizes power error.

Operation of the RF control system in a pulsed magnetic field environment requires ground loop isolation between RF systems in the machine area and control room for safety and noise reduction. A four channel high speed isolation amplifier based on the Cheele CIM1100 isolator module (120MHz bandwidth) separates machine area and control room grounds. Impedance matching of input and output lines to the isolator module with OP amp buffers reduces the effective isolator bandwidth and introduces additional phase shift between input and output signals. The isolator achieves minimal signal distortion at 10MHz, 19ns rise time, and 24ns delay between input and output.

Forward power from the klystron is stabilized by an AGC system using a PID control loop. P_F measured by the diode detector is monitored by a PID control system through the isolation amplifier. The measured power is compared to a reference value to generate an error signal; the summed value of time integrated, time differentiated, and scaled (proportional), error signals are used to drive the VVA through a channel of the isolation amplifier. An OP-amp buffer connected to the VVA and tuned to have a frequency response inverse to that of the VVA extends the attenuator bandwidth to 2MHz. When tuned to the highest stable gain, negative feedback from the PID control system acts to null out fluctuations in RF power.

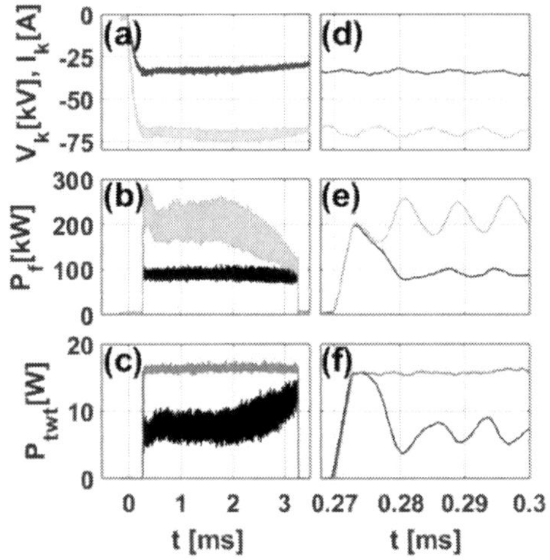

Fig 7. Voltage (light blue) and current (dark blue) supplied to klystron (a) exhibit ripple and droop over the high voltage pulse. Plotted in (b), the resulting change in klystron gain induces ripple and droop in RF power (green) which is stabilized with feedback control (black). Plotted in (c), the feedback controller modulates the output from the TWT (black) driving the klystron in inverse to the ripple and droop to stabilize RF power. TWT output power without feedback is shown in magenta. Zoomed values of (a,b,c) are plotted in (d,e,f). Voltage ripple (d, light blue) and power (e, green) are phase locked to with a constant phase offset. Settling time of feedback controller (e, black) is ~10μs after an initial power spike caused by integrator windup.

Feedback control of RF power with the AGC system achieves a significant reduction in RF power droop and ripple. RF power from the klystron is plotted in **Fig 7** (b) with the AGC enabled (black) and disabled (green). In the absence of AGC feedback, noise (e) and droop (b) in power (green) are observed during the pulse corresponding to ripple (d) and droop (a) in voltage (light blue). With the AGC feedback enabled, RF power from the klystron is leveled at ~100kW, and noise is reduced; the inverse of the noise and droop is modulated onto the TWT power feeding the input of the klystron (c,f black). During the leading edge of the RF pulse, the AGC exhibits a slightly underdamped response with a settling time of ~10μs as plotted in (e, black). Initial power overshoot on the leading edge is due to integrator windup as the AGC system remains active when the PIN diode switch inhibits RF power; although it is not detrimental to operation of the system, future systems will address this issue with an integrator reset when RF power is off.

2.4 Klystron protection system

Damage to klystron cathodes due to arcing is prevented with use of a snubber and crowbar circuit to limit arc fault energy. A resistive capacitive (RL) snubber constructed with a 50ohm resistor in parallel with a 400uH inductor is placed in series with the klystron to limit fault current during an arc. The snubber provides additional protection to the power supply by preventing a HV pulse from being reflected on the coaxial cable during an arc. A triggerable spark gap in parallel with the klystron crowbars voltage during a fault, shunting current away from the cathode. The triggered gap is designed to fire at overvoltage conditions determined by gap spacing, self triggering with high dI/dt conditions, or remote triggering if an imminent arc is predicted. The self-triggering system uses a 3mH inductor in series with the klystron connected across a trigger electrode. High di/dt during a klystron arc triggers the spark gap by providing a high voltage difference across the trigger electrode. Current flow through the trigger electrode is prevented by a 5nF series capacitor. Prior to a klystron arc, gain drops off ~50μs prior to arc initiation. The klystron protection system uses an RF power comparator to remotely trigger the crowbar spark gap if the P_F/P_{TWT} power ratio falls below a set threshold.

2.5 Antenna arcing and coupling

Identification of good and bad shots is required to select valid experimental data and prevent sustained operation at settings resulting in antenna arcing. Although the presence of x-ray emission indicates RF heating within the plasma, the absence of x-ray emission alone is not sufficient to determine if power is coupled into the plasma, or dissipated into an arc within the antenna structure, particularly during the exploration of new parameter space. Analyzing of coupling and fluctuations in reflected power and phase between forward and reflected power are instrumental in determining successful operation of the antenna system.

Phase between forward (P_F) and reflected (P_R) power is recovered through digital synchronous demodulation (DSD) [18] of down-converted IF signals shown in **Fig 6**. An IF frequency at 455kHz is generated by mixing the P_F and P_R signals with a local oscillator (LO) locked at 455kHz above the RF generator frequency. RF signals are ground loop isolated on both shield and center coaxial connections and are high-pass filtered at 800MHz to reduce noise pickup. A MITEQ DMX0418L double balanced mixer down-converts P_F and P_R signals to 455kHz. Since LO and RF signals are phase locked, the difference in phase between P_F and P_R IF signals provide the phase difference at the RF frequency. The phase diagnostic is useful for edge density fluctuation measurements and identification of antenna arcing.

Fluctuations in edge density in the RFP vary the location of the R and L cutoff surfaces and edge density scale length, these fluctuations appear on the phase and reflected power measurement. During an arc, plasma within the waveguide or pillbox window structure presents a load with reduced reflected power fluctuations and phase fluctuations compared to power coupled into the plasma. An example arc is plotted in **Fig 8**, occurring at ~17.5ms; reflected power fluctuations (a, blue) and phase fluctuations (b) are shown to decrease substantially. Simultaneously, termination of x-ray bremsstrahlung from heated electrons (c) measured with a limiter and target probe [19] is observed. Although reflected power shows an initial dip and subsequent climb, this characteristic is not uniformly observed during all arcs. Reflected power can remain low following an arc; in all cases, high frequency fluctuations in reflected power, and HXR emission, similar to those before 17.5ms, uniformly cease.

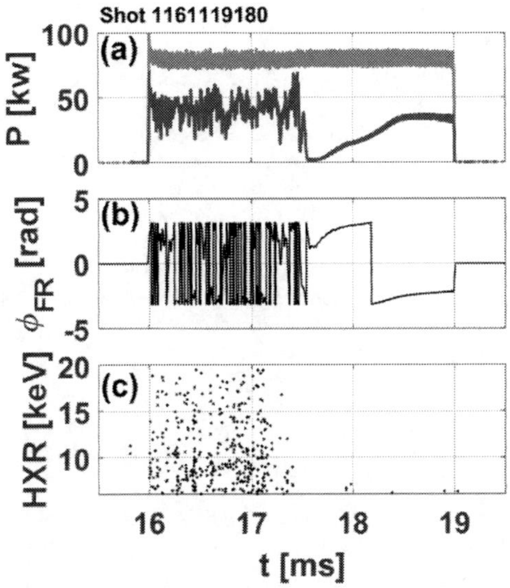

Fig 8. Characteristic changes in reflected (blue) power and phase due to arcing occurring at the ~17.5ms during an 80kW forward power (red) RF pulse. Substantial reduction in reflected power fluctuations (a) and phase (wrapped to $\pm\pi$) fluctuations (b) are observed with simultaneous termination of target bremsstrahlung x-rays (c) from heated electrons.

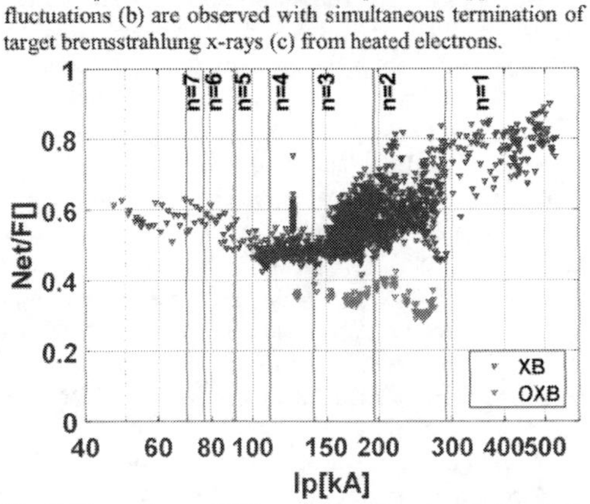

Fig 9. Net vs forward RF power for X-mode (blue) and O-mode(red) launches over a range of plasma currents corresponding to 1st-11th harmonic absorption. Boundaries between harmonics are demoted with vertical red lines.

Coupling from the waveguide antenna is a function of plasma edge conditions [11]; sharp jumps in coupling are not observed when crossing harmonic boundaries. Coupling from the waveguide antenna is plotted in **Fig 9** during X mode (blue) and O mode (red) launch; coupling for XB conversion is typically 50-80%, while OXB coupling is typically 30-40%. Lower coupling in the OXB case is expected as optimal launch for OXB occurs at an oblique angle; the lack of a phased array antenna in this experiment mandates a radial launch.

3 Conclusions

RF heating in the RFP has been observed using the EBW on MST. Conditioning of a surplus radar klystron extended achievable pulse width from 18μs to 3ms at a reduced output power; arc damage to dispenser cathodes is recoverable through high temperature conditioning, and preventable with arc detection and LR snubber circuits. Reduction in antenna porthole diameter allows heating at lower harmonics and improves radial accessibility. Feedback control stabilizes RF power from the klystron and reduced voltage ripple induced noise. Coupling in XB conversion is typically 50-80% of forward power. The current configuration favors XB to OXB heating due to a radial launch. Addition of a local limiter to optimize edge density gradient may improve coupling in XB conversion heating, however ceramic covers were shown to limit power handling of the antenna; a coaxial type limiter surrounding the antenna is recommended. This material is based upon work supported by the U.S. Department of Energy Office of Science, Office of Fusion Energy Sciences program under Award No. DE-FC02-05ER54814.

References

[1] J. Preinhaelter, V. Kopecky, J. Plasma Physics. 10, 1 (1973)
[2] H. Sugai, Phys. Rev. Lett. 47, 1899 (1981)
[3] H. P. Laqua, et al. Phys. Rev. Lett. 78, 3467 (1997)
[4] H. P. Laqua, et al. Phys. Rev. Lett. 90, 075003 (2003)
[5] A. Mueck, et al. Phys. Rev. Lett 98,175004 (2007)
[6] T. Maekawa, et al. Phys. Rev. Lett. 86, 3783 (2001)
[7] V. Shevchenko, et al. Phys. Rev. Lett. 89, 265005 (2002)
[8] A. K. Ram, S. D. Schultz Phys. Of Plas. 7,4084 (2000)
[9] S. Shiraiwa, et al. Phys. Rev. Lett. 96, 185003 (2006)
[10] P.K. Chattopadhyay, et al. Phys. Plasmas 9, 752 (2002)
[11] M. Cengher, J.K. Anderson, V. Svidzinski, C.B. Forest. Nuc. Fusion. 46, 521, (2006)
[12] C.B. Forest, P.K. Chattopadhyay, R.W. Harvey, and A.P. Smirnov, Phys. Plasmas 7, 1352 (2000)
[13] R.N. Dexter, et al. Fus. Sci. and Tech. 19,131 (1991)
[14] R. Bilato, et al. Nuc.Fusion. 49, 075020 (2009)
[15] A. H. Seltzman, J. K. Anderson, S. J. Diem, J. A. Goetz, and C. B. Forest Phys. Rev. Lett. 119, 185001 (2017)
[16] V. Pericoli et.al. Nuc. Fusion 45, 1085 (2005)
[17] P. J. Fimognari, et al. Plasma. Phys. Cont. Fusion. 52, 095002 (2010)
[18] D. W. Choi, et al. Rev. of Sci. Inst. 57.8 (1986)
[19] A. H. Seltzman, J. K. Anderson, A. M. DuBois, A. Almagri, and C. B. Forest. Rev. Sci. Instrum. 87, 11E329 (2016)

EPJ Web of Conferences **203**, 02011 (2019) https://doi.org/10.1051/epjconf/201920302011
EC20

Mode analysis limitations of ECE-I & ECE measurements at the plasma edge

B. Vanovac[1,*], *S.S.* Denk[2], *E.* Wolfrum[2], *M.* Willensdorfer[2], *W.* Suttrop[2], *R.* Fischer[2], *N.C.* Luhmann Jr[3], and the ASDEX Upgrade Team

[1]DIFFER - Dutch Institute for Fundamental Energy Research, De Zaale 20, 5612 AJ Eindhoven, the Netherlands
[2]Max-Planck-Institut für Plasmaphysik, D-85748 Garching, Germany
[3]Department of Applied Science, the University of California at Davis, Davis, CA 95616, USA

Abstract. Interpretation of the Electron Cyclotron Emission (ECE) and ECE Imaging (ECE-I) measurements concerning the measurement position and the radiation temperature (T_{rad}) at the plasma edge is associated with significant uncertainty. Various limitations such as low and high-density limits, relativistic and Doppler shift-broadening, mode overlap were identified in the past. Here, we analyse the influence of density profile variation onto the radiation temperature in H-mode plasmas at ASDEX Upgrade tokamak. We show that, in the region of steep gradients, the variation of the density profile leads to an outward-shift of the measurement position, towards lower T_{rad}. The analysis is extended towards the study of the contribution of the ordinary mode (O-mode) emission to the measured T_{rad} at the plasma edge. With this analysis, we show the qualitative agreement between experimental observations and modeling when taking into account the O-mode contribution to the ECE measurements. All of these effects are assessed via the Electron Cyclotron Radiation transport forward model that is now routinely used to model the response of the ECE-I and ECE systems in different plasma scenarios.

1 Introduction

The Electron Cyclotron Emission (ECE) diagnostic is widely used to measure the electron temperature (T_e) and electron temperature fluctuations (δT_e) [1]. The electron cyclotron emission imaging (ECE-I) diagnostic [2] works on the same principle, with the difference that it has multiple lines of sight (LOS), which allow measurements of δT_e in a poloidal plane. At ASDEX Upgrade, both diagnostics measure the extraordinary mode (X-mode) microwave intensity, which can be converted into a radiation temperature T_{rad}. The calibrated T_{rad} can, in many cases, be directly related to T_e at the measuring position defined by the strength of the toroidal magnetic field, only. The measurement position defined in this way is called 'cold' resonance position. However, in specific plasma scenarios and plasma regions, the microwave radiation does not always originate at the cold resonance. This ambiguity is of particular importance for the measurements at the plasma edge. Various types of limitation of the ECE radiometry at the plasma edge such as low and high-density limits, relativistic and Doppler broadening effects are reviewed in [3]. Shine-through in ECE measurements at ASDEX Upgrade is studied in detail in [4, 5] and for the DIII-D ECE-I system in [6]. The ECE-I diagnostic at ASDEX Upgrade has an oblique observation angle, and hence an enhanced Doppler shift of the emission frequency. It also suffers from even higher relativistic down-shift of the respective frequencies due to an oblique angle [7] which makes the scrape-off layer (SOL) non-accessible. However, it has been shown that strong perturbations such as density filaments in the SOL can be detected but the reconstruction of an accurate T_e profile from T_{rad} is, in general, not possible [8]. In [9] the slow variations in density measured with Lithium Beam Emission Spectroscopy (Li-BES)[10] are found to influence the changes in the T_{rad} signal measured by ECE-I and ECE diagnostics in the pedestal. From here it was concluded that the knowledge of the electron density (n_e) plays a vital role in the interpretation of the ECE-I in this region. The same effect has been studied for the marginally optically thick plasmas at KSTAR [11].

In this work, we study the response of the ECE-I and ECE systems to the variations of n_e profile in the optically thick ($\tau > 3$) region. To probe the emission volume for each channel and interpret the radiation temperature T_{rad} we use the electron cyclotron radiation transport model. In this way, we acquire the detailed knowledge on the position of the channels and T_{rad} associated with it. This information is essential for the interpretation of measured variations in T_{rad}. Since the ordinary mode (O-mode) contribution to the ECE radiation is not perfectly blocked by the polarization filter at the ECE antennas in ASDEX Upgrade, the radiation transport model was extended to take this effect into account. In order to test the modeling of the O-mode contribution, we examine a case where the O-mode contribution to the measured T_{rad} becomes significant.

In Section 2 we introduce the radiation transport forward model as the primary tool for studies performed in this manuscript. The results on the diagnostics response to the density variation are shown in Section 3. Furthermore, in

*e-mail: branka.vanovac@ipp.mpg.de

Section 4, we show the particular case where the O-mode contribution is significant.

2 Electron Cyclotron Radiation transport forward model

At ASDEX Upgrade we routinely use the radiation transport model for interpreting measurements of ECE and ECE-I diagnostics [12, 13]. The radiation transport model accounts for the geometry of the detection systems, and radiation transport effects. For a specific frequency, ray tracing is performed, and the radiation transport equation is solved backwards until the ray reaches the antenna. This model is able to treat both X-mode and O-mode polarization. The radiation transport model delivers the information on the location of the emitted radiation associated to a certain channel and the T_{rad} associated with that channel. The location of the emitted radiation is determined as the peak of the birth place distribution of the observed intensity (D_ω) (see figure 1 and figure 2), defined in [12]. We label this peak as the 'warm' resonance position. This 'warm' resonance can differ from the 'cold' resonance position. The cold resonance ($\omega = neB/m_e$, where n is the harmonic of the cyclotron emission) is determined only by the strength of the magnetic field. The characteristic positions for a single ECE-I midplane channel of frequency f = 110.27 GHz are shown in figure 1. The warm resonance (peak in D_ω) is shifted outwards in comparison to the cold resonance (black dashed line). The measured electron emission does not originate from a single location but rather from a broad portion of the plasma determined by the width of D_ω. The most important broadening effects in the case of the ECE-I system are instrumental and Doppler broadening. The Doppler shift-broadening originates from the oblique observation angle of the ECE-I diagnostics. The outward shift towards lower temperatures

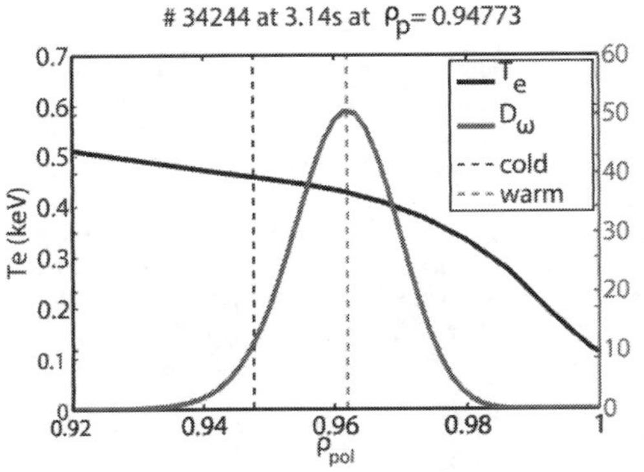

Figure 1. # 34244 at t = 3.14 s: Midplane edge channel with frequency f = 110.4 GHz of the ECE-I diagnostic. The cold resonance position at ρ_{pol} = 0.947, is determined by the magnetic field strength, and is shown as black dashed line. A birthplace distribution of the observed intensity (D_ω) is shown in blue. The peak of the D_ω represents a warm resonance of a channel. The T_e profile is shown in a solid black line.

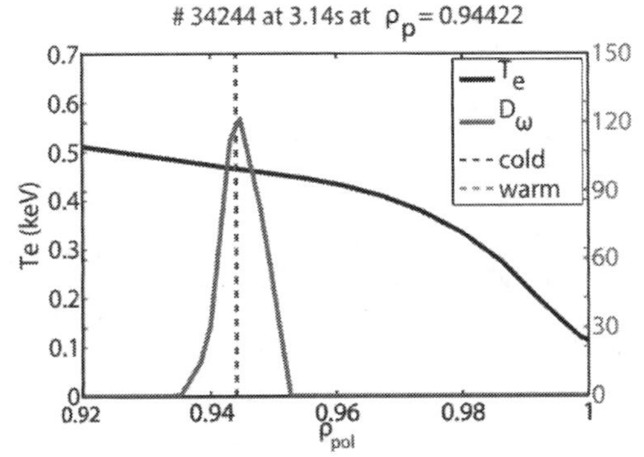

Figure 2. # 34244 at t = 3.14 s: Midplane edge channel of f = 110.27 GHz of the ECE diagnostic. The cold resonance position, determined by the magnetic field strength, is shown as black dashed line. A birthplace distribution of the observed intensity (D_ω) is shown in blue. The peak of the D_ω represents a warm resonance for a respective channel. The T_e profile is shown in a solid black line.

and lower magnetic field compared to the cold resonance (black dashed line) is about 0.5 cm. This is proximately half of the radial resolution for this channel (1.1 cm for this channel).

The ECE system, on the other side, has negligible Doppler shift-broadening (see figure 2) as the observation angle is perpendicular to the magnetic field lines. The radial resolution of this ECE channel, determined by the width of D_ω, is about 5 mm. However, due to a small divergence of the antenna pattern, a slight Doppler shift can also be observed for the case of the ECE system.

3 Variation of the density profile

We apply the radiation transport model to the case of the discharge # 34244 at 3.14 s. This discharge features strong inter-ELM activity in the low-frequency range measured as n_e and T_e fluctuations simultaneously. To simulate the experiment, a background density profile is modified and used as the input for the Electron Cyclotron Emission Forward Model (ECFM). A modified profile consists of the background profile with 10% Gaussian perturbation added on top of it. The forward modeling of the electron cyclotron radiation is performed for all channels of both diagnostics. The T_e profile and the equilibrium mapping are kept the same for the two cases. The n_e and T_e profiles are shown in figure 3. The profiles are obtained with Integrated Data Analysis (IDA) within the framework of Bayesian probability theory [14]. The background profile is displayed as a solid black line in figure 3. With added Gaussian perturbation on top of the background profile, we obtain the modulated n_e profile displayed in figure 3 as a black dashed line. The position of the largest change in density profile corresponds to the position of the ECE and ECE-I channels that measure the mode activity in the experiment. In the following, we use the notation as case 1 and case 2. For case 1 we use the background density

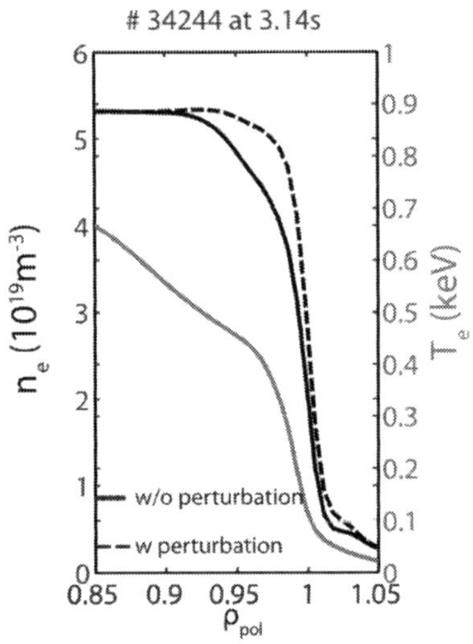

Figure 3. #34244 at 3.14 s: Input profiles of T_e (red) and n_e (black) used to model the response of the ECE-I and ECE channels to the presence of the density modulation. Background density profile is shown as a black solid line while the perturbed density profile is displayed as black dashed line. The temperature profile is kept the same for the two cases.

profile (black solid line in figure 3), and for the case 2 we use a density profile with added perturbation (shown as dashed black line in figure 3). Figure 4(a) shows the synthetic $T_{\rm rad}$ for all channels of the ECE-I system for the two cases shown in figure 3. The channels are mapped onto warm resonances, and they align very well with the profile of the electron temperature inside the confined region. The channels with the cold resonances outside the separatrix obey a relativistic frequency downshift, and are excluded from the analysis. The optical depth (τ) for the two cases is shown in figure 4(b). The effect of modified density is reflected in the form of a displaced profile of the optical depth (blue profile in figure 4(b)). The background profile is shown in gray. In case 1 the channel position is at $\rho_{\rm pol,1}$ = 0.978. It corresponds to the major radius position of $R_{\rm warm,1}$ =2.138 m. In case 2, the same channel has its warm resonance at $\rho_{\rm pol,2}$ = 0.981 with $R_{\rm warm,2}$ = 2.14 m. The $T_{\rm rad,1}$ =347 eV in case 1, and $T_{\rm rad,2}$ = 323 eV in case 2. For ΔR_{warm} of 2 mm the ΔT_{rad} is ~20 eV. This makes 3 % change in the $T_{\rm rad}$ and is comparable to the sensitivity level of the present ECE-I system. The shift of 2 mm is within the radial resolution of this ECE-I channel (11.4 mm). The result of the modeling for the profile ECE system is shown in figure 5. For case 1, the position of the channel is $\rho_{\rm pol,1}$ = 0.977 with major radius position at $R_{\rm warm,1}$ = 2.135 m. The warm resonance of the same channel for case 2 is $\rho_{\rm pol,2}$ = 0.980 and $R_{\rm warm,2}$ = 2.137 m. The $T_{\rm rad,1}$ = 350 eV, and $T_{\rm rad,2}$ = 315 eV. The ΔR_{warm} is 2 mm, and is the same as for the ECE-I system. However, ΔT_{rad} is 35 eV and is higher than that of the ECE-I channel. This makes about 5 % of T_e at the channel position. The radial resolution of this ECE channel is ~ 6 mm thus,

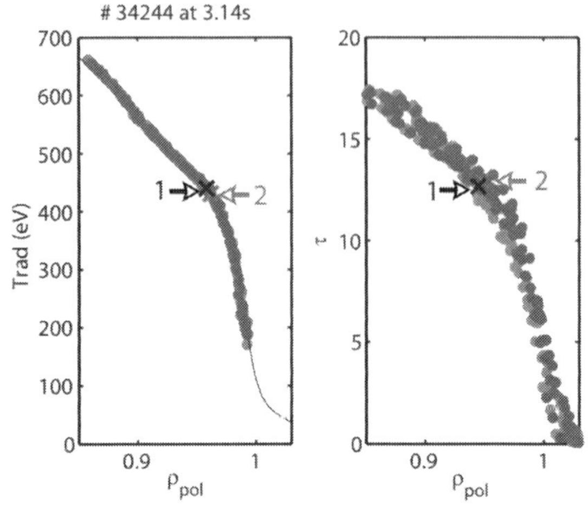

Figure 4. a) Modeled $T_{\rm rad}$ profile of the ECE-I system shown against warm resonances. The dark gray circles correspond to a case without density perturbation with input density profile shown as a solid black line in figure 3. The blue circles correspond to a case where the perturbed density profile, shown as a dashed black line in figure 3, was used. The channel analyzed in this manuscript is labeled as 1 for the case without perturbation, and 2 with perturbation present. b) Optical depth for ECE-I channels for case 1 and 2, marked as gray and blue circles, respectively.

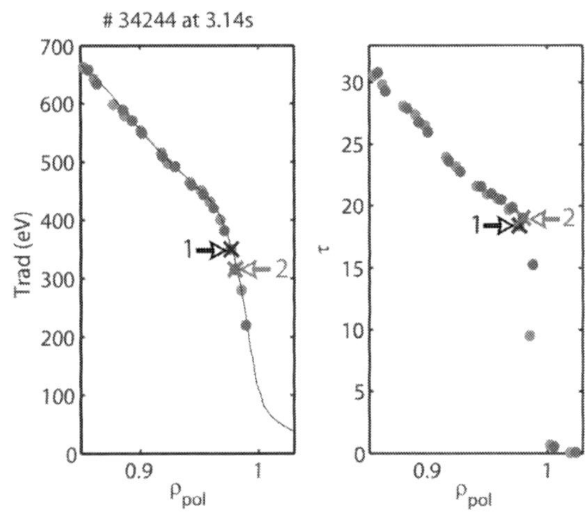

Figure 5. a) Modeled $T_{\rm rad}$ profile of the ECE system shown against warm resonances of respective channels. The dark gray circles correspond to a case with background density profile (displayed as black solid line in figure 3), and the blue circles correspond to a case with added Gaussian (dashed black profile in figure 3). The channel analyzed in this manuscript is labeled as 1 for the case without perturbation, and 2 with perturbation present. b) Optical depth for ECE channels for case 1 and 2, marked as gray and blue circles, respectively.

the radial shift of 2 mm is half of the radial resolution of this diagnostic. Therefore, in both diagnostics, variation of the density profile (no T_e modulation in the modeling) leads to the modulation of measured T_{rad}, visible in a single channel in the optically thick pedestal region. An increase in n_e leads to a decrease of T_e, hence if n_e and T_e are modulated in phase, then the ECE underestimates the measured amplitude. Since the density fluctuations during the inter-ELM modes are nearly almost always present, it is necessary to perform the measurements of the density fluctuations simultaneously. As shown, the effect on the channels that should be sensitive to T_e only (the optical depth is high) is non-negligible and falls within the sensitivity of both ECE-I and ECE diagnostics.

4 O-mode emission

To measure the X2 (extraordinary polarization $E_{wave} \perp B_t$, second harmonic) radiation only, it is necessary to prevent the coexistent O2 (ordinary polarization $E_{wave} \parallel B_t$) contribution via proper orientation of the polarizing grid. The polarizing grid is not always perfectly aligned (pitch angle changes with plasma safety factor q), therefore some of the O-mode leaks into the waveguides, and some of the X2 radiation is missing. This way the O-mode contributes to measured radiation temperatures. To study the significance of the O-mode contribution, the radiation transport model described in [12] has been extended. While the formalism is unchanged for X- and O-mode individually, the calculation of the mixture of O- and X-mode was added. As the polarizing wire grid is aligned toroidally, the poloidal magnetic field introduces an inherent misalignment of the polarization filter. This misalignment is superimposed by the small toroidal and poloidal inclination of the lines of sight. We calculate the O-mode contribution by projecting the polarization vector $\vec{e}$ of each mode onto the polarization vector of the filter $\vec{p}$. The polarization vector of the microwave is calculated from the cold plasma dispersion relation [15, 16] using the magnetic field vector and the electron density n_e at the separatrix position. This is the common way of calculating the optimal polarization for electron cyclotron resonance heating (ECRH)[17] therefore, also it applies to the ECE case, because of reciprocity. However, it should be noted that the Faraday effect and the Cotton-Mutton effect can change the polarization of the wave while it traverses the SOL[18]. Nevertheless, in this work, any change of wave polarization in the SOL is neglected. Since the intensity depends on the square of the electric field, $T_{rad,mod}$ is composed of the X- and O-mode according to:

$$T_{rad,mod} = (\vec{e}_X \cdot \vec{p})^2\, T_{rad,mod,X} + (\vec{e}_O \cdot \vec{p})^2\, T_{rad,mod,O} \quad (1)$$

With this formalism synthetic spectra containing a mixture of X-mode and O-mode can be computed. Because the optical depth for the O-mode contribution is usually a lot lower than for X-mode, the birthplace distribution is significantly different and can be much wider for O-mode electron cyclotron emission. This formalism, integrated into the radiation transport model, is applied to one case where the X-mode polarized ECE of the second harmonic is in cut-off for the large fraction of the low-field side (LFS) radius. In figure 6 we show the radial domain of the fundamental (f_c) and second harmonic ($2f_c$) cyclotron frequency for the magnetic field configuration of this discharge # 32934 at t = 3.298 s with an on axis $B_t = -1.8$ T. One probing frequency of the ECE radiometer (f_{ECE}) is shown as a horizontal solid line. Assumed O-mode ray propagation is depicted in the lower plot. If we were to measure T_{rad} via the X2 mode it would not be possible due to its cut-off at most of the LFS. Note that $f_R > 2f_c$ for the large fraction of the LFS hence, the X2 cyclotron emission is non-accessible. This includes the cold resonance positions of all ECE measurements in the frequency range f_{ECE} = 86 to 98 GHz. For clarity, note that the vertical solid black line corresponds to the magnetic axis. Although the X-mode electron cyclotron emission is in cut-off for this discharge, the measured T_{rad}, shown as green circles in figure 7, is significant and is in the range of 200 to 700 eV. This intensity originates from the second harmonic O-mode emission, which is not affected by cut-off because $f_p < 2f_c$. Synthetic and measured ECE spectra for the discharge # 32934 at t = 3.298 s are compared in figure 7. The ECE measurements are shown as green dots and the synthetic ECE spectrum as red triangles. Depicted are n_e and T_e-profiles. The kinetic profiles were computed with the integrated data analysis technique[14]. On the same figure, we also show the polarizer efficiency $(\vec{e}_X \cdot \vec{p})^2$ as purple triangles. The polarizer efficiency is

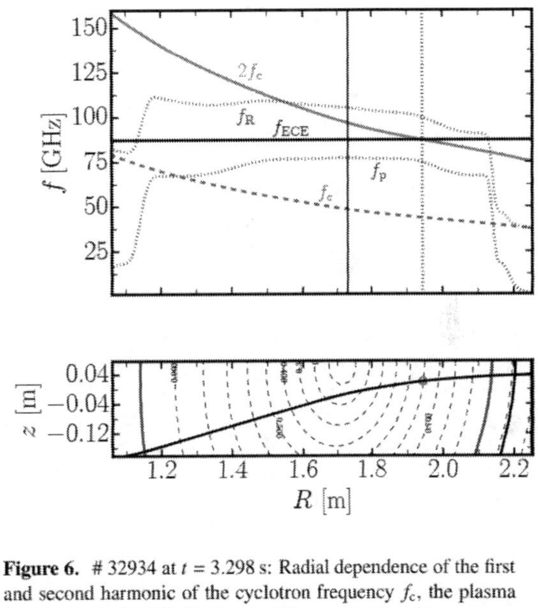

Figure 6. # 32934 at t = 3.298 s: Radial dependence of the first and second harmonic of the cyclotron frequency f_c, the plasma frequency f_p, the right hand cut-off frequency f_R and a measurement frequency of 86 GHz. The solid vertical line indicates the position of the magnetic axis. The bottom plot shows the calculated Ordinary mode (O-mode) ray propagation for f_{ECE} = 86 GHz in the (R,z) plane.

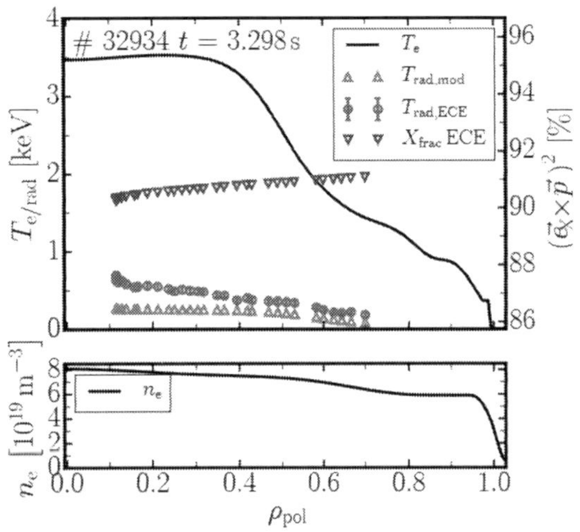

Figure 7. # 32934 at $t = 3.298$ s: T_e (top) and n_e (bottom) profiles versus normalized poloidal radius (black lines). Radiation temperatures $T_{rad,ECE}$ (measured, green circles) is compared to $T_{rad,mod}$ (modelled with ECFM, red triangles), with the polarizing efficiency $(\vec{e}_X \cdot \vec{p})^2$ of 91 % shown as purple triangles.

about 91 %. 100 % of polarizing efficiency would correspond to a perfect X-mode polarization. As can be seen, the synthetic $T_{rad,mod}$ and measured $T_{rad,ECE}$ do not match perfectly. There are two possible explanations for the observed discrepancy. The first is the influence of wall reflections due to the low optical depth $\tau_\omega \approx 0.05$ of the O2 emission. At the stellarator TJK, full wave calculations were performed to asses which parts of the plasma can contribute to ECE measurements at extremely low optical depth[19]. They found that radiation from any point in the machine is capable of reaching their ECE antenna, because of the wall reflections. The model of infinite reflections between two plane mirrors, which is employed in the radiation transport model [12], is not suitable in this case. Furthermore, the reflections of the metallic surface of the vessel can cause mode conversion between O- and X-mode. The model assumes that $T_{rad,mod,X} = 0$ due to the cut-off. It means that any mode conversion from O to X-mode could greatly increase the modeled T_{rad} due to the much larger acceptance of X-mode radiation of the polarizing beam splitter. Since only 91 % of the O-mode radiation is rejected in the current model, even small changes in the polarization state of radiation could have a significant impact on the observed T_{rad}. Unlike the influence of wall reflections, this effect is independent on the optical depth and could also affect scenarios in which the O-mode is optically thick. Additional effects, that were omitted in this work, include polarization change in the SOL and X-mode emission reflected from the cut-off layer back to the antenna.

5 Conclusions

The radiation transport model enables us to model the response of the ECE-I and ECE systems to different plasma scenarios at the ASDEX Upgrade tokamak. Of special interest are H-mode plasmas with pronounced density fluctuation at the plasma edge region during the inter-ELM periods. We have modeled such a scenario and shown the effect of the density at the pedestal top on an ECE-I and ECE channels both being shifted towards lower magnetic fields and lower radiation temperatures. The effective change in the radiation temperature seen by one channel due to the density is within the sensitivity for both ECE-I and ECE systems and might, therefore, be interpreted as real temperature variation. This effect can have an impact when analyzing the cross phase between n_e and T_e [20]. It is not possible to disentangle n_e and T_e fluctuations with ECE-I and ECE measurements only, however, the measurement of n_e fluctuations allows a quantification of their contribution to T_{rad} via radiation transport model. Furthermore, we have shown that in the particular plasma scenario, where X-mode polarization is in cut-off, the measured radiation temperature is significant. This opens up the possibility to treat the O-mode contribution to the radiation temperatures. However, with the current radiation transport model it is not possible to quantitatively analyze ECE of the second harmonic O-mode measured at ASDEX Upgrade. Nevertheless, it is important to note that even when the X-mode is not in cut-off, T_{rad} is underestimated by 5-10 % due to O-mode contributions.

6 Acknowledgments

This work has been carried out within the framework of the EUROfusion Consortium and has received funding from the Euratom research and training programme 2014-2018 under grant agreement No 633053. The views and opinions expressed herein do not necessarily reflect those of the European Commission.

References

[1] H.J. Hartfuss, T. Geist, M. Hirsch, Plasma Physics and Controlled Fusion **39**, 1693 (1997)

[2] I.G.J. Classen, C.W. Domier, N.C.L. Jr., A.V. Bogomolov, W. Suttrop, J.E. Boom, B.J. Tobias, A.J.H. Donné, Review of Scientific Instruments **85**, 11D833 (2014)

[3] W. Suttrop, A. Peeters, IPP report (1996)

[4] S.K. Rathgeber, L. Barrera, T. Eich, R. Fischer, B. Nold, W. Suttrop, M. Willensdorfer, E. Wolfrum, the ASDEX Upgrade Team, Plasma Physics and Controlled Fusion **55**, 025004 (2013)

[5] S.S. Denk, R. Fischer, O. Maj, E. Poli, M. Willensdorfer, J. Stober, U. Stroth, W. Suttrop et al., *Shine-through in electron cyclotron emission measurements*, in *44th EPS Conference on Plasma Physics* (2017), proceedings of the 44th EPS Conference on Plasma Physics

[6] B.J. Tobias, M.E. Austin, J.E. Boom, K.H. Burrell, I.G.J. Classen, C.W. Domier, N.C.L. Jr., R. Nazikian, P.B. Snyder, Review of Scientific Instruments **83**, 10E329 (2012), `https://doi.org/10.1063/1.4733742`

[7] M. Willensdorfer, S.S. Denk, E. Strumberger, W. Suttrop, B. Vanovac, D. Brida, M. Cavedon, I. Classen, M. Dunne, S. Fietz et al., Plasma Physics and Controlled Fusion **58**, 114004 (2016)

[8] B. Vanovac, I. Classen, S. Denk, E. Wolfrum, M. Hoelzl, A. Lessig, F. Orain, N.C.J. Luhmann et al., *ELM filaments on ASDEX Upgrade: ECEI observations and modelling*, in *43th EPS Conference on Plasma Physics* (2016), proceedings of the 43th EPS Conference on Plasma Physics

[9] B. Vanovac, E. Wolfrum, S.S. Denk, F. Mink, F.M. Laggner, G. Birkenmeier, M. Willensdorfer, E. Viezzer, M. Hoelzl, S.J. Freethy et al., Plasma Physics and Controlled Fusion **60**, 045002 (2018)

[10] M. Willensdorfer, E. Wolfrum, R. Fischer, J. Schweinzer, M. Sertoli, B. Sieglin, G. Veres, F. Aumayr, the ASDEX Upgrade Team, Review of Scientific Instruments **83**, 023501 (2012)

[11] M.K. Ayub, G.S. Yun, W. Lee, H.K. Park, Journal of the Korean Physical Society **70**, 268 (2017)

[12] S.S. Denk et al., *Radiation transport modelling for the interpretation of oblique ECE measurements*, in *19th Joint workshop (EC-19) on Electron Cyclotron Emission (ECE) and Electron cyclotron Resonance Heating (ECRH)* (EPJ, 2016)

[13] S.S. Denk, R. Fischer, H.M. Smith, P. Helander, O. Maj, E. Poli, J. Stober, U. Stroth, W. Suttrop, E. Westerhof et al., Plasma Physics and Controlled Fusion **60**, 105010 (2018)

[14] R. Fischer, C.J. Fuchs, B. Kurzan, W. Suttrop, E. Wolfrum, the ASDEX Upgrade Team, Fusion Science and Technology **58**, 675 (2010)

[15] M. Bornatici, R. Cano, O.D. Barbieri, F. Engelmann, Nuclear Fusion **23**, 1153 (1983)

[16] F. Albajar, N. Bertelli, M. Bornatici, F. Engelmann, Plasma Physics and Controlled Fusion **49**, 15 (2007)

[17] R. Prater, Physics of Plasmas **11**, 2349 (2004), `https://doi.org/10.1063/1.1690762`

[18] TSUJIMURA et al., Plasma and Fusion Research **1** (2016)

[19] G. Sichardt, E. Holzhauer, A. Köhn, M. Ramisch, T. Hirth, *Electron cyclotron emission measurements at the optically thin plasmas of the stellarator TJ-K*, in *44th EPS Conference on Plasma Physics* (2017)

[20] S.J. Freethy, G.D. Conway, I. Classen, A.J. Creely, T. Happel, A. Köhn, B. Vanovac, A.E. White, Review of Scientific Instruments **87**, 11E102 (2016)

Research activities and progress on the long pulse ECRH launcher for EAST

Xiaojie Wang[1,*], *Fukun* Liu[1], *Wei* Wei[2], *Yunying* Tang[1], *Bo* Li[1], *Liyuan* Zhang[1], *Handong* Xu[1], *Dajun* Wu[1], *Jiafang* Shan[1], *Huaichuan* Hu[1], *Jian* Wang[1], *Zege* Wu[1], *Wendong* Ma[1], *Jian* Zhang[1], *Miaohui* Li[1], *Yang* Zhang[1], *Yong* Liu[1] and EAST team

[1]Institute of Plasma Physics Chinese Academy of Sciences, No.350 Shushan lake Road, 230031, Hefei, China
[2]Hefei University of Technology, No.193 Tunxi Road, 230009, Hefei, China

Abstract. A long pulse Electron Cyclotron Resonance Heating (ECRH) system is developed on EAST tokamak for plasma heating and current profile tailoring. The ECRH system is designed to operate at 140GHz and to inject 4MW CW power. With respect to the physical objectives of the newly built ECRH system, a quasi-optical launcher is designed to inject 4MW continuous wave into plasma through an equatorial port. Gaussian beams delivered from evacuated corrugation waveguides will be focused and reflected by high thermal conductive metal mirrors, and then steered by using push-rod steering mechanism with entire scanning range of ±25° toroidally and over 30° poloidally in plasma cross section. The mirrors are carefully designed with mega watts power handling capability and optimum optical characteristics. The performance of steering mechanism has been tested before installation, an open-loop control system for ECRH launcher has been implemented for required mirror movement and proper polarization between plasma discharges. This paper will present the overall design and progress of the launcher, along with the performance in EAST campaigns. Considerations and possible upgrade of the design features relevant to long pulse operation are discussed.

1 Introduction

Electron Cyclotron Resonance Heating (ECRH) scheme has many advantages in magnetic fusion devices, e.g. perfect power coupling with the plasma, highly localized power deposition [1-4] and good theoretical understanding of the wave-particle interaction physics. Besides, due to the technological advantage of launching structure, ECRH will support steady state operation for next step devices, such as ITER [5] and DEMO [6].

As one of the most promising external auxiliary heating and current drive methods, ECRH provides large flexibility and complies with diverse physics demands. It was developed and studied in most tokamaks such as DIII-D [7], ASDEX [8], HL-2A [9] and the stellarators as well, like W7-X [10] and LHD [11]. A long pulse ECRH program is launched on EAST under the support of National Magnetic Confinement Fusion Science Programs of China since 2011. It will be used for central heating and current drive applications as well as off-axis control of magnetohydrodynamic (MHD) instabilities such as Neoclassical Tearing Modes (NTMs), sawtooth, and Edge Localized Mode (ELM). The EAST ECRH system is planning to deliver a total installed power of 4 MW into plasma for either in stationary operation mode with a pulse length up to 1000 sec or a modulation mode with a frequency up to 1 kHz [12, 13]. Considering the operational range of toroidal magnetic field on EAST and the frequency of a commercially available gyrotron, the new system uses 140 GHz to match the second harmonic of the electron cyclotron resonant frequency.

The ECH power generated from 4 gyrotrons will be transmitted by 4 individual evacuated waveguides and fed to plasma by a front steering quasi-optical launcher from the low field side (LFS) via an allocated equatorial port (M-p) (as shown in Fig.1). Therefore, the launcher should be compatible with the requirements of long pulse power handling and flexible control of localized power deposition. These design goals present a number of challenges. This paper describes the overall design and progress of the launcher.

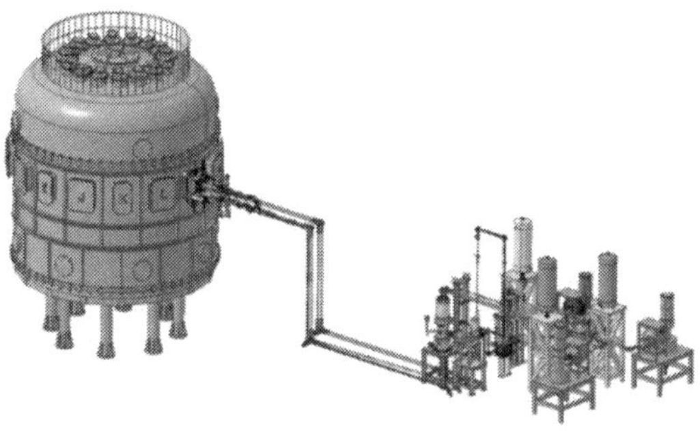

Fig. 1. Schematic of 140GHz/ 4MW long pulse ECRH system on EAST

* Corresponding author: xjiew@ipp.ac.cn

2 Design of launcher

2.1 Layout design

To investigate the operational range of the ECRH system on EAST, numerical simulations of ECCD have been carried out by using the ray tracing code TORAY. For the given equilibrium parameters of EAST, the deposition of EC power was greatly dependent on the scanning of launcher position, and both the poloidal and toroidal steering angles. It is found that a launch from Z = 30 cm (the vertical position) is the best scheme for core localized heating and the suppression of neoclassical tearing modes [14, 15].

Taken into account the available port dimensions (518 mm in width and 960 mm in height), the beam launch position is chosen to locate at 30 cm vertically away from the equatorial plane and at R = 3 m with radial injection. The driven current density and its radial deposition location contours, as a function of the launch angles, are presented in figure 2. The range of toroidal injection angle from -25° to +25° is suitable to obtain higher ECCD efficiency in co-direction or ctr- direction, meanwhile, the beam spans 30° poloidally, a wide range (0.03 a ~ 0.8 a) of plasma areas of interest is covered.

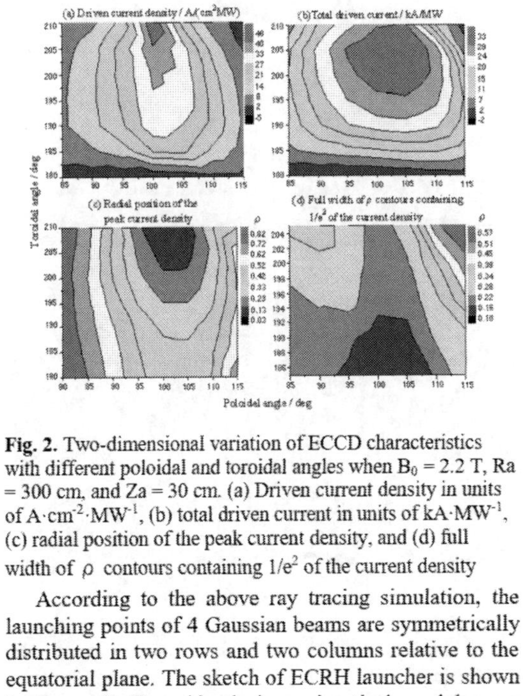

Fig. 2. Two-dimensional variation of ECCD characteristics with different poloidal and toroidal angles when B_0 = 2.2 T, Ra = 300 cm, and Za = 30 cm. (a) Driven current density in units of $A{\cdot}cm^{-2}{\cdot}MW^{-1}$, (b) total driven current in units of $kA{\cdot}MW^{-1}$, (c) radial position of the peak current density, and (d) full width of ρ contours containing $1/e^2$ of the current density

According to the above ray tracing simulation, the launching points of 4 Gaussian beams are symmetrically distributed in two rows and two columns relative to the equatorial plane. The sketch of ECRH launcher is shown in figure 3. Four identical quasi-optical modules are designed for the beams propagating, each contains a 63.5 mm-i.d stainless steel (SSL) corrugated waveguide with interiors nickel plated, a set of quasi-optical arrangement formed by a fixed-focusing and a rotatable-flat mirror. Each beam emitted from the SSL waveguides is obliquely incident on a focusing mirror, and then guided into the flat steerable mirror, which purely changes the beam angle in the poloidal and toroidal directions. All the subassemblies will be integrated to the main frame, an external shield structure is mounted for protection against contamination of lithium (Li) wall coatings.

2.2 Mirrors design

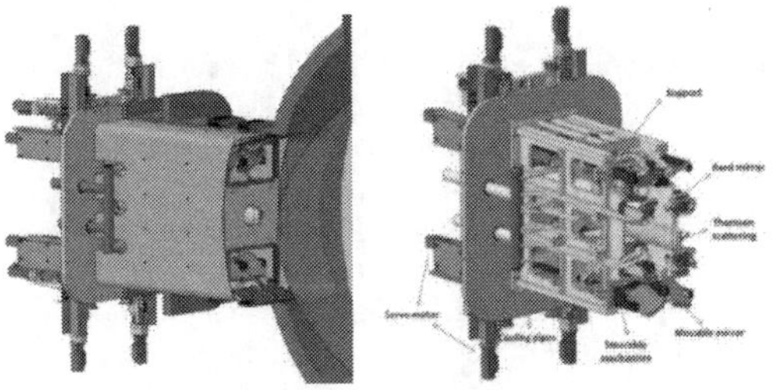

Fig. 3. (Left) Illustration of the 4MW front steering equatorial launcher design, (Right) Configuration of the launcher

The focusing mirror, at angle of 45° with respect to the incident beam, allows the possibility of changing the beam size for the higher localized current drive. Spherical focusing mirror was chosen for simplicity, the focal length of the fixed mirror has been optimized in a reference absorption layer in the plasma. The evolutions of the beam waist by varying the focal length under given distance from the SSL waveguide termination to focusing mirror are displayed in figure 4, The radius of curvature of the spherical mirror is optimized to be 1400 mm, the focusing mirror is located at 240 mm away from the waveguide output and the beam radius is about 31 mm at the resonance layer (about 1200mm). Considering the optimized results of the beam propagation properties and the spatial limitation of the port duct, the dimensions of the focusing mirror are 127 mm × 90 mm and mobile mirror sizes are170 mm × 113 mm, which are chosen to ensure that at least 99.99% Gaussian beam power intercepted by reflecting surfaces of mirrors.

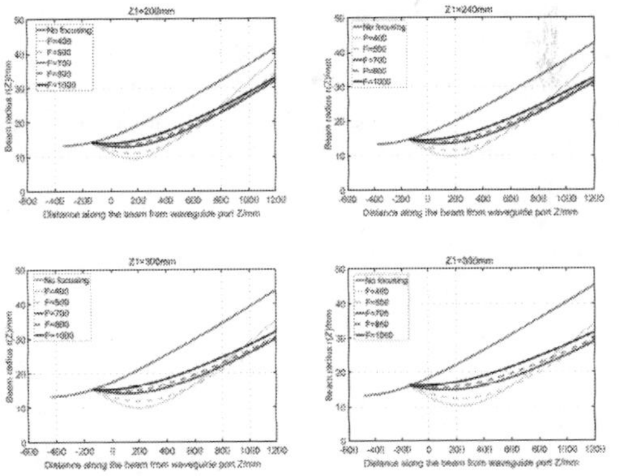

Fig. 4. The beam waist along the propagation trajectory, as a function of shaping mirror focal length. The horizontal -140 mm and 0 mm refer to the positions of the focusing mirror and steering mirror, respectively.

The mirrors should be actively cooled to withstand the heat loads during 1 MW long pulse operation. In the initial design, all possible radiation power of EAST future full power operation are estimated for the calculation, therefore, there is big margin of the total heat load. 3 D models of mirrors are set for thermal stress analysis, the choices of material and structure of mirrors were guided by thermal calculations and estimations of heat loads on the reflective surfaces [16]. The oxygen-free copper (OFC) is adopted as mirror surface and body for requiring higher thermal conductivity, the parameters of internal cooling channels are carefully optimized to provide sufficient heat exchange with the smooth surface. The heat load on the movable mirror (as shown in figure 5) is higher than on the fixed focus mirror, the cooling channels are cut directly into the copper layer and form several S-shaped loops going from the center to the edge of the mirror. A sheath of stainless steel is backside brazed for higher mechanical robustness. Finite element thermal calculations predict that a maximum temperature is about 296°C located at the center of the mirror surface and the peak stress is 178.1 MPa at the corner of cooling channels, the peak temperature and stress are allowable.

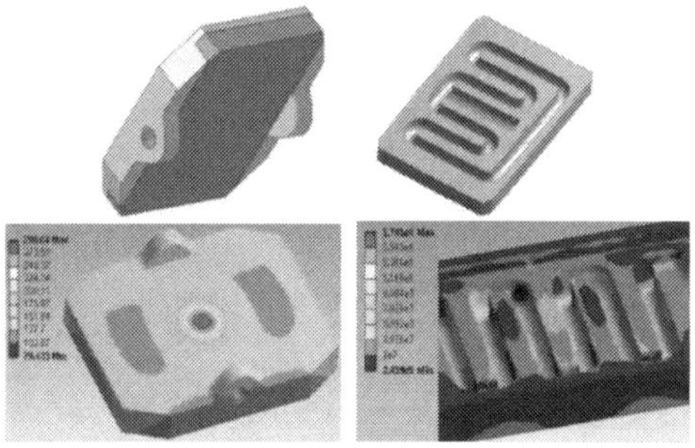

Fig. 5. (Upper) The cooling design of the mobile mirror, (Lower) the temperature and the stress distributions of the mirror.

The uniform thickness is 20 mm for the flat mobile mirror, the choice of the metal material and increased mass are often accompanied by high electromagnetic forces. The distribution of magnetic field on the mirror, induced current density, force and torque are computed by MAXWELL software and the structure stress and deformation of mirror are analyzed in the case of plasma current quench with exponential decay [17]. The result of the electromagnetic transient analysis indicates that the mirror could work safely under the strong electromagnetic environment. A maximum deformation as seen from figure 6, locates on the top of mirror is 0.0225 mm which is very small with respect to the size of mirror. However, to decrease the induced eddy current density, force and torque on the edge of the mirror, design modification of mirrors is under investigating, such as further optimizing the cooling design to decrease the heat volume with high conduction, trying a tapered thickness mirror to reduce electromagnetic force.

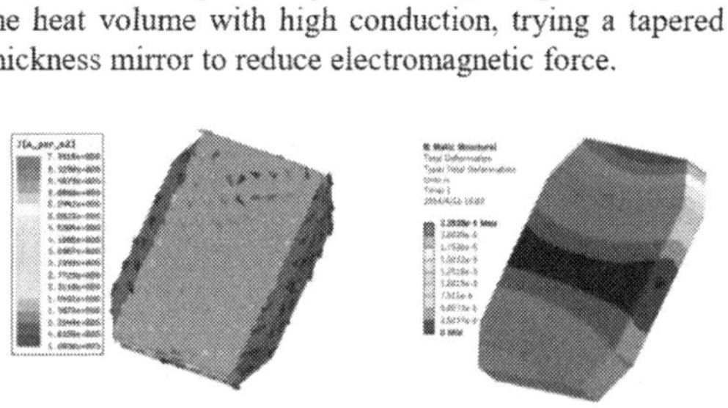

Fig. 6. The distribution of induced eddy current density and the maximum deformation locates on the top of mirror

2.3 Steering mechanism and performance

A pair of steering mechanisms for the mobile mirror enables extensive control of the toroidal and poloidal injection angles for the beam. As shown in figure 7, the two directional steering motions are respectively actuated by horizontal motor and vertical motor, the angle steering is achieved by a push-rod drive [18]. The poloidal injection degree could be changed independently by pushing or retracting the rod, no coupling happens during poloidal movement when toroidal is fixed. The toroidal movement is realized through the rotation of rod around its axis, the connecting rod outside the vessel transmits the up-and-down motion to the crankshaft, which changes it into rotary motion inside the vacuum box. A slight poloidal motion will occur when scanning the mirror toroidally, the poloidal angle has to be corrected accordingly. All rods are connected via welded bellow feed-throughs to servo-motors situated outside the port flange.

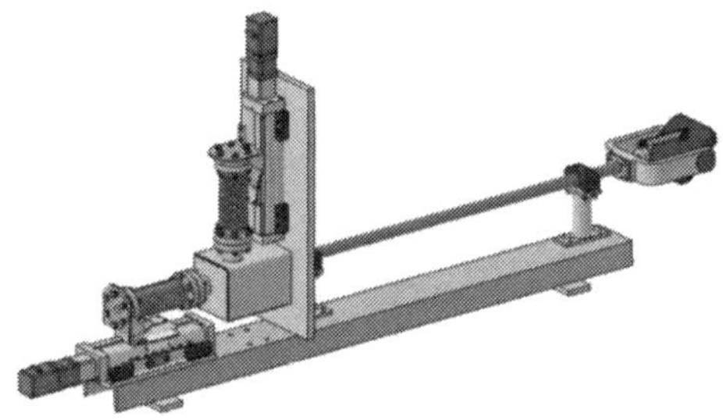

Fig. 7. 3D view of the steering mechanism design

Fig. 8. Optical calibration of the launcher

The two directional steering angles are accomplished via independent actuator rods. To determine the transformation functions of beam angles and displacements of pushrods, the configuration of the space linkage driving mechanism has been mathematically analyzed by using the direction of the normal to the steerable mirror and the geometry of reflection [19]. The sophisticated mathematical relationships between the displacements of linear driving sliders on both directions and the corresponding shift angles of reflected beam was derived, the kinematic analysis was programmed by Matlab and integrated in the launcher programmable logic controller (PLC) control system. Laser calibrations have been implemented to test the steering performances and the alignment of the components (as shown in figure 8).

The graph in figure 9 displays the scanning map and the test results on the front projective plane. The sector diagram is obtained based on the correlation analysis of steering structure, in which the longitude lines represent the calculated trajectories of poloidal angles with several given toroidal angles, while the latitude lines denote the toroidal scanning. During the test, the theoretical positions of the randomly selected injection angles were marked with stars. The positions of laser points were marked with triangular symbols, the purple one and the light blue one represent the back and forth moving directions, respectively. It is obvious that the tracking routes of laser points are in good agreement with the dotted theoretical lines, hence, the steering feasibility and validity of the expressions were confirmed, it can provide a reliable mathematical model for the accurate control of beam steering. Whereas, the deviation seems increasing when the mirror oriented at its steepest angle relative to the incident beam, it is particularly sensitive to the toroidal angle. The maximum deflection of the angle is about 0.5 deg. One of the factors which affect the accuracy is that the fit clearance and the elastic deformation of the slender shaft were ignored in the mechanical kinematic analysis, however, the exiting tolerance of all the machining parts certainly led to the positioning bias. The torque caused by the plastic deformation of the flexible metal cooling pipe also gives rise to the backlash of the rotary motion. Moreover, the beam is assumed to ideally inject on the center of the mirrors, a small displacement of the mirror arrangement will consequentially result in undesired shift of the beam deposition on the target area.

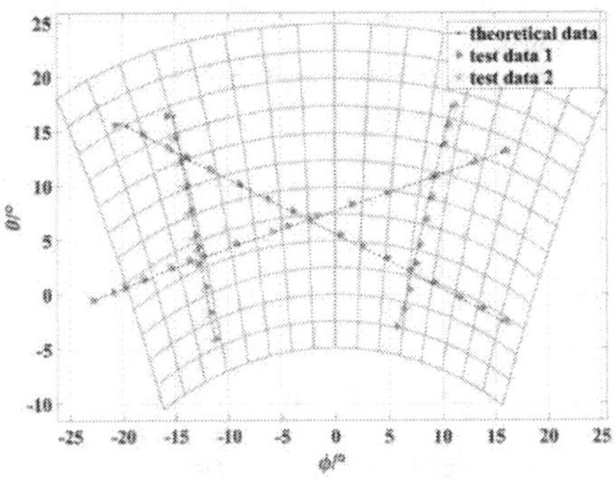

Fig. 9. Scanning map of EAST ECRH launcher on the projective target (distance L from the launching point is 2m)

To procure a launcher that fulfills the requirements of EAST in terms of performance and system reliability, Tests activities including the steering optics, actively cooling capability, the thermomechanical and fatigue aspects of the design should be carried out during the R&D phase for design validation prior to installation in EAST port. In particular, the vacuum seal performance of launcher with pressurized cooling pipes was rigorously inspected in a testing chaember under vacuum and 200◦C baking conditions for the sake of security operation.

Errors in the positioning of the mirrors and the deformations of the port plug structure will also influence the launch points and injection angles, therefore, validation test of optics after installation was performed to inspect the impact of possible misallignment.

2.4 Shutter design

In EAST campaigns, lithium (Li) powder evaporation was used as a routine wall conditioning technique to access the H-mode during plasma discharges. In order to protect the mirrors surface and steering mechanism against coating contamination by deposited Li powder, a tight mandatory metal shield is designed. The critical part of the shield structure is the movable shutter in front of the beam propagating route. Figure 10 shows the schematic drawing of the shutter, two pieces of SSL plates are used and pneumatic actuated by cylinder, a long driveshaft with a crank links to the plate, the linear motion of the rod generates the plate rotary along its fixed edge. Considering the required angular range of plate motion, the travel distance of the rear slider was carefully designed and positioned. The movable performance was tested using a prototype under the bulk lithum coating experiment during R&D, the shutter can be smoothly open and close after lithum coating.

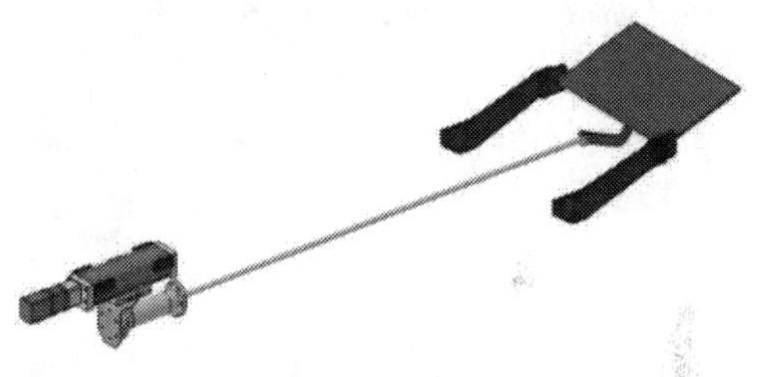

Fig.10. Schematic drawing of the shutter

3 Progress of launcher

3.1 Current status

The fabrication of the launcher was in two steps, the lower two channels of the launcher were developed and installed on EAST in 2014 without power. The lower right module of the launcher for power injection has been in service since 2015 spring campaign till now. The launcher with all 4 modules was built in last year, it is presently installed on EAST as shown in figure 11. In the coming experiment, the second gyrotron from GYCOM is tested and ready for operation, another launcher module will be employed and await its initial operation.

Fig.11. Photos of the launcher plug-in

3.2 Engineering commissioning

The commissioning of launcher in situ was done during the initial operation of 1 MW ECRH system powered by #1 gyrotron. Figure 12 indicates the experimental result of power scanning capability of the first launcher under a modulated operation at frequency of 10Hz/1s. Nine sets of beam angles varied in a sequence of discharge shots, the deposit locations were measured by ECE diagnostics. The obtained scanning range is accessible from core to boundary close to 0.8. During the test, no obvious hot spots were found in the high filed wall which confirmed that polarizations and injection angles were as expected [20]. Comparing the measured deposition positions with the predicted value, most of them agree well except some offset around 0.3, the experiment will be further done to identify the position precision. Till now, the launcher is stably operated under 100 s pulse length with 500kW injection power. The influence on sawtooth period and tearing mode stabilizing were investigated due to the flexible steering capability [21].

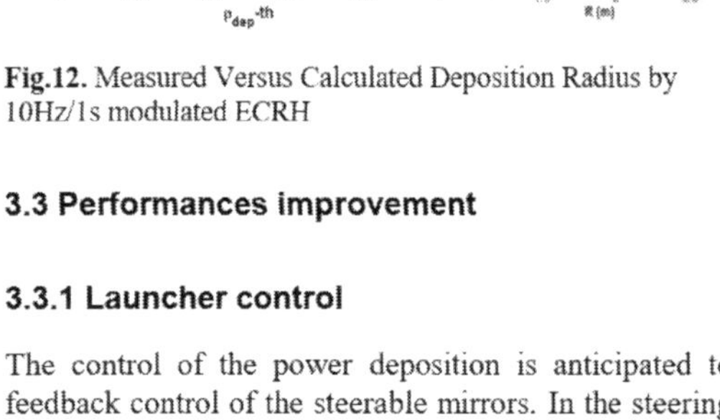

Fig.12. Measured Versus Calculated Deposition Radius by 10Hz/1s modulated ECRH

3.3 Performances improvement

3.3.1 Launcher control

The control of the power deposition is anticipated to feedback control of the steerable mirrors. In the steering calibration test, the velocity and the response time of the steering mechanism are primarily tested, the obtained velocity is about 5°/50ms in poloidal. The direct current servomotor has an adequate ratio of torque to inertia, but for the purpose of fast scanning, more work should be done to optimize the mechanical structure to reduce the friction of the pivot joints. The divation result from the mechanical tolorenance and backlash needs to be investigated and appropriately compensated in the algorithm. In addition, the mirror mounting structures and the dynamic tracks of cooling bellows will be further optimized to facilitate the position adjustment during the optical calibration, and minimize the slight angular misalignment. Hence, the position accuracy and the steering velocity need to be further determined.

The launcher steering is remote controled by a control loop based on PLC system. At present, the beam deposition is feed forward controlled, the mirror angles are adjusted shot by shot during discharge experiment. The PLC system works for mirror movement control as well as monitoring the general status of the motors. The aiming process was not fully automated, yet, a database for the objective ECCD deposition position is off-line computed by TORAY. The target depositions could be achieved by manually setting injection angles on the control screen, the steering commands are dispatched through the PLC to the steering mirror drives, the encoders transform them into the motion parameters and trigger the motor to implement it.

Fast poloidal steering is important for rapid tracking of instabilities. However, at the moment, the response time of polarization control is nearly 1 sec, which is not fast enough for real time operation. In order to meet the requirement of NTM suppression, a fast poloidal scan within small range ($\pm 4°$ around the initial position of mirror) was developed to sweep the ECCD in a small step for a better alignment to the island with no need of polorization change, the angular resolution of the mirror rotation is desired to be better than 1°. The representative fast scan was tested in 2017 EAST campaign, the achieved angular velocity obtained from the response in plasma is 2° /200ms.

In the near term plan, a mock-up shown in figure 7 will be fabricated and actuated with a control modular (in figure 13), extensive test will be excuted for further analysis and optimization of the steering performance. Improvement of the control system by integrating real time calculations of RF deposition using fast real time EFITs and kinetic profiles using an accurate diffraction model will be part of future work.

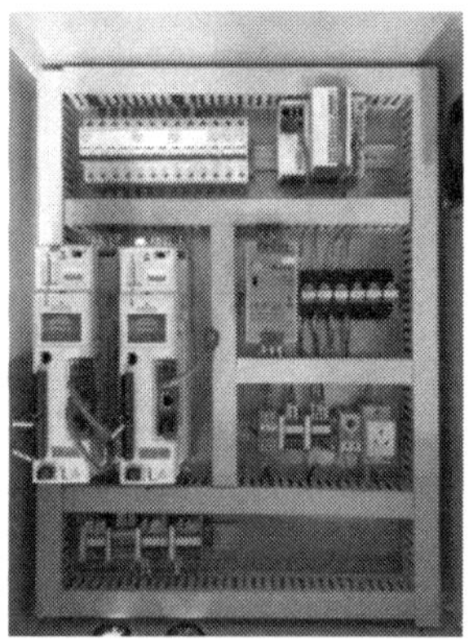

Fig. 13. Control modular of the steering mirror

3.3.2 Mirror replacement

After each campaign with lithum wall conditionning, the deposition of dust or lithium could be found on the surface of the mirrors, especially the steering mirror (as shown in figure 14), which are need to be replaced or cleaned. To remove the mirror, the flexible cooling tube connected with the mirror should be cut off, and the steering mechanism should be taken apart. After new mirror mounted, the mechanism and the cooling should be re-assembled, the leakage test as well as the optical and mechanical test should be repeated to ensure the reliability of the launcher. The tricky issue encountered in the burdensome maintenance process is the copper mirror surface protection. Now, a novel design concept of detachable mirror surface is ongoing to simplify the mirror replacement work.

Fig. 14. Contaminations on the steering mirror surface

4 Summary

A long pulse quasi-optical launcher rated for 4 MW / 1000 sec power injection has been developed on EAST ECRH system, the design is overviewed in this paper. Efforts have been made on the design of the most critical elements such as steering mechanism of the mirror and shutter, flexible cooling connection for mobile mirror, a series steering mechanism test and optical calibration of the rotation angle. Gaussian beams delivered from evacuated corrugation waveguides will be focused and reflected into designated plasma area by actively cooled metal mirrors, with entire scanning range of ±25° toroidally and over 30° poloidally. The mirrors are carefully designed to satisfy the desired physical requirements. Characteristics of beam zooming and propagation has been optimized, the thermal-mechanical evaluation has been performed by finite element analysis software. Test activities including beam steering, cooling and leakage inspection have been done before installation. Based on the commissioning results, Attentions are paid on steering performance improvement and development of the open-loop control system towards real time feedback control ability.

Acknowledgement

This work supported by the National Magnetic Confinement Nuclear Fusion Power Development Special Fund of China (Grant Nos. 2011GB102000, 2012GB103000 and 2015GB102004). The authors greatly appreciate all kind person from the following laboratories: GA, PPPL, Tore Supra and NIFS for the fruitful discussions.

References

1. H. Zohm, Fusion Sci. Technol. **52** (2007) 134-144.
2. M. Bornatici, et al., Nucl. Fusion **23** (1983) 1159
3. V. Erckmann, Plasma Phys. Controlled Fusion **36** (1994) 1869
4. R. Prater, Phys. Plasmas **11** (2004) 2349.
5. T. Omoria, Fusion Eng. Des. **86** (2011) 951-954
6. E. Poli, Nucl. Fusion **53** (2013) 013011
7. J. Lohr, J Infrared Milli Terahz Waves **32** (2011) 253-273
8. J. Stober, EPJ Web of Conferences 201232020 (2012)
9. X. R. Duan, Nucl. Fusion **50** (2010) 095011
10. V. Erckmann, et al., Fusion Sci. Technol. **52** (2007) 291-312
11. Yoshimura, et al., Fusion Sci. Technol. **58** (2010) 551
12. X. J. Wang, et al., Fusion Eng. Des. **96-97** (2015)
13. H. D. Xu, et al., Plasma Sci. Technol. **18** (2016) 442
14. W. Wei, et al., Chin. Phys. B **23** (2014) 055201
15. X.G. Wang, et al., Phys. Scr. **86** (2012) 065501
16. Y. Tang, et al., Fusion Eng. Des. **89** (2014) 847
17. Y. Tang, et al., J Fusion Energy **34** (2014) 422-426
18. R Ellis, et al., Fusion Engineering **1-4** (2011)
19. L. Zhang, et al., Machine Des. Research **31** (2015) 5
20. D. J. Wu, et al., J Fusion Energy **33** (2014) 634-639
21. Y. Yuan, et al., Phys. Plasmas **23** (2016) 062503

ECCD operations in the second experimental campaign at W7-X

M. Zanini, H.P. Laqua, T. Stange , C. Brandt , M. Hirsch, U. Höfel, N. Marushchenko, U. Neuner, K. Rahbarnia, J. Schilling, H. Thomsen, R.C. Wolf and the W7-X Team

Max-Planck-Institute for Plasma Physics, Wendelsteinstr. 1,17491 Greifswald, Germany

Abstract. In the Wendelstein 7-X stellarator, up to 7MW of power are delivered to the plasma by an electron cyclotron resonance heating system consisting of ten 140 GHz gyrotrons [1]. Due to the flexible front steering mirror of each beam line, the power deposition can be varied over the whole plasma radius and is optionally combinable with additional current drive. This flexibility, together with small toroidal currents in the stellarator, makes W7-X a perfect testbed for electron cyclotron current drive (ECCD) experiments, which have been successfully accomplished during the first two experimental campaigns OP1.1 and OP1.2a.
Long discharges (lasting up to 30s) have been performed in OP1.2a, thus allowing the study of the current drive time evolution and the possibility to compensate the bootstrap current.
ECCD efficiency has been studied using different power deposition profiles combined with a variation of the injection angles in relation to the magnetic field.
During ECCD experiments, saw-tooth-like oscillations have been observed. Depending on the driven current density, ECCD can significantly modify the rotational transform (iota) profile, which can locally reach low order rational, thus triggering plasma instabilities.
Different current density profiles have been tested, in order to try to understand the main trigger parameter for the instabilities. In particular, effects caused by current density gradient have been investigated producing both co- and counter-current drive at different radial positions: the total current drive is negligible, but a strong current gradient arises by driving currents in opposite directions.
In this work an overview of ECCD operations in OP1.2a is given and first results, comparing different diagnostics, are presented. An initial 1-D model, coupled with the ray tracer TRAVIS, is developed, in order to have an estimation of current diffusion times and the radial position where a low order rational crosses the disturbed iota profile.

1. Introduction

The superconductive optimized stellarator Wendelstein 7-X (W7-X) [1] generates the rotational transform by means of five-period external coils, therefore toroidal current is not needed to confine the plasma. During the first two operation campaigns (OP1.1 and OP1.2a) plasma start-up and heating have been obtained using Electron Cyclotron Resonance Heating (ECRH), with up to 10 gyrotrons at 140 GHz, able to operate both in O2 and X2 mode, for a maximum delivered power of 7MW [2]. ECRH is a versatile tool to deposit power in the plasma, since the resonance region is extremely localised and it can be varied by steering the launcher mirror. If oblique injection takes place, Electron Cyclotron Current Drive (ECCD) is generated [3]; the generation of ECCD is well localised too and it can be used to locally modify the plasma rotational transform and shear or to compensate bootstrap current. Nevertheless, during ECCD experiments in abovementioned experimental campaigns, fast collapses of electron temperature have been observed [1] [4] to present strong similarity with well-known saw-tooth- [3] [5]crashes in tokamaks [6].

A short overview of the experimental setup will be presented, and then a remarkable ECCD shot will be introduced, in order to highlight the main issues related to this kind of collapses. In the end, the 1-D model for current evolution, under development, will be presented.

2. Experimental Setup

50 superconductive non-planar coils and 20 planar coils generate W7-X magnetic field [1]. It is possible not only to change the current intensity on coils in order to change the magnetic field strength, but also to change the current ratio between different coils, in order to generate different magnetic configurations, with different rotational transforms. During OP1.1 and OP1.2, different magnetic configurations have been employed, with a magnetic axis rotational transform between 0.7 and 1.04 and a LCFS

rotational transform between 0.95 and 1.2. The toroidal magnetic field on the axis is between 2.46T and 2.52T. The effective minor radius of the plasma depends on the magnetic configuration and plasma parameters, but generally, it is between 0.49m and 0.55m. The major radius is 5m, hence the aspect ratio of the stellarator is about 10.

The magnetic field lines have a discrete periodicity in the toroidal direction and the machine can be subdivided into five modules. ECRH launchers are placed on module 1 and module 5, where electromagnetic beams are injected into the plasma on the proximity of bean-shaped planes (ϕ = 0° and ϕ = -72°). Radiative electron temperature is measured by means of 32 Electron Cyclotron Emission (ECE) channels, with a sampling rate of 1 MHz.

Thomson Scattering provided electron temperature and density profiles, with a sampling rate of 10 Hz.
Different Rogowski coils provided toroidal current measurements.

The plasma heating has been provided by ECRH only, which also provided current drive, in both co- and counter-direction, i.e. increasing or decreasing the rotational transform. ECRH power deposition and ECCD are simulated using the ray-tracer code TRAVIS, developed at IPP [5].

3. ECCD-driven temperature collapses

During ECCD experiments, fast and repetitive crashes of electron temperature in ECE and soft X-rays camera have been detected. Since this phenomenon appeared during ECCD operations only, it has been hypothesised that the rotational transform reached a low order rational value, where an MHD instability was triggered. An overview can be find in [4].

From a comparison between the reference discharge (without ECCD), plotted in Figure 1 and a characteristic shot with current drive (Figure 2), it is possible to notice that temperature collapses (second timetrace in each plot, in magenta) are present in the second discharge only. This consideration leads to the hypothesis there crashes are caused by a modification of the rotational transform profile, caused by localised current drive.
In this work the effort will be focused on a plasma ECCD discharge. An example is provided by discharge 20171206.028 (Figure 2 and Figure 3). This discharge was performed using standard magnetic configuration (axis rotational transform = 0.85) and an inverted magnetic field. In this discharge, three gyrotrons heated the plasma, for a total power of 1800 kW delivered into it. One gyrotron injected power in perpendicular direction, while the other two with oblique injection ($\phi \approx$ -11°), thus driving current in co-direction. The discharge is about 13s long and it is possible to notice that the temporal distance between two adjacent crashes is not constant. Moreover, two different types of crash can be recognized. The first type has higher frequency, but a smaller amplitude with respect to the second one, whose collapse effects have been detected also on the diamagnetic energy. The discharge ended abruptly, with a plasma current of -14kA.
The time interval between strong adjacent crashes changes and increases as the discharge evolves. Total power into the plasma and the density do not change, therefore the other parameter that can influence the time necessary to trigger an instability is the current profile, but since the power deposition is unchanged, one can hypothesise that, after a crash, the current redistributes in such a way to have an accumulation close to the axis. In this way, the rotational transform decreases after each crash and it takes more time for the ECCD to reach a resonance. This hypothesis has still to be proven.

Due to the similarity with saw-tooth crashes, which are related to m=1, n=1 mode, it has been assumed that ι =1 is reached, where a magnetic island is formed and whose collapse modifies, in very short time, the electron temperature inside the affected volume.

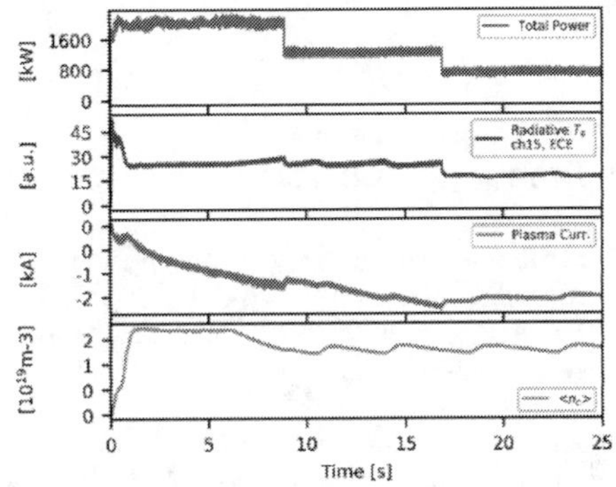

Figure 1: Reference discharge without current drive. Temperature and plasma current changes are due to modifications of the total injected power (blue timetrace).

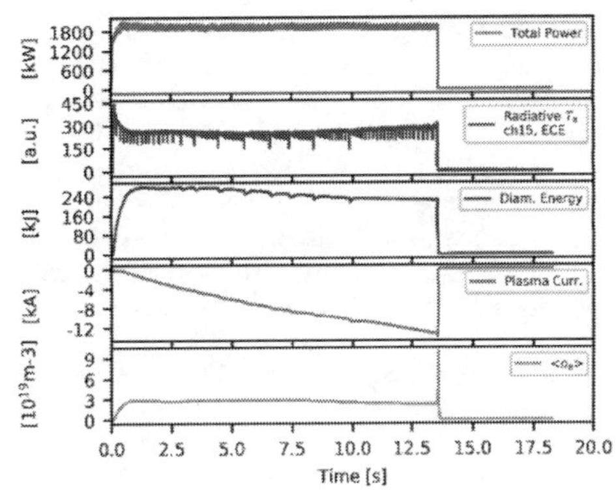

Figure 2: ECCD discharge example. It is possible to notice repetitive temperature crashes on the ECE channel (in

purple). Instability effects are visible also on the diamagnetic energy signal (green).

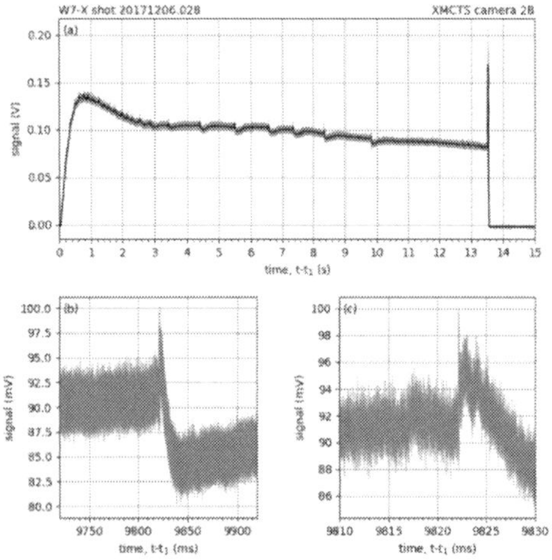

Figure 3 : Soft X rays (XMCTS) data for the analysed discharge. In figure (a) the time traces are displayed. It reproduces main feature of ECE in Figure 3. Plots (b) and (c) are crash zooms and are similar to ECE in Figure 4.

4. Precursor activity

Most of the ECE channels display precursor activity before crashes.

The comparison and correlation between different channels is still not complete yet, but it can be noticed that precursors on channel 10 and 15 have the same frequency (around 6.1 KHz), but opposite phase (180°). The electron temperature detected by these two channels drops abruptly, with a typical time of the order of 10-50 μs. On the other hand, the channel 18 does not display a relevant precursor activity, probably too small to be detected by the actual Fourier analysis routine. A temperature spike can be seen; the temperature increases abruptly and returns to the previous value in the timescale for the temperature collapse of the other two channels.

This behaviour is slightly different with respect to saw-tooth in tokamaks, where channels display a symmetric behaviour with respect to centre.

In further analyses, the radius shift caused by diamagnetic shift and Shafranov shift will be taken into account. In this work, only the cold resonance position of ECE channels has been considered.

After the crash, a strong MHD activity, whose frequency is of about 170 kHz, has been measured by Mirnov coils (in Figure 4 only one Mirnov coil channel has been plotted). The activity starts 0.5 ms after the temperature collapse and it is 0.2 ms long.

Since these oscillations are detected after the crash, this could indicate reconnective phenomena that are taking place after the collapse. A further investigation with all Mirnov coil channels will be performed in order to obtain more information.

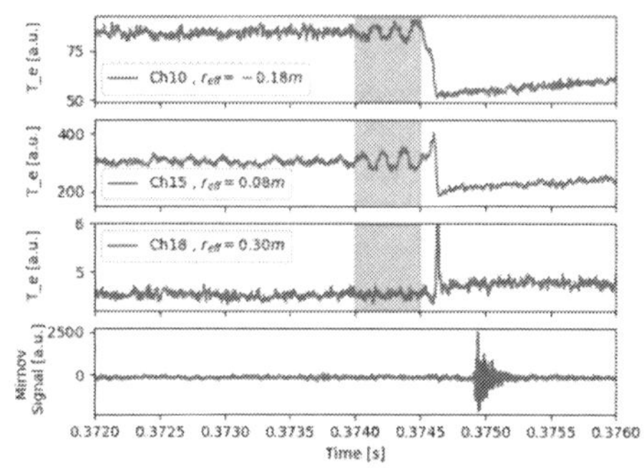

Figure 4 : ECE (first three time traces) and Mirnov coil signal (fourth time trace) before a temperature collapse. Strong oscillation are visible in channel 10 and 15 (red box), while weak oscillation are slightly visible on channel 18.

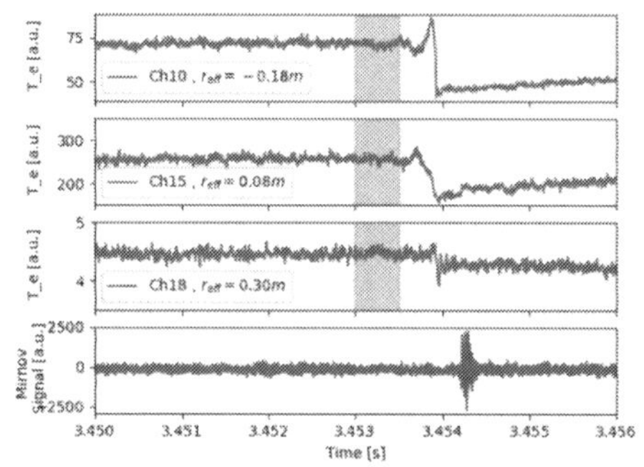

Figure 5 : Three ECE channels and Mirnov coil signal related to a typical crash of Figure 7. Main features are different with respect to Figure 4, especially with respect to precursor activity (comparing red boxes in this figure and Figure 5, it is possible to notice that the strength of precursor activity is different.)

5. Crash evolution in time

Long discharges at W7-X are an excellent tool to study evolution of saw-tooth-like crashes.

The plasma discharge here presented ended abruptly: since similar global collapses have been observed in other discharges when the current reached a threshold value (about -14 kA), it is crucial to understand whether smaller collapses play a role in the development of this big collapse. It was previously mentioned that the behaviour of collapses displays different behaviour over time, even if total ECRH power and density have not been changed. This can be noticed in Figure 6 and Figure 7, where the interval from 0s to 3s and the interval from 3s to 6s of ECE channels are respectively plotted.

In these plots, normalised ECE time traces are displayed. In the first plot, where the first three seconds of the discharge are displayed, it is possible to notice that, at the same channel, crashes have almost the same amplitude. The central collapse region is included between channel 8 and 16; temperature decreases during the collapse in the central region, while increases in the outer one. An example of crash in this region is given in Figure 4. The crash behaviour slightly changes, since two different types of crash are distinguishable. The weaker one, with higher frequency and a stronger one with smaller frequency. The central region, where the temperature decreases, is slightly expanded in Figure 7 with respect to Figure 6, probably indicating that the inversion radius for this kind of instabilities increases over time. Another feature that is not clear yet is represented by the fact that precursor activity, as shown in Figure 4 is stronger at the beginning of the discharge (such as in the time interval plotted in Figure 6) and becomes weaker (Figure 5) as the discharge develops (interval represented in Figure 7). In fact, in Figure 5 a typical crash for this second time interval is displayed. Precursor activity is not recognisable. A characterization of different crash behaviours is going on, in order to better classify and understand this phenomenon.

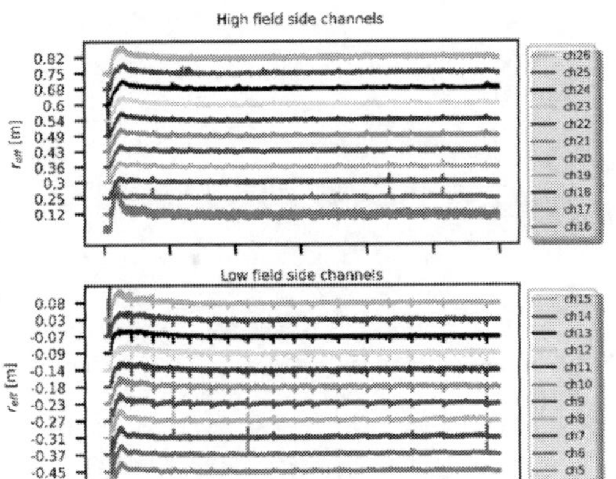

Figure 6 : discharge interval from 0s to 3s. ECE signals are normalised with respect to their average value. Cold resonance position of channels is provided. ECE cold resonance radii are represented on the y-axis

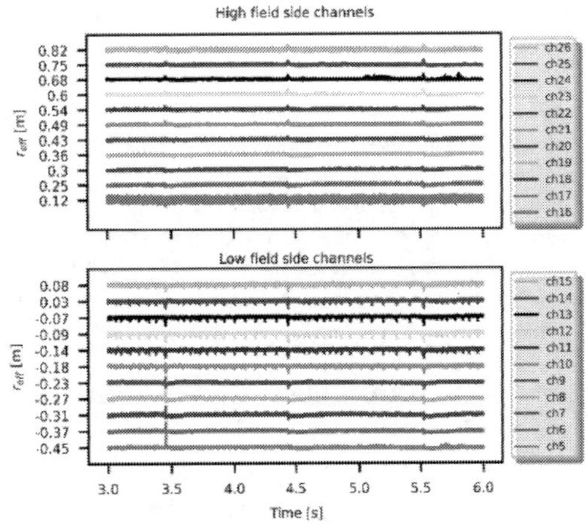

Figure 7 : discharge interval from 3s to 6s. ECE signals are normalised with respect to their average value. . Cold resonance position of channels is provided. . ECE cold resonance radii are represented on the y-axis.

6. Discharge termination

Thanks to long discharges conducted in OP1.2a, it has been possible to investigate saw-tooth-like instabilities for rather long shots.

The maximum toroidal current that has been achieved is about -14kA; after this level a global crash takes place and the global confinement is lost.

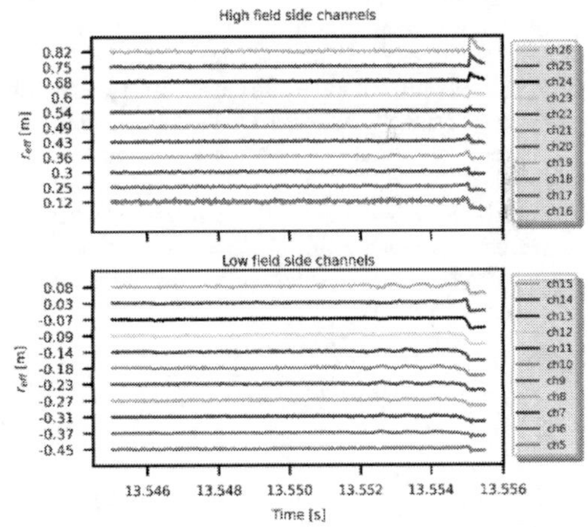

Figure 8 : Time traces zoom on ending part of the discharge. ECE signal are normalised with respect to their average value. LFS channels display temperature decrease, HFS channels display a temperature increase. . ECE cold resonance radii are represented on the y-axis.

In Figure 8 the last microseconds of the discharge are plotted. It is possible to notice that LFS channels have a small precursor activity, followed by a temperature decrease that will lead to discharge termination. Precursor in channel 15 are out phase with respect to precursor in channel 6.

On the other hand, HFS channels do not display any precursor, but they experience a temporary temperature increase, before the discharge ending. The first interpretation consists in a global plasma displacement from the LFS to HFS leading to the termination of the discharge. The main reason is not clear yet, but since it has not been possible to reach a current higher than -14kA, this phenomenon limits global plasma performances, so a deep investigation has to be conducted using soft-x-rays and magnetic diagnostics.

7. 1-D current evolution model

Electron cyclotron current drive is well localised along the effective radius, therefore the estimation of the current profile is crucial to understand and model the rotational transform time evolution.

So far, no direct measurements of plasma current profile are available; therefore, toroidal plasma current can be simulated only. In order to evaluate whether the generated ECCD can be responsible of temperature crashing by means of localised rotational transform modifications, the following model has been developed.

When ECCD is generated inside the plasma (for sake of simplicity, we assume the current to be constant along the same flux surface and we assume topological modification of the original magnetic configuration to be negligible), a counter induced current is induced by the plasma itself to compensate the change of magnetic flux (Lenz's Law). This shielding current will then diffuse, so a net current will rise, modifying the rotational transform. In W7-X the R/L time, the typical time for the current to reach a stationary state is estimated to be around 30s, but toroidal current effects can be perceived on faster timescales. Assuming that the ECCD profile does not change in time and using a cylindrical geometry, since the current drive has toroidal direction only, the shielding current evolution is described by the following equation:

$$\frac{\partial j(r,t)}{\partial t} = \frac{1}{r}\frac{\partial}{\partial r}\left[r\frac{\partial}{\partial r}\big(\eta(r)j(r,t)\big)\right] \qquad 1)$$

Where μ_0 is the magnetic permeability of vacuum, η is the plasma resistivity, r is the radius of the cylinder and j(r,t) is the counter-induced current density.
The rotational transform is then affected by the net current and can be decomposed into two terms.
The first term corresponds to the vacuum iota, which is given by external coils only, while the second term is given by the plasma current.

Considering initially a cylindrical geometry, the second term depends on the total current and position. The relation is the following:

$$\iota \propto {}^{I(r)}/_{r^2} \qquad 2)$$

In order to explain the repetitiveness of these crashes, it is legitimate to suppose that magnetic surfaces are destroyed during a crash, so the current and the temperature redistribute. Since magnetic surfaces are not existing and temperature decreased, ECCD becomes negligible. Then, according to the saw-tooth crash theory, magnetic surfaces are restored, temperature reaches a stationary value again, current drive is newly driven and, again, a counter induced current shields it.

Since the current drive is generally run between 10% and 30% of the effective radius, during low-beta and relatively low density (about $2\cdot10^{19}$ m^{-3}), the bootstrap current in this region is not expected to provide a relevant contribution to rotational transform modification, therefore effects of ECCD and of the relative shielding current only will be considered. Data collected by Thomson Scattering are used to estimate the electronic temperature and density, which have to be taken into account for current drive calculations.
The ECCD is simulated using TRAVIS, as displayed in Figure 9:

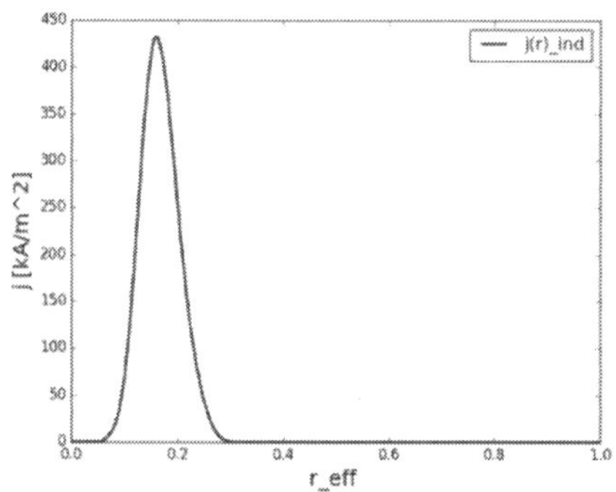

Figure 9: ECCD profile, simulated with TRAVIS, as a function of the normalised radius.

Assuming that, at t=0 s, the ECCD and the shielding current compensate each other. This second component diffuses and a net current arises, as it can be seen on Figure 10, where the evolution of the two components sum is plotted at different time. In this initial version of the code, non-linear changes of the toroidal flux and topology alterations are neglected, even if it is foreseen to take them in account in future developments. The sum of current-free rotational transform and the contribution yielded by the toroidal current in Figure 10 are plotted in Figure 11. The resonant values is reached on a timescale of 100ms or 200ms, which is compatible with the typical time between two crashes.

This simplified 1-D model demonstrates the possibility for the toroidal current to make iota reach a resonant value. Nevertheless, improvements are required in order to take into account contribution provided by inner current redistribution after the crash.

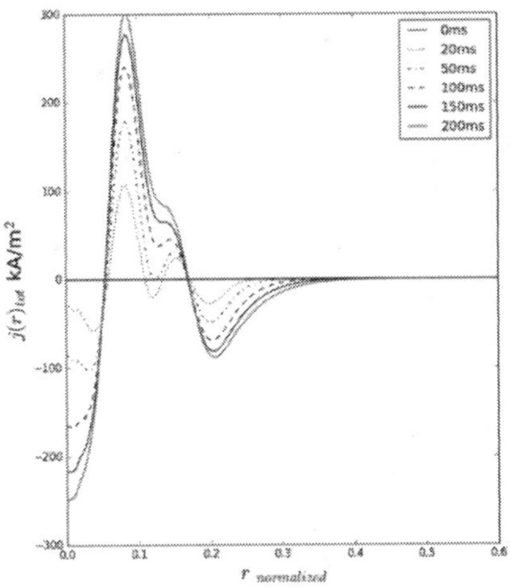

Figure 10: Sum of ECCD and shielding current at different time-steps. At t=0s, the two components compensate each other, therefore the resulting current is null. As the time increases, the shielding current diffuses and a net current arises.

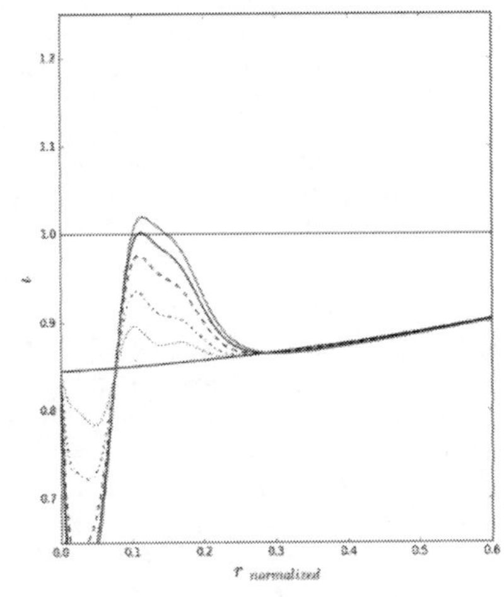

Figure 11: Current changes modify rotational transform, which is plotted for different times as a function of the effective radius.

8. Conclusions

In this work, an overview of a relatively long ECCD driven crash, performed during the second experimental campaign of W7-X, has been presented.

This phenomenon has been noticed during electron cyclotron current drive experiments only, which can be related to current driven MHD instabilities.

A preliminary Fourier analysis of ECE signal displays growing oscillations before a crash, whose frequency is between 1 and 10 KHz, while Mirnov coil signals show a strong activity after the crash, which may indicate reconnective phenomena taking place. Temperature crashes display different behaviour over time: a preliminary analysis suggested that the inversion radius increases. The discharge terminated when a maximum current value was reached; further analyses will be carried out to understand whether this is a threshold value for toroidal current in W7-X.

Since a proper mode analysis is still going on, a 1-D model has been developed, in order to understand whether rotational transform changes, induced by localised current, can reach a resonant value.

This work has been carried out within the framework of the EUROfusion Consortium and has received funding from the Euratom research and training programme 2014-2018 under grant agreement number 633053. The views and opinions expressed herein do not necessarily reflect those of the European Commission.

References

[1] R. Wolf and al., "Major results from the first plasma campaign of the Wendelstein 7-X stellarator," *Nuclear Fusion 57*, 2017.

[2] T. Stange and al., *EPJ Web of conferences*, vol. 157, no. 02008, 2017.

[3] N. Fisch and A. Boozer, "Creating an asymmetric plasma resistance with waves," *Phys. Rev. Lett.*, vol. 45, 1980.

[4] H. Laqua and al., "Saw tooth like collapses by ECCD at the Wendelstein7-X Stellarator," *submitted to Phys. Rev. Lett.*, 2018.

[5] N. Marushchenko and al., "Ray-tracing code TRAVIS for ECR heating, EC current drive and ECE diagnostic," *Computer Physics Communications*, vol. 185, 2014.

[6] J. Wesson, Tokamaks, Oxford University Press, 1987.

Advances in turbulence measurements using new Correlation ECE and nT-phase diagnostics at ASDEX Upgrade

Simon J. Freethy[1,2], *Tobias* Görler[1], *Alex J.* Creely[2], *Garrard D.* Conway[1], *Severin S.* Denk[1,3], *Tim* Happel[1], *Pascale* Henniquin[4], *Christian* Koenen[5], *Anne E.* White[2], and *The ASDEX Upgrade* team[1]

[1]Max Plank Institute for Plasma Physics, 85748 Garching, Germany
[2]Plasma Science and Fusion Center, Massachusetts Institute of Technology, Cambridge, MA 02139, USA
[3]Physik-Department E28, Technische Universität München, 85748 Garching, Germany
[4]Laboratoire de Physique des Plasmas, Ecole Polytechnique, 91128, Palaiseau cedex, France
[5]Lehrstuhl für Hochfrequenztechnik, Technische Universität München, Arcisstr. 21, 80333 München

Abstract. Guided by predictions from nonlinear gyrokinetic simulations, two new turbulence diagnostics were designed and installed at ASDEX Upgrade (AUG) to probe the fundamentals of ion-scale turbulent electron heat transport. The first, a 30-channel correlation ECE (CECE) radiometer (105-128 GHz, 2nd harmonic X-mode), introduces a novel channel comb arrangement. This allows measurements of high radial resolution profiles ($0.5 < r/a < 0.8$) of low-k ($k_\theta \rho_s < 0.3$) temperature fluctuation amplitudes, frequency spectra and radial correlation length profiles in unprecedented detail. The second diagnostic is formed by the addition of two W-band and one V-band X-mode reflectometers on the same line of sight as the CECE to enable measurements of the phase angle between turbulent density and temperature fluctuations. Historically, the radial alignment between reflectometer and radiometer has been a challenge due to the requirement that alignment is achieved within a radial correlation length (< 5 − 10 mm). This challenge is significantly alleviated by using the CECE channel comb arrangement and the maximal coherence between reflectometer and radiometer can be unambiguously captured. Measurements of these quantities have been made in an AUG L-mode plasma, at the same radial location and have provided simultaneous quantitative constraints on realistic gyrokinetic simulations [Physics of Plasmas **25**, 055903 (2018)] using the gyrokinetic code GENE. Here we present diagnostic detail for this study.

1 Introduction

Understanding the turbulent driven heat flux in a tokamak remains one of the key goals of fusion research. Anomalous transport up to two orders of magnitude above what one would expect from neoclassical theory is observed and this is now understood to be caused by turbulent fluctuations in the plasma density, temperature and potential, originating from drift-wave like instabilities which grow and non-linearly saturate [1, 2]. In order to study in detail the turbulence giving rise to electron heat transport, the ASDEX Upgrade (AUG) correlation ECE (CECE) diagnostic was significantly upgraded, introducing a channel comb arrangement. This new diagnostic then measures high radial resolution fluctuation amplitude, $\delta T_{e\perp}/T_e$ profiles. The AUG CECE channel comb also provides the first measurements of high radial resolution profiles of $L_r(T_{e\perp})$ using the CECE channel comb, showing the proportionality of $L_r(T_{e\perp})$ to the ion sound gyroradius, ρ_s. Measurements of the radial correlation length of the temperature perturbations, $L_r(T_{e\perp})$, have been reported previously [3, 4], but are nonetheless rare and the relationship to ρ_s never explicitly demonstrated.

The heat flux driven by turbulent fluctuations depends not only on the amplitudes of the fluctuating quantities, but on their relative coherencies and phase angles [5]. Experimental measurements of these cross-field quantities are rare due to the difficulty in attaining them, but offer extremely valuable information with which to constrain high fidelity models of turbulence. For example, the phase angle between temperature and density fluctuations α_{nT} has been measured in the core plasma by the combination of a reflectometer and radiometer [5, 6]. We follow this approach here, combining two W-band (75-110 GHz) and one V band (50-75 GHz) X-mode reflectometers along the same line of sight as the AUG CECE. We find that the CECE channel comb significantly improves the reliability of the radial alignment between radiometer and reflectometer. All quantities have been measured simultaneously at the same radius in dominantly electron heated L-mode plasma and have been compared directly to non-linear ion scale GK simulations [7]. It is found that both electron and ion heat flux can be matched, along with α_{nT} and $L_r(T_{e\perp})$, however $\delta T_{e\perp}/T_e$ is higher in the simulations than measured.

2 Correlation ECE diagnostic

In order to investigate the ion-scale electron heat transport at ASDEX Upgrade, a Correlation ECE receiver was

https://doi.org/10.1051/epjconf/201920303001

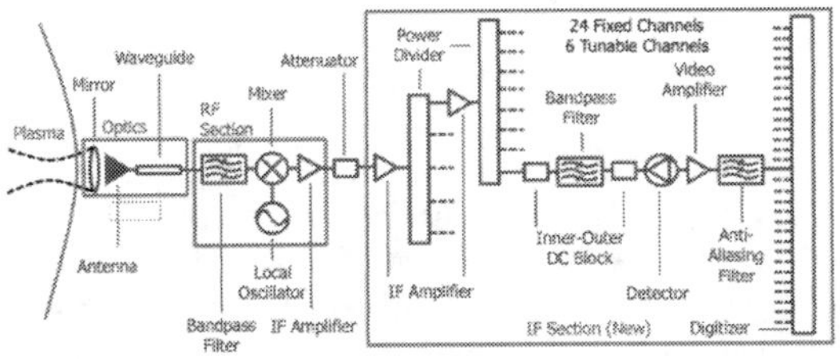

Figure 1. Reproduced from [8]. Block diagram showing the layout of the new AUG CECE diagnostic.

designed and built which employs both a high channel number and a novel channel comb arrangement. Three interchangable radio frequency (RF) sections employ tunable oscillators in the range 105.5-109.1 GHz, 110.2-113.8 GHz and 113.5-117 GHz, which in turn feed a filter bank of 24 fixed channels covering the range 4-8 GHz in steps of either 100 or 200 MHz, as shown in Figure 1. This results in a radial spacing on the order of the ECE linewidth ensuring maximal spatial resolution of between 2-4mm. Four additional tunable YiG filters increase the frequency range up to 14 GHz. This configuration was guided by ECE modelling using a radiation transport model [9] and existing non-linear gyrokinetic (GK) simulations for predictions of the radial correlation length of the perpendicular electron temperature fluctuations $L_r(T_{e\perp})$.

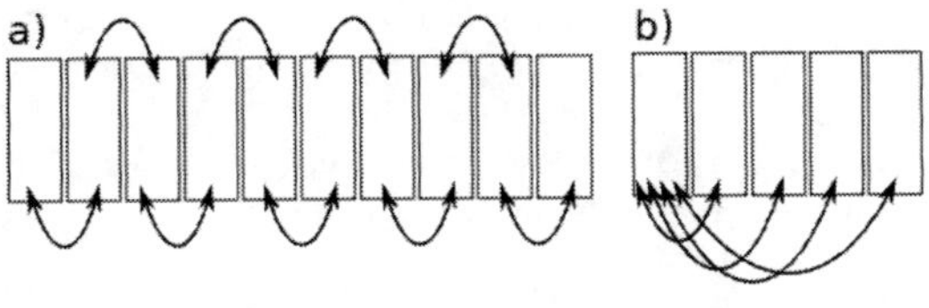

Figure 2. Schematic showing example channel combinations for high radial resolution $\delta T_e/T_e$ profiles (a) and correlation length measurements (b). Reproduced from [7].

The novel channel comb allows the measurement of high resolution $\delta T_{e\perp}/T_e$ and $L_r(T_{e\perp})$ profiles. $\delta T_{e\perp}/T_e$ is calculated from a nearest neighbour cross-correlation as shown in Figure 2 (a). The correlation length is calculated from cross-correlations of increasing distance as shown in Figure 2 (b). In this case, the reference channel can be changed and the process repeated, allowing a correlation length profile to be built up. Figure 3 shows $\delta T_{e\perp}/T_e$ and $L_r(T_{e\perp})$ profiles for a medium density L-mode plasma with $T_e > T_i$. It has previously been reported for density fluctuations that $L_r(n_e) \simeq 5 - 10\rho_s$ [10]. We can use this data to confirm for the first time a similar trend for the electron temperature fluctuations. In our case $L_r(T_{e\perp}) = 8.5\rho_s$, as shown in Figure 3 (b).

Non linear gyrokinetic results for these quantities at $\rho_{\rm pol} = 0.86$ produced using the GENE gyrokinetic code [11] and described in a previous publication [7] are also shown in the figure, denoted by the black stars. As you can see, reasonable agreement between GENE and the experiment is found, although the code slightly over predicts $\delta T_{e,\perp}/T_e$.

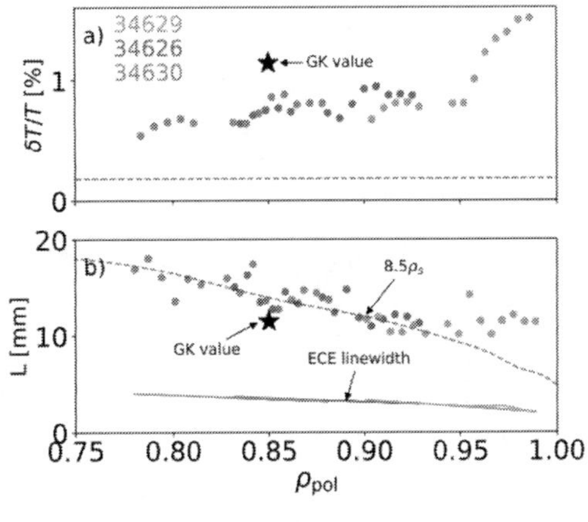

Figure 3. High radial resolution $\delta T_{e,\perp}/T_e$ profile (a) and correlation length profile, $L_r(T_{e\perp})$, (b) as measured by the AUG CECE diagnostic. $L_r(T_{e\perp})$ is shown to agree well with $8.5\rho_s$ in this case, with a jump at $\rho_{\rm pol} = 0.95$. The ECE linewidth $w_{\rm ECE}$ and gyrokinetic values are also shown for comparison.

3 nT cross-phase diagnostic

The cross-phase angles between the fluctuating fields are important quantities for the validation of gyrokinetic models, since they determine the magnitude of the heat flux along with the fluctuation amplitudes. The most experimentally accessible quantity for core fluctuations is the phase angle between temperature and density fluctuations. This quantity has been shown to vary when experimentally varying the T_e normalised gradient scale length $1/L_{Te}$, which is a drive for the trapped electron (TEM) and electon temperature gradient (ETG) modes [12]. This work indicates that α_{nT} is an experimental measure of the balance of TEM and ITG structure present in the turbulence. α_{nT} has been measured previously by using a combination of a reflectometer for the density fluctuations and a radiometer for the temperature fluctuations, which share the same line of sight to the plasma [5, 6] and a similar approach is taken here. As can be seen from Figure 4, the CECE is installed on the same oversized waveguide as W and V band X-mode reflectometers. They are coupled via a 3dB wire-grid splitter and share the antenna and focussing mirror, giving both diagnostics the same line of sight. Then, if radial alignment can be achieved by choosing the reflectometer frequency appropriately, cross-correlation of radiometer and reflectometer signals may be used to calculate α_{nT}.

A significant problem in the operation of a combination reflectometer radiometer α_{nT} diagnostic is the stringent radial alignment constraint. The frequency of the reflectometer should be chosen with sufficient accuracy such

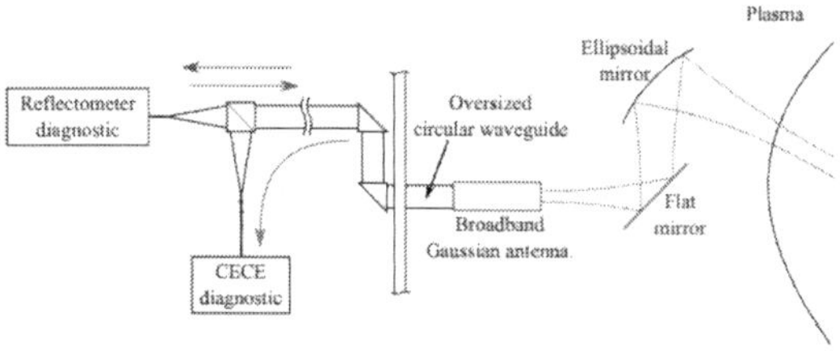

Figure 4. Adapted from [13]. A schematic of the CECE waveguide and optical setup. The CECE is set up to use the same line of sight as a reflectometer, coupled via a 3dB wire-grid splitter.

that the cut-off position aligns well within a turbulent radial correlation length of an ECE radiometer channel, i.e. < 5 mm. Within the uncertainties of the experiment, the chances of aligning a single reflectometer and radiometer channel first time are small. Figure 5 shows the X mode cut-off and cyclotron harmonics for the target L-mode plasma. One can see that a 20% uncertainty in the density gives rise to a significant uncertainty in the cut-off position. One way of alleviating this issue is to use the CECE channel comb to continuously cover a larger radial range, and one can easily estimate that a 5GHz comb would be required in this case. This is sufficient to achieve radial alignment using only a single plasma discharge, as we may also take advantage of 3 simultaneous reflectometer frequencies, thus significantly improving the reliability, productivity and usefulness of the diagnostic.

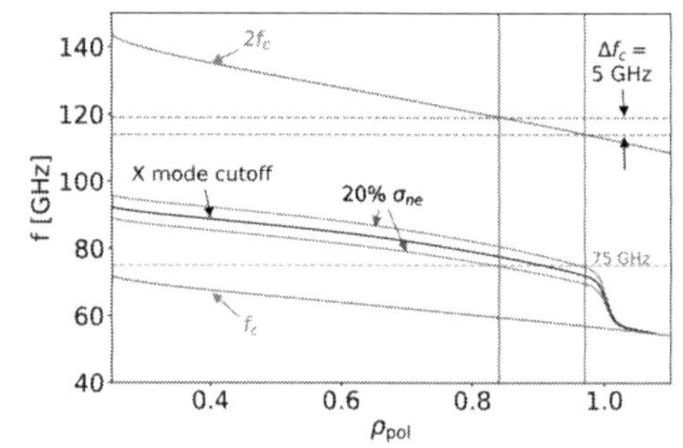

Figure 5. The first two cyclotron harmonics and the X-mode cut-off frequency profile. A 20% error in the density profile leads to a significant uncertainty in the radial position of the reflectometer. For this case, a CECE channel comb with a bandwidth of 5GHz should be sufficient to capture the coherence between reflectometer and radiometer.

The chosen technique is to illuminate the plasma at a fixed frequency, and the resulting amplitude fluctuations in the return waves are then correlated with each of the CECE channels in turn. It is then possible to build up a profile of the coherence between reflectometer and radiometer (refl.-ECE) as shown in Figure 6 (a) and (b). These panels show the coherence averaged between 0-100 kHz for refl.-ECE (black squares) and between radiometer channels (ECE-ECE, orange triangles) with respect to a reference channel chosen to line up with the peak refl.-ECE coherence. The width of the latter is dominated by the $L_r(T_{e\perp})$ as shown in section 2. One can see that the refl.-ECE coherence function is asymmetric, being broader towards higher radii compared to the ECE-ECE coherence function. This suggests that the reflectometer scattering carries information about density fluctuations away from the cut-off location and is thus correlated with fluctuations more than a radial correlation length away from the cut-off.

Figure 6 (c) and (d) show the cross-phase averaged from 10-40 kHz for refl.-ECE correlations (black squares), ECE-ECE correlations (orange triangles). One can see that there is a slope in the average cross-phase present both in ECE-ECE and refl.-ECE correlations and this is due to a finite radial time delay of the turbulent structures. This is considered to originate from the finite turbulent eddy tilt angle, as there is no poloidal projection to this line of sight. Both refl.-ECE and ECE-ECE correlations show approximately the same phase behaviour with radius.

The advantage of the channel comb is then clear in the presence of such a behaviour in the cross-phase, as the ECE-ECE correlations are required to correct the phase trend in the refl.-ECE correlations for a robust measurement. Further, the measurement is only valid at peak coherence and without knowledge of the linear trend in the phase, the tolerance on radial alignment between ECE and reflectometer position is only a small fraction of the radial correlation length (< $\pm 0.4 L_r$ or < ± 2-4 mm), despite the fact that the reflectometer is sensitive to fluctuations over a broader radial range than the ECE. This degree of alignment can only be achieved with a channel comb in either ECE or reflectometer, or by painstakingly repeating the measurement with minute adjustments to achieve the same result.

4 Conclusions

A new Correlation ECE (CECE) and density temperature phase angle, α_{nT}, diagnostic has been designed and built for ASDEX upgrade. The system uses 24 non-overlapping frequency channels arranged in a comb pattern, which when the finite channel bandwidth, Doppler broadening, relativistic broadening and radiation transport are considered, give near continuous radial coverage of a small radial region. This diagnostic gives detailed measurements of the radial correlation length of the temperature fluctuations $L_r(T_{e\perp})$ and confirms for the first time the proportionality of $L_r(T_{e\perp})$ to the local thermal ion gyroradius ρ_s. The addition of 2 W-band and 1 V-band X-mode reflectometers along the same line of sight allows the measurement of the phase angle between density and temperature fluctuations α_{nT}. The channel comb significantly improves the reliability of the diagnostic, and it has been shown that the condition of radial alignment can be more stringent than previously thought due to a linear trend in the cross-phase as a function of the separation between reflectometer cut-off and ECE channel position. High fidelity, ion-scale, non-linear gyrokinetic simulations performed agree well with the experimentally inferred electron and

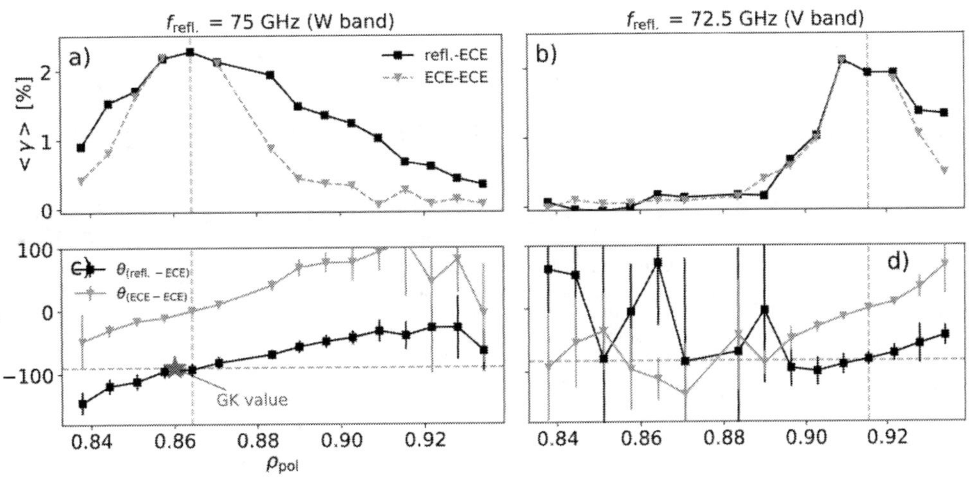

Figure 6. (a) and (b): Coherence between ECE channels (orange triangles) and ECE and reflectometer amplitude (black squares) averaged between 0-100 kHz for reflectometer probing frequency of 75 GHz and 72.5 GHz. The ECE coherence is scaled by 0.4. (c) and (d): The cross phases averaged between 10-40 kHz between the ECE channels (orange triangles), ECE and reflectometer amplitude (black squares).

ion heat fluxes, while simultaneously matching the measured α_{nT} and $L_r(T_{e\perp})$. Temperature fluctuation amplitudes are slightly above those measured in the plasma.

Acknowledgements

This work is supported by the US DOE under grants DE-SC0006419 and DE-SC0017381, and was performed in the framework of the Helmholtz Virtual Institute on Plasma Dynamical Processes and Turbulence Studies using Advanced Microwave Diagnostics. It has also been carried out within the framework of the EUROfusion Consortium and has received funding from the Euratom research and training programme 2014-2018 under grant agreement No 633053. The views and opinions expressed herein do not necessarily reflect those of the European Commission. The gyrokinetic simulations have been performed at the MARCONI supercomputer at CINECA and the MPCDF supercomputing facilities.

References

[1] J. Connor, H. Wilson, Plasma Physics and Controlled Fusion **36**, 719 (1994)

[2] W. Horton, Reviews of Modern Physics **71**, 735 (1999)

[3] L. Porte, S. Coda, T. Goodman, A. Pochelon, V. Udintsev, V. Vuille, EPJ Web of Conferences **32**, 03007 (2012)

[4] M. Fontana, L. Porte, S. Coda, O. Sauter, Nuclear Fusion **58**, 024002 (2018)

[5] A. White, W. Peebles, T. Rhodes, C. Holland, G. Wang, L. Schmitz, T. Carter, J. Hillesheim, E. Doyle, L. Zeng et al., Physics of Plasmas **17**, 056103 (2010)

[6] M. Häse, M. Hirsch, H.J. Hartfuss, Review of Scientific Instruments **70**, 1014 (1999)

[7] S.J. Freethy, T. Görler, A.J. Creely, G.D. Conway, S.S. Denk, T. Happel, C. Koenen, P. Hennequin, A.E. White, Physics of Plasmas **25**, 055903 (2018)

[8] A.J. Creely, S.J. Freethy, W.M. Burke, G.D. Conway, R. Leccacorvi, W.C. Parkin, D.R. Terry, A.E. White, Review of Scientific Instruments **89**, 053503 (2018)

[9] S. Denk, R. Fischer, O. Maj, E. Poli, J. Stober, U. Stroth, B. Vanovac, W. Suttrop, M. Willensdorfer, EPJ Web of Conferences **147**, 02002 (2017)

[10] T.L. Rhodes, J.N. Leboeuf, R.D. Sydora, R. Groebner, E.J. Doyle, G.R. McKee, W.A. Peebles, C.L. Rettig, L. Zeng, G. Wang, Phys. Plasmas **9**, 2141 (2002)

[11] F. Jenko, W. Dorland, Plasma Physics and Controlled Fusion **43**, A141 (2001)

[12] J. Hillesheim, J.. DeBoo, W. Peebles, T. Carter, G. Wang, T. Rhodes, L. Schmitz, G. McKee, Z. Yan, G. Staebler et al., Physics of Plasmas **20**, 056115 (2013)

[13] S.J. Freethy, G.D. Conway, I. Classen, A.J. Creely, T. Happel, A. Köhn, B. Vanovac, A.E. White, Review of Scientific Instruments **87**, 11E102 (2016)

Design and development of the ITER CTS diagnostic

Søren B. Korsholm[1,*], *Bruno* Gonçalves[2], *Heidi* E. Gutierrez[1], *Elsa* Henriques[4], *Virginia* Infante[4], *Thomas* Jensen[1], *Martin* Jessen[1], *Esben* B. Klinkby[3], *Axel* W. Larsen[1], *Frank* Leipold[1], *André* Lopes[2], *Raul* Luis[2], *Volker* Naulin[1], *Stefan* K. Nielsen[1], *Erik* Nonbøl[3], *Jesper* Rasmussen[1], *Mirko* Salewski[1], *Morten* Stejner[1], *Arianna* Taormina[1], *Alberto* Vale[2], *Catarina* Vidal[2], *Laura* Sanchez[5], *Raul* M. Ballester[5], and *Victor* Udintsev[6],

[1]Technical University of Denmark, Department of Physics, 2800 Kgs. Lyngby, Denmark
[2]Instituto de Plasmas e Fusão Nuclear, Instituto Superior Técnico, Univ. Lisboa, Portugal
[3]Technical University of Denmark, Center for Nuclear Technologies, 4000 Roskilde, Denmark
[4]IDMEC, Instituto Superior Técnico, Univ. Lisboa, Portugal
[5]Fusion for Energy, 08019 Barcelona, Spain
[6]ITER Organisation, 13115 Saint Paul Lez Durance, France

Abstract. The Collective Thomson Scattering (CTS) diagnostic will be a primary diagnostic for measuring the dynamics of the confined fusion born alpha particles in ITER and will be the only diagnostic for alphas below 1.7 MeV [1]. The probe beam of the CTS diagnostic comes from a 60 GHz 1 MW gyrotron operated in a ~100 Hz modulation sequence. In the plasma, the probing beam will be scattered off fluctuations primarily due to the dynamics of the ions. Seven fixed receiver mirrors will pick up scattered radiation (the CTS signal) from seven measurement volumes along the probe beam covering the cross section of the plasma. The diagnostic is planned to provide a temporal resolution of ~100 ms and a spatial resolution of ~a/4 in the core and ~a/20 near the plasma edge where a = 2.0 m is the nominal minor radius of ITER. The front-end quasi-optics will be installed in an equatorial port plug (EPP#12). A particular challenge will be to pass the probing beam through the fundamental electron cyclotron resonance, which is located in the port plug (R=10.3 m) for the nominal magnetic field B_t = 5.3 T. Hence, particular mitigation actions against arcing have to be applied. The status of the design and specific challenges will be discussed.

1 Introduction

The primary aim of the next step fusion experiment ITER is to demonstrate – in an integrated system – the production of net power from fusing deuterium and tritium. The fusion born alpha particles (birth energy of 3.5 MeV) will have much higher relative densities than in present machines and will have a substantial effect on the overall dynamics of the ITER fusion plasma. Therefore, diagnosing the dynamics of the fast alpha particles are important, and at the same time challenging. Only a few ITER diagnostics can measure the confined fast alpha dynamics. One of these is the ITER Collective Thomson Scattering (CTS) diagnostic which has the primary role for the fast ion energy spectrum. Recent analysis shows that for alpha energies below 1.7 MeV, the CTS diagnostic will be the only diagnostic on ITER [1].

The work towards having a fast ion CTS diagnostic on ITER dates back more than 20 years. Driven by the need for a better diagnostic coverage of the fast ion dynamics in ITER, CTS was included as an enabled diagnostic (more on the implications of 'enabled' later) in the updated ITER baseline in 2007. Scientifically, the inclusion was in part based on the feasibility study of a CTS diagnostic system for ITER performed at Risø National Laboratory (now DTU – Technical University of Denmark) [2, 3], and also on the advances of the CTS diagnostics on present machines. This article does not intend to give an overview of the present day CTS diagnostics and also not the fast ion CTS principle and analysis. For these, please refer to [4, 5].

This paper intends to give a brief overview of the ITER CTS diagnostic, and more particularly the development of the in-vessel front-end components.

2 Overview of the ITER CTS Diagnostic

The ITER CTS diagnostic – as proposed in the feasibility study [2, 3] – operates in the 60 GHz frequency range, which for the nominal ITER magnetic field of B_0 = 5.3 T is below the fundamental EC resonance in the plasma. The probing beam injected in X-mode is generated by a ~1 MW 60 GHz gyrotron. The frequency range of the scattered radiation will be from 55 to 65 GHz. The expected temporal resolution will be 100 ms and the spatial resolution will be in the order of a/4 in the center, where a = 2.0 m is the nominal minor radius of ITER. The spatial resolution varies with plasma scenario and location of measurement volume. For the

* Corresponding author: sbko@fysik.dtu.dk

reference scenario (Scenario 1 – ITER Baseline Scenario) the resolution will vary from ~a/2 on the high field side to ~a/30 on the low field side near the edge. The CTS system will have seven measurement volumes in the plasma.

The ITER CTS diagnostic is an 'enabled' diagnostic in the ITER baseline. The consequence is that primarily the in-vessel components and captive components that cannot be retrofitted after first plasma are part of the credited construction baseline. Fusion for Energy (F4E) will be responsible for delivering the in-vessel front-end components, as well as the functional specifications of the rest of the system. The responsibility for the design and delivery of the ex-vessel components (including the vacuum vessel windows) of the CTS diagnostic is under ITER IO. The different components of the CTS diagnostic are presented in the schematic in Figure 1. The present paper will focus on the design and development on the front-end components, i.e. the in-vessel components (in the white box in the left of the figure).

The in-vessel CTS components will be installed in a separate drawer #3 in equatorial port plug #12. A drawer consists of a diagnostic shield module (DSM) and a diagnostic first wall (DFW). The gyrotron will be installed in a future enclosure in the assembly hall and the matching optics unit (MOU) box will include a polarizer set to ensure X-mode injection in the plasma. Further, it will be connected to the port plug by a ~110 m transmission line consisting of circular corrugated waveguide and mitre bends. At the closure plate, the gyrotron radiation is coupled through a Ø110 mm inner diameter (ID) CVD (Chemical Vapour Deposition) diamond window by a set of two ex-vessel coupling mirrors. The ex-vessel transmission of nine identical receiver lines will be through circular corrugated waveguides connecting the ~37 meters to the diagnostic building where the receiver electronics will be located.

The number of receiver lines was limited by two factors: possible space allocation along the path between the port cell and the diagnostic building – limit nine waveguides of ID 63.5 mm – and the space allocation on the closure plate. Each vacuum window takes up space for the actual window, the window assembly, and the zones for installation and maintenance. In case an individual window was used for each receiver transmission line, this would have meant that no more than five receiver windows would fit – and similarly only five transmission lines could leave the portplug. Hence, it was proposed to put three large (Ø160 mm ID) windows that could be used commonly by several receiver lines. The quasi-optical paths, designed at DTU, could fit three quasi-optical lines through each window assembly using a set of two in-vessel and two ex-vessel coupling mirrors for each line. In this way the maximum number of transmission lines between the diagnostic building and the vacuum vessel was set to nine. The geometry of the closure plate is illustrated in Figure 2.

The current status of the CTS diagnostic development process is that the Conceptual Design Review was conducted and passed by ITER IO in 2013. The F4E framework partnership agreement contract with the DTU/IST consortium was initiated in 2014 and the Intermediate Design Review (or Pre Annex B Design Review) was conducted and passed in 2017. Currently, the design of the in-vessel system is in the preparatory

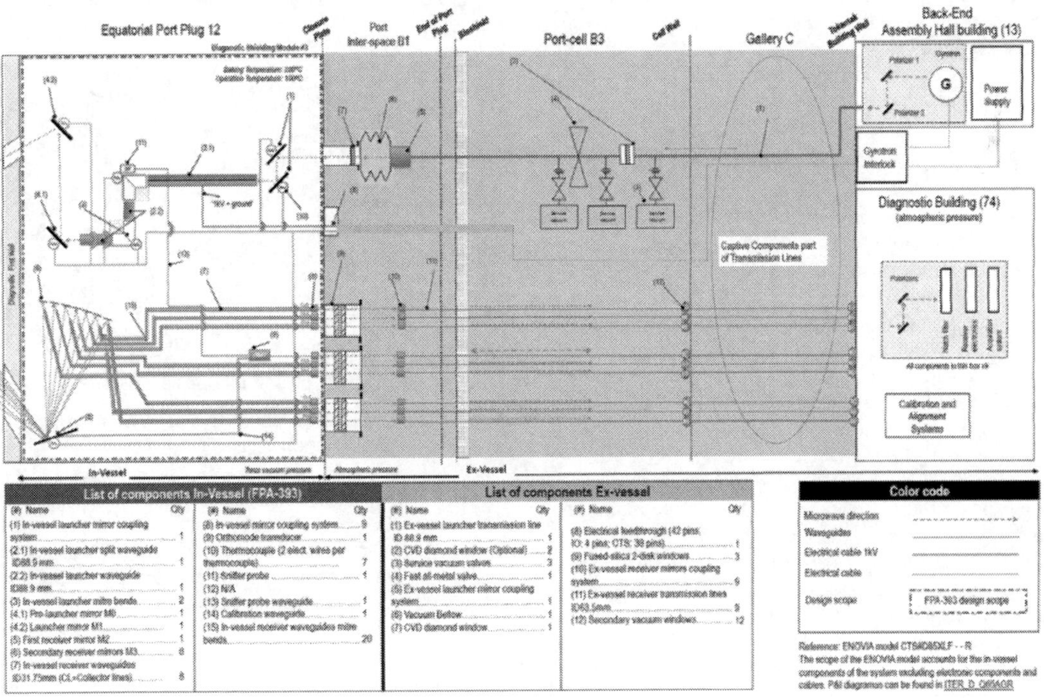

Fig. 1. Schematic for the ITER CTS diagnostic.

design phase, and the Preliminary Design Review for the full system is expected within 12 months.

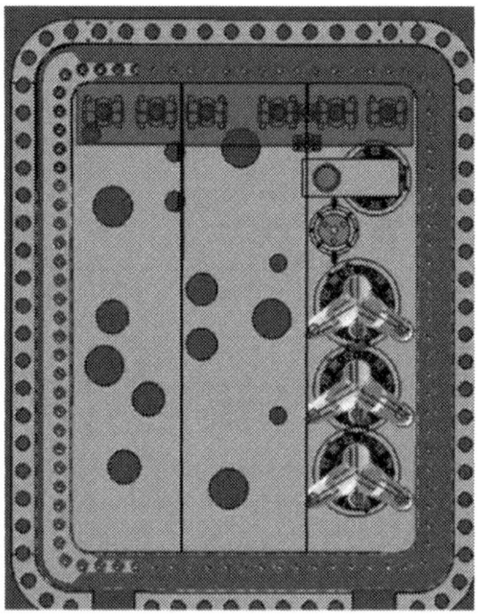

Fig. 2. CAD drawing of the closure plate of equatorial port plug #12 seen from outside the vacuum vessel. DSM #3 is to the right. The launcher line window (Ø110mm CVD) is in the top and below are the three receiver windows (Ø160mm fused silica double disk).

Prior to the Intermediate Design Review, a series of analysis and reports were performed and presented for the system design at that time. This included e.g. diagnostic performance analysis, critical component analysis and design, prototype needs analysis and plans, Reliability, Availability, Maintainability and Inspectability (RAMI) analysis, thermomechanical analysis (to be included in structural integrity analysis), remote handling analysis, neutronics analysis Some of the analyses have been published, like e.g. the RAMI analysis [6] and the neutronics modelling results [7].

3 CTS front-end component design

The in-vessel parts of the ITER CTS diagnostic are exclusively located in drawer #3 of the remote handling and first plasma equatorial port plug #12. As any in-vessel ITER system the design is restricted by a long list of interfaces and requirements; and in addition some more internal challenges. A full account of these is given in the documentation for the Intermediate Design Review. In this paper we shall only briefly describe some of the major issues.

An overall challenge for the design has been the requirement to have no moveable parts, while ensuring diagnostic performance and still having some margin or flexibility to different plasma scenarios. At the same time the quasi-optical sizes of components (at 60GHz) and required apertures are important for diagnostic performance, while large cut-outs in the DSM and the DFW are challenging from a neutronics and thermomechanical point of view. Particularly the cut-out for the launcher beam in the DFW is a compromise between diagnostic performance (larger cut-out could focus the beam better in the plasma and increase spatial resolution) and (neutronics) shutdown dose rate at the closure plate.

As mentioned in the section above, the limitations at the closure plate (and beyond) limited the number of transmission lines from the vessel. The number of channels to the diagnostic building is nine, and they have been allocated with seven actual receiver lines, one passive view, and one line for a calibration/sniffer probe. Present experiments on e.g. ASDEX Upgrade [8] have shown that events in the plasma contribute to the background signal in the CTS measurements, in a way that is not properly captured by the normal ECE background subtraction. Therefore, the introduction of a passive view can help diagnosing these parts of the signal and improve the data analysis significantly. The passive view is a receiver line as the other seven receiver lines, with the important exception that there is no overlap between the gyrotron probing beam and the passive view beam. At the same time, for the passive view, the beam geometry and the plasma of view should be as similar as possible to the active receiver views.
The seven receivers observe the gyrotron probe beam at seven different radial locations. In this way, the CTS diagnostic can provide a radial profile of the fast ion measurements despite not having any moveable parts. As it can be seen in Figure 3, all seven receiver beams (as well as the passive view) share the common plasma facing receiver mirror in the bottom of the drawer. From there each receiver line has a separate secondary receiver mirror that couples the received signal into circular corrugated waveguides of Ø31.75 mm. The transmission lines continue to the mirror coupling system at near the closure plate via waveguides and three mitre bends per line.

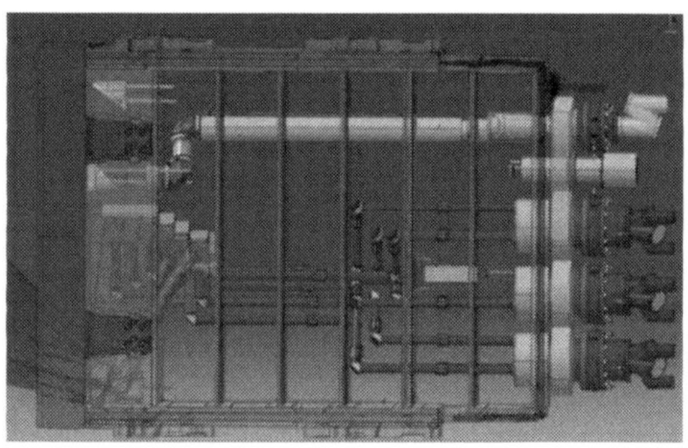

Fig. 3. CAD drawings of ITER CTS diagnostic in drawer #3 of equatorial port plug #12.

The last of the nine transmission lines are reserved for the double use as a gyrotron transmission line sniffer probe during diagnostic operation and as a receiver calibration source transmission line during maintenance. This can be done as the two functionalities will never be done simultaneously and in this way we saved one transmission line through the closure plate. It has been enabled by an in-vessel passive switching system using directional couplers and orthomode transducers ensuring sufficient isolation (>30 dB) between the two uses. The sniffer probe will be installed in the first in-vessel mitre bend (seen in the direction of the gyrotron probe beam) and it will observe the long launcher waveguide to be described below. The calibration waveguide will be connected to a small hole in the center of the common receiver mirror and will emit microwaves in a uniform pattern that will be picked up by the individual secondary receiver mirrors in order to calibrate the receiver transmission lines. The calibration source will be located in the diagnostic building and can be controlled with broadband signal or sweeping signal and can naturally also be varied in power level as appropriate.

The primary challenge of the ITER CTS diagnostic in the port plug is the transmission of the gyrotron probe beam through the fundamental electron cyclotron resonance, which is located inside the port plug for most of the proposed operating main magnetic fields of ITER. For the nominal magnetic field of 5.3 T the fundamental resonance for 60 GHz (B=2.14 T) is at a major radius of R=10.4 m. The location of the resonance will naturally move with different operating magnetic fields (e.g. for B_0=3.5T the resonance is at R=9.6 m).

In order to limit the risk of stray light and in order to limit the required cut-outs in the DSM, the design solution has been to use waveguide rather than quasi-optical transmission via mirrors through the DSM. The chosen inner diameter of the circular corrugated waveguide is 88.9 mm. Transmitting a 1 MW probe beam through this waveguide and passing the fundamental resonance at an uncontrolled ambient pressure constitutes a genuine risk for creating a plasma breakdown inside the waveguide. This would hamper the diagnostic capability and in the worst case lead to partial or full absorption of the 1 MW beam in the waveguide. For this reason, mitigation mechanisms have to be implemented. Several mitigation actions have been considered. Presently, the most developed mitigation action is adapting a split biased waveguide. This solution has previously been used at DITE and DIII-D. In short, the launcher waveguide is split along its axis into two halves. They are isolated from each other (and the surroundings), and one half is grounded, while the other is connected to a bias voltage of +1 kV. In case the gyrotron probe beam ionises a neutral atom or accelerates a free electron, the sweep time (time from waveguide center to the biased wall) is much shorter than the gyration time. In this way, the risk of a breakdown should be mitigated. As a side effect, the biased waveguide will act as a diagnostic for such electron sweeping events; as such an event will generate a current.

Some of the quasi-optical components have been prototyped and tested in the DTU microwave laboratory in order to verify the quasi optical codes and mirror designs. The prototypes of the mirrors were produced by 3D plastic printing at DTU. Following this, the mirror surfaces were covered by metal foil. This method has been used and tested before on prototyping of quasi optical components.

An example can be seen in Figure 4 and 5 below. This prototype was designed to test the receiver transmission line coupling system at the closure plate. The tests comprised of verification of the mirror and beam shapes – verifying the codes and design – and confirmation that the in-vessel cylinder around the mirrors (mitigating stray light) and the vacuum window assembly "tunnel" do not give rise to interference.

The results of the prototype tests (see example in Figure 6) verify that the designed quasi-optical components (in particular the mirrors) for the ITER CTS transmission line are performing as expected and have an acceptable correspondence to the modelling. The next step prototypes include 3D plastic printing of circular corrugated waveguides. These will then have a conducting coating applied by either conducting spray paint or other external coating methods.

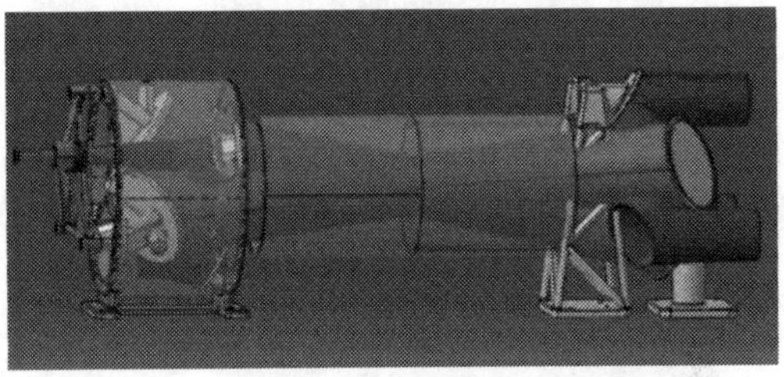

Fig. 4. CAD drawings of the ITER CTS prototypes of the closure plate coupling system for three receiver beams. The green duct represents the 400 mm long vacuum window assembly.

Fig. 5. The prototype testing setup in the DTU microwave lab of the receiver line coupling mirror units as presented in the CAD model in Figure 4.

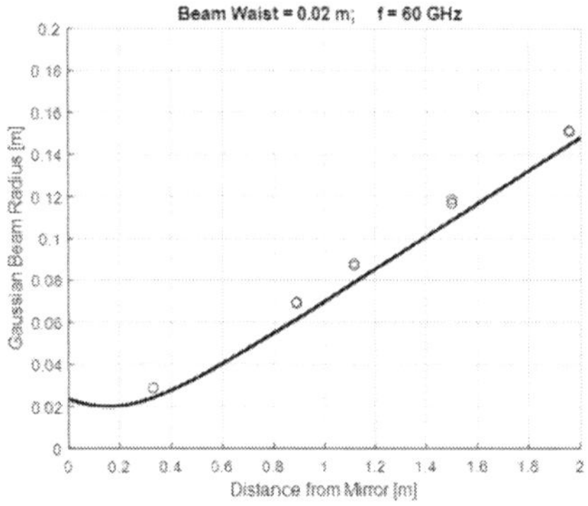

Fig. 6. Results of the prototype testing of the receiver transmission line coupling mirrors at the closure plate (shown in Figure 4 and 5). The solid black line shows the calculated (desired) beam radius and the red (blue) dots show the measurements without (with) the vacuum window assembly "tunnel". (1.0 x Gaussian width). The entrance of the waveguide is located at the position of the beam waist (0.157 m from the mirror).

4 Measurement capabilities

The present design of the ITER CTS diagnostic as presented above does not have any moveable parts, and thus the paths of the seven primary receiver beams are fixed. The probing frequency of 60 GHz was chosen as a compromise between the acceptable background ECE levels, maximum obtainable signals, and spatial resolution. For 60 GHz the O-mode cut-off is at the edge of the plasma, hence X-mode injection is required. However, the X-mode R-cut-off is quite close to the receiver frequency range, so depending on the plasma scenario (primarily the density profile) and the angle of injection (to the density gradient), refraction of the microwave beams can be significant. This is illustrated by the ray tracing results in Figure 7. It is seen that refraction is most significant for the outermost receiver beams having the largest poloidal angle to the density gradient. The positive side effect of this is a more perpendicular incidence on the probing beam, and hence an increased spatial resolution in the edge, i.e. ~a/20 – a/30, where a is the plasma minor radius. However, the fast ion density for these scattering volumes is presumably very low. The refraction in the poloidal plane will thus be different for various plasma scenarios and density profiles, and optimisation of the fixed mirror geometry will have to be based on a few reference scenarios and optimised for some robustness.

The probe and receiver beams do also have a toroidal injection angle in order to have a resolved angle $\varphi=\angle(k^{\delta},B)$ that is at least 5 degrees from 90 degrees in order to avoid interference in the CTS spectrum from plasma waves propagating perpendicularly to the magnetic field. φ is the angle between the main magnetic field B and the wave vector of the resolved fluctuations k^{δ}. The (slightly) toroidal injection angles do also give rise to refraction in the toroidal direction, which will depend on plasma scenarios and density profiles. However, the toroidal refraction is similar for the probe and receiver beams, and hence the overlaps between them should not be lost.

In the following we define the spatial resolution as the extent of the region from which 75% of the scattered signal originates.

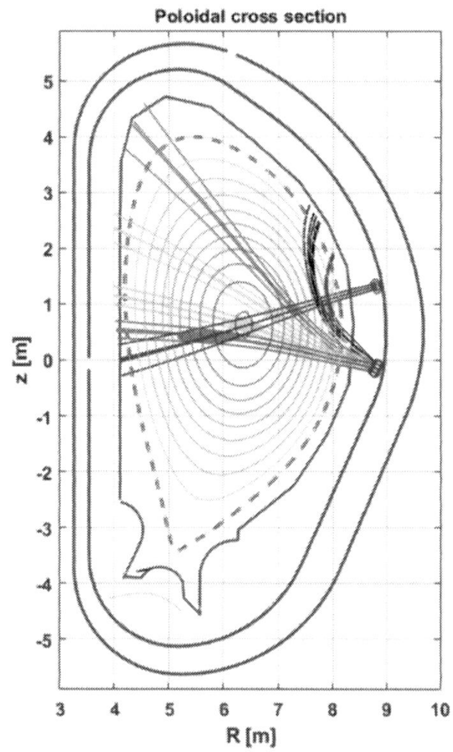

Fig. 7. Poloidal view of the ray tracing results for the ITER CTS probe and the seven receiver beams. The equilibrium is for Scenario 1 (reference scenario) and has been provided by F4E. The electron density and temperature profiles for the scenario are presented in Figure 8.

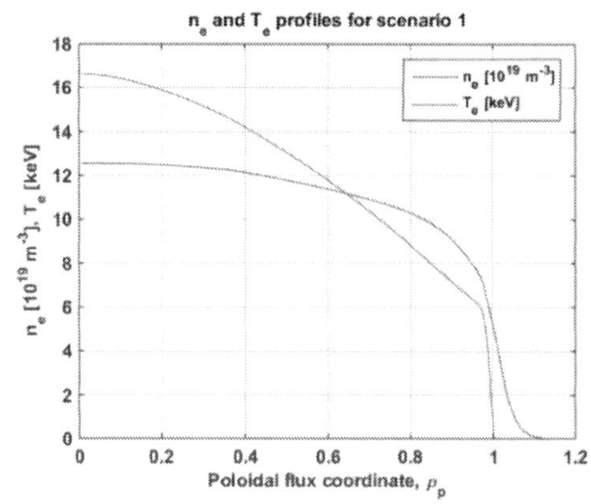

Fig. 8. Electron density and temperature profiles for the provided Scenario 1 (ITER Baseline Scenario) that was used for the ray tracing presented in Figure 7.

For the presently provided reference Scenario 1[a], the ITER Baseline Scenario (see Figure 8), the spatial resolution for the receiver observing the high field side is ~a/2 and for the center of the plasma it is ~a/4, and then increasing towards ~a/30 near the low field side plasma edge.

The study of the potential measurement capabilities will be reported in a separate publication, and the extent of the study is too comprehensive to fully review here. However, the conclusion is that the ITER CTS diagnostic can provide measurements of the projected 1D fast-ion velocity density function of the confined alphas and fast ions, as well as the alpha fast ion density (under isotropy assumptions) with a temporal resolution of 100 ms and within the required statistical uncertainty of 20%. A series of fitting tests with synthetic data has also shown that the present CTS system setup and analysis codes can reproduce the true alpha density used in the tests with only small discrepancies.

CTS has also been assigned a supplementary role for measuring the bulk ion temperature profile as well as the n_H/n_D and the fuel ion ratio n_T/n_D in the plasma core. As presented in [9, 10] the two latter will require a rather perpendicular scattering geometry, i.e. $\varphi=\angle(k^\delta,B)$~near 90 degrees. As the CTS mirror geometry is fixed, and deliberately off-perpendicular as described above, the ideal setup for fuel ion and isotope ratio measurements cannot be achieved. Thus these measurements are not design driving for the ITER CTS diagnostic. However, for a given plasma scenario and scattering spectrum, it may be possible to infer some information relating to the fuel ion ratio, as there will be spectral content containing information on this even with the off-perpendicular geometry. However, integrated data analysis accounting for other measurements may be required. In the analysis of the CTS spectrum with the aim of getting the information on the fast ion dynamics, the analysis will additionally provide some information on the bulk ion temperature and the ion rotation velocity, and may thus contribute to the determination of these via integrated data analysis.

The performance analysis results presented briefly above are based on the (conservative) assumption of a background ECE level of 100 eV. Since the CTS frequency range is far below the (cold) fundamental resonance the estimation of ECE background levels are not straightforward. Some modelling results predict virtually 0 eV, but high temperatures, multiple reflections of the beams on the vessel walls and Doppler shift of the resonance will give rise to a non-zero ECE level. While still searching for a good method/code for prediction of the level, we have assumed 100 eV.

Presently, no official full data sets of ITER plasma scenarios, plasma parameters and profiles, fast ion densities etc. exist. We have assembled the required information from sources at ITER IO and F4E. We follow closely the development of the official and traceable sets of ITER plasma scenarios and parameters performed in the IMAS project at ITER IO, and will update our analysis and predictions based on this data.

[a] See last paragraph of Section 4 concerning lack of traceability of plasma scenario data

5 conclusions

The present paper has intended to give an overview of the present status of the development of the ITER CTS diagnostic. It is in no way exhaustive of the extensive work done on a long range of topics for both the in-vessel and the ex-vessel parts of the system. Specific publications on these are planned and underway.

We have presented that the ITER CTS diagnostic is in the preparatory design phase, and the Preliminary Design Review for the system is expected within 12 months.

The diagnostic will be installed in drawer #3 of Equatorial Port Plug #12 and will consist of robust non-moveable quasi-optical parts. Based on modelling, the diagnostic is expected to fulfil its primary role for measuring the fast alpha and fast ion dynamics in seven measurement volumes in the plasma.

Acknowledgement

The work leading to this publication has been funded partially by Fusion for Energy under the Framework Partnership Agreement F4E-FPA-393. This publication reflects the views only of the author, and Fusion for Energy cannot be held responsible for any use which may be made of the information contained therein. The views and opinions expressed herein do not necessarily reflect those of the ITER Organization.

References

1. M. Salewski, et al., Alpha-particle velocity-space diagnostic in ITER, submitted to Nucl. Fusion, (2018)
2. H. Bindslev, F. Meo and S.B. Korsholm, ITER Fast Ion Collective Thomson Scattering, Feasibility study and conceptual design, EFDA Report (2003)
3. H. Bindslev, et al, Rev. Sci. Instrum, **75**, 3598 (2004)
4. S.B. Korsholm et al, Nucl. Instr. Meth. Phys. Res. A, **623**, 677 (2010)
5. S.K. Nielsen, et al, Phys Scr, **92**, 024001 (2016)
6. V. Infante, et al, Fus. Eng Design, **123**, 663 (2017) (http://dx.doi.org/10.1016/j.fusengdes.2017.05.001)
7. A. Lopes, et al, Neutronics analysis of the ITER Collective Thomson Scattering System, submitted to Fus. Eng. Des. (2018)
8. S.K. Nielsen, et al, Plasma Phys. Control. Fusion **57** 035009 (2015)
9. S.B. Korsholm et al, Phys. Rev. Lett. **106** 165004 (2011) (http://dx.doi.org/10.1103/PhysRevLett.106.165004)
10. M. Stejner et al, Plasma Phys. Control. Fusion **53** 065020 (2011)

Progress in ITER ECE Diagnostic Design and Integration

V.S. Udintsev[1,*], S. Danani[2], G. Taylor[3], T. Giacomin[1], J. Guirao[1], S. Pak[1], S. Hughes[1], L. Worth[1], G. Vayakis[1], M.J. Walsh[1], M. Schneider[1], H.K.B. Pandya[2], R. Kumar[2], V. Kumar[2], S. Jha[2], S. Thomas[4], S. B. Padasalagi[2], S. Kumar[2], P. E. Phillips[4], W. L. Rowan[4], M.E. Austin[4], A. Khodak[3], R. Feder[3], H. Neilson[3], A. Basile[3], A. E. Hubbard[5], A. Saxena[6], C. Nazare[6], P. Maquet[7], N. Gimbert[7]

[1] ITER Organization, Route de Vinon-sur-Verdon, CS 90 046, 13067 St. Paul Lez Durance Cedex, France
[2] ITER-India, Institute for Plasma Research, Bhat, Gandhinagar-382 428, India
[3] Princeton Plasma Physics Laboratory, Princeton, NJ 08543, USA
[4] Institute for Fusion Studies, University of Texas at Austin, TX 78712, USA
[5] Plasma Science and Fusion Center, MIT, Cambridge, MA 02139, USA
[6] Fircroft France, 4 rue du Bailliage, 78000 Versailles, France
[7] Bertin Technologies, 155 rue Louis Armand, CS 30495, 13593 Aix-en-Provence Cedex 3, France

Abstract. The ITER Electron Cyclotron Emission (ECE) diagnostic is progressing towards its Preliminary Design Review (PDR). In parallel, the diagnostic integration in the Equatorial Port is ongoing. Port Integration has to address the structural integrity to withstand various loads, maintenance and the safety aspects of ECE diagnostic. The ITER ECE system includes radial and oblique lines-of-sight. Recently, a successful peer-review of the in-port plug Hot Calibration Source has taken place and its performance and integration feasibility has been demonstrated. Four 45-meter long low-loss transmission lines are designed to transmit mm-wave power in the frequency range of 70- 1000 GHz in both X- and O-mode polarization from the port plug to the ECE instrumentation room in the diagnostic building. Prototype transmission lines are being tested [1]. A prototype polarizing Martin-Puplett type Fourier Transform Spectrometer (FTS) operating in the frequency range 70-1000 GHz, has a fast scanning mechanism and a cryo-cooled dual-channel THz detector system. Its performance has been tested as per ITER requirements. Assessment of the instrumentation and control requirements, functional and non-functional requirements, operation procedures, plant automation are ongoing for the PDR. The current status of the diagnostic, together with integration activities, is presented.

1 Introduction

The Electron Cyclotron Emission (ECE) diagnostic design in ITER has to address the specifications defined in the project measurement requirements. Several requirements are specified with different spatial, temporal and spectral resolution and accuracy. Even for the same measurement, the electron temperature profile, the ECE measurement requirements can be different for advanced machine control and for physics. The relativistic downshift at high temperatures, which in ITER is substantial, will strongly influence the access to the plasma for ECE. Integrated performance for the temperature measurement is, however, adequate when all electron temperature (T_e) systems (ECE and two Thomson scattering systems) are considered together [2]. A prioritization has been given to the core electron temperature measurement and to Neoclassical Tearing Modes (NTMs) detection [3, 4]. For the temperature perturbation caused by NTMs, ECE is the only diagnostic that can deliver this measurement. There is a functional requirement for the broadband measurement system to measure the total radiated power due to ECE. The specified frequency for the broadband measuring system is from 70 GHz up to 1 THz. The design should take into account the physics and technical constraints that may affect, on one hand, the layout of the transmission line and the back-end instrumentation and, on the other hand, the performance of the system for priority measurements, such as the core electron temperature profile and the NTM amplitude and localization.

In particular, ECE integration into the ITER complex is a very important task that requires interfaces with other systems, like port plug, buildings and remote handling. The latter is particularly important as the port plugs will have to be remotely handled during installation/removal, including refurbishment, environmental and functional tests in the Hot Cell Facility (HCF) outside the tokamak. Another very important design driver for the diagnostic is occupational safety for workers who will have to perform hands-on maintenance operations. Finally, various parts of ITER ECE systems will experience different load conditions, arising from electromagnetic events (during disruption), thermal and neutronic heating, and seismic events.

* Corresponding author: victor.udintsev@iter.org

This paper updates the status of ITER ECE design and discusses current and future activities.

2 ITER ECE requirements

As stated in the ITER Project Requirements, the measurements shall be delivered by the suitable set of diagnostics with certain spatial and temporal resolution and accuracy. Clearly, depending on the physical principles of each diagnostic, some systems are more appropriate for a particular measurement than others. ECE has a primary role (which means that it is well suited to the measurement) in the delivery of the core electron temperature profile and detection of NTMs (δT/Te; 100 ms integration time). The system has a backup role (which means it provides similar data to primary role, but has some limitations) in the measurement of pedestal in the H-mode, Edge Localized Modes (ELMs) and L-H mode transition indicator. The ECE diagnostic has a supplementary role (which means that diagnostic validates or calibrates the measurement, but is not complete in itself) in the delivery of plasma energy, radiated power, runaway electrons, edge electron temperature profile and high frequency instabilities (i.e. TAEs δT/T).

Based on the measurement requirements, a prioritization analysis and flow-down has been performed using a limited set of criteria, to establish diagnostic measurement specifications. Prioritization is important in establishing a focus for development of a system design as well as accommodating constraints set by physics and integration limitations. It should be noted that assessments were performed for both the early lower performance operating phase, as well as for operation during the planned Q = 5 to10 burning plasma phase. Table 1 gives an overview of specifications for primary role measurements after the completion of requirements flow-down.

The spatial resolution required for electron temperature profiles is specified at 67 mm (or a/30); however, depending on the plasma parameters and physical principles, some diagnostics that measure Te may provide better or worse values. For core electron temperature profile measurement by ECE, the high temperatures foreseen in ITER (30 – 40 keV) will affect the spatial resolution, due to the large relativistic and Doppler broadening, which will lead to a widening of the radial extension from which a given frequency in the ECE spectrum is emitted. It should be noted that the target of a/30 would be possible only in regions with a low enough value of Te (typically r/a > 0.5 for Te(0) = 30 keV on ITER). The relativistic downshift at the high temperatures in ITER is substantial and will strongly limit the access to the plasma. It is important to remark that the high temperature scenarios considered for ITER (Te(0) up to 40 keV) will degrade the spatial resolution of Te due to a further widening of the emission layer. Further methods to improve spatial resolution of the diagnostic by deconvolution will be investigated in coming years as a part of diagnostic performance assessment. It is also advisable to get some support from ECE diagnostics in operating fusion devices to testify the applicability of methods to enhance the spatial resolution at higher Te.

Table 1. Expanded table of ITER ECE diagnostic specifications for primary role requirements.

Measur-t	Range	Time/ Freq. resolution	Spatial resolution
Core Te	0.5 - 15 keV	10 ms/ 1 μs	(a/30)/ 2-5 cm, increases with Te
	15 – 40 keV		(a/30)/ 5-13 cm, increases with Te
NTM δT / Te	(0.1 - 5) x 1E-2; ΔT/T > 1% sensitivity decreasing with bandwidth	0.1 -10 kHz/ 0.1 – 1000 kHz	(m,n) < (2,1), (3,2). dr=5cm/ 4-10 cm, depends on Te

ECE will play a key role in the targeting of the Electron Cyclotron Heating (ECH) to control neoclassical tearing modes (NTMs). The ECH deposition widths for the NTM stabilization are in the range of 30 – 70 mm. An island with the width of w > 20 mm is detectable by ITER ECE (Fig. 1). Thus, assumption for ECE channel separation of 20 mm or smaller at NTM radii is adequate for the mode detection and localization. The response time of the feedback system once the mode is tracked is in the order of tens of ms. It is anticipated that it will take about 20 ms to begin moving the ECH steering mirror and then some time to lock onto the mode. The total response time might be 30 – 50 ms or even longer to lock onto the mode and begin suppressing it with Electron Cyclotron Current Drive (ECCD), depending on how far the mode rational surface is away from where the mirror is pointing at the time. The growth time of the NTM is of the order of 100 ms, and the target resolution of the diagnostic expressed in terms of sampling rate could be in the range of 0.1 – 1000 kHz. Response to the NTM in ECE measurements is limited by the intrinsic resolution of the instrument that includes the relativistic broadening and the antenna pattern. Also, the signal-to-noise ratio limits the minimum island size to about 2 cm for a dT_e/T_e ~1%. The latency needed for the NTM feedback is estimated to be 1 ms which is sufficient to match the response time of the feedback system. To improve the NTM detection and make the ECH feedback more effective, correlation with another ECE line-of-sight or with magnetic diagnostics will be used to increase the sensitivity of the measurements. For islands as small as 10 mm and located closer to the core, the measurement of ΔTe will be more challenging, but Fast Fourier Transform (FFT) techniques can most likely be

 https://doi.org/10.1051/epjconf/201920303003

employed to enhance the signals relative to the noise. This is a technique that has been proven to be efficient in other tokamaks where NTM detection and stabilization is routinely performed [5].

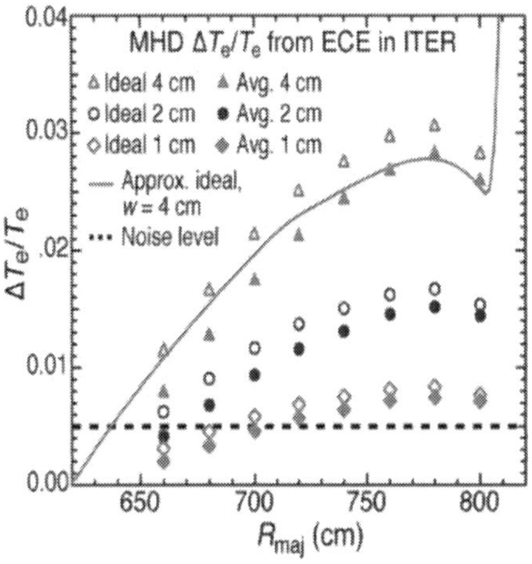

Fig. 1. Expected **ΔTe/Te** from rotating MHD islands. Correlation technique could be used to get even better resolution. Note: ITER plasma major radius is 6.2 m.

Details of backup and supplementary measurements and their specifications for ITER ECE are given in [3] and are not being repeated here.

3 Diagnostic instrumentation

To enable measurements discussed in the previous section, the following design and instrumentation are being designed:

- •Two lines-of-sight from Equatorial Port #9; one perpendicular and another oblique at ~10 degrees (used for redundancy in primary measurements and also for non-thermal electron population detection);
- •Frequency coverage: 70 GHz – 1000 GHz; both X- and O-modes; ensured by polarization splitters near the port plug
- •One 2X-mode radiometer, 220 – 340 GHz, and one 1O-mode radiometer, 122 – 230 GHz -> @ Bt = 5.3 T;
- •Two, high throughput, reciprocating Martin-Puplett Fourier Transform Spectrometers, 70-1000 GHz, a frequency resolution ≤ 5 GHz, instrument transmittance > 70 % and a scanning repetition rate ≤ 20 ms.

Figure 2 shows the prototype Polarizing Fourier Transform Spectrometer (FTS) developed and tested at ITER-India Lab and supplied by the Blue Sky Spectroscopy. The tests are ongoing; parameters already achieved are quite close to the measurement specifications for FTS instruments planned to be used in ITER during operation.

Fig. 2. Prototype Polarizing Fourier Transform Spectrometer being tested at ITER-India Lab.

4 Operations at 1.8 T

There are several stages for ITER operations currently assumed. Recently, the ITER research plan [6] has been modified to include the option of operation at 1.8 T. The exploitation of this option offers the possibility of access to H-mode operation at lower injected powers, implying that H-modes (and ELM control) may be studied at an early stage of the research program where the available levels of heating power will be lower than the baseline of 73 MW. ITER ECE diagnostic performance has been assessed for compatibility with 1.8 T operations. As an input, the data was provided for He- H-modes with a magnetic equilibrium, EQDSK file representing an ITER plasma at 1.8T along with electron density (ne) and Te profiles coming from the METIS 0.5D transport code [7] and from the ASTRA code for the L-mode [8]. The GENRAY ray tracing code was used to simulate ECE. Figure 3 shows a plot of ECE frequencies, multiple harmonics, versus major radius for the 1.8 T case. The low frequency radiometer (115-220 GHz) would be able to measure the low-field-side Te profile with 3rd harmonic frequencies. However, the best coverage in terms of spatial resolution would be for the 2nd harmonic X-mode, in the frequency range of 75 – 120 GHz. This range is currently not covered by a radiometer instrument in the diagnostic plan and has to be developed. For information, current radiometers cover low frequencies (1-O mode at 5.5 T at 122 – 230 GHz) and high frequencies (2X-mode at 5.3T at 220 – 340 GHz). Calculations also indicate very high optical thickness, which means that it is straightforward to obtain Te. The conclusion is that the low frequency radiometer as such has limited capability to measure at 1.8 T. Although 3rd harmonic X-mode profile measurement is possible with current instruments, a better coverage with better spatial resolution can be ensured with the 2nd harmonic X-mode ECE. If the system will be needed to operate at 1.8 T magnetic field, a radiometer instrument for 2nd harmonic X-mode (75 – 120 GHz) will have to be developed to ensure Te measurements with better resolution, including the pedestal region. The 75-120 GHz coverage would be implemented by replacing the RF section of the 220-340 GHz radiometer, using the same IF and video bands as the 220-340 GHz radiometer.

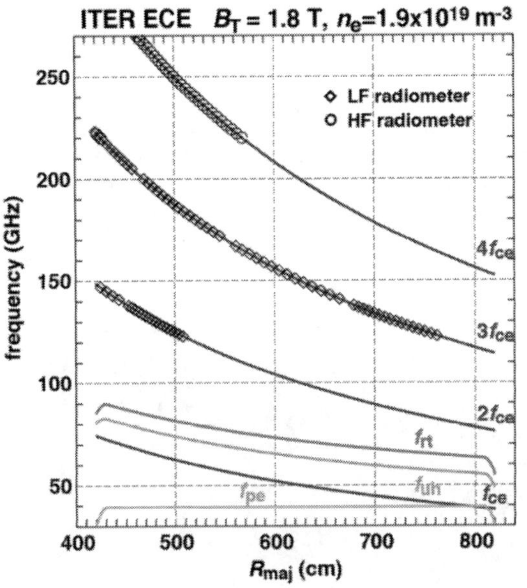

Fig. 3. Simulated frequencies and radial coverage by presently designed radiometers for 1.8 T ECE operation assuming recent METIS 0.5D kinetic profiles for Te and ne.

5 Integration in the tokamak infrastructure

ECE diagnostic integration in the tokamak is interfacing with port infrastructure, buildings and services, such as vacuum, electrical and remote handling. The general layout of ECE system is shown in Fig. 4. Many components are located in the inner, Port Interspace and outer, Cell, areas. These two areas of the Port Cell are radiation zones with substantially different (~ factor of 10) exposure levels but common atmosphere. The design of the diagnostic has to make a provision for minimizing worker's presence there for maintenance operations.

Various parts of ITER ECE systems will experience different load conditions, arising from electromagnetic events (during disruption), thermal and neutronic heating, and seismic events. There are other accidental loading cases, such as fire hazard in the port cell or gallery, which have to be taken into account during design, too. Finally, ECE in ITER has to be protected from microwave stray radiation by the ECH installation (170 GHz) and by other millimeter wave emission, like Collective Thomson scattering, planned to operate at 60 GHz.

The diagnostic front-end is being currently integrated in Equatorial Port #9. Integration is compatible with the latest modular Diagnostic Shield Module (DSM) design that is proven to provide the required neutron shielding performance and fulfil maintenance requirements. The ECE front end is fixed on the vertical plates as shown in Fig. 5 and is surrounded by Boron Carbide (B4C) blocks for neutron absorption, which reduces the Shutdown Dose Rate (SDDR) on the port closure plate and in the Port Interspace - Port Cell. Work is continuing together with integration of the port itself with other tenant diagnostics in the same port.

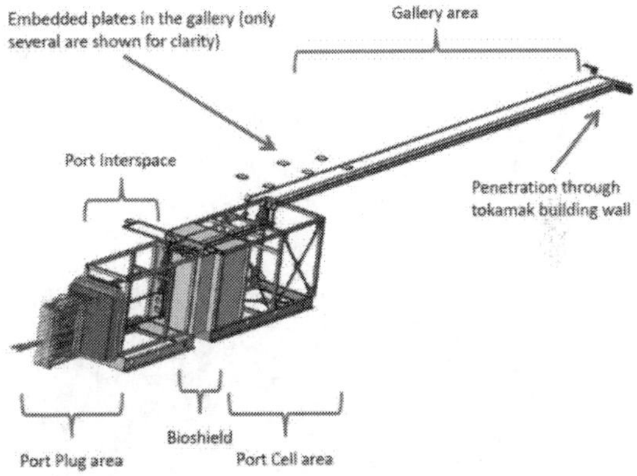

Fig. 4. Schematic layout of the ITER ECE diagnostic from port plug to the tokamak building boundary.

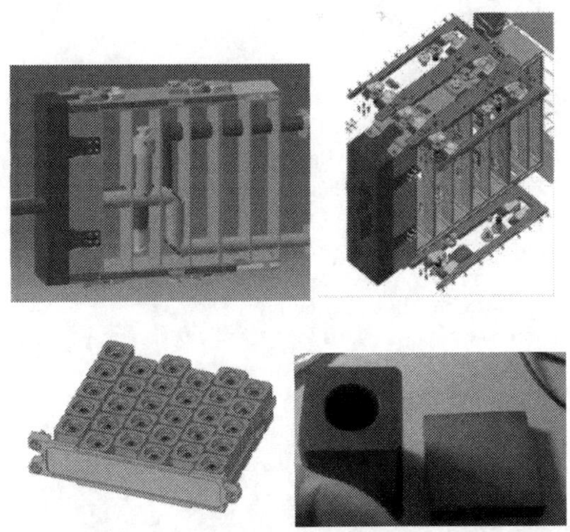

Fig. 5. Integration of ECE front-end in the modular DSM (top row; different colors represent two different lines-of-sight); B4C shielding trays and single units to stop neutrons (bottom row).

In-port plug ECE Hot Calibration Source design [4] successfully passed its Peer Review in 2017; some tests, like steam/ water leak event compatibility, are still ongoing. The heated Silicon Carbide (SiC) emitter has reached the temperature > 700 C in the test prototype, and outgassing tests in vacuum have demonstrated compliance with ITER requirements. Figure 6 shows the prototype test layout provided by the US ECE Team. The design of the shutter/ switchable mirror to orient the ECE beam towards the plasma or towards the Hot Calibration source is still being developed; a few concepts ranging from mechanical rod actuators to a piezoelectric motor were considered. More development work is expected in the coming months for the design and integration of the shutter mirror; also, integration of lessons learned about ECE shutters operation from existing machines would be beneficial.

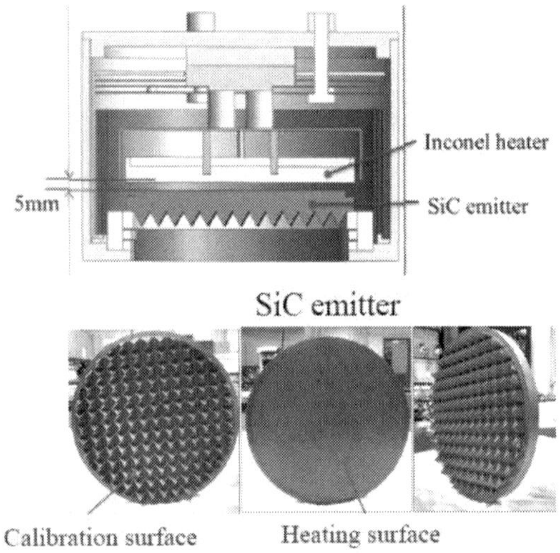

Fig. 6. Hot Calibration Source basic design (top) and SiC emitter (bottom).

The development of the primary diagnostic window assemblies for ECE (Fig. 7) is progressing towards the Final Design. Lessons learned during the Preliminary Design phase indicate that windows design must be updated to be fully compliant with structural and performance needs and to fulfil safety requirements. It is important to ensure that no additional loads or stresses on the primary windows are introduced by the compliance seal box located on the atmospheric side of the port closure plate next to the window assemblies, but not reacting on windows mechanically.

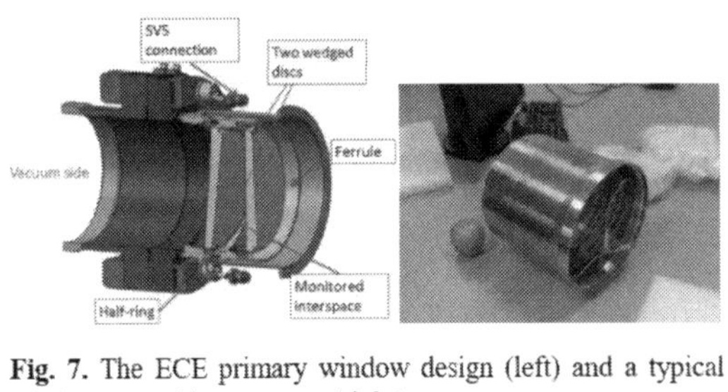

Fig. 7. The ECE primary window design (left) and a typical window assembly prototype (right).

The design is being considered to avoid excessive loading of the window assembly. In parallel, several prototyping/ Research and Development (R&D) activities are being undertaken to demonstrate the feasibility of novel window assemblies design in ITER, or to validate specific features of the design.

Transmission lines outside the port plug are integrated in the port interspace and port cell support structures and routed through the gallery and building penetrations to the dedicated area in the diagnostic building where back-end instrumentation is installed. The transmission lines are then safety important as they contribute to the tritium confinement as a secondary barrier of the tokamak infrastructure. Waveguides are attached to the supports connected to embedded plates in the ceiling of the building. The design of supports is entering its final phase, and it is being in close collaboration with ITER construction team; the completion of installation of captive supports is a priority task to be completed to ensure that the building' infrastructure is ready for the First Plasma operations in ITER, even if ECE is not meant to be operational for the First Plasma. Figure 8 shows the layout of the ECE transmission lines and their supports in the tokamak building complex.

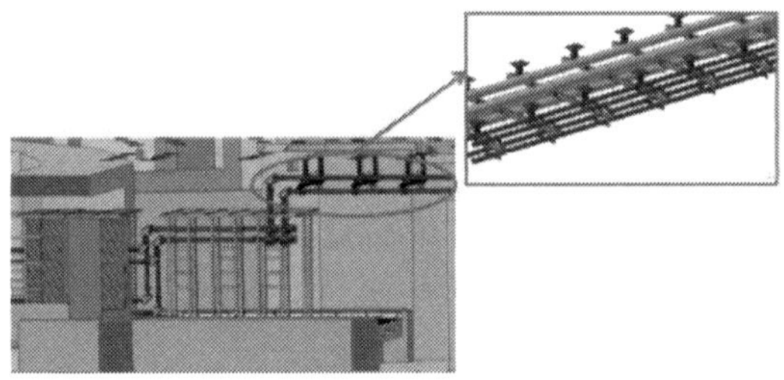

Fig. 8. Ex-Port Plug transmission lines layout and zoon into the waveguide supports attached to the ceiling of the tokamak building.

6 Ongoing R&D activities on transmission lines characterization

Currently, ex-port plug transmission lines (TLs) are being prototyped by the ITER India team, in collaboration with ITER Central Team and US Team. Three types of transmission lines, including: 10 m long smooth-wall waveguides, 2 m long corrugated waveguides and 4 m long dielectric waveguides, are being tested (separately) as an assembly including mitre bends and using FTS instrumentation. The tests are being carried out both in air and in vacuum. Results are important for the engineering solution for transmission line layout: currently, ITER ECE foresees using fully evacuated (at rough vacuum) smooth-wall transmission lines with safety valves at building boundaries to ensure confinement function, to minimize mm-wave absorption on water vapour in air, and to avoid Bragg-scattering effects that would be present if corrugated waveguides are used. However, if the tests demonstrate that ECE over the entire range from 70 GHz to 1000 GHz is still usable for all planned measurements to meet requirements for the transmission line in air and with corrugated transmission lines, it would be beneficial to use passive secondary windows at the building boundaries, instead of valves with safety control function. This would also avoid excessive loading of the primary window assemblies at the closure plate due to the absence of the rough vacuum on the glass material from the interspace side.

The test setup is shown in Fig. 9. The first results are presented in [1]. It is planned to complete these studies in 2018 and to make the final decision on the transmission line layout for the Preliminary Design of ECE ex-port plug part.

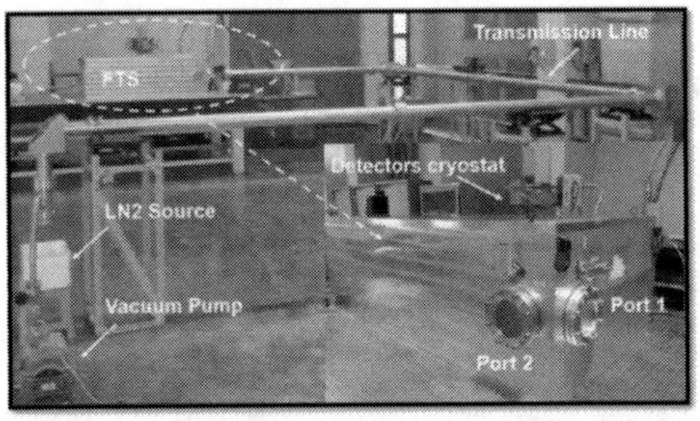

Fig. 9. Transmission line mock-up at ITER India [1].

7 Summary

Development of the ITER ECE is progressing well towards its Preliminary Design Review. The R&D and Peer Review of the Hot Calibration Source have been completed; however, important design work on shutter design remains open. Integration work related to port infrastructure is ongoing; integration in the tokamak buildings is also important in order to meet construction need-dates for ECE transmission line captive supports. R&D to characterize the transmission lines outside the port plug and to finalize its layout is ongoing to ensure the maximum diagnostic performance while complying with safety requirements. Finally, future work to improve the spatial resolution of the system at higher Te by applying numerical methods and support from ECE diagnostics in operating fusion devices to test the applicability of methods to enhance the spatial resolution at higher Te would be very helpful.

References

1. R. Kumar *et al.*, "Comparative studies of various types of transmission lines in the frequency range 70 GHz -1 THz for ITER ECE diagnostic", this Workshop.
2. V.S. Udintsev *et al.*, Fusion Engineering and Design, Volume 88 (6) (2013)
3. V.S. Udintsev *et al.*, "Exploring measurement capabilities of ITER ECE system." *Proc. FEC (San Diego) IAEA-CN-197/ITR/P5-41* (2012)
4. G. Taylor *et al.*, in Proc. of EC-19 Workshop, Ahmedabad, India (2016)
5. F. Felici *et al.*, Integrated real-time control of MHD instabilities using multi-beam ECRH/ECCD systems on TCV, *Nuclear Fusion*, *52*(7):074001 (2012)
6. ITER Technical Report ITR-18-003
7. J.F. Artaud *et al., in Proc. of the 32nd EPS Conf. on Control. Fusion and Plasma Phys.*, ECA Vol. 29C, P1.035 (2005)
8. G. Pereverzev and P. N. Yushmanov, Max-Planck-Institut für Plasmaphysik, Report IPP 5/98, February 2002

Design of the Synthetic Aperture Microwave Imager Upgrade for measurement of the edge current density on MAST-U

J.O. Allen[1,*] and *C. H.* Vincent[2] *R. G. L.* Vann[1]

[1]York Plasma Institute, Department of Physics University of York, York, YO10 5DD, UK
[2]Centre for Advanced Instrumentation, Department of Physics Durham, DH1 3LE, UK

Abstract. The Synthetic Aperture Microwave Imager (SAMI) has demonstrated the feasibility of 2D Doppler backscattering for measurement of the edge magnetic pitch angle on MAST and NSTX-U. The aim of SAMI-Upgrade (SAMI-U) is to build on this methodology to produce higher quality pitch angle data simultaneously in multiple spatial locations, enabling calculation of the edge current density. This movement from proof of principle to production quality necessitates several alterations to the design. There will be a fourfold increase in the number of antennas, as minimising the sidelobe level is key to ensuring maximum resolution in the reconstructed Doppler backscattered power map. SAMI-U will actively probe the plasma with two frequencies at the same time. These correspond to two different backscattering locations in the edge plasma which allows the edge current density to be calculated from the measured magnetic field vector. Dual-polarised sinuous antennas will be used in the array as they are planar and broadband. Polarisation separation is necessary for differentiation between the O- and X-mode cut off surfaces, as their locations can be separated by up to a few centimetres. Due to spatial constraints many of the components will be placed on a PCB behind each antenna. FPGAs will be used to stream the high data throughput, over 16 GB s^{-1}, into PC memory.

1 Introduction

For future tokamaks, such as ITER, the energy released by the largest edge-localised mode (ELM) instabilities will critically damage components [1, 2], therefore understanding the ELM mechanism in order to suppress or mitigate their effects is important for future devices. Several diagnostic methods have been employed so far to measure the edge current density. These include the Zeeman effect [3, 4], the motional Stark effect (MSE) [5] and beam emission polarimetry [6]. However, these are all indirect methods which do not necessarily work at all times throughout a plasma shot. Consequently a diagnostic to deliver direct and continuous edge current density measurements is desirable. By making high quality direct measurements of the edge magnetic pitch angle at multiple radial locations, SAMI-U will produce a profile of the edge current density throughout the plasma lifetime on MAST-U.

2 Summary of SAMI results

SAMI was used to make passive measurements, of plasma thermal emission, as well as active measurements, when probing the plasma with a broad microwave beam, of the plasma edge. It was installed on MAST to passively investigate electron Bernstein wave (EBW) mode conversion windows [7, 8] and additionally microwave bursts during ELMs [9]. In active probing mode SAMI has been used to record the edge magnetic pitch angle on MAST at multiple radial locations in the plasma [10, 11]. The system was subsequently moved to NSTX-U where the pitch angle measurements were repeated. These results demonstrated the feasibility of 2D DBS for measuring the magnetic pitch angle and motivate the construction of a production quality diagnostic, the SAMI-Upgrade.

3 System overview

A block circuit diagram of the SAMI-U radio frequency (RF) and digitisation system can be seen in figure 1. The antenna feed is connected to a balun on the RF PCB module, highlighted in green, which is situated perpendicular to the antenna PCB. The signal is amplified before passing through a power divider into IQ mixers at the fundamental. These two mixers combine the antenna signals with two local oscillator signals in the channel frequency range 10 – 40 GHz. The IF is band pass filtered between 1 MHz, to block leakage from the LO signal, and 25 MHz, to prevent aliasing; this is under half of the ADC sampling rate of 65 MS s^{-1}. The upper frequency of the band pass filter is kept below the Nyquist frequency to allow for the non-negligible frequency roll off of the filter.

Coaxial cables connect the RF modules to the digitisation system where first the signal is converted by an ADC. The digital output is processed by an FPGA. SAMI-U will produce over 16 GB s^{-1} of data throughout a plasma shot, which is processed into Ethernet packets and sent over multiple 10 Gbps fibre links to the data storage machine.

*e-mail: j.allen@york.ac.uk

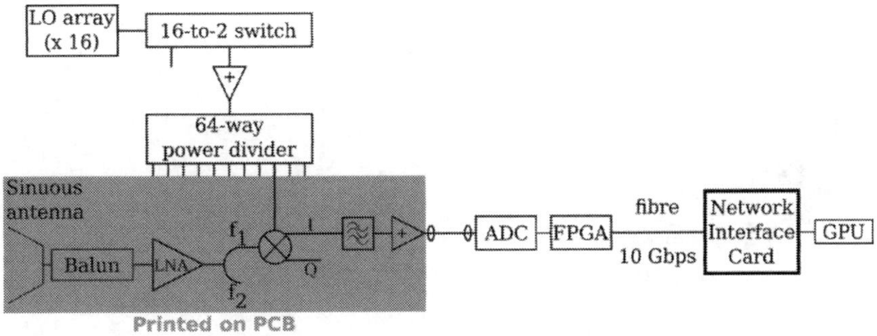

Figure 1. A block circuit diagram of the circuitry for one of two antenna polarisations in the SAMI-U array. Repeating sections have been omitted. The components in the green section will be printed or surface mounted on PCB which will feed the IF into the digitisation system.

Using this method SAMI-U can accommodate planned increases in pulse length on MAST-U and is only limited by the amount of RAM available in the storage machine. GPUs will be used in future for real-time calculation of the edge current density which could allow SAMI-U to become part of the set of feedback control diagnostics.

4 The sinuous antenna array

There are several constraints affecting the SAMI-U array design, the most restrictive of which is the 160 mm port window diameter which sets the maximum array diameter. This limits the number of antennas of a given size that can be contained while retaining independent baselines. In order to cover as much of the pedestal as possible, the antennas should be wide-band to encompass frequencies over at least one octave in the frequency range 10 – 40 GHz. To enable separation of O- and X-mode radiation they must be dual-polarisation. A planar structure would protect against spurious reflections within the array.

The sinuous antenna is the top candidate when considering all of these requirements and furthermore it is especially compact, a diameter of 13 mm is required to reach a low end frequency of 10 GHz. Its dual-polarisation log-periodic structure is shown in figure 2 with an example symmetric feed for both pairs of arms in the centre. The arms interleave to enable the dual-polarisation capability and each polarisation will be printed on either side of a PCB. Due to the top frequency of 40 GHz, the central area in which to extract the signals is small. Therefore, designing the geometry of this feed network is the biggest challenge when working with sinuous antennas at these frequencies.

SAMI-U will be situated at the midplane out of vacuum behind a port window of diameter 160 mm and its antenna array has a 150 mm diameter to retain an antenna field of view of ±40° in both the horizontal and vertical directions, accounting for the window thickness. Each of the 32 antenna elements requires a circular area of diameter 15 mm, including 2 mm of substrate free from copper surrounding the antenna arms. The antennas may not overlap one another and must lie wholly within the array PCB.

Figure 2. A dual-polarisation sinuous antenna, vertical polarisation in green and horizontal polarisation in purple. The log-periodic structure of the sinuous antenna offers a relatively small diameter for the low end frequency compared with other antennas and enables two polarisations to interleave. An array of sinuous antennas will not suffer from internal reflections due to their planar shape. At the desired maximum frequency, 40 GHz, the central feed region becomes very small making the design of the feed network the most complex part of the design. The feed design is omitted here so as not to prejudice future publications.

When arranging the original SAMI array, a simulated annealing optimisation method was used to minimise the sidelobe level of the array response to a point source [12]. However, the SAMI-U element placement will be chosen from a set of random placements as the benefit of using an optimisation algorithm in this case is unlikely to be worth the computational expense. Figure 3 shows an example antenna arrangement. The response of this array to a point source is displayed in figure 4.

All the sidelobes are below −18 dB making uncertainty due to these spurious signals sufficiently small in relation to the real backscattered signal. This is significantly below the maximum sidelobe of over −7 dB in the comparable response of the existing SAMI system, displayed in figure 5.

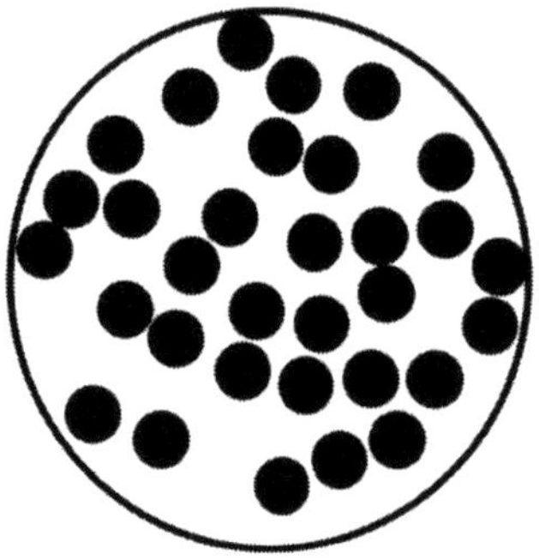

Figure 3. An example SAMI-U array layout. Ideally the layout should be maximally random in order to maximise the number of independent baselines and thus give the maximum amount of unique information in Fourier space. This is a random antenna arrangement with the lowest sidelobe level from a set of 10,000 candidate arrangements. The computational expense of optimising the placement of the 32 elements is likely to be prohibitive.

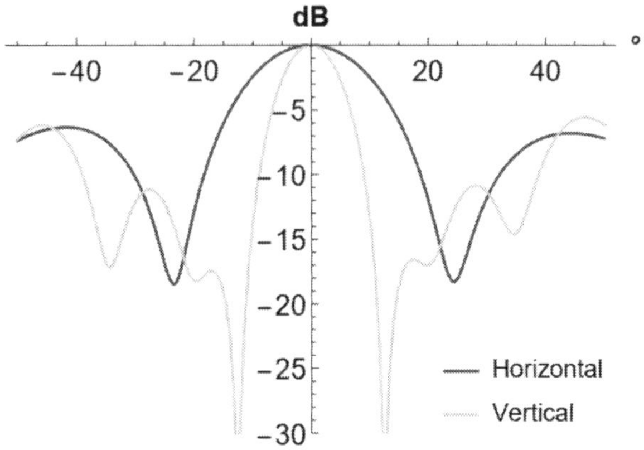

Figure 5. Response of the SAMI array to a point source at $(\theta, \phi) = (0, 0)$ at frequency 16 GHz. The maximum sidelobe is above -7 dB in both the horizontal (blue) and vertical (green) planes.

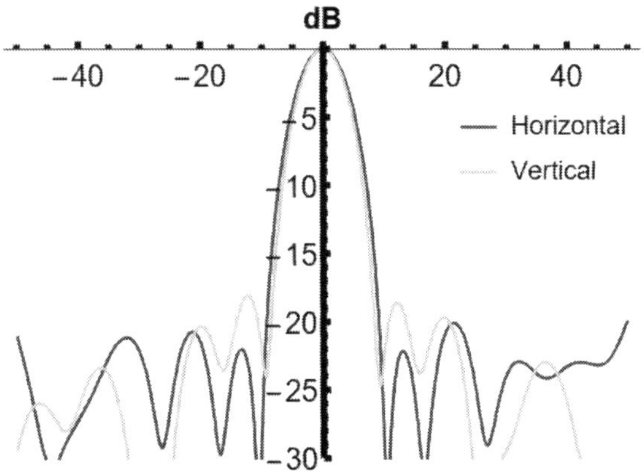

Figure 4. Response of the SAMI-U candidate array in figure 3 to a point source at $(\theta, \phi) = (0, 0)$ at frequency 16 GHz. The maximum sidelobe is below -18 dB in both the horizontal (blue) and vertical (green) planes.

5 Conclusion

Components for the SAMI-U diagnostic are being designed and built with the view to installing the completed system on MAST-U in time for the first experimental campaign. This device will extend the capability of the SAMI diagnostic to measure magnetic pitch angle at multiple radial locations to give the edge current density profile. The evolution of these current density profiles will aid understanding of edge instabilities including ELMs.

6 Acknowledgements

The authors would like to thank the Fusion CDT, EUROfusion and the EPSRC for supporting this research. This work was funded by the EPSRC under grant EP/L01663X/1.

References

[1] A. Loarte, G. Saibene, R. Sartori, D. Campbell, M. Becoulet, L. Horton, T. Eich, A. Herrmann, G. Matthews, N. Asakura et al., Plasma Physics and Controlled Fusion **45**, 1549 (2003)
[2] J.W. Connor, A. Kirk, H.R. Wilson, S. Benkadda, AIP Conference Proceedings **1013**, 174 (2008)
[3] D.M. Thomas, A.W. Leonard, H.W. Mueller, Review of Scientific Instruments **75**, 4109 (2004)
[4] D.M. Thomas, A.W. Leonard, R.J. Groebner, T.H. Osborne, T.A. Casper, P.B. Snyder, L.L. Lao, Physics of Plasmas **12**, 1 (2005)
[5] M.F.M. De Bock, J. Citrin, S. Saarelma, D. Temple, N.J. Conway, A. Kirk, H. Meyer, C.A. Michael, Plasma Physics and Controlled Fusion **54**, 025001 (2012)
[6] E. Viezzer, R. Dux, M. Dunne, Review of Scientific Instruments **87**, 11E528 (2016)
[7] V.F. Shevchenko, R.G.L. Vann, S.J. Freethy, B.K. Huang, Journal of Instrumentation **7**, P10016 (2012)
[8] S.J. Freethy, B.K. Huang, V.F. Shevchenko, R.G.L. Vann, Plasma Physics and Controlled Fusion **55**, 124010 (2013)
[9] S.J. Freethy, V.F. Shevchenko, B.K. Huang, R.G.L. Vann, *Localised microwave bursts during ELMs on MAST*, in *EPJ Web of Conferences* (EDP Sciences, 2015), Vol. 87, p. 03008
[10] D.A. Thomas, K.J. Brunner, S.J. Freethy, B.K. Huang, V.F. Shevchenko, R.G.L. Vann, Nuclear Fusion **56**, 026013 (2016)
[11] R.G.L. Vann (2018), 20th Joint Workshop on Electron Cyclotron Emission (ECE) and Electron Cyclotron Resonance Heating (ECRH)

 https://doi.org/10.1051/epjconf/201920303004

[12] S.J. Freethy, V.F. Shevchenko, R.G.L. Vann, IEEE Transactions on Antennas and Propagation **60**, 5442 (2012)

Investigation of Optically Grey Electron Cyclotron Harmonics in Wendelstein 7-X

*Neha Chaudhary**, *Johan W. Oosterbeek*, *Matthias Hirsch*, *Udo Höfel*, *Robert C. Wolf*, and *W7-X Team*

Max-Planck-Institut für Plasmaphysik, 17491 Greifswald, Germany

Abstract. For a magnetic field of 2.5T, the electron cyclotron harmonics are spectrally well separated in W7-X as it has a large aspect ratio. Because of this advantage from the geometry of W7-X stellarator, it is easier to scan these higher harmonics (70,140,210.....GHz) compared to tokamaks with small aspect ratio. For confinement reasons, W7-X is planned to work at high plasma densities applying O2 electron cyclotron resonance heating (ECRH). For such plasmas, which already have been demonstrated in experimental campaign OP1.2a of W7-X, electron cyclotron emission (ECE) from second harmonic extraordinary mode (X2) is in cutoff. In that case, optically grey higher harmonics provide the only access to ECE signal and hence electron temperature profiles. A Michelson interferometer will be used in the next operational campaign of W7-X for broadband (50-500 GHz) ECE scan and that will be compared to modeling of ECE emission at different plasma parameters from radiation transport calculations (TRAVIS). The strong stray radiation from non-absorbed ECRH is a major concern in scanning these higher harmonics using an interferometer because this stray radiation will dominate the spectra thus masking the ECE signal. For this purpose, a multimode notch filter design based on a multilayer dielectric structure is constructed to attenuate the stray radiation at 140GHz by 50dB.

1 Introduction

In W7-X for various plasma conditions, the X2 mode of ECE is optically thick and hence it can be taken as a blackbody emission representing an electron temperature [1]. At present, the X2 mode is scanned using a heterodyne radiometer [2] consisting of 32 channels distributed over a frequency range of 120-160 GHz. In OP 1.2a of W7-X, the X2 mode was in cut-off at a plasma density of $1.2\times 20\ m^{-3}$ [3]. This demonstrates that there is no direct access to X mode of ECE using the radiometer at high plasma densities. W7-X achieved the highest triple product (0.68×10^{20} $keVm^{-3}s$) [4] for stellarators in OP1.2a. And it is further intended to work at high plasma density considering different modes of electron cyclotron resonance heating [4]. In such scenarios, ECE from optically grey higher harmonics has the potential to be used as the diagnostic tool to get the electron temperature profiles in magnetic confinement devices. To explore diagnostic capability, a Michelson Interferometer [5] will be used to scan these harmonics over a broadband (50-500 GHz) in the next operational campaign OP1.2b of the W7-X. The current work focuses on how to tackle the problem of stray radiation [6] from multiple reflections of non-absorbed ECRH at 140 GHz in the W7-X. A similar situation occurs while scanning the ECE from the radiometer. A Bragg notch filter [2] with notch bandwidth of several 100 MHz is being used to attenuate this stray radiation by approximately 40 dB. However, a monomode Bragg notch filter integrated into a waveguide can be used only for a limited frequency range as it will introduce notches at extra frequencies as well and will not work for the broad spectral range of Michelson interferometer. Therefore, a multimode notch filter is designed based on a multiple-layer dielectric structure [7] with dielectric layers of high and low refractive indices. The design and corresponding transmission characteristics of the filter are given in following sections.

2 Michelson Interferometer as an ECE diagnostic

A Michelson interferometer will scan the broad spectral range of electron cyclotron harmonics emitted in the W7-X. It is planned to do a cross-calibration of the interferometer with the standard ECE radiometer diagnostic alongside an independent calibration with the hot source. The Michelson Interferometer has a temporal resolution of 23 ms and a spectral resolution of approximately 11 GHz corresponding to mechanically limited mirror excursion of 15mm. The Fourier transform of interferograms will provide the spectral information of ECE. These spectra consisting of higher harmonics of ECE will be explored as an option to get the electron temperature profiles. For further analysis, it is necessary to calculate the broadband ECE spectra from TRAVIS [8]. Hence, one can model electron temperature profiles and plasma density profiles from the microwave radiation transport code TRAVIS. An example from TRAVIS calculations showing higher harmonics of X and O mode ECE with X2 mode in cut-off is given below in Fig. 1.

* Corresponding author: neha.chaudhary@ipp.mpg.de

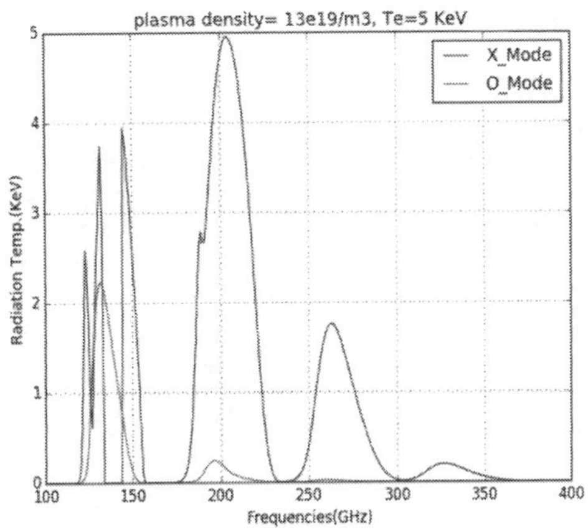

FIG. 1. Higher harmonics of ECE calculated from TRAVIS with parabolic electron temperature and plasma density profiles as input parameters

3 Multimode Notch Filter

Notch filters are generally designed for single cavity applications and for the narrow operational spectral range. Strong stray radiation from non-absorbed ECRH will enter the X2 mode of ECE and it will dominate the ECE in spectra scanned using the interferometer. For this specific application, a notch filter is required for a narrow band stop at 140 GHz to attenuate this stray radiation. Consequently, ECE spectral features will be well defined with the filter.

3.1. Design and transmission of the filter

The design of the filter consists of the structure made up of multiple layers of dielectric with different permittivity. For microwave applications with wavelength (λ) in millimeter range, dielectric plates with thickness of the order of λ can be directly assembled together to construct the filter.

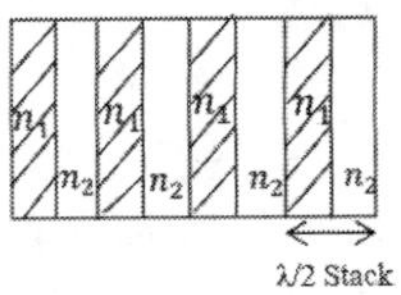

FIG. 2. Schematic of the multimode notch filter with a multilayer structure

The design proposed here for multimode notch filter is consisting of 'x' number of stacks where one stack is made up of two λ/4 thick dielectric plates with the refractive indices n_1 and n_2 respectively resulting in a λ/2 stack as shown in Fig. 2. Transverse reflection coefficients ρ_i for such a structure with $2x + 1$ number of interfaces are given by,

$$\rho_{Ti} = \frac{n_{i-1} - n_i}{n_{i-1} + n_i}, \quad i = 1,2, \ldots \ldots, 2x + 1$$

The transmission of this λ/2 stack structure vanishes at odd multiples of frequency (f) corresponding to λ [7]. The theoretical transmission spectrum of the filter for normal incidence as a function of frequency normalized to 140 GHz is given below,

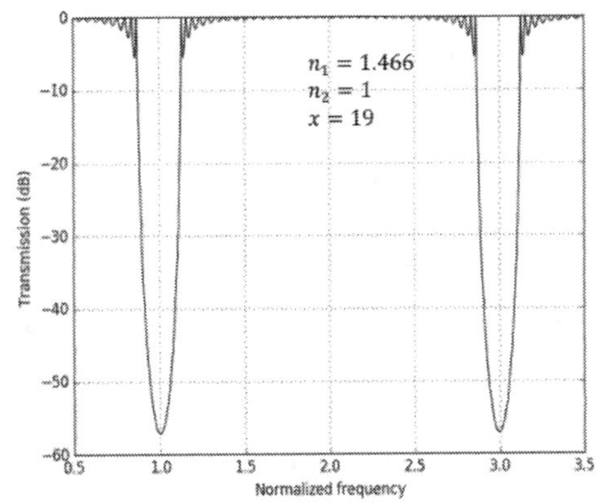

FIG. 3. Theoretical transmission spectrum of the multimode filter

From the theoretical study of the transmission characteristics of such a structure, it is evident that the total number of plates has a direct effect on attenuation at the notch frequency. Hence, depending on the magnitude of attenuation required at the notch frequency one can adjust this parameter accordingly.

3.2 Effect of design parameters on transmission characteristics

3.2.1 Dielectric constant and number of plates

The first transmission minimum in filter characteristic in Fig. 3 is defined as the notch of the filter with corresponding frequency as notch frequency and attenuation in transmission as notch depth. The transmission characteristics of such a structure depend on various physical parameters of the filter such as refractive indices, the thickness of plates, the total number of stacks and angle of incidence θ. The relationship between notch bandwidth Δf and these parameters is given by [7] ,

$$\frac{\Delta f}{f} = \frac{1}{45}\left[\sin^{-1}\left|\frac{n_1 - n_2}{n_1 + n_2}\right| \text{degrees}\right]$$

For the specific purpose of attenuating stray radiation from nonabsorbed ECRH, a large ratio of notch depth to width is required. This can be achieved by having high refractive indices of two dielectric plates in the stack. Also, the notch depth is directly proportional to the total number of λ/2 stacks assembled together in filter design as shown in Fig.4.

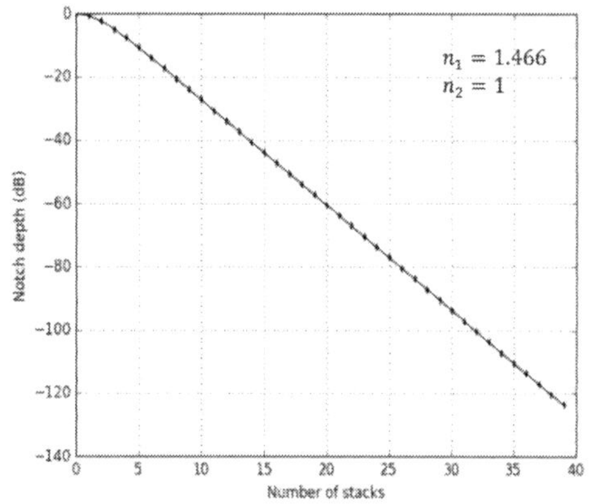

FIG. 4. Variation of notch depth as a function of the total number of stacks

3.2.2 Oblique angle of incidence

For oblique incidence, optical path length traveled by incident rays is equal to optical path length for normal incidence multiplied by the cosine of the angle of transmittance. Depending on the polarization of the incident field, the refractive index has to be replaced by a function of the angle of incidence using Fresnel formula in the calculations [7]. Notch frequency for oblique incidence shifts towards the higher values. This property of the filter can be used in fine-tuning of the notch frequency. Also a small variation in thickness (d_1 and d_2) of either of the plates in the stack can lead to a shift in the notch frequency as shown in Fig. 5.

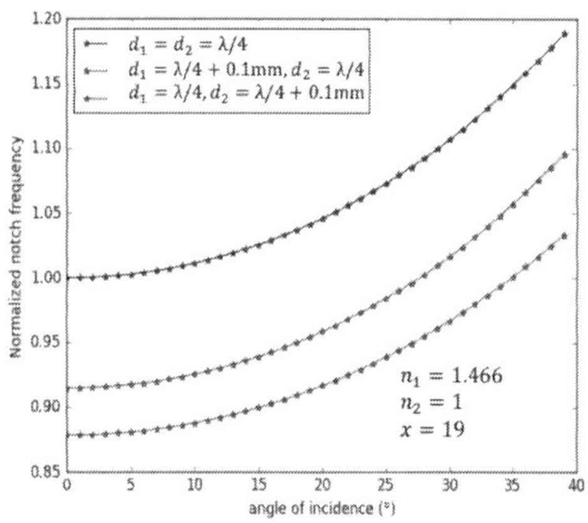

FIG. 5. Notch frequency variation with angle of incidence and thickness of dielectric plates

From the above-shown plot, it can be observed that notch frequency has a strong response to a small variation (0.1 mm) in the thickness of plates.

3.3 Experimental results

A first prototype design test was done with easily available material PTFE which also has low insertion loss [9]. The stack elements for the filter were PTFE (2.1mm thick) and air gap (2.9mm) introduced by square metal rings (12×12 cm) resulting in one 5λ/2 stack with λ corresponding to 2.32mm and 10 of such 5λ/2 stacks were assembled together to construct the filter. The experimental transmission characteristic of this filter was in good agreement with the theoretical prediction. This led to design the final filter, which will be used with the Michelson to attenuate the stray radiation at 140 GHz.

The final filter was constructed with a 3λ/2 stack with λ corresponding to 2.2mm with each stack consisting of PTFE (1.1mm thick) and air-gap (1.65mm). The thicknesses of dielectric plates are chosen to be 3λ/4 instead of λ/4 because of manufacturing limitations. The circular cross-section in metal rings for introducing air gap was optimized for the input of the Michelson. PTFE and air were chosen as stack elements because they have a small value of $n_1 - n_2$ satisfying the theoretical condition for small Δf . The total number of plates were chosen from the studies of stray radiation at the location of Michelson interferometer [6]. The theoretical broadband transmission spectrum of such a filter design can be calculated easily and it has four notches at odd multiples of $f/3$ in the spectral range of 50-500 GHz.

The experimental setup as shown in Fig.6 consisted of a broadband source (120-160 GHz) as input, tapers, and a network analyzer to process the output. Spherical mirrors were used to make input beam parallel with beam waist equal to half of the filter's input cross-section. A rotating platform (optimized for the filter dimensions) was used for the fine-tuning by varying angle of incidences.

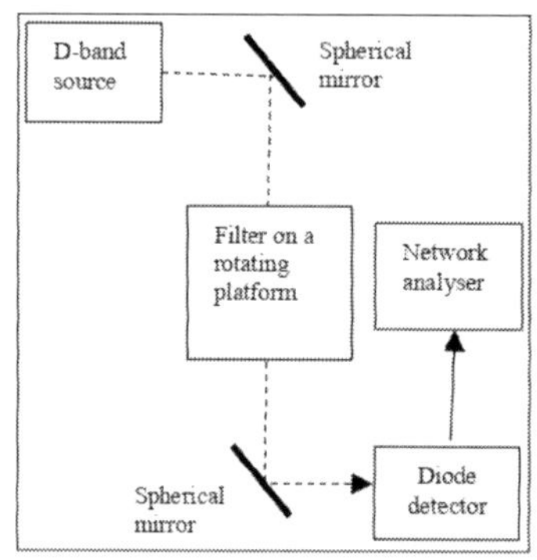

FIG. 6. Schematic of experimental setup

The filter was designed to have notch at slightly lower frequency than 140 GHz and later tune it to 140 GHz by changing the angle of incidence (rotating the filter). This design specificity has an additional advantage other than the tuning of frequency that the reflected power at 140 GHz will not re-enter the input transmission line to the Michelson thus preventing the multiple reflections. The experimental transmission of filter as shown in Fig.7 gives that at 40° with respect to normal incidence there is a 4 GHz wide notch with attenuation of approximately 50dB at 140 GHz.

Experimentally, the notch frequency shifts with a change in the angle of incidence (aoi) as predicted by theory. The experimental results in the frequency range (120-

160GHz) agree with the predicted results. Due to laboratory limitations, the filter was not tested for higher frequency ranges.

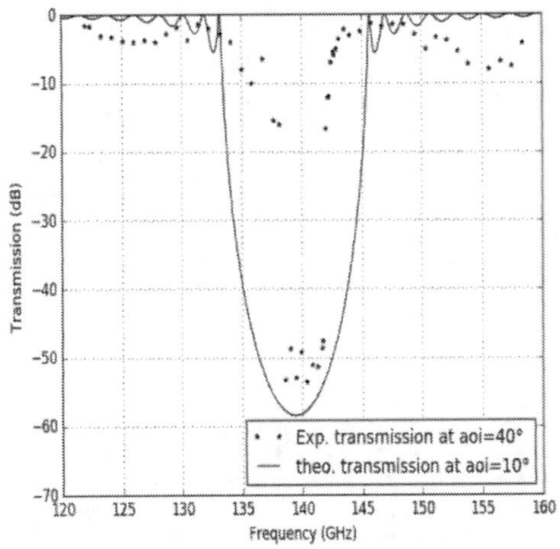

FIG. 7. Experimental transmission characteristics of the notch filter

4 Discussions

Theoretically, a predicted shift in the notch frequency was observed with the change in angle of incidence. But the notch at 140 GHz was predicted at an angle of incidence of 10° opposed to what is observed experimentally at 40°. The explanation for this could be that one 1.1mm thick PTFE plate was assembled by putting 1mm and 0.1mm thick PTFE plates together. The same process was used to create 1.65mm wide air gap by assembling metal rings of 1.5, 0.1 and 0.05mm together. This could have led to a change in the total optical thickness of the structure. Hence, we see the notch frequency of 140 GHz at a higher angle of incidence.

The experimentally observed notch width is smaller compared to the predicted one -- this could suggest that there are further multiple reflections in the structure that we do not understand. The insertion loss at the theoretically transparent part of transmission spectrum is relatively high and hence a proper calibration of the filter will be required at these frequencies.

5 Acknowledgment

This work has been carried out within the framework of the EUROfusion Consortium and has received funding from the Euratom research and training programme 2014-2018 under grant agreement No 633053. The views and opinions expressed herein do not necessarily reflect those of the European Commission.

6. References

[1] Hartfuss Hans-Jürgen, Geist Thomas, Fusion Plasma Diagnostics with mm-Waves (Wiley-VCH, 2013).
[2] Hirsch M. et al., Proceedings of the 20th Joint Workshop on Electron Cyclotron Emission and Electron Cyclotron Resonance heating, May 14-17, 2018, Greifswald, Germany.
[3] Brunner K J. et al., Proceedings of the 20th Joint Workshop on Electron Cyclotron Emission and Electron Cyclotron Resonance heating, May 14-17, 2018, Greifswald, Germany.
[4] Wolf R C. et al., Electron-cyclotron-resonance heating in Wendelstein 7-X: A versatile heating and current-drive method and a tool for in-depth physics studies, submitted to Plasma Phys. Control. Fusion 2018.
[5] Austin M E. et al., Rev. Sci. Instrum. **68** ,480 (1997).
[6] Oosterbeek W J. et al., Proceedings of the 20th Joint Workshop on Electron Cyclotron Emission and Electron Cyclotron Resonance heating, May 14-17, 2018, Greifswald, Germany.
[7] Young L. et al., IEEE Tranactions On Microwave theory And Techniques VOL. MTT-14, NO.2.
[8] Maruschenko N B. et al., Computer physcics Communications 185(2014)165-176.
[9] Lamb W J. et al., International Journal of Infrared and Millimeter Waves, December 1996, Volume 17, Issue 12, pp 1997-2034.

Runaway electron diagnostics for the COMPASS tokamak using EC emission

Michal Farnik[1,2,*], *Jakub* Urban[1], *Jaromir* Zajac[1], *Ondrej* Bogar[1,3], *Ondrej* Ficker[1,2], *Eva* Macusova[1], *Jan* Mlynar[1,2], *Jaroslav* Cerovsky[1,2], *Mikita* Varavin[1,2], *Vladimir* Weinzettl[1], and *Martin* Hron[1]

[1]Institute of Plasma Physics of the CAS, Za Slovankou 1782/3, 182 00 Prague 8, Czech Rep.
[2]Faculty of Nuclear Sciences and Physical Engineering, Czech Technical University, Brehova 7, 115 19 Prague, Czech Rep.
[3]Faculty of Mathematics, Physics and Informatics, Comenius University in Bratislava, Mlynska dolina 6280, 842 48 Bratislava, Slovakia

Abstract. An electron cyclotron emission (ECE) diagnostic of suprathermal electrons was utilised for runaway electron (RE) experiments purposes in the COMPASS tokamak. Our vertical ECE (V-ECE) system consists of a 16-channel heterodyne radiometer and an E-band horn antenna with a 76.5-88 GHz frequency range front-end. Simulations used for the design of the diagnostic showed a possibility of detecting the emission of low-energy (50-140 keV) runaway electrons. We realized measurements with both extraordinary (X-) and ordinary (O-) mode linear polarizations. The amplitudes of the X-mode and O-mode signals are similar, which can be explained by depolarised reflected radiation. V-ECE measurements in low-density flattop discharges and in discharges with massive gas injections of high-Z elements show correlations with other RE diagnostics. Our results are in the agreement with the principles of the primary runaway generation mechanisms.

1 Introduction

Electron cyclotron emission (ECE) diagnostic is usually used for the electron temperature (T_e) profile and fluctuation measurements in tokamak devices. However, the electron cyclotron emission measured vertically along the line of a constant magnetic field can yield information about the electron velocity distribution function and its evolution during a discharge [1]. This could be used for the direct detection of the runaway electron presence during runaway experiments performed in the COMPASS tokamak [2]. Ray-tracing simulations were used to design and interpret a measured signals. This motivated us to install a vertical ECE diagnostic in the COMPASS tokamak.

The article is structured as follows. In the second section the suprathermal ECE is introduced. Afterwards, in the third section, the experimental setup with simulations are presented. Finally, in the fourth section, first experimental results are given. Conclusions are drawn at the end.

2 Suprathermal Electron Cyclotron Emission

Every electron moving in an external magnetic field emits electromagnetic radiation at the cyclotron frequency. The finiteness of the Larmor radius implies that the electromagnetic field generated by an electron gyrating with the cyclotron frequency is not simply harmonic and higher harmonics $n\omega_c$ emerge. Thanks to the relativistic effects, distribution function and the Doppler effect, the emitted frequencies have a wide spectrum. The frequency at which an electron moving with a velocity **v** is able to emit or absorb the waves having a wave vector **k**, through the n-th harmonic contribution, is given by

$$\omega = n \cdot \omega_{ce}(B, v) + k_{\|}v_{\|} = n \cdot \omega_{ce}(B)/\gamma + k_{\|}v_{\|}. \quad (1)$$

The relativistic effects are represented by the Lorenz gamma factor. The second term on the right hand side denotes the longitudinal Doppler effect, the index $_{\|}$ indicates the component parallel to **B**.

It is possible to restrain the influence of the varying magnetic field in (1) via a vertically positioned antenna (**V-ECE**). The toroidal magnetic induction value becomes approximately constant along the line of sight. Thus, the resonance condition depends only on the velocity. In such a situation, the emission at a particular frequency (ω) for a parallel optical index $N_{\|}$ and a particular cyclotron harmonic (n) occurs only from the electrons satisfying the resonance condition

$$\frac{\omega(c - N_{\|}v_{\|})}{n \cdot \omega_{ce}c} = \sqrt{1 - \left(\frac{v_{\|}}{c}\right)^2 - \left(\frac{v_{\perp}}{c}\right)^2}. \quad (2)$$

This condition gives us a relationship between the frequency and the electron energy where only the emission from certain velocity phase space ($v_{\|}$, $v_{\perp}$) fulfils the condition (2). As a result, one can diagnose the density and the velocity anisotropy of the electrons as a function of the energy at the expense of returning to a chord-averaged measurement.

From the solution of the radiation transfer equation we calculate the optical depth (τ). To secure the vital practical

*e-mail: farnik@ipp.cas.cz

https://doi.org/10.1051/epjconf/201920303006

requirement, avoiding the detection of multiple reflections from the wall, a high optical depth ($\tau \gg 1$) is needed. [3]

3 Experimental Setup

The ECE/EBW heterodyne radiometer was utilized in the past on COMPASS for the measurement of electron Bernstein mode waves (EBW) [4]. The radiometer consists of a separated front-end and a 16-channel intermediate frequency receiver with the bandwidth 1.5–15 GHz. Each channel has a second mixer stage, a digitally controlled built-in attenuator 0–30.5 dB in 0.5 steps, a power detector and a DC amplifier, which provides a linear output voltage of 0–10 V / DC–1 MHz and the bandwidth 850 MHz.

The system, shown on the left in Figure 1, contains front-ends with the Ka-band (26.5–40 GHz) or the E-band divided into two subbands (60–74.5 GHz/76.5–90 GHz). The receiver uses a vertical perpendicular vacuum chamber port. Horn antenna with a teflon lens is placed behind a silicon port window, as it is shown in Figure 2 and it is connected to the radiometer over a ~2.5 m waveguide. [4][5]

The cutoff frequencies increase with the electron plasma frequency ω_{pe}, which is dependent on the electron density $\omega_{pe} \propto \sqrt{n_e}$. It is favourable to choose the front-end frequencies as high as possible to be able to measure in the highest possible plasma densities. The electron cut-off densities were calculated for $B = 1.15$ T at the magnetic axis of the COMPASS tokamak. In the case of the E2-band frequencies (76.5–90 GHz), the minimum of the cutoff density was $n_{eX} = 4.21 \cdot 10^{19}$ m^{-3} for the X-mode and $n_{eX} = 7.26 \cdot 10^{19}$ m^{-3} for the O-mode. If the density value is well below the shown numbers, the measured wave is not limited by the cutoff area anywhere during its propagation through the whole plasma. This condition is fulfilled in the vast majority of the discharges within the runaway electron dedicated campaigns in COMPASS.

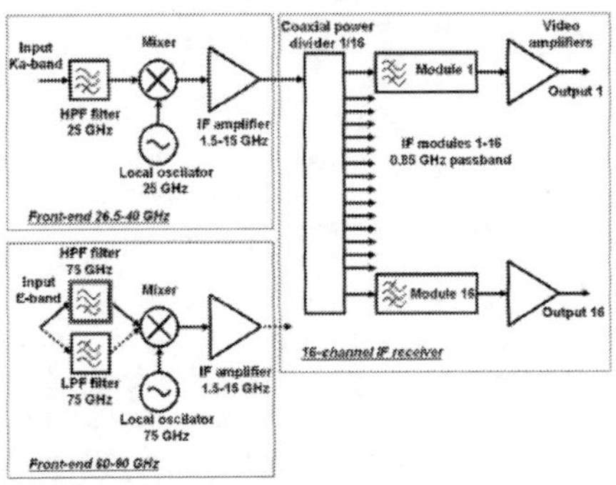

Figure 1. *Upper:* Block scheme of the ECE/EBW radiometer, taken from [4]

3.1 Energy Spectrum

The electron kinetic energies as well as the velocity phase space which fulfil the resonance condition can be determined from the range of frequencies with the use of

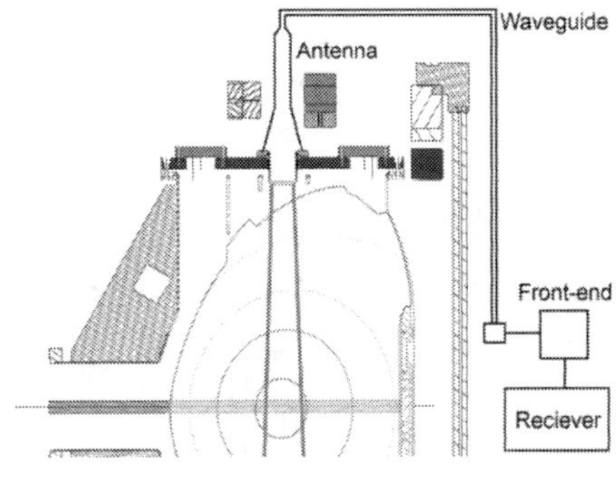

Figure 2. Scheme of the V-ECE measurement setup.

(2). As it was mentioned earlier, the emitted cyclotron frequency is down-shifted in connection with reaching relativistic velocities. Along the line of constant magnetic field $B = 1.15$ T the cyclotron frequency has value $\nu_{ce} = 32.2$ GHz. This means the E2-band could detect the down-shifted radiation from the third and higher harmonic cyclotron frequency. Figure 3 shows the velocity phase space for the resonance condition and the kinetic energy for the 3rd harmonic frequency of the E2-band.

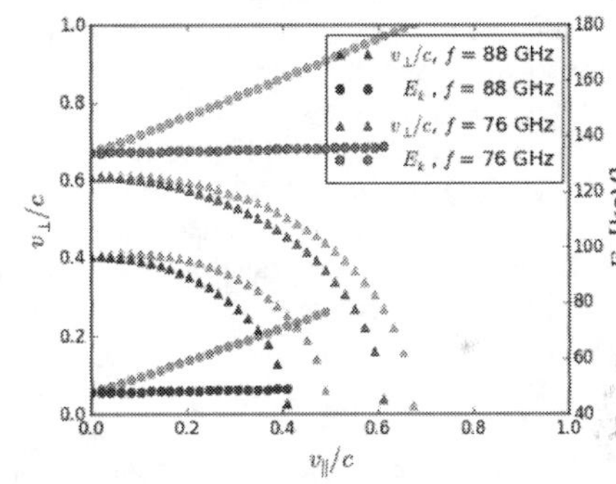

Figure 3. Solution of the resonance condition for the velocities (△) and the electron kinetic energy E_k (•) relevant to the plotted point of the phase space for the 3rd harmonic for the boundary E2-band frequencies. *Rich colour:* $N_\parallel = 0.005$; *Pale colour:* $N_\parallel = 0.1$.

The later given ray-tracing simulations yield the value $N_\parallel = 0.005$. Higher $N_\parallel$ causes resonance with higher energy electrons. Then, we are able to determine the electron energies in resonance with the E2-band frequencies which are ~ **50 – 140 keV**. Therefore, it could possible to investigate the low-energy REs with the parallel velocities above 0.4 c. We can consider the electrons in this phase space "runaway" because of their resonant $v_\parallel$ (assuming $v_\perp < 0.1$ c) is higher than the maximum critical velocity v_c in the V-ECE experiments ($n_{e,max} = 4.2 \cdot 10^{19}$ m^{-3}, $U_{loop} = 0.7$ V, $\ln\Lambda = 13$) which is $v_c = 0.37$ c. It corresponds to the kinetic energy $E_{ck} = 36$ keV. The electrons with velocities higher than the critical velocity are further accelerated by the electric field. They are called the runaway electrons (RE). [6]

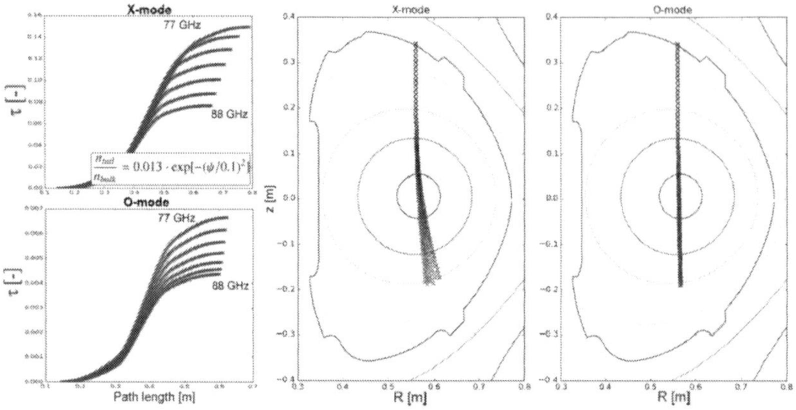

Figure 4. *Left:* Values of the optical depth τ dependent on the ray path length s for both O-mode and X-mode for E2-band frequencies. *Center and right:* Poloidal view of the trajectories of O-mode and X-mode rays in the COMPASS tokamak for odd E2-band frequencies, $B = 1.15$ T.

3.2 Ray-tracing Simulations

The SPECE code [7] was used for the analysis of the electron cyclotron emission from suprathermal electrons. This code solves the radiation transfer equation along the ray trajectories in a tokamak equilibrium. It uses the cold dispersion relation for propagation, while the absorption and emission coefficients are obtained as a solution of the relevant fully relativistic dispersion relation, which is valid at high electron temperatures.

In a Maxwellian plasma, there is a low number of the resonant electrons, therefore neither absorption nor emission occurs in the centre along the line of constant magnetic field B_0 in the measured frequency range 76.5 – 88.3 GHz. It is necessary to add a non-thermal component into the simulation to increase the absorption coefficient and hence the optical depth. The SPECE can use bi-Maxwellian plasma with added non-thermal component. In our simulations these parameters were used: velocity at the end of the plateau $u_{max} = 0.4$ c, temperature $T_{tail} = 60$ keV distribution function in space and normalised magnetic flux coordinate in Figure 4.

Values of the optical depth for the X-mode are 20 times higher than for the O-mode (see Figure 4). During the evaluation of various simulations this difference between values of the optical depth stayed almost the same. As it was mentioned earlier, the goal is to prevent the detection of waves reflected from the tokamak wall. In the case of $\tau \gg 1$ the reflections are negligible. Unfortunately, the parameters of the plasma and the available radiometer on COMPASS do not provide this suitable condition, which complicates the interpretation of experimental data.

4 First Experimental Results

Firstly, it has to be verified that the suprathermal electrons are the source of the measured emission. Due to the possible detection of a reflected emission, the thermal emission could be measured from the 2nd harmonic (HFS) and the 3rd harmonic (LFS) during the ordinary discharges with $B = 1.15$ T. In these regions in Figure 5 the non-relativistic electrons fulfil the cold electron cyclotron resonance condition i. e. (1) for $\gamma \rightarrow 1$.

No dependence of the V-ECE signal on the electron temperature has been found in any of the approximately 300 studied discharges. In the discharges with higher magnetic field $B = 1.5$ T, where the region of the 2nd harmonic thermal emission shifts towards the LFS into the exact centre of the tokamak vessel right in the line of V-ECE diagnostic line of sight. The V-ECE signals in density ramp-up scenario from the standard and high-B discharge are shown in Figure 6.

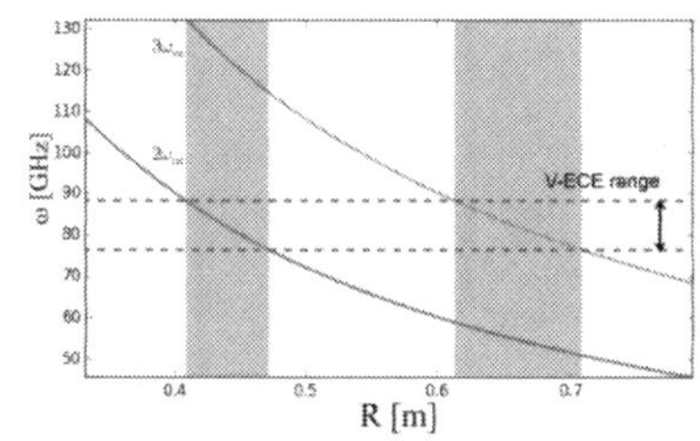

Figure 5. EC frequency profiles for $B = 1.15$ T and regions of thermal resonances of the 2nd harmonic (red) and the 3rd harmonic (green) inside the COMPASS plasma.

We do not see any significant differences between V-ECE signals in these two scenarios. Even on the individual channels we do not observe any conspicuous differences. This indicates that the signal comes from the low-energy runaway electrons and does no correlate with electron temperature. The emission intensity from the REs is probably much higher than the thermal emission intensity.

The V-ECE measurements proved to be a good diagnostic of the low-energy runaways and useful tool for detection of the initial runaway seed. A runaway seed significantly influences the evolution of the RE presence within a discharge. These connections are shown in Figure 7. In the yellow rectangle, the difference between a shot with a lower initial RE seed and a shot with a larger initial RE

https://doi.org/10.1051/epjconf/201920303006

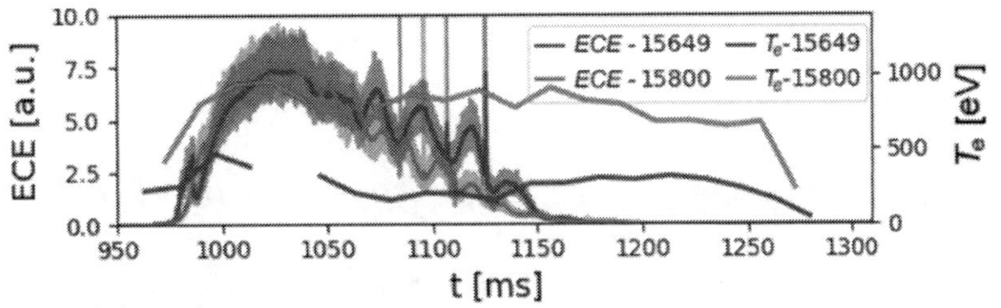

Figure 6. Comparison of the V-ECE measurement and Thomson scattering T_e at the resonance region of the ordinary B = 1.15 T (#15649) and high-field B = 1.50 T (#15800) discharge. O-mode, channel 1 = 76.5 GHz: (O,ch1). *Pale colour*: raw signal; *Rich colour*: smoothed signal. t > 1150: overdense ⇒ cutoff.

seed is highlighted. The primary generation in the shot 14642 is lower due to the higher electron density. Because of the missing whole runaway primary generation, we do not measure strong V-ECE signal neither the signal leap during the discharge. The HXR scintillator measures only the signal during the plasma termination at the end of the discharge (not due to REs). The case of the shot 14643 is the same as it was discussed earlier, it is discharge with the density ramp-down and a large RE population. It is worth mentioning that initial RE losses are not detected by the HXR scintillator. That could imply the confinement of a vast majority of the generated runaways in the initial seed.

By normalising and shifting the HXR signal, we get the confinement time with respect to the initial RE seed and the primary generation. This is shown in Figure 8. The low-energy runaway population is to some extent terminated as the 24ms delayed detection of the HXRs shows us. The REs are accelerated and collide with the tokamak first wall.

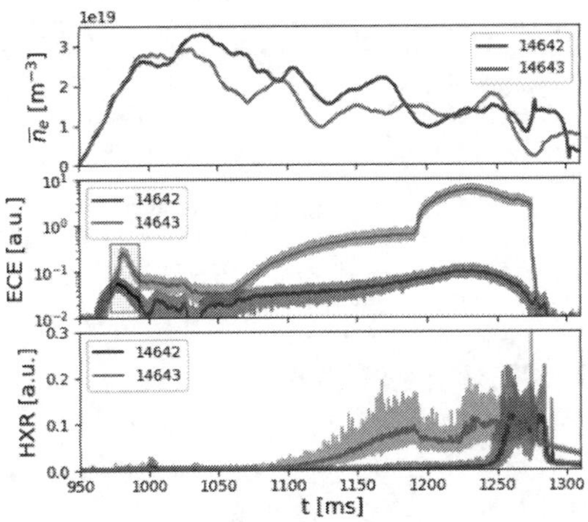

Figure 7. Comparison of the electron density, V-ECE and HXR measurement from the shots 14642 and 14643. Yellow rectangle highlights the runaway seed. (O,ch1)

5 Conclusion

The final design of the radiometer was utilized for runaway electron measurements in COMPASS. The V-ECE measurement provides information about low-energy runaway electron presence especially about the primary generation at the beginning and in the flat-top phases of the discharge. V-ECE can be used as a diagnostic in the RE campaigns.

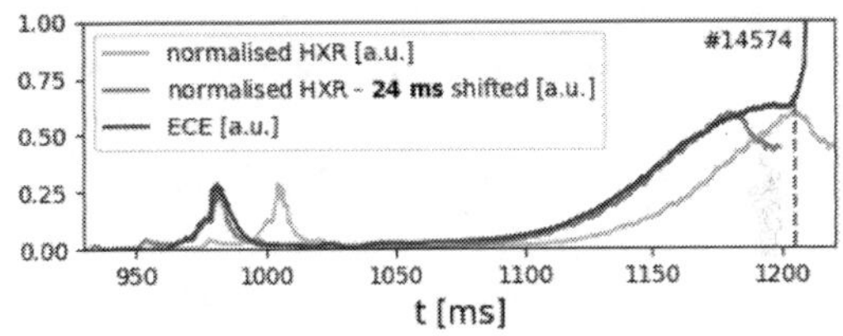

Figure 8. Smoothed HXR and V-ECE signal for the shot 14574. (X,ch1)

Acknowledgements

The author acknowledges Daniela Farina and Lorenzo Figini for providing the SPECE code. This work was supported by the Czech Science Foundation Project No. GA18-02482S.

References

[1] K. Kato, I.H. Hutchinson, Physical Review Letters **56**, 340 (1986)

[2] O. Ficker et al., Nuclear Fusion **57**, 076002 (2017)

[3] I.H. Hutchinson, *Principles of Plasma Diagnostics* (Cambridge University Press, New York, 2002), ISBN 0521803896

[4] J. Zajac, J. Preinhaelter, J. Urban et al., *EBE/ECE Radiometry on COMPASS Tokamak—Design and First Measurements*, in *AIP Conference Proceedings* (2009), Vol. 1187, pp. 473–476

[5] J. Zajac, J. Preinhaelter, J. Urban et al., Review of Scientific Instruments **81** (2010)

[6] H. Dreicer, Physical Review **115**, 238 (1959)

[7] D. Farina, L. Figini, P. Platania, C. Sozzi, *SPECE: A code for electron cyclotron emission in tokamaks*, in *AIP Conference Proceedings* (2008), Vol. 988, pp. 128–131

 https://doi.org/10.1051/epjconf/201920303007

ECE Diagnostic for the initial Operation of Wendelstein 7-X

Matthias Hirsch[1], *Udo* Höfel[1], *Johan Willem* Oosterbeek[1], *Neha* Chaudhary[1], *Joachim* Geiger[1], *Hans-Jürgen* Hartfuss[1], *Walter* Kasparek[3], *Nikolai* Marushchenko[1], *Boudewijn* van Milligen[2], *Burkhard* Plaum[3], *Torsten* Stange[1], *Jakob* Svensson[1], *Hayato* Tsuchiya[4], *Dietmar* Wagner[1], *Gavin* McWeir[1], *Robert* Wolf[1] and the W7-X Team

[1] Max-Planck-Institut für Plasmaphysik, Greifswald/Garching, Germany
[2] Laboratorio Nacional de Fusion, CIEMAT, Avda. Complutense 40, 28040 Madrid, Spain
[3] Inst. für Grenzflächenverfahrenstechnik und Plasmatechnologie, Univ. Stuttgart, Germany
[4] National Institute for Fusion Science, Toki, Japan

Abstract. The ECE diagnostic at W7-X in its standard mode of operation measures in X2 mode polarization with a 32 channel radiometer in the frequency band around 140 GHz for central magnetic field 2.5T. The radiometer is calibrated by a noise source and the overall system absolutely calibrated by means of a hot-cold source placed outside the torus in front of a Gaussian telescope optics with identical geometry and transmission line as it is installed for the measurements in the plasma vessel. The system is supplemented with a 16 channel zoom device with 4 GHz span for higher frequency resolution at a suitable radial range and a Michelson interferometer for the characterization of higher harmonics sharing the same line of sight.

1 Introduction

Wendelstein 7-X is a second generation Stellarator with optimized physics properties of its three-dimensional confining magnetic field along the HELIAS concept [1]. Aim of the first operational phase, OP1, split into essentially three campaigns, is to prove the success of this optimization and develop an Electron Cyclotron Resonance heated (ECRH) high-density scenario with island divertor edge and a high-beta core as a basis for steady-state operation. Instead of the later high heatflux divertor and fully cooled in-vessel components, a geometrically identical but inertially cooled test-divertor unit is used. Long-pulse high-density operation relies on the flexible ECRH system [2] with total power ~ 7 MW provided to the plasma by 10 cw-gyrotron units and optical transmission lines and by microwave absorption using mutipass O2 mode launch [3] instead of X2 launch as used at lower density.

During the first test operation, OP1.1, dedicated to integrated commissioning, tests of components and diagnostics, the device was equipped with five uncooled inboard carbon limiters only, which restricted the magnetic configuration and the overall energy input for a single experiment scenario to avoid uncontrolled overheating [4], [5]. Test-divertor operation started September to December 2017 (OP1.2a) and has been completed July to Oct. 2018 (OP1.2b). Due to still uncooled first wall and divertor elements the overall energy input during an experiment scenario was restricted to < 200 MJ, resulting in detached high performance plasmas with a pulse duration of 32s @ 5MW O2-heating and feedback controlled density ~ 0.9 10^{20} m^{-3} not limited by plasma performance. The first operation of the NBI system with P_{NBI}= 3.4 MW (Hydrogen injection 55 keV) and maximum technically available duration of a few seconds was performed to explore direct ion-heating, deep fueling (resulting in peaked density profiles with n(0)= 2.0 10^{20} m^{-3} and fast ion confinement and for the commissioning of beam-related diagnostics as well. Long-pulse operation is scheduled early 2021 (OP2) after the high-heatflux divertor and fully cooled in-vessel components have been installed.

The Electron Cyclotron Emission Diagnostic (ECE) has been operated throughout these experiments with the task to monitor ECR heating and measure electron heat transport [6],[7]. This contribution summarizes the technical status and characterization of the ECE diagnostic as a basis for an overview on typical results and the status of their evaluation.

2 ECE diagnostic for W7-X

2.1. Sightline

Due to the large aspect ratio of W7-X the cyclotron harmonics are well separated and the cold resonances up to the 4th harmonic do not overlap. The full 2nd harmonic at 2.5T operation is covered with a 32 channel heterodyne radiometer in the frequency band 126 GHz to 162 GHz with a single mixer only. Sightline location and geometry use the manifold of the 3D magnetic topology of the stellarator magnetic field to optimize the conditions for the ECE measurement [8],[9]: For

* Corresponding author: matthias.hirsch@ipp.mpg.de

localized temperature measurements the toroidal location is selected such that the local magnetic field strength increases Tokomak-like with the distance from the observation optics towards the inner side of the torus. Simultaneously the magnetic field on the axis is different from the 2.52 T at the ECRH launching ports to avoid that access to the plasma center is hampered by 140 GHz stray-radiation of unabsorbed ECRH beams [10]. This radiation is particularly strong in cases of O2-heating where single pass absorption is 70% only. To alleviate this the toroidal location of the ECE is chosen nearly opposite to the ECRH launching plane. Notch filters have been developed as described below for the radiometer and the Michelson interferometer. However, temporary perturbations occur in individual ECE channels due to spurious sidebands in particular gyrotron modes of operation. They easily can be identified and need to be suppressed for data evaluation. At the selected slightly elongated local plasma cross section the flux surfaces are compressed by about a factor of two with respect to an equivalent circular cross section where the average minor radius is a~50cm (Fig. 1). For a maximum spatial resolution of ECE, the variation of B in the probed emitting volume needs to be minimized. This is achieved by a small emitting volume defined by a narrow Gaussian beam characteristic focused moderately at the plasma center. The 1/e width of its power distribution is $w_0 < 23$ mm all along the pathlength of about 55 cm where the sightline crosses the confinement region. The temperature variation within a probed volume defined by Gauss beam and bandwidth of a given radiometer channel is minimized by a sightline widely perpendicular to the crossed flux-surfaces and thus to the fieldlines. The criterion coincides with the requirement that magnetic field components *along* the sightline should be minimized to reduce broadening by Doppler shift. Furthermore a sightline that is widely perpendicular to the flux surfaces and thus to the density gradient reduces beam refraction during high-density operation. Due to helical excursions of the plasma axis - and thus in contrast to a tokamak - the required condition is not straightforward achieved in a stellarator. For the selected diagnostic position it results in a tilt of 6 deg and -10deg of the sightline axis to the vertical and horizontal plane respectively [9].

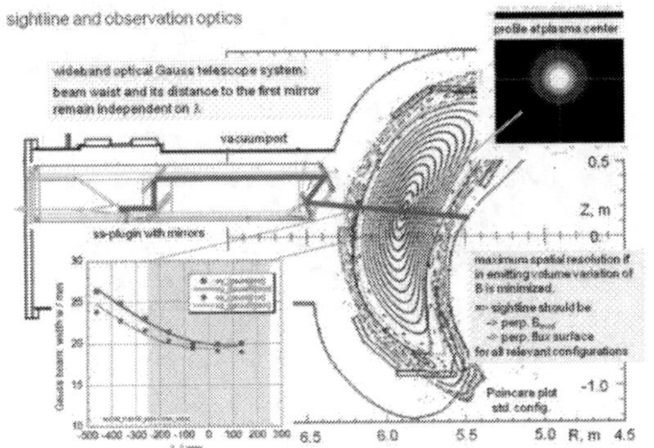

Fig. 1: Cut through the W7-X plasma and -vessel at the poloidal plane of the ECE sightline and the Gaussian telescope optics plugin in the diagnostic port.

The selected ECE sightline is realized by intrinsically broadband Gaussian beam telescope optics (Fig. 1), consisting of 2 elliptical and 2 plain stainless steel mirrors - the latter for beam folding and -steering - and an adapted broadband horn antenna altogether assembled in a common stainless-steel frame attached to one of the major port flanges. For later long-pulse operation the optics is already prepared to be water cooled by a meander structure integrated in the front mirrors and cooling pipes along the support frame.

Due to the low magnetic shear of W7-X, the pitch angle of the magnetic field varies only moderately along this ECE sightline - e.g. by 6 deg for W7-X standard magnetic field configuration resulting in rather well defined conditions for ordinary O- and extraordinary X-mode polarization. At the plasma axis the magnetic field is tilted by 13 deg with respect to the horizontal direction. During the first campaign X-mode polarization was selected already in the vessel by a microwave horn with rectangular fundamental mode exit. For OP1.2 a conical 70mm long extremely wideband horn has been developed by means of an optimizer [11]. The free space Gauss mode TEM10 is converted to a TE11 at its circular 4 mm diameter exit which allows for both X- and O-mode measurements selecting the desired polarization outside the torus. A 4 mm diameter circular Cu waveguide then leads the radiation to the broadband vacuum window consisting of 100 μm thick Viton-sealed mica sheet in a 4 mm circular aperture. The calculated mode purity of the horn exceeds 97% for frequencies 140 to 500 GHz and remains > 93% down to 50GHz thus reducing cross polarization coupling. For high frequencies (> 250 GHz) the calculated horn characteristics approach those of a truncated waveguide. In the horn the nominal beam waist has been selected to increase with ~ sqrt(λ) which guarantees frequency independent wavefront curvature and thus broadband capabilitiy of the telescope arrangement. The nominal beam waist location in the horn varies only moderately keeping the location of the beam waist on the other side reasonably well at the plasma center. So far calculated and measured waists and waist locations have been compared for frequencies 5GHz < f < 200 GHz and agree well.

2.2 Transmission line and notch filter

An oversized transmission line which essentially consists of uncorrugated circular Cu waveguides with inner diameter d= 28mm guides the microwave signal from the torus through a dog-leg structure in the radiation shield to the radiometer outside the torus hall (Fig. 3). For this path 11 mitre bends are required plus one polarization tuner that consists of additional 3 mirrors to adapt the selected polarization to the input of the radiometer. Right outside the window the TE11 mode is launched to the oversized waveguide and finally reconverted at the radiometer by two 0.39 m long d= 28mm to d= 4mm tapers on both sides of the transmission line. In the transmission line the TE11 mode has moderate losses, and is rather insensitive to perpendicular alignment inaccuracies (e.g. radial displacements) at the connection between subsequent Cu-tubes. However, care has to be taken to avoid any tilts in the waveguide (at flanges, inaccurate mitre bends etc.), which lead to the excitation of spurious unwanted modes. The total length is 23.7m including the two tapers. To minimize crosstalk X- and O-mode are split already right outside the torus by a wire grid in the oversized waveguide and the corresponding O-mode transmission line runs in parallel to the X-mode one. The overall attenuation of this transmission line in the spectral range of the radiometer 124 to 162 GHz measured with a noise source (Fig. 3) increases from 8 to 10 dB mainly resulting from losses at the mitre bends which essentially act like gaps in the oversized waveguides with a distance of transmission line diameter. In contrast cross polarization coupling is as low as ~1% only.

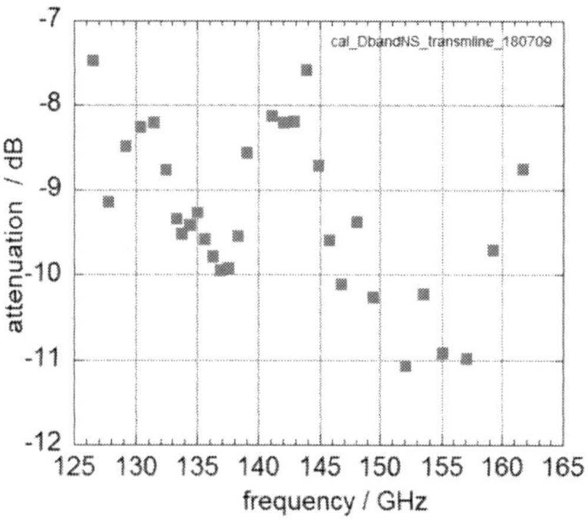

Fig. 2: Insertion loss of the transmission line determined by differential measurement with a noise source and the radiometer cannels used for detection.

Fig. 3 shows an overview of the transmission line system. Besides the calibration optics discussed in 2.4 below it shows a second plasma observation from the opposite "high field side". It views along the same line of sight but reversely and has been prepared to allow for studies of optically grey emission expected there which gives access to the electron distribution function [8]. This high-field side observation may be selected by a switch in the oversized transmission line. The antenna consists of a 10 mm aperture stainless steel horn integrated in a tile of a water-cooled baffle-like heatshield selecting X-mode radiation. The horn aperture is slightly retracted into the surface of the plasma facing graphite tile which itself prolongs the flare angle of the horn. Optics or larger horn dimensions are impossible to realize due to geometrical restrictions and the expected convective heatloads of < 300 kW/m^2. Calculated maximum temperatures reach 450 °C near the horn itself which is just acceptable for the CuCrZr heat-sink. Fundamental mode and slightly oversized rectangular waveguides hidden behind the heat shield lead the signals to the next port and the respective 4 mm diameter mica vacuum window at its end.

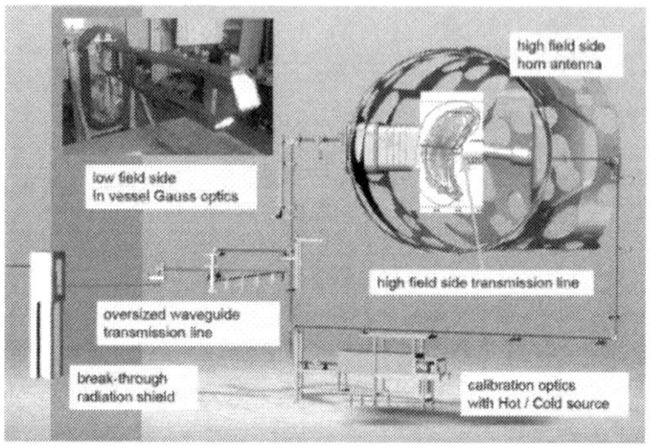

Fig. 3: Scheme of the overall transmission line system. The additional transmission line for O-mode radiation running in parallel from the low field side observation port to the radiometer has been omitted for clarity.

The frequency band of ECRH stray-radiation is 139.9 to 140.4 GHz resulting from the various individual long-pulse gyrotrons and taking into account their individual frequency downchirp of ~ 300 kHz as they are switched on. It is cut out of the spectrum directly in front of the radiometer by a waveguide Bragg reflection notch filter [12] with depth > 40 dB (this value being limited by the available measurement equipment) and an insertion loss of about 5 dB outside this frequency band (Fig. 4).

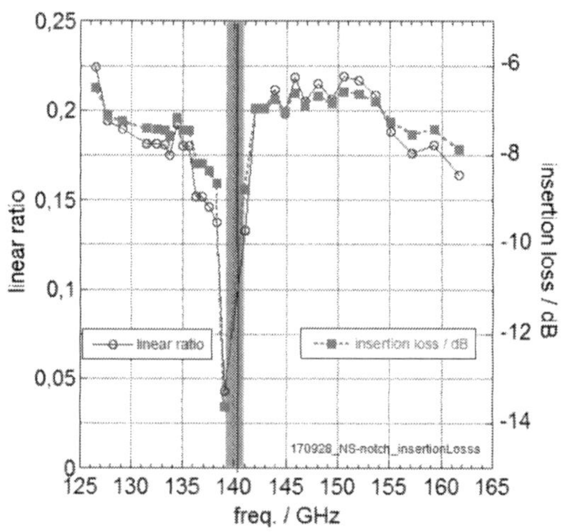

Fig. 4: Insertion loss of the Bragg reflection notch filter measured with a noise source and the radiometer channels. About 2 dB of the measured values account for the tapers and waveguides used for the measurement itself. The overall insertion loss is 5 dB in the probed frequency band. Ch16 and 17 are already at the edge of the notch and thus difficult to calibrate.

2.3 Radiometer

The radiometer uses a single broadband mixer and an LO of 122 GHz, stabilized by a phase locked loop and a quartz oscillator for down conversion to an intermediate frequency range 2 to 40 GHz (Fig. 5). Already in the mixer itself the IF signal is split into the bands of the subsequent filterbanks 2 - 18 GHz and 18 - 40 GHz with 16 channels each [6[,[14]. The bandwidth of the individual filters (0.25 - 1.4 GHz) corresponding to a radial resolution between 0.5 to 1 cm is adapted to the radial resolution of the ECE emission, as calculated from the optical depth at the expected plasma conditions for standard plasma profiles [9]. The detector diode output after pre-amplification is launched and filtered via differential amplifiers (Tektronix AM502), to the subsequent DAQ system. Its analogue input stage expands the nominal range of ± 5 V to the capabilities of the 2 Ms/s, 16 bit DAQ (ATCA-MIMO-ISOL based on an FPGA) which allows to study fast mesoscale events also and simultaneously limits the measured frequency range according Nyquist theorem by a low pass filter with a 3dB compression at 550 kHz.

The overall noise level of the radiometer is determined by the broadband mixer and the subsequent broadband video amplifiers, the noise of the IF system itself is negligible. In each of the two IF-systems the gain can be adapted by calibrated attenuators such that the diodes are operated in their linear range. The fixed gain of the broadband video pre-amplifiers has been set such that linearity is guaranteed if the output does not exceed 1V which is monitored during operation and attenuation is adapted if necessary.

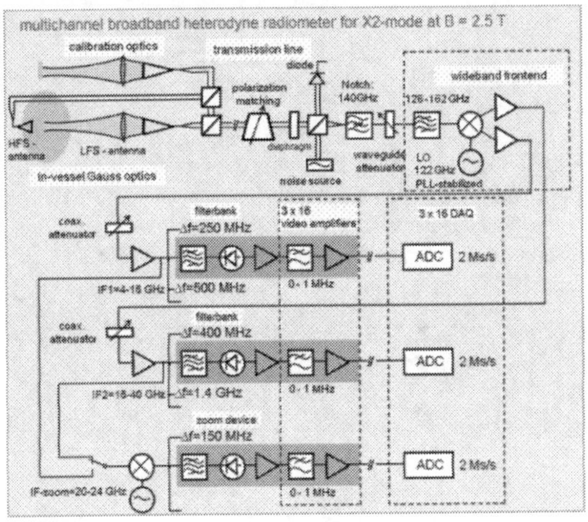

Fig. 5: Scheme of the 32ch radiometer and the 16ch zoom device.

A characteristic of the two IF-systems is the rather large variety of individual channel sensitivity which varies by up to a factor of 20 generally decreasing towards the higher frequencies (Fig. 6). For the individual channels the signals are levelled by attenuators in front of the diodes such that for all channels of a filterbank linearity is guaranteed up to the same level of microwave input power as proven with the noise source

For higher frequency resolution at a suitable radial range the radiometer is supplemented with a "zoom" device in parallel to the standard filterbank [15] with 16 channels (bandwidth 150 MHz) spread over an IF band of 4 GHz only (20 - 24GHz). It uses a tunable LO and a wideband mixer to select any part of the original 2 to 40 GHz IF-band. In magnetic radial coordinates this frequency span of 4 GHz covered with 16 channels corresponds to a radial range of Δr ~ 6 cm at the High-Field Side (HFS) or ~15cm at the Low Field Side (LFS), respectively.

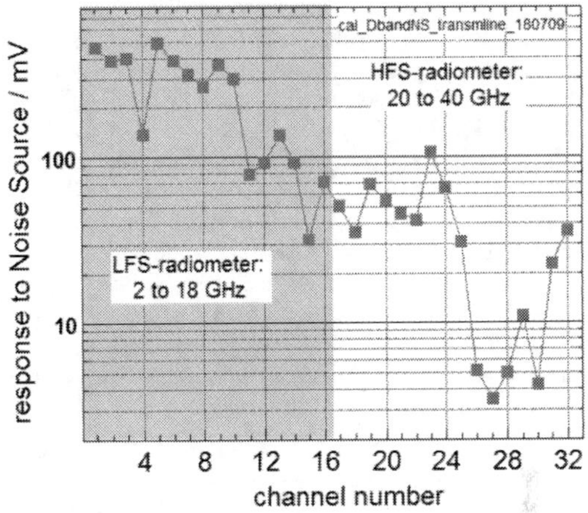

Fig. 6: Individual response of the radiometer channels to the chopped Noise Source.

2.4 Calibration

For an absolute overall calibration of the diagnostic a second identical Gaussian optical system has been built as a twin outside the torus (Fig. 3). This calibration unit includes identical waveguide components, mica window, polarization splitter and a geometrically identical transmission line, however, with a hot-cold calibration source chopping between LN_2 temperature and room temperature in front of it. Laboratory measurements with this arrangement showed that the two optical systems can be relative calibrated relative to each other with an accuracy of about 5% provided spurious reflections in the transmission path are carefully avoided. A detailed analysis of this hot/cold calibration process including a straightforward Bayesian error analysis is given in Höfel et al. [16].

Calibration signals and signals measured during plasma operation differ by ~ 5 orders of magnitude. To cope with that, the gain of the detection chain is increased by reducing the attenuation of the calibrated attenuators in each of the two IF branches 2 to 18 GHz and 20 to 40 GHz. The hot/cold modulation of the signal is completely buried in the radiometer noise and conditional averaging is therefore mandatory. The DC offset of each diode resulting from detected frontend noise is balanced in the AM502 video amplifiers and the resulting AC signal is amplified by a factor 1000 to avoid bit-noise in the ADC. Overall the ADC including its analogue input stage yield a bit-to-voltage conversion factor is 5513 [bit/V] with an accuracy of ± 0.4%. However, as the diagnostic calibration is done with the same data acquisition during experiment and calibration possible calibration errors of the DAQ modules cancel in the final determination of radiation temperatures. The video amplifier gain shows an accuracy of ± 1.5% over the full three decades. The DC offset used during calibration to balance the finite DC component of the signal resulting from the frontend noise does not hamper this linearity. Worst cases exceeding this value of ± 1.5% range up to a non-linearity of 2.7% for a few individual channels.

Conditional averaging over 1 - 2 hours uses the signal from the rotating mirror (rotation speed ~ 3 Hz) that serves as a chopper between hot and cold for reference. Finally, this absolute calibration of the ECE channels is directly hampered by the temporal stability of all components in the signal chain, which may depend on thermal drifts. To minimize drifts the DAQ system temperature is controlled. As a monitor for drifts and to detect possible individual channel failure during operation, a D-band (120 to 160 GHz) noise source can be selected by waveguide switch in the oversized transmission line right before the notch filter. It provides a relative calibration correction with respect to the last Hot/Cold calibration that needs to be taken into account.

Due to the large variation of channel sensitivity some individual channels with low signal-to-noise ratio during the hot/cold calibration are recalibrated by means of their neighbor channels using the ECE emission of selected plasma discharges as broadband noise sources. To avoid perturbation with possible ECR heating effects only time windows significantly after ECRH has been switched off are used for this relative calibration step.

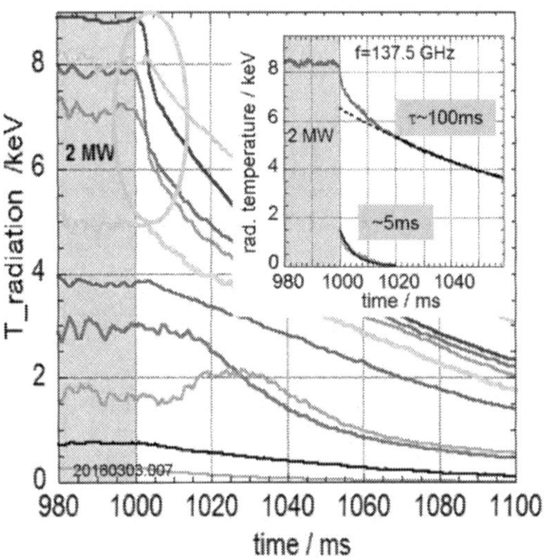

Fig. 7: Time traces from the ECE Low-Field Side right after 2 MW ECRH in a stationary low density plasma (n_e= 1.5 10^{19} m^{-3}) has been terminated. The inlay shows exponential fits to the two timescales for a selected central LFS channel. Note that the High Field Side channel (f= 141.1 GHz) plotted in light green displays the energy confinement time only.

3 ECE time traces, spectra and temperature profiles

Besides X2 blackbody emission the ECE radiation temperature spectra T_{rad} (f) show features that display EC emission from hot core electrons for frequencies where the plasma is optically grey (Fig. 8). Moreover, in low density plasmas with small optical depth the ECE radiation temperatures on the low field side near the center ($f < f_{ECRH}$ = 140 GHz) tend to be increased due to shine-through from the hot ECRH-heated core region right behind. These non-thermal emission contributions can be distinguished from local blackbody emission by ECRH power steps downward, where their decay is faster than that of the thermal spectrum (order of 5 ms in contrast to order confinement time ~100 to 150 ms) as it displays local changes in the core electron distribution functions rather than temperature changes (Fig. 7).

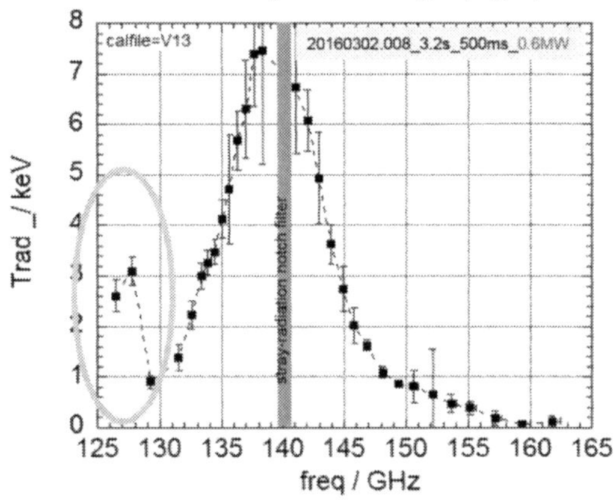

Fig. 8: The radiation temperature spectrum T_{rad} (f) at its Low-Field-Side shows downshifted emission from hot core-electrons as well. The example is a low density discharge with low with heating P_{ECRH}= 0.6 MW.

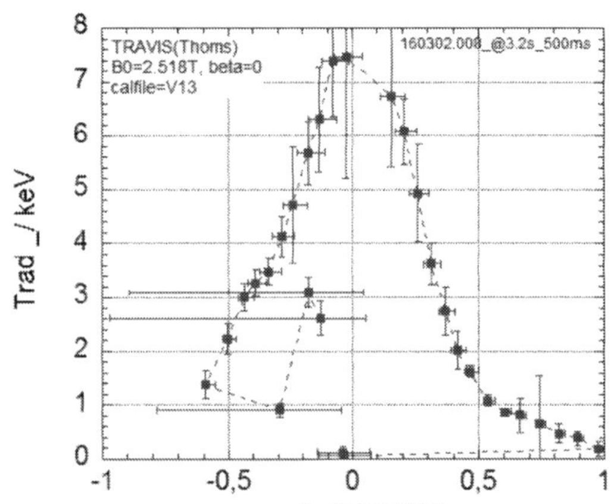

Fig. 9: Electron temperature profile derived from the spectrum above using the beam tracing and radiation transport code TRAVIS [18] with the density profile from Thomson scattering as input and the plasma equilibrium from VMEC [19] using β~0. For many discharge conditions the outermost channels on the

High-Field-Side already display downshifted core emission from the 3rd EC harmonic as seen by their fast decay at ECRH power switch down as confirmed by TRAVIS.

Electron temperature T_e profiles are derived using ray tracing of the radiation transport along the line of sight with the TRAVIS code, equilibrium calculations from VMEC and the Bayesian framework Minerva (Fig. 9). The latter allows for a rigorous error analysis and the integration with the information obtained from other diagnostics as well. A detail description of this process is in preparation by Höfel [17].

4 Dynamic phenomena

Stationary and dynamic transport studies were conducted together with the flexible ECRH system with its option for on- and off-axis heating, power steps and power modulation. Heatwave propagation experiments, with on- and off-axis power modulation of the gyrotrons, allow for the determination of the ECRH power deposition profiles as well as for dynamic electron transport studies [20].

The ECE radiometer also displays spontaneous radially propagating electron temperature perturbations. They have been used as tracer to display electron heat transport applying transfer entropy as an advanced technique that detects causal information transfer [21]. The zoom device with its radially close channels also allowed for first temperature fluctuation studies using radial correlation techniques [22].

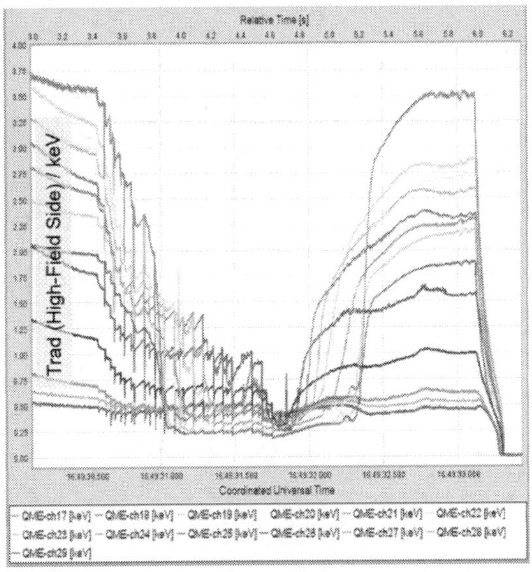

Fig. 10: Time traces of selected High-Field-Side radiometer channels during a discharge with a series of pellets driving the central ECE channels to cut-off. After the pellets the ECE signals recover channel for channel as the density is decreasing again.

The radiometer also provides MHD mode activity [23],[24] such as a nearly coherent activity with frequency around 7 to 10 kHz is visible in ECE channels located ~ r/a= 0.5 in several scenarios possibly related to an ι = 4/5 rational surface and edge related crashes.

Individual events such as pellet injection [25] and ECCD-driven current induced crash events [26],[26] have been investigated as well. Fig. 10 shows time traces during a density increase supported by a series of pellets. As a consequence X2 cut-off is reached for the core channels. ECE emission is recovered - one channel after the other -as the density falls again.

5 Outline: ECE diagnostic for high-density steady state operation of W7-X

For the desired high-density (overdense) steady-state scenarios of W7-X at $n > n_{co}(X2) = 1.2\ 10^{20}\ m^{-3}$ heated by O2 ECRH, where X2 ECE fails, a continuous core temperature measurements needs to be developed. As an option the diagnostic capability of ECE at optically grey harmonics such as X3 is studied. Another alternative would be Electron Bernstein Wave Emission. As a first step a Michelson interferometer has been operated in OP1.2b which shares the same line-of sight as the X2-radiometer [28]. Strong emission of the 3rd and weaker at the 4th harmonic are observed [29]. For high performance plasma conditions the X3 emission is close to blackbody radiation in particular on its low frequency side as expected from TRAVIS calculations. In order to derive absolute temperatures an integrated data analysis using the frequent Thomson profiles and interferometry signals is being prepared.

The broadband spectra derived with the Michelson interferometer in both X- and O-mode also support the interpretation of the 2nd harmonic radiometer results as e.g. the intensity of the O2 emission is affected by multipass effects resulting from reflection at the torus backwall. High heat loads there hamper the realization of an anti-reflection structure. The fact that in the stellarator this backwall is toroidally and poloidally tilted with respect to the ECE sightline complicates the interpretation of the multipass effects. A concern is that back-reflected microwave components which may be partially depolarized pollute the X2 measurements.

A critical issue of the Michelson interferometer is the suppression of ECRH stray-radiation which in O2 heating scenarios at high densities is significantly increased with respect to X2 heating scenarios due to the reduced single pass absorption [28]. Therefore a quasi-optical 140 GHz notch filter with wide passband has been developed which allows to measure the full ECE spectrum 50 to 500 GHz [29]. Gyrotron sidebands are a further issue to be taken into account under these diagnostic conditions.

For long-pulse radiometer operation drifts of the radiometer sensitivity and thus of the calibration may become crucial. A more careful temperature stabilization of the radiometer therefore is mandatory. Also the option to recalibrate the system during a discharge by switching temporarily to a noise-source needs to be tested.

6 Acknowledgement

This work has been carried out within the framework of the EURO fusion consortium and has received funding from the EURATOM research and training programme 2014-2018 under grant agreement No 633053. The views and opinions expressed herein do not necessarily reflect those of the European Commission

References

[1] Grieger G. et al. 1992 *Phys. Fluids* **B 4** (7) 2081

[2] Laqua H. et al. 2018 *this workshop*, Wolf R. et al. 2018 Electron-cyclotron resonance heating in Wendelstein 7-X: A versatile heating and current drive method and a tool for in depth physics studies, submitted to *Plasma Phys. Contr. Fusion*

[3] Brunner K. J. et al. 2018 *this workshop*, Stange, T. et al. 2018 First Demonstration of Magnetically Confined high Temperature Plasmas beyond the X2 cut-off density sustained by O2-heating only, submitted to *Phys Rev. Lett.*

[4] Hirsch M. et al. 2017 *Nucl. Fusion* **57** 086010

[5] Dinklage A. et al. 2018 *Magnetic Configuration effects on the Wendelstein 7-X Stellarator*, accepted for publ. *Nature Physics*

[6] Fuchs Ch. and Hartfuss H.-J. 2001 *Fus. Eng. Design* **53** 451-456

[7] Hirsch M. et al. 2016 *Proc. of the 43rd EPS conference on Plasma Physics* (EPS2016), P4.007

[8] Maruschenko N. B. et al. 2006 *Fus. Sci. Technol.* **50** 395

[9] Schmuck S. et al. 2009 *Fusion Eng. and Design* **84** 1739-1743

[10] Oosterbeek J. W. et al. 2018 *this workshop*

[11] Plaum B. 2018 *J of Infrared Milli Terahz Waves* 39:984-995

[12] Wagner D. et al. 2011 *J Infrared Milli Terahz Waves* **32** 1424–1433

[13] Schmuck S. et al. *36th EPS Conference on Plasma Phys. Sofia* ECA Vol.33E, P-4.212

[14] Hirsch M. et al. 2015 Proceedings of Science, *1st EPS Conference on Plasma Diagnostics ECPD 14-17 April 2015, Villa Mondragone, Frascati Italy.*

[15] Fuchs Ch. and Hartfuss H.-J. 2001 *Rev. Sci Instrum* **72** (1) 383

[16] Höfel U. et al. *Bayesian Modelling of Microwave Radiometer Calibration on the example of the Wendelstein 7-X Electron Cyclotron Emission diagnostic* to be published.

[17] Höfel U. et al. *Bayesian Modelling of the ECE diagnostic at W7-X* to be published.

[18] Marushchenko N. B., Turkin Y. and Maassberg H. 2014 *Computer Physics Communications* **185** 165 – 176 ISSN 0010-4655

[19] Hirshman S. P. and Whitson J. C. 1983 *The Physics of Fluids* **26** 3553–3568

[20] Höfel U. et al. 2016 *Proc. of the 43rd EPS conference on Plasma Physics* (EPS2016) P4.008

[21] v. Milligen B. et al. 2016 *Nucl. Fusion* **58** 076002

[22] Weir G. et al. 2016 *Proc. of the 43rd EPS conference on Plasma Physics* (EPS2016) P4.009

[23] Rahbarnia K. et al. 2016 *Proc. of the 43rd EPS conference on Plasma Physics* (EPS2016) P4.011

[24] Thomsen H. et al. 2016 *Proc. of the 43rd EPS conference on Plasma Physics* (EPS2016) P4.010

[25] Bozhenkov S. et al. 2018 *Proc. of the IAEA conference, Ahmedabad* EX/P8-8

[26] Laqua H. et al. 2018 Observation of core electron temperature collapses during ECCD experiments at the Wendelstein7-X Stellarator, submitted to *Phys Rev. Lett.*

[27] Zanini M et al. *this workshop*

[28] Oosterbeek H., Chaudhary N. et al. 2018 *Broad Band ECE Measurements in ECW heated Plasmas*, SOFT 2018 submitted to *Fusion Eng. and Design*

[29] Chaudhary N. et al. 2018 *this workshop*

Overview of the ECE measurements on EAST

Yong Liu[1,*], *Hailin* Zhao[1], *Tianfu* Zhou[1], *Xiang* Liu[1], *Zeying* Zhu[1], *Xiang* Han[1], *Stefan* Schmuck[2], *John* Fessey[2], *Paul* Trimble[2], *C.W.* Domier[3], *N.C.* Luhmann, Jr.[3], *Ang* Ti[1], *Erzhong* Li[1], *Bili* Ling[1], *Liqun* Hu[1], *Xi* Feng[4], *Ahdi* Liu[4], *W.L.* Rowan[5], *He* Huang[5], *P.E.* Phillips[5], and *Lorenzo* Figini[6]

[1]Institute of Plasma Physics, Chinese Academy of Sciences, Hefei 230031, China
[2]CCFE, Culham Science Centre, Abingdon, Oxon, OX14 3DB, UK
[3]Department of Applied Science, University of California at Davis, Davis, California 95616, USA
[4]University of Science and Technology of China, Hefei 230026, China
[5]Institute for Fusion Studies, University of Texas at Austin, TX 78712, USA
[6]CNR, Istituto di Fisica del Plasma, Milano 20125, Italy

Abstract. Radiometer systems and a Michelson interferometer, have been operated routinely to detect the electron cyclotron emission (ECE) from EAST plasmas for diagnosing the local electron temperature. A common quasi-optical antenna placed inside the vacuum vessel is employed to collect and focus the plasma emission, and the line of sight is along a radial chord. All of the systems are located in a diagnostic room where the plasma emission is transmitted by overmoded corrugated waveguide. In-situ absolute intensity calibration has been carried out for both the radiometer systems and the Michelson interferometer independently, to ensure that the ECE diagnostic provides an independent electron temperature measurement. In order to diagnose the small-amplitude electron temperature fluctuation, a correlation ECE (CECE) diagnostic has been designed and commissioned recently. The CECE diagnostic employs an independent antenna system which has improved poloidal resolution. A synthetic diagnostic is realized by using the simulation code SPECE to interpret the ECE data in plasmas with non-Maxwellian distribution, and preliminary results imply that the ECE data could be still useful as a localized measurement in plasmas with non-thermal electrons, such as the LHW-heated plasmas on EAST.

1 Introduction

Electron cyclotron emission (ECE) measurements have been widely employed in tokamak plasmas since the 1970s as a powerful diagnostic [1, 2]. Generally, ECE provides the temporal evolution of the local electron temperature with good spatial and temporal resolutions. In addition, information about the electron density, the electron velocity distribution, and the wall reflectivity can also be inferred from the entire ECE spectrum [3, 4].

On the experimental advanced superconducting tokamak (EAST), the ECE diagnostic has been operated since the first plasma of EAST and was improved continuously since then. There are a few radiometer systems [5, 6] and a Michelson interferometer [7] for conventional ECE measurement to diagnose the electron temperature profiles and electron temperature fluctuation driven by macroscopic instabilities. A correlation ECE (CECE) diagnostic has been developed to detect the turbulence associated small-amplitude electron temperature fluctuation. Synthetic diagnostic is studied by using the forward modeling code SPECE, and this facilitates the explanation of ECE data in plasmas with non-Maxwellian distribution. The conventional ECE diagnostics on EAST will be described in Section 2, and Section 3 introduces the CECE diagnostic. Results of synthetic diagnostic are given in Section 4. Section 5 is the summary.

*e-mail: liuyong@ipp.ac.cn

2 Conventional ECE diagnostic

2.1 Antenna and transmission line

Since 2014, a quasi-optical (QO) antenna has been adopted for collecting and focusing EAST plasma emission along a horizontal line of sight. As shown in Figure 1, the antenna comprises 3 reflectors: an ellipsoidal mirror and two flat mirrors. The ellipsoidal mirror can be rotated facing either the plasma or a hot-load calibration source. The ellipsoidal mirror is located 2000 mm away from the plasma center, and the distance to the waveguide is 1000 mm. The ellipsoidal mirror has an effective focal length of 900 mm, and forms a beam size of roughly 40 mm for a toroidal magnetic field of 2.3 T.

A single flat window with thickness of 15 mm was used in the year 2014, and this results in parallel-plate interference effects and consequently transmission ripple. Even though the in-situ absolute intensity calibration has been carried out, the transmission ripple has influence on the measurement and it is difficult to evaluate this effect quantitatively. In order to reduce the transmission ripple, the flat window was replaced with a double wedge window

 https://doi.org/10.1051/epjconf/201920303008

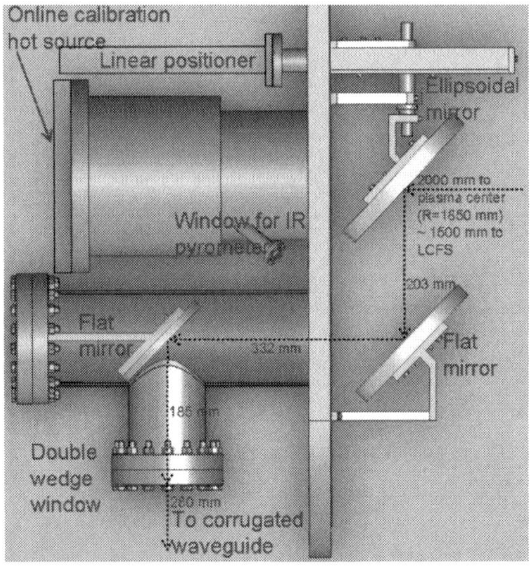

Figure 1. Quasi-optical antenna for the ECE diagnostic on EAST.

in 2015. Figure 2 shows the simulation results of the transmission for the single flat window and the double wedge window. The parameters of the double wedge window are: the wedge angle Ψ is 12.5 degree, the thickness of the thin side d_1 is 10 mm, and the distance between both windows d_w is 1 mm. The material for both the flat window and the double wedge window is high density polyethylene with dielectric constant ϵ_r of 2.3.

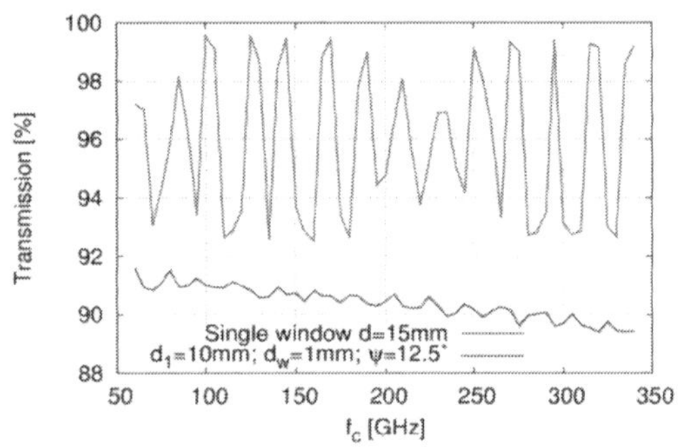

Figure 2. Comparison of the transmission for a single flat window and a double wedge window.

A wire grid polarizer is placed between the window and the waveguide to select the extraordinary mode (X-mode) polarization. Low-loss corrugated waveguides with 63.5 mm-id is used. The transmission line (TL) comprises roughly 45 meters of waveguides, and 8 miter bends. Because some bends are not oriented with identical or orthogonal planes of incidence, the linear polarization is rotated by roughly 63.15 degrees after transmission through the 8 miter bends.

In the diagnostic room, a notch filter is used to prevent the measurement systems from being damaged by electron cyclotron heating (ECH) stray radiation and then the plasma emission is divided into 3 paths by using two wire grid polarizers (see Figure 3). The power ratio for the three paths is 5:4:1. The first two paths are for the radiometer systems, and the remaining one is for the Michelson interferometer.

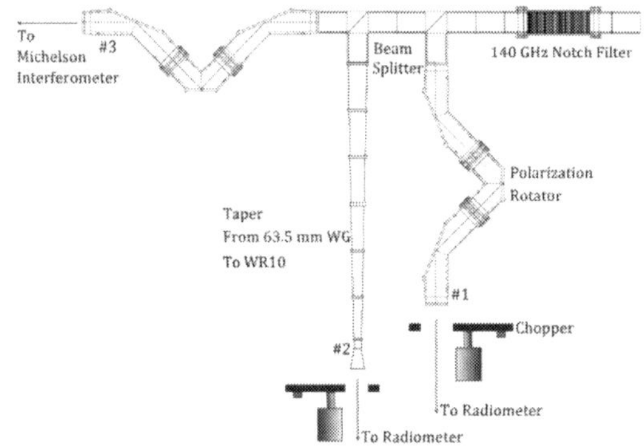

Figure 3. The layout of the transmission line (TL) in the diagnostic room.

2.2 ECE measurement systems

2.2.1 Radiometer systems

A 16-channel radiometer system which has a frequency coverage of 98.5-126 GHz has been operated since the first plasma of EAST in 2006 [5]. In 2012, a 32-channel radiometer system was designed and commissioned to measure the plasma emission in the frequency range of 104-168 GHz [6], and it uses the #1 output of the TL. This 32-channel radiometer has been upgraded in 2015, and the frequency interval for the lower frequency range upto 120 GHz is reduced to 1 GHz from 2 GHz by introducing 8 channels in the intermediate frequency (IF) part. In order to extend the frequency coverage, an independent 8-channel radiometer system covering a frequency range of 97-104 GHz (local oscillator of frequency 90 GHz) was built in 2015, and it uses the #2 output of the TL. With the combination of these radiometer systems, they have a frequency coverage capable of measuring the second harmonic ECE for plasmas with toroidal B_t higher than 1.75 T.

The new system has been fully characterized in the laboratory, and the test focuses on the performance of the IF bandwidth and the linearity. The results of the linearity test (see Figure 4 (a) and 9c)) indicate a good linearity in the range from 0 to 2 V. The sensitivity of each channel (including the RF part) has been adjusted to ensure that the system output in plasma measurements is within this dynamic range. The frequency response of the heterodyne radiometer electronics is shown in Figure 4 (b) and (d), and the result shows that each channel is responsive to only a single band of radio frequency (RF) frequencies.

The RF bandwidth f_{RF} of the radiometer systems varies from roughly 250 MHz to 500 MHz, and the video bandwidth f_V of the systems is upto 400 kHz. This results in a sensitivity of 4-5.7 % ($\approx \sqrt{\frac{2f_V}{f_{RF}}}$). This is adequate for most applications of ECE measurements such as diagnosing the electron temperature profile and the electron temperature fluctuation driven by MHD instabilities. However, it is not adequate to detect the small-amplitude, high-frequency fluctuations (with typical amplitudes around 1%) driven by micro-instabilities.

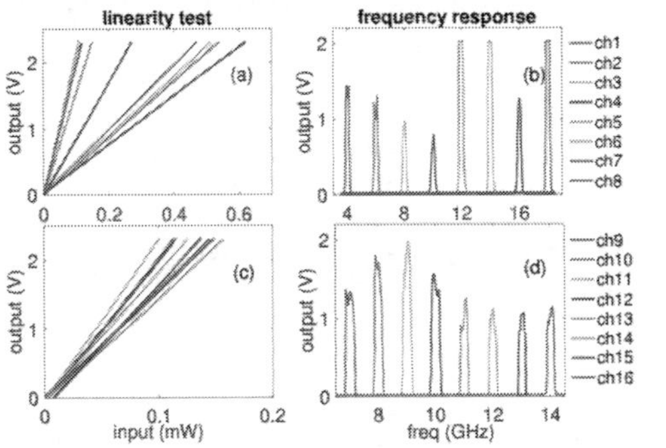

Figure 4. Tests of IF linearity ((a) and (c)) and frequency response ((b) and (d)) for the new 16 channels. The IF central frequency of the top is from 4 to 18 GHz, with 2 GHz separation, while the bottom is from 7 to 14 GHz, with 1 GHz separation.

2.2.2 *Michelson interferometer*

One of JET's Michelson interferometers was loaned to EAST in 2012, and it has been operated routinely sine the year 2014 [7]. This Michelson interferometer is one of the original systems developed by SPECAC in the 1980s for the ECE measurements in magnetically confined plasmas. This system had not been in operation since the late 1990s, and it has been refurbished when the loan agreement was signed in late 2011.

The system has a spectral resoluton of 2.7 GHz, and a temporal resolution of roughly 33 ms. The instrument is able to measure the ECE spectra from 80 GHz to 350 GHz, and this is mainly limited by the transmission line and the in-situ calibration. This frequency range typically covers the first to fourth ECE harmonic ranges for EAST magnetic field configuration.

2.2.3 *In-situ intensity calibration*

In-situ absolute intensity calibration has been carried out independently for the radiometer systems and the Michelson interferometer since 2014 [8]. Two types of hot sources have been developed for the calibration on EAST. One of them (#1) is a copy of the JET hot source [9], and the other one (#2) is the ITER prototype developed by the Institute for Fusion Studies [10]. Source #1 operates at atmosphere, while source #2 needs to be operated in vacuum. Normally, an in-situ calibration was carried out before an experimental campaign starts by using the hot source #1.

A digital coherent averaging technique and lock-in amplifier are separately used to measure the weak signal (a few tens μV), and both of the methods require the input signal to be modulated by an optical chopper. The lock-in amplifier gives RMS (root mean square) value of the modulated signal, while the coherent averaging technique results in the peak-to-peak value. The coherent averaging technique is more efficient because one can analyze all channels simultaneously.

Following the in-situ intensity calibration, the ECE diagnostic on EAST provides independent measurements of the electron temperature profiles and electron temperature fluctuation. Figure 5 shows the measurement of a few electron temperature profiles. The left figure of Figure 5 is for a discharge at a toroidal magnetic field of 1.8 T. The black line is the electron temperature measured by a Michelson interferometer [7], which can give the profile of the whole plasma region with a temporal resolution of roughly 33 ms, the red closed dots are the electron temperature measured by the existing 32-channel heterodyne radiometer, and the blue closed dots are the electron temperature measured by the newly upgraded 16-channel heterodyne radiometer. We can see that at such low magnetic field, the upgraded heterodyne radiometer can give the high field side electron temperature profile. The right figure of Figure 5 is the electron temperature profile for a discharge at a toroidal magnetic field of 2.47 T, and the result demonstrates that the radial separation is decreased with the new 48-channel heterodyne radiometer, from ~4 cm to ~2 cm.

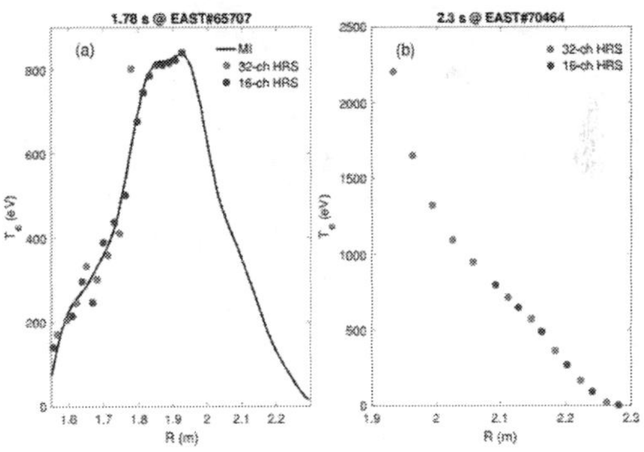

Figure 5. The electron temperature profiles at toroidal magnetic field of 1.80 T (left) and 2.47 T (right).

3 Correlation ECE diagnostic

In order to detect the small-amplitude electron temperature fluctuation associated with the anomalous transport, a CECE diagnostic has been designed since 2016. An independent QO antenna was designed, which has an improved poloidal resolution (roughly 15 mm for the frequency range of the CECE diagnostic). This is essential for the CECE diagnostic, and it determines the maximum wavenumber of the detected fluctuation. As illustrated in Figure 6, the QO antenna comprises a flat mirror and an ellipsoidal mirror. The ellipsoidal mirror is located 800 mm away from the plasma center, and the distance to the waveguide is 616 mm. The ellipsoidal mirror has an effective focal length of 390.6 mm. This antenna is integrated for both the CECE diagnostic and an existing Doppler backscatter diagnostic, and the flat mirror is rotatable in the poloidal direction. In contrast to the conventional ECE diagnostic, the heterodyne radiometer part of the CECE diagnostic is placed a few meters away from the diagnostic port.

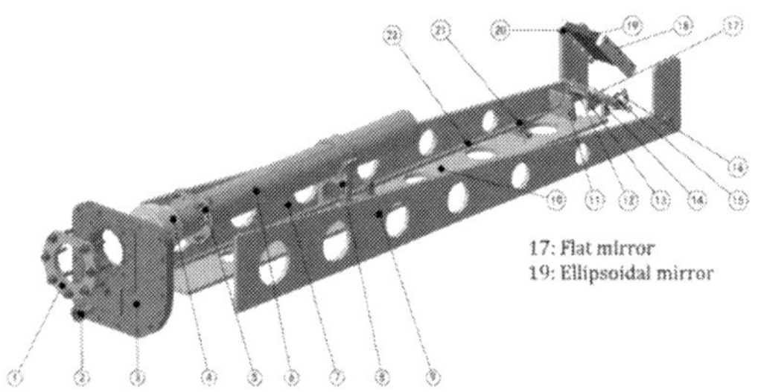

Figure 6. QO antenna of the CECE diagnostic.

The heterodyne detection part (see Figure 7) is similar to a conventional radiometer system, and a unique aspect is the application of YIG filters. The YIG filters (OMNIYIG INC) are of tunable central frequency in the frequency range of 4-18 GHz, and a 3 dB bandwidth of 100-250 MHz (central frequency dependent). This feature is very important for facilitating the radial correlation. The sensitivity of this CECE diagnostic ($\sqrt{\frac{1}{\sqrt{N}}\frac{2f_V}{f_{RF}}}$, where N is the number of data points for correlation) can reach around 0.2-0.3% for a correlation analysis using 10^6 data points.

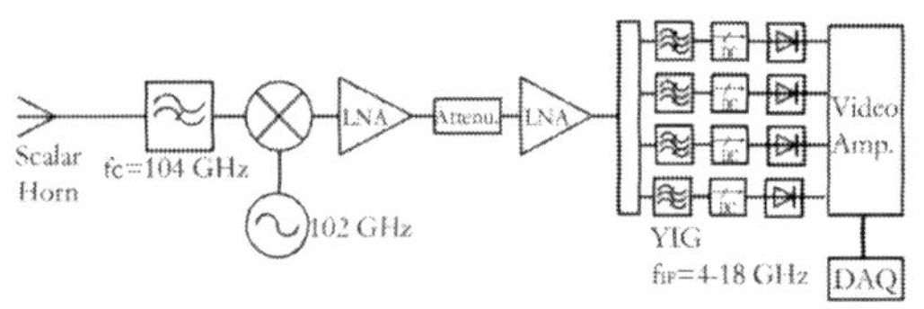

Figure 7. The heterodyne detection part of the CECE diagnostic.

This system has been commissioned in the EAST experimental campaign carried out in 2017. Preliminary results indicate that the broadband electron temperature fluctuations are detected in some cases. More tests will be conducted to confirm the observations in future.

4 Synthetic diagnostic

In order to interpret the ECE data in plasmas with non-Maxwellian distribution, the ECE ray-tracing code SPECE [11], developed as a support tool for the oblique ECE diagnostic at JET, is adopted for simulating the ECE spectrum in terms of radiative temperature on EAST. The input data for SPECE is described as follows: (i) the magnetic equilibrium is from EFIT, (ii) the electron temperature profile and the electron density profile are respectively provided by the Thomson scattering diagnostic and the reflectometry system, (iii) the LOS is described based on the real ECE diagnostic configuration. With the input data, the code solves the radiation transport equation along the ray trajectories in a tokamak plasma. Ray trajectories are computed making use of the cold dispersion relation, while the absorption and emission coefficients are obtained solving the relevant fully relativistic dispersion relation valid at high electron temperature. The actual antenna pattern is simulated by means of a multi-rays calculation, and the spatial resolution of the ECE measurements is computed by means of an algorithm that takes properly into account the emission along each ray of the beam.

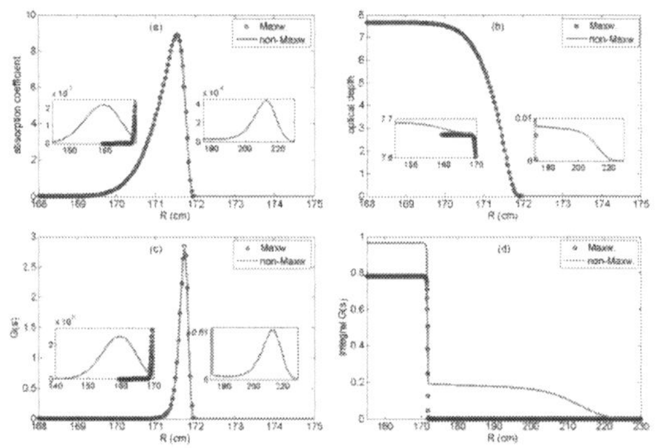

Figure 8. Example illustrating the spatial localization of 110 GHz ECE for plasmas (B_0 = 1.8 T @ 1.85 m) separately with Maxwellian and non-Maxwellian distribution functions: (a) absorption coefficient $\alpha(s)$, (b) optical depth $\tau(s)$, (c) $G(s) = \frac{\omega^2}{8\pi^3 c^2} T_{rad}(s)\alpha(s)e^{-\tau(s)}$, (d) $\int G(s)ds$. Here, s is along the major radius.

Figure 8 illustrates the spatial localization of 110 GHz (corresponds to the second harmonic ECE for the case with Maxwellian distribution) for plasmas (B_0 = 1.8 T @ 1.85 m) separately with Maxwellian and non-Maxwellian distribution functions. As the results (see Figure 8 (a)) show, two additional resonance layers will appear in separately HFS and LFS for the case with non-thermal electrons. The resonance layer in HFS is due to the relativistic down-shift of the second harmonic, while the resonance layer in the LFS is due to the relativistic down-shift of the third harmonic. Propagating from HFS to LFS, the emission in HFS will be absorbed completely for frequencies with high optical depth, and eventually the received intensity from LFS (the line of sight for a conventional ECE diagnostic) is mainly attributed to two emission layers (see Figure 8 (c)). In the rest of this paper, these two emission layers are named as the thermal emission layer and the non-thermal emission layer. As indicated in Figure 8 (c), the thermal emission layer for the non-Maxwellian case is close to that for the Maxwellian case. The reason is that the optical depth of the non-thermal emission is small (as shown in the small panel at the right side of the Figure 8 (b)), and consequently the absorption of the thermal emission layer is negligible when the emission propagates through the non-thermal resonance layer. Even though the non-thermal emission layer is much broader than the thermal emission layer, the intensity is much weaker.

In order to study the influence of the non-thermal electrons on the spatial localization of ECE measurement, a parameter $R = \int G_{thermal}(s)ds / \int G(s)ds$ is defined to describe the contribution of the thermal emission layer (local electron temperature) to the received intensity by an actual ECE diagnostic located in LFS, where the numerator is the integral for the thermal emission layer. The other parameter is the localization of the non-thermal emission layer, and it is described by the position of the peak value and the FWHM of the non-thermal emission layer.

To qualitatively study the contribution of the thermal emission layer to the total intensity and the localization of the non-thermal emission layer, some of the parameters describing the non-thermal electrons are scanned. The simulation results are summarized in Figures 9 and 10. The input data such as the electron temperature profile, the electron density profile, and the magnetic equilibrium for the simulations are from measurements. The main parameters are: central electron density $n_{e0} = 3.5\times10^{19}$ m^{-3}, central electron temperature T_{e0} = 1100 eV, central magnetic field strength B_0 = 1.8 T, the major radius $R_0 \approx$ 1.85 m, the minor radius $a \approx 0.45$ m.

As illustrated in Figure 8 (c), for the frequency range with associated high optical depth and the non-thermal resonance layer at the HFS is behind the thermal resonance layer, the emission is absorbed completely and will not be detected at the LFS. This corresponds to radial range with major radius of roughly 1.6-2.1 m in Figures 9 and 10. Because the non-thermal emission layer is attributed to the third harmonic emission, it is obvious that the influence is stronger at the HFS and the influenced raidal range goes to LFS deeper as the energy of the non-thermal electrons increase (Figures 9 (d)). For a set of reasonable parameters ($\eta_0 = 0.001 - 0.002$, $\psi_0 = 0.4$, $\Delta\psi = 0.3$, $T_{tail} = 25 - 30$ keV, these parameters give a good agreement between the simulated and the experimental spectra [12]), the intensity attributed to the thermal emission layer is larger than 70% for the radial range 1.7-2.1 m. For the radial range with major radius larger than 2.1 m, the optical depth becomes smaller. Hence, the emission from the non-thermal resonance layer at the HFS is not absorbed completely and contributes to the observed intensity. This explains why there is a drop for most cases in Figure 9. As the location of the non-thermal electrons (ψ_0) changes from plasma core to plasma edge, the number of resonant electrons becomes smaller and the contribution of the non-thermal electrons is smaller (see Figure 9 (b)).

As Figure 10 shows, the localization of the non-thermal emission layer (position and width) is strongly dependent of the radial position and radial distribution of the non-thermal electrons, and the dependence on the fractional density and tail energy is weak.

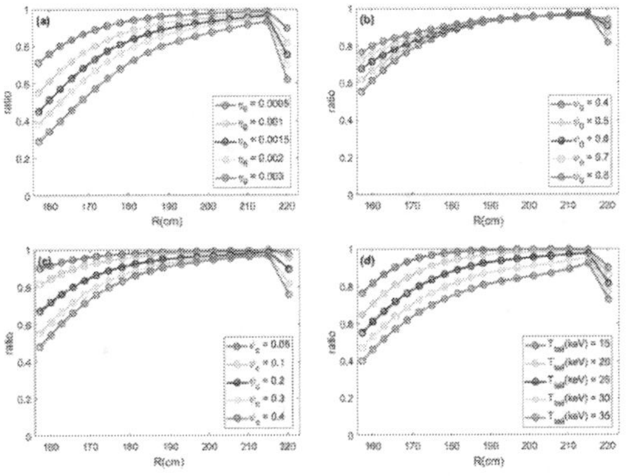

Figure 9. The ratio between intensity attributed to the thermal emission layer and the total intensity for scan of (a) the fractional density (ψ_0 = 0.4, $\Delta\psi$ = 0.3, T_{tail} = 25 keV, $u_{\parallel,max}$ = 0.4), (b) the radial position (η_0 = 0.001, $\Delta\psi$ = 0.3, T_{tail} = 25 keV, $u_{\parallel,max}$ = 0.4), (c) the radial distribution (η_0 = 0.001, ψ_0 = 0.4, T_{tail} = 25 keV, $u_{\parallel,max}$ = 0.4), and (d) the tail energy (η_0 = 0.001, ψ_0 = 0.4, $\Delta\psi$ = 0.3, $u_{\parallel,max}$ = 0.4).

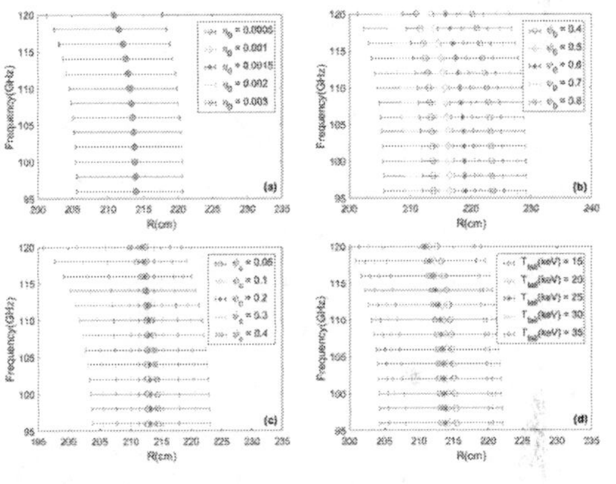

Figure 10. The localization (position defined at the peak value, width defined as the FWHM) of the non-thermal emission layer for scan of (a) the fractional density (ψ_0 = 0.4, $\Delta\psi$ = 0.3, T_{tail} = 25 keV, $u_{\parallel,max}$ = 0.4), (b) the radial position (η_0 = 0.001, $\Delta\psi$ = 0.3, T_{tail} = 25 keV, $u_{\parallel,max}$ = 0.4), (c) the radial distribution (η_0 = 0.001, ψ_0 = 0.4, T_{tail} = 25 keV, $u_{\parallel,max}$ = 0.4), and (d) the tail energy (η_0 = 0.001, ψ_0 = 0.4, $\Delta\psi$ = 0.3, $u_{\parallel,max}$ = 0.4).

5 Summary

A complete set of measurement systems, including a number of radiometer systems and a Michelson interferometer, have been developed and operated on EAST to detect the plasma emission in the ECE frequency range. Two independent QO antennas were adopted to collect the plasma emission for the conventional ECE diagnostic and the CECE diagnostic. In-situ intensity calibration has been carried out to ensure independent electron temperature measurement since 2014. Following the calibration, the ECE diagnostic on EAST provides the information of electron temperature profiles and fluctuation with fairly good spatial and temporal resolutions. Synthetic diagnostic using forward modeling is used to interpret the ECE data in plasmas with non-Maxwellian distribution, and preliminary results imply that the ECE data could be still useful as a localized measurement in plasmas with non-thermal electrons, such as the LHW-heated plasmas on EAST.

Acknowledgements

This work was supported by the National Natural Science Foundation of China under Grant No. 11405211, and the Innovative Program of Development Foundation of Hefei Center for Physical Science and Technology. This work was also partly supported by National Magnetic Confinement Fusion Science Program of China under Contract Nos. 2015GB101003 and 2015GB103002.

References

[1] A.E. Costley, TFR Group, Phy. Rev. Lett. **38**, 1477 (1977)

[2] I.H. Hutchinson, D.D. Komm, Nuclear Fusion **17**, 1077 (1977)

[3] L. Figini, S. Garavaglia, E. De La Luna, D. Farina, P. Platania, A. Simonetto, C. Sozzi, JET-EFDA Contributors, Rev. Sci. Instrum. **81**, 10D937 (2010)

[4] S. Schmuck, J. Fessey, J.E. Boom, L. Meneses, P. Abreu, E. Belonohy, I. Lupelli, Rev. Sci. Instrum. **87**, 093506 (2016)

[5] A. Ti, B.L. Ling, Q.S. Fei, X. Gao, Int. J. Infrared Milli. Waves **28**, 243 (2007)

[6] X. Han, X. Liu, Y. Liu, C.W. Domier, N.C. Luhmann, Jr., E.Z. Li, L.Q. Hu, X. Gao, Rev. Sci. Instrum. **85**, 073506 (2014)

[7] Y. Liu, S. Schmuck, H.L. Zhao, J. Fessey, P. Trimble, X. Liu, Z.Y. Zhu, Q. Zang, L.Q. Hu, Plasma Science and Technology **18**, 1148 (2016)

[8] X. Liu, H.L. Zhao, Y. Liu, E.Z. Li, X. Han, C.W. Domier, N.C. Luhmann, Jr., A. Ti, L.Q. Hu, X.D. Zhang, Rev. Sci. Instrum. **85**, 093508 (2014)

[9] S. Schmuck, J. Fessey, T. Gerbaud, B. Alper, M.N.A. Beurskens, E. de la Luna, A. Sirinelli, M. Zerbini, JET-EFDA Contributors, Rev. Sci. Instrum. **83**, 125101 (2012)

[10] A. Ouroua, J. Beno, A. Bryant, A. Khodak, P. Phillips, W. Rowan, G. Taylor, D. Weeks, IEEE TRANSACTIONS ON PLASMA SCIENCE **46**, 1239 (2018)

[11] D. Farina, L. Figini, P. Platania, C. Sozzi, AIP Conf. Proc. **988**, 128 (2008)

[12] T. Zhou, Y. Liu, A. Ti, L. Figini, H. Zhao, Z. Zhu, B. Ling, Fusion Science and Technology **74**, 154 (2018)

Intense intermittent radiation at the plasma frequency on EAST

Yong Liu[1], *Tianfu* Zhou[1], *Yemin* Hu[1,*], *Hailin* Zhao[1], *Zeying* Zhu[1], *Xiang* Liu[1], *Bili* Ling[1], *Ruijie* Zhou[1], and *Tao* Zhang[1]

[1]Institute of Plasma Physics, Chinese Academy of Sciences, Hefei 230031, China

Abstract. Intense intermittent radiation has been observed regularly in EAST by using a Michelson interferometer and a Q-band radiometer system. The radiation bursts are related to a fast process with characterization time of a few microseconds. An electron density window exists for the occurrence of the bursts, and the upper electron density threshold is dependent of the toroidal magnetic field. The frequency of the emission f is at the plasma frequency, and the frequency bandwidth Δf is very narrow (~1.5 MHz FWHM, $\Delta f/f \sim 3 \times 10^{-5}$). Fine structure of the spectrum with multi-peaks have been observed, and the frequency interval is around 3 MHz. Numerical simulation results of cavity modes indicate that the frequencies of these modes are close to the central plasma frequency, and the departure from the central plasma frequency becomes larger for higher order modes.

1 Introduction

Electron cyclotron emission (ECE) measurement has been widely employed in tokamak plasmas since the 1970s as a powerful diagnostic. Along with the development of the diagnostic, some of the characteristics of the measured spectra were not as expected as the ECE theory predicts. Under certain plasma conditions, additional peak besides the ECE harmonics, occurs at the electron plasma frequency ω_{pe0} corresponding to the measured value of electron density n_{e0} [1, 2]. This quiescent emission has been interpreted as the consequence of a Cherenkov resonance with the runaway electrons [3–5]. Fluctuating ω_{pe} emission with narrow bandwidth was also observed and studied on Alcator [6, 7]: the emission typically lasts 5-10 μs, and it is very intense with radiation temperatures exceed 10^5 eV. The spectral width of the emission is in the range of 400-800 kHz. Similar fluctuating emission was observed as well on DIII-D [8]. Two individual theories (cavity mode[9, 10], and triple-wave resonance model[11]) have been proposed to interpret the fluctuating ω_{pe} emission. Both of the theories involve two stages. The first stage is the generation of a tail with a bump of the electron velocity distribution parallel to the magnetic field. Under certain plasma conditions, the Maxwellian electron distribution begins to grow a runaway electron tail in the presence of a longitudinal electric field. This distribution function with a tail becomes unstable in a strong magnetic field owing to the anomalous Doppler effect [12]. Consequently, a bump is generated on the tail as a result of the instability due to the anomalous Doppler effect. The physics of the first stage is well accepted [13], and adopted by both of the theories as the origin of the free energy. However, the two theories have difference in explaining the origin of the ω_{pe} emission. The first theory assumes that the cavity mode is excited, while the other theory proposed that the ω_{pe} emission is caused by a triple-wave resonance instability. A recent study [14] on the nonlinear kinetic simulations of the anomalous Doppler instability of suprathermal electrons in plasmas reveals that it exhibits a spectral feature which may correspond to the fluctuating ω_{pe} emission.

Another unexpected feature is the intense bursts in the ECE frequency range correlated with edge-localized modes (ELMs), and this phenomenon has been reported widely in JET[15], TFTR[16, 17], DIII-D[18], and MAST[19]. Recently this phenomenon is explained by the action of the anomalous Doppler instability which enables fast collective radiative relaxation, in the perpendicular direction, of electrons accelerated in the parallel direction by inductive electric fields generated by the initial ELM instability [20]. Besides, intense millimeter wave bursts corresponding to the upper hybrid frequency near the low field plasma edge were observed in plasmas well conditioned with lithium in the Tokamak Fusion Test Reactor [21].

Besides the observations in laboratory plasmas, different types of emission bursts at the plasma frequency have been detected in astrophysical plasmas ever since the 1950s[22–24]. This has been explained as a multi-stage mechanism with the first stage being generation of Langmuir waves through a streaming instability driven by a positive gradient $df(v_\parallel)/dv_\parallel > 0$, and subsequent stages involving partial conversion of the Langmuir turbulence into escaping radiation at the plasma frequency [25–27]. Despite many successes of this theory, the question is still open because this theory still has difficulty in explaining the type I bursts [28].

All of the existing theories/models for explaining the bursts of emission at plasma frequency in either the tokamak plasmas or the astrophysical plasmas, rely on the assumption of the non-Maxwellian electron velocity distri-

*e-mail: yeminhu@ipp.ac.cn

bution. Provided the mechanism is well understood, this phenomenon can be used to obtain the information on the energetic tail parameters, and this is of serious concern with regard to safe operation of tokamaks. On the other hand, the bursts are in the frequency range of millimeter-wave, and they are so intense that this could be one of the reasons that causes the malfunction of the millimeter-wave components of reflectometry diagnostics on some tokamak machines [29].

On the experimental advanced superconducting tokamak (EAST), bursts at both the plasma frequency (hereinafter ω_{pe} bursts) and ECE frequency were observed, and the former is the subject of the present study. Even though Lithium wall conditioning is used a lot on EAST, the phenomenon which is discussed in this paper is not like that on TFTR and was also observed in plasmas with Silicon wall conditioning. In the following, the section 2 describes the essential diagnostics relevant to the study, the section 3 presents the phenomenology of the ω_{pe} bursts on EAST, the numerical results of cavity mode are given in the section 4, and the section 5 includes the conclusions of the present work and prospects the future study.

2 Diagnostics relevant to the study

2.1 Conventional ECE diagnostics

Common quasi-optical antenna placed inside the vacuum vessel is adopted to collect and focus the plasma emission, and the line of sight is along a radial chord. A long transmission line transmits the emission to the diagnostic room. A wire grid polarizer is placed between the window and the transmission line to select the extraordinary mode (X-mode) polarization. The transmission line is composed of roughly 45 meters of corrugated waveguide and 8 miter bends. Even though the transmission line is optimized in the frequency range of 60-300 GHz, estimation of the theoretical losses indicates that the transmission line still has a good performance even for the frequency range down to 30 GHz (The attenuation per 10 meters of waveguide is 23.8%, and the loss per bend is 3.4 % for the frequency of 30 GHz [30]. This results in a transmission loss of roughly 78% for the transmission line on EAST). Different systems, including a few radiometer systems [31] and a Michelson interferometer [32], have been operated routinely for ECE measurement in EAST. The radiometer system adopts the conventional millimeter-wave heterodyne technique: the plasma emission is downconverted by mixers, and the frequency selection is accomplished by individual bandpass filters leading to detectors. The Michelson interferometer produces interferograms by splitting a input beam so that one beam strikes a fixed mirror and the other a movable mirror. Fourier transform of the interferogram reveals the spectrum of the input light assuming the light is stable during the scan. In-situ absolute intensity calibration has been carried out for both the radiometer system and the Michelson interferometer independently, to ensure that the ECE measurement provides independent absolute electron temperature profiles.

Table 1 lists the spectral and temporal specifications of the ECE diagnostic. The frequency range of the radiometer systems spreads from 97 to 168 GHz, and it is realized by using 48 separate channels. The frequency interval is 1 GHz in the frequency range of 97-121 GHz, while it is 2 GHz in the remaining frequency range. The frequency coverage of the Michelson interferometer is mainly limited by the specification of the transmission line.

System name	Freq. coverage	Spectral	Temporal
Radiometer	97-168 GHz	0.2/0.5 GHz	2.5 μs
Michelson interf.	<30-300 GHz	2.8 GHz	30 ms

Table 1: The spectral and temporal specifications of the ECE diagnostic on EAST.

2.2 Q-band radiometer system

The bursts have been observed regularly ever since the Michelson interferometer was commissioned in 2014 But the bursts are not observed by the radiometer systems. The radiometer could detect either the second harmonic or the third harmonic ECE as the toroidal magnetic field varies. Since the Michelson interferometer and the radiometer systems share the same transmission line, it is believed that the frequency range of the bursts is not in either the second or the third ECE harmonics.

In order to demonstrate the rough frequency range of the bursts, a 90 GHz dichroic plate which acts as a high pass filter was placed in front of the input aperture of the Michelson interferometer. The comparison of the results for discharges with similar plasma parameters rules out the possibility of higher ECE harmonics. Based on the observations on other tokamak machines, it is very likely that the bursts observed on EAST are ω_{pe} bursts. Thereafter, a Q-band radiometer system was built to distinguish the frequency of the bursts and consequently characterize the bursts in 2016. This system shares the same transmission line with the conventional ECE diagnostics. A local oscillator (LO) with frequency of 34 GHz was employed to downconvert the plasma emission. After amplification, the intermediate frequency (IF) signal is divided into 8 channels. One channel is detected directly, while the remaining channels are filtered by band pass filters (some YIG filters with adjustable central frequency are used). In order to obtain the frequency spectra, one channel of the Q-band radiometer system is fed to a real-time spectrum analyzer (realized by the combination of a 26.5 GHz PXI vector signal analyzer PXIe-5668 and a PXI high-speed data storage module HDD-8261 provided by National Instruments) for a few dedicated discharges.

3 The phenomenology of the ω_{pe} bursts

3.1 Occurrence

The ω_{pe} bursts have been observed regularly in ohmic plasmas and electron cyclotron wave heated plasmas, but

 https://doi.org/10.1051/epjconf/201920303009

it rarely happens in lower hybrid wave heated plasmas. There exists an electron density window for the occurrence of the ω_{pe} bursts (as illustrated in Figure 1), and the upper threshold is strongly dependent of the toroidal magnetic field. The occurrence of ω_{pe} bursts always accompany with enhancement of radiation in the energy range from a few hundreds keV to a few MeV (see Figure 1 (d)), and this indicates the presence of run away electrons. But the presence of run away electrons is not sufficient, there are cases no bursts were observed even though the run away electrons exist.

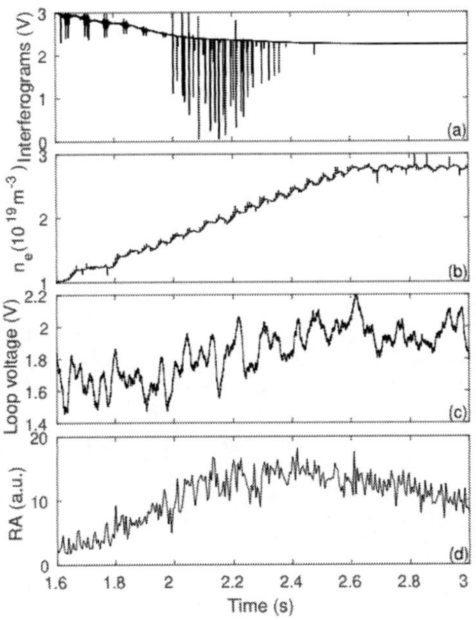

Figure 1: A typical example (EAST #72621) shows the existence of electron density threshold for the occurrence of the ω_{pe} bursts: (a) interferograms measured by the Michelson interferometer, (b) line averaged electron density along a vertical chord close to the plasma center, (c) plasma loop voltage, (d) γ ray counts.

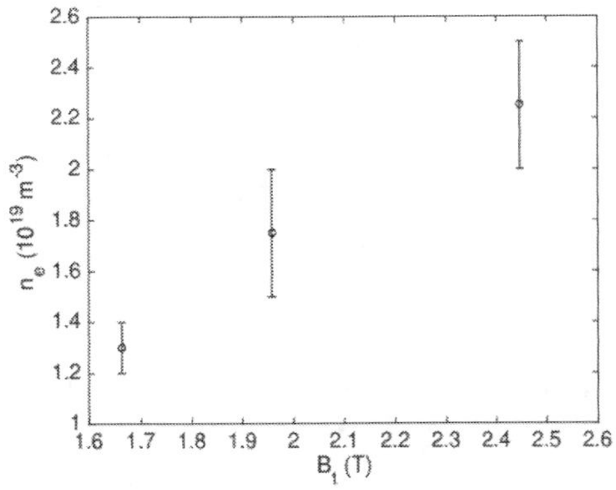

Figure 2: The upper electron density threshold (line averaged electron density along a chord close to the plasma center) for the occurrence of the ω_{pe} bursts versus the toroidal magnetic field at major radius of 1.85 m.

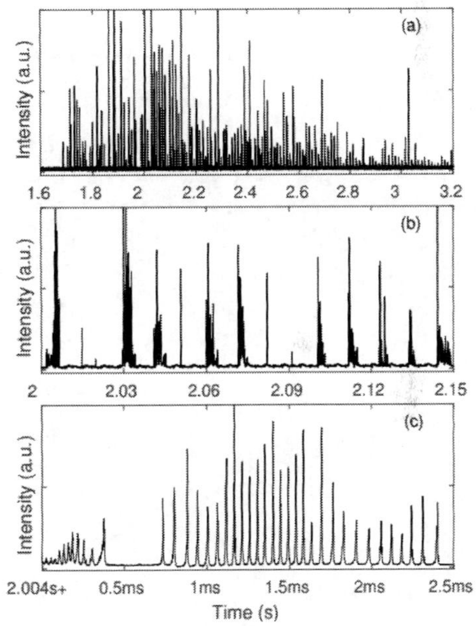

Figure 3: A typical example (EAST #70273) shows the pattern of the occurrence of the ω_{pe} bursts: (a) time scale of 1.6 s, (b) time scale of 150 ms, (c) time scale of 2.5 ms.

In order to obtain the threshold, a number of discharges for both with and without ω_{pe} bursts are analyzed statistically. The threshold is determined to be the average value for two discharges with closest electron density on the map of the ω_{pe} bursts occurrence. Figure 2 shows the trend of the upper electron density threshold (line averaged electron density along a chord close to the plasma center) versus the toroidal magnetic field at major radius of 1.85 m.

A typical example illustrating the pattern of the occurrence of the ω_{pe} bursts is shown in Figure 3. Each individual ω_{pe} burst has a timescale of a few μs (see Figure 3 (c)), and the frequency of the occurrence of ω_{pe} bursts can reach upto a few kHz within a timescale of a few milliseconds followed by a quiescent period of the order of ten milliseconds (see Figure 3 (b)).

3.2 Frequency and fine structure

Figure 4 illustrates a typical measurement (EAST #70444) of the Q-band radiometer system during an electron density ramp up phase. It is clear that the frequency of the bursts changes as the electron density ramps up. The plasma frequencies calculated from the electron density profiles measured by a microwave reflectometer and the frequency of the bursts for a few time slices are given in

https://doi.org/10.1051/epjconf/201920303009

the Figure 5, and the frequency of the ω_{pe} bursts is close to the central plasma frequency.

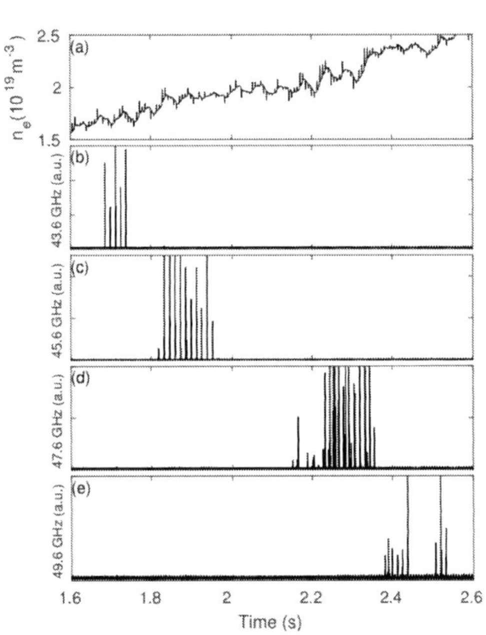

Figure 4: The frequency of the ω_{pe} bursts changes as the electron density ramps up (EAST #70444): (a) line averaged electron density along a vertical chord close to the plasma center, (b)-(e) results of the Q-band radiometer system.

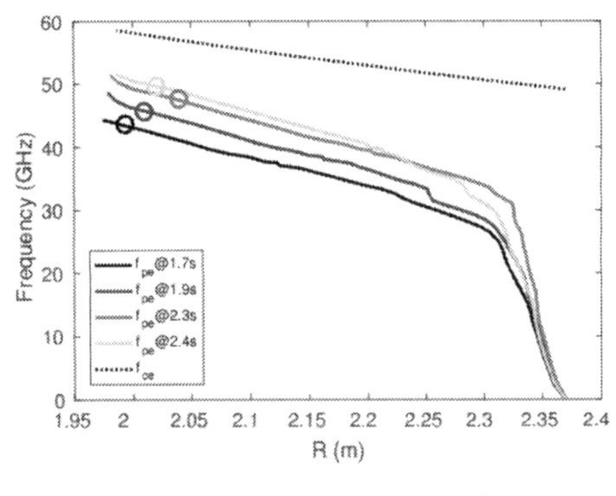

Figure 5: Comparison of the plasma frequency (solid lines) and the ω_{pe} bursts frequency (circles, the position is determined assuming the frequency is ω_{pe}) (EAST #70444, magnetic axis ~1.9 m). The electron density is measured by a microwave reflectometer.

The precise frequency of the ω_{pe} bursts has been determined from the measurement of a real-time spectrum analyzer (using a carrier wave of 7.5 GHz, and sampling the data with a rate of 200 MS/s). Frequency chirping with a rate of magnitude of MHz/μs has been observed, and an

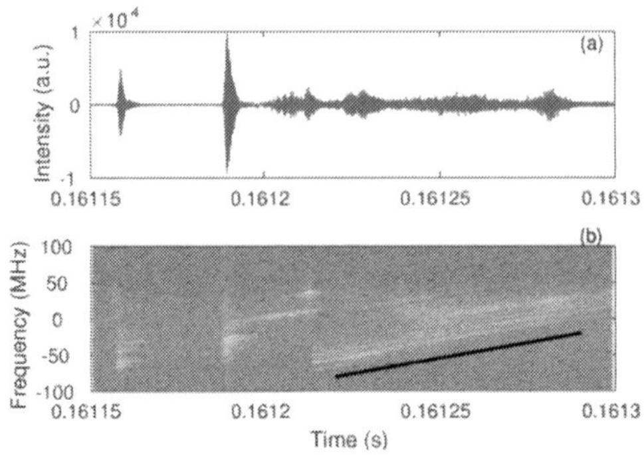

Figure 6: The results for a group of ω_{pe} bursts measured by the real-time spectrum analyzer: (a) raw data, (b) spectrogram obtained using FFT, and the black solid line indicates a sweeping rate of 0.86 MHz/μs. The frequency of the emission is the summation of the frequency illustrated in the figure, the carrier wave frequency of the real-time spectrum analyzer (7.5 GHz), and the LO frequency of the radiometer (34 GHz).

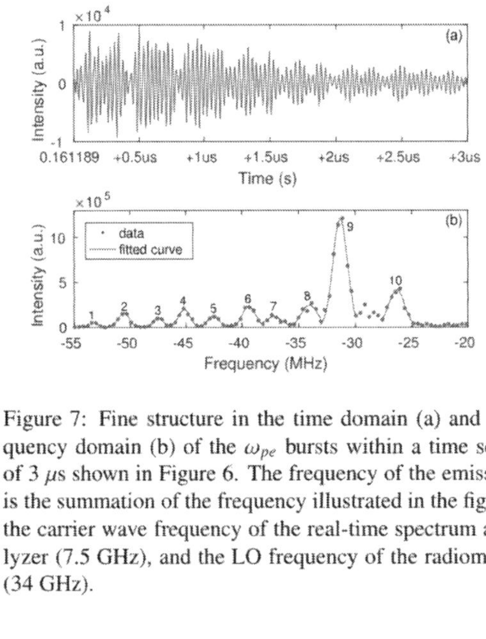

Figure 7: Fine structure in the time domain (a) and frequency domain (b) of the ω_{pe} bursts within a time scale of 3 μs shown in Figure 6. The frequency of the emission is the summation of the frequency illustrated in the figure, the carrier wave frequency of the real-time spectrum analyzer (7.5 GHz), and the LO frequency of the radiometer (34 GHz).

example is illustrated in Figure 6. Figure 7 shows the raw data and the spectrum within a time scale of 3 μs, and it shows a clear structure with multiple peaks. Table 2 summarizes the results of Gaussian fitting of the peaks in Figure 7(b). The peak separation is of roughly 3 MHz, and the FWHM of the peaks is about 1-1.7 MHz.

4 Numerical simulation of cavity mode

As mentioned in section 1, some theoretical work has been carried out to explain the process of ω_{pe} burst formation. In this work we are following the theory of I. H.

Peak	Position (MHz)	FWHM (MHz)	FWHM/freq.
#1	-53.28	0.9449	2.28e-5
#2	-50.57	1.1616	2.80e-5
#3	-47.47	1.0625	2.56e-5
#4	-45.16	1.2512	3.02e-5
#5	-42.51	1.2573	3.03e-5
#6	-39.39	1.3835	3.34e-5
#7	-37.18	1.7101	4.12e-5
#8	-34.05	1.7334	4.18e-5
#9	-31.33	1.5054	3.63e-5
#10	-26.35	1.6701	4.03e-5

Table 2: Results of Gaussian fitting for the fine structure of the ω_{pe} bursts shown in Figure 7(b), and the frequency of the emission is the summation of the number in the column "Position", the carrier wave frequency of the real-time spectrum analyzer (7.5 GHz), and the LO frequency of the radiometer (34 GHz).

Hutchinson and R.F. Gandy et.al. [9], and modeling the general cavity mode structure for waves near ω_{pe} in the EAST plasma by numerical method. The numerical analysis model is to solve Maxwell's equations

$$\nabla \times \boldsymbol{E} = -\frac{\partial \boldsymbol{B}}{\partial t} \quad (1)$$

$$\nabla \times \boldsymbol{B} = \mu_0 \boldsymbol{j} + \frac{1}{c^2}\frac{\partial \boldsymbol{E}}{\partial t} \quad (2)$$

Here, in order to be convenient, we take the same expression and symbols as that in Ref.[9] and [10]. Under the conditions of EAST parameters, the model we adopted for the plasma response is also that of a plasma of cold electrons and stationary ions. So we followed a cold radially inhomogeneous plasma and ignored thermal corrections. For the time dependence of the plasma response $\propto \exp(-i\omega t)$ and a quasi-cylindrically symmetric with background magnetic field in the z direction, Maxwell's equations can be written out in coordinate system (r, θ, z) to seek solutions proportional to $\exp i(kz + m\theta)$. So Eqs.(1) and (2) can be rewritten as [9] Eqs (3)-(7).

where N is the longitudinal refractive index kc/ω, k_0 the vacuum k vector ω/c, and

$$D = \frac{\left(N^2 - \epsilon_\Lambda\right)^2}{\epsilon_\Lambda} + \frac{\epsilon_x^2}{\epsilon_\Lambda} \quad (8)$$

$$\epsilon = 1 - X, \quad (9)$$

$$\epsilon_x = -\frac{iXY}{Y^2 - 1}, \quad (10)$$

$$\epsilon_\Lambda = 1 + \frac{X}{Y^2 - 1}, \quad (11)$$

where $X = \omega_{pe}^2/\omega^2$, $Y = \Omega/\omega$, and ω, ω_{pe} and Ω are waves eigenfrequency, electron-plasma frequency and electron cyclotron frequency, respectively.

For Eqs.(3) and (4), we normalized E and B to E_0 and B_0 i.e. $E_{r,\theta,z} \rightarrow \frac{E_{r,\theta,z}}{E_{z0}}$ and $B_{r,\theta,z} \rightarrow \frac{B_{r,\theta,z}}{B_{z0}}$, and take the boundary conditions as $E_z|_{r=0} = B_z|_{r=0} = 1, E_z\Big|_{r=\infty} = B_z\Big|_{r=\infty} = 0$, and $\frac{dE_z}{dr}\Big|_{r=0} = \frac{dB_z}{dr}\Big|_{r=0} = 0$, where E_{z0} and B_{z0} are electric field and magnetic field in z direction at $r = 0$. For a monotonically decreasing plasma density profile, we found that the cavity mode structure and dispersion relations are not so sensitive to the specific profile by the later numerical results, so we chose the similar density profile as that in Ref.[9] and [10] as following

$$n_e = n_{edge} + \left(n_{e0} - n_{edge}\right)\left(1 - r^2\right), \quad (12)$$

where n_{edge} and n_{e0} are the density at the last close magnetic surface and magnetic axis, respectively. So we can solve the coupled normalized eigenmode E_z and B_z equations (3) and (4) with given longitudinal refractive index N, and then obtain the corresponding eigenfrequency ω. As following the analytical analysis of mode structure in Ref.[9], we definite the mode number n to be the number of the zero E_z with respect to r direction. In Fig. 8, the eigenmode structure of the E_z modes $(m, n) = (0, 140)$ for $N = 24, B_0 = 2.19$ T$, n_{e0} = 2.3 \times 10^{19}$ m^{-3} and aspect ratio $R_0 = 4$ are shown. The results indicate that E and B are both highly localized near the axis and the field at the plasma edge is negligible for the present plasma parameters. Besides, it is also found that $B/E \ll 1$, and this means that the present electromagnetic modes can be approximate to be quasi-static electric modes. Fig. 9 shows the dispersion relations for the E_z modes $(m, n) = (0,0), (0,1), (0,2), (1,0), (1,1)$ and $(1,2)$, respectively, with respect to the longitudinal refractive index N in a quasi-cylindrical EAST plasma. It is found that the frequencies of waves are close to central plasma frequency, and the deviation from the center plasma frequency becomes larger for higher order modes.

Fig. 10 shows the comparison between ω_{pe} and eigenmode frequency with (m, n)=(0, 0), (0, 140) and (0, 234) with respect to r. It is found that the eigenmode frequency became smaller for mode with higher n. This may explain why that the measured frequency deviates from the central plasma frequency (see Fig. 5).

Table 3 gives the mode frequency and adjacent frequency interval of Eigenmodes with different mode number $m = 0, 1$ and $n = 136 \sim 141$ for $N = 24$. It is found that the frequency intervals are in good agreement with the observed fine structure of the ω_{pe} burst frequency as showed in the Fig. 7.

5 Discussions and conclusions

The ω_{pe} bursts have been observed regularly in EAST. The individual burst has a time scale of a few microseconds. An electron density window exists for the occurrence of the related instability, and the upper electron density threshold is dependent of the toroidal magnetic field. The frequency of the emission f is at the plasma frequency, and the frequency bandwidth Δf is very narrow ($\sim$1.5 MHz FWHM, $\Delta f/f \sim 3 \times 10^{-5}$). Fine structure

$$\left[\frac{1}{r}\frac{d}{dr}\frac{d}{dr}\frac{r}{D}\left(N^2-\frac{\epsilon_\Lambda^2+\epsilon_x^2}{\epsilon_\Lambda}\right)-\frac{m^2}{Dr^2}\left(N^2-\frac{\epsilon_\Lambda^2+\epsilon_x^2}{\epsilon_\Lambda}\right)+\frac{im}{r}\left(\frac{d}{dr}\frac{N^2}{D}\frac{\epsilon_x}{\epsilon_\Lambda}\right)-k_0^2\epsilon\right]\frac{E_z}{c}$$
$$=\left\{\frac{1}{r}\frac{d}{dr}\frac{Nr}{D}\frac{\epsilon_x}{\epsilon_\Lambda}\frac{d}{dr}-\frac{m^2}{r^2}\frac{N}{D}\frac{\epsilon_x}{\epsilon_\Lambda}-\frac{im}{r}\left(\frac{d}{dr}\frac{N}{D}\left(\frac{N^2}{\epsilon_\Lambda}-1\right)\right)\right\}B_z, \tag{3}$$

$$\left[\frac{1}{r}\frac{d}{dr}\frac{r}{D}\left(\frac{N^2}{\epsilon_\Lambda}-1\right)\frac{d}{dr}-\frac{m^2}{r^2D}\left(\frac{N^2}{\epsilon_\Lambda}-1\right)+\frac{im}{r}\left(\frac{d}{dr}\frac{\epsilon_x}{D\epsilon_\Lambda}\right)-k_0^2\right]B_z$$
$$=\left[\frac{1}{r}\frac{d}{dr}\frac{Nr}{D}\frac{\epsilon_x}{\epsilon_\Lambda}\frac{d}{dr}-\frac{m^2}{r^2}\frac{N}{D}\frac{\epsilon_x}{\epsilon_\Lambda}-\frac{i}{r}\left(\frac{d}{dr}\frac{mN}{D}\left(\frac{N^2}{\epsilon_\Lambda}-1\right)\right)\right]\frac{E_z}{c}, \tag{4}$$

$$E_\theta=-\frac{mNc}{krD}\frac{\epsilon_x}{\epsilon_\Lambda}B_z+\frac{iN^2}{Dk}\frac{\epsilon_x}{\epsilon_\Lambda}\frac{dE_z}{dr}+\left(\frac{mN^2}{krD}E_z+\frac{iNc}{kD}\frac{dB_z}{dr}\right)\left(\frac{N^2}{\epsilon_\Lambda}-1\right), \tag{5}$$

$$B_\theta=\frac{m}{krD}\left[\frac{N^2}{\epsilon_\Lambda}\left(N^2-\frac{\epsilon_\Lambda^2+\epsilon_x^2}{\epsilon_\Lambda}\right)+\frac{N^2\epsilon_x^2}{\epsilon_\Lambda^2}\right]B_z+\frac{iN}{kcD}\left(\frac{\epsilon_\Lambda^2+\epsilon_x^2}{\epsilon_\Lambda}-N^2\right)\frac{dE_z}{dr}+\frac{mN^3}{krcD}\frac{\epsilon_x}{\epsilon_\Lambda}E_z+\frac{iN^2}{kD}\frac{\epsilon_x}{\epsilon_\Lambda}\frac{dB_z}{dr},$$

$$E_r=-\frac{c^2}{\omega\epsilon_\Lambda}\left(\frac{m}{r}B_z-kB_\theta\right)-\frac{\epsilon_x}{\epsilon_\Lambda}E_\theta, \tag{6}$$

$$B_r=\frac{m}{\omega r}E_z-\frac{kE_\theta}{\omega}. \tag{7}$$

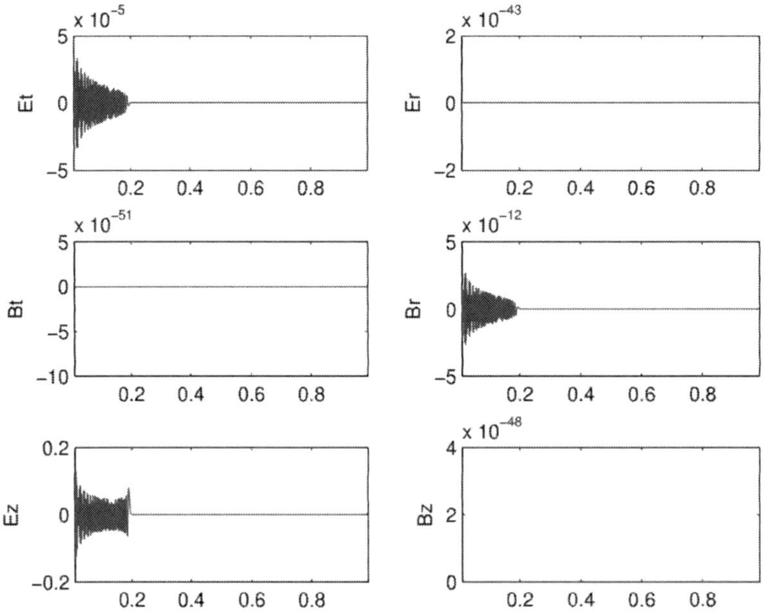

Figure 8: r, θ and z components of electric field E and magnetic field B with respect to the normalized minor radius r for cavity mode with $(m, n) = (0, 140)$, shows that the mode is localized in plasma center and it is quasi-static electric mode. The mode number n is defined to be the number of zero E_z with respect to r direction.

of the spectrum with multi-peaks have been observed, and the frequency interval is around 3 MHz.

In order to explain this phenomenon, the present work follows the cavity mode theory and calculates the cavity mode structure and dispersion relation numerically by solving scalarized Maxwell's equations in a quasi-cylindrical EAST plasma with monotonically decreasing density profile. The results have shown that (i) the lowest-order electromagnetic modes are approximate to be quasi-static electric modes, and the corresponding eigenmode

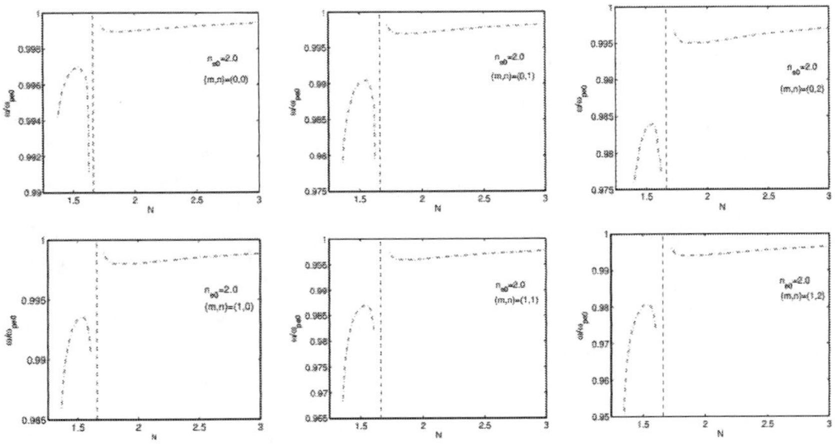

Figure 9: Cold plasma dispersion relation with respect to the longitudinal refractive index N for the first few modes.

n	136	137	138	139	140	141
$f(GHz)(m=0)$	42.24956	42.24381	42.23815	42.23238	42.22670	42.22093
$f(GHz)(m=1)$	42.24669	42.24066	42.23523	42.22955	42.22379	42.21809
δf_{01}(MHz)	2.86	3.15	2.91	2.83	2.91	2.83

Table 3: Frequency and adjacent frequency offset of eigenmode with different mode number m and n.

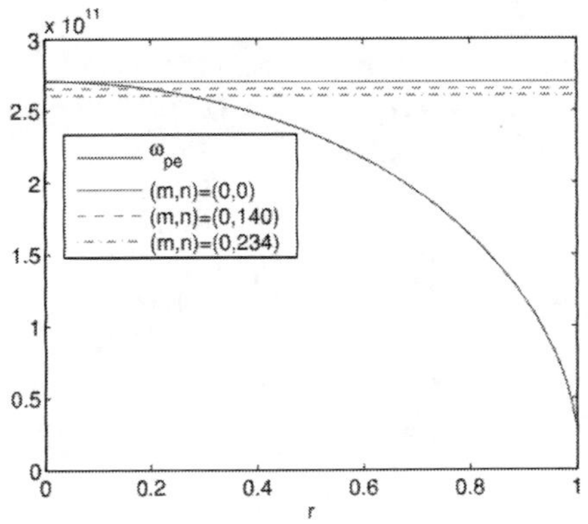

Figure 10: The comparison between ω_{pe} and eigenmode frequency with $(m,n) = (0,0), (0,140)$ and $(0,234)$ with respect to r.

frequencies are close to the central plasma frequency for a large range of longitudinal refractive index, (ii) the deviation of the eigenmode frequencies from the central plasma frequency becomes larger for higher order modes, (iii) the frequency interval for eigenmodes with high mode number ($m = 0, 1$ and $n = 136 \sim 141$) for $N = 24$ is roughly 3 MHz. To some extent, these characteristics of the cavity modes are in agreement with the observations of the ω_{pe} bursts.

However, the underlying physics of the time scale, the frequency bandwidth, the radiation transport, and the intensity of the observed the ω_{pe} bursts are not discussed in the present work. A further study by using kinetic simulations may reveal these questions.

Acknowledgements

One of the authors (Y. Liu) would like to express his thanks to Dr. Gary Taylor, Dr. Max Austin, Prof. Boris Breizman, and Prof. Ian Hutchinson for fruitful discussions. Y. Liu also thank Mr. Kewei Qian and Mr. Liyang Tian for implementing the real-time spectrum analyzer. This work was supported by the National Natural Science Foundation of China under Grant Nos. 11405211 and 11375234, and the Innovative Program of Development Foundation of Hefei Center for Physical Science and Technology. This work was also partly supported by National Magnetic Confinement Fusion Science Program of China under Contract Nos. 2015GB101003 and 2015GB103002. The authors would also like to acknowledge the ShenMa High Performance Computing Cluster at the Institute of Plasma Physics, Chinese Academy of Sciences.

References

[1] A.E. Costley, TFR Group, Phy. Rev. Lett. **38**, 1477 (1977)

[2] I.H. Hutchinson, D.D. Komm, Nuclear Fusion **17**, 1077 (1977)

[3] H.P. Freund, C.S. Wu, L.C. Lee, D. Dillenburg, Physics of Fluids **21**, 1502 (1978)
[4] H.P. Freund, L.C. Lee, C.S. Wu, Phys. Rev. Lett. **40**, 1563 (1978)
[5] K. Swartz, I.H. Hutchinson, K. Molvig, Physics of Fluids **24**, 1689 (1981)
[6] I.H. Hutchinson, S.E. Kissel, Physics of Fluids **23**, 1698 (1980)
[7] I.H. Hutchinson, S.E. Kissel, Physics of Fluids **26**, 310 (1983)
[8] Private communication with Dr. Max Austin from the University of Texas at Austin.
[9] I.H. Hutchinson, R.F. Gandy, Physics of Fluids **28**, 1090 (1984)
[10] R.F. Gandy, I.H. Hutchinson, D.H. Yates, Phys. Rev. Lett. **54**, 800 (1985)
[11] R.O. Dendy, C.N. Lashmore-davies, M.M. Shoucri, Nuclear Fusion **25**, 721 (1985)
[12] S.C. Luckhardt, A. Bers, V. Fuchs, M. Shoucri, The Physics of Fluids **30**, 2110 (1987)
[13] V.V. Parail, O.P. Pogutse, Nuclear Fusion **18**, 303 (1978)
[14] W.N. Lai, S.C. Chapman, R.O. Dendy, Physics of Plasmas **20**, 102122 (2013)
[15] L. Porte, D.V. Bartlett, D.J. Campbell, A.E. Costley, in *Proceedings of the 18th European Conference on Controlled Fusion and Plasma Physics* (European Physical Society, Petit-Lancy, Switzerland, 1991), Vol. 15C, Part IV, p. 357
[16] G. Taylor, C.E. Bush, E.D. Fredrickson, H.K. Park, A.T. Ramsey, Nuclear Fusion **32**, 1867 (1992)
[17] G. Taylor, E. Fredrickson, A. Janos, Rev. Sci. Instrum. **66**, 830 (1995)
[18] Fuchs Ch, M.E. Austin, Physics of Plasmas **8**, 1594 (2001)
[19] V.F. Shevchenko, R.G.L. Vann, S.J. Freethy, B.K. Huang, Journal of instruments **7**, P10016 (2012)
[20] S.J. Freethy, K.G. McClements, S.C. Chapman, R.O. Dendy, W.N. Lai, S.J.P. Pamela, V.F. Shevchenko, R.G.L. Vann, Phys. Rev. Lett. **114** (2015)
[21] G. Taylor, D.K. Mansfield, Nuclear Fusion **38**, 1819 (1998)
[22] J.P. Wild, L.L. McCready, Aust. J. Sci. Res. A Phys. Sci. **3**, 387 (1950)
[23] J.P. Wild, L.L. McCready, Aust. J. Sci. Res. A Phys. Sci. **3**, 399 (1950)
[24] J.P. Wild, L.L. McCready, Aust. J. Sci. Res. A Phys. Sci. **3**, 541 (1950)
[25] V.L. Ginzburg, V.V. Zheleznyakov, Sov. Astron. **2**, 653 (1958)
[26] V.V. Zheleznyakov, V.V. Zaitsev, Sov. Astron. **14**, 47 (1970)
[27] V.V. Zheleznyakov, V.V. Zaitsev, Sov. Astron. **14**, 250 (1970)
[28] D.B. Melrose, Rev. Mod. Plasma Phys. **1:5** (2017)
[29] Private communication with technicians on EAST, and DIII-D.
[30] Private communication with John Doane from General Atomics.
[31] X. Han, X. Liu, Y. Liu, C.W. Domier, N.C. Luhmann, Jr., E.Z. Li, L.Q. Hu, X. Gao, Rev. Sci. Instrum. **85**, 073506 (2014)
[32] Y. Liu, S. Schmuck, H.L. Zhao, J. Fessey, P. Trimble, X. Liu, Z.Y. Zhu, Q. Zang, L.Q. Hu, Plasma Science and Technology **18**, 1148 (2016)

Assessment of ECH stray radiation levels at the W7-X Michelson Interferometer and Profile Reflectometer

Johan. W. Oosterbeek[1,*], *Neha* Chaudhary[1], *Matthias* Hirsch[1], *Udo* Höfel[1], *Robert. C.* Wolf[1], and the W7-X team[1]

[1]Max-Planck-Institut für Plasmaphysik, Teilinstitut Greifswald, Wendelsteinstraße 1, 17491 Greifswald, Germany

Abstract. Electron Cyclotron Heating and Electron Cyclotron Current Drive are key components for heating and control in magnetically confined fusion plasmas. The high power microwaves are not always completely absorbed leading to stray radiation [1], [2]. At W7-X, the total injected microwave power can be up to 7.5 MW @140 GHz while the entire Electron Cyclotron Emission picked-up by an observer at the edge of the plasma is a fraction of a mW. In the situation of a Michelson Interferometer, the principle measurement is the entire ECE spectrum. Thus, any stray radiation is bound to enter the spectrum. In this work initial stray radiation measurements without filters at the location of two microwave receivers - the Michelson Interferometer and the Profile Reflectometer - are discussed. The data is used to dimension a notch filter to be used with the broad band Michelson Interferometer.

1 Introduction

A Michelson Interferometer is being prepared for W7-X operations from campaign OP1.2b onwards, scheduled July - October 2018. The objective of the work reported here is to find both the absolute power at the input of the instrument as well as the ratio of ECH to ECE power to assess the impact of stray radiation. In support of this, stray radiation measurements at the location of Michelson Interferometer as well as at the location of the Profile Reflectometer have been carried out during the previous W7-X experimental campaign (OP1.2a). Spurious gyrotron mode have not been included in this work.

2 Instruments and waveguide system

Both the Michelson Interferometer and the Profile Reflectometer use in-vessel horn antennae. The signals are relayed from the vessel by means of oversized waveguide to the Diagnostic Hall. A sketch is given in Fig. 1.

2.1 Michelson Interferometer

The Michelson Interferometer is designed to operate from $50 \cdots 300$ GHz. The in-vessel antenna of the ECE diagnostic is located in module 4 (port 'AEE41'). It is constructed by a quasi-optic section of mirrors that project the collected beam onto a horn antenna with a diameter of 36 mm. The output of the horn is tapered to 4 mm circular waveguide and subsequently passes a single disk mica vacuum window. On passing the window the waveguide is tapered to 28 mm smooth walled circular waveguide and split into X- and O-mode by a wire grid. These waveguides then continue to the diagnostic hall where, by means of switches, X-mode is coupled to the radiometer and O-mode to the Michelson Interferometer. The overall waveguide length between vessel and instrument is ≈ 22 m. The losses of waveguide up to the radiometer have been measured in advance of OP1.2a using a 140 GHz source and are 13.5 +/- 0.5 dB (waveguide relay in TE_{11} mode). The losses starting at the input of the wire grid through the vacuum windows, tapers and up to the horn antenna are estimated to be 3 dB, resulting in an estimated overall transmission line loss of 16 dB.

*e-mail: hans.oosterbeek@ipp.mpg.de

2.2 Profile Reflectometer

The edge Density Profile Reflectometer is situated in module 2 (port 'AEA21') and is still under development. In the case of the profile reflectometer the in-vessel antennas consists of two horns with diameter 62 mm, located such that their beams intersect in the edge region of the plasma. The upper horn was used in the experiment. The horn guides the power into a section of circular waveguide that tapers to 4 mm diameter in order to pass the mica vacuum window. Following the window the waveguide is tapered up to 28 mm smooth walled circular waveguide and continues to the diagnostic hall. While the overall waveguide length is comparable to the ECE-system, the losses are lower. This can possibly be explained by fewer components in the Transmission Line (e.g. no X/O splitter, no polarisation tuner). The losses in the Tx-Line from window to instrument were measured, similarly as in Sec. 2.1, and are 8 +/- 0.5 dB. The loss from vacuum window up to the in-vessel horn aperture is estimated at 2 dB, such that the overall loss of the Tx-Line is set to 10 dB.

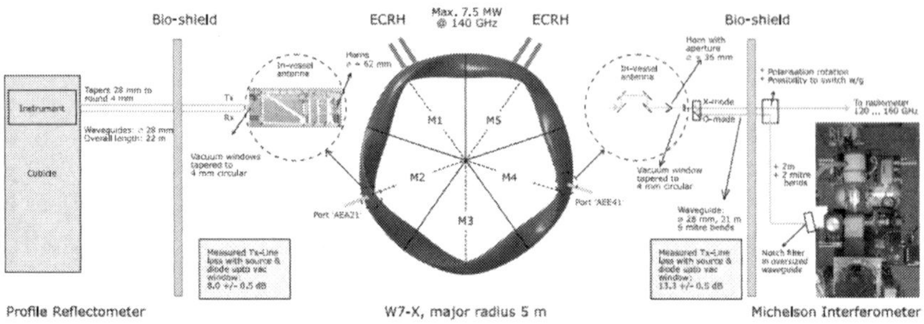

Figure 1: Michelson and Profile Reflectometer waveguide systems at W7-X

2.3 Stray radiation probe

The stray radiation measurements were carried out close to the entrance of the instruments, i.e. directly behind the tapers from circular 28 mm diameter to fundamental. For the stray radiation measurements the 28 mm circular waveguide was tapered to D-band (110 ··· 170 GHz). The signal was - by means of a variable attenuator - terminated on a single mode Schottky detector diode. The detectors were calibrated with a 140 GHz source using an isolator and a variable attenuator. With the attenuator set to 0 dB, the output power was measured as 0.6 mW, corresponding to 10 V at the ADC. The combination of detector diode and pre-amplifier resulted in a sensitivity of the sniffer probe of 17 V/mW at low power. The lowest power level that could be detected without additional video conditioning was around - 35 dB below 0.6 mW (i.e. 0.2 μW). In the Michelson configuration the probe measured O-mode and in the Profile Reflectometer configuration the probe measured X-mode.

3 Expected power levels

3.1 ECH

In-vessel stray radiation levels are obtained from [3]. For over-dense plasmas (O2-heating) stray radiation levels as measured by bolometers are of the order of 10 kWm^{-2} for module 2 (Reflectometer) and 4 kWm^{-2} for module 4 (Michelson). This data is for injection of 5 MW ECH. For X2 heated plasmas, optically thick and not in cut-off, stray radiation will predominantly originate from the ECH launchers and mismatches in polarisation launch, and these levels will be lower. In the presence of a plasma a more or less isotropic and homogeneous power density is found in the volume between the plasma surface and the vessel components. In such case, the stray radiation is present under all angles and polarisations and the absolute ECH power coupled into the horn is obtained by multiplication of the physical antenna aperture with the power density but reduced by the antenna directivity. In this assessment the antenna directivity of the horns is taken to be 20 dB, compared to the ideal situation of an omni-directional sniffer probe at 0 dB. The power that reaches the detector is subsequently reduced by i) rejection of modes, and ii) attenuation of the waveguide for stray radiation. The reduction of modes is here taken to be the ratio of all modes that can propagate to those reaching the detector, i.e. one mode in this case (single mode detector). An estimate considering all possible modes in circular oversized waveguide is given by: $N = 0.85\left(\frac{f}{f_0}\right)^2$ [4], with f_0 the fundamental cut-off frequency in the waveguide (here TE_{11}). For the Michelson horn antenna one finds of the order of 700 modes while for the Reflectometer horn one finds of the order of 2000 modes. Expressed in decibels the ratios multi-mode at horn aperture to single mode at detector are -28 dB resp. -33 dB. The attenuation of the waveguide for stray radiation is here taken to be the same as that measured with the RF-source and diode in section 2.

3.2 ECE

The ECE power is emitted in two modes (X- and O-mode). The circular in-vessel antenna has been arranged such that the antenna pattern is completely illuminated by the ECE in either of these two modes. Under such condition, and with the plasma optically thick, the power coupled into the waveguide is ≈ kTB per mode using the Rayleigh-Jeans approximation for black body radiation, with k Boltzmann's constant, T the electron Temperature and B the bandwidth [5]. The TRAVIS code [6] was used to verify the ECE spectrum for typical temperatures (5 keV) and densities ($5 \cdot 10^{19}$m^{-3}), see Fig. 2. It is seen that the power in the 2-nd harmonic O-mode is less the that in the 2-nd harmonic X-mode. As the Michelson probe measures in O2-mode, less power then based on kTB may be expected. When calculating the power over the entire spectrum, one should take the particular electron temperature profile into account as the central electron temperature would result in an over-estimate. In this estimate a parabolic electron temperature profile has been assumed which results in a reduction of the overall power by a factor of 1.5 with respect to a 'top-hat' profile.

3.3 Estimate

Table 1 gives the expected power as found using the data in sections 3.1 and 3.2.

 https://doi.org/10.1051/epjconf/201920303010

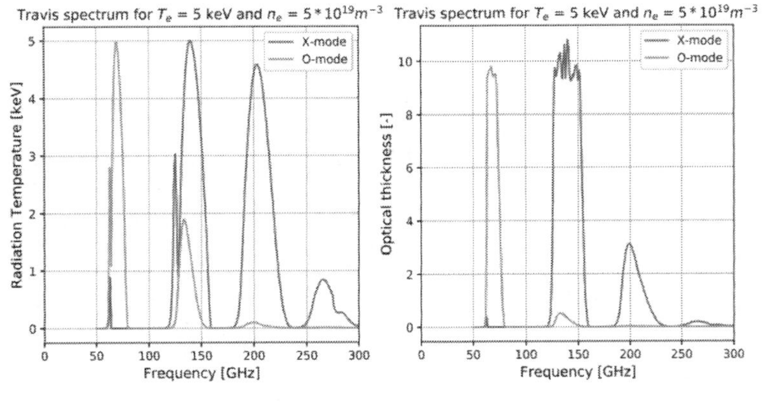

Figure 2: Data from Travis, left: radiation spectrum and right: optical depth.

	Power Density [Wm^{-2}] / 5MW	System	Ant. dia. [mm]	P_ant. [W]	P_{ECE} (kTB) [μW]	ECH/ECE @ Ant. [dB]	ECH/ECE on Det. [dB]
Module 2	12000	PR	61	0.35	32	40	7
Module 4	4000	MI	36	0.04	32	31	3

Table 1: Expected power levels. The column P_{ant} gives the maximum ECH power coupled into the antennae: the power density multiplied with the antenna aperture but reduced by the antenna directivity. PR stands for Profile Reflectometer and MI for Michelson. The column P_{ECE} is the ECE power over D-Band, taking a central electron temperature of 5 keV and assuming a parabolic profile. Next the power ratio ECH to ECE in-vessel is calculated. The last column gives this ratio in front of the Michelson: here the stray radiation is strongly reduced due to the transition from oversized horn antenna to single mode detector.

4 Experiment

Availability of waveguides - in demand for other diagnostic data - restricted the measurements to a number of days towards the end of campaign. Nevertheless useful data was obtained but in X2 heated plasmas only. It was found that RF-attenuation before the diode could be reduced to below a factor 10 before the ECH signal during *start-up* approached 1 mW on the detector. During the *plasma phase* the stray radiation power was mostly too low to detect but note that only X2-heated plasmas were available. Fig. 3 shows a panel with time traces, the left hand column shows the overall pulse, the right hand column a zoom during start-up. The first row (a & b) shows signals for the total ECRH power, the second row (c & d) the Michelson stray radiation probe ('MI-probe'), and the third row (e & f) the Profile Reflectometer stray radiation probe ('PR-probe'). In this discharge the central electron temperature was around 5 keV. The central density was varied, but between 1 to 4 seconds it was in the range of $5{\cdot}10^{19}$electrons m^{-3}. The attenuation in front of the MI-probe was 4 +/- 1 dB while for the PR-probe it was set to 0 dB. It is seen that the stray radiation probes, on the 10 V scale of the ADC, only show signal during the plasma start-up, i.e. when ECRH power is injected but there is no absorption by the plasma yet.

The experimental program was not dedicated to stray radiation measurements. However, polarisation scans with the probes themselves were carried out and the W7-X program required various ECH launcher angles. An initial assessment shows no large deviation from the presented data, however, a systematic analysis of launcher angles and polarisation - requiring additional measurements - is planned.

During start-up the data shows a large standing wave pattern of stray radiation that varied with launching angle and polarisation. But the amplitude rarely clipped the limit of the ADC, and if it did so, only just. During the plasma phase, limited stray radiation (below the levels at start-up) is observed during a number of pulses where the density is too high or too low.

The relatively low waveguide attenuation for the PR-probe, the 0 dB setting of the attenuator and possibly the X-mode measurement opposed to O-mode, made it possibly to see a finger-print of the ECE during the plasma-phase of the discharge on the PR-probe. This is shown in Fig. 4, now with a *vertical* zoom on the PR-probe and as a reference a central radiometer EC-channel (# 14, 138.085 $\cdots$ 138.435 GHz).

5 Discussion

Both probes show strong ECH stray radiation during plasma start-up. As the power fluctuates wildly, and is made up out of standing waves, it is not possible to give an accurate estimate of the power. But the signal very rarely

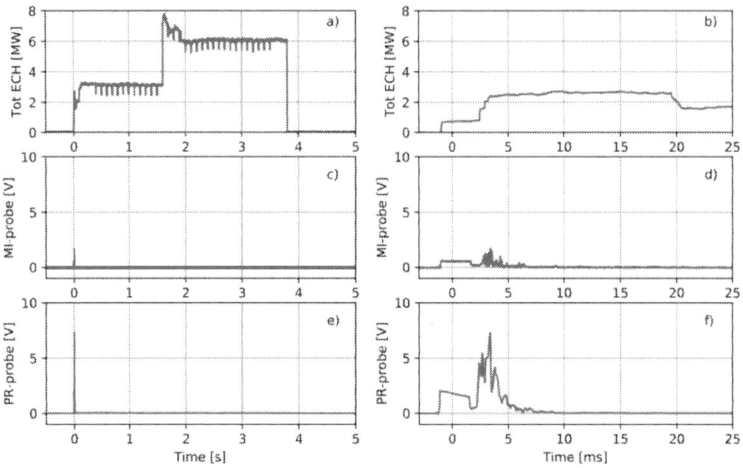

Figure 3: Data for # 171207009. Traces in the left hand column show the overall discharge, traces in the right hand column a zoom on plasma start-up. Top: ECRH total power, center: the Profile Reflectometer stray radiation probe and lower the Michelson Interferometer stray radiation probe.

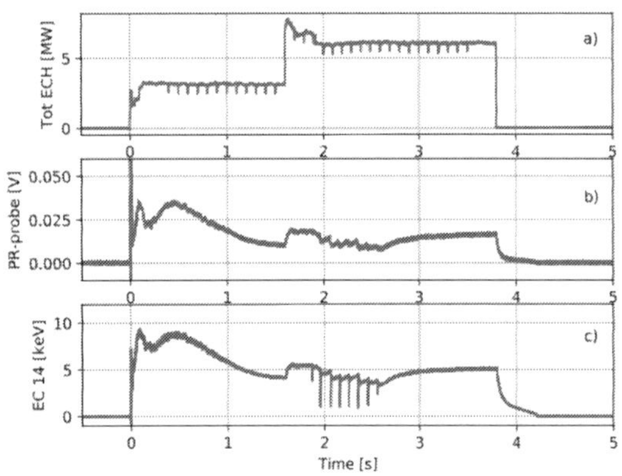

Figure 4: ECE finger-print on Profile Reflectometer stray radiation probe (b) using the data as in Fig. 3. For comparison a central ECE channel is shown (c). After ECH switch-off (@ 3.8 s) the signal from the probe falls off exponentially with the confinement time showing that no large ECH stray radiation power component seemed present.

saturated and is limited to 1 mW in the single mode measured. The Michelson is multi-mode but with the number of modes limited by the $\approx$ 10 modes that the 4 mm section of transmission line in the vacuum window can support (see Sec. 3.1). The total power reaching the Michelson is then expected to be at maximum 10 times larger, i.e. at maximum 10 mW during start-up. Note that this is based on the *estimates* of modes, and the assumption that the total power is spread equally over these modes. As the power at start-up is expected (and observed) to be far in excess of stray radiation during the plasma phase of the experiment, the notch filter maximum ECH power is set at 10 mW. The ECE losses from vessel to Michelson are 16 dB, or $\approx$ a factor 40. If one demands the ECH power to be reduced to the overall ECE power at the input of the Michelson one finds a ratio of 10 mW/ 1 μW = 40 dB. Such a ratio is useful as the detector will remain in the same dynamic range during start-up and during the plasma phase.

During the plasma phase the expected ratio ECH to overal ECE for a single mode detector is given in the last column of Table 1. For the Michelson, exposed to 10 modes, the ratio is 10 dB higher. When rounded the ECH power in O2-heated plasma is expected $\approx$ 15 dB more then the overall ECE power. In the experiment, **but X2 mode heating opposed to O2-mode**, the ratio ECH to overall ECE appears to be below unity.

To validate the ECE component a consistency check can be made. Referring to Fig. 4 the ECE power scales to around 20 mV in the ADC. This signal is the ECE response over 120 $\cdots$ 170 GHz in X-mode (Reflectometer probe) and as such represent a 'spatially averaged' ECE power. From the detector scaling of 17V / mW it follows that 20 mV measured with the ADC corresponds to $\approx$ 1.2 μW on the detector. Applying the overall transmission line attenuation of 16 dB between the instrument and the in-vessel antenna, the power at the in-vessel antenna is 16 dB more, or 48 μW. In Table 1 it is estimated to be $\approx$ 32 μW. . From this it is concluded that the expected ($\approx$ 30 μW) and measured ECE power ($\approx$ 50 μW) are consistent given the margins on such an ECE power measurement.

The probe on the Michelson does not show an ECE signal. But the losses in the Michelson transmission line are higher by 6 + 4 (attenuator) = 10 dB. This reduces the 1.2 μW as measured by the Profile Reflectometer probe to 0.12 μW, indeed at the limit of the probe.

6 Conclusions and future work

Measurement complemented by modeling suggest that the maximum stray radiation power during ECH start-up, at the input of the Michelson, will not exceed 10 mW in the specific W7-X configuration. A notch at 40 dB brings the signal in the same order of magnitude as the overall expected ECE power. Expected stray radiation levels during the plasma phase in O2-heated plasmas are of the order of 15 dB above the ECE levels. A notch filter at 40 dB would cope with this. This margin is comportable given uncertainties such as i) extrapolation from power measured with a single mode detector to power measured in multi-mode with the Michelson, ii) assumtions on the difference in directivity of a sniffer probe compared to a ECE-horn, and iii) a limited data set in ECH heated plasmas as of yet, and excluding O2-heated data. A careful measurement campaing with the Michelson, with the notch filter in place, is required to verify the results.

The authors are grateful for the discussions with the W7-X ECH team, Dr. Walter Kasparek and Prof. Manfred Thumm during this work.

This work has been carried out within the framework of the EUROfusion Consortium and has received funding from the Euratom research and training programme 2014-2018 under grant agreement No 633053. The views and opinions expressed herein do not necessarily reflect those of the European Commission.

References

[1] M. Moseev et al., Review of Scientific Instruments, 87(8), 1-9. [083505]

[2] S. Marsen et al., Nuclear Fusion, Volume 57, Number 8

[3] S. Marsen, Poster Session during the 20th Joint Workshop on Electron Cyclotron Emission (ECE) and Electron Cyclotron Resonance Heating (ECRH), May 14-17, 2018, Greifswald, Germany.

[4] Nikitin et al., Proceedings of the 2011 IEEE International Symposium on Antennas and Propagation (APSURSI), 3-8 July 2011, Spokane, WA, USA

[5] Hartfuss, Hans-Jürgen, *Fusion Plasma Diagnostics with mm-Waves* (Wiley-VCH, 2013)

[6] N.B. Marushchenko, Computer Physics Communications 185 (2014) 165–176

[7] N. Chaudhary et al., Proceedings of the 20th Joint Workshop on Electron Cyclotron Emission (ECE) and Electron Cyclotron Resonance Heating (ECRH), May 14-17, 2018, Greifswald, Germany.

Integration of an Electron Cyclotron Imaging diagnostic system on the WEST tokamak

Roland Sabot[1,*], *Yoonbum* Nam[1], *Cyril* Brun[1], *Didier* Elbèze[1], *Frédéric* Faisse[1], *Laurent* Gargiulo[1], *Minwoo* Kim[2], *Woochang* Lee[3], *Philippe* Lotte[1], *Hyeon K.* Park[2], *Gunsu* Yun[4] and *WEST team*

[1]CEA, IRFM, F-13108 Saint Paul-lez-Durance, France.
[2]Ulsan National Institute of Science and Technology, Ulsan 44919, Korea
[3]National Fusion Research Institute, Daejeon 34133, Korea
[4]Department of Physics, Pohang University of Science and Technology, Pohang 37673, Korea

Abstract. An Electron Cyclotron Emission Imaging (ECEI) diagnostic system for the WEST tokamak has been developed under the UNIST-WEST collaboration. This diagnostic system is designed to overcome accessibility and thermomechanical constraints for long pulse operation. The first O-mode channel will be installed in the first trimester of 2019 to probe the low field side (LFS) of the WEST plasma. Two large metallic reflective mirrors are installed inside the duct which is being used for maintenance access. They are suspended on a rail to facilitate mirror manipulation. The ex-vessel optical system (lens, detection array, etc.) is housed in a compact optical enclosure that fits in a tight free space between the port flange and tokamak access lobby. The design emphasized reproducibility of the precise alignment between in-vessel mirrors and optical enclosure since the both elements must be removed during shutdown period for maintenance access. The overall optical system was fully tested at UNIST last year. The test results demonstrated that the imaging optics can full access at any radial position on the LFS. The 2D beam pattern measurements were consistent with the design values.

1 Introduction

WEST (Tungsten (W) Environment in Steady-state Tokamak) is transformation of the Tore Supra tokamak plasma in an X-point divertor configuration. WEST aims to test ITER divertor technology for high heat and particle fluence during long pulse operation [1]. Internal coils were installed to obtain flexible equilibrium: lower, upper or double X-point configuration. All carbon plasma facing components (PFC) were replaced with tungsten surface (W monoblock and carbon tiles coated with tungsten). The stainless steel water-cooled panels protecting the vacuum vessel were refurbished to cope with the new elongated configuration. The first WEST plasma was commissioned in 2016 and the divertor plasma configuration was obtained in 2017.

Control of magnetohydrodynamic (MHD) instabilities and metallic impurity transport is an essential step to sustain long discharges. Microwave imaging instrument offering a 2-D visualisation of the MHD as well as turbulence phenomena inside the plasma gives a unique insight of instability characteristics. Electron Cyclotron Emission imaging (ECEI) diagnostic system, first developed in the 90's, images the temperature fluctuations [2,3]. It has proven extremely powerful for understanding the physics of sawtooth crash [4,5], the structure of ELMs [6] or fast particles induced instabilities [7].

An ECEI system developed through the UNIST-WEST collaboration will be installed during the first 2019 trimester. The ECEI will complement the set of diagnostics dedicated for fluctuation measurements [8]. 2D images of the temperature and density fluctuations induced by MHD instabilities could thus be obtained simultaneously [9].

This paper describes innovative approaches to integrate the ECEI optics on WEST. Although the WEST diagnostic relies on the ECEI principles, introduction of the large reflective optics to be positioned inside the man-access port, which induce new constraints on the optics design compared to standard ECEI. The solution of both the in-vessel mirrors and compact ex-vessel optics was proposed to overcome the constraints in WEST will be described in part 2. The integrated optical system was fully tested last year at UNIST, Korea; the laboratory test results along with the alignment procedure will be presented in part 3. The article concludes that although the ECEI implantation on large steady-state tokamaks is challenging, innovative solutions can overcome the integration issues to get the unrivalled information provided by 2D images of temperature fluctuations.

2 WEST ECEI characteristics

2.1. Implantation in the man access port

The ECE diagnostics is based on emission of the electrons around the magnetic field lines to measure the local electron temperature [10]. In the ECEI system, the single

* Corresponding author: roland.sabot@cea.fr

antenna of the conventional 1D ECE is replaced by an array of antennas equivalent to the layers of radiometers stacked vertically [3]. To guide the emission beam into the plasma to the order of 1-2 cm vertical and radial direction, the ECEI system needs a large access window and large optics to focus the emission beam.

Having a large vacuum window on WEST is very difficult due to the accumulative plasma radiation during long pulse discharge. Moreover, diagnostic ports were almost fully occupied. The only option was the maintenance access port where no diagnostic were implemented because of the tangential view to the plasma and a flange at several meters from the plasma edge. However, these drawbacks shadow the radiation fluxes and thus soften the thermomechanical constraints on the large window.

A preliminary design by Y. Nam et al [11] demonstrated a feasibility of ECEI system using an O-mode with the in-vessel mirrors at midway to refocus and redirect the emission beam (see fig. 1).

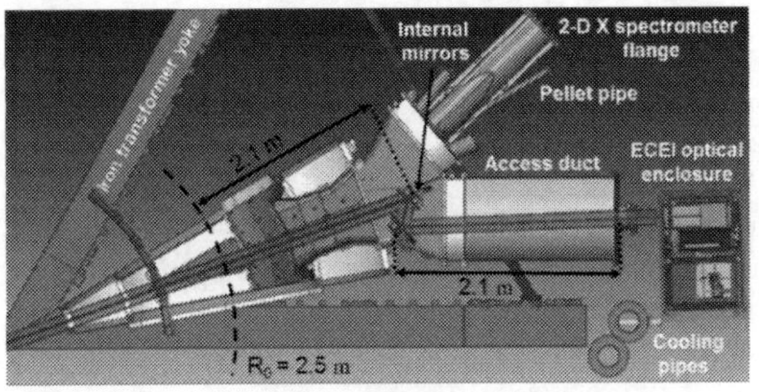

Fig. 1. ECEI beam path (purple colour). Two in-vessel mirrors are installed in the Y duct of the Q5A port shared by the maintenance access (bottom), the pellet injector and the 2D X spectrometer diagnostic (top). The ECEI beam is show in purple with the line of sight in blue.

Implementation of the current design in WEST presents several issues. Most ECEI diagnostics are located close to the window and benefit from the radial view. Here, the optics, usually composite lens, can be installed outside the vacuum vessel. In contrast, the in-vessel mirrors would be the first optical elements of the WEST ECEI system. The tight space for the ex-vessel optics made the integration of the focusing lens and detection array challenging. The available space between the flange and the access lobby was of the order of 1 m^2 (fig 2).

Interference was another concern for the design because the ECEI diagnostic system would block the maintenance access for maintenance. The ex-vessel optics had to be removed to connect the access lobby (few meters long) to the torus during maintenance period. Also the in-vessel mirrors should be designed so that they can be displaced for the passage of maintenance crew. Moreover, both operations should have a marginal impact (few hours delay) on the first entrance inside the vessel.

Repositioning of the optics after each operation should also be guaranteed to avoid re-alignment effort after the maintenance period.

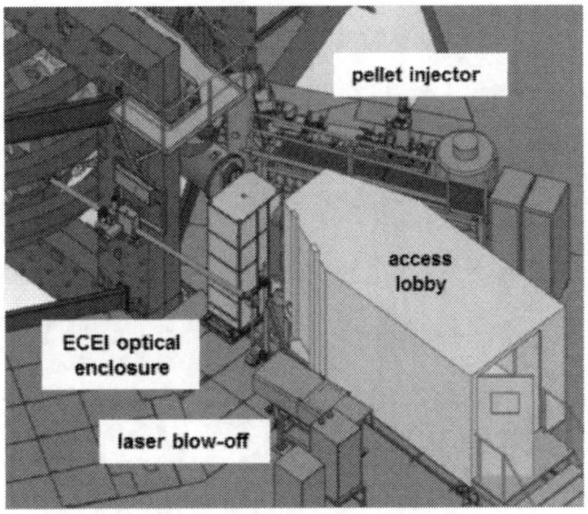

Fig. 2. View of WEST Q5A sector with the main elements showing the ECEI optical enclose, the access lobby and the pellet injector. The laser blow-off system is connected to the neighbouring sector. The X-ray imaging crystal spectrometer (not shown) is now installed on the right side of the pellet injector.

2.2 In-vessel optics

2.2.1 In-vessel-mirrors

The in-vessel mirrors are implemented in the Y duct at about ~2.5 m from the plasma centre (fig 1). The design sought to maximize the mirror height (H~59 cm) to perverse good beam properties at all vertical position. The width was restricted to 20 cm (external width 27 cm) to avoid self-shadowing. The first mirror from the plasma has a horizontal curvature (R=4.0 m) and the second one is vertically curved (R=3.5 m) (see fig 3a).

During the maintenance, the mirrors will remain inside the vacuum vessel to avoid the radiation and safety controls imposed on all elements exiting from the vacuum vessel. The mirrors on the rail track will be moved on the right to provide the path for crew to the vessel (see fig 3b). To facilitate mirror manipulation, the mirrors are suspended on the rails. The rails are similar to those installed in the WEST Articulated Inspection Arm (AIA) [12]. They can withstand machine baking and they have shown reliable operation under vacuum. Although the mirrors will not be manipulated under the vacuum, reliability of manipulation is a key safety issue.

To preserve the alignment after manipulation, the mirrors are mounted on a shaft with two pivot points. The top one frees the rotation and the displacement along the rail (see fig 4a). Reproducibility of the repositioning is guaranteed with a pin. Fine orientation of tuning (±2°) and lateral position (±2 cm) is done using the second pivot point at the centre of the mirror (see fig 4b).

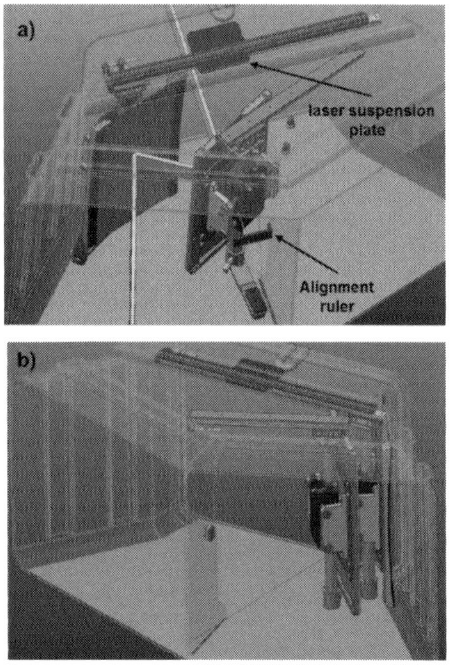

Fig. 3. Mirror positions during plasma operation (a) and during maintenance (b). Elements for alignment are also show (see 3.2)

The mirrors are attached to the middle of the shaft to ensure free and symmetric dilation during long pulses. The plasma radiation can heat the first mirror. Thermomechanical analysis predicts that the temperature at the centre of the first mirror would rise up to 430°C at the end of the reference WEST long pulse (1000s, 9 MW of RF heating). The thermal expansion remains small and the optical quality of the microwave beam is not affected. Water cooling of the first mirror is thus not required.

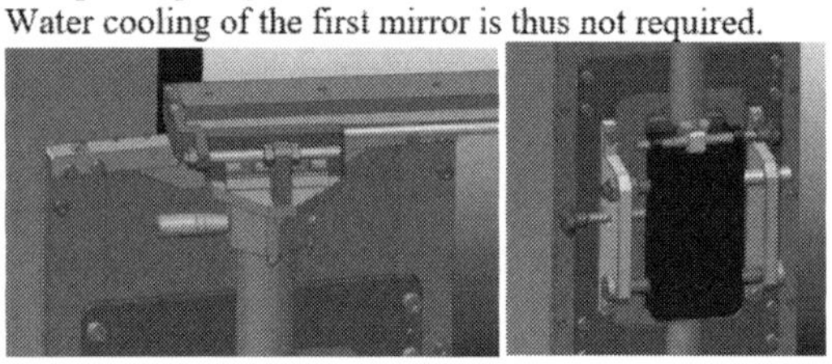

Fig. 4. Details view of the top pivot point (a) and fine tuning mechanisms (b).

2.2.2 Angle of sight

The magnetic ripple field in WEST reaches over 2% at the outer edge of the plasma. To minimize the broadening of the ECE channel due to the toroidal extension of the beam, a sight angle as close as possible to the middle of the port (lowest ripple variation) was desirable. The ECEI sight angle could not be lower than 1° from the middle to avoid shadowing diagnostics installed in the port. As shown in the figure 3, the ripple-induced channel overlapping at 1° is negligible even at the edge. A second sight angle at 1.2° can be used to widen further the pellet camera view.

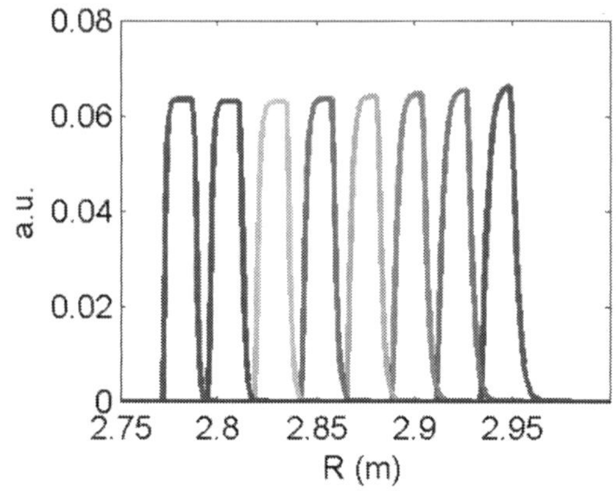

Fig. 5. Channel overlapping at the edge due to the ripple for the reference 1° sight angle. The toroidal waist of the ECEI beam is 6 cm and the channel separation is ΔF=900 MHz.

2.3 External optics

2.3.1 Optical enclosure

The housing for external optics and detection system had to fit in a limited space between the port and access lobby. Benefiting from the height of the WEST equatorial plane (~2.5 m above the ground floor), the imaging optical components are vertically aligned inside the compact (0.8 m x 1.2 m x 2.5m) enclosure.

The lens at the optical enclosure entrance will first refocus the beam to the detectors (see fig 6). The 45° mirror will reflect the beam downward. Dichroic plates will reject image frequencies. The polarization is then rotated to match with the antenna pattern of the detector array. The focal position of the beam in the plasma will be set by the position-controlled lens. The last lens focuses the elongated beam to the detector array made of two rows of 12 mini-lenses and mixers [13]. Extra effort was invested in design to make the imaging and local oscillator as compact as possible. In particular, a couple of aspheric lenses was used to shorten the optical path length of the LO optics to ~0.5 m [14]. The design includes provision for a beam splitter for high field side (HFS) view. Notch filter will be inserted if the ECRH gyrotrons at 118 GHz resume operation.

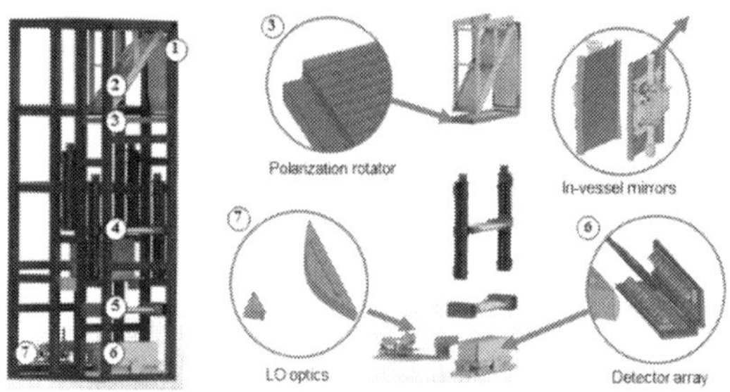

Fig. 6. Optical enclosure: ① entrance lens, ② 45° mirror, ③ polarization rotator, ④ focusing lens (moveable), ⑤ collimation lens, ⑥ detector array box ⑦ LO optics.

https://doi.org/10.1051/epjconf/201920303011

All elements were delivered to CEA in April 2018. Due to its height, the optical enclosure will be lifted by the torus hall crane and it will be placed in a dedicated area inside the torus hall during maintenance period.

2.3.2 Optical enclosure supporting frame

To ensure precise alignment after each reinstallation, the optical tower will be mounted on a supporting frame that enables to adjust, the lateral position, the orientation, the height and tilt of the optical enclosure (see fig 7). The supporting frame will be removed with the optical enclosure during maintenance period.

The optical enclosure will rest on dedicated plates fixed to the metallic beams below the ground floor. Socket ball like feet will guarantee the required reproducible positioning of the supporting frame after each manipulation.

Fig. 7. Optical enclosure supporting frame. Screws and bolt allow tuning the position, orientation, height and tilt of the enclosure with respect to the ECEI beam.

2.4 Control and acquisition cubicles

Three cubicles in the basement below the ground floor house the IF electronics, the control and data acquisition system. The cubicles are assembled side by side. They are shielded against the electromagnetic and radiofrequency noises. The power supply for both channel will be installed in the central cubicle and each side cubicle will house the electronics and acquisition of one channel. 24 low loss 20 m long cables (True Blue 290 model from Teledyne) will connect the optical enclosure to the cubicles.

The diagnostic is operated by an commercial personal computer running LINUX OS connected to the WEST control and data acquisition network. Diverse devices in the cubicles or in the optical enclosure such as the LO frequency synthesizer, the step motors etc. are connected to the PC though a local network. The WEST timing network provides the time-stamping and event triggers for starting the acquisition.

The 192 channels are recorded by 6 acquisition units with 32 channels each at a sampling rate up to 2Ms/s (ACQ132CPCI from D-tAcq), similar to KSTAR acquisition [15]. Up to 10 s at 500 kS/s can be recorded. The acquisition can be split in several triggers for synchronized measurements with other fast acquisition diagnostics during long discharges. To measure the background noise, the first trigger will start before plasma.

2.5. Frequency range and plasma accessibility

The first channel of the diagnostic will operate in the W-band covering the frequency range of 85 – 113 GHz. At the WEST nominal magnetic field (B=3.7 T), the diagnostic will fully cover the low field side (LFS) (see fig. 8). As the core MHD is a primary concern, the ECEI can probe the region beyond the plasma centre (from R=2.3 at B=3.7T).

The diagnostic probes 8 radial positions with 900 MHz separation. The 7.2 GHz IF bandwidth provides a radial coverage around 18 cm in the plasma centre and 24 cm at the edge. The diagnostic has 24 vertical channels to provide 8x24 pixel images of electron temperature fluctuations.

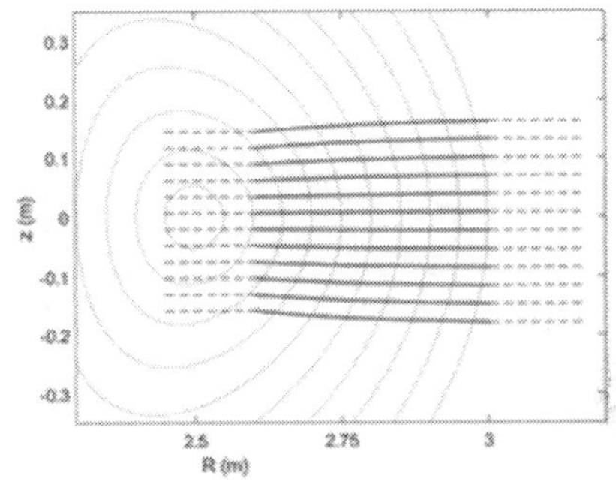

Fig. 8. First (LFS) channel accessibility at WEST nominal magnetic field (3.7 T).

A second channel at higher frequencies (100 – 130 GHz) will be installed later to cover the inner side (R=2 to R=2.6 m) with some overlapping with the LFS channel in the plasma core.

3 Optics properties and alignment

3.1 Imaging optics test

3.1.1 Experimental set-up

The metallic mirrors were sent to UNIST, Korea in 2017 for integration test of the whole imaging optics. The beam path has been modified for laboratory test to facilitate measurements (fig. 9). These modifications have negligible effect on the beam property:

- the vertical 45° mirror in the optical enclosure is not used and the optical elements are horizontally aligned
- the angle of the horizontally curved mirror have been adjusted to make the beam parallel.
- the beam source is placed at the detector box position, then the beam pattern at the plasma position is measured with a detector (reversed beam path).

The tests were performed at 90 GHz using a frequency synthesizer and an active frequency multiplier (sextupler). Corrugated horn antenna and focusing lenses shaped beams with a 4.7° divergence close to the 5° divergence of the ECEI detection lens. More details on the test can be found in reference [16].

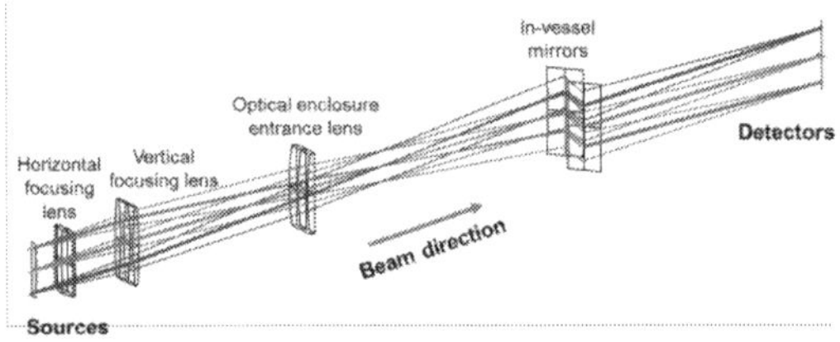

Fig. 9. Setup for integrated optics test for beam properties in laboratory

3.1.2 Focusing capability test

The focal position in the plasma is controlled by the vertical focusing lens: moving the lens toward the top optical enclosure lens moves the focus to the plasma edge. The beam pattern of the center channels (#12, 13) along the plasma position has been measured at the focus plane at the edge, at the mid-radius or at the plasma centre.

The imaging optics can focus on any radial position on LFS by controlling the vertical focusing lens. (2.5 ~ 3.0m). The size at the focal position is around 1.6 cm and divergence is small enough for the beam size to be almost constant over the extension of the probing zone (24 cm at the edge) as shown in figure 10.

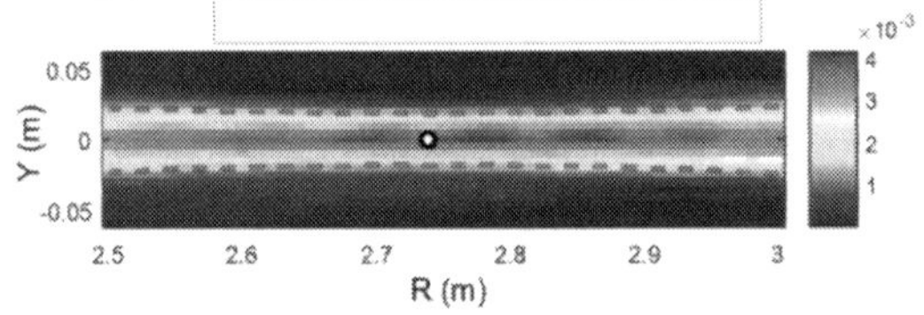

Fig. 10. Beam pattern (vertical direction) along the major radius. The white spot shows the position of the focal plane.

3.1.3 Beam pattern of the different channels

Beam pattern along the vertical direction was also measured for the same focal position: plasma edge, mid-radius and plasma centre. Results were compared to the design values (see fig. 11 for the mid-radius focus). The measured beam patterns are in good agreement with the design values. This demonstrates that the optical components are fabricated within the design precision.

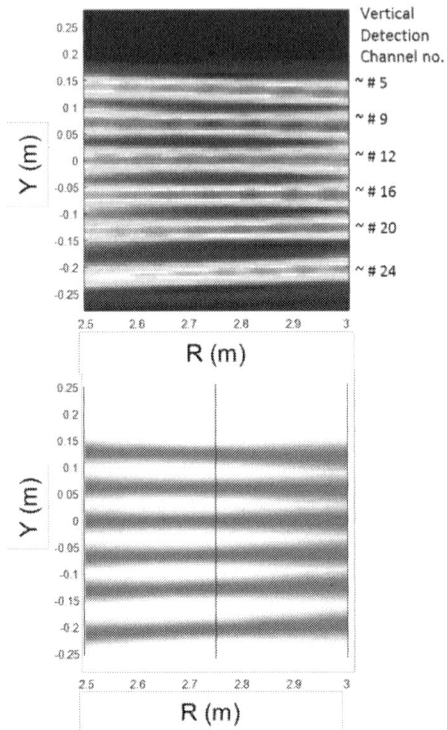

Fig. 11. Beam pattern (vertical direction) along the major radius. The white spot shows the position of the focal plane.

3.2 Alignment procedure

The WEST safety rule requires that all entry to the vacuum vessel must go through the access lobby. Hence, the alignment must be split in two phases: first, the in-vessel mirror alignment performed with the access lobby, then the optical enclosure after removing the access lobby.

The alignment will be performed with a cross laser (deviation less than 2 mm at 10 m) to get both the equatorial and vertical planes. The laser will be suspended on the innermost rail. The first step is to align the laser with reference markers (horizontal and vertical planes) on the inner wall and back of the Y duct (see fig 12). We will move the first mirror (M1) to get the laser beam at the mirror centre. The second mirror (M2) will then be set to its reference position and the angle of the first mirror will be adjusted so the laser beam points to the center of the M2 mirror. The M2 mirror will then oriented to get the laser beam on markers showing the position of the vacuum window middle inside the access lobby. The in-vessel alignment step is completed and the access lobby can be removed.

The height and position of the optical enclosure will be adjusted to get the laser beam at the centre of the 45° mirrors. The lenses have to be removed because they are not transparent to the laser beam. By tuning the optical enclosure orientation and tilt angle, the detector array can be aligned with the 2D laser beam. Each lens starting from the bottom one will be installed and its position will be checked. The alignment ends after the installation of the last lens at the top of the optical enclosure. However, the alignment procedure will not yet be completed. The optical enclose must be removed and the access lobby must be connected. Then we can enter the vacuum vessel to remove the 2D cross laser.

During the first diagnostic alignment, we foresee to check that moving the mirrors back and forth or lifting up down the optical enclosure do not alter the diagnostic alignment.

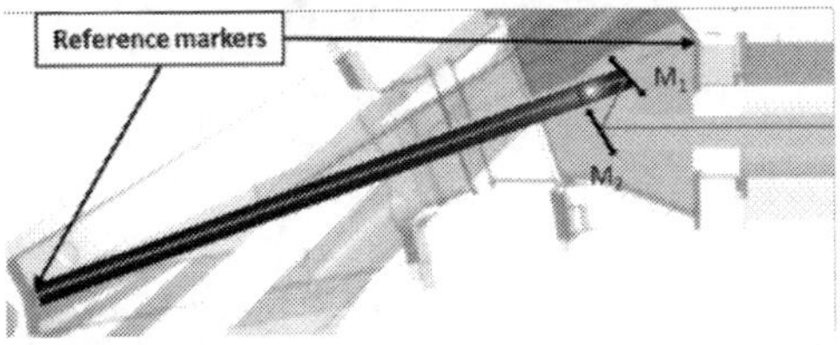

Fig. 12. Alignment set-up with a cross laser suspended on the innermost rail. Alignment order: laser, M1 position M2 position, M1 angle, M2 angle.

4 Conclusion

An optical systems for ECEI system that that can accommodate the hostile environment and constraint of long pulse discharge was designed and built for WEST. Innovate solutions have been developed to resolve the safety issue associated with the in-vessel access and tight space issues. Integration test demonstrated that the performances of the fabricated optics confirmed the precision of the design values. The ECEI system is now in a final assembly phase at Cadarache and will be installed on WEST during the 2018-19 winter shutdown.

The installation of the ECEI diagnostic on the next generation of devices would be much more challenging (thermomechanical stresses, vacuum window and radiation issues). However, the WEST ECEI demonstrates that some of these issues can be overcome in order to obtain the unrivalled value of 2-D images of temperature fluctuations for MHD and turbulence studies.

This diagnostic development was supported by a grant from the Région Sud Provence-Alpes-Côte d'Azur of France under Contract DEB 15-1376 and by the NRF of Korea under Contract Nos. NRF-2014M1A7A1A03029881 and NRF-2014M1A7A1A03029865.

References

1. J. Bucalossi, A. Argouarch, V. Basiuk, O. Baulaigue, P. Bayetti, M. Bécoulet, B. Bertrand, S. Brémond, P. Cara, M. Chantant, Y. Corre, X. Courtois, L. Doceul, A. Ekedahl, F. Faisse, M. Firdaouss, J. Garcia, L. Gargiulo, C. Gil, C. Grisolia, J. Gunn, S. Hacquin, P. Hertout, G. Huysmans, F. Imbeaux, G. Jiolat, M. Joanny, L. Jourd'heuil, M. Jouve, A. Kukushkin, M. Lipa, S. Lisgo, T. Loarer, P. Maget, R. Magne, Y. Marandet, A. Martinez, D. Mazon, O. Meyer, M. Missirlian, P. Monier-Garbet, P. Moreau, E. Nardon, S. Panayotis, B. Pégourié, R. A. Pitts, C. Portafaix, M. Richou, R. Sabot, A. Saille, F. Saint-Laurent, F. Samaille, A. Simonin, and E. Tsitrone, Fusion Engineering and Design **86**, 684 (2011).

2. R. P. Hsia, W. R. Geck, S. Cheng, W. -M. Zhang, C. W. Domier, and N. C. Luhmann, Review of Scientific Instruments **66**, 834 (1995).

3. H. Park, C. C. Chang, B. H. Deng, C. W. Domier, A. J. H. Donné, K. Kawahata, C. Liang, X. P. Liang, H. J. Lu, N. C. Luhmann, A. Mase, H. Matsuura, E. Mazzucato, A. Miura, K. Mizuno, T. Munsat, Y. Nagayama, M. J. van de Pol, J. Wang, Z. G. Xia, and W.-K. Zhang, Review of Scientific Instruments **74**, 4239 (2003).

4. TEXTOR team, H. K. Park, N. C. Luhmann, A. J. H. Donné, I. G. J. Classen, C. W. Domier, E. Mazzucato, T. Munsat, M. J. van de Pol, and Z. Xia, Phys. Rev. Lett. **96**, 195003 (2006).

5. TEXTOR Team, H. K. Park, A. J. H. Donné, N. C. Luhmann, I. G. J. Classen, C. W. Domier, E. Mazzucato, T. Munsat, M. J. van de Pol, and Z. Xia, Phys. Rev. Lett. **96**, 195004 (2006).

6. G. S. Yun, W. Lee, M. J. Choi, J. Lee, H. K. Park, B. Tobias, C. W. Domier, N. C. Luhmann, A. J. H. Donné, J. H. Lee, and KSTAR Team, Physical Review Letters **107**, (2011).

7. B. J. Tobias, I. G. J. Classen, C. W. Domier, W. W. Heidbrink, N. C. Luhmann, R. Nazikian, H. K. Park, D. A. Spong, and M. A. Van Zeeland, Physical Review Letters **106**, (2011).

8. R. Sabot, P. Hennequin, and L. Colas, Fusion Science and Technology **56**, 1253 (2009).

9. R. Sabot, D. Elbèze, W. Lee, Y. Nam, H. Park, J. Shen, G. Yun, M. Choi, J.-C. Giacalone, T. Nicolas, C. Bottereau, F. Clairet, P. Lotte, and D. Molina, Comptes Rendus Physique **17**, 1018 (2016).

10. I. H. Hutchinson, Cambridge Core (2002).

11. Y. Nam, W. Lee, G. S. Yun, H. K. Park, D. Elbeze, J. L. Segui, R. Sabot, and J. P. Chenevoix, Review of Scientific Instruments **83**, 10E318 (2012).

12. J.-C. Hatchressian, V. Bruno, L. Gargiulo, D. Keller, Y. Perrot, P. Bayetti, J.-J. Cordier, J.-P. Friconneau, J. D. Palmer, and F. Samaille, J. Phys.: Conf. Ser. **100**, 062031 (2008).

13. B. Tobias, X. Kong, T. Liang, A. Spear, C. W. Domier, N. C. Luhmann, I. G. J. Classen, J. E. Boom, M. J. van de Pol, R. Jaspers, A. J. H. Donné, H. K. Park, and T. Munsat, Review of Scientific Instruments **80**, 093502 (2009).

14. Y. B. Nam, D. J. Lee, J. Lee, C. Kim, G. S. Yun, W. Lee, and H. K. Park, Review of Scientific Instruments **87**, 11E130 (2016).

15. J. B. Kim, W. Lee, G. S. Yun, H. K. Park, C. W. Domier, and N. C. Luhmann, Review of Scientific Instruments **81**, 10D931 (2010).

16. Y. Nam, in *45th European Physical Society Conference on Plasma Physics* (Prague, 2018).

Developments for collective Thomson scattering equipment with a sub-THz gyrotron in LHD

Teruo Saito[1,*], *Yoshinori* Tatematsu[1], *Yuusuke* Yamaguchi[1], *Masafumi* Fukunari[1], *Takumi* Hirobe[1], *Ryushi* Shinbayashi[1], *Shunsuke* Tanaka[1], *Kunizo* Ohkubo[1], *Shin* Kubo[2], *Takashi* Shimozuma[2], *Kenji* Tanaka[2] and *Masaki* Nishiura[3]

[1]Research Center for Development of Far-Infrared Region, University of Fukui, 3-9-1 Bunkyo, Fukui city, 910-8507, Japan
[2]National Institute for Fusion Science, National Institute of Natural Sciences, 322-6 Oroshi cho, Toki city, 509-5292, Japan
[3]Graduate School of Frontier Sciences, The University of Tokyo, 5-1-5 Kashiwanoha, Kashiwa city, 277-8561, Japan

Abstract. Plan of collective Thomson scattering (CTS) experiment for the Large Helical Device (LHD) in NIFS with a 303 GHz gyrotron is under way. Use of a sub-THz gyrotron expands the CTS-applicable region of plasma parameters. In LHD, sub-THz CTS can be applied to the high density operation region, plasmas with impurity hole, etc. Moreover, sub-THz CTS is expected to be free from ECE noise. Its "collective" use with 77 GHz and 154 GHz CTS will compose a powerful diagnostic system. A high power sub-THz gyrotron with a frequency of 303 GHz has been developed. Its maximum power is 320 kW. It oscillates in pulse mode and the maximum pulse width is around 100 μs, which is sufficient for use in CTS experiments. A whispering gallery mode $TE_{22,2}$ was adopted for this gyrotron to avoid mode competition. Careful frequency measurement has proved purely single mode oscillation of the $TE_{22,2}$ mode including turn-on and turn-off phases of the oscillation pulse. This is consistent with mode competition calculations taking account of a finite voltage rise time. A low loss transmission line is necessary for CTS. We have two possibilities. One is a new line with 1.25 inch corrugated waveguides that are optimized for the 300 GHz band. Transmission test with the 303 GHz gyrotron has been carried out and a sufficiently low loss coefficient has been confirmed. The other is to use an existing line with 3.5 inch corrugated waveguides for lower frequencies such as 77 GHz and 154 GHz. Transmission test has been carried out with the 303 GHz gyrotron and a sufficiently low loss coefficient has been confirmed also for 3.5 inch corrugated waveguides. An existing line with 3.5 inch corrugated waveguides will be used in the initial phase of 303 GHz CTS experiment.

1 Introduction

CTS diagnostics using gyrotrons for electron heating around 100 GHz have been developed [1, 2]. Sub-THz waves are promising for CTS because they do not suffer from refraction and absorption in plasmas and noise due to strong electron cyclotron emissions (ECEs). The power source for CTS should be more than 100 kW to realize a suitable signal-to-noise (SN) ratio [3]. Then, a high power sub-THz gyrotron has been developed for use in CTS diagnostics in LHD.

A second harmonic gyrotron was initially developed. It realized single mode oscillation approaching 100 kW at 389 GHz. However, a competing fundamental harmonic mode prevented further increase in power [4]. We subsequently developed a fundamental harmonic prototype gyrotron with a whispering gallery mode (WGM) and used it to verify the design concept for stable single-mode oscillation [5]. The same design concept to use WGM mode was adopted in the design of a practical gyrotron and fabricated for use in CTS diagnostics in the LHD.

The designed operating frequency of the practical gyrotron is 303.3 GHz, which corresponds to the minimum for the ECE for standard LHD operation [6]. The cutoff density corresponding to 303 GHz is about 10^{21} m^{-3}. This gyrotron can be used in the high density operation mode of LHD.

The oscillation mode is the $TE_{22,2}$ mode, which is a moderately high order WGM similar to the $TE_{14,2}$ mode used in the prototype gyrotron. WGM mode is suitable to satisfy requirements for mode competition avoidance and low cavity surface ohmic loss simultaneously [5]. WGM modes were used in the initial stage of the development of gyrotrons for fusion plasma heating and stable oscillations without mode competition were shown [7, 8]. Volume modes were then used to reduce the ohmic loss for the continuous wave (cw) 1 MW class gyrotrons required for the International Thermonuclear Experimental Reactor (ITER) [9]. Because cw oscillation is not required in CTS doagnostics, the 303 GHz gyrotron is operated in pulse mode. WGM modes are promising for complete suppression of spurious modes, including the turn-on and turn-off phases of each oscillation pulse.

* Corresponding author: saito_2729@fir.u-fukui.ac.jp

A low loss transmission line is necessary for realization of CTS diagnostics because the gyrotron is located far away from LHD. The transmission line will be tens of meters long. A corrugated waveguide (CW) system is a prospective candidate of a low loss transmission line. Two types of CW are now considered. One is a new line with 1.25 inch CW optimized for 300 GHz band. The other is a 3.5 inch CW line already installed in LHD for electron heating with lower frequency gyrotrons.

We have first evaluated the transmission efficiency of the 1.25 inch CW with the 303 GHz gyrotron power. Five 2.085 m long waveguides were prepared for this test [10, 11]. An attenuation coefficient of 0.02 – 0.04 dB/m sufficiently small for use in CTS was obtained. However, theoreticaliy calculated attenuation coefficients are smaller than 0.001 dB/m [12, 13]. Experimental errors in the measurement of the very small attenuation might cause the discrepancy. In addition, absorption by water vapor of the order of 0.01 dB/m should also be considered.

Since installtion of a new line with the 1.25 inch CW is difficult for the time being, an exsiting line with 3.5 inch CW will be used in the initial stage of 303 GHz CTS. Then, we have measured the transmisson efficiency with spare waveguides.

2 Design, fabrication and test of 303 GHz gyrotron

It was necessary to use WGM higher than that of the prototype gyrotron to extend the pulse width and the duty ratio up to 10%. Preliminary design consideration was carried out to find candidate modes of the practical gyrotron. Figure 1 plots the stationary oscillation power obtained with a gyrotron design code [14] including the competition with neighboring modes. This figure indicates that WGM modes are not always appropriate for a high power oscillation. The $TE_{19,2}$ and $TE_{20,2}$ modes subject to competition with neighboring modes. The $TE_{22,2}$ mode was finally adopted for the practical gyrotron [15]. The same mode was also employed in [16 - 18]. Oscillation characteristics of the $TE_{22,2}$ mode were examined in detail with the gyrotron design code and stable single mode oscillation was expected. A power of more than 300 kW was predicted for a beam voltage V_k of 65 kV, a beam current I_B of 15 A, and a velocity pitch factor α of 1.2. Ohmic loss on the cavity surface is safely smaller than 2 kW/cm^2 up to the duty ration of 10%.

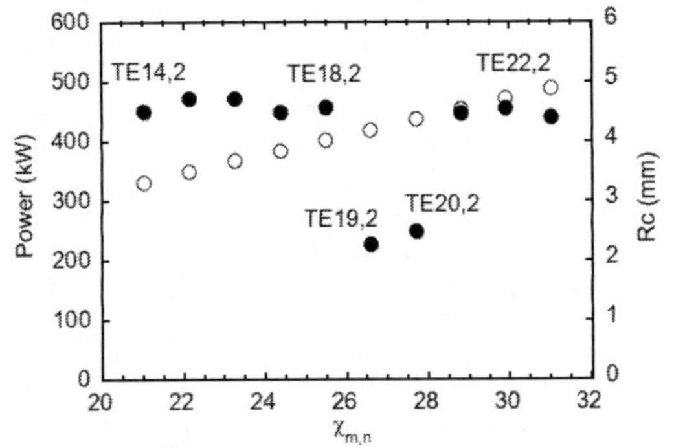

Fig. 1. Power calculation of candidate modes (solid circles). The beam voltage and beam current were set at 65 kV and 20 A, respectively. Possible competing modes are included in the calculation. The cavity radius corresponding to the oscillation frequency of 303.3 GHz is also shown by open circles.

An electron gun was newly designed and optimized for the $TE_{22,2}$ mode by using the same design principle as reported in Ref. [19]. The design predicted electron beams of high quality with the expected α values and sufficiently small velocity spreads up to $I_B = 20$ A [20].

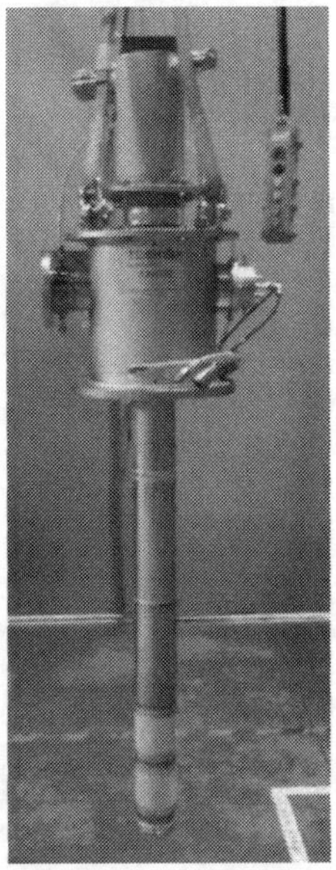

Fig. 2. Picture of the fabricated practical gyrotron. A Gaussian beam is radiated through the vacuum window at the left side.

Figure 2 is a picture of the fabricated gyrotron. This gyrotron is mounted on a liquid He-free 12 T SC magnet. The diameter of the room temperature bore of SC is 100 mm. An internal mode converter composed of a helical-cut Vlasov type launcher and four mirrors is installed into a rather narrow room. The vacuum window is made of a c-axis cut crystal sapphire disk. Its thickness is optimized for 303.3 GHz.

Oscillation tests have been carried out and a maximum power higher than 320 kW for $V_k = 65$ kV and $I_B = 15$ A was obtained [20, 21]. The oscillation efficiency was higher than 30%. Figure 3 shows the experimentally obtained power and efficiency as functions of the beam current. The calculated values are also plotted. In the calculation, α values were evaluated for each experimental parameters with the E-GUN code. The experimental results are essentially consistent with calculations with α ranging from 1.1 to 1.3, although the experimentally obtained powers are slightly lower than the calculated values. In this calculation instant voltage rise is assumed and interaction between the operating

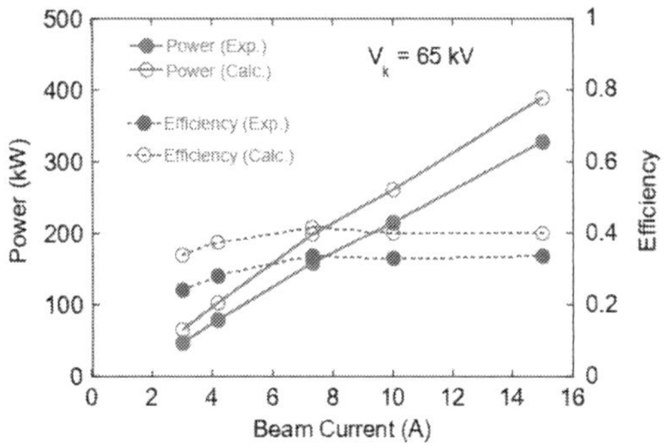

Fig. 3. Experimental and calculated values of the power and the efficiency are plotted. The TE21,2 mode is included in the calculation of the power and efficiency.

$TE_{22,2}$ mode and the $TE_{21,2}$ mode is included. Although many other modes exist in between the two modes, interaction between the two modes is particularly strong.

The radiation pattern was measured using an infrared camera as the increasing temperature profile on a polyvinyl chloride (PVC) plate located in front of the vacuum window. A Gaussian like radiation pattern similar to that expected from the mode converter design calculation was obtained. The radiation pattern was somewhat elliptic and a small fraction of higher order modes was observed [20].

Measurement of the oscillation frequency using a Fabry-Perot interferometer verified single-mode oscillation of the $TE_{22,2}$ mode. The oscillation frequency was measured more accurately using a heterodyne receiver system. The measured frequency was 303.3 GHz when V_k = 62 kV and the magnetic field at the cavity B = 11.59 T. This frequency is equal to the designed $TE_{22,2}$ mode frequency to within an accuracy of the order of 10 MHz.

Frequency control is very important for application to CTS because a narrow band notch filter is used. A notch filter is necessary to protect the highly sensitive receiver system from the gyrotron-frequency wave stemming from scattering off the vacuum vessel. The oscillation

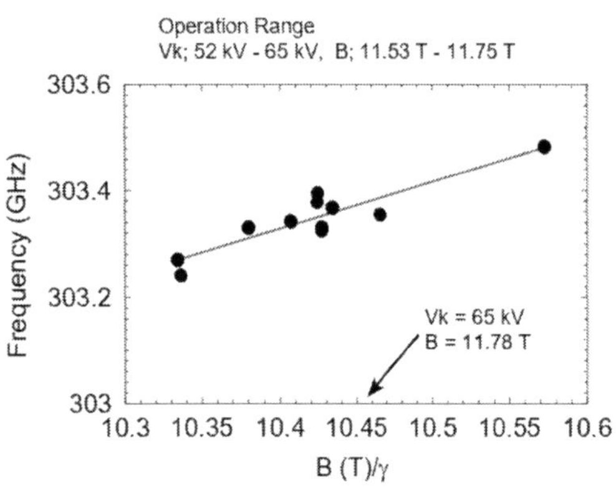

Fig. 4. Frequency variation as a function of a ratio of the magnetic fireld strength at the cavity B and the relativistic factor of the electrn beam γ.

frequency slightly varies with the operation parameters such as V_k and B. We can control the oscillation frequency at around 303.3 GHz by appropriately changing V_k and B. Figure 4 indicates controllability of the frequency. The frequency variation with B and V_k was less than ±100 MHz. The center frequency of the notch filter is 303.3 GHz and the full notch width is 0.5 GHz. The observed frequency variation around 303.3 GHz is safely smaller than the rejection width. The expected frequency width of the scattered signal is much wider than the notch width [3].

The pulse width was extended up to 100 μs. Relevant waveforms are plotted in Fig. 5. The output power is approximately 200 kW for this case. The oscillation signal was measured using a pyroelectric detector in the time integration mode. The linear increase in this signal

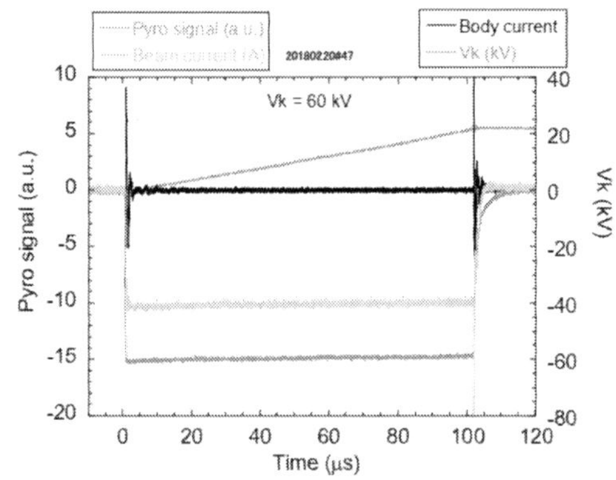

Fig. 5. Waveforms of an 100 μs oscillation. The beam voltage and beam current were set at 60 kV and 10 A, respectively.

indicates very stable power through the pulse width. This pulse width is sufficient for use in CTS diagnostics.

The frequency spectrum measured with the heterodyne receiver system was very narrow and stable across the pulse width. The frequency variation in one pulse is smaller than 10 MHz and the pulse to pulse difference is less than 20 MHz [15]. Parasitic modes were searched for over a wide frequency band, ranging from 290 GHz to 312 GHz, using the same heterodyne receiver system. No oscillation other than the $TE_{22,2}$ mode was observed throughout the pulse width, including during the turn-on and the turn-off phases. The dynamic range of measurement of about 40 dB is larger than the power range of the expected scattered signal [3].

The voltage of the modulation anode of the triode electron gun was set with a simple resistive divider. Stable and no-parasitic-mode oscillation without sophisticated voltage control of the modulation anode is likely due to a very fast rise time of the beam voltage. The beam voltage attains to 65 kV in about 0.25 μs. A fast semiconductor switch realizes this short rise time. Suppression of parasitic modes for a voltage rise time shorter than 1 μs was predicted [22]. A theoretical study just corresponding to the present gyrotron also predicts single mode oscillation of the $TE_{22,2}$ mode [23]. The frequency mismatch of the $TE_{22,2}$ mode is zero or even negative when the beam voltage starts to increase.

Analysis with a non-stationary axial field structure shows that a backward wave of the same transverse mode as the design mode is first soft-self excited and gradually transits to the high power gyrotron mode. This was predicted in [24].

3 Transmission test of 3.5 inch corrugated waveguides [25]

For CTS in LHD, it is necessary to transmit the gyrotron power to LHD with a low loss transmission line composed of corrugated waveguides over about 60 m. At the first stage of the 300 GHz CTS experiment, an existing 3.5 inch CW line is used for power transmission. Since this waveguide is not optimized for 300 GHz but for lower frequency gyrotrons for electron heating. Thus, transmission test is necessary. The corrugation parameters of the 3.5 inch CW are shown in Table I. The Bragg frequency of this waveguide is 187 GHz.

Table 1. Corrugation parameters of 3.5 in waveguide.

Waveguide diameter : a	88.9 mm
Corrugation depth: d	0.60 mm
Corrugation width: w	0.60 mm
Corrugation Pitch: p	0.80 mm

The output beam from the gyrotron is not axisymmetric and expands along propagation. Using a Matching Optics Unit (MOU), the output beam from the gyrotron was shaped into a Gaussian beam that couples to the HE_{11} mode of the waveguide with a high coupling coefficient [26]. Figure 6 indicates setup of MOU. It is composed of two elliptic mirrors to tailor the beam radiated from the gyrotron to a Gaussian beam with the waist at the waveguide entrance position and two plane mirrors to adjust the beam direction.

The beam shape near the design waist position was

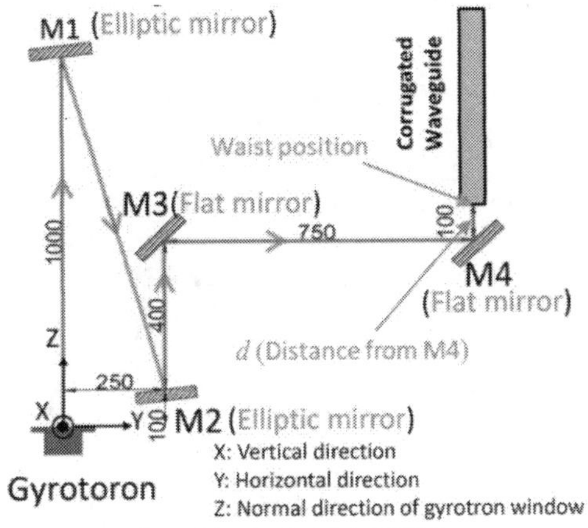

Fig. 6. Setup of Matching Optics Unit (MOU).

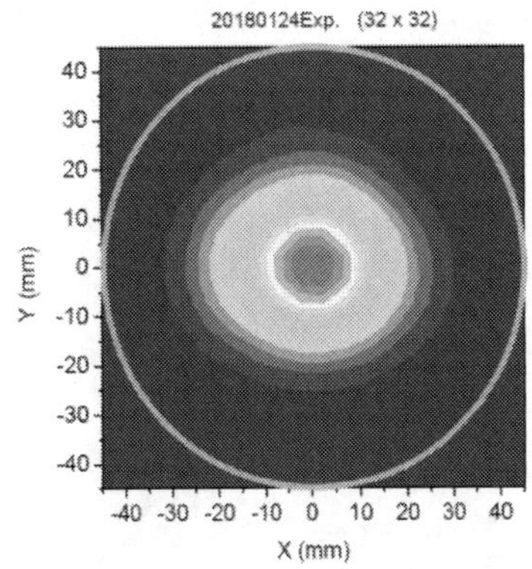

Fig. 7. Beam shape at the entrance position of 3.5 corrugated waveguide.

slightly elliptical. This is possibly because the gyrotron beam contains higher order modes. Figure 7 represents the beam shape at the waveguide entrance position measured as the distribution of temperature increase on a PVC plate. The green circle represents the waveguide cross section.

Figure 8 is a picture of the transmission test. Ten one-meter long waveguides were connected straight on a tight frame. The relative transmitted power was

Fig. 8. Picture of transmission test.

evaluated at every 2 m distance by integrating the distribution of the temperature increase on a PVC plate over the waveguide cross section. The PVC plate was located just at the out pout position for each distance measurement.

Figure 9 plots the relative power as a function of the transmission distance along 3.5 inch CW. Although measurements includes some error, attenuation is very low probably smaller than 0.01 dB/m. Correspondingly, the transmission loss is smaller than 0.6 dB for 60 m.

Figure 9 represents the total power transmitted to each measurement position. It includes attenuation of higher order modes excited in waveguides. In fact, owing to excitation of higher order modes, very complicated structures were observed at each measurement point.

Then, nonlinear mode content analysis was carried out with the same method as shown in [27]. In this

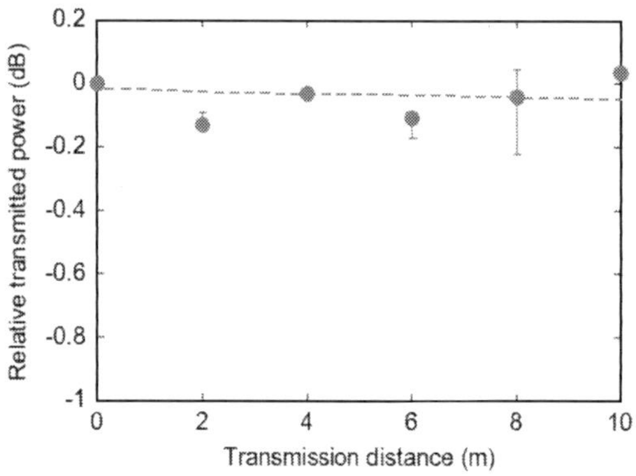

Fig. 9. Total power transmitted to each measurement position.

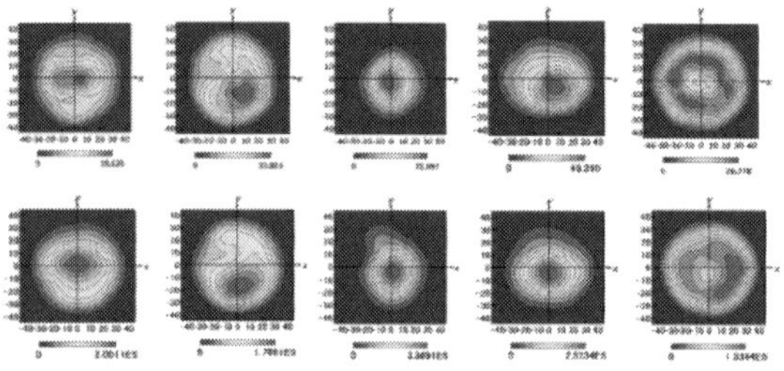

Fig. 10. Comparison between observed pattern of power distribution at each measured position (upper row) and the composed power distribution with calculated power rfactions and phases (lower row).

analysis, 19 higher order modes were taken into account. Figure 10 represents the result of the analysis. Calculated power distributions as the composition of the HE_{11} mode and higher order modes well reproduces the measured power distribution. This composition takes account of relative phases between modes. The power fraction of HE_{11} mode was about 82%. Main higher order modes were co polarizing HE_{21} mode and HE_{12} mode with power fractions around 4%.

The transmission test shows that the existing 3.5 inch waveguide line can be used for 300 GHz CTS experiment. In the CTS experiment, more axisymmetric beam will be coupled into the waveguide.

4 Summary

Sub THz CTS experiment is under preparation with a pulse gyrotron in the 300 GHz band with collaboration between FIR UF and NIFS.

The design concept of the practical gyrotron is the same as that used in the prototype gyrotron. Higher mode TE22,2 was adopted to reduce Ohmic loss on the cavity surface. Stable single mode oscillation has been confirmed and the oscillation power and frequency are very stable over the pulse width. No parasitic mode was observed during the turn-on and turn-off phases. This is probably due to a finite voltage rise time. This gyrotron can be used in CTS diagnostics in LHD.

For power Transmission, an existing transmission line with 3.5 inch corrugated waveguides is used at the first stage. Although this waveguide is not optimized for 300 GHz band, transmission test with the gyrotron have shown very low transmission loss.

Sub THz CTS experiment will start soon. Its "collective" use with 77 GHz and 154 GHz CTS will compose a powerful diagnostic system in LHD.

References

1. Bindslev et al., Phys. Rev. Lett. **97**, 205005 (2006).
2. M. Nishiura et al., Nucl. Fusion **54**, 023006 (2014).
3. T. Saito et al., Journal of Phys.: Conference Series **227**, 012013 (2010).
4. T. Saito et al., Phys. Plasmas **19**, 063106 (2012).
5. Y. Yamaguchi et al., Nucl. Fusion, **55**, 013002 (2015).
6. T. Notake et al., Rev. Sci. Instrum. **79**, 10E732 (2008).
7. K. E. Kreischer et al., Phys. Fluids **B2**, 640 (1990).
8. A. Kasugai et al., Fusion Eng. Design **26**, 281 (1995).
9. K. Sakamoto et al., Nucl. Fusion **49**, 095019 (2009).
10. T. Saito et al., IRMMW-THz 2016, H5P.16.03, Copenhagen, Sep. 25-30 (2016).
11. T. Hirobe et al., Plasma Conference 2017, 21P-53, Himeji, Nov. 20-24 (2017).
12. J. Doane, *Infrared and Millimeter Waves*, Ed. K. J. Button, Academic Press, vol.13, pp. 123-170, 1985.
13. E. A. Nanni et al., J. Infrared Milli Terahz Waves **33**, 695 (2012).
14. Y. Tatematsu et al. J Infrared Milli Terahz Waves **33**, 292 (2012).
15. T. Saito et al., IRMMW-THz 2017, RA2.1, Cancun, Aug 27 – Sept 1 (2017).
16. K. Felch et al., Conf. Digest of 17th Int. Conf. Infrared and Millimeter Waves, p.184, Pasadena, Dec. 14-17 (1992).
17. K. Sakamoto et al., ibid, p.188, Pasadena, Dec. 14-17 (1992).
18. K. Sakamoto et al., Phys. Rev. Lett. **73**, 3532 (1994).
19. Y. Yamaguchi et al., Phys. Plasmas **19**, 113113 (2012).
20. Y. Yamaguchi et al., J. Instrumentation **10**, C10002 (2015).
21. T. Saito et al., Plasma Fusion Res. **12**, 1206013 (2017).
22. G. S. Nusinovich et al., Phys. Plasmas **15**, 103101 (2008).
23. O. Dumbrajs et al., Phys. Plasmas **23**, 023106 (2016).
24. G. S. Nusinovich et al., Phys. Rev. Lett. **96**, 125101 (2006).
25. R. Shinbayashi et al., Plasma Conference 2017, 21P-55, Himeji, Nov. 20-24 (2017).
26. J. P. Crenn, Appl. Optics **23**, 3428 (1984).
27. K. Ohkubo et al., J Infrared Milli Terahz Waves **38**, 853 (2017).

Development of a correlation ECE radiometer for electron temperature fluctuation measurements in Heliotron J

Gavin M. Weir[1], *Kazunobu* Nagasaki[2,*], *Jinxiang* Zhu[3], *Maoyuan* Luo[4], *Hiroyuki* Okada[2], *Takashi* Minami[2], *Shinichiro* Kado[2], *Shinji* Kobayashi[2], *Satoshi* Yamamoto[2], *Shinsuke* Ohshima[2], *Shigeru* Konoshima[2], *Yuji* Nakamura[4], *Akihiro* Ishizawa[4], *Xiang-xun* Lu[4], *Linge* Zang[5], *Nikolai* Marushchenko[1] and *Yasuo* Yoshimura[6], and *Tohru* Mizuuchi[2]

[1]Max-Planck Institute for Plasma Physics, Greifswald, Germany
[2]Institute of Advanced Energy, Kyoto University, Gokasho, Uji, Kyoto, Japan
[3]Zhejiang University, Hangzhou, China
[4]Graduate School of Energy Science, Kyoto University, Uji, Kyoto, Japan
[5]Southwestern Institute of Physics, Chengdu, China
[6]National Institute for Fusion Science, Toki, Japan

Abstract. A radial correlation ECE radiometer diagnostic has been developed for electron temperature fluctuation measurements in the helical-axis heliotron device, Heliotron J. The radiometer consists of two heterodyne detection systems. One system scans the frequency of a local oscillator from 52 to 64 GHz with a single intermediate frequency filter, and the second system has a fixed frequency, 56 GHz local oscillator with four intermediate frequency filters. This frequency range covers measurement positions spanning from the plasma core to the half radius. Laboratory tests indicate that each system has narrow intermediate frequency bandwidth and high-sensitivity over a large dynamic range. During plasma experiments with NBI heating, radiation temperature fluctuation measured by the CECE radiometer decrease with increasing ECCD commensurate with previous measurements of energetic particle driven modes on Heliotron J.

1 Introduction

Turbulence in density and temperature is believed to govern energy and particle confinement in magnetically confined fusion plasmas. Several theories suggest that turbulent electrostatic and/or electromagnetic fluctuations are responsible for this anomalous transport. Drift wave turbulence has been experimentally and theoretically studied, and Trapped Electron Modes (TEMs) have been found to lead to experimentally relevant electron heat transport [1-3] in tokamak and helical devices.

While plasma density fluctuations may be directly measured in the confinement region with microwave or optical diagnostics, electron temperature fluctuations are measured indirectly through decorrelation of Electron Cyclotron Emission (ECE) [4]. This is necessary because thermal noise dominates electron temperature fluctuations in the ECE spectrum and standard ECE radiometry techniques are not suitable for measuring high-frequency, small amplitude temperature fluctuations.

Progress has been made in extracting electron temperature fluctuations from the dominant thermal noise in tokamaks and helical devices. The first proof-of-principle measurements of core electron temperature fluctuations were reported from the W7-AS stellarator [5] and the TEXT-U tokamak [6-9], and electron temperature fluctuations might contribute significantly to the anomalous electron heat flux when they are correlated with potential fluctuations [10]. Electron temperature and plasma density fluctuations have been measured on the DIII-D and ASDEX Upgrade tokamaks, and the cross phase between the two has been found to be a sensitive indicator of drift wave driven turbulence when compared to nonlinear gyrokinetic simulations of ITG/TEM driven turbulence [11-14]. Correlation ECE radiometry is reviewed in Ref. 4, and four decorrelation schemes are suggested:

1. spatial decorrelation through multiple sightlines of the same plasma volume
2. spectral decorrelation of separate frequency bands along a single sightline
3. spatial decorrelation of disjoint measurement volumes
4. autocorrelation of radiation decorrelated in time

Method (2) is the most commonly used decorrelation scheme; however, method (1) was used on W7-AS stellarator, and (4) was used on the TEXT-U tokamak. A poloidal correlation ECE radiometer using method (3) has been developed for the W7-X stellarator [15].

Recently we have designed and installed a new correlation ECE radiometer system to measure core electron temperature fluctuations on the medium-sized stellarator/heliotron (S/H) device, Heliotron J. The CECE system consists of two separate heterodyne detection systems, and either spectral decorrelation along the same sightline or spatial decorrelation through two toroidally

* Corresponding author: nagasaki@iae.kyoto-u.ac.jp

separated measurement volumes may be used to decorrelate thermal noise in the ECE spectrum.

In this paper, we will concentrate on the spectral decorrelation method and report on the development of a correlation ECE (CECE) radiometer for temperature fluctuations and recent plasma experimental results in the helical-heliotron device, Heliotron J. The organization is as follows: The Heliotron J device and CECE radiometer are described in Sec. 2. Laboratory tests are presented in Sec. 3, and CECE measurements during plasma experiments with ECRH and NBI heating are presented in Sec. 4. A summary of the new CECE diagnostic and its first results is presented in Sec. 5

2 Heliotron J and the CECE diagnostic

2.1. The Heliotron J device

Heliotron J is a medium-sized plasma experimental helical device [16][17]. The average major radius is R=1.2 m, and the average minor radius is a=0.1–0.2 m depending on magnetic configuration. The rotational transform also changes with magnetic configuration, $\iota/2\pi$=0.3–0.8, and the maximum magnetic field strength on the magnetic axis is B=1.5 T. The magnetic coil system is composed of an L = 1/M = 4 helical coil, two types of toroidal coils, and three pairs of vertical coils. Here L and M denote the poloidal and toroidal pitch numbers, respectively. The magnetic configuration of Heliotron J can be changed significantly by varying the current ratios in each coil, allowing us to investigate the properties of global confinement, neoclassical transport, turbulent transport and MHD stability in a wide range of magnetic configurations.

Plasmas are produced and heated by second harmonic X-mode ECH at 70-GHz [18]. The maximum injected EC power is 380 kW, and the focused EC beam radius at $1/e^2$ power is 3 cm at the magnetic axis. This is much smaller than the plasma minor radius, a~17 cm, localizing the power deposition and driven current profiles. A steerable mirror allows the injection angle of the ECRH to varied in the toroidal and poloidal directions, and $N_\|$ sampled by the ECRH beam ranges from -0.05 to 0.6 for positive magnetic field, limited by the chamber port size. Here $N_\|$ is the parallel refractive index, determined by the angle between axial magnetic field and EC wave vector under vacuum condition, and positive magnetic field corresponds to the counter-clockwise direction. The EC launcher is positioned between the "straight" and "corner" sections of the non-axisymmetric vacuum vessel. For the extraordinary wave polarization (X-mode) the absorbed power is greater than 80%, and can be larger than 90 % under adequate conditions ($0.1 < N_\| < 0.6$). In this paper we report results from plasmas with ECRH and NBI heating, where the core electron temperature is measured to be approximately 1 keV and the plasma line-average density is $n_e=1.2\times10^{19}$ m^{-3}.

Figure 1 shows the poloidal cross-section of the Heliotron J plasma at the CECE measurement position. The trajectory of 70 GHz electron cyclotron waves is calculated using the TRAVIS code. The magnetic field of the Heliotron J device has a tokamak-like mod-B contour at the viewing port for the CECE radiometer, and second harmonic X-mode ECE between 52 and 72 GHz radiates from the plasma core to the edge region. The core optical depth is approximately 3 for a plasma with $n_e=1.2\times10^{19}$ m^{-3} and T_e=1.0 keV in Heliotron J, and the average emission from the core is optically thick. The antenna is located at the #11.5 low field side port of the Heliotron J vacuum chamber and is comprised of a pyramidal horn and focusing polytetrafluoroethylene lens with a beamwaist of approximately 2 cm. The detected ECE signal is transmitted to the radiometer with approximately 10 m of oversized rectangular waveguides (X-band). A 70 GHz notch filter (band stop) filter is used to protect the CECE diagnostic by suppressing stray ECRH power.

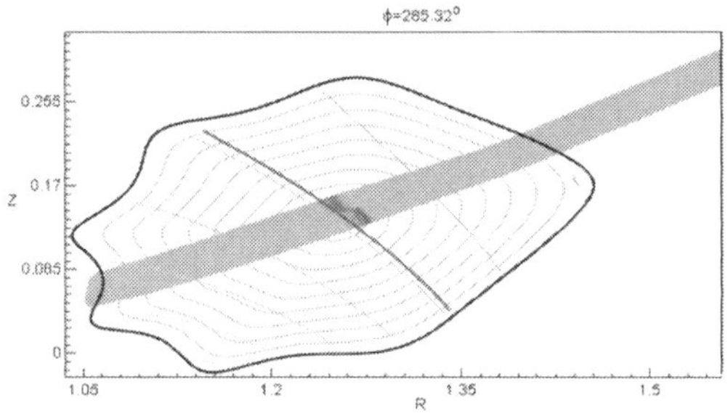

Fig. 1. Poloidal cross-section of Heliotron J at the CECE viewing location. The trajectory of 70 GHz EC waves is calculated using the TRAVIS code.

2.2 Correlation ECE radiometer

A single-sightline correlation electron cyclotron emission (CECE) radiometer has been developed and implemented on Heliotron J for studying bulk electron temperature fluctuations. A conventional multi-channel ECE radiometer is routinely used for electron temperature profile measurements [19], and the new radiometer shares some previously existing components from that system. The new radiometer consists of two detection systems. Figure 2 illustrates a schematic of the CECE radiometer. One detection system (CECE-RF) includes a variable frequency local oscillator (LO) that can be scanned from 52 to 64 GHz with high frequency resolution using a signal generator and a 4-times multiplier. Post-mixing, the down-converted intermediate frequency (IF) signal is bandpass filtered by an 8 GHz intermediate frequency filter with a bandwidth of B_{IF}=0.2 GHz. The filtered IF signal is detected with a crystal diode RF detector, and then amplified with an AC/DC coupling video amplifier. The other system (CECE-IF) has a 56 GHz fixed frequency LO, and the down-converted IF signal is divided into four lines with band pass filters at 4, 8, 12 and 16 GHz. The IF bandwidth of these filters has been measured to be between 0.1 and 0.2 GHz. This frequency range spans measurement positions from the core to the half radius on both sides of the magnetic axis. Post-detection, each channel is amplified by a 10 MHz bandwidth, +20 dB video amplifier and low-pass filtered by 500 KHz anti-aliasing filters. The signal from the

CECE radiometer back-end is then split into DC- and AC-coupled components and digitized with a 1 MHz sampling frequency for simultaneously measuring the mean electron temperature and electron temperature fluctuations respectively. The effective video bandwidth of each CECE channel is B_{vid}=0.5 MHz. Figure 3 shows a photo of the CECE radiometer. The signal from the plasma is split using a -3 dB, V-band directional coupler (WR-15) before being fed to each radiometer system.

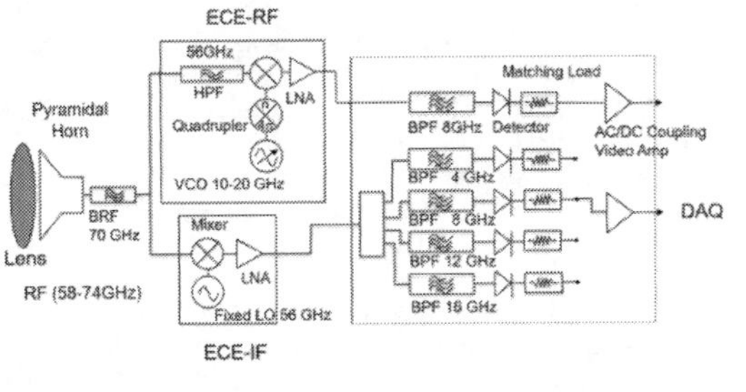

Fig. 2. Schematic view of correlation ECE radiometer in Heliotron J.

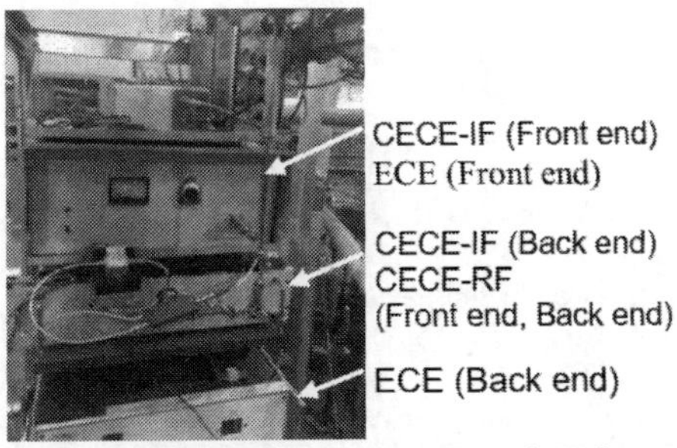

Fig. 3. Photo of CECE radiometer system in Heliotron J.

The sensitivity to electron temperature fluctuations is given by Equation (1),

$$\frac{\langle \tilde{T_e}^2 \rangle}{\langle T_e \rangle^2} \geq \frac{1}{\sqrt{N}} \frac{2B_{sig}}{B_{IF}}, \tag{1}$$

where N=$\sqrt{2B_{sig}\tau_i}$ is the number of unique samples and B_{sig} is the post-detection signal bandwidth. With a post-detection bandwidth limited by the video amplifiers, and an integration time of τ_i =0.1 sec at 1 MHz sampling frequency, the CECE radiometer is sensitive to electron temperature fluctuations above 0.4%. With a 2 cm beamwaist, the CECE diagnostic is sensitive to poloidal wavenumbers less than 1.6 cm^{-1}.

3 Laboratory tests

Pre-installation in the Heliotron J torus hall, the frequency conversion, linearity and bandwidth of the CECE radiometer were tested in the laboratory. Figure 4 shows the signal output from the CECE-RF channel with a 65 GHz Gunn oscillator applied as an input signal. The frequency of the variable local oscillator was scanned to measure the bandwidth of the CECE-RF channel, which is shown in Fig. 4. The 65 GHz input signal is downshifted by 8 GHz to 57.0 GHz, and the measured -3 dB bandwidth of the filter is 0.2 GHz.

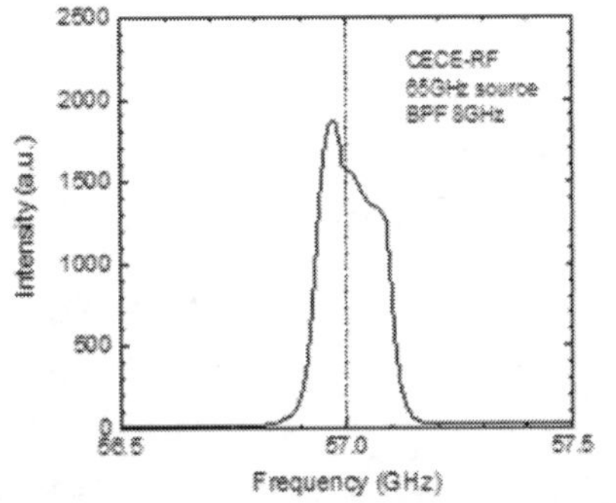

Fig. 4. Frequency conversion and bandwidth of the CECE-RF channel of the CECE radiometer. The LO was scanned to measure the CECE-RF response to a 65 GHz Gunn oscillator.

The intermediate frequency bandwidth and relative sensitivity of the CECE-IF channels was measured by applying an RF source (2-20 GHz) to the IF line of each channel. As shown in Figure 5, the measured bandwidth of each channel is between 0.1 and 0.2 GHz and the sensitivity of the radiometer decreases with higher IF frequency. Post-detection amplification is used to increase the signal measured at higher frequencies.

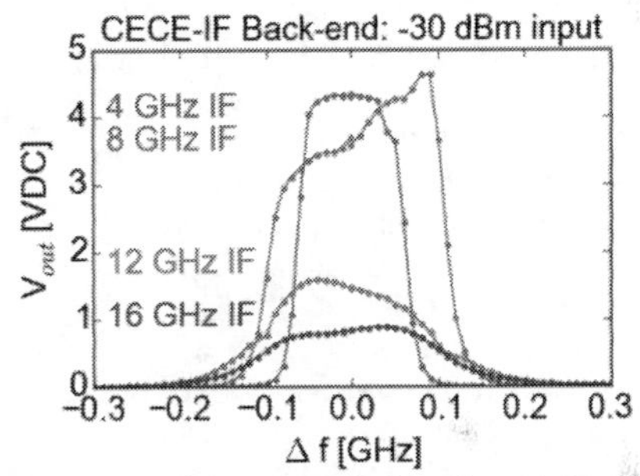

Fig. 5. Response of each CECE-IF channel to -30 dBm input.

Figure 6 show the linearity of the CECE-IF channel output voltage versus input power. The 8 GHz IF channel is used here to demonstrate the linearity of the radiometer, and it is representative of the other channels.

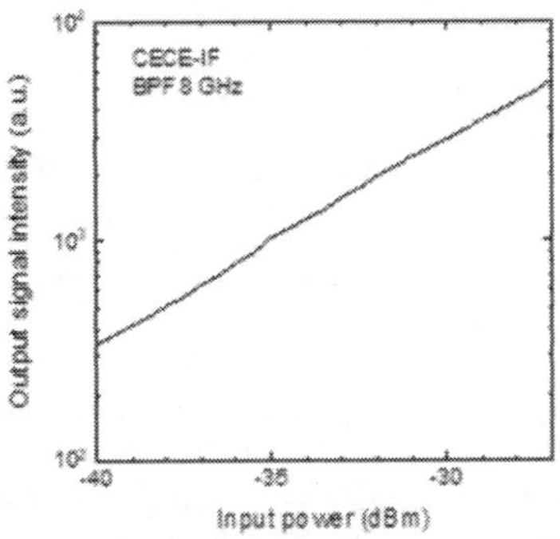

Fig. 6. Linearity of output voltage versus power applied to the 8 GHz IF line of the CECE-IF channel.

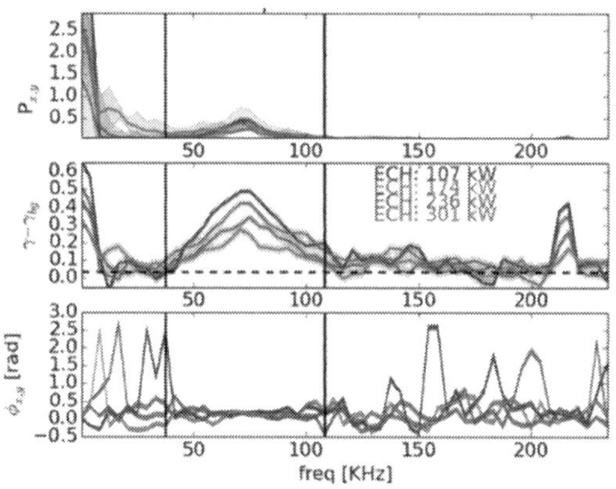

Fig. 7. Cross-power, coherence, and phase (top to bottom) between the CECE-RF and CECE-IF channels during an experiment with varying ECCD in Heliotron J.

4 Plasma experiments

The first measurements with the new radial CECE diagnostic were made in a plasma with both ECRH and NBI heating. Energetic particle modes (EPM) are excited by fast ions with NBI heating in Heliotron J, and these modes are stabilized by using counter-ECCD [20-21]. During this experiment, the CECE-RF channel was set to measure 68.3 GHz, and it was correlated over 90 ms with the 12 GHz CECE-IF channel, corresponding to a spectral separation of 0.3 GHz and a sensitivity limit of 0.41%. The cross-power, coherence, and cross-phase between the two channels are shown in Figure 7. The coherence between 40-110 KHz and 210-220 KHz measured by the CECE diagnostic decrease with increasing ECCD, consistent with the previously published results on EPM suppression in Heliotron J. The electron temperature fluctuation level is estimated by integrating the complex coherence [22], and the fluctuation level between 40-110 KHz drops from 1.8% with 107 kW of launched power to 1.1% with 301 kW of launched power. The fluctuation level of the 210-220 KHz component is below the sensitivity limit during this experiment; however, there is a low frequency component between 1-15 KHz that increases with ECCD power from 0.6% to 1.0%.

Separately, preliminary results in ECH plasmas indicate that the radial correlation length is on the order of a few mm, which is comparable to that of electron density fluctuations measured with a two-frequency reflectometer. Those results will be separately reported.

5 Summary

A new correlation ECE radiometer has been designed and installed on the Heliotron J device to measure electron temperature fluctuations. The diagnostic measures second harmonic X-mode ECE radiation between 56-72 GHz, corresponding to r/a<0.5. The system is comprised of two separate heterodyne radiometers. One radiometer has a single intermediate frequency channel and a variable local oscillator frequency, and the other has four intermediate frequency channels with a fixed local oscillator frequency. Laboratory tests indicate that the system has high spectral resolution and high-sensitivity over a large dynamic range. The first correlation ECE measurements on Heliotron J are consistent with expectations from previous experiments on Heliotron J, and the fluctuation levels measured by the CECE diagnostic are larger than the sensitivity limit of the CECE radiometer.

The authors are grateful for the Heliotron J staff for operating plasma experiments. This work was carried out with the support from the auspices of the Collaboration Program of the Laboratory for Complex Energy Processes, IAE, Kyoto Univ., the NIFS Collaborative Res. Program (NIFS10KUHL030), the NIFS/NINS project of Formation of International Network for Scientific Collaboration, JSPS Postdoctoral Fellowships for Research in Japan and JSPS Grant-in-Aid for Scientific Research (B) 18H01199. This work has been carried out within the framework of the EUROfusion Consortium and has received funding form the Euratom research and training programme 2014-2018 under grant agreement No 633053. The views and opinions expressed herein do not necessarily reflect those of the European Commission.

References

1. F. Ryter, *et al.*, Phys. Rev. Lett. **95**, 085001 (2005)
2. J. C. DeBoo, *et al.*, Phys. Plasmas **19**, 082518 (2012)
3. G. M. Weir, *et al.*, Phys. Plasmas **22**, 056107 (2015)
4. C. Watts, Fusion Sci. Technol. **52**, 176 (2007)
5. S. Sattler and H. J Hartfuss, "Evidence for Temperature Fluctuations in the W7-AS Stellarator", presented at Int. School of Plasma Physics, Varenna, Italy, 1993
6. R. F. Gandy, *et al.*, "Measurement of Electron Cyclotron Emission Fluctuations in TEXT", presented at Int. School of Plasma Physics, Varenna, Italy, 1993
7. C. Watts, *et al.*, Rev. Sci. Instrum. **66**, 451 (1995)
8. M. Kwon, *et al.*, Rev. Sci. Instrum. **63**, 4633 (1992)
9. G. Cima, *et al.*, Phys. Plasmas **2**, 720 (1995)
10. S. Sattler, H. J. Hartfuss adn W7-AS team, Phys. Rev. Lett. **72**, 653 (1994)
11. A. E. White, *et al.*, Phys. Plasmas **15**, 056116 (2008)
12. A. E. White, *et al.*, Phys. Plasmas **17**, 020701 (2010)
13. S. Freethy, *et al.*, Rev. Sci. Instrum. **87**, 11E102 (2016)
14. S. Freethy, *et al.*, Phys. Plasmas **25**, 055903 (2018)
15. G. M. Weir, *et al.*, this workshop
16. M. Wakatani, *et al.*, Nucl. Fusion **40**, 569 (2000)
17. T. Obiki, *et al.*, Nucl. Fusion **41**, 833 (2001)
18. K. Nagasaki, *et al.*, Contrib. Plasma Phys. **50**, 656 (2010)
19. K. Nagasaki, *et al.*, Fusion Eng. Design **34**, 463 (1997)
20. K. Nagasaki, *et al.*, Nucl. Fusion **53**, 113041 (2013)
21. S. Yamamoto, *et al.*, Nucl. Fusion **57**, 126065 (2017)
22. A. J. Creely, *et. al.*, Rev. Sci. Instrum. **89**, 053503 (2018)

Development of ECE/ECEI diagnostics and MHD-related studies on HL-2A tokamak

Z.C. Yang[1], M. Jiang[1,*], Z.B. Shi[1], X.T. Ding[1], N.C. Luhmann, Jr.[2], W.L. Zhong[1], W. Chen[1], P.W. Shi[1], Y. Xu[3], J. Wen[1], A.S. Liang[1], Y. Liu [1], Q.W. Yang[1] and HL-2A team

[1]Southwestern Institute of Physics, P. O. Box 432, Chengdu 610041, China.
[2]Department of Electrical and Computer Engineering, University of California, Davis, California 95616, USA
[3]Institute of Fusion Science, School of Physical Science and Technology, Southwest Jiaotong University Chengdu 610031, China

Abstract. A novel 60-channel electron cyclotron emission (ECE) radiometer has been designed and tested for the measurement of electron temperature profiles on the HL-2A tokamak. This system is based on the intermediate frequency division technique, and has the features of wide working frequency range (60−90 GHz) and high temporal-spatial resolution (3 μs, 1 cm), which covers almost the entire plasma region. Also, an electron cyclotron emission imaging (ECEI) system has been developed for studying two dimensional electron temperature fluctuations. It is comprised of several front-end quasi-optical lenses, a 24 channel heterodyne imaging array with a tunable RF frequency range spanning 60−135 GHz, and a set of back-end ECEI electronics that together generate two 24×8 array images of the 2nd harmonic X-mode electron cyclotron emission from the HL-2A plasma. The measurement region can be flexibly shifted due to two independent local oscillator sources, and the field of view can be adjusted easily by changing the position of the zoom lenses as well. The temporal resolution is about 2.5 μs and the achievable spatial resolution is 1 cm. The ECE/ECEI diagnostics have been demonstrated to be powerful tools to study MHD-related physics including the multi-scale interaction between macro-scale MHD and micro-scale turbulence on the HL-2A tokamak.

1 Introduction

Electron temperature (T_e) is one of the basic parameters in fusion plasma, and its measurements are essential for successful completion of the fusion device mission. At present, it can be diagnosed in various ways, including laser Thomson scattering, soft x-ray using silicon drift detector, Langmuir probes, and electron cyclotron emission (ECE). However, for Thomson scattering the time resolution is about several to tens of millisecond, and for the Langmuir probe, the measurement can only reach the edge region. The interpretation for silicon drift detector measurements are tricky since they depend on the electron temperature, density and impurity. In contrast, ECE diagnostic is an effective method to determine the T_e profile and T_e fluctuation $\tilde{T}_e$ with good temporal and spatial resolution, and therefore would become a routine tool to detect T_e evolution in ITER [1]. ECE electron temperature data will be important for evaluating the radial deposition profile of EC heating and the time evolution of the alpha particle heating. Furthermore, real-time ECE measurements will be used to identify magnetohydrodynamic (MHD) modes, such as neoclassical tearing modes that can often be deleterious to plasma performance.

Under the condition of optically thick emission, the 2nd harmonic of ECE can be considered as Blackbody emission, which is described by Planck's law in the Rayleigh-Jeans limit. Consequently, the intensity of ECE for the optically thick case is expressed by Eq.1 [2, 3].

$$I_{ECE} = \frac{\omega^2 T_e}{8\pi^2 c^2} = CT_e \tag{1}$$

By calibrating the coefficient C, the T_e profile can be obtained. The previous ECE radiometers on HL-2A tokamak had a finite working frequency range which limits the available measurement coverage in the plasma especially when the toroidal magnetic field changes. In addition, the spatial resolution was about 2−3 cm, which makes it difficult to identify the accurate locations of some events, such as ECH deposition, sawtooth inversion radius, and tearing modes. Also, the linear operating range needs to be extended to satisfy the higher heating power condition during the discharges. Hence, a novel 60 channel ECE radiometer has been developed to measure the T_e profile of the entire plasma region with a spatial resolution of 0.5−1 cm. The detailed introduction of this system will be given in Section 2. The traditional 1D ECE diagnostic measures the electron temperature locally at different radii along mid-plane of the device [4]. To satisfy the need for multi-dimensional MHD physics measurements, the 2D electron cyclotron emission imaging (ECEI) technique has been developed by extending the traditional ECE approach by incorporating large aperture quasi-optical lenses and vertically aligned antenna arrays. The detailed description of ECEI on HL-2A and the optical improvements will be pre-

*e-mail: jiangm@swip.ac.cn

sented in Section 3. In Section 4, two examples regarding MHD-related studies by using ECE/ECEI are presented. Last, a summary is presented in Section 5.

2 60-channel ECE with intermediate frequency division technique

FIG.1 shows the schematic of the newly developed 60-channel ECE diagnostic. Compared to the previous ECE radiometers, this system is based on the direct intermediate frequency (IF) filtering and detection technique by using several multiplexers instead of the second frequency down-converting (mixing) and power division technique [5]. Note that the power divider will introduce large insertion loss and is quite expensive especially at high frequency. This direct frequency filtering technique has three advantages: lower cost, wider working frequency range, and higher IF output power. The local oscillator (LO) source can be tunable between 50 and 90 GHz at a power level of 10 dBm, and the ECE signal from plasma is down converted to an IF signal with a frequency of 1−33 GHz. In the IF modules, the signal is divided by a 2-way resistive-type power divider, and then amplified by the wide-band low noise amplifier (LNA) separately. After that, the signals are split into 60 channels (2 to 31.5 GHz with a step of 0.5 GHz and bandwidth of 0.5 GHz) by eight-frequency multiplexers, and then detected, video amplified and low pass filtered with a bandwidth of 300 kHz.

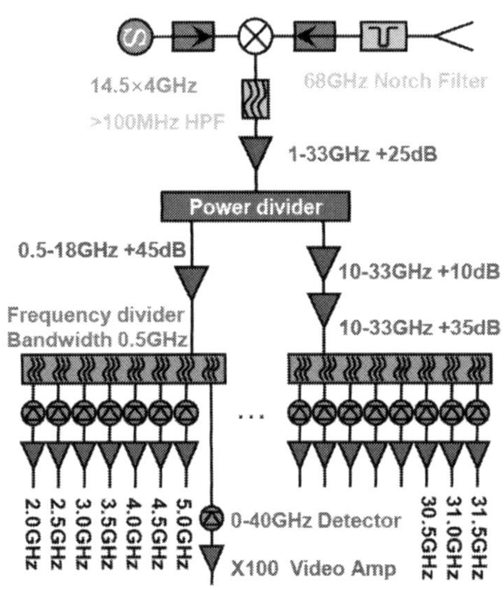

Figure 1. Schematic of the newly developed 60-channel ECE radiometer.

The system was tested in the laboratory before installation on the HL-2A tokamak. Figure 2(a) shows the frequency response of the LNA used in the IF module after the two-way power divider as illustrated in Fig.1. The LNA can amplify IF signals by 30-40 dB over the whole frequency band. The high IF output power improved the signal to noise ratio significantly. Shown in Fig.2(b) is the frequency response of the filter for each IF channel. The frequency interval between adjacent channels is 500 MHz and the bandwidth of the filters is also 500 MHz. The test result shows that there is almost no overlap between neighboring channels, and a relatively flat response is obtained over the full working band. The utilization of the narrow band filter guaranteed a high spatial resolution of T_e measurement, i.e., ~ 1 cm/0.5 cm in the B_t = 1.3/2.4 T operation. Fig. 2(c) shows the power response of a detector with f_{IF} = 23 GHz. The detector works over a very wide linear region with input IF power less than 1.1 μW and output power within -2 V−2.5 V. In this case, the detector output voltage is proportional to the electron temperature. Base on the system performance introduced above, the ECE system can provide reliable profiles even with megawatt electron cyclotron resonance heating (ECRH). For the routine discharge on the HL-2A tokamak, the measurement range will cover about 66 cm on the midplane, with a spatial resolution of 1.1 cm and temporal resolution of 3 μs.

3 ECEI on HL-2A and the optical optimization

Although the 1D ECE measurement can provide T_e profile with high temporal and spatial resolution, it cannot provide the global quantities of the instabilities in the plasma, due to the fact that the MHD physics is always multi-dimensional. To capture the 2D characteristic of the MHD mode, we have developed two 24×8 ECEI array under the collaboration with University of California, Davis. ECEI is an advanced extension of the traditional ECE radiometry with the same basic principle; however, it can image the plasma onto a vertically aligned antenna array by using a set of quasi-optical lenses with the heterodyne frequency capability yielding 2D images. The ECEI diagnostic has been proven to be a powerful tool for MHD physics through electron temperature fluctuation measurements, providing invaluable images on core/edge physics [6–11]. The ECEI instrument includes four subsystems, i.e., quasi-optical system, antenna array, microwave receivers and digitizers. The detailed introduction of each subsystem, its laboratory calibration, upgrade and preliminary experimental results are described in previous publications [12, 13]. According to different physical requirements, the system can work in narrow zoom pattern (higher spatial resolution) and wide zoom pattern (larger spatial coverage) by adjusting the relative positions of the zoom lenses. Using two separate LO sources, the measurement regions of the two 24×8 ECEI arrays can be shifted flexibly in the radial direction. It has 16 channels in the horizontal direction and 24 channels in the vertical direction, covering a rectangular area of 40−55 cm (vertical) × 12−17 cm (radial). The temporal and spatial resolutions are 2.5 μs and 1−3 cm, respectively.

In the laboratory calibration and plasma experiments, we found that the quasi-optical system had some limitations: (i) the focal area can only reach the core and low

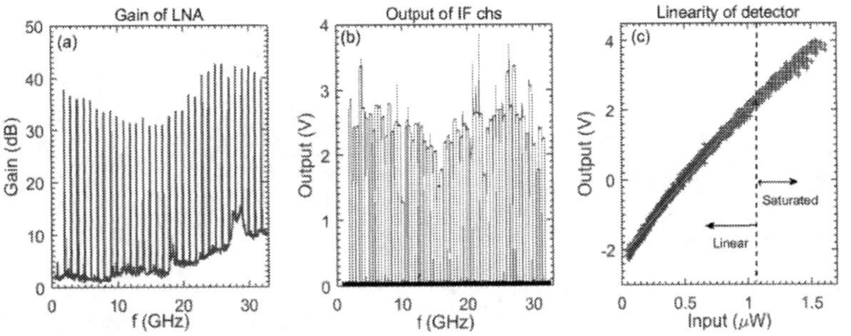

Figure 2. (a) the gain of the wide-band low noise amplifier in the IF module, (b) the frequency response of each IF channel, and (c) the power response of the detector.

field side (LFS) region of the plasma, but not the high field side (HFS) region; (ii) the focal plane is very curvy. The curved focal plane indicates that the edge vertical channels are out of focus, leading to a low signal to noise ratio of the signals. Consequently, last year, we have optimized the quasi-optical system to obtain better focal performance by changing the curvature radius of the focal lens, the surface type of the zoom lenses, and the relative distance between the lenses. Figure 3 shows the results of the Gaussian beam tracing of the radio frequency optics before and after optimization. The focal lens, zoom lenses, and the distance between the antenna and the focal lens (L) are marked in the figure. The zoom lenses are used to adjust the coverage of the vertical field of view. By moving the focal lens, the radial location of the focal plane can be shifted. It can be seen that before the modification of the optics, the focal plane is very curvy, while after the modification it becomes much flatter.

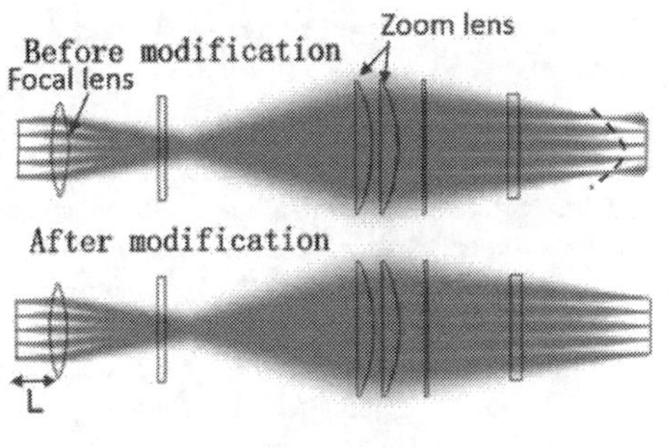

Figure 3. Gaussian beam tracing of the radio frequency optics before and after the optics modification.

Shown in Figure 4 are distributions of the focal planes with different L values under narrow zoom and wide zoom patterns, calculated from the real ray tracing. In the wide zoom pattern, the vertical field of view is larger than that in the narrow zoom pattern. It was found that in both narrow and wide zoom patterns, the field curvature of the focal plane (ΔR between the core channel and the edge channel) is within 6 cm. In addition, the focal area can reach the high field side plasma region. The optimized optics were fabricated, tested and finally installed on the HL-2A tokamak for the 2018 spring experimental campaign. The quality of the ECEI signals were highly improved in this experiment.

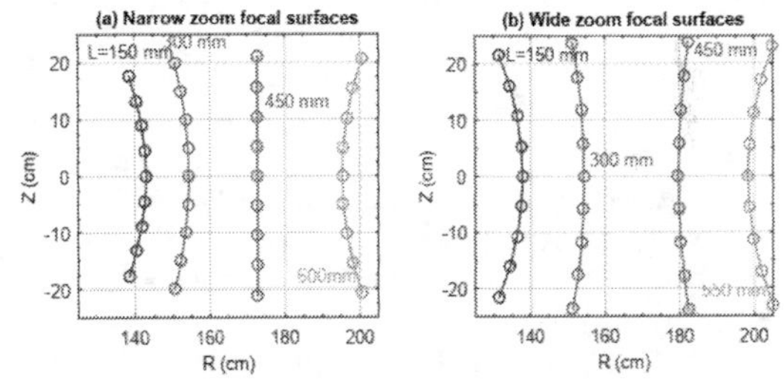

Figure 4. Focal plane with different positions of focal lens under narrow zoom and wide zoom patterns.

4 MHD-related studies by ECE/ECEI

4.1 Calibration of the ECE system

Normally, due to the different electronic response of each channel, the ECE measurement has to be calibrated to represent the T_e profile. As we know, the detection locations of ECE depend on the working frequency and magnetic field. Hence, the ECE radiometer can be relatively calibrated by changing the working frequency or magnetic field slightly in two similar discharges with the assumption of a stiff T_e profile. The absolute values of the T_e can be obtained through cross comparison with the core Thomson scattering data. The calibration method by hopping working frequency has been introduced in [5]. Here we show one example of the ECE calibration by shifting the magnetic field, which is the routine way to obtain T_e profiles in HL-2A. The calibrated T_e profiles are shown in figure 5(a). As we mentioned in Section 2, this ECE radiometer can work in a wider, linear input region compared with previous systems; it has been verified that even under high power ECRH the system can still work well. Fig. 5(b) indicates that with on-axis ECRH, the plasma center shifted outward by 5 cm and the core T_e was increased by 150% compared to the ohmic heating case.

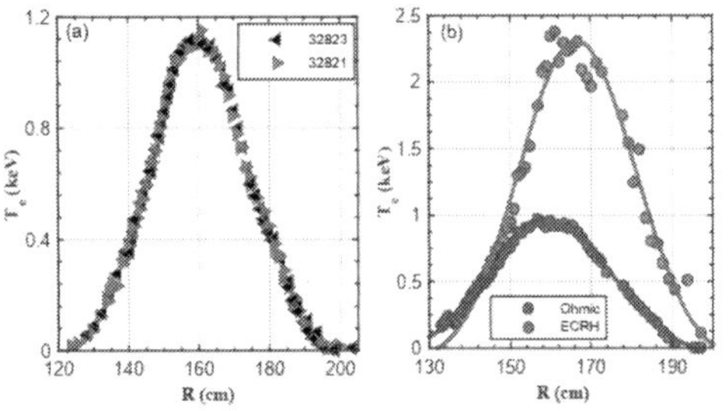

Figure 5. (a) Calibrated T_e profiles obtained by shifting the magnetic field in two similar discharges, and (b) T_e profiles under ohmic and ECRH heating conditions.

4.2 Analysis of radial mode structure

The newly developed ECE system basically covers 90% of the plasma along the midplane with a spatial resolution of ~1 cm. Consequently, it represents a good tool to detect the radial structure of MHD modes, such as the accurate location and width of the tearing mode (TM). In the experiment, the mode rotates past the ECE detector position, so ECE signal variations can be used to distinguish between kink and tearing modes. Plotted in figure 6 is a representative comparison of the features between the m/n=2/1 TM and the m/n=1/1 ideal kink mode. Fig. 6(a) illustrates the time history of the T_e located inside, but near the q=2 surface, and Fig. 6(e) shows T_e near the q=1 surface. The ideal kink mode produces sinusoidal temperature variation near the resonant surface (see Fig. 6(e)), while for the tearing mode, due to the flat profile near the O-point of the island, the T_e variation near the resonant surface will be deformed when the island structure passes by (see Fig. 6(a)), which is clearly illustrated by the cartoon in Reference [14]. For the position inside, but near the q=2 surface, the more flattened trough of the T_e variation indicates that the O-point of the island is across the probed location (blue vertical line), while the more peaked crest means the X-point is through there (red vertical line).

Figures 6(b), 6(c), 6(f) and 6(g) show the amplitude and phase of relative electron temperature variations ($\delta T_e/\bar{T}_e$) using fast Fourier transform analysis for TM and ideal kink mode, respectively. The shaded area illustrates the locations of the modes. The TM mode is located at 136 cm < R < 144 cm in the HFS and 180 cm < R < 188 cm in the LFS, which corresponds to the two peak areas in Fig. 6(b) and the flat T_e profile with the O-point of island passing by (blue circles in Fig. 6(d)). For the 1/1 ideal kink mode, the mode structure is encompassed by the q=1 surface, which is R=150 cm in the HFS and R=180 cm in the LFS. The T_e profile is not flattened for the kink mode (see Fig. 6(h)), demonstrating that there is no magnetic reconnection. In addition, for TM T_e perturbation takes on the same phase (2π) between the HFS and LFS (see Fig. 6(c)), consistent with the even behavior of m=2, whereas for the internal kink mode an opposite phase associated with odd structure was observed (see Fig. 6(g)), in agreement with m=1. It was also found that the core T_e profile becomes much flatter when there exists a TM compared to without TM cases, probably due to the impurity accumulation in the core plasma region.

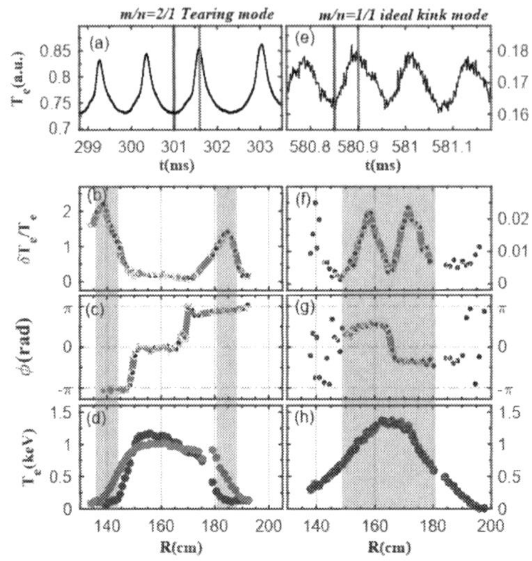

Figure 6. (a) time evolution of T_e inside but near the q=2 surface, (b-c) the amplitude and phase of the relative T_e variation $\delta T_e/\bar{T}_e$ with an m/n=2/1 tearing mode, (d) T_e profiles at two moments (O-/X-point passing-by) marked by the blue and red vertical lines in (a), (e) time evolution of T_e near the q=1 surface, (f-g) the amplitude and phase of the relative T_e variation $\delta T_e/\bar{T}_e$ with an m/n=1/1 ideal kink mode, and (h) T_e profiles at two moments marked by the blue and red vertical lines in (e). The locations of the mode were shaded in the figure.

4.3 Multi-scale interaction between TM and turbulence

In HL-2A, the ECEI diagnostic has been extensively used to detect the 2D structure of large scale MHD modes [10, 11, 15, 16], such as tearing mode, fishbone, and sawtooth crash. Figure 7 shows the mode structure of the m/n=2/1 tearing mode. The phase of T_e variations is opposite on two sides of the q=2 surface. The X-point and O-point of the island are indicated with the black X and O marks in the figures 7(a) and 7(b), respectively. Recently, multi-scale physics such as the interaction between magnetic islands and turbulence was found to play an important role in regulating the turbulent transport [15, 17–23], and therefore has become an important topic in fusion research. ECEI was demonstrated to be a powerful tool for analyzing such multi-scale physics. On the one hand, from the ECE images, the rotation frequency and direction, X-/O-point, the location and width of the island can be identified accurately. On the other hand, it can also provide information of high frequency T_e fluctuations, $\tilde{T}_e$.

Figure 8 is a typical example of the interaction between the magnetic island and T_e turbulence. In Fig. 8(a), the power spectrum of raw signals (passing through the island) measured by ECEI is plotted. The coherent peak at

$f \approx 4.3$ kHz indicates the modulation on the local electron temperature by the island rotation. To gain further insight into the effect on broadband turbulence, we analyzed the envelope of temperature fluctuations ($\xi_{\tilde{T}_e}(t)$) in the frequency range of 20-100 kHz (see shaded areas), i.e., $\xi_{\tilde{T}_e}(t) = (\frac{1}{2\pi}\int_{20\ kHz}^{100\ kHz} \tilde{T}_e(f)e^{i2\pi ft}df)^2$, where $\tilde{T}_e(f)$ is the Fourier transform of $\tilde{T}_e(t)$. The frequency spectra of $\xi_{\tilde{T}_e}(t)$ is shown in the inset of Figs. 8(a). The envelope spectrum exhibits clear peaks at the rotating TM frequency, signifying the modulation of broadband turbulence by rotating islands. Fig. 8(b) further display the time traces of $\tilde{T}_e$ (integrated from 20–100 kHz), where the envelope of $\tilde{T}_e$ was calculated via the Hilbert transform and depicted by the red curves. The amplitude in $\tilde{T}_e$ is modulated during the rotation of the island, i.e. minimum at the O-point and maximum at X-point, which is consistent with gradient-driven turbulence. Furthermore, using two vertically adjacent ECEI channels, the poloidal flow velocity could be deduced from the vertical pattern velocity. Thus, the poloidal flow shear in the island region can be investigated. It was reported on KSTAR that the combined effects of the T_e gradient and the poloidal flow shear determine the two-dimensional distribution of the T_e turbulence [22].

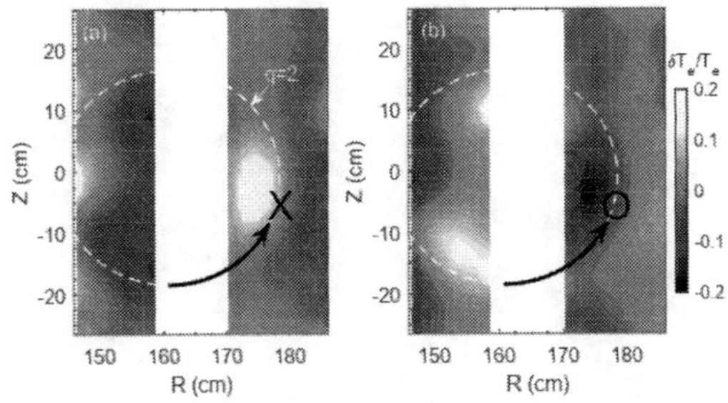

Figure 7. 2D mode structure of the m/n=2/1 tearing mode measured by ECEI. The white dashed curves denote the q=2 surface, and the X-point and O-point of the island are indicated with the black X and O marks, respectively.

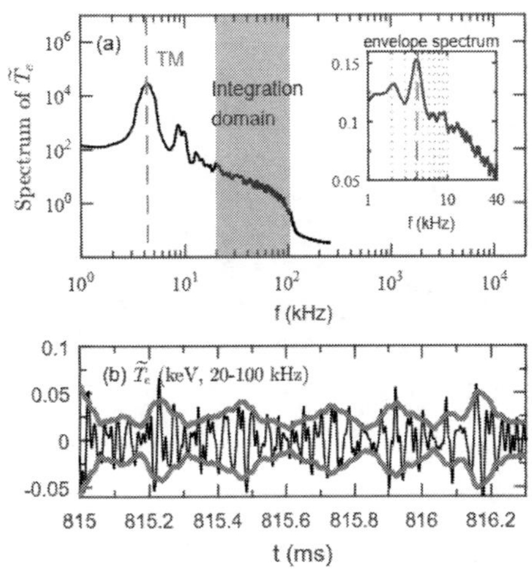

Figure 8. (a) Power spectrum of $\tilde{T}_e$ measured at R=182.14 cm and Z=3.46 cm. The inset in (a) shows the amplitude spectrum of the corresponding envelope signal, $\xi_{\tilde{T}_e}(t)$ in the frequency range Δf=20-100 kHz (shaded in (a)). (b) Time histories of $\tilde{T}_e$ (integrated from 20–100 kHz)and the envelopes of $\tilde{T}_e$ depicted by the black and red curves, respectively.

5 Summary

A novel 60-channel ECE radiometer has been developed for measuring T_e profile and T_e variations on the HL-2A tokamak. It is based on the direct intermediate frequency division technique by utilizing several multiplexers which ensures high IF output power, and hence high signal to noise ratio of the signals. The measurements of this system can cover almost 90% of the plasma along the midplane, and the temporal and spatial resolutions are 3 μs and 1 cm for the routine B_t=1.3 T operation. Two 24×8 ECEI arrays have been developed by incorporating quasi-optical lenses and vertical aligned antenna array for two dimensional $\tilde{T}_e$ measurements. The rectangular coverage of the two arrays can be flexibly shifted due to two independent LO sources, and the field of view can be adjusted easily by changing the position of the zoom lenses as well. The time and maximum spatial resolutions are 2.5 μs and 1 cm, respectively. The optics of ECEI has been modified to cover broader plasma region and improve the imaging quality. The broad coverage and high spatial resolution of the ECE make it a powerful tool to study MHD-related physics, such as distinguishing the tearing mode from the kink mode, the location and width of magnetic island. In addition, multi-scale interaction between TM and high frequency turbulent $\tilde{T}_e$ has been investigated by ECEI diagnostic.

This work is partially supported by the National Natural Science Foundation of China under grant No. 11705051, 11475057, 11775069 and 11705049, partially supported by National Key R&D Program of China (Grant Nos. 2017YFE0301201 and 2017YFE0301203), and partially supported by Sichuan Science and Technology Program (Grant No. 2018RZ0123).

References

[1] G. Taylor, et al., EPJ Web of Conferences **87**, 03002 (2005)
[2] M. Bornatici, et al., Nucl. Fusion **23**, 1153 (1983)
[3] I. H. Bornatici, Principles of Plasma Diagnostics (Cambridge University Press, Cambridge,1987), Chapter 5.
[4] H. J. Hartfuss, T. Geist and M.Hirsch, Plasma Phys. Controlled Fusion **39**, 1693 (1997)
[5] Z.B. Shi, et al., Rev. Sci. Instrum. **85**, 023510 (2014)
[6] H.K. Park, et al., Phys. Rev. Lett. **96**, 195003 (2006)
[7] I.G.J. Classen, et al., Phys. Rev. Lett. **98**, 035001 (2007)

[8] I.G.J. Classen, et al., Plasma Phys. Controlled Fusion **53**, 124018 (2011)
[9] J. Lee, et al., Phys. Rev. Lett. **117**, 075001 (2016)
[10] M. Jiang, et al., Phys. Plasmas **24**, 022110 (2017)
[11] M. Jiang, et al., Nucl. Fusion **58**, 026002 (2018)
[12] M. Jiang, et al., Rev. Sci. Instrum. **84**, 113501 (2013)
[13] M. Jiang, et al., Rev. Sci. Instrum. **85**, 076107 (2015)
[14] V. Igochine et al., Nucl. Fusion **57** 036015 (2017)
[15] W. Chen et al., Nucl. Fusion **57** 114003 (2017)
[16] L.M. Yu et al., Nucl. Fusion **57** 036023 (2017)
[17] Bardóczi L. *et al* Phys. Rev. Lett. **116** 215001 (2016).
[18] Bardóczi L. *et al* Phys. Plasmas **24** 056106 (2017).
[19] Ji X. Q. *et al* Sci. Rep. **6** 32697 (2016).
[20] Ida K. *et al* Sci. Rep. **5** 16165 (2015).
[21] Zhao K. J. *et al* Nucl. Fusion **57** 126006 (2017).
[22] Choi M. J. *et al* Nucl. Fusion **57** 126058 (2017).
[23] Sun P. J. *et al* Nucl. Fusion **58** 016003 (2018).

Design and Performance of Microwave Components for ECH and ECE Applications at General Atomics

James Anderson*, *John* Doane, *Charles* Moeller, *Howard* Grunloh, *Raymond* O'Neill, *Michael* Brookman, *Matthew* Smiley, and *David* Su

General Atomics, San Diego, CA, 92121, USA

Abstract. This paper provides an overview of high power components for the application of Electron Cyclotron Heating transmission lines, and broadband devices for Electron Cyclotron Emission detection systems. The unique fabrication and assembly challenges are discussed, particularly in the context of ITER. The ITER ECH system will require robust, vacuum-compatible components such as polarizers, dummy loads, and switches that are sufficiently cooled to withstand 1 MW for 3,600 seconds. These elements, along with overmoded corrugated waveguide, are necessary to form transmission lines with efficiencies of 90%, and 90% transmitted HE_{11} mode purity. Recent high power test results are summarized and scaled from the 63.5 mm internal diameter design to the 50 mm diameter version that will be used for ITER. Elements designed for Electron Cyclotron Emission detection and reflectometry systems are discussed, such as frequency filters and polarization rotators. The large frequency operating range of corrugated waveguide is exploited for such applications. The application of additive manufacturing technology towards both low and high power components is considered as a promising new area of development.

1 Overview

Microwave technology for fusion devices has evolved considerably over the past several decades. Experience gained from the practical application of electron cyclotron heating (ECH) systems such as at DIII-D, KSTAR, JT-60, NIFS, and EAST has led to steady improvements along with numerous innovations. While development of components has advanced through sometimes painful trial and error, reliability has gradually increased with lessons learned during the accumulation of operational hours on EC systems. As the ITER project looms, a robust microwave industry, spurred on by shared knowledge and recurring interaction, is more essential than ever.

General Atomics (GA) began design and production of high power, high frequency transmission line components in the early 1990s in support of the DIII-D ECH program for the US DOE. Specified for 110 GHz at 1 MW, transmission lines of 31.75 mm ID corrugated waveguide were fabricated, installed and operated at DIII-D. Currently DIII-D's ECH systems include eight 90-meter transmission lines. GA has since provided numerous full-transmission line sets of components to a number of tokamak devices over the world. These components include corrugated waveguide, miter bends, polarizers, power monitors, waveguide switches, dummy loads, expansion units, pumping tees, tapers, DC-breaks, and gate valves.

Transmission lines for fusion applications can be divided into high power such as Electron Cyclotron Heating and Current Drive, and low power such as Electron Cyclotron Emission diagnostic systems. As will be discussed in this review of microwave components and update on the state-of-the-art, each application has its own set of unique design criteria. A final section on additive manufacturing discusses a tantalizing future area of exploration.

2 Review of ECH Components

Applications such as ECH, where power levels can exceed 1 MW, present formidable challenges for component design and fabrication. As microwave output from gyrotron sources has increased both in terms of peak power and pulse length, the design emphasis has been placed firmly on maximizing transmission efficiency and minimizing mode conversion. Modern gyrotrons for ITER's ECH system, for example, are intended to produce 1 MW for over 3600 s [1]. Factory acceptance tests of these gyrotrons have demonstrated 1 MW for 300 s [2], and 0.9 MW for 500 s [3]. Under such conditions, scaled measurements from QST's gyrotron facility in Naka, Japan [4], suggest that over 512 W/m of heating will be dissipated for the HE_{11} mode supported in 6061-T6 aluminium alloy corrugated waveguide of 50 mm diameter at 170 GHz far from a miter bend. Within the approximately 1 m decay length for higher order modes at a miter bend, power absorption increases by 4 to 5 times. Therefore even though the HE_{11} mode is very low loss overall, the absorbed power is high enough to warrant active cooling of components.

* Corresponding author: andersonjp@fusion.gat.com

Furthermore, ECH transmission lines generally require a high purity output mode in order to reduce added loading from stray radiation. For ITER, the output mode purity must be greater than 90% HE_{11} [5]. (For reference, the ECH transmission lines at DIII-D are generally around 90% HE_{11} [6].) As a result of these strict requirements, as well as demands from other high purity ECH systems, manufacturing tolerances on microwave components have become extremely tight. This creates a significant challenge in terms of fabrication, not to mention maintaining adequate quality control processes.

GA is currently ramping up its production capabilities for the considerable amount of waveguide components necessary for ITER. As part of this process, GA has supplied 63.5 mm and 50 mm 170 GHz corrugated waveguide to several high power test facilities. GA has also designed, built, and tested many prototype components that may be used for the ITER ECH transmission line.

2.1. Corrugated Waveguide

Fabricating extended sections of corrugated waveguide over half a meter in length generally requires special tooling. At GA, equipment has been developed to produce corrugated waveguide in lengths up to 2.1 m, although techniques have been developed to make even longer waveguide assemblies. The corrugation machines are adaptable enough to accommodate diameters from 31.75 mm up to 88.9 mm in materials such as aluminium, copper, and stainless steel. Corrugation of a full 2.1m section can take as little as 30 minutes, although in practice this step is just one of many, adding up to multiple man-hours.

2.1.1 Ex-vessel Waveguide

As described in [7], the 4.2 m water-cooled aluminium corrugated waveguide assembly meets ITER tolerances. In 2015, this component was installed and tested at the QST high power gyrotron facility at 450 kW over 300 seconds. The results revealed double the expected power absorption compared to previous measurements, although this was due to unexpected reflections elsewhere in the transmission line [8].

The waveguide was made in 63.5 mm diameter rather than the 50 mm diameter in ITER's current specification, yet the experiments suggested the waveguide is suitable for an ITER ECH transmission line. Work is underway to model and analyse a scaled water-cooled corrugated waveguide design at 50 mm diameter. Generally the attenuation losses scale as the inverse of the radius cubed, so the expected absorption becomes roughly twice that measured previously [4]. However, this increase is not expected to impact the approach to active waveguide cooling.

Uncooled regions where joint couplings are necessary are important to analyze. Couplings can be designed to have significant heat capacity as well as much higher thermal conductance than regular waveguide because of the larger cross section. If the coupling is about 20 cm long and has an OD of about 10 cm, its heat capacity plus the heat capacity of the waveguide is about 2850 J/K. If the waveguide absorbs 100 W in the 20 cm length, then in a 3600 second pulse with radial conduction to the coupling, the average temperature rise of the coupling and waveguide underneath would be about 126° C. Axial conduction through the neighboring walls to the nearby cooling channels will reduce this heating further, so there should be no need to cool the couplings.

2.1.2 In-vessel Waveguide

Inside a fusion device, the environment generally becomes quite severe, not to mention unpredictable and chaotic. Ultra-high vacuum conditions coupled with the presence of energetic particles creates a different set of limitations on transmission line design, both in terms of cooling approach and material selection. In addition, the region within the bio-shield is not typically accessible, so reliability becomes an important factor, along with the development of proper installation and assembly procedures.

Informed by experience from operating ECH launcher components at DIII-D, General Atomics has developed a number of recent in-vessel waveguide components for tokamaks, including JT-60SA. At KSTAR, the launcher region was instrumented with a number of sensors. The measurements indicated thermal equilibrium was reached for the in-vessel water-cooled launcher mirrors during a 60.5 s gyrotron pulse at 800 kW [9], an encouraging result for long-pulse operation.

A 1.5 m stainless steel corrugated waveguide was built by GA as a prototype for launcher components at JT-60SA (Figure 1). A stainless steel sheath was welded to the outside to form a water-cooled jacket. The corrugations were plated with nickel in a sulfamate bath to reduce the resistivity at the metal surfaces. As revealed by looking at a sample with a scanning electron microscope (SEM), the process produced a uniform coating of nickel (Figure 2). The thickness ranged from 5 to 10 microns, several multiples of the skin depth. According to [10], the resistivity of electrolytic nickel plating is 8.6×10^{-6} ohm-cm, compared to 7.4e-5 ohm-cm for 316 L stainless steel alloy. Since the attenuation is proportional to the square root of the resistivity, the amount of power absorption is reduced by a factor of 2.9 by applying the relatively inexpensive plating process.

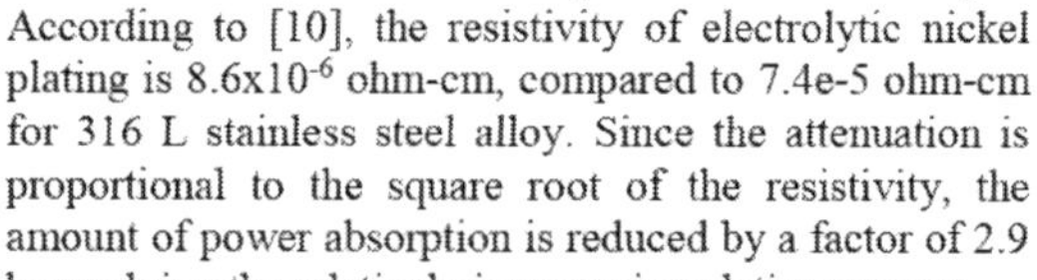

Fig. 1. A water-cooled stainless steel corrugated waveguide for JT-60SA.

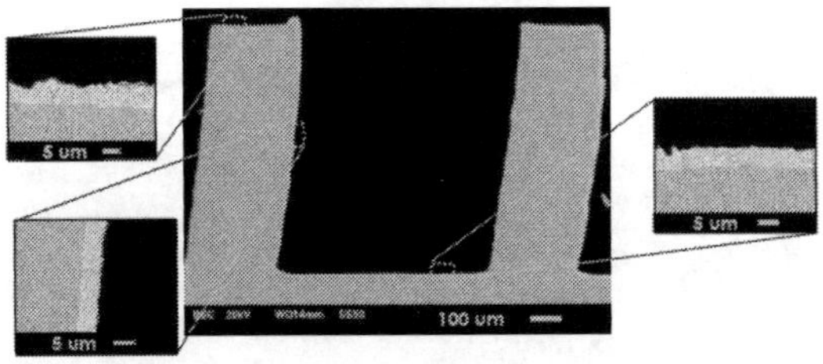

Fig. 2. SEM images of nickel-plated stainless steel corrugations. The images show the uniformity of the plating process. The slight distortion of the corrugations occurred when the sample was cut.

A transmission line was constructed to test ECH components for the JT-60SA tokamak at QST. Recent experiments were conducted comparing power absorption in aluminium waveguide with both plated and unplated stainless steel waveguide using 20 s pulses from a 110 GHz gyrotron at 500 kW [11]. The measured temperature rise of a section of unplated stainless steel waveguide was around 2.5 times that of an aluminium section, which matches the theoretical value when accounting for the heat capacity of the materials. In a similar comparison, the nickel plated section had a slightly lower temperature rise than that of aluminium. The theoretical ratio is 0.9.

For the ITER reactor, in-vessel components are subject to a number of requirements which place constraints on design of corrugated waveguide intended for the equatorial and upper ECH launchers. Among these constraints are welds, which must be full-penetration at water/vacuum boundaries, and subject to 100 percent volumetric inspection using non-destructive examination (NDE) techniques. GA has contributed to the development of a conceptual design for a 2 m long, 50 mm inner diameter equatorial launcher corrugated waveguide, as shown in Figure 3. The component is nickel-plated stainless steel, with a welded outer water jacket. To improve the cooling uniformity around the circumference of the waveguide, two input channels are used with two return channels. The water connections are accessible at the non-vacuum side of the component. The number of fabrication steps make the assembly challenging, but within the capability of modern machining and welding techniques. Analysis of the design using ANSYS shows the water cooling approach is acceptable for 1.2 MW transmission at steady state conditions (Figure 4).

Fig. 3. Conceptual design of equatorial ECH launcher waveguide for ITER.

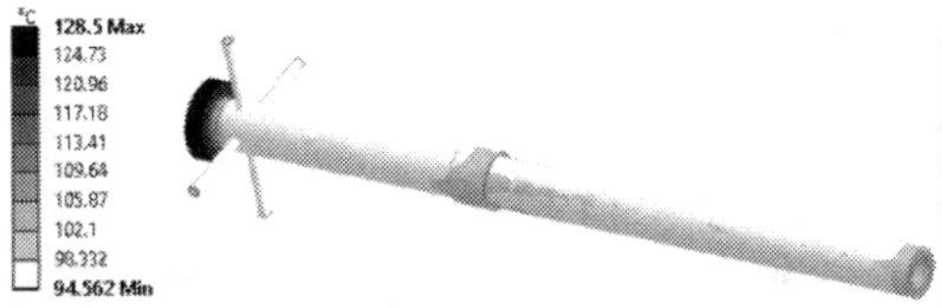

Fig. 4. Temperature distribution of the outer surface of the conceptual equatorial launcher waveguide. The most significant heating occurs at the non-cooled flange at the non-vacuum side, where the temperature rise is around 33° C. The water inlet temperature is 94.5° C.

2.2. Polarizers

Miter bends with rotatable grooved mirrors are often used for polarization control in ECH systems. Two miter bends (a linear polarization rotator and a circular polarizer) can produce the full range of polarization states, although it is possible to achieve some amount of polarization control with a single grooved mirror albeit with higher losses [12]. The design of a motorized 63.5 mm diameter polarizer miter pair for ITER is presented in [13], along with a generalized theory for predicting the ohmic losses of such components. Prototypes of these polarizers were tested under high power conditions (~350 kW over 500 s) at QST. Low power measurements and finite element analysis has shown the effectiveness of the design [14].

Adjusting to updated requirements, GA has developed an ITER polarizer design for 50 mm diameter. A steady-state analysis of the mirror design for 1.2 MW incident power at an input flow rate of 20 liters / min. indicates the cooling will be sufficient to limit the temperature rise at the mirror surface to around 120° C, even at the groove orientation with highest losses. The maximum amount of distortion is under 30 μm over the surface.

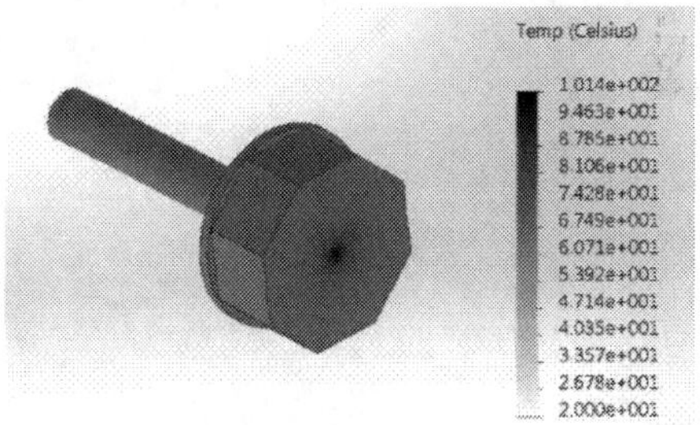

Fig. 5. Temperature distribution for 50-mm polarizer.

A ferromagnetic fluid-based seal is used to enable smooth, rapid rotational motion of the mirror driven by a motor external to the vacuum. This type of seal, commonly found in the semiconductor industry, is considered to be highly reliable. To reduce impact of failure, it is possible to design a small secondary interspace between the seal and the primary vacuum. The interspace can be monitored for leaks, and may even provide an automatic isolation seal triggered by a pressure differential.

Since many ECH systems are now multi-frequency, there is an increased interest in broadband polarizers. For example, KSTAR has gyrotrons which produce power at 105 and 140 GHz. By applying the theory as discussed in [13], it can be shown that GA's polarizer pairs used for that system (originally designed for 140 GHz) have stable operating points over the full range of polarizations at 105 GHz. A recent set of polarizers produced for the upgrade at TCV are designed to operate at 82.6, 84, and 118 GHz. (These polarizers also required fast setting of mirror position, with a full 180° rotation of the mirror in 0.2 s.) Finally, there is recent consideration of making the ITER installation a dual-band system. Calculations show that the GA polarizer pair for ITER will produce the full polarization spectrum at both 104 and 170 GHz.

2.3 Dummy Loads

High power microwave dummy loads are essential for conditioning, testing, and commissioning gyrotron sources. The dummy load system developed at GA typically consists of three elements: a downtaper to 31.75 mm diameter, a waveguide attenuator, which absorbs most of the power, and a compact tank load which dissipates the remaining power. As described more fully in [15], the attenuator consists of a section of varying corrugation depths either in copper or CuCrZr copper alloy that converts the HE_{11} mode to a surface wave, which has high attenuation. The advantage of this approach is that it has very low reflected power, relatively low water flow requirements, short time constants, no welds, and no rotating parts. The system has been used successfully at DIII-D, KSTAR, and EAST. Recently a prototype system has been built to handle 1.4 MW continuous power at 170 GHz for ITER (Figure 6). The waveguide attenuator was designed to absorb 85% of the power (1.2 MW) with the rest of the power (200 kW) collected in the tank. Around 80% absorption was measured in the waveguide attenuator during cold test. Although designed to transition from 63.5 mm waveguide, the system is easy to adapt to a 50 mm transmission line by replacing the taper in Figure 6(a).

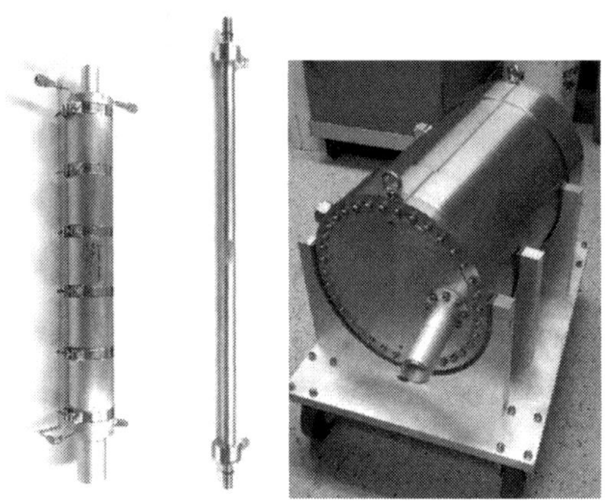

(a) (b) (c)

Fig. 6. The downtaper (a), attenuator (b), and tank (c) elements of the dummy load system designed for ITER conditions.

During gyrotron conditioning, the load system installed at the EAST facility withstood 630 kW of power at 1 s. The pulse width was limited during the time since the attenuator was absorbing less of the total power than expected (60%, rather than 80%), raising concerns about overheating the tank load. Calculations of the load geometry based on ray tracing conclude that the power rating of the tank load is about 500 kW CW in the most favourable input polarization orientation, and 250 kW CW in the orthogonal orientation. Therefore it should be possible to condition the gyrotron using long pulses despite the 60% absorption. The lower-than-expected absorption of power in the waveguide attenuator could be due to low HE_{11} mode purity. While measurements showed that the output mode purity from the gyrotron was around 90%, it is possible that the HE_{11} content is smaller at the input of the load due to any misalignments in the transmission line between the gyrotron and load.

2.4 Switches

Waveguide switches are used in applications where microwave power is necessary at multiple locations. Alternatively they enable a facility with several gyrotrons to dump power into a single load. For example, a switch such as the one shown in Figure 7(a) can be used to switch ECH power between the equatorial and upper launchers at ITER. This motorized switch, which employs a rotating mirror, is designed for 1.4 MW at steady state. It is fairly compact to accommodate 300 mm spacing between waveguide centerlines of adjacent switches, and has few parts compared to other types of waveguide switches, which improves reliability. In addition, it has low consumption of water for cooling, ≤ 0.1 kg/s with <0.6 MPa total pressure drop. The switching time has been demonstrated to be as low as 0.6 seconds, which may be useful in certain mode suppression scenarios for ITER. (Another type of rotary switch, developed for the ITER Low Field Side Reflectometer, has switching speeds < 0.1 second). While it was originally designed for 63.5 mm diameter, the ECH unit is easily scaled to 50 mm since the heating conditions of the flat mirror will be less than that for the grooved mirrors of the polarizers discussed in the previous section.

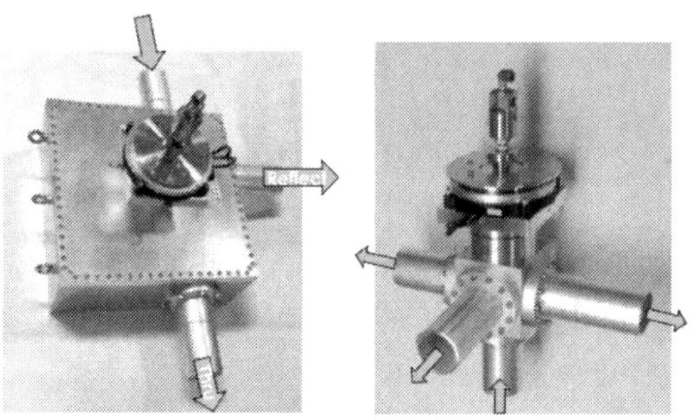

(a) (b)

Fig. 7. In-plane (a) and out-of-plane (b) motorized rotary waveguide switches for high power applications.

The rotary switch shown in Figure 7(a) has input and output ports in the same plane. An alternate configuration has output ports out of plane from the input port, such as the four port switch built for the TCV upgrade (Figure 7(b)). This particular approach allows for up to four different switching states. For vacuum applications, the assembly incorporates the same ferromagnetic fluid feedthrough as the polarizer. Overall, the rotary waveguide switches have proven to be robust, reliable, and inexpensive.

2.5 Transmission Line Diagnostics

Microwave power at the megawatt level has the potential to cause significant damage, in addition to posing an occupational safety hazard. It is not difficult to find instances of ECH power destroying sensitive equipment or disrupting delicate systems [16]. Such cases underline the importance of properly monitoring high power transmission lines. Moreover, it is often useful to have the ability to determine key transmission parameters such as power, polarization, frequency, and mode purity. This information may be used to determine problems with components in a microwave system, particularly the gyrotron. Often the signals from these devices are employed as safety interlocks.

GA has developed a number of different types of diagnostic components for corrugated transmission lines. As described in [17], two different diagnostics have been built and tested for measuring power in ECH transmission lines for ITER. A new power monitor miter bend using a CuCrZr mirror machined with small sampling holes was instrumented with a pair of diode detectors to indicate forward and reverse power of a 170 GHz signal produced by a gyrotron at the QST test facility. The power monitor miter bend provides instantaneous accurate power levels, as well as signal directionality. The signals from the sampling ports can also be used to determine frequency. An alternative is the calorimetric miter bend, which uses a CuCrZr mirror that is thermally isolated from the body of the miter bend. By measuring the temperature rise of the cooling water passing through the mirror, it is possible to determine the power in the transmission line. Such a device is not expensive. However, since it requires thermal equilibrium, it does not provide an instantaneous power measurement.

2.5.1 Inline Power and Polarization Monitor

A more simplified version of the power monitor miter bend is the inline power monitor, such as the one modelled in Figure 8. A calorimetry-based inline power monitor has been used at DIII-D, but since it was not placed far enough from miter bends and other sources of mode conversion, it has had limited success in providing accurate power measurements.

An updated inline power monitor, based on a gap-type pumpout port, uses diode detectors to provide instantaneous signals. The small gap allows a small amount of signal to leak out to the sampling ports. In its simplest form, a diode detector at a single sampling port can be used as an indication of power. For a high mode purity system, the signal from this inline power monitor may provide a calibrated power measurement, assuming it is placed far enough from sources of mode conversion. Two opposing ports may be instrumented with diode detectors at 90° orientations. In such a configuration it is possible to use the device as a polarization discriminator. By instrumenting all four ports with detectors, it may be possible to collect information on the mode structure of the travelling wave, making for an effective technique for monitoring misalignments in the transmission line.

Fig. 8. Inline power monitor with multiple ports for polarization discrimination.

A lower cost option is the inline power monitor based on a single-hole design, rather than the gap-type approach. This version provides a monitor port with a coupling of around -70 dB, which is large enough for a diode detector to measure assuming 1 MW incident power. Although the device does not allow for instantaneous polarization comparison, it may be used in applications where a simple indicator of power (or frequency) is all that is required.

2.5.2 Mode-Sampling Miter Bend

For more detailed analysis of the mode structure in an overmoded transmission line, GA has designed a mode-sampling miter bend [18]. Sometimes referred to as a "leaky" miter bend, the mirror of this device contains many small signal-coupling holes. Polarization discrimination is enabled with a transition down to a standard waveguide orthomode transducer at the back of the leaky mirror. Alternatively, several mode couplers can be used, each coupler with its own diode detector. In this configuration it is possible to instantaneously collect calibrated power measurements of the HE_{11} mode, as well as the LP_{11} odd and even modes commonly associated with tilt and offset misalignments. A version of this mode sampling miter bend was built and instrumented for the ECH transmission lines at DIII-D for 110 GHz. A prototype has also been built for 170 GHz which may be used as a diagnostic on an ITER high power 63.5 mm diameter transmission line test bed. This leaky mirror design requires approximately 5,000 holes through the 1 mm thick plate, which must be leak tight.

3 Review of Diagnostic Components

Overmoded transmission lines play an important role in many low power applications as well, although the design of these components face a different set of challenges than those intended for high power applications such as ECH. For example, plasma diagnostics like Electron Cyclotron Emission (ECE) detection systems and reflectometers require very wide operating bands. The emphasis is not on heat management or maintaining strict vacuum requirements, but rather on optimizing signal quality over a range of frequencies. Low loss is still essential.

3.1. Frequency Filters

The ECH system at KSTAR is dual-frequency, with gyrotrons producing power at 105 GHz and 140 GHz. The ECE detection system is designed to operate over a band which encompasses those frequencies. GA has built several notch filters to protect the diagnostic's sensitive electronics. The notch filters are made using a series of biaxial polystyrene sheets separated by thin spacer disks in a corrugated waveguide assembly. The assembly for two notches, as shown in Figure 9, is around 760 mm in length. The measured response showing the two notches is also shown. A similar notch filter was delivered to EAST, although with a single notch at 140 GHz.

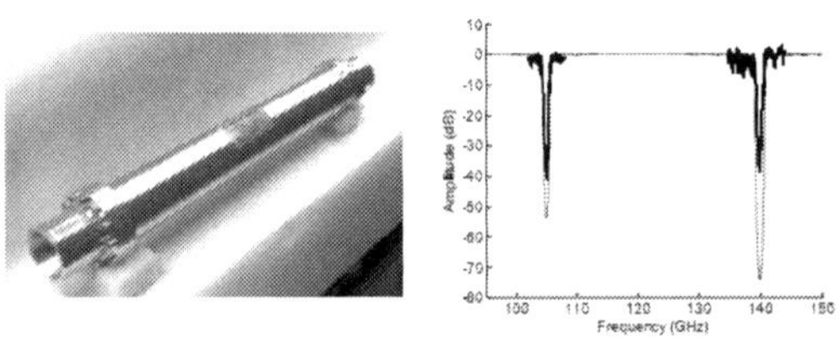

(a) (b)

Fig. 9. Dual-notch filter for KSTAR ECE diagnostic in 63.5 mm corrugated waveguide (a). Measured frequency response near 105 GHz and 140 GHz (b).

A different type of frequency filter is planned as a protection element for the Low Field Side Reflectometer for ITER. For this system, which will operate from 30 to 165 GHz, it is necessary to have a low pass filter at 170 GHz. A diffraction grating was designed to replace the mirror of a miter bend (Figure 10). For frequencies below 170 GHz, the signal will reflect off the mirror at an angle equal to the angle of incidence. For higher frequencies, the power is reflected back into the input waveguide where it is attenuated within about a meter from the mirror. The grating is polarization sensitive, so two miter bends in orthogonal planes are necessary to protect the diagnostic for all possible incoming field orientations.

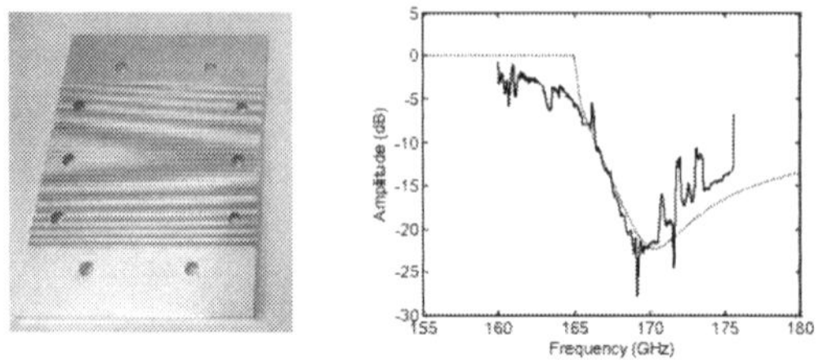

(a) (b)

Fig. 10. Diffraction grating reflector for 63.5 mm miter bend (a). Measured and theoretical frequency response near 170 GHz (b).

3.2. Polarization Rotators

Diagnostics using microwave signals typically rely on information gained from specific polarizations. In emission detectors and reflectometers, it is therefore advantageous to rotate linearly polarized signals, either on transmit or receive. One means of doing this is to use a three mirror polarization rotator configuration, such as shown Figure 11 for the EAST ECE system. In such an arrangement it is possible to achieve a full 90° rotation of polarization by mechanically rotating the assembly 45° around the axis of the input and output waveguide. A motorized version of the three-mirror polarization rotator using rotary joints has been designed for the ITER Low Field Side Reflectometer (LFSR) system.

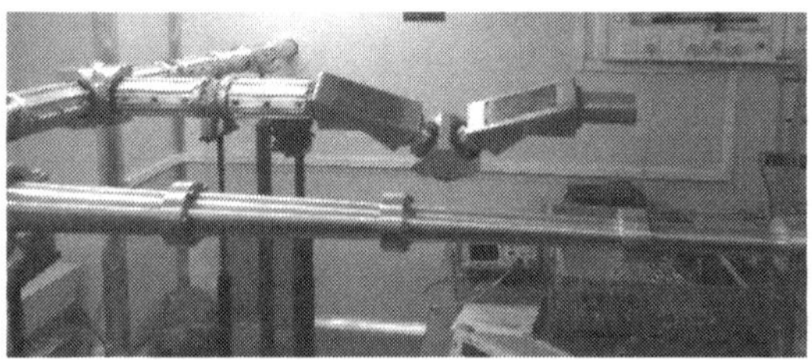

Fig. 11. Polarization rotators for ECE diagnostic at EAST.

3.3. Broadband Waveguide

Microwave diagnostics such as radiometers and reflectometers operate with frequencies in the range of the electron cyclotron resonance and right hand cutoff, which can vary by 2-3 octaves in a tokamak plasma due to dependencies on density and magnetic field. Examples of recent or planned reflectometers are the 40-120 GHz reflectometer for EAST, the 15-75 GHz Plasma Position Reflectometer for ITER, and the 30-165 GHz LFSR for ITER. Such applications are ideal for corrugated waveguide, which maintains low ohmic loss in the HE_{11} mode over large frequency ranges. The calculated loss for the three diagnostic waveguide types is shown in Fig. 12 in 6061-T6 aluminium alloy (assuming $4x10^{-6}$ ohm-cm resistivity and 30% roughness enhancement for surface resistance).

 https://doi.org/10.1051/epjconf/201920304001

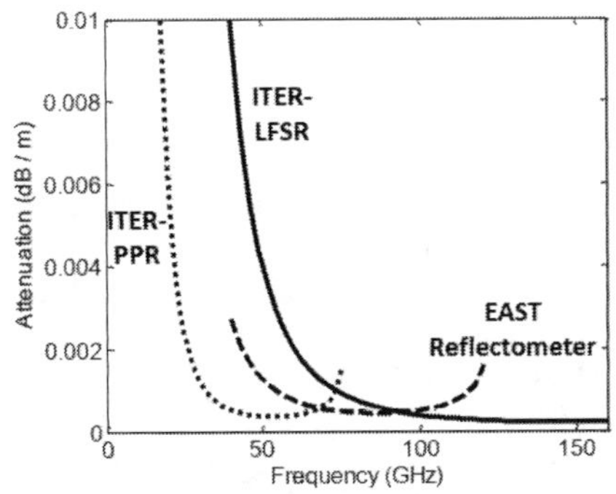

Fig. 12. Theoretical HE_{11} attenuation in 6061-T6 aluminium for ITER PPR waveguide (dotted), EAST reflectometer waveguide (dashed), and ITER LFSR waveguide (solid).

4 Additive Manufacturing

As the technology has advanced, additive manufacturing has played an increasing role in design and fabrication of microwave components. Although the process is primarily associated with plastic, the range of materials has expanded to include many different metals, including stainless steel, aluminium, and copper. While it is possible to procure metal-plated plastic microwave components from commercial sources specializing in additive manufacturing [19], examples may also be found of direct printing metallic waveguide structures using laser sintering [20]. At GA, additive manufacturing has been used to explore means of lowering fabrication costs, accelerating development of new components, and enhancing performance [21]. The research is divided into two areas: low power and high power microwave components.

4.1. Low Power Components

Many low power microwave applications make use of overmoded components which are fairly compact, such as elements used to transition to corrugated waveguide from fundamental rectangular waveguide. In applications where the volume of the part is within the build size of an additive manufacturing machine, it may be advantageous to print the part, since the turn-around time for a printed part is generally short compared to a machined or electroformed part. With additive manufacturing, it is therefore possible to establish a rapid prototyping development cycle for optimizing a design quickly at relatively low cost.

Direct Metal Laser Sintering (DMLS) has been found useful in making tapered corrugated horns with output apertures as large as 63.5 mm in diameter. While the individual layer thickness of parts produced from current DMLS machines is within the +/-25 μm tolerance resulting from a conventional machining approach, the resolution of a geometry produced on a given layer of printed metal is around 350 μm. As a result, the maximum frequency for such components is limited to around 100 GHz. For higher frequency components, plastic additive manufacturing processes such as PolyJet (15 μm layer thickness, 40 μm X/Y resolution) or projection stereolithography (SLA) (5 μm layer thickness, 25 μm X/Y resolution) are more effective. Commonly used electroless copper and nickel plating processes of plastic parts have been proven for coating the fine corrugations, even in difficult-to-access regions such as the throat of a corrugated horn. The metal plating produced by these processes is not as electrically conductive as electrolytic plating, but the applications for such low power components are usually not focused on minimizing ohmic loss.

A recent example of a 3D printed corrugated component is shown in Figure 13. The part, which represents a section of a transition from WR-6 to 63.5 mm diameter corrugated waveguide, was made by copper plating a corrugated geometry via SLA. Low power measurements of the assembly using the printed part was compared to the assembly using a machined version (in two sections). The resulting HE_{11} mode purity in both instances was above 97%.

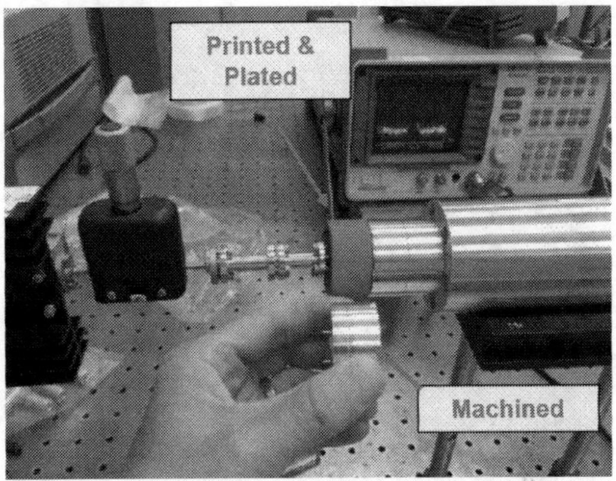

Fig. 13. Metal-plated corrugated section made via additive manufacturing used in assembly to transition from WR-6 to 63.5 mm corrugated waveguide at 140 GHz.

4.2. High Power Components

In the realm of high power microwaves, plastic parts are typically avoided, and metal printing processes such as DMLS are the default technology. The focus is no longer on rapid prototyping, but fabricating robust vacuum-qualified parts which can withstand high temperatures. One advantage that additive manufacturing offers in this scenario is its ability to incorporate conformal cooling channels. While it is possible to provide some amount of complexity to cooling in conventionally machined high power parts, the fabrication process often requires advanced assembly techniques such as brazing or electron beam welding which add technical risk. Additive manufacturing has an established capability of producing internal channels which are consistent with the geometry of the part, and can provide cooling to crucial regions that may be otherwise difficult to access.

https://doi.org/10.1051/epjconf/201920304001

A flat copper mirror, such as one used in a typical overmoded miter bend, is currently being investigated at GA (Figure 14). Two sample mirrors representative of those used in a 63.5 mm diameter waveguide miter have been produced in commercially available DMLS copper alloy materials (C18200 CuCr and CuNi2SiCr). The mirror blocks have identical spiral-shaped cooling structures. This channel geometry is theoretically more efficient in terms of heat transfer than the straight cooling channels enabled by conventional machines for a miter mirror. The surfaces were polished to near mirror-quality (around 8 microinch Ra). No leaks were discovered when vacuum pumping on the mirror surfaces to $1x10^{-9}$ Pa/m^3-s. A static water pressure test was also applied to the cooling channels with 0.7 MPa. Nevertheless, a hot isostatic press (HIP) process was applied to further minimize porosity of the sintered copper alloy.

(a) (b)

Fig. 14. 3D-printed copper mirror for a 63.5 mm miter bend (a), and X-ray image of spiral-shaped internal cooling channel (b).

Following the initial tests to verify the process, a finalized mirror made via DMLS is being planned which may be incorporated into the miter bend housing of a high power transmission line for testing.

This work was funded through internal research and development at General Atomics. The authors would like to acknowledge K. Sakamoto, K. Takahashi, Y. Oda, T. Kobayashi, R. Ikeda (QST), Y. Uemoto (MHI), M. Joung (NFRI), G. Wang (UCLA), L. Yong (ASIPP), T. Goodman (EPFL-SPC), M. Henderson (ITER-IO).

References

1. M. Henderson, "Status of the ITER EC system," presented at US/EU/JA Workshop on RF Heating Tech., Santa Monica, CA, USA, Sep. 4-7, 2017.
2. K. Takahashi *et al.*, "Development of ITER gyrotrons and equatorial launcher," presented at US/EU/JA Workshop on RF Heating Tech., Santa Monica, CA, USA, Sep. 4-7, 2017.
3. T. Goodman *et al.*, "FALCON test facility," presented at US/EU/JA Workshop on RF Heating Tech., Santa Monica, CA, USA, Sep. 4-7, 2017.
4. J.P. Anderson *et al.*, "Heating effects in overmoded corrugated waveguide for ITER," J. Infr. MMW-THz Waves, **37**, 1, 2016.
5. G. Hanson, "ITER ECH transmission line system update," presented at US ECH Tech. Community Meeting, San Diego, USA, Feb. 6, 2017.
6. J. Lohr *et al.*, "Performance history and upgrades for the DIII-D gyrotron complex," presented at US ECH Tech. Community Meeting, Oak Ridge, TN, USA, May 8, 2014.
7. J.P. Anderson, *et al.*, "High power testing of water-cooled waveguide for ITER-like ECH transmission lines," *Nucl. Fusion*, **57**, 5, 2017.
8. W. Bin *et al.*, "Status of the high power matched load activity at IFP-CNR," presented at US/EU/JA Workshop on RF Heating Tech., Santa Monica, CA, USA, Sep. 4-7, 2017.
9. M. Joung *et al.*, "KSTAR ECH status and plan for high performance and steady-state operation," presented at 22nd Conf. on RF Power in Plasmas, Aix-en-Provence, France, May 30 – June 2, 2017.
10. W.H. Safranek, *Properties of Electrodeposited Metals and Alloys*, 2nd Ed., Amer. Electroplaters Soc., Dec. 1986.
11. T. Kobayashi *et al.*, "Progress in ECRF system development and preparation for JT-60SA," presented at US/EU/JA Workshop on RF Tech., Santa Monica, CA, USA, Sep. 4-7, 2017.
12. M. Saigusa *et al.*, "Thermal analysis and ohmic loss estimation of polarizer for ITER ECCD system," *Fusion Eng. Des.*, **74**, 2005.
13. J. Doane, H. Grunloh, W. Martin, and W. Wu, *Fusion Eng. and Design*, **102**, 2016, pp. 99-107.
14. S. Jawla *et al.*, "MIT program update," presented at ECH Tech. Community Meeting, San Diego, USA, Feb. 6, 2017.
15. J.L. Doane, "Tests of compact dummy loads designed for 1 MW CW," *Proc. Of the 12th Joint Workshop on ECE and ECH, Aix-en-Provence, France, 13-16 May 2002*, World Scientific, 2003.
16. J. Lohr *et al.*, "Operating a multiple gyrotron system," presented at US/EU/JA Workshop on RF Heating Tech., Sedona, AZ, USA, Sep. 22-24, 2014.
17. J. Doane, *et al.*, "Power monitor miter bends for high-power microwave transmission," *Fusion Eng. And Design*, **93**, 2015.
18. C. Moeller *et al.*, "Status of mode-selective ECH miter bend power monitor," presented at US/EU/JA Workshop on RF Heating Tech., Santa Monica, CA, USA, Sep. 4-7, 2017.
19. www.swissto12.com
20. T. Horn *et al.*, "Additively manufactured WR-10 copper waveguide," presented at Int'l Vacuum Elect. Conf. (IVEC), Monterey, CA, USA, April 24-26, 2018.
21. J.P. Anderson *et al.*, "Additive manufacturing techniques for millimetre-wave components," presented at 41st Int'l. Conf. on Infrared, Millimeter and THz Waves, Copenhagen, Sep. 25-30, 2016.

Recent progress of the development of a long pulse 140GHz ECRH system on EAST

Handong Xu[1,*], *Xiaojie Wang*[1], *Jian Zhang*[1], *Fukun Liu*[1], *Yiyun Huang*[1], *Jiafang Shan*[1], *Weiye Xu*[1], *Miaohui Li*[1], *J.Lohr*[2], *Y.A.Gorelov*[2], *J.P.Anderson*[2], *Yang Zhang*[1], *Dajun Wu*[1], *Huaichuan Hu*[1], *Yong Yang*[1], *Jianqiang Feng*[1], *Yunying Tang*[1], *Bo Li*[1], *Wendong Ma*[1], *Zege Wu*[1], *Jian Wang*[1], *Liyuan Zhang*[1], *FeiGuo*[1], *Haozhang Sun*[1], *Xinsheng Yan*[1] *and EAST Team*[1]

[1]Institute of Plasma Physics, Chinese Academy of Sciences, Hefei 230031, China
[2]General Atomics, San Diego, CA 92186-5608, USA

Abstract.A long pulse ECRH system with a goal of 140GHz 4MW 100~1000s has been developed to meet the requirement of steady-state operation on EAST. Gycom gyrotrons are employed in the No.1 and No.3 systems, CPI gyrotrons are used in the No.2 and No.4 systems. The development of the two Gycom gyrotron systems has been finished. The first short pulse EC wave injection has been demonstrated successfully during the EAST 2015 Spring campaign. In the commissioning and operation towards steady-state operation, 0.4MW 100s has been injected to plasma successfully by using the No.1 system, 4.7keV 102s L-mode and 102s H-mode plasma have been achieved on EAST with the help of ECRH. Recently, a longest pulse of 0.55MW 1000s has been obtained based on calorimetric dummy load measurements on the No.3 gyrotron. The No.2 gyrotron also has been installed and partially tested, 500kW 80s has been demonstrated in the dummy load. The remaining No.4 gyrotron will be ready to test in 2018 or 2019. The whole 4MW system will be completed within two years. The 400s fully non-inductive H-mode operation would be expected in the next four years in the condition of fully tungsten diverter on EAST.

1 . Introduction

The superconducting tokamak EAST [1] has been in operation since 2006; the main auxiliary systems, including neutral beam injection (NBI) [2], lower hybrid current drive (LHCD) [3] and ion cyclotron resonance heating (ICRH) [4] also have been installed. A long pulse ECRH system, which has been designed to meet the requirement of steady-state operation in EAST, has been planned since 2011. The general preliminary design of the system was completed in 2012, and all sub-systems have been under development since 2012. The system is mainly composed of four 140GHz gyrotron systems, 4 ITER-like transmission lines, 4 independent channel launchers and corresponding power supplies, water cooling, control & interlock system etc. Each gyrotron is expected to deliver a maximum power of 1MW and be operated at 100s~1000s pulse lengths. Gycom gyrotrons are employed in the No.1 and No.3 systems, CPI gyrotrons are used in the No.2 and No.4 systems. Two gyrotron systems were installed in 2014 and the first short pulse EC wave injection has been demonstrated successfully during the EAST 2015 Spring campaign [5]. Significant plasma heating and MHD instability suppression effects were observed. In addition, high confinement (H-mode) discharges triggered by ECRH were obtained.

2 . A long pulse ECRH system on EAST

2.1 Overview of the system

Considering the development progress of gyrotrons[6] around the world, two kinds of gyrotron were considered, those being the 170GHz gyrotron developed for ITER and the 140GHz gyrotron, which is used widely in ASDEX-U, HL-2A, W7-X, etc. The normal operation range of toroidal magnetic fields is about 2.0T~3.0T on EAST tokamak. In order to fit most experiments of EAST in a wide range of magnetic fields, the 140GHz gyrotron was adopted finally such that the EC wave would be coupled to plasma as an X2 mode.

According to the overall arrangement of EAST ports, an equatorial M port was reserved for the ECRH system; correspondingly, the gyrotron system could be located in an adjacent building to the EAST hall. Considering the space limitations of the M port and gyrotron building, 4 sets of gyrotrons were designed. The system layout design is optimized due to limited space in EAST and the gyrotron building. The final layout is shown in Fig.1. The distance between two gyrotrons is about 4.5 m to prevent mutual magnetic disturbance, and the measured stray magnetic field coming from the EAST tokamak

* Corresponding author: xhd@ipp.ac.cn

and its poloidal coils power supply is lower than 2 gauss in the horizontal direction and less than 5 Gauss in the vertical direction, which satisfies the requirement of gyrotrons. The length of the transmission line is 30 m approximately. The system is composed of 4gyrotron systems, 4 waveguide transmission lines consisting of various evacuated corrugated waveguide components, 4 independent channel launchers and corresponding diversified power supplies, water & oil cooling system, control & inter-lock system, auxiliary vacuum system, etc.[7]. Two cathode power supplies are employed to power the gyrotrons, with each cathode power supply feeding two gyrotrons, and an independent body power supply and dummy water load configured for every gyrotron.

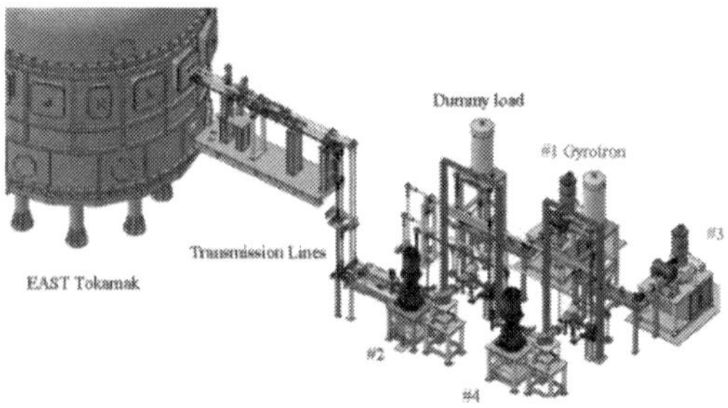

Fig. 1. The layout of gyrotron systems and transmission lines.

2.2 Gyrotrons

The gyrotrons come from two manufacturers. Two tubes, numbers 1 and 3, are made by Gycom Ltd., Russia, and the other two are provided by CPI, USA. Their main parameters are shown in table 1. They are all designed to operate at 140GHz with a power goal of 900kW~1MW and pulse length goal of 100s~1000s. Diamond windows and Gaussian output beams are employed in all gyrotrons. The differences between the two kinds of gyrotron focus on the cavity mode, the cathode and filament, cathode voltage, efficiency, ion pump, matching optic unit (MOU) and collector power limit. A two-mirror MOU is used with the Gycom gyrotrons, and a single mirror MOU is adopted for the CPI gyrotrons. An oil cooling system is required to improve the insulation of the electron gun and remove the heat from the filament or ceramic in all gyrotrons. A series of parameters of 1MW 1s, 900kW 10s, 800kW 95s and 650kW 753s have been demonstrated successfully on the No.1 gyrotron system. Recently, a longest pulse of 0.55MW 1000s has been obtained based on calorimetric dummy load measurements [8] on the No.3 gyrotron. The No.2 gyrotron also has been installed and partially tested, 500kW 80s has been demonstrated in the dummy load. The remaining No.4 gyrotron will be ready to test in 2018 or 2019.

Table 1. Main parameters of gyrotrons.

	No.1	No.3	No.2 and No.4
Manufacturer	Gycom		CPI
Type	Diode		Diode
Cavity mode	TE22,8		TE28,7
Max. output power	1MW		900kW
Pulse Length	100~1000s		100~1000s
Output beam	Gaussian beam		Gaussian beam
Collector type	Depressed		Depressed
Collector power limit	1.0 MW	1.3MW	1.8MW
Cathode voltage	-46kV		-59kV
Beam current	42		40
Body voltage	+25kV		+22kV
Filament power	1000~1200W		200~250W
MOU	Two mirrors		Single mirror
SC magnet	Liquid helium	Helium free	Helium free

2.3 Transmission lines and launcher

An ITER-Like TL system using a 63.5mm diameter evacuated waveguide system was adopted for EAST. As shown in Fig. 2, each TL is a chain consisting of corrugated waveguides, waveguide switch, dummy load, miter bends, polarizers, DC break, bellows, CVD window, and gate valve. The function of each component is similar to the ITER design [9]. The total power loss of each TL is about 10% ~ 15%. When the approximately 10% power dissipated in the MOU and internal loads is taken into account, the remaining 75%~80% fraction of the generated gyrotron power can be injected into EAST.

A launcher with four independent channels was developed for EAST, with the four channels designed in a symmetrical arrangement relative to equatorial plane as well as the M port of EAST tokamak. Each channel uses a fixed focus mirror and a steering mirror, and a section of stainless steel waveguide is adopted to transmit millimeter waves to the focus mirror in the vacuum of EAST. The launcher has a main feature of active cooling in all mirrors which will ensure the availability of CW operation. Through the optimization study of the launcher design [10], the launcher parameters were decided as follows: the poloidal sweep range is 65° ~ 95°, the toroidal range is 165° ~ 205°, and the definition of these angles is consistent with the GA-TORAY code [11]. The lower two channels of launcher have been installed in EAST prior to the 2015 Spring campaign.

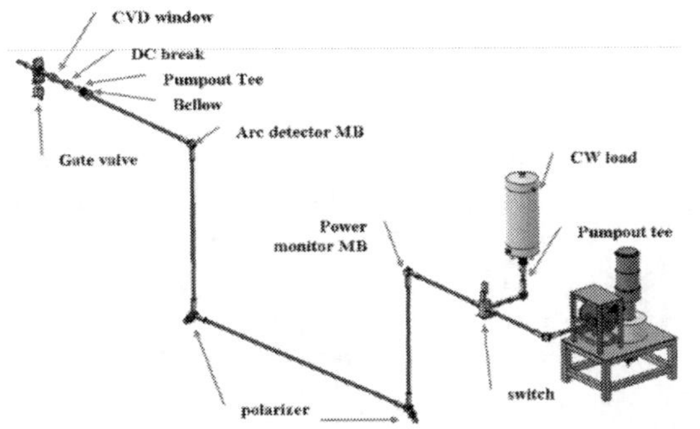

Fig. 2. The layout of transmission lines.

2.4 High voltage power supplies

The four gyrotrons will be fed by two sets of cathode power supplies, with each power supply capable of delivering -60kV/80A in continuous wave (CW) mode, and also -70kV/100A for 100s. The architecture of the cathode high voltage power supply is shown in Fig. 3. A pulse step modulator (PSM) type is used in the cathode HVPS. It has a capability of fast modulation up to 1 kHz, which is intended for the potential application of MHD suppression by modulated ECRH. It has a fast response time of less than 10 microseconds once a fault signal is received, and the maximum energy dissipated in a gyrotron is less than 10J in an arc event. The body HVPS has an output capability of +30kV/100mA in CW mode, with the response time less than 5 microseconds and the dissipated energy is also less than 10J in an arc event.

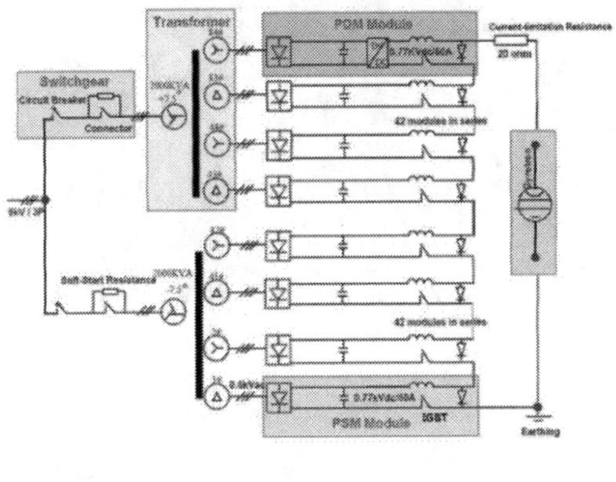

Fig. 3. The architecture of the cathode high voltage power supply.

2.5 Control & Interlock system

A control & interlock system has been developed to ensure the safe operation of the whole system, which consists of several sub-systems, each sub-system having specific functions. The architecture of the control & interlock system is shown in Fig. 4. A central control unit is mainly applied to control cathode and body HVPS according to the turn-on and turn-off sequence of a gyrotron, and also to identify the gyrotron silence fault. The control unit is designed using flexible FPGA technology. A fast response interlock sub-system is developed to detect arc events inside gyrotron or waveguide as well as other over-current faults[12], and send a fault signal to switch off cathode and body HVPS within one microsecond. A RF protection sub-system is developed to monitor the forward and the backward RF power [13]. A PLC sub-system is employed to monitor the general status of the whole system including all electric signals, water and oil cooling parameters and vacuum status etc. It is also used to control mechanical movements of launcher, waveguide switches, and gate valves. A few PXI computer systems are developed for data acquisition [7], polarizer control [14] as well as setting parameters of all power supplies.

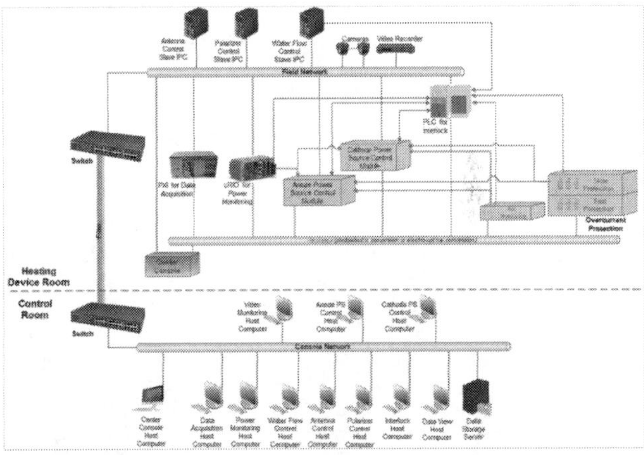

Fig. 4. The architecture of the control & interlock system.

2.6 Auxiliary system

A vacuum pumping system is designed to produce the required vacuum of transmission lines, which should be better than 1×10^{-3}Pa in the long pulse operation. Three pumping ports, which are shown in Fig.2, are reserved in each TL channel. They are located at MOU, water load and a port close to the CVD window on the launcher side. In order to handle the heat load in the long pulse operation of system, a large water cooling system was specially built for the ECRH system; all components of the system should be cooled by water. The water cooling system has a maximum heat exchange capability of 8MW in the CW mode when the environmental wet bulb temperature is less than 28℃ and a required water temperature of less than 35℃.

3 . Commissioning results on dummy load

3.1 Alignment commissioning of gyrotron

The expected low loss mode of the transmission line is the HE11 mode, and the purity of the mode is very important for the EC wave. In the initial installation of No.2 gyrotron system, a large vertical tilt and offset resulted in a lower fraction of HE11 content, which was measured by infrared camera and calculated based on phase retrieval analysis [15]. A very careful adjustment by shimming gyrotron structure has been carried out to improve the alignment. Finally the tilt was reduced from 0.39 deg to 0.2deg, and the offset was reduced by about 3 mm shown in Fig. 5. Fig. 6 shows the phase front of the back-propagated beam at the waveguide mouth is seen to be extremely flat after adjustment, the signature of a high quality Gaussian beam. Analysis shows the HE11 purity was increased from 67.5% to 89.5%, which is similar to results of the DIII-D ECRH system [16].

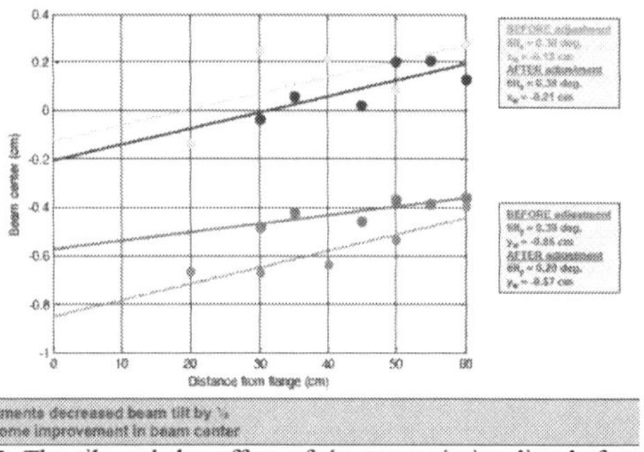

Fig. 5. The tilt and the offset of the transmission line before adjustment and after adjustment.

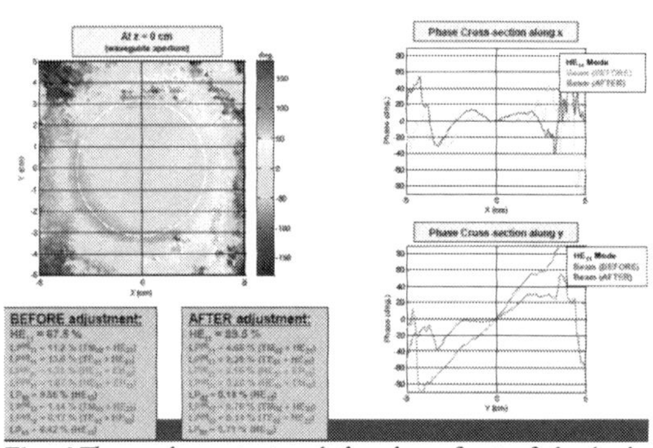

Fig. 6. The mode content and the phase front of the back-propagated beam at the waveguide mouth before adjustment and after adjustment.

3.2 Long pulse commissioning on dummy load

Long pulse tests into a dummy load have been carried out for No.1, No.2, and No. 3 gyrotron systems in 2015~2018.

A series of parameters of 1MW 1s, 800kW 95s and 650kW 753s have been demonstrated successfully on the No.1 gyrotron system. These results were obtained on corresponding shots without filament compensation. Fig. 7 shows waveforms of the 650W 753s pulse, the pulse was stopped by an over-current fault.

The best results of 500kW 80s, 800kW 300ms have been achieved successfully on No.2 gyrotron system in ASIPP, and 460kW 1000s has been obtained in CPI factory testing. Fig.8 shows waveforms of the 500kW 80s pulse with filament compensation.

A series of parameters of 780kW 20s, 630kW 100s and 550kW 1000s have been demonstrated successfully on the No.3 gyrotron system. Fig. 9 shows waveforms of the 550kW 1000s pulse with filament compensation.

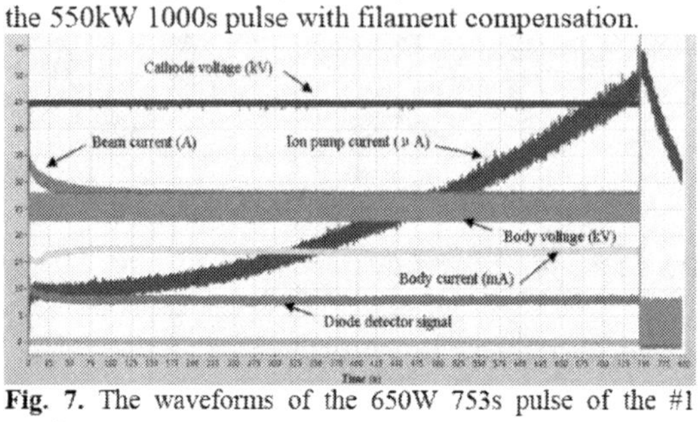

Fig. 7. The waveforms of the 650W 753s pulse of the #1 gyrotron.

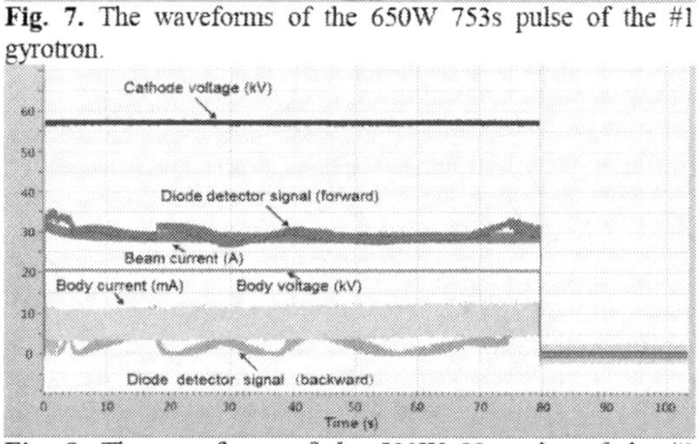

Fig. 8. The waveforms of the 500W 80s pulse of the #2 gyrotron.

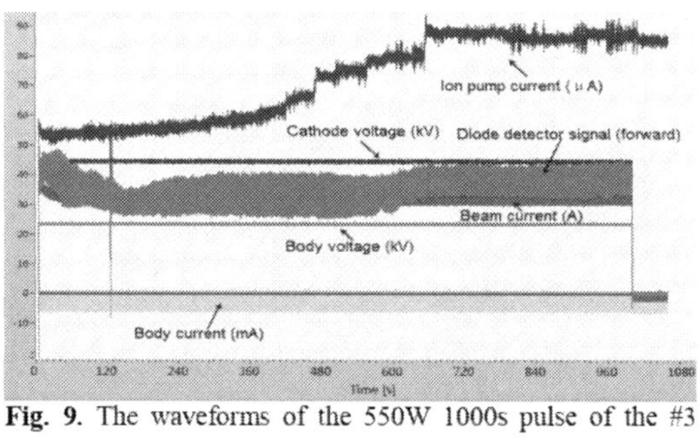

Fig. 9. The waveforms of the 550W 1000s pulse of the #3 gyrotron.

4.Experimental results on EAST

4.1 Verification of polarizer control

The EC waves can propagate in the plasma in the ordinary (O) and extraordinary (X) modes. The toroidal field on EAST is normally in the range of 2.0~2.6T. For 140GHz waves, the interactions happen at the second harmonic electron cyclotron resonance in plasma. Because the ordinary mode is not well absorbed at higher harmonic resonances, we choose the second harmonic extraordinary(X2) as the operating mode. The wave coupling in plasma is determined by the polarization of injected EC wave.

The plasma absorption efficiency is affected by different polarizations. Fig. 10 shows this effect. When the polarization is not right, the EC waves can not be well absorbed by the plasma. The EC waves directly hit the first wall of tokamak. The obvious bright spots can be seen through an infrared camera.

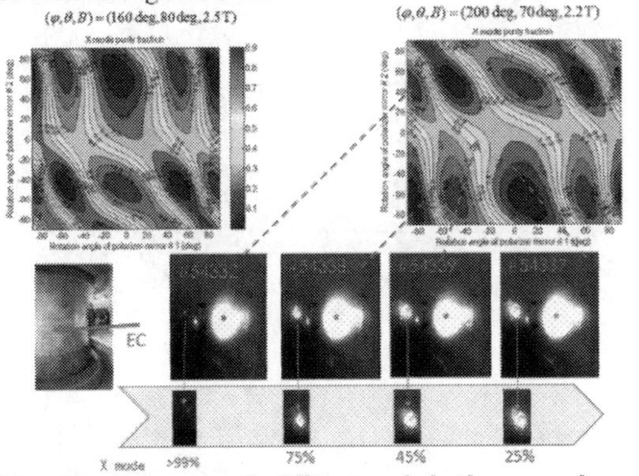

Fig. 10. Influence of different polarization on plasma absorption efficiency.

4.2 First short pulse EC wave injection

The No.1 system was applied in EAST experiments from 2015. The first short pulse EC wave injection has been demonstrated successfully during the EAST 2015 Spring campaign. Significant plasma heating and MHD instability suppression effects were observed.Fig. 11 and Fig. 12 show a typical waveform of ECRH power injected into Ohmic target plasma (plasma current I_p = 400 kA, line-averaged density $n_{e\_av} = 2.2 \times 10^{19} / m^3$). In can be seen that, when the EC wave with P_{EC} = 400 kW turned on, the plasma stored energy (W_{MHD}) and the central electron temperature (T_{e0}) were increased by 26 kJ and 1 keV respectively, compared with the Ohmic heating phase, which suggests good plasma heating effect was obtained by the EC wave.

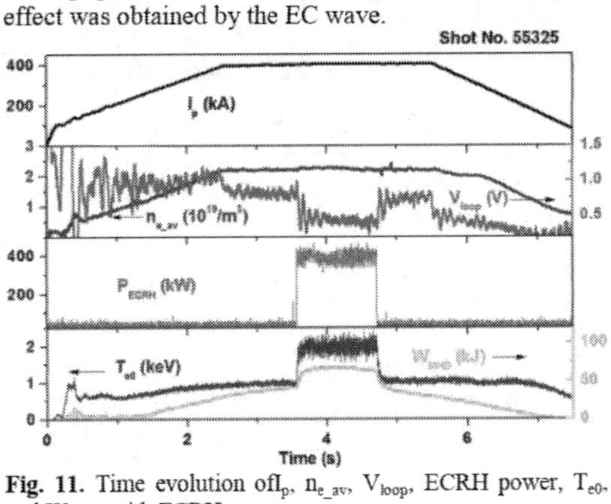

Fig. 11. Time evolution ofI_p, n_{e_av}, V_{loop}, ECRH power, T_{e0}, and W_{MHD} with ECRH.

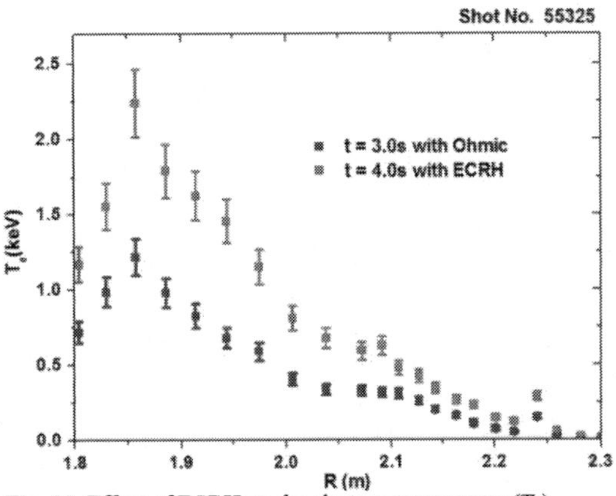

Fig. 12. Effect of ECRH on the electron temperature (T_e).

4.3 Application in 100 s high T_e operation

The long pulse operation also has been attempted to be applied during the EAST 2016 Spring campaign. A fully non-inductive 102s plasma discharge has been sustained successfully by the synergy of LHCD and ECRH[17]. Fig. 13 shows the typical waveforms of ECRH power injected into LHCD target plasma (plasma current I_p = 400 kA, line-averaged density $n_{e\_av} = 1.5\times 10^{19} / m^3$), the loop voltage is maintained at zero. In order to guarantee the EC power absorbed in the central region, the toroidal injection angle and poloidal angle were set to be 180 ° and 75 ° respectively. The magnetic field on the axis was about 2.5 T, corresponding to the value for the cold 2nd harmonic resonance. Fig. 14 shows the central electron temperature is close to 5keV, which is measured by Thomson scattering diagnostics.

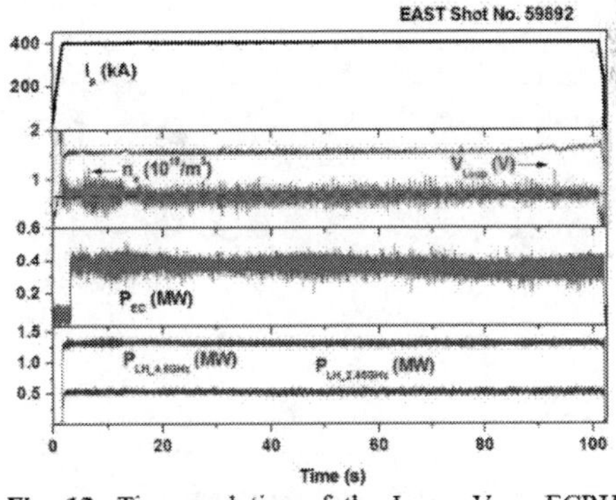

Fig. 13. Time evolution of the I_p, n_e, V_{loop}, ECRH power, LHCD (4.6 GHz) power, and LHCD (2.45 GHz) power for the 102s high T_e operation of EAST.

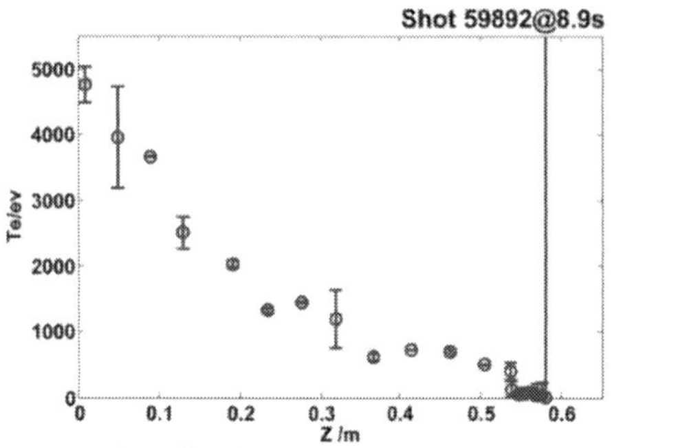

Fig. 14. The radial distribution of electron temperature for the 102s high T_e operation on EAST.

4.4 Application in H-mode experiments

The long pulse H-mode operation has been attempted to be applied during the EAST 2016~2017 campaign. The minute-long steady-state H-mode discharges on EAST have been realized in the EAST 2016 Autumn campaign [18]. A steady-state H-mode 101s plasma discharge has been sustained successfully by the synergy of LHCD, ICRF, and ECRH in the EAST 2017 Summer campaign.

Fig. 15 shows the typical waveforms with ECRH power injected into target plasma (Shot #73999, I_p=0.4MA, B_t=2.5T, PRF=3.0MW, ne=3.0x10^{19}/m3, T_e=4.0KeV, H_{98y2}=1.1, USN).

EAST successfully implements the H-mode with small disturbance amplitude edge local mode (ELMs) under conditions of low momentum injection of pure RF heating, tungsten diverters. A fully non-inductive steady-state high-performance plasma with a constrained improvement factor of H_{98y2} greater than 1.1 was obtained, and a soft landing of the plasma operation was achieved through precise control of the advanced magnetic configuration.

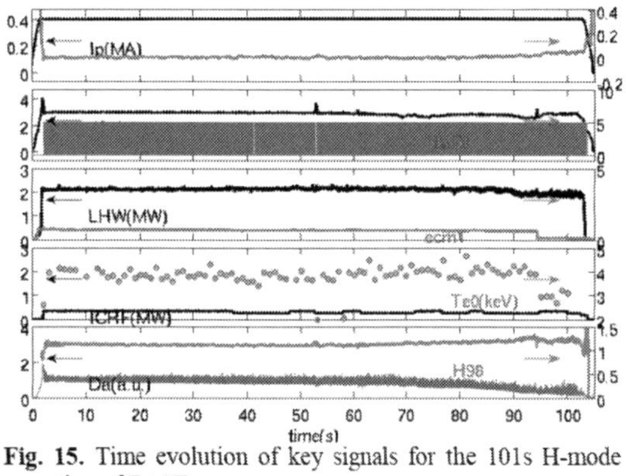

Fig. 15. Time evolution of key signals for the 101s H-mode operation of EAST.

The performance degradation in an electron heating dominant H-mode plasma after ECRH termination in EAST was found [19].

As shown in Fig. 16, the stored energy (W_{MHD}) drops after ECRH termination. The reason for the energy confinement degradation may be the shift of LHW power deposition from a central region with lower diffusivity to a peripheral region with larger diffusivity after ECRH termination[19]. The feedback mechanisms are shown in Fig. 17.

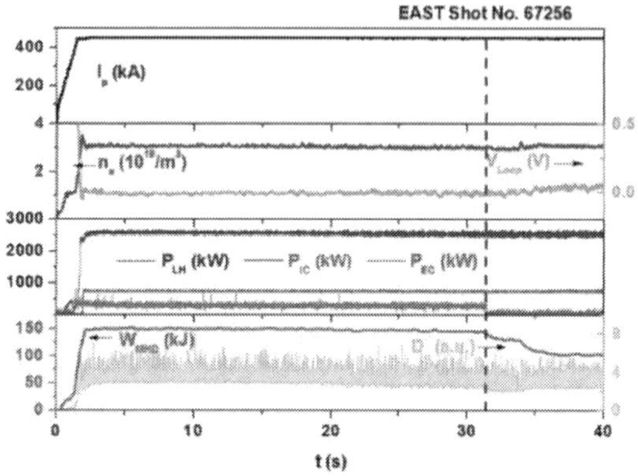

Fig. 16. Time evolution of key signals of the shot # 67256.

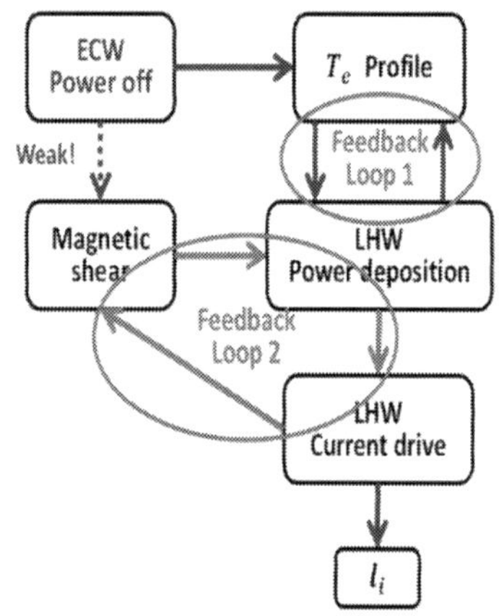

Fig. 17. Feedback mechanisms based on parametric dependences of LHW deposition.

4.5 Low loop voltage start-up experiments

We have tried to do the low loop voltage start-up experiments. A experiment results are shown in Fig. 18, With the pre-ionization by ECR and LHW, the minimum central toroidal electric field can be reduced to ~0.16V/m.

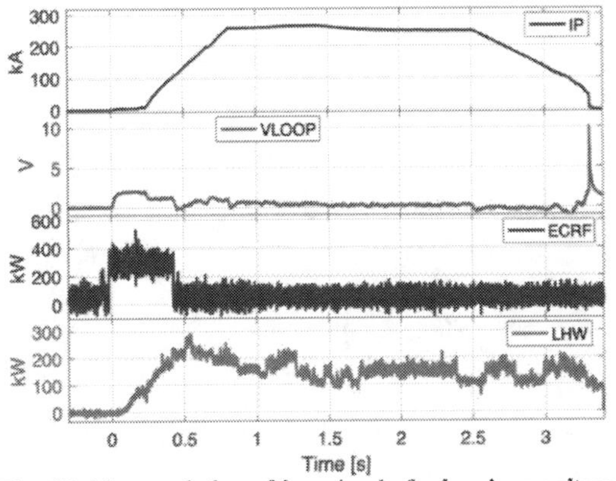

Fig. 18. Time evolution of key signals for low loop voltage start-up experiments.

4.6 Operation reliability of No.1 system

The gyrotron reliability is an important issue for long pulse plasma operation. The first short pulse EC wave injection using No. 1 gyrotron has been demonstrated successfully during the EAST 2015 Spring campaign. The discharge success rate of this gyrotron is good in the 2015 experiments and in the beginning of the 2016 experiments. Then, the reliability decreased in the first half of 2016 experiments which is shown in Fig.19. There may be some damage inside the gyrotron. In order to solve this problem, we tried to make low-power, improve conditioning and optimize the operation region. Now, the reliability of No. 1 gyrotron is good for 300 kW~500 kW output power.

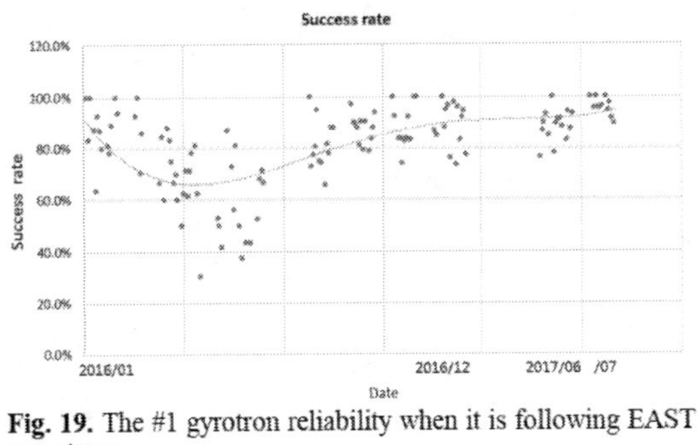

Fig. 19. The #1 gyrotron reliability when it is following EAST experiment.

5. Future plan

In order to explore high β_N(>2) H-mode operation shown in Fig.20 on EAST, #1 system will be upgraded to dual frequency (140/105GHz) mode to fit low B_T (2.5T→1.9T) operation.

For the purpose of upgrading the No.1 gyrotron to dual frequency operation, the collector, mirrors system and electron gun of #1 gyrotron will be upgraded, slight modification of transmission line is required. The parts that need to be upgraded are marked in red fonts in Fig. 21. It is expected to deliver a power of 0.6MW/100s~1000s at 140GHz & 105GHz.

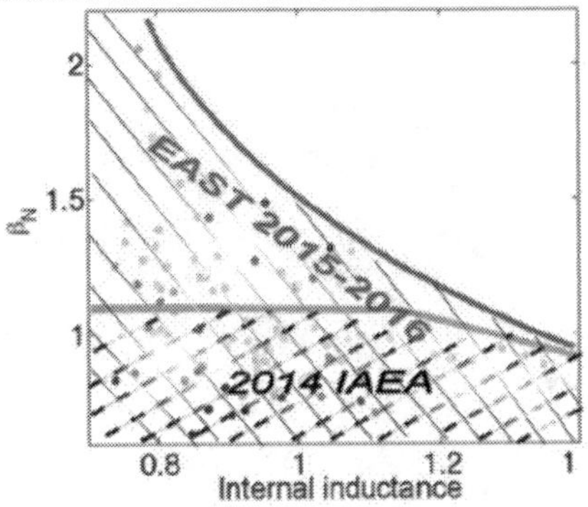

Fig. 20. The β_N in the relationship with internal inductance.

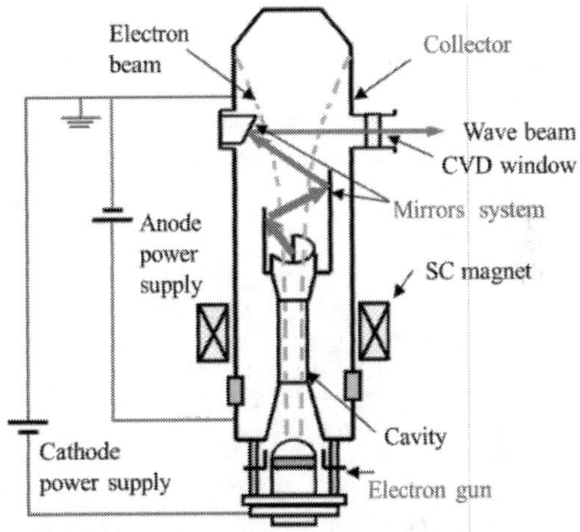

Fig. 21. The composition of the gyrotron, where the parts marked in red fonts are parts that need to be upgraded.

6. Conclusions

A long pulse ECRH system with a goal of 4MW 100~1000s has been developed for EAST, which comprises four 140GHz gyrotron systems. The 3MW system has been installed and partially tested, and long pulse operation of the system has been demonstrated successfully during the EAST 2016~2017 campaign. The remaining No. 4 CPI gyrotron will be ready to start testing in 2018 or 2019, and the whole system will be completed within two years. The 400s fully non-inductive H-mode operation would be expected in the next four years in the condition of fully tungsten diverter on EAST.

Acknowledgements

This work was supported by the National Key R&D Program of China(Grant Nos.2017YFE0300401 and 2016YFA0400600).The authors would like to thank Gycom& CPI experts for their great effort in the development of gyrotrons. We also thank all colleagues from GA, PPPL, NIFS, CEA and SWIP for their help in the design and commissioning of EAST ECRH system.

References

1. B. Wan, Nucl Fusion **49**, 104011 (2009)
2. C. Hu and N. B. I. Team, Plasma Science and Technology **14**, 567 (2012)
3. L. Zhao, J. Shan, F. Liu, H. Jia, M. Wang, L. Liu, X. Wang, and H. Xu, Plasma Science and Technology **12**, 118 (2010)
4. Y. P. Zhao *et al.*, Fusion Eng Des **89**, 2642 (2014)
5. H. Xu *et al.*, Plasma Science and Technology **18**, 442 (2016)
6. K. Sakamoto, Fusion Science & Technology **52**, 145 (2007)
7. W. Xu, H. Xu, F. Liu, F. Hou, and Z. Wu, Fusion Eng Des **113**, 119 (2016)
8. W. Xu, H. Xu, F. Liu, J. Wang, X. Wang, and Y. Hou, Plasma Science and Technology **19**, 105602 (2017)
9. F. Gandini *et al.*, Fusion Sci Technol **59**, 709 (2011)
10. X. Wang *et al.*, Fusion Engineering & Design **96-97**, 181 (2015)
11. W. Wei, B.-J. Ding, X.-J. Zhang, X.-J. Wang, M.-H. Li, E.-H. Kong, and L. Zhang, Chinese Phys B **23**, 055201 (2014)
12. W. Xu, H. Xu, J. Feng, Y. Yang, H. Hu, and F. Liu, IET Science, Measurement & Technology, Online, DOI: 10.1049/iet-smt.2017.0214, (2018)
13. W. Xu, H. Xu, F. Liu, Y. Tang, Z. Wu, X. Wang, J. Wang, and J. Feng, IEEE Access **4**, 5809 (2016)
14. D. J. Wu *et al.*, J Fusion Energ **33**, 634 (2014)
15. J. P. Anderson, M. A. Shapiro, R. J. Temkin, and D. R. Denison, Ieee T Microw Theory **50**, 1526 (2002)
16. Y. A. Gorelov, J. Lohr, A. Torrezan, J. P. Anderson, D. Ponce, and M. Cengher, in *APS Meeting*2014)
17. J. Qian *et al.*, Plasma Science and Technology **18**, 457 (2016)
18 X. Gong *et al.*, Plasma Science and Technology **19**, 032001 (2017)
19. H. Du *et al.*, Nucl Fusion **58**, 066011 (2018)

Studies towards an upgraded 1.5 MW gyrotron for W7-X

Konstantinos A. Avramidis[1,*], *Tobias* Ruess[1], *Felix* Mentgen[1,4], *Jianbo* Jin[1], *Dietmar* Wagner[2], *Gerd* Gantenbein[1], *Stefan* Illy[1], *Zisis C.* Ioannidis[1], *Heinrich P.* Laqua[3], *Ioannis Gr.* Pagonakis[1], *Tomasz* Rzesnicki[1], *Manfred* Thumm[1], *Robert C.* Wolf[3], and *John* Jelonnek[1]

[1]Karlsruhe Institute of Technology, 76131 Karlsruhe, Germany
[2]Max Planck Institute for Plasma Physics, 85748 Garching, Germany
[3]Max Planck Institute for Plasma Physics, 17491 Greifswald, Germany
[4]*Present affiliation:* European Space Agency - ESTEC, 2200 AG Noordwijk, The Netherlands

Abstract. Studies towards a 1.5 MW, 140 GHz CW gyrotron, with the capability of MW-class operation also at 175 GHz, are ongoing at Karlsruhe Institute of Technology in view of a possible future upgrade of the ECRH system of the stellarator W7-X. The upgrade of the existing 1.0 MW, 140 GHz European gyrotron for W7-X has been chosen as a development path. Detailed designs of the cavity, the non-linear uptaper, and the quasi-optical launcher for the upgraded gyrotron have been obtained and have been validated numerically. In parallel, a mode generator, intended for low-power tests of the quasi-optical mode converter system of the upgraded gyrotron, has been designed, manufactured, and successfully tested.

1 Introduction

The stellarator Wendelstein 7-X (W7-X) is equipped with a steady-state (1800 s) Electron Cyclotron Resonance Heating (ECRH) system consisting of ten 1 MW gyrotrons, which operate at 140 GHz [1]. The available electron cyclotron heating and current drive power in the plasma ranges from 7 to 9 MW [2, 3]. Hence, W7-X is using the world's largest ECRH system today. The ECRH is used for plasma start-up, heating and current drive, and plasma vessel conditioning.

The possibility of even higher ECRH power in the future is under discussion. In addition, MW-class microwave power at a frequency around 175 GHz would be attractive for Collective Thomson Scattering (CTS) diagnostics. Motivated by these considerations, studies towards an upgraded 140 GHz, 1.5 MW Continuous Wave (CW) gyrotron, with the option for MW-class operation also at 175 GHz, have been initiated at Karlsruhe Institute of Technology [4]. Several possibilities have been investigated, using successful existing European gyrotron designs as a starting point. It has been identified that the most promising development path, with respect to risk and cost, would be the upgrade of the existing W7-X $TE_{28,8}$-mode gyrotron in order to operate in the $TE_{28,10}$ mode at 140 GHz and in the $TE_{36,12}$ mode at 175 GHz.

Following this outcome, detailed designs for the gyrotron cavity and non-linear uptaper as well as for the launcher of the quasi-optical mode converter system of the gyrotron have been obtained and validated numerically. The designs fulfil the specifications and require the smallest changes with respect to the existing W7-X gyrotron layout. A first assessment of the necessary modifications for the rest of the gyrotron components has also been made. In support to the design work, the components of a mode generator, intended for low-power tests of the quasi-optical mode converter system, have been designed and manufactured. The mode generator is based on the obtained cavity design and is using a coaxial cavity with perforated wall. It has been assembled and experimentally tested, delivering very good results.

This paper is organized as follows: In Section 2, the selection of the operating modes for the upgraded gyrotron, driven by particular specifications and constrains, is described. The designs for the cavity, non-linear uptaper, and launcher are given in Section 3, together with their numerical validation. In Section 4, the design and assembly of the mode generator as well as the results from its experimental testing are presented. Finally, Section 5 gives an outlook to the next steps towards the realisation of the upgraded gyrotron.

2 Operating mode selection

The selection of the operating modes and the design parameters for the upgraded gyrotron for W7-X has primarily been driven by the following list of requirements:

(i) CW operation at 140 GHz with a power of 1.5 MW
(ii) CW MW-class operation at 175 GHz
(iii) Minimization of risk and cost of development
(iv) Use of the existing gyrotron power supplies at W7-X

For the first requirement, the order of the operating TE mode for 140 GHz operation should be high enough to ensure acceptable Ohmic loading of the wall of the

* Corresponding author: konstantinos.avramidis@kit.edu

gyrotron cavity. The limit for the wall loading is 2 kW/cm^2.

The simultaneous fulfilment of the first and second requirement imposes two additional criteria for the selection of the pair of modes for 140 GHz and 175 GHz operation: (a) The ratio of the eigenvalues of the two modes needs to be as close to 175/140 as possible, in order to take full advantage of the corresponding $\lambda/2$ resonances of the output window of the gyrotron and minimize the window reflection. (b) The caustic radii $R_{caustic}$ of the two modes must be very similar, in order to be possible for the electron gun and the quasi-optical launcher to support both modes.

To address the third requirement, the upgraded gyrotron should be based on already proven long-pulse gyrotron designs and the deviation from these designs must be as small as possible. With respect to this, the European $TE_{28,8}$-mode 140 GHz, 1 MW CW gyrotron for W7-X [5] and the European $TE_{32,9}$-mode 170 GHz, 1 MW CW gyrotron for ITER [6] have been considered as starting points.

Under the aforementioned considerations, the most promising mode-pairs for the upgraded gyrotron have been identified and are shown in Table 1.

Table 1. Possible modes for the upgraded W7-X gyrotron.

Starting point	W7-X gyrotron	ITER gyrotron
	140 GHz operation	
Operating mode	$TE_{28,10}$	$TE_{32,9}$
Electron beam radius (mm)	10.0	11.4
	175 GHz operation	
Operating mode	$TE_{36,12}$	$TE_{40,11}$
Electron beam radius (mm)	10.2	11.4
Window reflection (%)	0.10	0.15
Difference in $R_{caustic}$ (%)	2.64	0.26

According to Table 1, if the existing 140 GHz $TE_{28,8}$-mode gyrotron for W7-X is used as starting point for the upgrade, then it should be modified for operation with the $TE_{28,10}$ mode, requiring a 10 mm electron beam radius at the cavity. The beam radius of the existing $TE_{28,8}$-mode gyrotron is also 10 mm [5]. This implies that the upgraded gyrotron could use practically the same cathode and beam tunnel as those of the existing W7-X gyrotron. However, this would not be the case if the existing 170 GHz $TE_{32,9}$-mode gyrotron for ITER, which uses a beam radius of 9.44 mm [6], was used as starting point. According to Table 1, the required beam radius for the upgraded gyrotron would then be significantly larger (11.4 mm); thus, a significantly different cathode and beam tunnel should probably be used for the upgraded gyrotron. From this observation, it was concluded that it is preferable, in terms of cost and risk, to use the existing W7-X gyrotron as the basis for the upgraded W7-X gyrotron.

Finally, with respect to the fourth requirement, the capabilities of the existing high-voltage power supplies at W7-X allow for a ~3.3 MW electron beam with cathode voltage up to 65 kV and preferred acceleration voltage in the range of 70-85 kV. The nominal 1 MW operation of the existing $TE_{28,8}$-mode gyrotron calls for an accelerating voltage V_{acc} = 81 kV, a beam current I_b = 40 A, and an electron velocity ratio α = 1.3 [5]. To support an output power of 1.5 MW, the accelerating voltage and the current should be increased accordingly. Since the margin for accelerating voltage increase is not very large, operating points in the vicinity of V_{acc} ~ 80 kV, I_b ~ 55 A, α = 1.3 have been considered for the upgraded 1.5 MW gyrotron, in order to be compatible with the existing power supplies at W7-X.

3 Design of components

3.1 Cavity and non-linear uptaper

A cavity and a non-linear uptaper have been designed for the upgraded gyrotron, to allow for 1.5 MW of output power with the $TE_{28,10}$ mode at 140 GHz. To keep the similarity with the existing W7-X gyrotron, the total length of the cavity and uptaper has been kept the same. In addition, the radius at the cavity entrance has been also kept the same to permit the use of a beam tunnel similar to that of the existing W7-X gyrotron. The radius at the cavity midsection is 22.83 mm, which is 2.35 mm larger than that of the midsection of the existing $TE_{28,8}$-mode cavity, to secure oscillation of the higher-order $TE_{28,10}$ mode at 140 GHz. A schematic of the cavity and uptaper design is shown in Fig. 1.

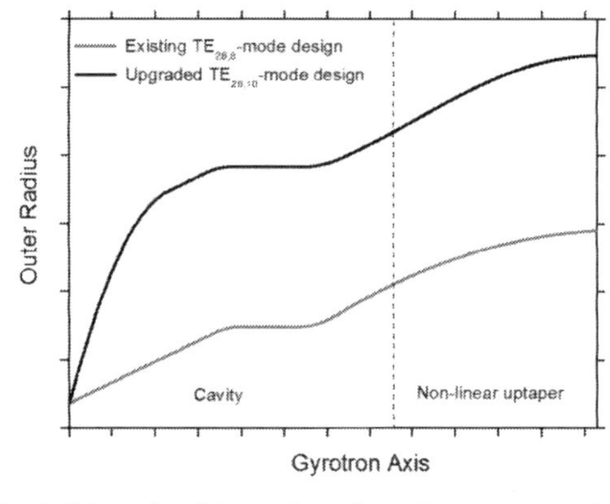

Fig. 1. Schematic of the cavity and non-linear uptaper of the existing W7-X gyrotron and of the upgraded W7-X gyrotron.

The performance of the non-linear uptaper, in terms of mode conversion, has been validated numerically. The calculated mode conversion is minor, resulting in 99.87% transmission of the $TE_{28,10}$ mode and in 99.81% transmission of the $TE_{36,12}$ mode at the end of the non-linear uptaper.

The performance of the cavity and the non-linear uptaper, in terms of beam-wave interaction, has also been numerically validated using the code-package EURIDICE [7]. The diode start-up is simulated by increasing the beam voltage V_b from 35 kV to 85 kV. The beam voltage is proportional to the electron kinetic energy ($V_b = E_{kin}/e$) and, during start-up, it follows the increase of the accelerating voltage applied by the power supplies. During the voltage increase in the simulation, the electron velocity ratio is changing adiabatically and the beam current obeys the Schottky formula. A large number of competing modes, with good resonance and coupling with the electron beam, are interacting simultaneously. A realistic magnetic field profile is considered, which is based on the profile of an existing magnet, used to test the gyrotrons for W7-X. In addition, relatively large spreads in the electron energy (1 % rms) and in the velocity ratio α (20 % rms) are assumed. A typical spread of two Larmor radii in the electron guiding-centres is also taken into account. All these considerations make the simulations as realistic as possible with respect to this stage of development.

The operation at 140 GHz, assuming a maximum magnetic field of 5.55 T, is illustrated in Fig. 2. During start-up, a series of modes is excited before the excitation of the operating $TE_{28,10}$ mode at around $V_b = 64$ kV. Then, the operating mode reaches the nominal operating point ($V_b = 78.5$ kV, $I_b = 56$ A, $\alpha = 1.3$) where it delivers 1.68 MW of microwave power at the end of the non-linear uptaper with an interaction efficiency of 39.5 %. Assuming the typical 5% additional losses until the gyrotron window, this corresponds to 1.6 MW of power at the window. There is a stability margin of 2 kV above the nominal point before mode loss. It should be noted here that the recent developments in the control of gyrotron operation [8] show very promising results regarding the feasibility of operation close to the stability limit of the mode.

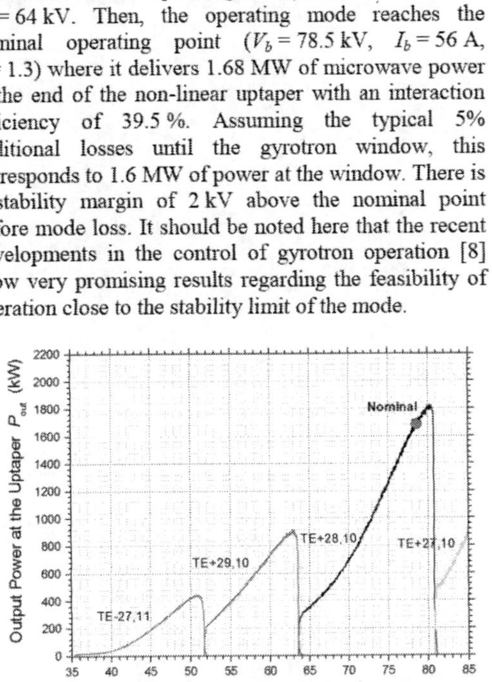

Fig. 2. Multi-mode simulation (36 competing modes) of the diode start-up for gyrotron operation at 140 GHz. A negative azimuthal index denotes a counter-rotating mode.

For the operation at 175 GHz, a maximum magnetic field of 7.05 T is assumed. Since an electron gun of the diode type is intended to be used for the upgraded gyrotron, there is some challenge in obtaining the appropriate values for the electron velocity ratio α both at a magnetic field of 5.55 T and at a magnetic field of 7.05 T, which is ~ 30 % higher. If the same electron energy is assumed in both cases, then for $\alpha = 1.3$ at 5.55 T, a first estimation of the achievable velocity ratio at 7.05 T is $\alpha \sim 0.8$ only. However, with a careful electron gun design and a magnet that permits some control over the amplitude and angle of the magnetic field vector in the gun region, higher values for α could be feasible [9]. For this reason, at this stage of studies, two different cases for the operation at 175 GHz have been examined: $\alpha \sim 0.8$ and $\alpha \sim 1.1$. The simulation results for operation at 175 GHz in the $TE_{36,12}$ mode are shown in Fig. 3. It turns out that stable MW-class operation is possible for both cases. In the case of $\alpha = 1.1$, the nominal operating voltage is not chosen on the grounds of a 2 kV stability margin but is limited to a lower value to secure a cavity wall loading of 2 kW/cm^2. Since operation at 175 GHz is required for CTS diagnostics, the spectral purity of the gyrotron signal is of paramount importance and, therefore, studies on the spectral purity of the present design are planned. Nevertheless, previous studies on a gyrotron for 60 GHz CTS [10] indicated that a spectral purity of ~100 dB within ±5 GHz of the central frequency can be expected.

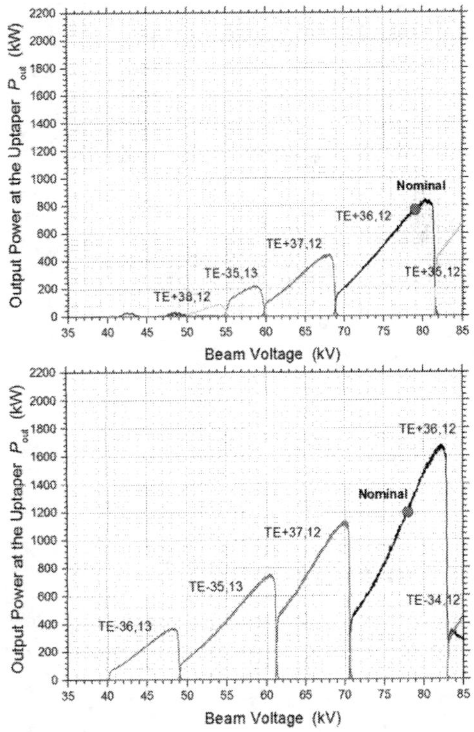

Fig. 3. Multi-mode simulation (46 competing modes) of the diode start-up for gyrotron operation at 175 GHz, assuming $\alpha = 0.8$ (top) and $\alpha = 1.1$ (bottom). A negative azimuthal index denotes a counter-rotating mode.

The operating parameters and the calculated performance of the upgraded gyrotron at 140 GHz and 175 GHz are summarized in Table 2. The cavity radius has been chosen in a way to give an oscillating frequency of 140.4 GHz for the $TE_{28,10}$ mode, in order to anticipate a frequency down-shift due to the cavity thermal expansion during long-pulse operation [11].

Table 2. Operating parameters and calculated performance of the upgraded W7-X gyrotron.

	140 GHz operation	175 GHz operation
	Operating parameters	
Operating mode	$TE_{28,10}$	$TE_{36,12}$
Electron velocity ratio α	1.30	0.82/1.13
Electron beam radius (mm)	10.0	10.2/10.3
Maximum magnetic field (T)	5.55	7.05/7.00
Beam voltage (kV)	78.5	79.0/78.0
Accelerating voltage[1] (kV)	82.1	81.8/81.0
Beam current (A)	55.7	55.8/55.6
	Calculated performance	
Frequency (GHz)	140.4	175.8
Power at end of uptaper (MW)	1.68	0.78/1.22
Interaction efficiency (%)	39.5	17.6/28.0
Ohmic wall loading[2] (kW/cm^2)	1.8	1.1/2.0
Power at window[3] (MW)	1.60	0.74/1.16
Total efficiency w/o depressed collector[1-3](%)	35.0	16.2/25.8

[1]Assuming a 60 % neutralization of the beam space charge [11]
[2]Assuming an effective conductivity of 1.73×10^7 S/m
[3]Assuming 5 % internal losses between uptaper and window

3.2 Quasi-optical launcher

A quasi-optical launcher has been designed for the upgraded gyrotron, taking into account the possibility for operation both at 140 GHz and at 175 GHz. The initial radius and taper angle of the launcher wall are compatible with the end of the non-linear uptaper in Fig. 1. The overall length of the launcher has been kept similar to that of the existing launcher of the $TE_{28,8}$-mode gyrotron at W7-X. The launcher is of the hybrid type [12] and the design of its perturbed inner wall is shown in Fig. 4. In the figure, the axis z of the cylindrical coordinate system coincides with the gyrotron axis. The azimuthal coordinate is denoted by α. The calculated performance of the launcher is very good. Assuming operation with the $TE_{28,10}$ mode at 140.0 GHz, a Gaussian microwave beam with a Gaussian mode content of 97.4% at the launcher aperture is generated. For operation with the $TE_{36,12}$ mode at 175.4 GHz, the Gaussian mode content of the generated microwave beam at the launcher aperture is 95.3%. The contour plots of the normalised amplitude of the electric field on the launcher surface are shown in Fig. 5. It should be noted that the two frequencies used in the calculations are the expected operating frequencies after the thermal expansion of the cavity. The Gaussian mode content in both cases is expected to be further increased with the help of the gyrotron mirror system, which will be designed at a next stage.

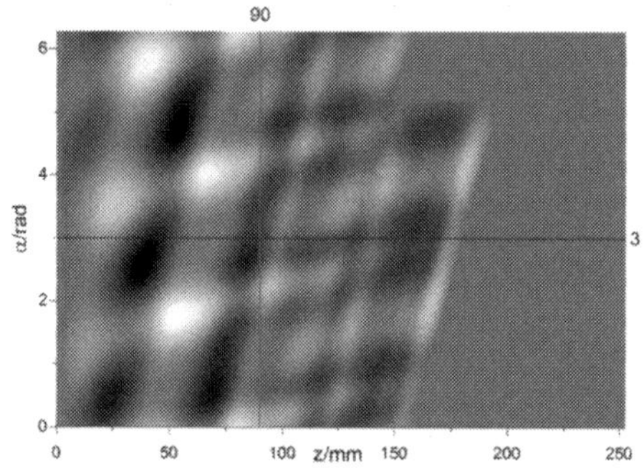

Fig. 4. Contour plot of wall perturbations of the designed hybrid-type quasi optical launcher for the upgraded gyrotron.

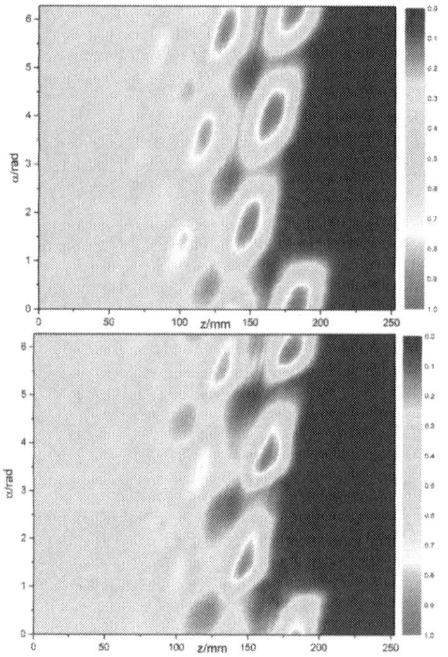

Fig. 5. Contours of the normalised amplitude of the electric field on the launcher surface for 140.0 GHz operation (top) and for 175.4 GHz operation (bottom).

4 Mode generator

A mode generator is necessary for low-power tests of the quasi-optical mode converter system of the gyrotron (launcher & mirrors). Based on the cavity and non-linear uptaper of Fig. 1, the components for a mode generator, namely the cavity and the uptaper, have been designed and manufactured. The assembled mode generator is shown in Fig. 6.

Fig. 6. Mode generator set-up.

The high-frequency wave is generated by a network analyser and is transmitted, using a horn antenna, as a spherical wave. The spherical wave is converted into a plane wave with a lens system (component at the left hand side in Fig. 6). A quasi-parabolic mirror is required to focus the wave on the caustic radius of the rotating mode to be excited in the cavity of the mode generator. The cavity (gold-coloured component near the centre of Fig. 6) has been designed using a scattering matrix code [13]. For easier excitation of the mode, a coaxial cavity is used. The cavity wall is perforated so that the wave can propagate from the quasi-parabolic mirror into the cavity [14].

The excited mode in the cavity needs to have a high purity, in order to resemble the gyrotron mode and make the low-power tests of the quasi-optical mode converter system reliable. In order to check the purity of mode excitation in the cavity, a non-linear uptaper is mounted after the cavity, to reach a diameter of 70 mm for better resolution in the measurements. (For the low-power test of the gyrotron launcher this uptaper will be removed and replaced by the launcher.) The excited mode is measured using a pick-up open-ended rectangular waveguide, which scans stepwise the plane in front of the non-linear uptaper. The measured signal is evaluated using the network analyser.

The frequency of the $TE_{28,10}$ mode and of the $TE_{36,12}$ mode in the mode generator cavity was found to be 140.155 GHz and 175.992 GHz, respectively. These frequencies are somewhat higher than the nominal frequencies because there is some uncertainty in the resonant frequencies in the cavity design because of the perforations. The diffractive quality factor of the cavity was measured to be 2600 and is in good agreement with the simulation. The mode pattern (electric field amplitude – vertical polarization) has been acquired using stepwise scanning with the pick-up waveguide. The resolution on the plane of measurement is 0.2 mm × 0.2 mm. Figure 7 shows the pattern of the $TE_{28,10}$ mode at 140.155 GHz and of the $TE_{36,12}$ mode at 175.992 GHz. The plots are normalized to the highest amplitude and values below -21 dB are clipped. The results are in very good agreement with the simulations. The azimuthal nodes are not clearly seen because the excited mode is by intention a rotating mode, as is the mode in the gyrotron cavity. Since only one polarization is measured, the first ring at the caustic radius is not closed. For the same reason, the rings are blurred in the area which is 45° inclined with respect to the linear polarization. This is a typical effect corresponding to the measurement setup [15]. The purity of mode excitation is good. The measured amount of the counter-rotating mode at the caustic radius is 0.88 % for the $TE_{28,10}$ mode and 0.42 % for the $TE_{36,12}$ mode. This means that the mode generator performs very well and is ready to be used for low-power measurements of the quasi-optical mode converter system of the upgraded gyrotron when available.

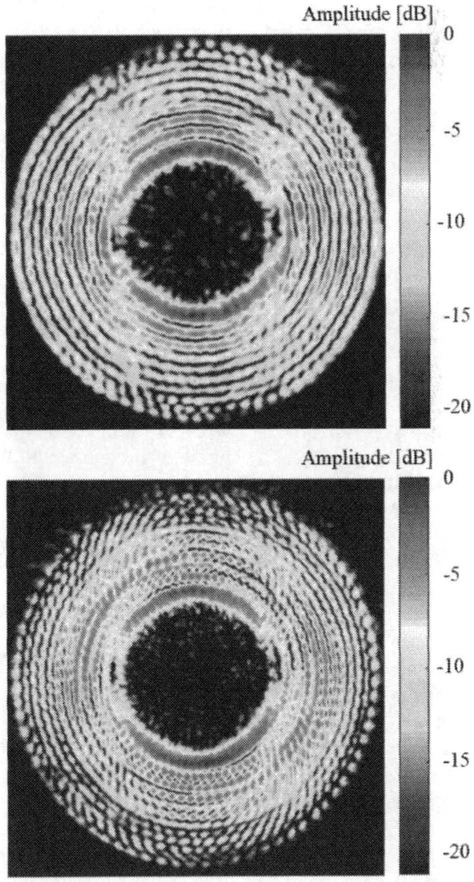

Fig. 7. Measured normalized pattern (electric field amplitude – vertical polarization) of the $TE_{28,10}$ mode at 140.155 GHz (top) and of the $TE_{36,12}$ mode at 175.992 GHz (bottom). The plots refer to a 70 mm × 70 mm domain.

5 Outlook and discussion

As next step, the design of further components for the upgraded gyrotron is planned, as described below.

(i) Diode electron gun. The existing cathode of the W7-X gyrotron could be used (maybe with some minor modifications), given that both the existing and the upgraded gyrotron require the same electron beam radius. However, the anode needs to be redesigned, in order to ensure optimum electron beam parameters at the required operating current of ~56 A, which is significantly higher than the current of 40 A used in the existing gyrotron. The capability of the gun to provide reasonable beam parameters also at ~30 % higher magnetic field (for 175 GHz operation) should also be taken into account.

(ii) Beam tunnel. Given again that the upgraded and the existing W7-X gyrotron require the same electron beam radius, the stacked beam tunnel of the existing gyrotron could, in principle, be also used for the upgrade. A minor redesign of the corrugations of the copper rings should be made, in view of the operation at 175 GHz. However, although the existing beam tunnel has shown satisfactory suppression of parasitic modes when operating with a 40 A beam current [16], it is not guaranteed that this would still be the case for operation at 56 A. A proposal would be to test the beam tunnel performance at higher beam current using one of the existing W7-X gyrotrons. Even short-pulse operation can clarify whether parasitic mode excitation is present.

(iii) Mirror system. The positions of the three mirrors inside the mirror box after the launcher in the upgraded gyrotron are not expected to be much different compared to the existing W7-X gyrotron. However, the surface and shape of the mirrors have to be redesigned. Given that the difference in caustic radius between the $TE_{28,10}$ mode and the $TE_{36,12}$ mode is 2.64 %, some difference in the direction of the microwave output beam for 140 GHz and 175 GHz operation is expected. This should be considered in the orientation of the third mirror before the gyrotron window. The optimum solution would be a steerable third mirror.

(iv) Window. The operating frequencies of the upgraded gyrotron have been chosen in a way that the existing diamond window of the W7-X gyrotron (with a thickness of 1.8 mm) can be used for the upgrade.

(v) Collector. The collector of the upgraded gyrotron would need to withstand 50 % more power compared to the existing collector. This requires an assessment of the existing collector in terms of operation at higher power levels and, if its capability is limited, a redesign of the collector with respect to size, shape, and electron beam sweeping options.

The design of all the components for the upgraded gyrotron will use well-established ideas as a starting point. However, in the course of the studies, the possibility of incorporation of innovative, newly verified concepts will also be assessed. These include improvements in the cooling of cavity and collector, different technologies for the emitter ring, and different materials for the beam tunnel.

References

1. M. Thumm *et al.*, IEEE Trans. Plasma Sci. **36**, 341-355 (2008)
2. R. C. Wolf *et al.*, Nucl. Fusion **57**, 102020 (2017)
3. T. Stange *et al.*, EPJ Web of Conf. **157**, 02008 (2017)
4. F. Mentgen, *Design studies towards a 140 GHz, 1.5 MW CW gyrotron (in view of an upgrade of the ECRH system at the W7-X stellarator)*, Master Thesis, Karlsruhe Institute of Technology, IHM, (2016)
5. M. Thumm *et al.*, IEEE Trans. Plasma Sci. **35**, 143-153 (2007)
6. Z. C. Ioannidis *et al.*, IEEE Trans. Electron Devices **64**, 3885-3892 (2017)
7. K. A. Avramides, I. Gr. Pagonakis, C. T. Iatrou, and J. L. Vomvoridis, EPJ Web of Conf. **32**, 04016 (2012)
8. F. Wilde *et al.*, *20th Joint Workshop on Electron Cyclotron Emission and Electron Cyclotron Resonance Heating*, 14-17 May 2018, Greifswald, Germany.
9. I. Gr. Pagonakis, K. A. Avramidis, G. Gantenbein, T. Rzesnicki, A. Samartsev, J. Jelonnek, Phys. Plasmas **24**, 033102 (2017)
10. R. Lankau, *Studies Towards a 60 GHz, 1 MW gyrotron for Collective Thomson Scattering applications*, Bachelor Thesis, Karlsruhe Institute of Technology, IHM, (2016).
11. A. Schlaich, C. Wu, I. Pagonakis, K. Avramidis, S. Illy, G. Gantenbein, J. Jelonnek, M. Thumm, Journal of Infrared Millimeter and THz Waves **36**, 797-818 (2015)
12. J. Jin, M. Thumm, G. Gantenbein, J. Jelonnek, IEEE Trans. Microwave Theory and Techniques **65**, 699-706 (2017)
13. D. Wagner *et al.*, Int. J. Infrared and Millimeter Waves **19**, 185-194 (1998)
14. M. Pereyaslavets *et al.*, Int. J. Electronics **82**, 107-116 (2010)
15. M. Losert *et al.*, *German Microwave Conference*, 16-18 March 2015, Nuremberg, Germany, proc. 256-259 (2015)
16. G. Gantenbein *et al.*, IEEE Trans. Plasma Sci. **38**, 1168-1177 (2010)

Adaptation of a Dual-Frequency 104/140 GHz Gyrotron for Operation at 175 GHz

Stephen Cauffman[1,*], *Monica* Blank[1], *Philipp* Borchard[1], and *Kevin* Felch[1]

[1]Communications and Power Industries, 607 Hansen Way, Palo Alto, California, 94304, USA

Abstract. A dual-frequency gyrotron capable of operation in the $TE_{28,7}$ cavity interaction mode at 140 GHz, or in the $TE_{22,5}$ mode at 104 GHz, has been developed for use in electron cyclotron heating in the W7-X stellarator at IPP Greifswald. The gyrotron incorporates an internal converter design that has been numerically optimized to convert either of the two operating modes into a high-quality Gaussian output beam. During short-pulse factory testing, the gyrotron produced 900 kW at 140 GHz, and 520 kW at 104 GHz. After delivery to IPP, the gyrotron was conditioned to long-pulse operation at 140 GHz, demonstrating 30-minute pulses at several power levels up to 811 kW, and producing ten consecutive ten-minute pulses at 811 kW as well. After long-pulse capabilities were demonstrated at 140 GHz, IPP requested an analysis of the feasibility of operating the gyrotron (without internal modification) in an additional mode with a frequency near 175 GHz. Several potential interaction modes were evaluated to determine the required operating parameters for excitation of these modes, and to assess the expected interaction efficiency, output power, internal diffraction losses, and output beam quality. The most promising modes appear to be the $TE_{33,9}$ (173 GHz) and the $TE_{34,9}$ (176 GHz), which should generate 400-500 kW of RF in a suitable magnet capable of producing the necessary 7.1 T field required for operation at these higher frequencies. Because the existing gyrotron's internal converter was not optimized for these modes, however, internal losses are expected to be higher than usual (up to 7%), and the output beam pattern will require external phase-correction in order to produce a Gaussian beam. A feasibility analysis for such external phase correction has been performed, demonstrating that a high-quality beam can be recovered using numerically synthesized external phase-correcting mirror surfaces.

1 Introduction

Gyrotrons are vacuum electron devices that make use of the cyclotron resonance maser instability to convert DC power into millimetre-wave radiation. An annular beam of helically propagating electrons travels along the flux lines produced by a solenoid magnet, and the electron beam's rotational energy (the energy of motion transverse to the magnetic field) is used to excite a resonant transverse electric eigenmode in a cylindrical interaction cavity. Operation at high average power (~1 MW) has been demonstrated in numerous gyrotrons at frequencies up to 170 GHz, using highly-overmoded interaction circuits. (For a comprehensive summary of the current state-of-the-art, see Reference 1.) The primary applications for such megawatt-class gyrotron oscillators are fusion plasma heating, current-drive, and plasma instability suppression. Localized electron cyclotron heating of fusion plasmas makes use of millimetre-wave beams injected with a frequency matching the electron cyclotron frequency for the magnetic field at the desired heating location in the plasma. While many gyrotrons are designed for operation at one specific frequency, some designs are optimized to support operation in several different eigenmodes.

Communications & Power Industries (CPI) has designed and fabricated a 2-frequency gyrotron capable of operation at either 140 GHz or 104 GHz, for use in plasma heating in the W7-X fusion stellarator at IPP Greifswald. Although the gyrotron was originally intended to support operation at these two frequencies, in order to allow for compatibility with stellarator plasmas generated at less than full field, the need for operation at the lower frequency did not materialize. Preliminary factory tests of the gyrotron confirmed proper operation at both frequencies, but subsequent tests focused solely on 140 GHz operation. After testing of the gyrotron, IPP requested an analysis of the feasibility of operating the gyrotron at a higher frequency near 175 GHz, for possible use in plasma diagnostics. CPI performed analyses to determine possible modes of operation, and to evaluate the expected performance in those modes.

This paper summarizes (a) the design of the gyrotron, (b) the demonstrated operation of the gyrotron during factory testing and during acceptance testing at the customer site, and (c) the results of the feasibility study evaluating the possibility of operation at ~175 GHz.

* Corresponding author: steve.cauffman@cpii.com

2 Gyrotron Design

Figure 1 shows the gyrotron, installed after delivery to IPP Greifswald. A schematic layout of the gyrotron is shown in Figure 2. The gyrotron employs a single-anode electron gun with a thermionic cathode, to produce an annular electron beam. The gyrotron is designed for compatibility with the magnetic field produced by a superconducting magnet which governs the propagation of the electron beam from the electron gun, through the interaction circuit, and into the collector, which dissipates the power remaining in the electron beam after the extraction of RF power. The collector voltage is depressed, relative to the interaction circuit, to improve overall device efficiency and reduce thermal loads. An internal converter consisting of a tapered, dimpled-wall launcher and a set of three phase-correcting and beam steering mirrors transforms the interaction mode into a freely propagating Gaussian TEM_{00} beam, which exits the gyrotron in the radial direction, through a vacuum window made of CVD diamond.

Fig. 1. The VGT-8141A gyrotron, installed at the customer site.

The gyrotron design is an adaptation of a previously demonstrated single-frequency design, which was developed to produce millimetre-wave radiation at 140 GHz, via excitation of the $TE_{28,7}$ mode [2]. To reduce technical risk, the pre-existing single-frequency 140 GHz gyrotron design was minimally modified, in order to allow for operation at 104 GHz as well. The electron gun, the interaction circuit, the output window, and the collector were kept unchanged. A slight modification to the beam tunnel, near the entrance to the interaction circuit, was necessary to ensure adequate electron beam clearance when operating at the slightly larger beam size required for excitation of the 104 GHz $TE_{22,5}$ mode. The modified internal converter was numerically optimized to maximize Gaussian content at both operating frequencies, while ensuring low diffraction losses inside the gyrotron, keeping the positions of the launcher and mirrors similar to those in the original design, and minimizing the angular divergence of the two possible output beams.

The gun and cavity were optimized for performance at the 140 GHz operating frequency, employing the $TE_{28,7}$ interaction mode, with simulations indicating that an 80 kV accelerating voltage and a beam current of 40 A should yield over 900 kW of output power, for a peak magnetic field value of 5.54 T. Simulations indicated that the 104 GHz $TE_{22,5}$ mode could be excited with reasonable efficiency, yielding an expected output power over 500 kW, by reducing the magnetic field to 3.98 T, modifying the relative currents in the superconducting magnet coils to yield the required beam radius for excitation of the $TE_{22,5}$ mode in the interaction circuit, and reducing the accelerating voltage to 60 kV to yield a suitable beam pitch factor ($v_\perp/v_\parallel$) for optimal output power.

The simulated performance of the dual-frequency internal converter is shown in Figure 3. The launcher converts either operating mode into a high-quality Gaussian beam, which is further shaped by numerically synthesized mirrors, which direct the beam to the output window. The two beams have slightly different sizes and waist positions, but are well centered on the window aperture, and have a differential angular divergence of less than 1.5 degrees.

3 Gyrotron Test Results

The nominal operating conditions and demonstrated performance of the dual-frequency gyrotron are summarized in Table 1. Factory testing was limited by the capabilities of the test set, which can operate at full beam voltage and current for short pulses (5 ms), or at beam currents up to 25A for longer pulses. Furthermore, although tests were performed to confirm operation at both frequencies, the tests at 104 GHz were limited in scope, because at the time of testing, the customer no longer anticipated a need for operation at the lower frequency. Conditioning of the gyrotron to full power and pulse length at 140 GHz was performed after delivery of the gyrotron to the customer site.

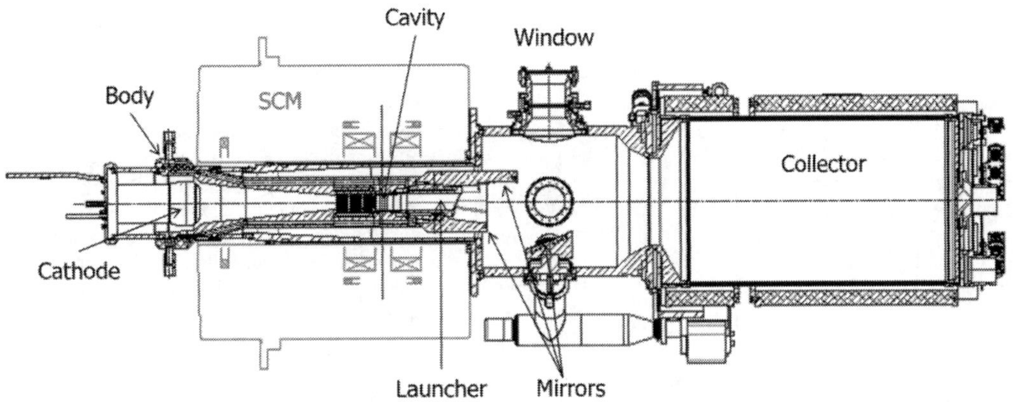

Fig. 2. A schematic of the VGT-8141A gyrotron.

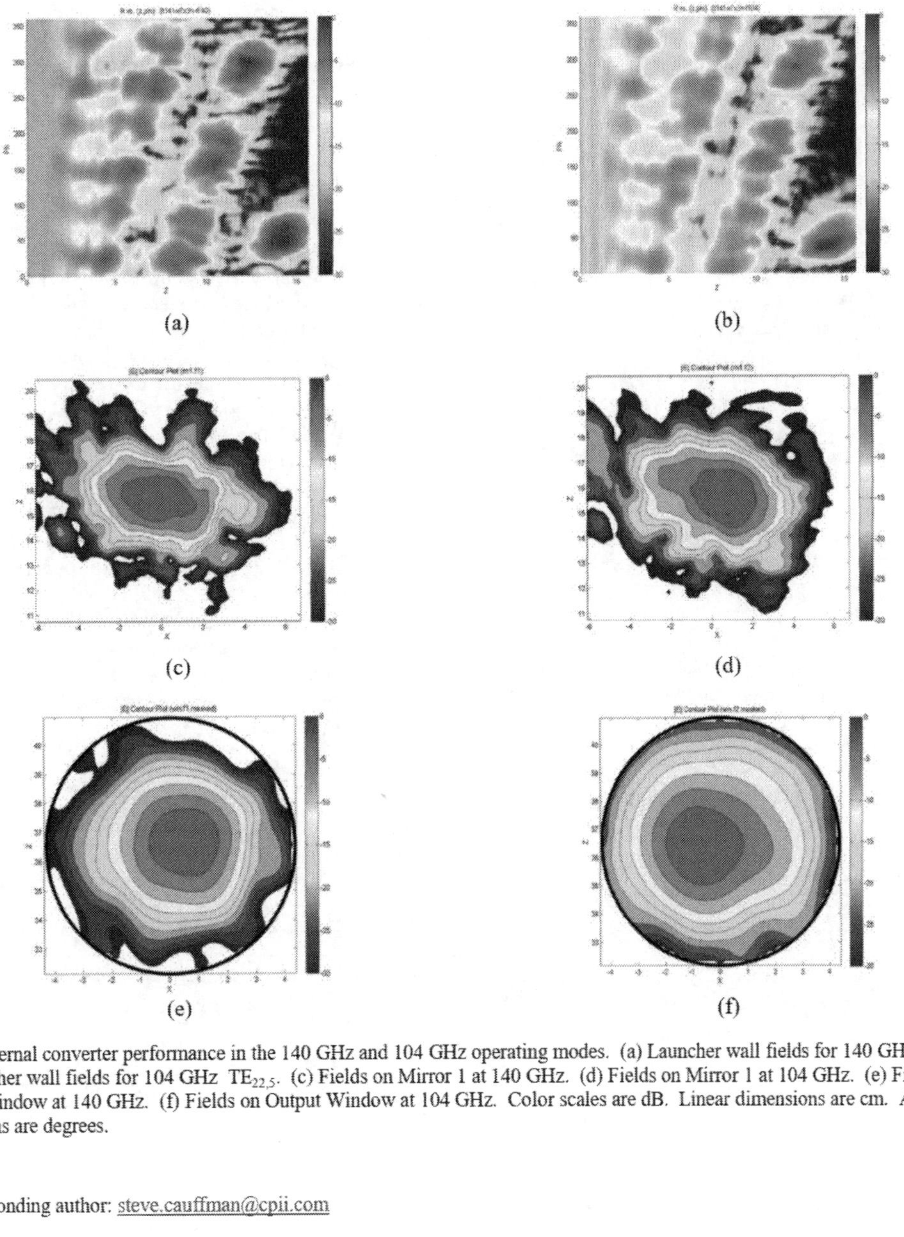

Fig. 3. Internal converter performance in the 140 GHz and 104 GHz operating modes. (a) Launcher wall fields for 140 GHz $TE_{28,7}$. (b) Launcher wall fields for 104 GHz $TE_{22,5}$. (c) Fields on Mirror 1 at 140 GHz. (d) Fields on Mirror 1 at 104 GHz. (e) Fields on Output Window at 140 GHz. (f) Fields on Output Window at 104 GHz. Color scales are dB. Linear dimensions are cm. Angular dimensions are degrees.

* Corresponding author: steve.cauffman@cpii.com

Table 1. Gyrotron Test Results.

Freq (GHz):	**140**		**104**	
Mode	$TE_{28,7}$		$TE_{22,5}$	
Pulse Len	20 ms	30 min	5 ms	100 ms
Acc. Voltage (kV)	84.4	84.3	60	60
Dep. Voltage (kV)	23.4	23.3	15	15
Beam Current (A)	46	39	40	25
Output Power (kW)	900	811	520	300
Test Location	IPP	IPP	CPI*	CPI**
Notes: *No 104 GHz tests at IPP. **No CW tests at 104 GHz. Demonstrated: 10 consecutive 10-minute pulses at 811 kW, no faults. 30-minute pulses at 300 kW, 500 kW, 811 kW.				

The gyrotron demonstrated 30-minute pulses at several power levels up to 811 kW and produced ten consecutive 10-minute pulses at 811 kW during final acceptance testing. Short-pulse testing yielded power levels over 900 kW, but stable operation for longer pulses was obtained by lowering the power level by about 10%.

4 Feasibility of 175 GHz Operation

After final testing of the 104/140 GHz gyrotron, IPP requested an analysis of the feasibility of operation at a higher frequency near 175 GHz (a value chosen, in part, to ensure compatibility with the existing gyrotron's single-disc output window thickness, which should be an integral number of half-wavelengths thick, to minimize reflection). CPI performed feasibility analyses to determine whether higher-frequency modes could be excited with reasonable efficiency, and to evaluate the quality of the generated output beam under such conditions.

The first challenge for operation at higher frequency is the need for a higher magnetic field in the interaction region. The magnet for the existing system is not capable of operation at the required ~7 T. However, an existing magnet design (model VYW-8170) for a similarly sized 170 GHz gyrotron (model VGT-8170) was found to be capable of generating the necessary field, and was also confirmed to be geometrically compatible with the VGT-8141A gyrotron, with slight modification to external mounting hardware.

Electron gun and cavity interaction simulations for the existing VGT-8141A gyrotron geometry were performed using the magnetic fields produced by the VGT-8170 magnet, to assess expected performance if the gyrotron were to be installed in the higher-field magnet. Several candidate modes in the frequency range of interest were identified and found to be capable of excitation in the existing gyrotron. The most promising candidates included the $TE_{31,10}$ (176.6 GHz), the $TE_{33,9}$ (173.0 GHz), and the $TE_{34,9}$ (176.0 GHz). Optimal operating parameters for these modes were found by varying the magnet coil currents and accelerating voltage, while assuming a maximum beam current of 40A. These modes and frequencies were then used as inputs to simulations of the internal converter system, to evaluate the expected internal diffraction losses and the output beam quality if the gyrotron were to be operated in these modes. A summary of the simulated performance in these modes is provided in Table 2.

The internal losses for the $TE_{31,10}$ were found to be too high (~20%) for long-pulse operation. The internal losses for the $TE_{33,9}$ and $TE_{34,9}$ are somewhat high (6-7%), but within the capabilities of the gyrotron's internal RF-absorbing water-cooled loads. The output beam quality for each candidate mode near 175 GHz is extremely poor. The dimpled-wall launcher design that was optimized for two specific modes at 140 GHz and 104 GHz employs a complex pattern of radial variations to yield excellent beam quality with a very short launcher length (about 7 helical RF bounces). For the expected input modes at 104 GHz and 140 GHz, these wall perturbations yield a mixture of modes that results in a high-quality Gaussian beam, but when the launcher is operated in other modes, these perturbations yield an undesirable mode mixture resulting in poor output beam quality. Although it would be possible to design a longer launcher with better performance over a wider range of modes using more gradual evolution of the mode mixture, this was not one of the original design criteria for the VGT-8141A.

Table 2. Simulated performance of the VGT-8141A gun, cavity, and internal converter.

Frequency (GHz)	140	104	175.6	173.0	176.0
Mode	$TE_{28,7}$	$TE_{22,5}$	$TE_{31,10}$	$TE_{33,9}$	$TE_{34,9}$
B_o (T)	5.54	3.98	6.76	6.96	7.10
Voltage (kV)	80	60	62	85	85
Current (A)	40	40	40	40	40
P_{cavity} (kW) [MAGY]	975	727	678	799	619
Diffraction Loss (%)	1.1%	2.2%	19.6%	6.7%	6.3%
Field Pattern at Window	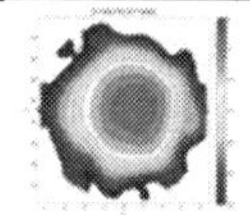	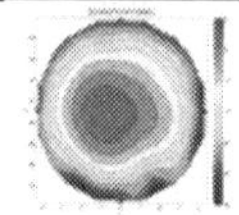	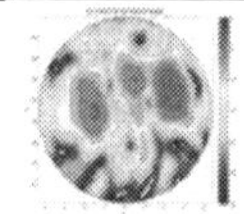	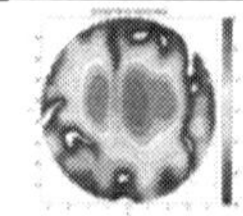	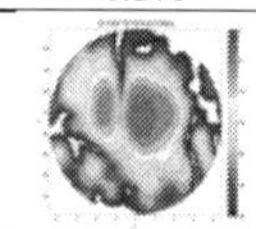

* Corresponding author: steve.cauffman@cpii.com

Although not as elegant as having a high-quality beam exiting the gyrotron window, external phase-correcting mirrors can be synthesized to correct the output beam pattern, converting it back into a useable high-quality Gaussian beam. To establish the feasibility this approach, synthesized mirrors were designed to convert the predicted poor-quality 176 GHz output beam into the desired Gaussian beam. Mirror positions were chosen based on the layout of IPP's external mirror system for directing and refocussing the output beam. The synthesized surfaces for the first two external mirrors yield a 98% match to the desired Gaussian beam at the transmission system's polarizer. The required sizes of the mirrors may be slightly larger than the existing layout allows for, however, so an adjustment of the mirror positions, by a few cm, may be necessary to ensure adequate clearance. Figure 4 shows the modelled layout of the external mirror system and the resultant RF beam pattern.

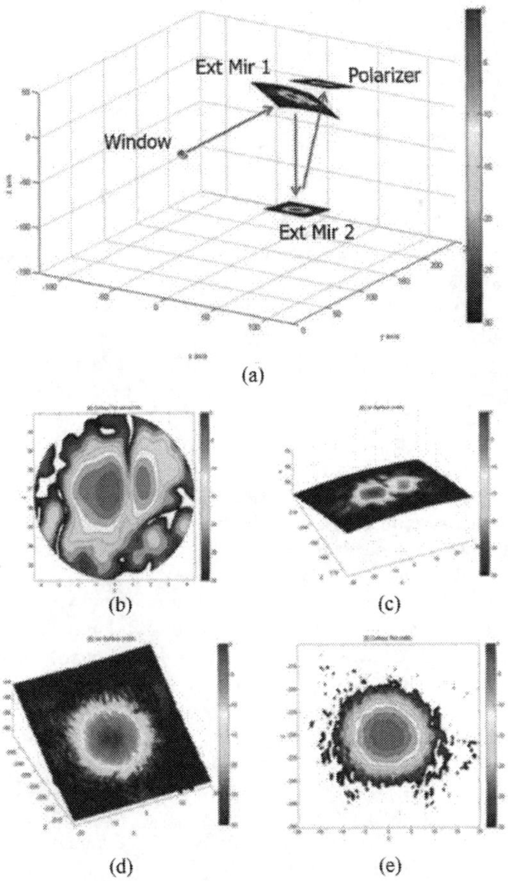

Fig. 4. External phase correction of 176 GHz radiation pattern using numerically synthesized mirror surfaces improves Gaussian beam coupling from 24% to 98%. Shown above: (a) Simulation geometry; (b) Field pattern at output window; (c) Field pattern at external mirror 1; (d) Field pattern at external mirror 2; (e) Field pattern at polarizer plane.

5 Summary

The VGT-8141A is a dual-frequency gyrotron capable of operation at either 140 GHz or 104 GHz. It has demonstrated short pulse operation at both frequencies (yielding over 900 kW at 140 GHz and over 500 kW at 104 GHz), and has demonstrated the ability to operate reliably at pulse lengths up to 30 minutes, for power levels up to 811 kW at 140 GHz.

The VGT-8141A gyrotron design is an adaptation of a previously demonstrated single-frequency gyrotron (the 140 GHz VGT-8141), with an internal converter that was re-optimized to support operation at 104 GHz while still producing a high-quality output beam.

Although the gyrotron was not designed for operation in other modes at other frequencies, it could be operated in either the 173 GHz $TE_{33,9}$ mode or the 176 GHz $TE_{34,9}$ mode, provided that a compatible 7.1 T superconducting magnet is available. Operation in such modes will yield higher internal RF losses (6-7%), and will result in a poor-quality output beam which will require external correction using synthesized mirror surfaces to convert the output radiation pattern into a TEM_{00} Gaussian beam.

References

1. M. Thumm, "State-of-the-Art of High Power Gyro-Devices and Free Electron Masers," KIT Scientific Reports 7750, Update 2017, DOI: 10.5445/KSP/1000081551.
2. M. Blank, P. Borchard, S. Cauffman, K. Felch, "Development and Demonstration of a 900 kW, 140 GHz gyrotron, " 2015 IEEE International Vacuum Electronics Conference (IVEC), Beijing, 2015.

* Corresponding author: steve.cauffman@cpii.com

Recent Status and Future Prospects of Coaxial-Cavity Gyrotron Development at KIT

Stefan Illy[1], *Konstantinos A.* Avramidis[1], *Gerd* Gantenbein[1], *Zisis* Ioannidis[1], *Jianbo* Jin[1], *Parth C.* Kalaria[1], *Ioannis Gr.* Pagonakis[1], *Sebastian* Ruess[1,2], *Tobias* Ruess[1], *Tomasz* Rzesnicki[1], *Manfred* Thumm[1,2], and *John* Jelonnek[1,2]

[1]*IHM, Karlsruhe Institute of Technology (KIT), Kaiserstr. 12, 76131 Karlsruhe, Germany*
[2]*IHE, Karlsruhe Institute of Technology (KIT), Kaiserstr. 12, 76131 Karlsruhe, Germany*

Abstract. Coaxial-cavity gyrotrons are microwave sources that can extend the possible power levels of hollow cavity gyrotrons significantly and make them attractive for future fusion experiments and power plants. KIT already demonstrated operation of a modular short-pulse 170 GHz coaxial-cavity gyrotron with an output power of 2.2 MW, operating in the $TE_{34,19}$ mode. Today's focus of KIT is the verification of this technology at longer pulses, which will prove the long-pulse capabilities also. At the moment, the current KIT prototype has been extended with cooling capabilities for all critical, highly loaded components of the tube and two new, different electron guns are available for operation. This paper will give an overview about the recent status and the future prospects connected to the coaxial-cavity gyrotron development at KIT, including both experimental and theoretical activities.

1 Introduction

At KIT, the activities in gyrotron research and development shall lead to a gyrotron design that will fulfil the increasing requirements for future nuclear fusion experiments and power plants, like DEMO [1], the fusion demonstration power plant that will follow ITER. Today, the requirements for the initial gyrotron installation at ITER are an RF output power capability of 1 MW (3600 s) at an operating frequency of 170 GHz for each single gyrotron [2]. For a future DEMO an RF output power level of minimum 2 MW CW at even higher operating frequencies up to 240 GHz is under consideration [3,4,5]. It is expected that the coaxial-cavity gyrotron [6,7] technology will allow an operation at those output power levels. Compared to the conventional hollow-cavity gyrotron technology as used for e.g. Wendelstein7-X (W7-X) and ITER today, the coaxial-cavity gyrotron technology allows the effective reduction of the voltage depression due to the space charge effect of the electron beam and the reduction of the mode competition due to the mode selectivity of the carefully chosen longitudinally corrugated inner conductor at high operating frequencies. The key challenges in the construction and operation of a coaxial-cavity gyrotron are the design of the coaxial insert in terms of beam-wave interaction, appropriate cooling of the insert and an accurate alignment of this relatively long structure. First experimental results obtained with the 170 GHz 2 MW short-pulse coaxial-cavity pre-prototype gyrotron (Fig. 1) at pulse length of a few milliseconds (ms) have shown the potential of the coaxial-cavity concept in the multi-MW operation regime. Nonetheless, until today, the coaxial-cavity gyrotron technology has not been verified at longer pulses above a few ms. Therefore, a focus of KIT is the verification of this technology at pulse lengths up to one second which will prove the long-pulse capabilities for longer pulses also. For the very first time, a coaxial cavity gyrotron will be operated and investigated in the experiment in long pulse operation.

At the moment, the current KIT prototype has been extended with cooling capabilities for all critical, highly loaded components of the tube [8]; a new triode electron gun equipped with non-emissive coating at the emitter edges [9] has been procured and a new inverse triode gun has been manufactured in-house [10]; both electron guns should increase the electron beam quality and therefore the overall performance and operation stability of the tube.

Intense theoretical design activities to support the manufacturing have been performed and are still on-going to assist during the experimental activities. This includes the design and optimization of different kinds of Magnetron Injection Guns (MIGs) [11], the simulation of the gyrotron interaction [12,13] and quasi-optical output launcher [14,15] (focusing also on a future dual frequency operation at 170 GHz / 204 GHz). In addition, this also includes multi-physics simulations of advanced cooling concepts for the outer cavity wall of the resonator [16] (including the manufacturing of mock-ups), fluid-dynamics simulations of the cooling structure of the coaxial insert [17] and investigations of enhanced longitudinal sweeping concepts for the collector [18,19].

* Corresponding author: stefan.illy@kit.edu

This paper is organized as follows: In Section 2, the current status and recent major results of the coaxial-cavity gyrotron development at KIT are presented; in Section 3, the design and manufacturing of the key components for the extension to longer pulses are described. Section 4 then presents the planned experimental activities in the near future.

2 Current Experimental Status

During the experimental short-pulse test campaign, the already existing KIT 170 GHz 2 MW coaxial-cavity modular pre-prototype did show an excellent and very stable performance. The tube was operated at an acceleration voltage of up to $U_c = 93$ kV and an electron beam current of $I_b = 80$ A. At the nominal operating parameters, listed in Table 1 ($U_c = 90$ kV, $I_b = 75$ A), an RF output power of 2 MW up to 10 ms pulse length has been achieved. An electronic (interaction) efficiency between the electron beam and the electromagnetic field in the cavity of ~30 % was achieved at nominal operating parameters [20]. During the experiments the magnetic field of the superconducting magnet was set to 6.90 T. The acceleration voltage was raised up to ~93 kV. At the same time, the electron beam current was set to 80 A. That allowed to generate a world-record in RF output power of 2.2 MW at 170 GHz with an efficiency close to 30 %. The calorimetrically measured RF output power and electronic efficiency versus cathode voltage is shown in Fig. 2. The achieved results are in excellent agreement with the multi-mode simulations performed at KIT [20].

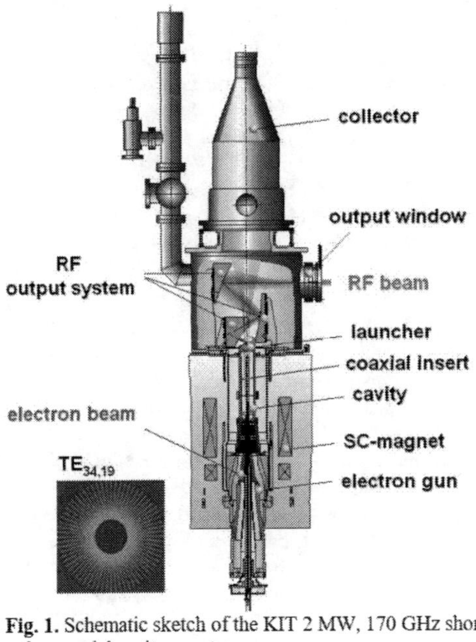

Fig. 1. Schematic sketch of the KIT 2 MW, 170 GHz short-pulse coaxial cavity gyrotron.

The experimental results were performed with an improved design of the launcher [14] with smoothed inner wall perturbations. That launcher is the major part of the quasi-optical system which converts the very high-order cavity mode into a Gaussian beam. The efficiency of the mode converter had been verified in low power measurements before implementing that launcher into the pre-prototype. The measured results are in excellent agreement with the expectations from numerical analysis. The fundamental Gaussian mode content had been extracted from the measured patterns and was calculated to be around 96 %, which corresponds very well to the simulated value of 96.3 %. In addition, at high output power, the profile of the RF beam had been measured with an infrared camera (IR-camera) (Fig. 3 [20]) at several different distances from the output window. The analysis of the data verified again the very good quality of the generated RF beam and it is in accordance with the low power measurements. The amount of the RF stray radiation losses inside the gyrotron tube had been obtained by measuring the RF power radiated through the relief window with a very sensitive ballistic calorimeter. Finally the stray radiation losses with the new launcher are significantly reduced from 7 % (previous experiments) down to 4 % [20].

Table 1. Design parameters for the coaxial-cavity short pulse gyrotron.

Operating cavity mode	$TE_{34,19}$
Frequency (f)	170 GHz
RF output power (P_{out})	2 MW
Beam current (I_b)	75 A
Accelerating voltage (U_c)	90 kV
Velocity ratio (pitch factor) (α)	~1.3
Electron beam radius	10.0 mm
Cavity magnetic field (B_{cav})	6.87 T
Overall efficiency (with depressed collector)	50 %
Pulse length (operation phase 1)	150 ms
Pulse length (operation phase 2)	1000 ms

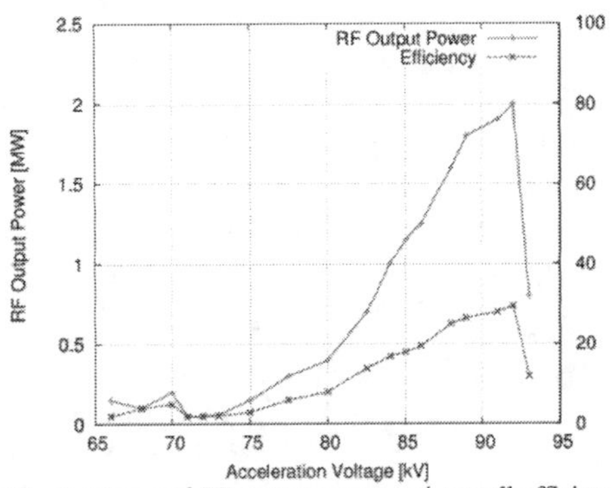

Fig. 2. Measured RF output power and overall efficiency as a function of the accelerating voltage U_c (obtained at $B_{cav} = 6.87$ T and I_b ~75 A, operation without depressed collector).

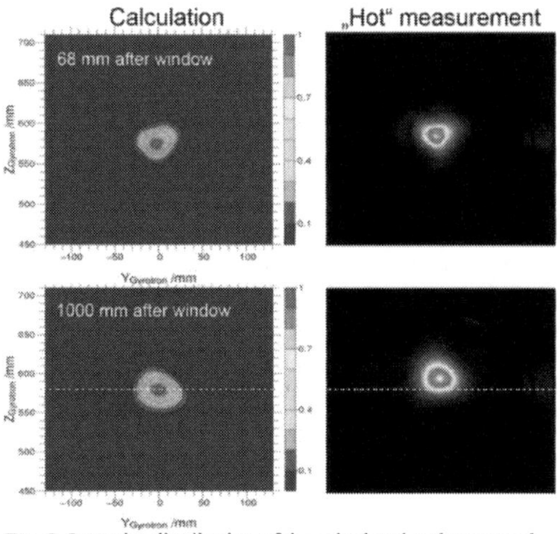

Fig. 3. Intensity distribution of the calculated and measured (IR-image) gyrotron RF output beam in 68 mm and 1000 mm distance from the window [21] (qualitative comparison).

Fig. 4. Water-cooled subcomponents of the KIT 170 GHz 2 MW coaxial cavity gyrotron for pulse lengths up to 1 s.

3 Design and Manufacturing of the Coaxial-Cavity Longer Pulse Pre-Prototype Gyrotron

In order to increase the pulse length from the ms-range up to the range of one second, the main components of the 2 MW coaxial-cavity gyrotron such as the beam tunnel, the cavity, the launcher, the mirrors, the chemical vapor deposited (CVD) diamond RF output window and the collector have to be equipped with an active cooling system. The corresponding design activity was starting in 2015. The already manufactured main components of the tube are shown in Fig. 4. One of the main project requirements is to keep the modularity of the gyrotron. Therefore, an independent cooling system for each component is considered. This allows the monitoring of the internal losses in each gyrotron component and of the final energy balance of the tube during longer-pulse operation. In the following, the advanced key components of the gyrotron are described.

3.1 Inverse Magnetron Injection Gun

A major step towards higher output power and operating frequency is the Inverse Magnetron Injection Gun (IMIG) [10]. An IMIG has been designed and already manufactured for the KIT 2 MW coaxial-cavity gyrotron [22]. It is expected that this gun leads to a more stable operation at even higher output power due to the avoidance of potential wells and secondary electrons. Due to the triode configuration of the IMIG, including a separate modulation anode, the IMIG has the flexibility to be operated at multiple frequencies such as 170 GHz and 204 GHz. Because of the possibility of larger emitter radius the gyrotron can be operated at a larger operating beam current and therefore at a higher RF output power. The requested electron beam parameters are presented in Table 1 together with the operation parameters of the tube. The simulated parameters of the IMIG indicate a high electron beam quality [10], which promises a reliable gyrotron operation at nominal operating parameters. Additionally, the IMIG is the first MIG which fulfils the design criteria for preventing the generation of trapped electrons as presented in [23]. In the existing IMIG design the emitter radius is set to 62 mm with an emitter thickness of 5 mm. The emitter has an angle of 25° with respect to the z-axis of the tube. In order to keep the temperature in the neighbouring regions of the emitter low at both edges, isolation gaps with a width of 200 μm are designed. As a result, the electrons, emitted from the emitter edges have a very high pitch factor, which is related to the field enhancement at the emitter edges. In order to mitigate this effect, the emitter is pushed into the cathode by 70 μm to reduce the local electric field at the edges.

In order to minimize mechanical shifts and misalignments caused by thermal effects, the thermo-mechanical behaviour of the IMIG was investigated by the use of COMSOL. In comparison to the "conventional" MIG (Fig. 6), the cathode and emitter of the IMIG (Fig. 5) are placed at the outside of the whole structure. Hence, it is directly cooled by passing oil at the outer surface of the cathode elements. That is the major benefit if comparing to the standard MIG used in all conventional gyrotrons. The components of the cathode and anode are made of materials with high thermal conductivity, mainly CuCrZr (indicated in Fig. 5 using orange colour) and Molybdenum (Fig. 5, marked by dark blue colour). Therefore, the heat loading of the neighbouring emitter regions is directly guided outside of the MIG and absorbed by the oil flow. Compared to conventional MIGs the temperature of the loaded parts is significantly reduced by 40 % down to ~150 °C. Corresponding simulation results are presented in [24].

Another significant influence on the beam quality, hence on the generated output power and efficiency, is caused by an inhomogeneous temperature distribution at the surface of the emitter. Obviously an inhomogeneous temperature distribution leads to an inhomogeneous emission. The emission inhomogeneity can be experimentally investigated by measuring the Current-Voltage Characteristics as published in [25]. Measurements of the temperature distribution of the emitter of the inverse MIG did indicate an azimuthal sinusoidal temperature distribution. The expected nominal temperature was 1000°C whereas the measured temperature varied between 993°C and 1007°C. One explanation for the non-constant azimuthal temperature distribution is the fact that the ends of the heater filament are not overlapping each other.

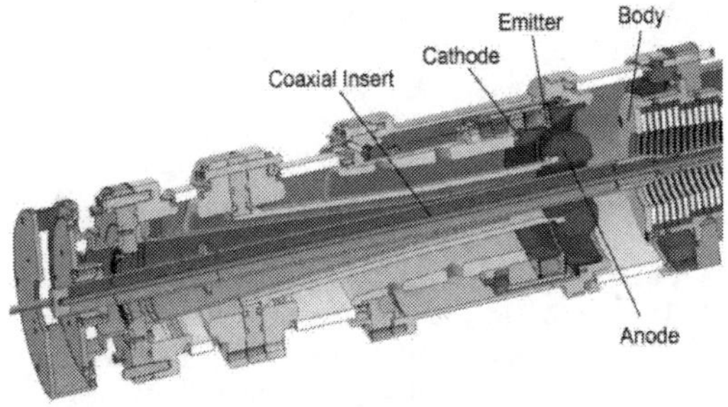

Fig. 5. Sketch of the already manufactured inverse magnetron injection gun (IMIG); the total length of the displayed section is approximately 700 mm.

3.2 Advanced Conventional Magnetron Injection Gun

The enhancement of the electric field at the emitter edges has a significant influence regarding the electron beam quality. Tolerance studies of the radial emitter position have confirmed those observations and have shown that manufacturing tolerances of some microns already become critical. A possible solution for this major issue is the implementation of an anti-emission coating at the edges of the emitter. For example, considering a misalignment of ±80 µm, the variation of the pitch factor measured over the emitter area is 83.3 % smaller compared to the variation using a conventional emitter [23]. A new conventional MIG with an advanced coated emitter for the 170 GHz 2 MW longer-pulse coaxial-cavity gyrotron, as shown in Fig. 6, has been designed by KIT and manufactured by Thales Electron Device (France) and is now available for experiments [11].

3.3 Beam Tunnel

The beam tunnel consists of alternating stacked copper rings and rings of lossy ceramics. Compared to the other components of the gyrotron, the thermal loading is relatively low; however, an active cooling is also considered for this component (see Fig. 4), which will allow to estimate the losses in the ceramic discs caused by RF stray radiation and spurious oscillations in the beam tunnel, if present.

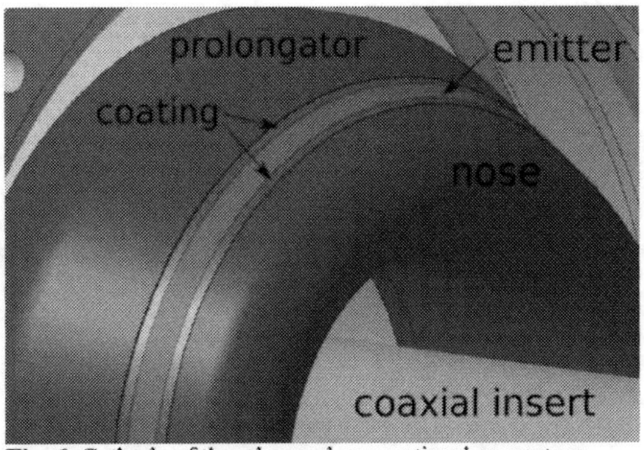

Fig. 6. Cathode of the advanced conventional magnetron injection gun (MIG) with an anti-emissive coating at the emitter edges (indicated in green).

3.4 Cavity

In the gyrotron cavity, the perpendicular kinetic energies of the individual electrons of the electron beam are transferred to the electromagnetic field. The expected total power loss in the cavity due to thermal wall loading is about 50 kW at nominal 2 MW gyrotron operation. Due to the high quality factor of the cavity one finds the peak thermal wall loading of approximately 2 kW/cm^2 in a very narrow region at the center of the cavity. It leads to a very high temperature gradient at the inner side of the cavity wall, which results in thermal stress, and, finally leads to a deformation of the cavity wall. Depending on that wall deformation, the operating frequency of the gyrotron is shifted and the quality factor of the cavity is changing significantly. Therefore, it is mandatory to implement effective water cooling. The chosen approach of the active cooling system is similar to the design of the cooling system for the launcher. However, due to the position and orientation, the implementation of a helix-type channel system is not necessary. The different temperatures and material properties of the coat and the outer cavity wall require also the implementation of a bellow for compensation of thermal expansion. The reduction of the water cooling gap (see Fig. 4, cavity) increases the flow velocity of the water and therefore the cooling capacity. The limitation of the cavity is a maximum temperature of approximately 350 °C. Based on this, the multi-physics software COMSOL predicts a maximum pulse length of approximately 1 s.

Additionally, the cavity uptaper is extended and optimized. The simulation results predict a transmission of 99.94 % of the $TE_{34,19}$-mode and a very low conversion to unwanted spurious modes.

3.5 Quasi-Optical Launcher

As stated before, the launcher together with the mirror system is responsible for the conversion of the cavity $TE_{34,19}$-mode into the fundamental Gaussian mode of the RF output beam [14]. Especially, at the tip of the launcher, where the complete RF-power is finally focused, the thermal loading of 0.4 kW/cm² is critical. In order to achieve a stable operation, it is necessary to cool that launcher tip in an efficient way. To achieve that, the water inlet and outlet of the launcher is connected at the bottom of the launcher. A helix cooling structure has been used to make that configuration possible. Due to that helix structure (see. Fig. 4) the channels have to be milled with a 5-axis milling cutter. In order to caulk the channels a stainless steel coat is imposed and brazed from the outside of the launcher. The introduction of a bellow is necessary as the launcher and the coat consist of different materials that have different thermal expansion coefficients. The already manufactured launcher shows a desired surface quality. The tolerances are in the range of below ±10 µm. The launcher performance has been successfully verified in a cold measurement test setup. Fig. 7 shows the radiated pattern which was sampled at a distance of 10 cm from the axis to the position of the first mirror. The measurement results correspond well to the simulation results and the measurements on the previous launchers. The simulated radiation pattern has been calculated with the full wave 3D vector analysis code SURF3D [26]. Technologically, soldered and brazed joints with a leakage rate of $<10^{-12}$ mbar l/s were achieved for the launcher parts.

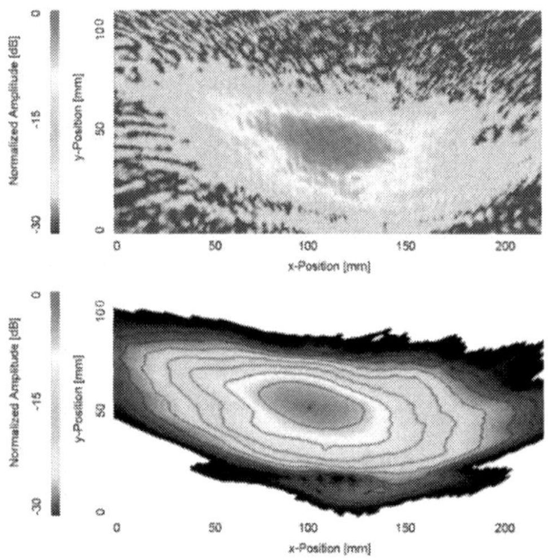

Fig. 7. Radiation pattern of the water-cooled launcher (top: cold measurement; bottom: corresponding simulation results).

4 Experimental Activities in the Near Future

During the last months, the assembly of the whole gyrotron has been finished and the tube has been equipped with a relatively old MIG for first verifications. The performed short-pulse measurements with pulse lengths ≤ 2.5 ms showed the same behaviour and output power levels as described in Section 2.

At the moment, the old MIG has been replaced by the new conventional triode type MIG with anti-emissive emitter edge coating (described in section 3.2) and new results will be expected during the next months. After finishing this measurement campaign, the experiments will be continued with the new IMIG.

5 Conclusion

KIT is pushing forward the coaxial-cavity gyrotron development by building up a modular longer-pulse 170 GHz, 2 MW pre-prototype targeting at pulse lengths of up to 1 s. The design principle and the manufacturing process of that pre-prototype have been presented here. All the subcomponents have been manufactured, brazed and successfully tested regarding vacuum tightness. Furthermore, the IMIG was manufactured at KIT with an expected high electron beam quality. Additionally, an advanced conventional MIG with coated emitter edges has been designed by KIT and manufactured at Thales Electron Devices. Both MIGs can be operated in hollow cavity as well as coaxial-cavity gyrotrons. The longer-pulse 2 MW coaxial-cavity gyrotron and the two MIGs will be tested during the next months.
Theoretical design activities are also on-going, focusing on possible two-frequency operation of the 2 MW coaxial cavity gyrotron at 170 GHz and 204 GHz [13].

This work has been carried out within the framework of the EUROfusion Consortium and has received funding from the Euratom research and training programme 2014-2018 under grant agreement No 633053. The views and opinions expressed herein do not necessarily reflect those of the European Commission. Part of the simulations was performed on the EUROfusion High Performance Computer (Marconi-Fusion).

References

1. G. Federici et al., 13th International Symposium on Fusion Nuclear Technology, 25-29 September 2017, Kyoto, Japan, to appear in Fus. Eng. Des.
2. J. Jelonnek et al., IEEE Trans. Plasma Sci., **42**, Issue 5, pp. 1135-1144. 2014, doi: 10.1109/TPS.2014.2301839.
3. J. Franck, *Systematic Study of Key Components for a Coaxial-Cavity Gyrotron for DEMO,* KIT Scientific Publishing, Karlsruhe, 2017.
4. J. Jelonnek et al., Fusion Engineering and Design, 2017, doi: 10.1016/j.fusengdes.2017.01.047.
5. H. Zohm et al., Nucl. Fusion **57**, 086002, 2017.
6. O. Dumbrajs, G.S. Nusinovich, IEEE Trans. on Plasma Science, **32**, 934-946, 2004.
7. B. Piosczyk et al., IEEE Trans. on Plasma Science, 25, 460-469 (1997).
8. S. Ruess et al., in Proc. EuMW Conference 2017, Nuernberg, Germany, Oct. 2017, EuMC40-5.
9. I. Gr. Pagonakis et al., Physics of Plasmas, **23**, 083103 (2016); doi: 10.1063/1.4959113
10. S. Ruess et al., IEEE Transactions on Electron Devices, **63**, p. 2104-2109, 2016,

DOI:10.1109/TED.2016.2540298.

11. I. Pagonakis et al., 20th Joint Workshop on Electron Cyclotron Emission (ECE) and Electron Cyclotron Resonance Heating (ECRH), May 2018, Greifswald, Germany
12. T. Ruess et al., German Microwave Conference (GeMiC 2018), 12 – 14 March 2018, Freiburg, Germany, Conference proceedings (online) DOI: 10.23919/GEMIC.2018.8335085.
13. T. Ruess et al., 20th Joint Workshop on Electron Cyclotron Emission (ECE) and Electron Cyclotron Resonance Heating (ECRH), May 2018, Greifswald, Germany
14. J. Jin et al., IEEE Transactions on Microwave Theory and Techniques, **57**, No. 7, pp. 1661-1668, 2009.
15. J. Jin et al., 20th Joint Workshop on Electron Cyclotron Emission (ECE) and Electron Cyclotron Resonance Heating (ECRH), May 2018, Greifswald, Germany
16. K. A. Avramidis et al., IEEE Trans. Electron Devices, January 2018 (online) DOI: 10.1109/TED.2017.2782365.
17. P.C. Kalaria et al., International Vacuum Electronics Conference (IVEC 2018), Monterey, California, 24-26 April 2018.
18. L. Jackowski, *Advanced Vertical Collector Sweeping for High Power Gyrotrons*, Bachelor Thesis, Oct. 2017, KIT, Karlsruhe
19. W. Kdous, *Technical Design of an Advanced Vertical Collector Sweeping System for High Power Gyrotrons*, Bachelor Thesis, April 2018, KIT, Karlsruhe
20. T. Rzesnicki et al., IEEE Transactions on Plasma Science, VOL. 38, NO. 6, June 2010.
21. T. Rzesnicki et al., 34th International Conference on Infrared, Millimeter, and Terahertz Waves, Busan, 2009, doi: 10.1109/ICIMW.2009.5324919
22. T. Rzesnicki et al, in Proc. 38th IRMMW-THz, Mainz, Germany 2013, We5-3.
23. I. Gr. Pagonakis et al., Physics of Plasma, **23**, Page 023105, 2016
24. S. Ruess et al., Vacuum Electronics Conference (IVEC) 2016 IEEE International, pp. 1-2, 2016.
25. J. Zhang et al., Proc. IVEC 2015, Beijing, China, 2015.
26. J. Neilson, Joint 29th Int. Conf. on IRMMW & 12th Int. Conf. on Terahertz Elect., Karlsruhe, Germany, 2004, pp. 667-668.

Report of recent experiments with the European 1 MW, 170 GHz CW and SP prototype gyrotrons for ITER

Zisis Ioannidis[1], *Tomasz* Rzesnicki[1], *Ferran* Albajar[2], *Stefano* Alberti[3], *Konstantinos* Avramidis[1], *William* Bin[4], *Tulio* Bonicelli[2], *Alex* Bruschi[4], *Ioannis* Chelis[5], *Francesco* Fanale[4], *Gerd* Gantenbein[1], *Virgille* Hermann[6], *Jean-Philippe* Hogge[3], *Stefan* Illy[1], *Jianbo* Jin[1], *John* Jelonnek[1], *Walter* Kasparek[7], *George* Latsas[5], *Carsten* Lechte[7], *François* Legrand[6], *Ioannis* Pagonakis[1], *Francisco* Sanchez[2], *Martin* Schmid[1], *Christian* Schlatter[3], *Manfred* Thumm[1], *Ioannis* Tigelis[5], *Minh Quang* Tran[3], *Anastasios* Zisis[5], *Andy* Zein[1]

[1]IHM, Karlsruhe Institute of Technology (KIT), Kaiserstr. 12, 76131 Karlsruhe, Germany
[2]European Joint Undertaking for ITER and the Development of Fusion Energy (F4E), Barcelona, E-08019, Spain
[3]Swiss Plasma Center (SPC), Ecole Polytechnique Fédérale de Lausanne, CH-1015 Lausanne, Switzerland
[4]Institute of Plasma Physics "P.Caldirola", National Research Council of Italy, Milan, Italy
[5]Faculty of Physics, National and Kapodistrian University of Athens, Zografou, GR-157 84, Athens, Greece
[6]Thales Electron Devices, 2 rue Marcel Dassault, Vélizy-Villacoublay, F-78141, France
[7]IGVP, University of Stuttgart, Pfaffenwaldring 31, 70569 Stuttgart, Germany

Abstract. The European 1 MW, 170 GHz industrial CW prototype gyrotron has been designed within EGYC (European GYrotron Consortium) in collaboration with the industrial partner Thales Electron Devices (TED) and under the coordination of Fusion for Energy (F4E). This is a conventional (hollow) cavity gyrotron that is based on the 1 MW, 170 GHz short-pulse (SP) modular gyrotron, which has been designed and manufactured by KIT in collaboration with TED. The SP prototype has been tested in multiple experimental campaigns since 2015 and the nominal cavity mode $TE_{32,9}$ is exited at 170.1 GHz, producing RF power above 1 MW with 35 % interaction efficiency. The first phase of the experiments with the CW industrial gyrotron was successfully completed at KIT in 2016, verifying most of the ITER specifications. Short pulses (<10ms) deliver RF power higher than 0.9 MW with a total efficiency of 26 % (in non-depressed collector operation). The Gaussian mode content of the RF beam is 97 %. Pulses with duration of 180 s (limited by the high-voltage power supply at KIT) produce power more than 0.8 MW with maximum efficiency 38 % (in depressed collector operation). In this work the achievements with the SP and the CW prototype gyrotrons are summarized.

1 Introduction

Europe is making significant effort for the development of high-frequency, high-power gyrotrons able to cover the electron cyclotron resonance heating (ECRH) and current drive (ECCD) needs of modern fusion devices, such as the ITER tokamak. In particular, the European GYrotron Consortium (EGYC) in cooperation with Thales Electron Devices (TED) and under the coordination of Fusion for Energy (F4E) is developing the EU 1 MW, 170 GHz (hollow-cavity) gyrotron.

According to the project planning of F4E, the development of the European prototype gyrotron for ITER is organized in two steps. First, a modular short-pulse prototype was developed by the Karslruhe Institute of Technology (KIT) in cooperation with TED in order to verify the scientific design of the gyrotrons key components with pulses in the ms range. Then one CW industrial prototype tube (with the same scientific design) was developed in order to verify the ITER requirements, in terms of output RF power, efficiency, RF beam quality and pulse length.

The basic scientific design of the 170 GHz, 1 MW modular prototype was successfully verified in multiple experimental campaigns between 2015 and 2017. Based on the SP gyrotron design, the manufacturing of the first industrial CW prototype gyrotron has been completed at TED in 2015. The testing of the CW industrial prototype is organized in two phases. For the first phase the tube was delivered to KIT in February 2016, where it was operated with the Oxford Instruments superconducting magnet. Due to a limitation of the power supply, the operation of the gyrotron at KIT is limited to 180 s pulses. For this reason, after the successful completion of the experiments at KIT, the CW prototype was delivered to the Swiss Plasma Center (SPC), Lausanne, where the tube is currently operated with a new cryogen-free superconducting magnet manufactured by Cryogenics. One of the key goals at SPC is to extent the pulse length up to one hour. In this work we summarize the key results from the recent experiments with the European

* Corresponding author: zisis.ioannidis@kit.edu

1 MW, 170 GHz CW and SP prototype gyrotrons for ITER.

2 Short-Pulse Prototype

Fig. 1 shows the modular short-pulse prototype gyrotron, which has a hollow-cavity (conventional gyrotron). This gyrotron has been tested in many experimental campaigns during the previous years. In the first experimental campaign that took place before the delivery of the magnetron injection gun (MIG), the tube was operated with the MIG that was initially designed for the modular KIT short-pulse pre-prototype 2 MW, 170 GHz coaxial-cavity gyrotron. Details for this campaign are presented in [1]. In this section we focus on the operation of the gyrotron with the nominal MIG that was delivered by TED.

Fig. 1. The EU 1MW, 170 GHz SP gyrotron.

2.1. Operation with the basic configuration

The basic (according to the initial scientific design) configuration of the short-pulse prototype gyrotron with the TED was tested at KIT in 2015 [2]. Taking advantage of the flexibility of the Oxford Instruments superconducting magnet, two different operating points have been defined, in terms of the accelerating voltage. The first one is the Low-Voltage Operating Point (LVOP) with accelerating voltage 76.7 kV and beam current 45 A and the second one is the High-Voltage Operating Point (HVOP) with accelerating voltage 85.1 kV and beam current 40 A.

From simulations, both operating points are equivalent and able to deliver output power 1 MW with practically the same efficiency. Fig. 2 presents a typical example of the output RF power with respect to the accelerating voltage. For both operating points the nominal mode $TE_{32,9}$ is easily excited at the frequency 170.1 GHz and it is possible to generate power higher than 1 MW with 30 % total efficiency in non-depressed collector operation, which corresponds to an electronic efficiency of 35 %, as expected from theory [3]. It is interesting to note that by significantly increasing the beam current to 63 A, more than 1.2 MW power was generated (in non-depressed collector operation) with an efficiency of approximately 30%.

The total efficiency of the gyrotron is further increased, when it is operated with a single-stage depressed collector (SDC). Fig. 3 presents the generated RF power with respect to the depression voltage for the HVOP. By applying 26 kV to the gyrotron body, it is possible to increase the efficiency to slightly higher than 40% without affecting the generated power significantly.

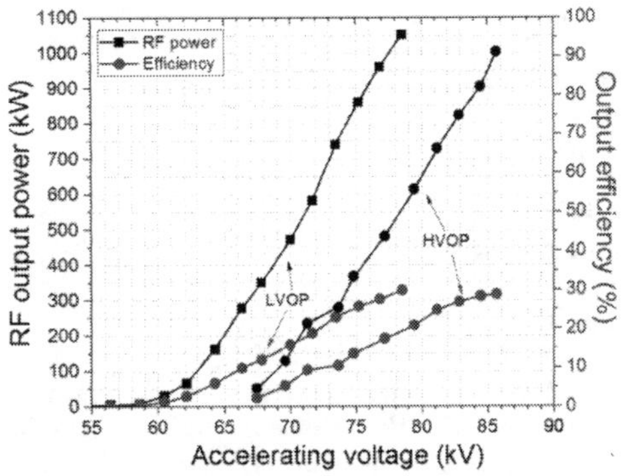

Fig. 2. RF power and corresponding efficiency (in non-depressed collector operation) with respect to the accelerating voltage for the LVOP ($V_{acc} \approx 78$ kV, $I_b \approx 45$ A) and the HVOP ($V_{acc} \approx 86$ kV, $I_b \approx 40$ A).

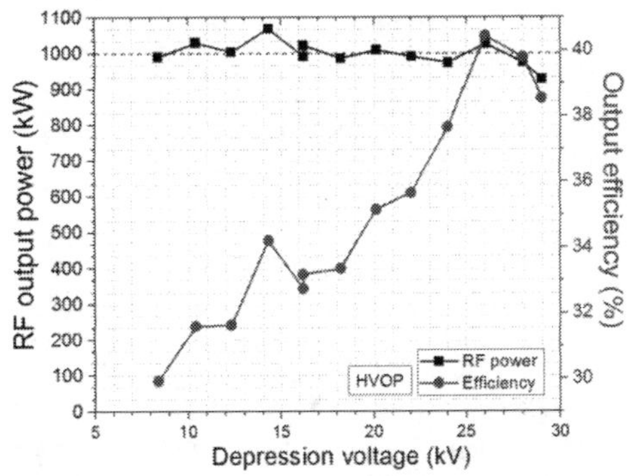

Fig. 3. RF power and corresponding total output efficiency (in depressed collector operation) with respect to the depression voltage for the HVOP ($V_{acc} \approx 86$ kV and beam current $I_b \approx 43$ A).

One of the key parts of the gyrotron that was also verified during this first campaign is the quasi-optical output coupler. The output RF beam of the gyrotron has been studied with an infrared camera. According to post processing of the experimental data the Gaussian content of the output beam was found to be almost 98 %, which meets very well the theoretical calculations.

2.2. Modified deceleration setups

According to the design/simulations, it should be possible to reach 50 % total efficiency by applying approximately 35 kV depression voltage. However, theoretical studies for the spent beam already predicted a limitation of the efficiency at lower values of the depression voltage [4], due reflection of part of the electrons in the area of the mirror box. In particular, the geometry and the internal arrangement of the mirror box results in an increase of the beam voltage depression and in turn to a significant drop of the kinetic energy of the electrons. For this reason, the maximum applicable deceleration voltage before electron reflections start is lower than initially expected. In order to investigate if it is possible to reach the ITER specifications in terms of total efficiency (larger than 50%), the modular SP was slightly modified and further tests were performed in depressed collector operation.

Fig. 4 (left) presents the first modification that was tested, which was to introduce in the mirror box a construction similar to the cooling pipes of the CW industrial prototype. Note that the SP prototype gyrotron is not equipped with a cooling system for the beam-tunnel, the cavity and the launcher, since this is not necessary for pulses in the ms range. On the contrary, in the CW prototype the pipes of the aforementioned cooling system pass through the mirror box. These pipes are located closer to the electron beam and they reduce the beam's voltage depression. Thus, an improvement of the gyrotron efficiency is expected.

Fig. 5 presents with red circles an example case (LVOP with output power approximately 850kW) of the total efficiency in depressed collector operation with respect to the depression voltage after the installation of the dummy cooling pipes. In the same figure the green triangles refer to the corresponding results of the initial gyrotron setup (without the dummy cooling pipes). Due to the installation of the cooling pipes mockup it was possible to increase the deceleration voltage roughly by 5 kV and to increase the total efficiency by 2 %.

The total efficiency can be further increased provided that the deceleration voltage is applied as close as possible to the collector. Fig. 4 (right) presents the second deceleration setup that was tested, where the SP gyrotron was operated using the isolated collector of the KIT 2 MW, 170 GHz coaxial-cavity short-pulse gyrotron. Using this collector and an additional isolating disc between the gyrotron and the magnet top plate it is possible to set the complete mirror box of the gyrotron to the body voltage and thus start the deceleration of the electrons very close to the collector. As shown in Fig. 5 with black squares, with this configuration it was possible to increase the deceleration voltage up to 39 kV and in this way to increase the total efficiency of the gyrotron to slightly higher than 44 %.

Based on the performance of the gyrotron when the complete mirror box is set on high-voltage, a metallic construction that may be installed inside the mirror box and can bring the body voltage closer to the collector was developed and will be tested [5].

Fig. 4. Upgrades on the SP gyrotron: internal pipes inside the mirror box (left) and gyrotron equipped with an isolated collector and cryostat insulating ring (right).

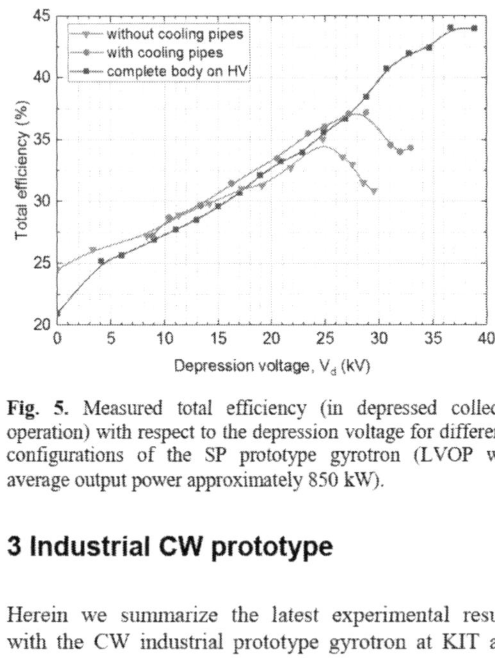

Fig. 5. Measured total efficiency (in depressed collector operation) with respect to the depression voltage for difference configurations of the SP prototype gyrotron (LVOP with average output power approximately 850 kW).

3 Industrial CW prototype

Herein we summarize the latest experimental results with the CW industrial prototype gyrotron at KIT and SPC. A detailed description of the results achieved at KIT is presented in [6]. The experiments at SPC are still ongoing.

3.1. Experiments at KIT

Fig. 6 shows the installation of the CW industrial prototype gyrotron in the KIT superconducting magnet during the 2016 experimental campaign. The KIT magnet is equipped with a set of dipole coils that were used for the optimization of the alignment of the magnetic field axis with the gyrotron cavity axis. The optimization procedure is based on the excitation circle method [7], according to which the beam is shifted by changing the dipole coil currents and the currents for which the nominal mode switches to one of its azimuthal competitors is recorded. Then the middle point of this area is used as the optimal one for operation.

A significant part of the experiments focused on the optimization of the gyrotron operation in the short-pulse regime. The gyrotron was operated for various magnetic field configurations and the magnetic field profile was optimized in terms of the field angle in the area of the emitter and the electron beam radius in the cavity [8]. In this way is was possible to generate 1 ms pulses with RF power higher than 930 KW and total efficiency 26 % (in non-depressed collector operation) The detailed characterization of the CW tube operated with short pulses is presented in [9].

Fig. 6. The EU 1MW, 170 GHz CW industrial prototype gyrotron installed at the KIT superconducting magnet.

Due to the excellent vacuum conditions of the tube the extension of the pulse length from the ms range to the second range was a straightforward procedure and after a few weeks conditioning it was possible to produce 180 s pulses with output power in the range of 800 kW. At KIT it is not possible to make longer pulses at this power level due to a limitation of the High-Voltage Power Supply. Focusing on the LVOP (due to the limited time of the experimental campaign), the gyrotron was operated with different magnetic field profiles. Fig. 7 presents the achieved output power with respect to the magnetic field angle φ_B in the emitter area and the beam radius R_b at the cavity. All the results correspond to pulses with the maximum possible accelerating voltage for each magnetic field profile, length at least 60 s and deceleration voltage in the 20-25 kV range. The corresponding total efficiency (in depressed collector operation) is shown in Fig. 8. The maximum RF output power is achieved when the gyrotron is operated with angle the φ_B = -3° and beam radius R_b = 9.50 mm. At this operating point the gyrotron delivers more than 800 kW with 36 % efficiency.

The RF output beam of the CW gyrotron was analyzed using the same method as for the SP prototype. In particular, the thermal footprint of the gyrotron output RF beam has been monitored on two different 200 mm diameter target plates, made of PVC and regular office printer paper, respectively. These plates were placed along the path of the RF beam at different distances between 180 mm and 1300 mm measured from the gyrotron output window. According to the measurements, the Gaussian content of the beam has been found to be at least 97 %, which fulfills the ITER specifications.

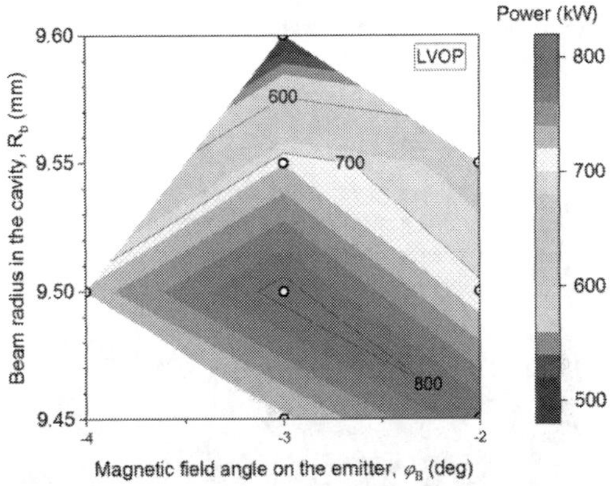

Fig. 7. RF power with respect to the magnetic field angle at the emitter and the radius of the electron beam in the cavity. Operation at the LVOP with long pulses (> 60 s).

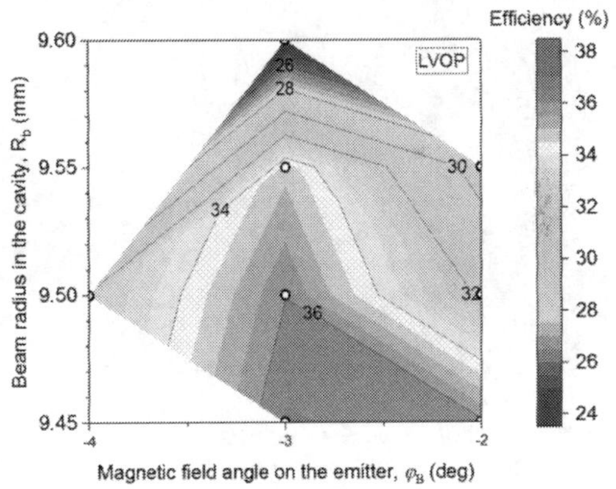

Fig. 8. Total efficiency with respect to the magnetic field angle at the emitter and the radius of the electron beam in the cavity. Operation at the LVOP.

3.2 Experiments at SPC

The industrial CW prototype gyrotron was transferred to the Swiss Plasma Center, EPFL Lausanne in the first quarter of 2017. Due to preparations on the test stand and the control system, the experiments with the gyrotron started at the end of 2017 and are still ongoing. Fig. 9 presents the gyrotron as it is currently installed in the SPC superconducting magnet. In the same figure the RF Coupling Unit (RFCU) and the RF load are also visible.

The superconducting magnet at SPC is not equipped with dipole coils and for this reason the gyrotron was installed in the magnet through an alignment XY-table that was designed and successfully tested at KIT for the SP gyrotron. The XY-table consists of two flat metallic plates that have low friction bearings between them. The bottom plate is mounted on the cryostat and the top plate on the gyrotron. The bearings allow the gyrotron to be mechanically shifted in the cryostat borehole in order to optimize the alignment of the cavity with respect to the magnetic field. A significant part of the experiments focused in the optimization of the position of the gyrotron with respect to the magnetic field axis. The method that has been followed is similar to the one with the dipoles coils and it is based on the recording of the excitation circle of the nominal mode. In particular for a specific accelerating voltage and beam current, the gyrotron was mechanically shifted on the x-y plane and the positions where the nominal mode $TE_{32,9}$ switched to the next azimuthal competitor $TE_{33,9}$ was recorded. Then the gyrotron was shifted to the middle point of the nominal mode's excitation area.

Fig. 9. The EU 1MW, 170 GHz CW industrial prototype gyrotron installed at the SPC superconducting magnet. The gyrotron is connected to the RF load using an RFCU under vacuum.

For the optimal alignment found, the gyrotron operation was first operated with short pulses in the range of 2 – 5 msec. A first coarse optimization of the gyrotron operation took place focusing on the LVOP and it was possible to reproduce the KIT results by achieving power slightly higher than 900 kW. Since the goal of the experiments at SPC was to optimize the operation of the gyrotron with pulses longer than those that are possible at KIT, the short-pulses campaign was relatively short and the short-pulse performance not fully optimized.

Currently the experiments focus on the extension of the pulse length towards 3600 seconds. For this reason, operating points able to provide power in the 500 kW – 800 kW range are selected and the pulse length is progressively increased in order to condition the tube as well as the RFCU and the RF load. In parallel, optimization of the test stand and the control system for long pulses is taking place. Up to now it was possible to make pulses longer than 200 seconds. The experiments were stopped in April and will be continued in the second half of 2018 with the goal to further optimize the operation of the gyrotron in terms of the output power and pulse length.

Acknowledgement

This work was supported by Fusion for Energy under Contract No. F4E-GRT-553 to the European Gyrotron Consortium (EGYC). EGYC is a collaboration among SPC, Switzerland; KIT, Germany; HELLAS, Greece; and IFP-CNR, Italy. The views expressed in this publication do not necessarily reflect the views of F4E or the European Commission.

References

1. I. Gr Pagonakis, et al., Status of the development of the EU 170 GHz/1 MW/CW gyrotron, Fusion Eng. Des. 96–97 (2015)
2. T. Rzesnicki et al., Recent experimental results of the European 1 MW, 170 GHz short-pulse gyrotron prototype for ITER, IRMMW-THz 2015, Proc. 3134554, Hong Kong, Aug. 23–28 (2015)
3. K. A. Avramidis et al., Simulations of the experimental operation of the EU 170 GHz, 1 MW short-pulse prototype gyrotron for ITER, IRMMW-THz 2016, Copenhagen, Denmark, Sep. 25–30 (2016)
4. I. Gr. Pagonakis et al., "Electron Beam Simulation in the Overall Gyrotron Geometry", Proc. 9th Int. Workshop "Strong Microwaves and Terahertz Waves: Sources and Applications", Nizhny Novgorod, Russia, 24-30 July (2014)
5. I. Pagonakis et al., Numerical Investigation on Spent Beam Deceleration Schemes for Depressed Collector of a High-Power Gyrotron, IEEE Trans. Electron. Devices, early access online (2018)
6. Z. C. Ioannidis et al., CW experiments with the EU 1-MW, 170-GHz industrial prototype gyrotron for ITER at KIT, IEEE Trans. Electron Devices, vol. 64, no. 9 (2017)
7. T. Rzesnicki et al., 2 MW Coaxial-Cavity Pre-Prototype Gyrotron for ITER -recent experiments with the modified gyrotron setup-, 37th International Conference on Infrared, Millimeter, and Terahertz Waves, Wollongong, NSW, Australia (2012)
8. I. Gr. Pagonakis et al., "Magnetic field profile analysis for gyrotron experimental investigation", Phys. Plasmas, vol. 24, no. 3 (2017)
9. T. Rzesnicki et al., Experimental Verification of the European 1 MW, 170 GHz Industrial CW Prototype Gyrotron for ITER, Fusion Eng. Des., Fusion Eng. Des., vol. 123 (2017)

Towards Advanced Fusion Gyrotrons: 2018 Update on Activities within EUROfusion

John Jelonnek[1,2*], *Gaetano* Aiello[3], *Stefano* Alberti[4], *Konstantinos* Avramidis[1], *Alex* Bruschi[5], *Ioannis* Chelis[6], *Thomas* Franke[7,8], *Gerd* Gantenbein[1], *Saul* Garavaglia[5], *Gustavo* Granucci[5], *Giovanni* Grossetti[3], *Stefan* Illy[1], *Zisis* Ioannidis[1], *Jianbo* Jin[1], *Parth* Kalaria[1], *George* Latsas[6], *Ioannis* Pagonakis[1], *Dimitrios* Peponis[6], *Tomasz* Rzesnicki[1], *Sebastian* Ruess[1,2], *Tobias* Ruess[1], *Theo* Scherer[3], *Martin* Schmid[1], *Dirk* Strauss[3], *Chuanren* Wu[1], *Manfred* Thumm[1,2], *Ioannis* Tigelis[6], *Minh Quang* Tran[4], *Fabian* Wilde[1,9], *Andy* Zein[1]

[1]IHM, [2]IHE, [3]IAM-AWP, Karlsruhe Institute of Technology (KIT), D-76131 Karlsruhe, Germany,
[4]Swiss Plasma Center (SPC), EPFL, CH-1015 Lausanne, Switzerland,
[5]Institute of Plasma Physics "P.Caldirola", National Research Council of Italy, Milan, Italy,
[6]National and Kapodistrian University of Athens, Faculty of Physics, Zografou, GR-157 84, Athens, Greece,
[7]EUROfusion Consortium, D-85748 Garching, Germany,
[8]Max-Planck-Institut für Plasmaphysik, D-85748 Garching, Germany,
[9]Max-Planck-Institut für Plasmaphysik, Teilinstitut Greifswald, D-17491 Greifswald, Germany

Abstract. During the ongoing pre-concept phase (2014 – 2020) for a possible future European DEMOnstration Fusion Power Plant (DEMO) the activities within EUROfusion WP HCD EC Gyrotron R&D and Advanced Developments are focusing on options for near-term solutions, and, at the same time, on long-term even more advanced options. The near-term target for DEMO is to realize pulsed operation. According to the current baseline it will probably use an EC system operating at 170 GHz and 204 GHz is being assessed, whereas the long-term target aims for steady-state operation and frequencies for current drive up to 240 GHz. Common targets for both are an RF output power per unit of significantly above 1 MW (target: 2 MW) and a total gyrotron efficiency of significantly higher than 60 %. Frequency step-tunability and multi-purpose/multi-frequency operation have to be considered. Those targets shall be achieved by considering the coaxial-cavity gyrotron technology and advanced technologies for key components (e.g. CVD diamond-disk Brewster angle window). Advanced simulation and test tools are complementing the research and developments. Gyrotron development is additionally supported by a significant investment into a new multi-megawatt long-pulse gyrotron test stand which is under final installation at KIT currently.

1 Introduction

In Europe, coordinated by the EUROfusion program [1,2] and within the EUROfusion Work Package Heating and Current Drive (WP HCD), the research and development is ongoing to find a suitable near-term solution for an Electron Cyclotron Heating and Current Drive (EC H&CD) system of a possible future European DEMOnstration Fusion Power Plant (DEMO). Start of operation is planned to be by 2050 [3]. The EC heating system should consider a minimum of 50 MW of heating power. In parallel, long-term, even more advanced options towards a real Fusion Power Plant (FPP) operating at steady state are under investigation. The timeframe during 2014 - 2020 is called the pre-concept phase of EU DEMO.

Within the WP HCD EC project a major focus is put on the development of a suitable high-power fusion gyrotron. Considering the baseline 2017 of EU DEMO, that gyrotron will have to operate at 170 GHz for heating and 204 GHz for localized plasma stabilization whereas the long-term target for a future FPP aims for frequencies up to 240 GHz (238 GHz) [4]. For both systems, an RF power per tube of significantly more than 1 MW (target: 2 MW) and a total gyrotron efficiency of higher than 60 % are assessed. RF beam steering into the plasma by frequency tuning might be an option. Therefore, the gyrotron must enable a frequency tuning in steps of around 2-3 GHz. A tuning bandwidth of approximately ±10 GHz is currently considered. Multi-purpose operation at multiples of the $\lambda/2$-resonance frequency of the RF gyrotron window, hence in leaps of about 30-40 GHz (e.g. at 136/170/204/238 GHz) might be considered for different scenarios such as plasma start-up, heating and plasma instability mitigation.

At the start of the EUROfusion program, both the advanced coaxial-cavity gyrotron technology, and, as a possible fall back solution, the conventional hollow-cavity gyrotron technology were under investigation [5]. The target was to achieve a fundamental understanding

* Corresponding author: john.jelonnek@kit.edu

on the efficiency and final operation stability at nominal operating conditions and at different parameter variations. A design strategy was developed that did help to find the optimum operating mode for the two different cavity topologies. Operating scenarios close to 2 MW for the coaxial-cavity technology and significantly above 1 MW for the conventional cavity technology were found in the theoretical analyses considering realistic targets for the thermal wall loading and the electron beam quality [6,7]. In the theoretical analyses it was found that the coaxial-cavity technology will offer a significant better potential with regards to the performance. The decision was done to focus on the coaxial-cavity technology. Concerning that, an additional experimental investigation using the unique KIT modular gyrotron technology at longer pulses (pulse lengths up to 100 ms (first stage) and up to 1 s (final stage)) is in preparation. It shall finally demonstrate the operation capabilities. In frame of frequency step-tunable RF sources, significant investments in advanced CVD diamond-disk Brewster-angle window technology are considered. Different promising concepts for multi-stage depressed collectors (MDCs) are under investigation. The new KIT teststand FULGOR is under final construction. It will support the long-pulse and CW tests of future DEMO gyrotrons operating up to 4 MW output power using dual-beam output (2x2 MW).

2 TOWARDS A LONG-PULSE COAXIAL CAVITY PRE-PROTOTYPE

Fig. 1. From left-to-right: The 170 GHz 2 MW short-pulse (ms) coaxial-cavity pre-prototype; the W7-X 140 GHz 1 MW CW series gyrotron; the ITER 170 GHz 1 MW short-pulse (ms) pre-prototype at the gyrotron test stand at Karlsruhe.

Figure 1 shows the two EU short-pulse pre-prototypes, the 170 GHz 2 MW short-pulse (ms) $TE_{34,19}$-mode coaxial-cavity tube (left) and the ITER 1 MW $TE_{32,9}$-mode short-pulse (ms) tube (right). In fig. 1 both tubes are surrounding the W7-X 140 GHz 1 MW CW series gyrotron mounted in the super-conducting magnet. The coaxial-cavity pre-prototype delivered already an RF output power of above 2.2 MW at short pulses, reliably in several experimental campaigns (ms-range) [8]. The verification at long pulses up to 1 s will be the final proof of concept and will demonstrate the applicability of the coaxial-cavity technology. The 1 s pulse length is considered to be the necessary time to achieve an operation close to steady state with regards to thermal expansion, beam neutralization and beam stability. The unique modularity of the KIT short-pulse pre-prototype shall be kept under all circumstances. The modular short-pulse pre-prototype is being upgraded in two steps. The target of the first step is to allow a reliable operation at 2 MW at pulse lengths up to 100 ms [9,10]. Figure 2 shows the finally upgraded pre-prototype before installation into the super-conducting magnet at KIT.

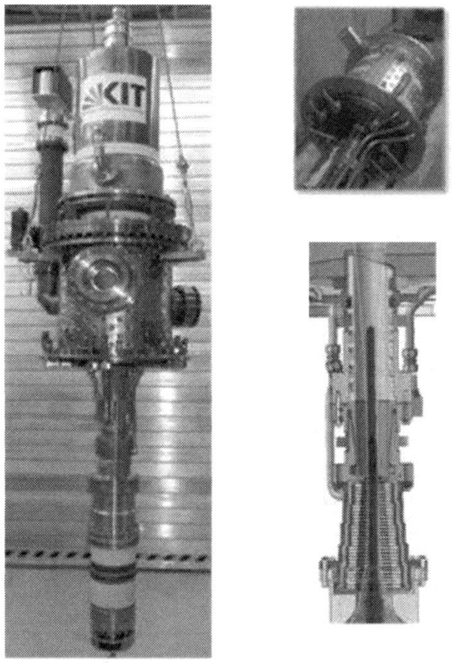

Fig. 2. Photo of the upgraded KIT 2 MW coaxial-cavity pre-prototype to operate at pulse lengths up to 100 ms (left). Upgrade of the mirror box showing the cooling system (upper right). Sketch of the key components which have been upgraded with advanced cooling systems (lower right).

This first stage includes the equipment of all thermally loaded key components of the short-pulse pre-prototype, namely beam tunnel, cavity, launcher and mirrors with advanced cooling structures. For the cavity and collector cooling different concepts are under investigation. The final second development step will include the introduction of a long-pulse collector and a CVD diamond disk RF output window. In [11] the latest developments and plans of the coaxial-cavity development are explained in more detail.

3 TOWARDS ADVANCED GYROTRON KEY COMPONENTS

The targets for a total gyrotron efficiency of higher than 60 % and the required frequency step-tunability are calling for the development of advanced key components within the EUROfusion program. The reduction of the manufacturing complexity and costs are additional drivers. In the following sections, four areas are briefly described which had been in the focus: new and advanced gun (MIG) designs, advanced cooling concepts for cavity and collector, novel concepts for multi-stage depressed collectors and the broadband CVD diamond disk Brewster-angle window.

3.1 Towards an advanced MIG design

Within the European gyrotron development program the design of the magnetron injection gun (MIG) is one of the major focus areas. The helical electron beam determines the proper excitation of the operating mode in the cavity, the maximum generated microwave power, and, finally, the gyrotron efficiency. Trapped electrons, manufacturing tolerances, surface roughness of the emitter ring, temperature inhomogeneity of the emitter surface and the electron beam neutralization are influencing the electron beam properties. [12] presents the research status on the influence of some of those factors on the electron beam. Several novelties developed are listed, e.g. detailed design criteria for the suppression of trapped electrons are shown. An emitter ring to minimize the influence of the manufacturing tolerances and edge effects on the beam quality is proposed. It bases on the finding, that the electron beam reacts very sensitive to the alignment of the emitter ring. A minor radial displacement due to manufacturing tolerances (~80 μm) is significantly influencing the beam properties already. A possible solution to this is the introduction of anti-emissive regions on the two end rims of the emitter ring [13]. A major step towards even higher output power and frequency will be an Inverse Magnetron Injection Gun (IMIG) as proposed in [14]. This novel IMIG is already manufactured for the KIT 2 MW coaxial-cavity gyrotron and waiting for tests. It is estimated that this type of inverse MIG will lead to more compact sizes and even more stable operation at even higher output power. Due to the triode configuration, the IMIG, originally designed for gyrotron operation at 170 GHz will also properly work at a frequency of 204 GHz. Because of the larger emitter radius, the gyrotron can be operated at a larger electron beam current and therefore at higher RF output power. The already performed inspection of the manufactured IMIG showed excellent surface conditions and small tolerances, which promises a homogenous electric field distribution in the emitter region and a high quality of the hollow electron beam. The placement of the emitter has been checked by using a precise optical laser measurement. Furthermore, proper welding and solder joints result in an excellent leakage rate <10^{-12} mbar/s. Figure 3 shows the two manufactured MIGs. The left part shows the MIG, produced by Thales, including an advanced emitter ring with non-emissive coatings. The right picture shows the inverse MIG, manufactured by KIT.

Fig. 3. Advanced coaxial MIGs prepared for tests with the KIT 2 MW 170 GHz coaxial-cavity gyrotron. Left: MIG including an advanced emitter ring with non-emissive coating; right: Inverse MIG with reversed configuration of emitter ring and anode.

3.2 Towards operation of a coaxial-cavity gyrotron at 170 GHz / 204 GHz / 238 GHz

In [6] the theoretical performance capability of a coaxial-cavity gyrotron operating at multiples of the λ/2-resonance frequency of the RF output window at up to 238 GHz was investigated by systematic theoretical analyses. It was shown that a coaxial-cavity gyrotron operating in a very high-order mode is capable to produce 2 MW RF output power. On the other hand, it would be much simpler, if one could operate a modified version of the existing 2 MW 170 GHz $TE_{34,19}$-mode pre-prototype at 204 GHz and at 238 GHz even. In [15] it is shown by simulations that the existing KIT $TE_{34,19}$-mode coaxial-cavity pre-prototype gyrotron is capable to operate in the new KIT FULGOR gyrotron test stand using a 10.5 T super-conducting magnet if the gyrotron construction will be slightly adapted. The modified gyrotron will be capable to operate as a multi-purpose / multi-frequency device. Without any other modifications it is expected that the gyrotron will deliver 1.7 MW (2.5 MW) at 204 GHz (170 GHz). The interaction efficiency is expected to be 31 % (37 %). Another modification of the cavity would raise the RF output power by ~0.4 MW. The interaction efficiency will increase by 1.3 % percent points at 204 GHz.

3.3 Towards a multi-stage depressed collector

Single-stage depressed collectors achieve collector efficiencies of about 60 % and are successfully used in high-power fusion gyrotrons, increasing the overall efficiency to above 50 %. The further increase of the overall gyrotron efficiency requires an efficient multistage depressed collector (MDC) system. For the same reason, e.g. space-TWTs are operating with MDCs very successfully. However, the design of a MDC for a high-power gyrotron operating at a magnetic field of several Tesla and a large hollow electron beam is not that trivial. Additionally, it requires the effective suppression of secondary electrons. A possible design is the so-called E×B drift concept as shown in fig. 4.

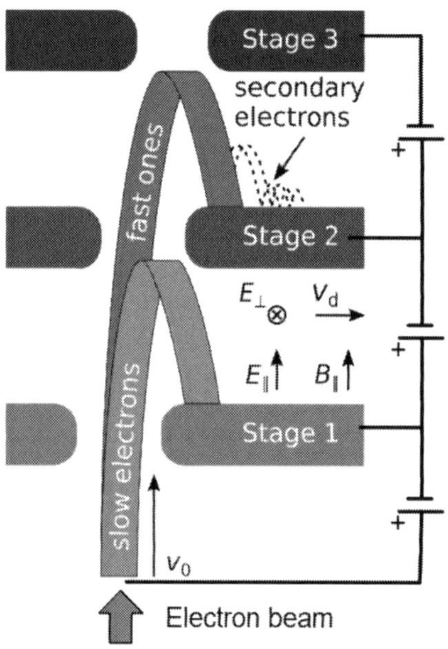

Fig. 4. Design principle of the E×B drift MDC concept.

Several theoretical design approaches based on the E×B drift have been recently published [16-20]. In [16] the E×B drift concept was proposed because of the efficient sorting of the magnetically confined gyrotron electron beam to the different electrodes. In that first investigation, a theoretical design was proposed for the demonstration of the E×B drift concept also. The collector efficiency was estimated to be of the order of 91 %. However, in that study an ideal situation with an infinite number of electrodes was considered due to limitations of the simulation tool used at that time. Figure 5 shows a so-called helical MDC which bases on the E×B concept as earlier proposed in [20]. In [21], advanced concepts of this type of MDC are numerically investigated with a full three dimensional simulation tool. MDCs with two, three and four stages have been optimized for a spent beam of a high power gyrotron. High efficiency has been demonstrated for a variety of realistic spent beam energy distributions with negligible reflected current.

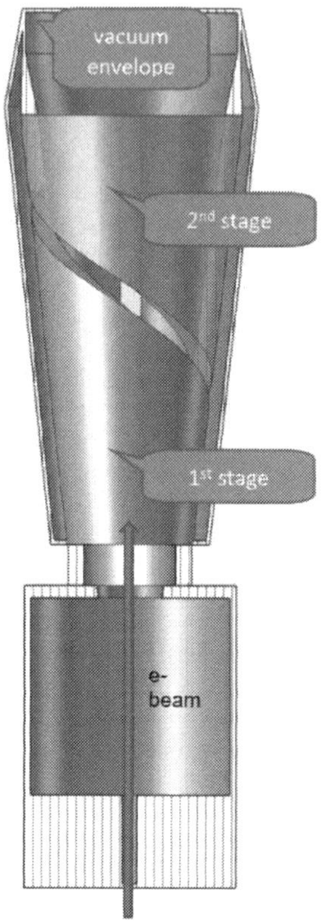

Fig. 5. Principle of a 2-stage collector with helical electrodes making use of the ExB drift concept according to [20].

3.4 Towards a broadband CVD diamond disk Brewster-angle window

It is assumed that future high-power transmission lines for fusion power plants will have to be capable to handle 2 MW of RF power at minimum. Hence, it is assumed that the waveguide diameter will be minimum 63.5 mm as originally foreseen for ITER. Figure 6 shows a sketch of such a waveguide with a CVD diamond Brewster-angle window. It follows that the large diameter of such a Brewster-angle window will have to be minimum 180 mm. State of the art is a maximum diameter of 140 mm.

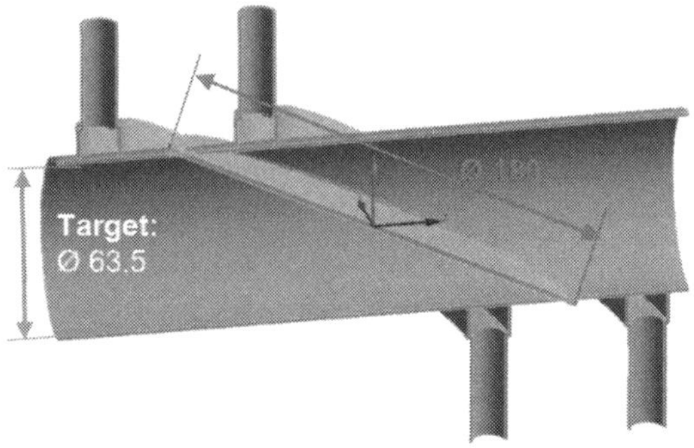

Fig. 6. Corrugated HE_{11} waveguide with future large diameter CVD diamond disk Brewster-angle window.

In 2017, at the industrial partner Diamond Materials, Freiburg, Germany [22], first attempts of directly growing a diamond disk with a large diameter of 180 mm using the microwave plasma chemical vapour deposition (CVD) process were done. These attempts resulted from the fact that the previous attempt of producing a large Brewster-angle diamond disk by joining diamond fragments using the hot filament method to overgrow the joint gap with diamond raised several drawbacks.
With the aim of growing diamond disks of 180 mm diameter, a development approach having several steps is applied. In this frame, up to November 2017, two growth experiments for parameter optimization have been successfully carried out. Each growth experiment run for approximately 350 hours resulting in rather homogeneous diamond wafers of good optical quality with a thickness of about 300 to 450 µm. Due to the very small thickness, the two wafers did not remain in one piece after dissolving the silicon substrate. Nevertheless, it was possible already to cut a 39 mm diameter disk to be used for loss tangent measurements at KIT. In 2018, growing experiments will be carried out to produce, first, a disk with a thickness of about 700 µm and then a disk of 2 mm thickness which represents the final target.

3.5 Towards smart control of gyrotrons

In future, smart control systems shall allow higher output powers of gyrotrons by reducing the necessary beam voltage safety margins [23]. Wendelstein 7-X (W7-X) uses ten 140 GHz gyrotrons for electron cyclotron resonance heating (ECRH). Approaching the practical maximum output power limit causes the excitation of unwanted parasitic and satellite modes which often results in a loss of the nominal operating mode. Therefore, the gyrotrons are operated with individual safety margins at reduced output powers to prevent a possible mode loss during operation. The power of those unwanted parasitic and satellite modes is converted to stray radiation. That radiation can be used as a precursor for mode loss indicating a potentially unstable operation state of the gyrotron. A dedicated diagnostic and control system could help to mitigate mode losses and to reduce necessary safety margins in operation. It would increase the maximum useable output power. In [24] the feasibility of a fast gyrotron controller with a real-time satellite mode activity feedback as a mode-loss precursor and an automated mode recovery is demonstrated.

4 FULGOR

KIT is currently funding a significant extension of its existing gyrotron test stand. That new test stand named FULGOR will allow the test of gyrotrons with up to 2x2 MW RF output power with dual-beam output [25].

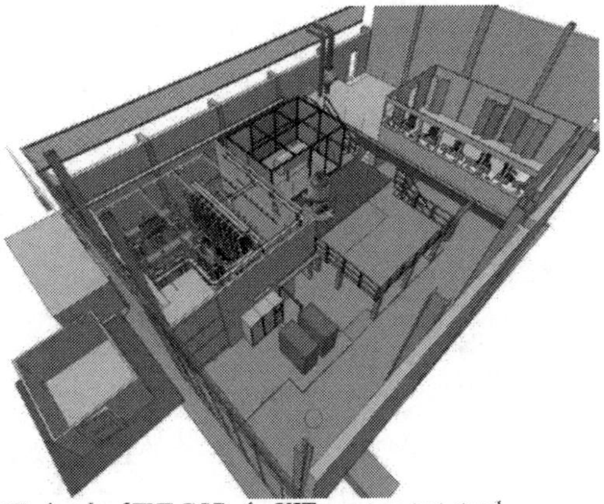

Fig. 8. 3D sketch of FULGOR, the KIT gyrotron test stand.

The installation of the main part, the 10 MW CW high voltage DC power supply (HVDCPS) has been completed in spring of 2018 by Ampegon, Turgi, Switzerland. The HVDCPS consists of an all solid-state modulator which is also capable of supplying gyrotrons with multi-stage depressed collector. While the effort for the development of the MDC is significant, the additional effort on the power supply is moderate. It is accompanied with a reduced noise level compared to a conventional pulse step modulator (PSM). This new PSM, which is the world largest today, consists of a cascade of so-called enhanced-PSM (EPSM) modules which provides a voltage of up to 90 kV. It can be tapped into at each intermediate connection point, thus giving considerable flexibility in selecting the voltage potential for each of the collector segments of a MDC. An additional option allows for 130 kV/120A non-depressed short pulse operation (all collector segments grounded) up to 5 ms every 2 s. The second major installation, will be a 10.5 T super-conducting magnet. The Call for Tender was launched in May 2017. After receiving competitive offers from three different companies and evaluating the proposals, the placement of the order was performed in November 2017. This SCM which is partially funded by EUROfusion will fulfill all requested specifications; the delivery is scheduled for the mid of 2019. It will provide a maximum magnetic field of 10.5 T at the center of the gyrotron cavity and will have a warm bore hole diameter of 261 mm. The allowed distance from the emitter to the cavity will be 400 mm. A dipole coil system for alignment of the electron beam will consist of 12 coils operating at full field with a possible shift of the beam at the cavity of 0.5 mm. The magnet will have a cryogen-free cooling system with two cooling heads. It does not contain a liquid-He-reservoir.

CONCLUSION

Within the EUROfusion Work Package Heating and Current Drive (WP HCD EC), the research and development is ongoing to find a suitable near-term solution for an Electron Cyclotron Heating and Current Drive system for the future EU DEMO. The development of a suitable high-power fusion gyrotron is one of the main research items. The coaxial-cavity gyrotron technology was chosen as the technology for future multi-megawatt gyrotrons operating at multiple frequencies. The existing KIT 170 GHz short-pulse pre-prototype has been upgraded to operate at longer pulses. It is under test currently. Using the new FULGOR test stand and the new 10.5 T magnet a modified version will be prepared to run at frequencies above 200 GHz. The gyrotron development is accompanied by the research on advanced key components, such as MIG, MDC and broadband RF window.

This work has been carried out within the framework of the EUROfusion Consortium and has received funding from the Euratom research and training programme 2014-2018 under grant agreement No 633053. The views and opinions expressed herein do not necessarily reflect those of the European Commission. A part of this work was carried out using the HELIOS supercomputer system at Computational Simulation Centre of International Fusion Energy Research Centre (IFERC-CSC), Aomori, Japan, under the Broader Approach collaboration between Euratom and Japan, implemented by Fusion for Energy and QST.

REFERENCES

1. EUROfusion, https://www.euro-fusion.org/, (2018)
2. G. Federici, C. Bachmann, L. Barucca, W. Biel, L. Boccaccini, R. Brown, C. Bustreo, S. Ciattaglia, F. Cismondi, M. Coleman, V. Corato, C. Day, E. Diegele, U. Fischer, T. Franke, C. Gliss, A. Ibarra, R. Kembleton, A. Loving, F. Maviglia, B. Meszaros, G. Pintsuk, N. Taylor, M. Q. Tran, C. Vorpahl, R. Wenninger, J. H. You, ISFNT13, (2017)
3. G. Federici, the PPPT PMU and Project Teams, http://www.firefusionpower.org/NAS_EU_DEMO_Federici.pdf, (2018)
4. J. Jelonnek, G. Aiello, S. Alberti, K. Avramidis, F. Braunmueller, A. Bruschi, J. Chelis, J. Franck, T. Franke, G. Gantenbein, S. Garavaglia, G. Granucci, G. Grossetti, S. Illy, Z. C. Ioannidis, J. Jin, P. Kalaria, G. P. Latsas, I. Gr. Pagonakis, T. Rzesnicki, S. Ruess, T. Scherer, M. Schmid, D. Strauss, C. Wu, I. Tigelis, M. Thumm, M. Q. Tran, Fus. Eng. Des., **123**, (2017)
5. J. Jelonnek, G. Aiello, S. Alberti, K. A. Avramidis, F. Braunmüller, J. Franck, T. Franke, G. Gantenbein, S. Garavaglia, G. Granucci, G. Grossetti, S. Illy, Z. Ioannidis, J. Jin, P. Kalaria, I. Gr. Pagonakis, T. Rzesnicki, S. Ruess, T. Scherer, D. Strauss, M. Thumm, I. Tigelis, M. Q. Tran, C. Wu, J. Zhang, EC-19 workshop, (2016)
6. J. Franck, KIT Scientific Publishing, http://dx.doi.org/10.5445/KSP/1000068000 (2017)
7. P. Kalaria, KIT Scientific Publishing, http://dx.doi.org/10.5445/KSP/1000073581 (2018)
8. T. Rzesnicki, B. Piosczyk, S. Kern, S. Illy, J Jin, A. Samartsev, A. Schlaich, M. Thumm, IEEE Trans. Plasma Sci., **38**, 6 (2010)
9. S. Ruess, G. Gantenbein, A. Heinzel, S. Illy, I. Gr. Pagonakis, T. Rzesnicki, M. Thumm, J. Weggen, A. Weisenburger, J. Jelonnek, EPJ Web of Conferences 149, 04015 (2017)
10. S. Ruess, K. A. Avramidis, M. Fuchs, G. Gantenbein, S. Illy, P. C. Kalaria, T. Kobarg, I. Gr. Pagonakis, T. Rzesnicki, M. Thumm, J. Weggen, J. Jelonnek, 47th EUMC, (2017)
11. S. Illy, EC-20 workshop, (2018)
12. I. Gr. Pagonakis, S. Alberti, K. Avramidis, F. Legrand, G. Gantenbein, J. Genoud, J.-P. Hogge, S. Illy, Z. Ioanidis1, P. Kalaria, B. Piosczyk, S. Ruess, T. Ruess, T. Rzesnicki, M. Thumm, I. Vomvoridis, J. Jelonnek, EC-20 workshop, (2018)
13. I. Gr. Pagonakis, S. Illy, M. Thumm, Phys. Plasmas, **23**, 83103 (2016)
14. S. Ruess, I. Gr. Pagonakis, G. Gantenbein, S. Illy, T. Kobarg, T. Rzesnicki, M. Thumm, J. Weggen, J. Jelonnek, IEEE Trans. Electron Devices, **63**, 5 (2016)
15. T. Ruess, K. Avramidis, G. Gantenbein, S. Illy, Z. Ioannidis, J. Jin, P. Kalaria, I. Gr. Pagonakis, S. Ruess, T. Rzesnicki, M. Thumm, J. Jelonnek, EC-20 workshop, (2018)
16. I. Gr. Pagonakis, J.-P. Hogge, S. Alberti, K. Avramidis, J. Vomvoridis, IEEE Trans. Plasma Science, **36**, 2 (2008)
17. O. Louksha, P. A. Trofimov, Tech. Phys. Lett. **41**, 9 (2015)
18. I. Pagonakis, C. Wu, S. Illy, J. Jelonnek, Phys. Plasmas, **23**, 4 (2016)
19. C. Wu, I. Gr. Pagonakis, G. Gantenbein, S. Illy, M. Thumm, J. Jelonnek, Phys. Plasmas, **24**, 4 (2017)
20. C. Wu, I. Gr. Pagonakis, K. Avramidis, G. Gantenbein, S. Illy, M. Thumm, J. Jelonnek, Phys. Plasmas, **25**, 033108 (2018)
21. B. Ell, I. Gr. Pagonakis, C. Wu, J. Jelonnek, EC-20 workshop, (2018)
22. Diamond Materials GmbH, 79108 Freiburg, http://www.diamond-materials.com/DE/index.htm
23. F. Wilde, H. P. Laqua, S. Marsen, T. Stange, K. Avramidis, G. Gantenbein, J. Jelonnek, S. Illy, I. Gr. Pagonakis, M. Thumm, Proc. IVEC, (2017)
24. F. Wilde, K. Avramidis, G. Gantenbein, S. Illy, J. Jelonnek, H. P. Laqua, S.Marsen, J. W. Oosterbeek, I. Gr. Pagonakis, T. Stange, M. Thumm, R. Wolf, the W7-X team, Proc. IVEC, (2017)
25. M. Schmid, M. Bader, T. Bourgeois, A. Epp, G. Gantenbein, M. Iten, J. Jelonnek, T. Kobarg, W. Leonhardt, D. Mellein, T. Rzesnicki, Fus. Eng. Des., **123**, (2017)

EPJ Web of Conferences **203**, 04008 (2019) https://doi.org/10.1051/epjconf/201920304008
EC20

Improved Simulation of Quasi-Optical Launchers for High Power Gyrotrons with Smoothing Algorithm

Jianbo Jin[1], *Gerd* Gantenbein[1], *Tomasz* Rzesnicki[1], *Manfred* Thumm[1,2], and *John* Jelonnek[1,2]

[1]IHM, [2]IHE, Karlsruhe Institute of Technology (KIT), Kaiserstr. 12, 76131 Karlsruhe, Germany

Abstract. Numerically optimized launchers for the quasi-optical output coupler of high power gyrotrons have been developed to provide RF beams with high Gaussian mode content. Generally, the profiles of numerically optimized launchers are quite complicated, so the improvement of their tolerance to fabrication errors is very important. In order to reduce the stray radiation generated by launchers with quite complicated wall profile, and also to reduce their sensitivity to fabrication errors, a method for the smoothing of the numerically optimized launcher wall has been developed at KIT, which is based on the spectrum reconstruction approach. As an example, a launcher designed for the KIT $TE_{34,19}$-mode, 2 MW CW coaxial-cavity gyrotron has been investigated.

1 Introduction

Gyrotrons are used as highest-power millimetre wave sources for Electron Cyclotron Resonance Heating (ECRH) of nuclear fusion plasmas. Those megawatt-level gyrotrons are usually operated in very high order cavity modes. Quasi-optical mode converters are employed to transform the short-wavelength and rotating asymmetric high-order cavity mode into a Gaussian-like distribution, which is directly usable for low-loss transmission in free space. The quasi-optical mode converter consists of a specific mode-converting waveguide/launcher and a mirror system. Oversized circular waveguide launchers with step-cut smooth wall were proposed by Vlasov et al. [1,2]. Launchers with harmonically deformed waveguide wall were originally proposed and demonstrated to improve the quality of the RF beams and decrease the diffraction loss of the power radiated from the launcher [3]. In 2006, by use of the Huygens' principle, a numerical method was developed for the synthesis of oversized mirror-line launchers for high power gyrotrons [4]. Later in 2009, a completely different numerical method was developed for the design of such mirror-line launchers at KIT [5]. Base on the integral Lorentz lemma, a systematic approach was proposed for the numerical optimization of multimode waveguides to convert a specific input mode into a different output field [6]. In a next step, a novel numerical method was developed at KIT for the synthesis of hybrid-type launchers [7]. The profiles of the numerically optimized launchers are relatively complicated. In order to smoothen the wall surface to satisfy fabrication requirements and to improve the tolerance of the launcher to fabrication errors and frequency shifts, it is very important to develop a method for smoothing of the surface of the launcher wall in the numerical synthesis procedure. A numerical method for the smoothing of wall surfaces has been developed based on the spectrum reconstruction method [8]. As an example, the mirror-line launcher developed for the KIT 170 GHz 2 MW $TE_{34,19}$-mode coaxial-cavity gyrotron has been investigated with and without wall smoothing in terms of the spectrum reconstruction method. In the case of the smoothened launcher wall, the minimum wall contour curvature radius increases from 1.5 mm (without smoothing) to 29 mm. The simulation results show that the Gaussian mode content of the RF beam at the launcher aperture only decreases from 96.26% (without smoothing) to 96%. The high power experimental results reveal that the Gaussian mode content of the RF beam is 96% at the output window, and the stray radiation is decreased from 7% (without smoothing) to 4% [9]. From the experimental results one can see that although the Gaussian mode content of the RF beam is slightly decreased when the wall surface is smoothed based on the spectrum reconstruction method, the stray radiation can be obviously reduced.

2 Numerically optimized launcher for the 170 GHz, $TE_{34,19}$-mode coaxial gyrotron

A mirror-line launcher has been designed for the $TE_{34,19}$-mode coaxial gyrotron at KIT [5]. The sketch of the launcher with a helical cut is shown in Fig. 1(a), where R, φ and L represent the radius, azimuthal angle and length of the launcher, respectively. In the procedure of the numerical optimization, the wall perturbations are checked in each step to see if the wall profile satisfies the requirement of the minimum curvature radius according to CNC milling machines. In the case that the curvature radius is smaller than the required minimum value, the perturbation of the wall at the sampling point will be modified by linear interpolation according to the neighbour points. If a perturbation value at a sampling point is modified, it will influence the curvature radius of its neighbour points, so the wall perturbations should be iteratively checked to be sure that the curvature radius at each sampling point is larger than the require minimum value. In the numerical synthesis procedure for the mirror-line launcher of the coaxial-cavity gyrotron, the wall perturbations were iteratively checked and modified by linear interpolation, so that the curvature radius at each mesh point should be larger than the minimum allowable curvature radius of 5 mm according to the tool used to machine the launcher. However, even though the launcher wall was iteratively interpolated 100 times in the synthesis procedure, the curvature radius at a very few mesh points was still as low as 1.8 mm. The wall profile of this non-smoothed mirror-line launcher for the coaxial gyrotron is shown in Fig. 1(b). The curvature radii are calculated via the second derivatives of this profile. Since the gyrotron cavity operates in a TE mode, we use the H_z-field component to show the field distribution in the launcher. The calculated H_z-field distribution on the launcher wall is plotted in Fig. 2. The Gaussian mode content η of the RF beam in the launcher is defined as the correlation coefficient of the H_z-field component u_z of the wave beam to an ideal Gaussian field distribution u_g at the aperture S of the launcher. The edges of the launcher aperture are shown by red lines in Fig. 2, Fig. 6 and Fig. 9.

$$\eta = \frac{\left|\int_S u_z^* \cdot u_g ds\right|^2}{\int_S |u_z|^2 ds \cdot \int_S \left|u_g\right|^2 ds} \quad (1)$$

The spectrum of the perturbed wall profile is shown in Fig. 3. As can be seen in Fig. 1(b), the minimum curvature radius in azimuthal direction is estimated as 1.8 mm and is 10 mm in longitudinal direction. The radius of the launcher is 32.5 mm.

A mirror system was also designed, manufactured and installed into the coaxial-cavity gyrotron together with the launcher shown in Fig. 1(b). The simulation result shows a Gaussian mode content of the RF beam at the launcher aperture of 96.26%. The high power measurement results reveal that the Gaussian mode content of the output beam is about 96%, and the internal stray radiation in the coaxial-cavity gyrotron is estimated to be 7%. From Fig. 1(b) one can see that there are many "ripples" on the launcher wall, and correspondingly, there are many small-amplitude high-frequency components in the spectrum of the wall profile as shown in Fig. 3. In order to depress the ripples, two methods have been studied to smoothen the perturbed launcher wall.

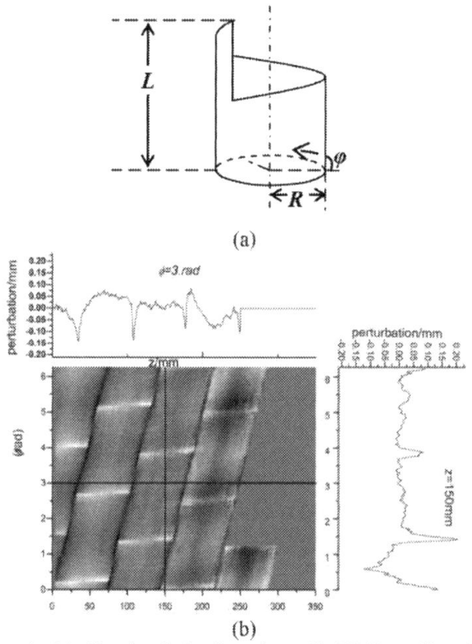

Fig. 1. (a) Sketch of the launcher, (b) Wall profile of the mirror-line launcher designed for the 170 GHz coaxial-cavity gyrotron.

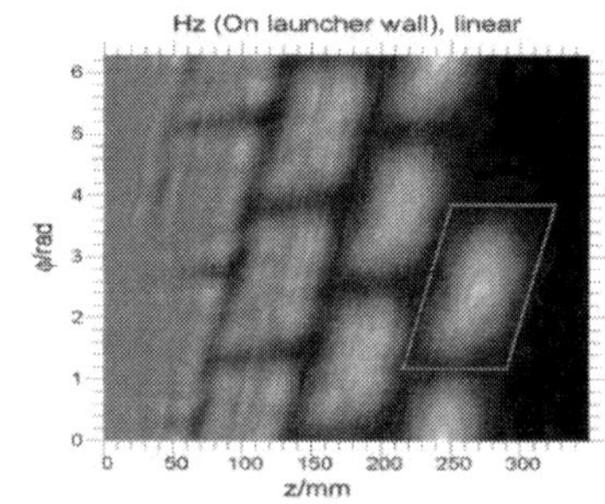

Fig. 2. H_z-field distribution on the wall of the mirror-line launcher (linear scale, 0~1).

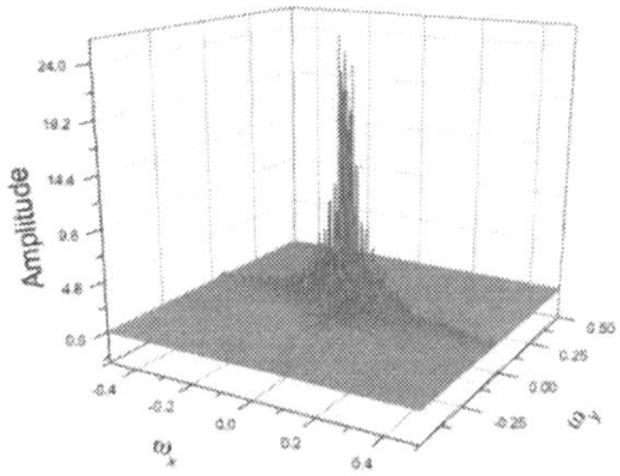

Fig. 3. Spectrum of the non-smoothened launcher wall profile.

3 Launcher wall smoothing by low-pass filtering

The principle of the low-pass filter is shown in Fig. 4. The cut-off frequencies for ω_x and ω_y are optimized to keep both the major part of the spectrum shown in Fig. 3 and high Gaussian mode content of the RF beam. The parameters of the low-pass filter for the mirror-line launcher developed for the $TE_{34,19}$-mode coaxial cavity gyrotron are shown in Fig. 4. In the synthesis procedure, the spectrum of the iteratively interpolated perturbations on the launcher wall was multiplied by the spectrum of the low-pass filter. The filtered spectrum of the profile of the launcher wall is plotted in Fig. 5.

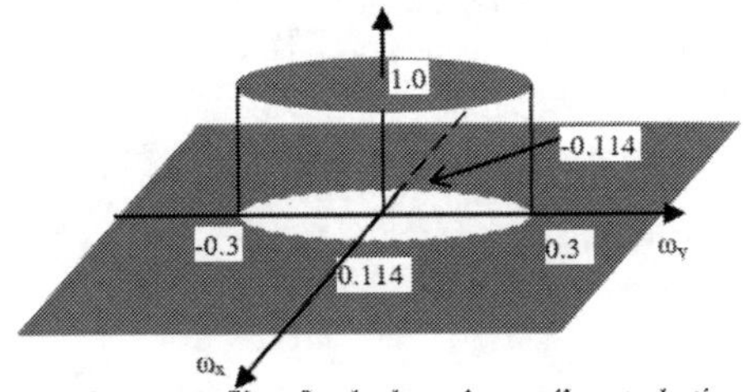

Fig. 4. Low-pass filter for the launcher wall perturbations.

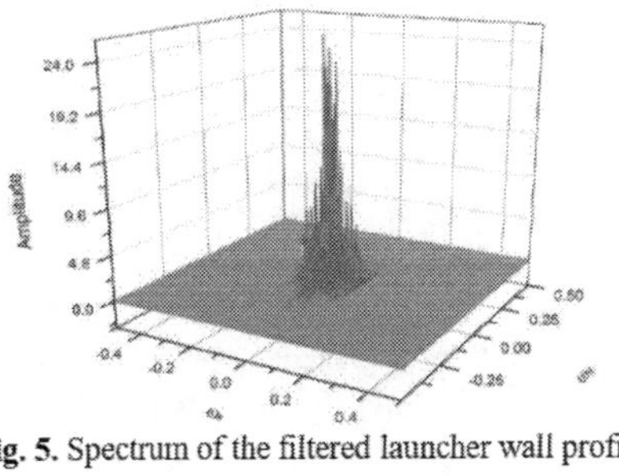

Fig. 5. Spectrum of the filtered launcher wall profile.

The wall profile was generated according to the filtered spectrum in Fig. 5. The simulation results show that the Gaussian mode content of the RF beam is increased a little bit to 96.27% at the launcher aperture as shown in Fig. 6. The wall profile is similar to that shown in Fig. 1(b), but the minimum curvature radii are increased to 6.4 mm and 67 mm in azimuthal direction and longitudinal direction, respectively. Comparing the field distribution shown Fig. 6 with that plotted in Fig. 2, the field contours still look somewhat irregular.

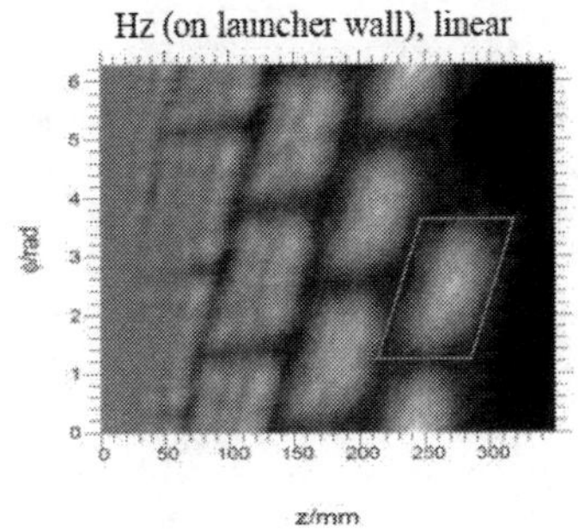

Fig. 6. H_z-field distribution on the wall of the mirror-line launcher smoothed using low-pass filtering (linear scale, 0~1).

4 Launcher wall smoothing by the spectrum reconstruction method

A spectrum reconstruction method has been developed to smoothen the launcher wall furthermore and to increase the minimum curvature radius. The method is described as follows [8]:

$$F_N(\omega_z,\omega_\varphi)=\begin{cases}F(\omega_z,\omega_\varphi) & when \quad |F(\omega_z,\omega_\varphi)|\geq A\\ 0 & when \quad |F(\omega_z,\omega_\varphi)|<A\end{cases} \tag{2}$$

Where

$$F(\omega_z,\omega_\varphi)=FFT(\Delta R(z,\varphi)) \tag{3}$$

$\Delta R(z,\varphi)$ is the wall deformation and A is the threshold of the amplitude of the spectrum used to select the relative strong components which give major contributions to the wall perturbation. FFT represents the Fast Fourier Transformation. Then the smoothed wall surface can be calculated as

$$\Delta R_N(z,\varphi)=FFT^{-1}(F_N(\omega_z,\omega_\varphi)) \tag{4}$$

The spectrum of the wall profile of the launcher smoothed on the basis of the spectrum reconstruction method is plotted in Fig. 7, where the threshold A=0.1 was used in the synthesis procedure. The small amplitude components in the spectrum have been cancelled as shown in Fig. 7. One also can see that the high-frequency components have disappeared in the spectrum and that there are fewer frequency components in the spectrum shown in Fig. 7 compared to those plotted in Figs. 5 and 3. The wall profile of this mirror-line launcher is shown in Fig. 8, the minimum curvature radius is 29 mm in azimuthal direction, which is just 3.5 mm smaller than the launcher radius. The minimum curvature in longitudinal direction is 107.8 mm. The wall profile shown in Fig. 8 is much smoother than that shown in Fig. 1 and the wall profile smoothed by low-pass filtering (Fig. 6).

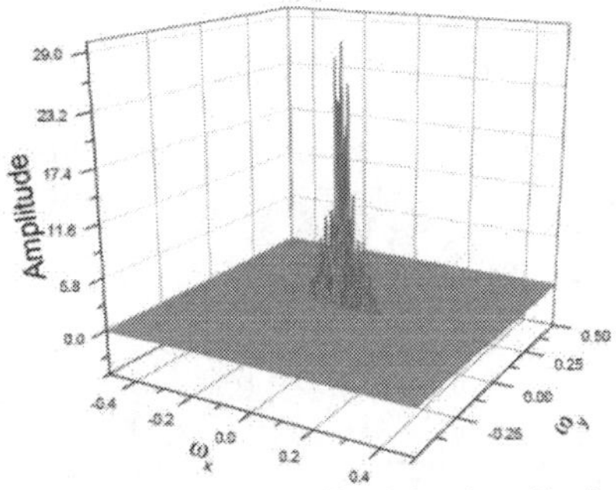

Fig. 7 Spectrum of the wall profile of the mirror-line launcher smoothed in terms of the spectrum reconstruction method.

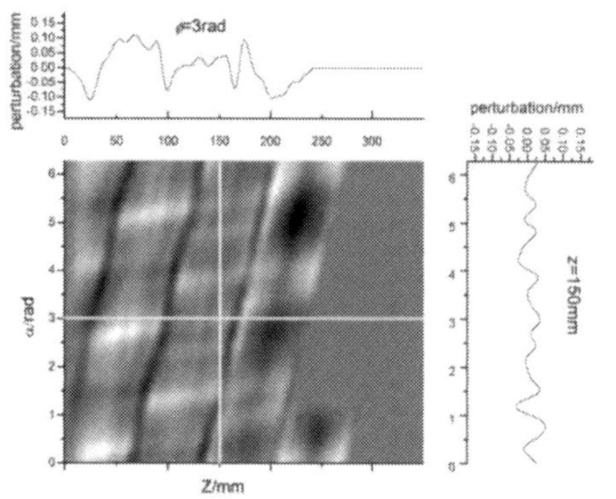

Fig. 8. Wall profile of the mirror-line launcher smoothed in terms of the spectrum reconstruction method.

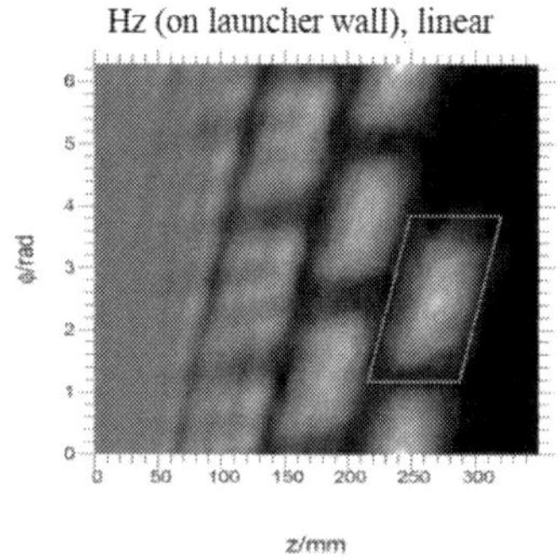

Fig. 9. H_z-field distribution on the wall of the mirror-line launcher smoothed in terms of the spectrum reconstruction method (linear scale, 0~1).

The field distribution on the improved launcher wall is shown in Fig. 9. The Gaussian mode content of the RF beam at the launcher aperture is estimated as 96.1% according to the simulation result. High power measurements shows that the Gaussian mode content at the output window is 96%, and the stray radiation was estimated by measurements to be 4% [9].

5 Summary and Outlook

Based on the spectrum reconstruction approach, a method for smoothing the launcher wall of numerically optimized high-power gyrotron quasi-optical output couplers has been developed at KIT. The curvature radii on the perturbed launcher wall have been remarkably increased. The improved launcher still provides an RF beam with high Gaussian mode content. The sensitivity of the improved launcher to fabrication errors and to frequency shifts will be investigated [8].

This work, supported by the European Communities under the contract of association between EURATOM and KIT, was carried out within the framework of the European Fusion Development Agreement. Part of this work was supported by Fusion for Energy under Grants F4E-2009-GRT-049 and within the European GYrotron Consortium (EGYC). The views expressed in this publication do not necessarily reflect the views of Fusion for Energy.

References

1. S.N. Vlasov and I.M. Orlova, Radiophysics & Quantum Electronics, Vol. 17, pp. 115-119 (1974).
2. S.N. Vlasov, et al., Radio Eng. & Electron Physics, Vol. 20, pp. 14-17 (1975).
3. G. G. Denisov, et al., Zapevalov, Int. J. Electron., Vol. 72, pp. 1079-1091 (1992).
4. A.V. Chirkov, et al., Radiophysics & Quantum Electronics, Vol. 49, No. 5, pp. 344-353 (2006).
5. J. Jin, et al., IEEE Trans. Microw. Theory Techn., Vol. 57, No. 7, pp. 1661-1668 (2009).
6. D. I. Sobolev and G. G. Denisov, IEEE Trans. Plasma Science, Vol. 38, No. 10, pp. 2825-2830 (2010).
7. J. Jin, et al., IEEE Trans. Microw. Theory Techn., Vol. 65, No. 3, pp. 699-706 (2017).
8. J. Jin, et al., "A method to improve the tolerance of quasi-optical launchers for high power gyrotrons," to be submitted to IEEE Trans. Plasma Science.
9. T. Rzesnicki, et al., "2 MW Coaxial-Cavity Pre-Prototype Gyrotron for ITER - recent experiments with the modified gyrotron setup -", Proc. IRMMW-THz 2012, Wollongong, Australia, DOI: 10.1109/IRMMW-THz.2012.6380413.

 https://doi.org/10.1051/epjconf/201920304009

Comparative studies of various types of transmission lines in the frequency range 70 GHz - 1 THz for ITER ECE diagnostic

Ravinder Kumar[1], *S.* Danani[1], *H.K. B.* Pandya[1, 2], *P.* Vaghashiya[1], *V.S.* Udintsev[3], *G.* Taylor[4], *M.E.* Austin[5], *Vinay* Kumar[1, 2]

[1]ITER-India, Institute for Plasma Research, Bhat, Gandhinagar-382428, India
[2]Homi Bhabha National Institute, Anushaktinagar, Mumbai-400094, India
[3]ITER Organization, Route de Vinon-sur-Verdon, CS 90 046, 13067 St. Paul Lez Durance Cedex, France
[4]Princeton Plasma Physics Laboratory, Princeton, NJ 08543, USA
[5]Institute for Fusion Studies, University of Texas at Austin, TX 78712, USA

Abstract. In ITER, an Electron Cyclotron Emission (ECE) diagnostic is planned to measure the electron temperature by measuring the cyclotron radiation in the frequency range of 70-1000 GHz. The cyclotron radiation is usually of low power and needs to be transported with low attenuation over a long distance of ~ 43 m, through a suitable transmission system. Pertaining to long distance, the transmission system will consist of straight waveguide sections, miter bends and waveguide joints. Low power, low loss transmission in a broadband frequency range over long distance makes the design of the transmission system challenging. To arrive at a suitable transmission system, attenuation measurements of three types of transmission lines (TLs) have been performed i.e. circular smooth walled, corrugated and dielectric coated waveguide. A polarizing Michelson interferometer based on Martin-Puplett design has been used to measure the spectrum from waveguide set ups and liquid nitrogen has been used as the black body radiation source. The measured spectrum shows atmospheric water vapour absorption lines in all types of TLs. The preliminary measurement shows that the attenuation of smooth walled waveguide is found to be comparable to corrugated waveguide up to ~ 600GHz and better than corrugated waveguide above 600 GHz for the chosen set of experimental conditions. Further, to avoid water absorption lines, a smooth walled TL is evacuated up to rough vacuum (~10^{-2}mbar) and it was observed that the attenuation is decreased and overall transmission is improved.

1 Introduction

In ITER, the electron cyclotron emission (ECE) diagnostic [1] will be used to determine the plasma electron temperature, plasma energy, ECE radiated power and runaway electrons in the frequency range 70 GHz - 1 THz by measuring the cyclotron radiation. The typical ECE system consists of front end optics including in-situ calibration sources, a set of transmission lines (TLs), and ECE radiation measurement instruments like radiometers and Michelson interferometers. However, one of the challenging requirements for the ITER ECE diagnostic is achieving low attenuation in the long TL (~43m length) for the broad frequency range. This is particularly challenging because of the low power (~ few nW) thermal radiation emitted from the in-situ calibration sources located in the port plug which needs to be measured by the ECE instruments that are located nearly ~ 43 m away in the diagnostic building. The TL will consist of straight waveguide sections, miter bends and waveguide joints. The attenuation of the TL will play an important role in determining the accuracy of measurements performed. Various investigations are underway to find a suitable TL system meeting this crucial requirement. Here, we have studied the performance of three types of circular TLs viz. smooth walled [2], corrugated [3] and dielectric coated circular waveguides. In addition, various TL components like 90° miter bends, waveguide joints, pump out units and straight waveguide sections are also characterized. The attenuation of TLs has been measured in the frequency range 70 GHz - 1 THz using a liquid-nitrogen-cooled 77 K black body source and Fourier-transform-based Michelson interferometer, and the attenuation is determined using the cutback technique. It is well known that the water vapour present in the air attenuates the millimetre waves at the water vapour absorption lines. In order to avoid absorption of the millimeter waves, the smooth walled TL system is evacuated and measurements are repeated. The TL performance is then compared in air and vacuum. The details of TL components, experimental set up and the results of the comparative studies of attenuation in three types of TLs will be discussed in this paper.

2 Transmission line and experimental set-up

Three types of circular TLs – smooth walled of inner diameter (ID) 72mm; corrugated, ID 88.9 mm; and dielectric coated, ID 69.9 mm, were characterized by measuring their attenuation in the broadband millimeter wave frequency range 70 GHz - 1THz. A broadband prototype Fourier Transform Spectrometer (70-1000

* Corresponding author: **ravinder.kumar@iter-india.org**

GHz) [4] has already been developed as per ITER ECE requirements and this FTS is used for all the prototype TL measurements. The FTS is Martin-Puplett polarizing type with linear scanning mirror to provide path length change of 15 mm resulting in a vacuum frequency resolution ~ 10 GHz and temporal resolution of 10 ms. To the best of our knowledge, all currently operating ECE interferometers are operated in air, and so this prototype FTS serves as a distinctive broadband measurement facility in vacuum, required for characterising the prototype TLs for ITER. Eccosorb [5] at liquid nitrogen temperature is used as a millimeter wave black body source for these measurements. The experimental set-up consists of the millimetre wave black body source, the FTS and the TL as shown in Figures 1 and 2. The typical TL consists of straight waveguide sections, miter bends and waveguide joints.

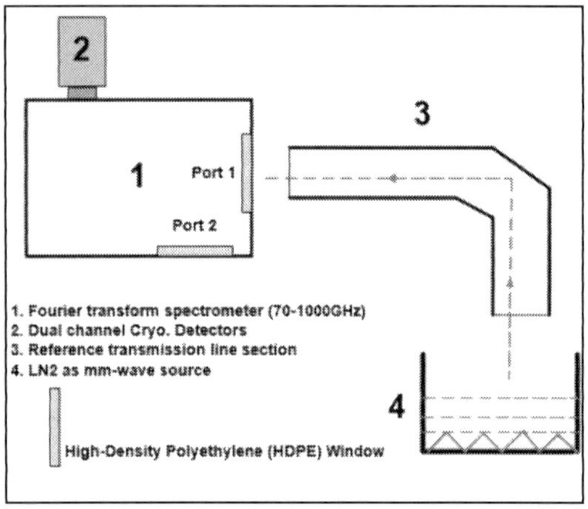

Fig. 1: Schematic of experimental set-up

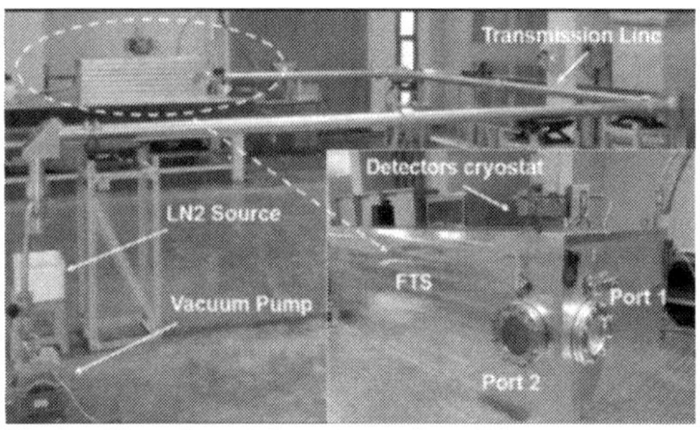

Fig.2: Experimental set-up for TL attenuation measurement

3 Results and Discussions

In this section, first the spectra measured for various transmission lines are shown. Then using the spectra, the transmission attenuation for different waveguides is derived and they are compared. As per availability of waveguide components, the attenuation is measured for 8 m of smooth walled TL, 4 m of dielectric coated and 1 m of corrugated waveguide.

3.1 TL measurements in air

Circular Smooth walled waveguide

Figure 3 shows the measured spectra for different lengths of smooth walled circular waveguide. There are water vapour absorption lines at 380, 448, 552, 752, 916 and 988 GHz. The broadness and depth of the lines increases with the length of the TL section.

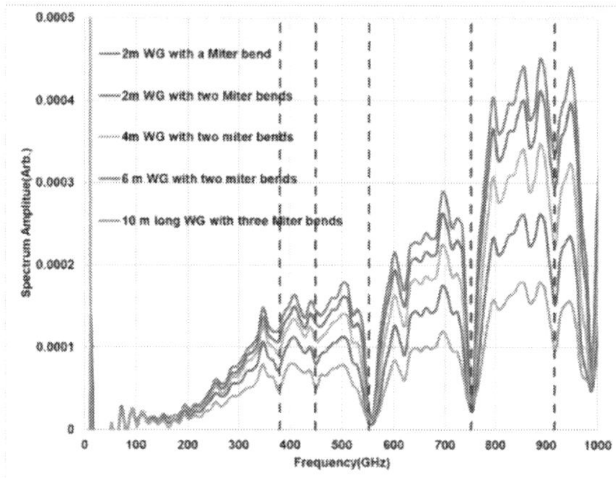

Fig. 3: Measured spectra for various TL sections

Corresponding to different TL lengths, the transmission attenuation is derived as shown in Figure 4.

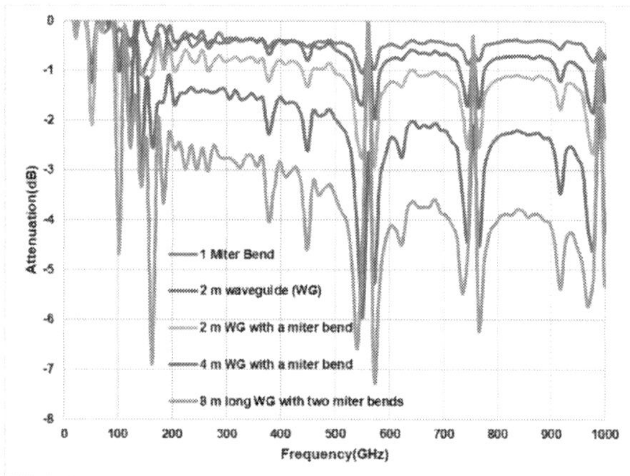

Fig. 4: Measured attenuation for various length of the smooth walled waveguide sections

Similarly, experiments were performed to measure the transmission attenuation for the dielectric coated and circular corrugated waveguides and the results are shown Figures 5 and 6 below.

Circular dielectric coated guide

https://doi.org/10.1051/epjconf/201920304009

The inner surface of the metallic waveguides is coated with optimal thickness of dielectric to minimize the transmission loss. The transmission loss is minimized as dielectric coated waveguide supports the hybrid mode, which has maximum power at the centre of waveguide and the electric fields are substantially vanished at the walls of the waveguide. Therefore, the dielectric thickness has to be chosen in such a way that the broadband millimetre wave transmits with a low loss in the desired frequency range.

The transmission attenuation is measured for 2 meters and 4 meters with 1 miter bend in a dielectric coated waveguide, and it is show in Figure 5

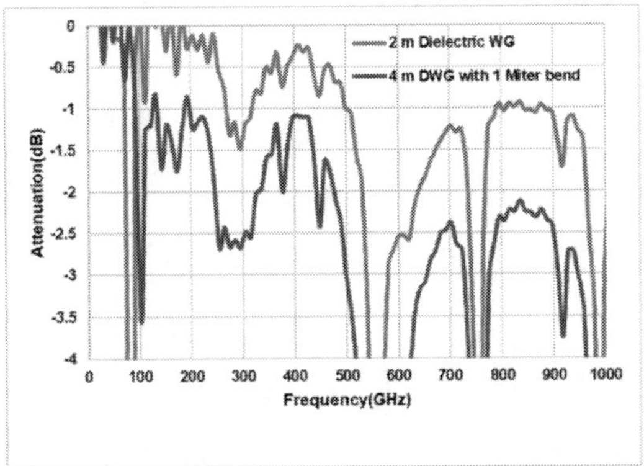

Fig. 5: Measured attenuation for various dielectric coated waveguide sections

Circular Corrugated waveguide

The 88.9 mm circular corrugated waveguide with a Bragg frequency of ~ 136 GHz is used for measuring the transmission attenuation in broadband millimetre wave frequency range 70 - 1000 GHz. The attenuation is measured in a 1-meter length waveguide section including a 90-degree bend, and is shown in Figure 6.

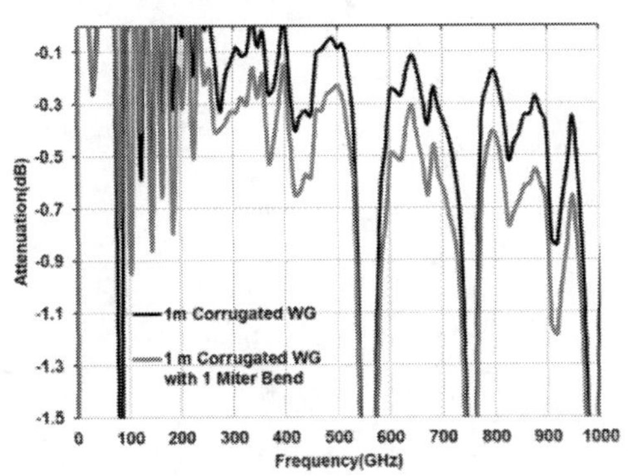

Fig. 6: Measured attenuation for circular corrugated Waveguide sections

3.2 Attenuation Comparisons

The transmission attenuation per unit length for each type of waveguide is derived and compared in Figure 7. It is observed that smooth walled circular waveguide has less attenuation at high frequency compared to Corrugated & dielectric coated waveguide.

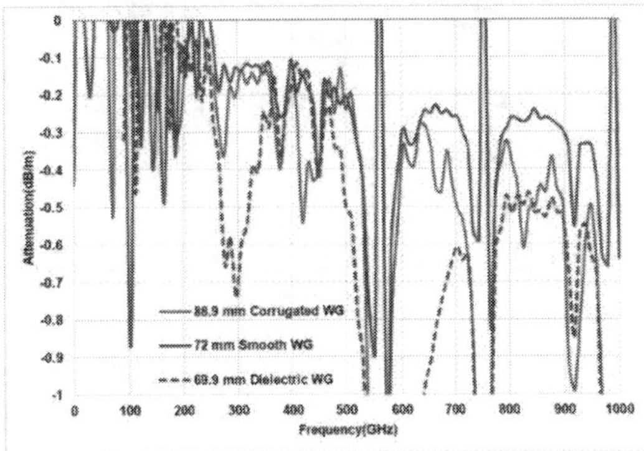

Fig. 7: Comparison of per meter transmission attenuation for various TLs

3.3 TL Measurements in a low Vacuum

In order to avoid the water absorption lines, the smooth walled TL is evacuated up to 10^{-2} mbar. The spectrum of a 10 m TL including 3 miter bends is measured in vacuum and compared with that of air TL. There is atmospheric path length i.e. ~ 550 mm in the transmission path as the millimeter wave source is not vacuum compatible. The comparison of measured spectra are shown in Figure 8. It is observed that there is some level of continuum absorption of the radiation in this frequency range by atmospheric gases.

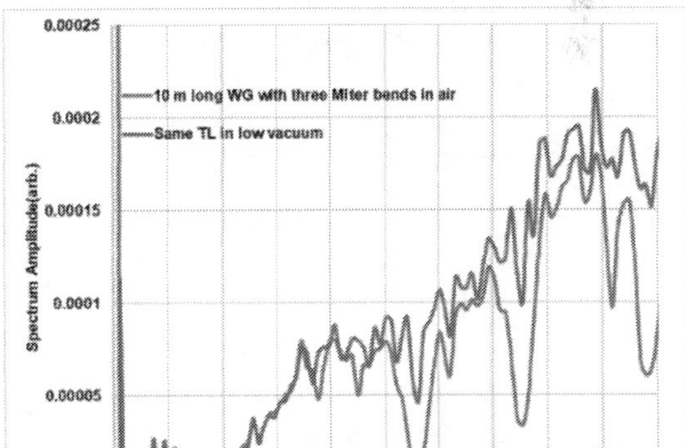

Fig. 8: Spectra of 10 m smooth TL in air and vacuum medium

The transmission attenuation is compared for the two cases and the preliminary result show that the attenuation is reduced in the case of evacuated TL and overall transmission is also improved.

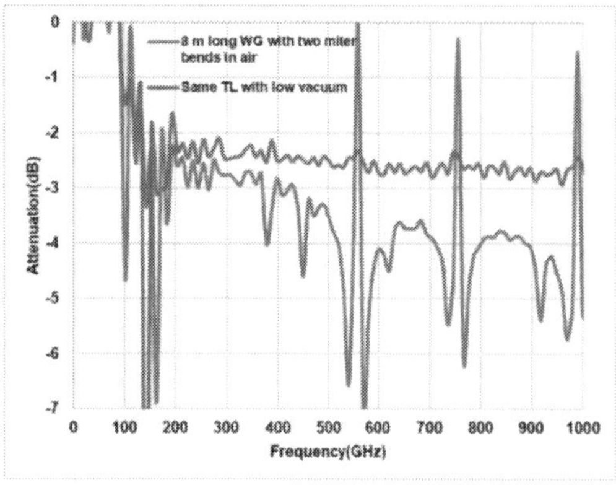

Fig. 9: Comparison of transmission attenuation in air and low vacuum

4 Conclusions

The transmission attenuations have been measured in three types of waveguides viz. circular smooth walled, corrugated and dielectric coated, using an FTS operating in the frequency range 70-1000 GHz. For each type of TL, attenuation is measured in air for a unit length of TL and compared. The preliminary measurements show that the attenuation of smooth walled waveguide is comparable to the corrugated waveguide up to ~ 600 GHz and it performed better as compared to corrugated guide above 600 GHz for the chosen set of experimental conditions as described in section 2. Water vapour absorption lines are found at certain frequencies with TLs operating in air. In case of evacuated TL, the attenuation is decreased and overall transmission is improved. It is also inferred from vacuum measurements that there is some level of continuum absorption of the radiation in this frequency range by atmospheric gases.

Acknowledgement

The author would like to thank Dr. W. Kasparek, University of Stuttgart, Stuttgart, Germany for providing dielectric coated waveguides used in the attenuation measurements.

References

1. V. S. Udintsev et al., "Progress in ITER ECE Diagnostic Design and integration," in this workshop, Griefswald, 2018.
2. H. B. Pandya, M.E. Austin, R.F. Eliis, Rev. Sci. Instrum., vol. 84, no. 103505, 2013.

3. Ravinder Kumar et al., "Fabrication and characterization of TL for ITER ECE Diagnostics," in National Symposium on Plasma Science & Technology (http://www.ipr.res.in/plasma-2017/), Gandhinagar, 2017.
4. H. B. Pandya et al., "Characterization of the Prototype Michelson Interferometer for the ITER ECE Diagnostic system," in National Symposium on Plasma Science & Technology, Gandhinagar, 2017.
5. http://www.eccosorb.com/

Disclaimer

The views and opinions expressed herein do not necessarily reflect those of the ITER Organization.

Simulation of Polarising and Reflector Gratings for High Power mm Waves

Carsten Lechte[1,*], *Walter* Kasparek[1], *Burkhard* Plaum[1], *Fritz* Leuterer[2], *Martin* Schubert[2], *Jörg* Stober[2], and *Dietmar* Wagner[2]

[1]Institut für Grenzflächenverfahrenstechnik und Plasmatechnologie IGVP, Universität Stuttgart, 70569 Stuttgart, Germany
[2]Max-Planck-Institut für Plasmaphysik, 85748 Garching, Germany

Abstract. High power mm waves for fusion plasma heating need to be elliptically polarised to ensure good absorption in the plasma. In some scenarios, electron cyclotron resonance heating (ECRH) at higher harmonics (X3 and O2) is used, but this has significant shine-through because of low single pass absorption. Grating reflectors at the inboard strike point form a holographic mirror that reflects the beam back into the plasma. This paper investigates the optical properties and ohmic losses of both the polariser and the reflectors with the 3D fullwave code IPF-FD3D. The reflection properties of a reflector for ASDEX Upgrade and the improved ohmic losses of a waveguide polariser were confirmed.

1 Introduction

Polarisers for multi-megawatt mm waves are a vital component in fusion plasma heating systems. The usual arrangement consists of 2 planar mirrors with straight gratings of depths $\lambda/4$ and $\lambda/8$ and periods much smaller than λ. This arrangement can transform any linearly polarised beam into any elliptical polarisation. Depending on the angle of the grooves to the incident plane, cross polarisation is produced in the desired quantity. The grooves themselves also introduce a phase shift between the E and H plane polarisations. It has been shown [1, 2] that the same effect can also be obtained with just two $\lambda/8$ gratings, which already have less ohmic loss than the deeper $\lambda/4$ grooves.

There are many possibilities for the shape of the grooves. For high-power applications, the main concerns are arcing and ohmic losses. The former is mitigated by rounded shapes, the latter by optimising the groove profiles. It has been shown by previous numerical investigations [3] that in general, wide grooves and thin ridges decrease the ohmic losses.

In electron cyclotron resonance heating (ECRH), advanced scenarios involving higher harmonics (X3 and O2) have significant shine-through because of reduced single-pass absorption. In the fusion experiments ASDEX Upgrade and Wendelstein 7-X, mirrors are placed at the inboard strike points which reflect the wave back into the plasma for a second absorption pass. The secondary strike point is situated on some robust section on the outboard side.

In both cases, the geometry requires 3D simulations to assess the optical qualities and the ohmic losses of the components. This is in contrast to the design stage of the components. For the polarisers, a simple mode analysis is sufficient to ascertain the polarisation parameters, and extensive 2D simulations of a single grating period were sufficient to determine the ohmic losses for certain scenarios, but not for arbitrary groove orientations relative to the plane of incidence.

In the case of the holographic reflector design, the optimisation decomposes the beam into many rays and uses a boundary element code to *independently* optimise a large number local gratings for each ray. A global optimisation is infeasible because of the large number of parameters arising from all the combinations of the local parameters. For the same reason, a 3D fullwave treatment is only feasible for verifying a small number of final results.

In the following section, the fullwave code IPF-FD3D is introduced. The polariser simulations are described in Sec. 3 and their results in Sec. 4. The reflectors are treated in Secs. 5 and 6, followed by the conclusions.

2 The Fullwave Code IPF-FD3D

The finite difference time domain code IPF-FD3D [4] can simulate wave propagation in arbitrary media and magnetised plasmas in 2D or 3D cartesian grids. Metallic components are modelled as unmagnetised high density plasmas. This method automatically forces the correct boundary conditions at the metallic surface, and the resulting plasma currents give a relative measure of the ohmic losses.

For the plasma case, the code solves the two Maxwell curl equations for $\vec{E}$ and $\vec{H}$ simultaneously with the cold electron equation of motion, which yields the plasma current $\vec{J}$. The plasma density is adjusted to a skin depth of

*e-mail: carsten.lechte@igvp.uni-stuttgart.de

1–2 grid cells.

$$\frac{\partial}{\partial t}E = \frac{1}{\varepsilon_0}\nabla\times H - \frac{1}{\varepsilon_0}J \quad (1)$$

$$\frac{\partial}{\partial t}J = \varepsilon_0\omega_{pe}^2 E - \omega_{ce}J\times\widehat{B_0} \quad (2)$$

$$\frac{\partial}{\partial t}H = -\frac{1}{\mu_0}\nabla\times E \quad (3)$$

where $\widehat{B_0}$ is the unity vector in the direction of the background magnetic field B_0, ω_{ce} the electron cyclotron frequency, and ω_{pe} the plasma frequency. No background magnetic field is present in the present simulations. All these quantities can be time and space dependent. The incident Gaussian beams are injected in the x-y plane.

3 Simulation of Polarising Gratings

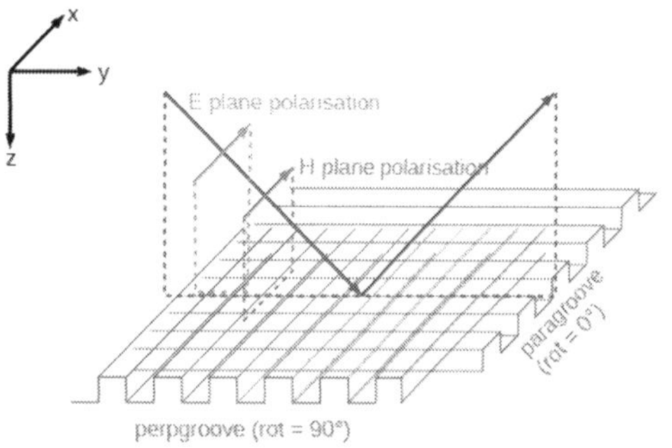

Figure 1. Geometry of the polariser simulations. The angle of incidence is always 45°. Simulations are done for E- and H-plane incident polarisation and the groove direction is measured with respect to the plane of incidence, with the 2 extreme cases indicated.

The geometric setup of the polariser simulations is given in Fig. 1. Since the polariser is incorporated into a mitrebend mirror, the angle of incidence is fixed at 45°. The direction of the grooves is chosen to effect the desired polarisation change. From the 2 incident polarisations (E- and H-plane), all elliptical input polarisations can be synthesised by complex linear combination of the complex 3D current distribution. The simulation was run with Gaussian beams with waist size of 4–6 λ. The spatial grid resolution was 186 points per wavelength, in order to accurately model the sub-mm shape of the grooves.

In addition to the conventional sinusoidal groove shape, the optimised profile "E8_4", developed at IGVP and shown in Fig. 2, was simulated. Basically, the groove width is increased and the ridge width is decreased, and finally the corners are rounded with an exponential function to avoid arcing. An additional constraint of high-power applications is the cooling of the ridges, which must therefore have a minimum width.

In Fig. 3, the electric field strength is shown in the plane perpendicular to the grooves.

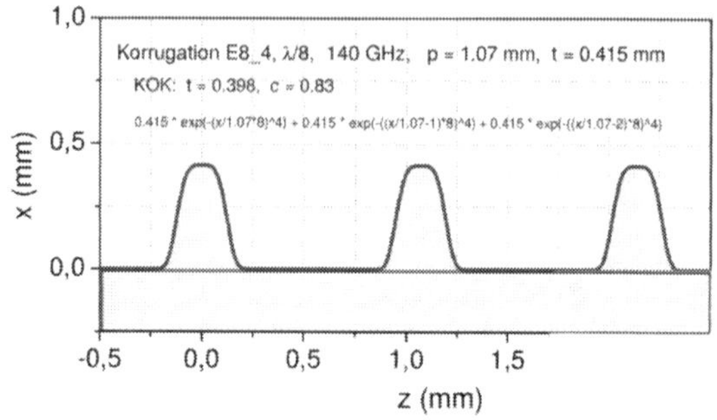

Figure 2. Side view of the optimised polariser profile "E8_4" with period $p = 1.07$ mm and ridge height $t = 415\,\mu$m. The equivalent rectangular groove parameters are reduced depth t and groove width c (second line).

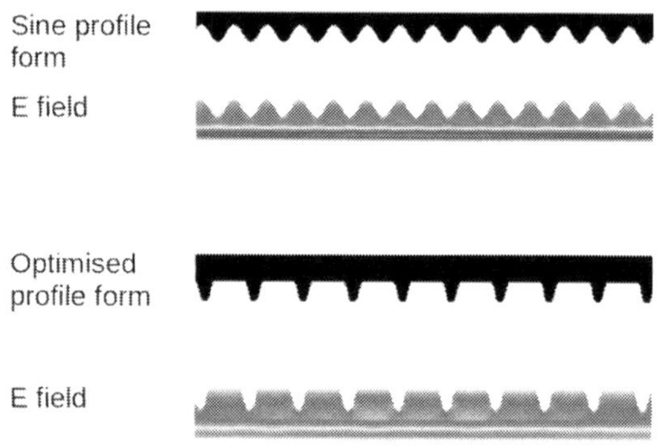

Figure 3. Electric fields in the polariser grooves for the sinusoidal and the optimised "E8_4" grating.

4 Results for Polarising Gratings

Fig. 4 shows the losses for $\lambda/8$ grooves perpendicular to the plane of incidence. They are normalised to the ohmic losses of perpendicular incidence on a plane mirror. The losses are maximal for E plane polarisation, where the wave can penetrate into the grooves and excites mostly the currents in the y and z directions. For H plane incidence, the E field is parallel to the grooves (x direction) and practically "sees" a slightly bumpy plane mirror. Consequently, the losses are caused only by currents in the x direction (J_x). For E plane incidence, the E field is equally distributed between E_y and E_z, and the losses for the sinusoid groove shape reflect this equipartition. In the optimised grating, which incidentally has more illuminated surfaces parallel to y (and x), more of the losses are carried by currents in the y direction.

Overall, the losses for polarisations near the E plane are reduced by about 20 % for the optimised groove profile. For the case of grooves parallel to the plane of incidence, not shown here, a similar improvement is seen near the H plane polarisation. The simulation results have been confirmed to within 15 % by measurements at ASDEX Upgrade [1, 2].

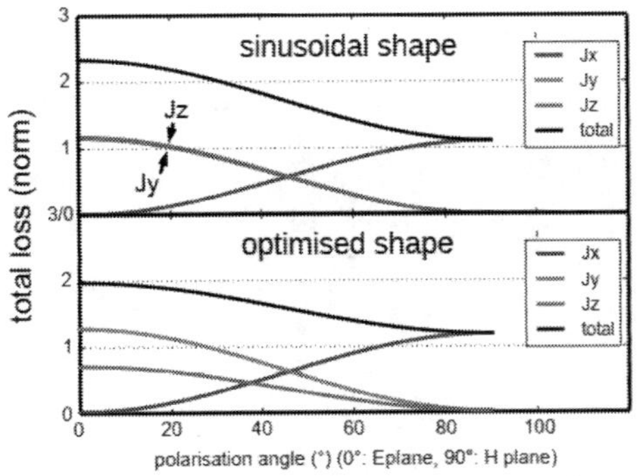

Figure 4. Losses of the 2 grating shapes for all linear polarisation angles illuminating grooves perpendicular to the plane of incidence. Losses are normalised to perpendicular incidence on a plane mirror.

5 Simulation of Holographic Gratings

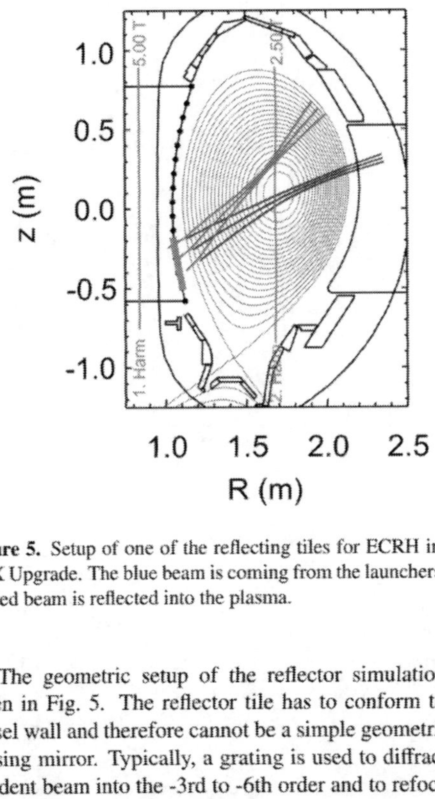

Figure 5. Setup of one of the reflecting tiles for ECRH in ASDEX Upgrade. The blue beam is coming from the launchers, and the red beam is reflected into the plasma.

The geometric setup of the reflector simulations is given in Fig. 5. The reflector tile has to conform to the vessel wall and therefore cannot be a simple geometric focussing mirror. Typically, a grating is used to diffract the incident beam into the -3rd to -6th order and to refocus it, preserving the incident polarisation. The term holographic mirror is also used for this component. The grating period is much larger than the one for the polarisers. The design of the grating shape is non-trivial because of the requirement to conserve the incident polarisation. The design process is detailed in the literature [5–7]. The finished tile was checked on the bench [8]. Previously designed tiles have been in operation at ASDEX Upgrade for years.

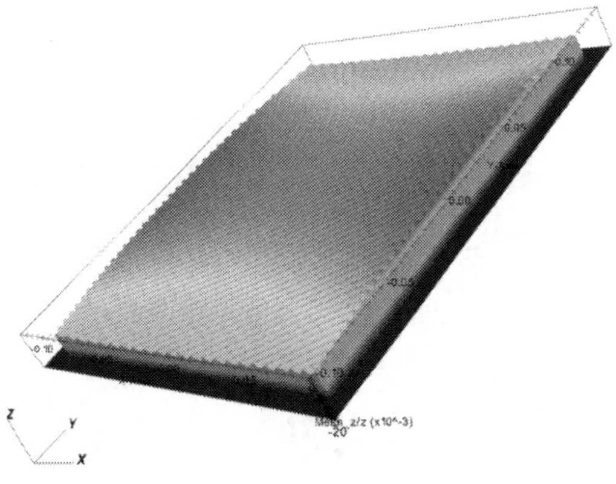

Figure 6. Shape of the holographic grating tile R7 for ASDEX Upgrade used in the simulation.

In addition, the properties are checked in a fullwave simulation using both the optimised grating shape (shown in Fig. 6) and the astigmatic elliptically polarised incident Gaussian beam. The simulation covers the full size of the grating tile with a spatial resolution of 20 points per λ. The incident wave is inserted into the grid in an x-y plane at z=5 mm. It radiates unidirectionally in the +z direction. The reflected wave can be observed in the x-y planes with z <5 mm.

6 Results for Holographic Gratings

The wave fields are shown in Fig. 7. The components E_y and H_x were chosen as representative and plotted in the x-z an y-z planes. Both slices go through the centre of the simulation region. The grating itself is the white shadow at the top of the plots. Positions with z >5 mm show the total field (incident + reflected), and only the reflected field is shown for z <5 mm.

The two top plots show the x-z plane. The incident beam is tilted in to $-x$ direction. It can be seen from the interference pattern and from the pure reflected field in the lower part of the plot that the reflected beam is nearly directly reflected into itself in x, as designed. The E_y component, being tangential to the groove surface, is 0 at the grating interface. The two bottom plots show the y-z plane. The incident beam tilts to the $+y$ direction, and the reflected beam has a much larger tilt, as designed.

The wave field is analysed with plane wave decomposition, shown in Fig. 8. It was taken in the total field region and therefore shows both beams. The incident beam at $k = (-15.0, +8.9)\,\mathrm{cm}^{-1}$ is the same as the design input beam within the simulation grid resolution (because it was put in that way). Its astigmatic nature is apparent in the elliptical shape of the peak.

The reflected beam is at $k = (+14.4, -14.4)\,\mathrm{cm}^{-1}$, which conforms to the design parameters to within 0.2 %. As sketched in Fig. 5, the beam should be reflected, but not directly back into itself. The peak's elliptical shape is stretched and rotated, but this in itself is not a problem

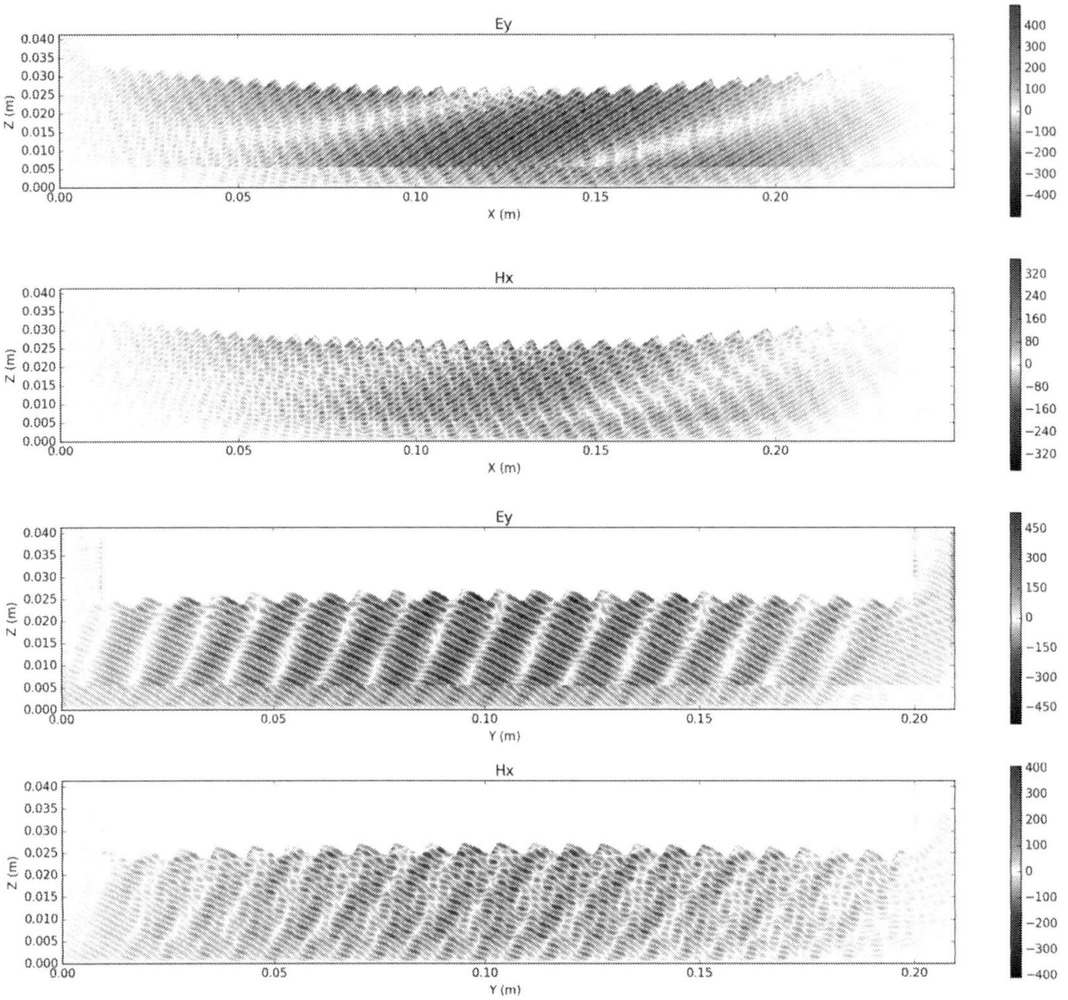

Figure 7. Wave E_y and H_x fields in the x-z (the two top plots) and y-z (two bottom plots) planes. The grating itself is the white shadow at the top of the plots. The incident wave is injected unidirectionally at z=5 mm.

for ECRH beams. The important properties of a heating beam are its (lack of) divergence and its polarisation. The reflected beam is not significantly divergent, which can be eye-balled from the extent of the peak, which is the same as the source beam's.

Even in the optimised grating, some percent of the power will inevitably go into unwanted diffraction orders, usually lower ones. One such order is just visible in the plot at $k = (4.6, -6.6)\,\mathrm{cm}^{-1}$. This is not problematic, since the power level is low and the beam will still hit the bulk plasma.

7 Conclusions

The application of the 3D fullwave code IPF-FD3D to two important beam forming components for ECRH has confirmed both the loss properties of polariser gratings, and the reflection properties of holographic mirrors gratings designed at IGVP.

For the polarisers, the 3D fullwave simulations have shown a significant decrease of ohmic losses, which was confirmed with high power measurements at ASDEX Upgrade. The optimised groove shape "E8_4" was shown to be advantageous compared to the standard sinusoidal shape.

The holographic grating reflectors for multi-pass ECRH are designed with a numerically intensive optimisation process that only looks at local groove profiles in 2D. For the finished product, a global 3D fullwave simulation is feasible. The imaging properties of the grating were confirmed to a high degree of accuracy. Results of the experimental verification of the manufactured reflecting tile are pending but are expected to support the result of the simulations.

The work was performed within the collaboration with the Max Planck Institute for Plasma Physics (IPP), Garching and Greifswald, Germany.

References

[1] D. Wagner, F. Leuterer, W. Kasparek, J. Stober, Journal of Infrared, Millimeter, and Terahertz Waves **38**, 191 (2017)

[2] F. Leuterer, D. Wagner, J. Stober, W. Kasparek, C. Lechte, ASDEX Upgrade Team, EPJ Web Conf. **149**, 03002 (2017)

[3] B. Plaum, E. Holzhauer, C. Lechte, Journal of Infrared, Millimeter and Terahertz Waves **32**, 482 (2011), 10.1007/s10762-011-9778-5

[4] C. Lechte, G.D. Conway, T. Görler, C. Tröster, the ASDEX Upgrade Team, Plasma Phys. Contr. Fusion **59**, 075006 (2017)

[5] B. Plaum, M. Schubert, A. Zeitler, W. Kasparek, C. Lechte, J. Stober, the ASDEX Upgrade Team, EPJ Web Conf. (2018), 20th Joint Workshop on Electron Cyclotron Emission and Electron Cyclotron Resonance Heating

[6] H. Höhnle, J. Stober, A. Herrmann, W. Kasparek, F. Leuterer, F. Monaco, R. Neu, D. Schmid-Lorch, H. Schütz, J. Schweinzer et al., Nucl. Fusion **51**, 083013 (2011)

[7] O. Mangold, Ph.D. thesis, Universität Stuttgart (2009)

[8] M. Schubert, J. Stober, A. Herrmann, F. Leuterer, F. Monaco, B. Petzold, E. Poli, S. Vorbrugg, D. Wagner, EPJ Web Conf. (2018), 20th Joint Workshop on Electron Cyclotron Emission and Electron Cyclotron Resonance Heating

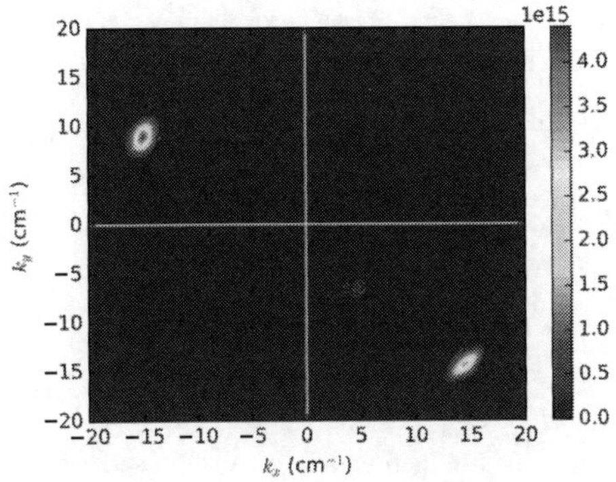

Figure 8. Power spectrum of the plane wave decomposition of incident (upper left) out reflected (lower right) wave. A spurious diffraction order is just visible at $k = (4.6, -6.6)\,\mathrm{cm}^{-1}$.

Overview on recent progress in magnetron injection gun theory and design for high power gyrotrons

Ioannis Pagonakis[1,*], *Stefano* Alberti[2], *Konstantinos* Avramidis[1], *Francois* Legrand[2,3], *Gerd* Gantenbein[1], *Jérémy* Genoud[2], *Jean-Philippe* Hogge[2], *Stefan* Illy[1], *Zisis* Ioannidis[1], *Parth* Kalaria[1], *Bernhard* Piosczyk[1], *Sebastian* Ruess[1], *Tobias* Ruess[1], *Tomasz* Rzesnicki[1], *Minh-Quang* Tran[2], *Trach-Minh* Tran†[2], *Manfred* Thumm[1], *Ioannis* Vomvoridis[4], and *John* Jelonnek[1]

[1]IHM, Karlsruhe Institute of Technology (KIT), D-76131 Karlsruhe, Germany
[2]Swiss Plasma Center (SPC), EPFL, CH-1015 Lausanne, Switzerland
[3]Thales Electron Devices (TED), Velizy, France
[4]National Technical University of Athens (NTUA), GR-15773 Athens, Greece

Abstract. The magnetron injection gun (MIG) is one of the most critical subcomponents in gyrotrons. The electron beam, which has the primary role on the gyrotron operation, is generated and configured at this part of the tube. The electron beam properties determine the excitation mode in the cavity, the power of the generated microwaves and the gyrotron efficiency. The operation of MIGs could be influenced by several factors such as trapped electrons, manufacturing tolerances, roughness of the emitter ring, emitter temperature inhomogeneity, electron beam neutralization effect, etc. The influence of many of these factors on the electron beam quality has been systematically investigated during the last years. Several novelties have been proposed in order to limit the influence of these factors on the gyrotron operation. In particular, new design criteria have been proposed for the suppression of electron trapping mechanisms, a new type of the emitter ring has been proposed to minimize the influence of the manufacturing tolerances and edge effects on the beam quality, alternative MIG design approaches have been proposed, etc. An overview of all these works will be presented here.

1 Introduction

The gyrotron is a millimeter and submillimeter wave source device, which is extensively used for plasma heating and current drive applications. The gyrotron operation is based on the electron cyclotron resonance maser (CRM) instability. A part of the transverse energy of the electron gyro-motion is converted to an electromagnetic wave in an open-ended cavity. The electrons emitted from a cylindrical emitter ring form an annular electron beam (HEB) under the influence of an externally applied static magnetic field. The beam quality at the entrance of the cavity plays an important role on the interaction efficiency η_e which ranges between 20% and 40%. For the generation of the electron beam a magnetron injection gun (MIG) is used. The geometry of the MIG subcomponents determines the quality of the HEB at the cavity. The beam electrons are produced from a heated emitter ring by temperature-limited thermionic emission. The emitter ring is part of the cathode, which is the main subcomponent of the MIG. The electrons are accelerated by the voltage U_a applied to the cathode and a second electrode, the anode. The convergent externally applied magnetic field guides the electrons towards the cavity, and generates the gyro-motion of the beam electrons. The MIG structure with a cathode and an anode is called a diode. More complex structures are also used with the introduction of a modulation anode at an intermediate potential, controlled by an independent power supply, increasing in this way the operational parameters. The MIG structure with a modulation anode is known as a triode. Even more complex MIG structures are also used for the coaxial cavity gyrotrons [1] due to the presence of the coaxial insert which is fixed at the rear part of the MIG. Theoretical models have been developed for the estimation of the beam electron trajectories in the MIG region where a strong electric field and convergent magnetic field are applied [2]. For more accurate calculations, electron optics codes have been developed, such as *EGUN* [3], *EPOSR* [4], *MICHELLE* [5], etc. In Europe (EU) three electron optics codes have been developed for gyrotron MIG study and design. These are (i) the *Ariadne* code [6] (finite element, parallel, three and two dimensional, electrostatic, tracking and self-consistent), (ii) the *ESRAY/ESPIC* code [7] (finite difference, two dimensional, tracking and PIC), and (iii) the *DAPHNE* code [8] (finite element, two dimensional, electrostatic, tracking and self-consistent).

In reality, the operation of the gun is not as ideal as it is in the numerical simulations. The quality of the generated electron beam could be influenced by many factors which are not usually taken into account during the design phase of the MIG. Such factors are the presence of trapped electrons, which can dramatically influence the gyrotron operation, the manufacturing tolerances and misalignments,

*e-mail: ioannis.pagonakis@kit.edu

 https://doi.org/10.1051/epjconf/201920304011

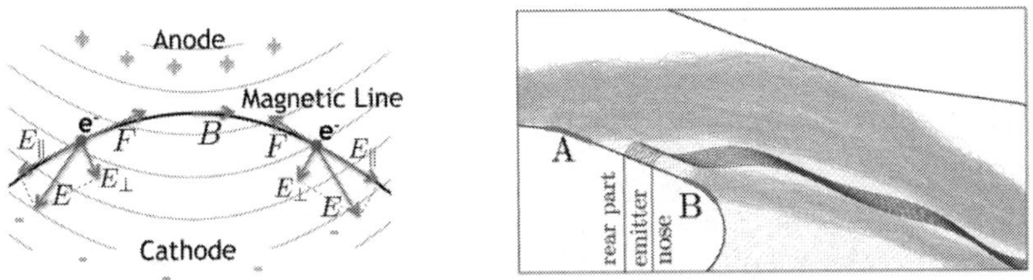

Figure 1. Equipotential lines are intersected by a magnetic field line formulating a magnetic potential well (left). Secondary electrons emitted from the region A and B are adiabatically trapped and they contribute to the generation of the harmful beam halo (right).

the emission inhomogeneity, etc. Many of these factors have been comprehensively investigated during the last years. An overview of the results of these studies will be presented here. In Sect. 2 the electron trapping mechanisms will be shortly discussed and the MIG design criteria defined for the suppression of the electron accumulation mechanisms will be presented. In Sect. 3 and Sect. 4 two important issues related to the emitter ring will be discussed. In particular, the influence of the manufacturing tolerances and the misalignment of the emitter in relation to its neighboring parts will be discussed in Sect. 3, while the effects of the azimuthal temperature non-uniformity of the emitter ring on the beam quality will be reported in Sect. 4. The paper is closing with a short summary.

2 Electron trapping mechanisms

During the experiments of the first industrial prototype of the EU 170 GHz/2 MW coaxial cavity gyrotron project for ITER [9], voltage stand-off problems even with cold emitter were observed in the presence of the nominal magnetic field with a high impedance power supply. Periodic discharges were observed for voltages around 60 kV, while for voltages more than 63 kV the leakage current exceeded the power supply current limitation. During short-pulse experiments with the electron beam, strong instabilities limited the operation of the gyrotron for accelerating voltages higher than 80–82 kV, although the nominal voltage was 90 kV.

The hypothesis of electron accumulation in the gun region was proposed for the explanation of this gyrotron behavior [10]. This hypothesis was also supported by the fact that the positions of many electron traces on the inner surface of the gyrotron components observed during the inspection of the tube after the end of the experiments, were correlated with the presence of trapped electrons. Two electron trapping mechanisms were investigated: (i) the presence of potential wells and (ii) the adiabatic trap of secondary electrons emitted from the cathode surface.

The potential well also known as Penning mechanism, can take place wherever a magnetic field line intersects the equipotential lines as shown at left of figure 1. The electrons guided by the magnetic field line are accelerated from one side and decelerated from the other side by a force $F = -eE_{\parallel}$ along the magnetic field line, where $E_{\parallel}$ is the component of the electric field parallel to the magnetic

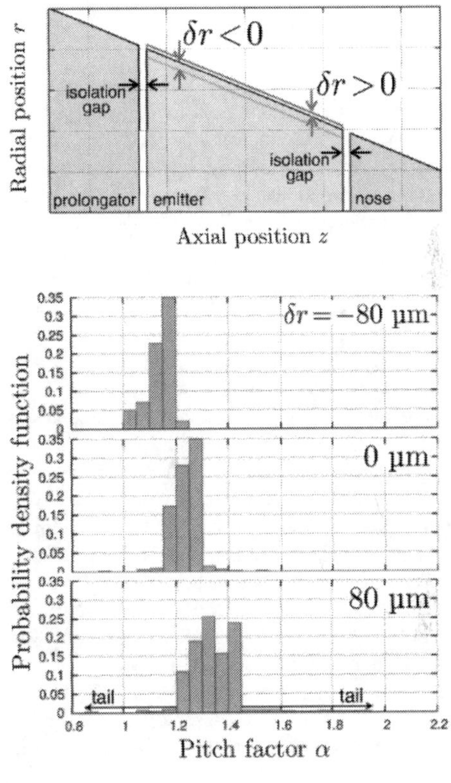

Figure 2. A small radial displacement of the emitter surface in relation to its neighboring parts (top) causes a significant variation of the pitch factor distribution (bottom).

field. Such type of potential well is usually formed at the rear part of the gun.

On the other hand, the adiabatic trap is a well known electron trapping mechanism in gyrotrons. It takes place in the region between the cathode and the cavity where the magnetic field lines converge. The transverse velocity of the beam electrons is increasing due to the conservation of the adiabatic invariant whereas the parallel velocity decreases in such a way that the total energy is conserved. As a result, some electrons can be reflected back by magnetic mirroring towards the cathode depending on their initial transverse velocity. On the other side, the decelerating

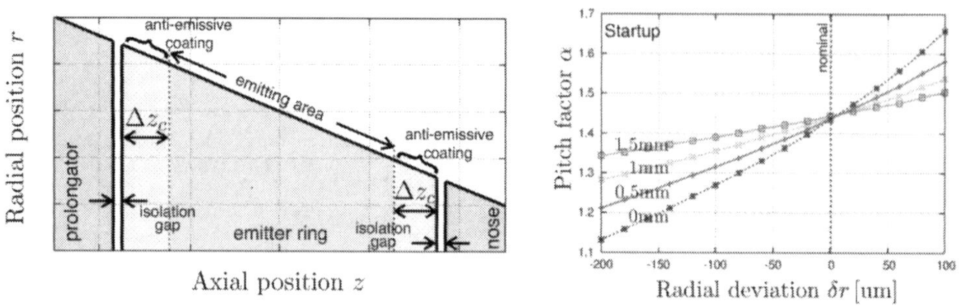

Figure 3. The new type of the emitter ring with coated edge rims (left). The variations of the average pitch factor α as a function of the radial deviation δr for several values of the coating length Δz_c (right).

field of the cathode reflects the electrons towards the cavity (electric mirror). It has been numerically shown that a minor number of adiabatically trapped electrons is adequate to initiate a harmful beam halo due to the adiabatic trap. These minor number of electrons bombard the cathode surface causing the emission of secondary electrons (see at right of figure 1). The properties of the secondary electrons emitted from the bombarded area of the cathode surface play a critical role in the formation of the beam halo.

Two design criteria have been determined for the suppression of the electron trapping mechanisms in MIG [11]: (i) the geometry of the MIG should be designed in such a way that the formation of potential wells is prevented, and (ii) no secondary electron emitted from the cathode surface is adiabatically trapped. These criteria have been considered for the design of the MIG of the refurbished first industrial prototype of the EU 170 GHz/2 MW coaxial cavity gyrotron project for ITER [12] and the MIG of the EU 170GHz/1 MW conventional cavity gyrotron for ITER [13]. The voltage standoff of both tubes was excellent, while during the operation of the gyrotrons no instability which could be correlated with the presence of trapped electrons was observed. The proposed design criteria were also considered in other MIGs designed in EU [14, 15].

Figure 4. A three-dimensional drawing and a photo of the cathode of the new MIG with the emitter ring with the coated edge rims.

3 Emitter ring tolerances

The influence of the emitter ring alignment versus its neighboring part (cathode nose and prolongator - see figure 2-top) on the beam quality has been numerically investigated. In this study the MIG design of the EU 170 GHz, 1 MW conventional cavity gyrotron for ITER [13] was considered.

An extreme sensitivity of the electron beam parameters in the emitter ring alignment was numerically identified. In particular, it has been shown that velocity spread and average pitch factor are significantly influenced by a minor emitter ring radial displacement. In figure 2 (bottom) the variation of the pitch factor distribution for a radial displacement (± 80 μm) is shown. The uncertainty on the pitch factor value could influence the mode excitation and the interaction efficiency at the gyrotron cavity. Furthermore, the formation of potentially harmful long tails in the pitch factor distribution is also predicted. This high sensitivity can potentially induce a significant spread in the performance of tubes with an identical scientific design.

A possible solution of this major issue for the gyrotron MIG designs was proposed by introducing anti-emissive regions on the two edge rims of the emitter ring [16], as shown in figure 3 (left). Using this idea, it was possible to significantly suppress that sensitivity as shown in figure 3 (right).

It should be note that for the 170 GH 2 MW 100 ms coaxial cavity gyrotron [17] a new MIG has been designed and manufactured which uses for first time in EU the new type of the emitter with coated edge rims (see figure 4). Experiments with the new MIG are expected in the near future.

4 Emission temperature inhomogeneity

The beam electrons are produced by thermionic emission from an emitter ring at around 900°C. The emission in gyrotrons takes place in the temperature-limited regime. For this type of emission, the current density J is defined as an exponential function of temperature (Richardson-Dushman equation with Schottky correction). Therefore, a small variation of temperature may dramatically influence the current density and cause a significant inhomogeneous emission.

The purpose of this study is to evaluate whether a typical specified tolerance of ±15°C for the temperature azimuthal inhomogeneity of the emitter ring has a negative impact on the gyrotron operation. For the definition of the current density inhomogeneity a similar procedure as in ref. [18] was followed. In particular, the current density was defined as $J(\varphi) = J_0(1 + \epsilon \cos(n\varphi))$, where ϵ is the amplitude and n is the harmonic of the inhomogeneity. Three dimensional simulations were performed for the EU 170 GHz, 1 MW conventional cavity gyrotron for ITER using different values of the harmonic number n. The inhomogeneity amplitude ϵ was set equal to 0.5. This value approximates an azimuthal current density inhomogeneity which corresponds to the temperature non-uniformity of the order of ±15°C.

The beam parameter which is mostly influenced by the current density inhomogeneity is the kinetic energy spread. As shown in figure 5, the kinetic energy of the electrons (left) and the kinetic energy distributions (right) are significantly different from the homogeneous emission case shown in black. The energy spread which in the ideal case is only 0.1%, at the worst case scenraio of $n = 1$ becomes 2%. However, preliminary mono-mode[19] and multi-mode[20] cavity simulations predict that the influence of such a higher energy spread does not have an important impact neither on the excitation and the stability of the operating mode nor on the interaction efficiency. More details on this work will be presented to the scientific community in the near future.

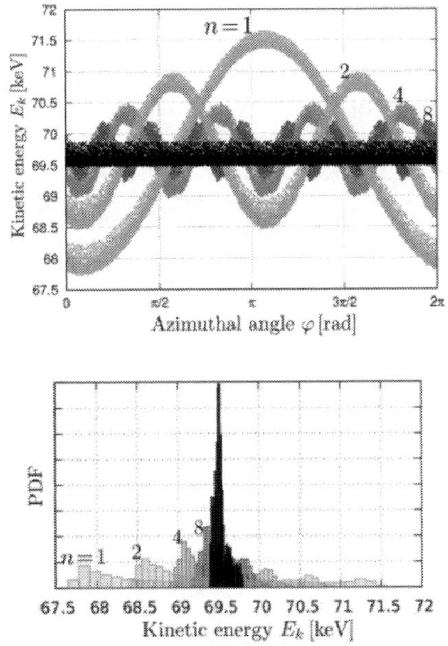

Figure 5. Kinetic energy of the beam electrons at top and the kinetic energy distribution for several values of inhomogeneity harmonic n at bottom.

5 Summary

Three important issues related to MIG theory and design were presented: (i) Two electron trapping mechanisms were discussed while design criteria were proposed for the suppression of the electron trapping mechanisms. (ii) The influence of a possible emitter ring radial displacement in relation to its neighboring parts on the beam quality was numerically investigated. The results show that such a radial displacement significantly influences the gyrotron operation. A new type of emitter ring with coated edge rims could suppress that sensitivity. (iii) The influence of the specified emitter ring temperature inhomogeneity on the beam quality was also shortly discussed.

Acknowledgments

Part of this work was supported by Fusion for Energy under Contract Nos. F4E-GRT-008, F4E-GRT-049, F4E-GRT-432, and F4E-GRT-553 to the European Gyrotron Consortium (EGYC). EGYC is collaboration among SPC, Switzerland; KIT, Germany; HELLAS, Greece; and IFP-CNR, Italy. In addition, a part of this work has been carried out within the framework of the EUROfusion Consortium and has received funding from the Euratom research and training program 2014-2018 under grant agreement No 633053. The views expressed in this publication do not necessarily reflect the views of F4E or the European Commission.

References

[1] B. Piosczyk, et al., IEEE Trans. Plasma Science **32**, 853 (2004).

[2] S.E. Tsimring, *Electron beams and microwave vacuum and electronics* (Wiley, New Jersey, 2007)

[3] W.B. Herrmannsfeldt, *EGUN - An Electron Optics and Gun Design Program* (Stanford Linear Accelerator Center, 1988)

[4] V.K.Lygin, V.N.Manuilov, Sh.E.Tsimring, Beams **96**, 385-388 (1996)

[5] J.J. Petillo, E.M. Nelson, et al. IEEE Trans. on Electron Devices **52**, 742-748 (2005)

[6] I.Gr. Pagonakis, J.L. Vomvoridis, IRMMW-THz, 657-658 (2004)

[7] S. Illy, J. Zhang, J. Jelonnek, Gyrotron electron gun and collector simulation with the ESRAY beam optics code, IEEE IVEC (2015)

[8] T.M. Tran, et all, EPS-APS (1994)

[9] J.P. Hogge, S. Alberti, et al., Fusion Science and Technology **55**, 204-210 (2009)

[10] I. Gr. Pagonakis,et al., Gun Design Criteria for the Refurbishment of the First Prototype of the EU 170GHz/2MW/CW Coaxial Cavity Gyrotron for ITER, IRMMW-THz (2009)

[11] I. Gr. Pagonakis, B. Piosczyk, et al., Physics of Plasmas **23**, 23105 (2016)

[12] S. Kern, et al., EPJ Web of Conference **32**, 04009 (2012)

[13] J. Jelonnek, F. Albajar, et al., IEEE Trans. on Plasma Science **42**, 1135 (2014)

[14] S. Ruess, I.Gr. Pagonakis, et al., IEEE Trans. on Electron Devices **63**, 2104-2109 (2016)

[15] J. Franck, I. et al., GeMiC, 260-263 (2015)

[16] I.Gr. Pagonakis, S. Illy, M. Thumm, Physics of Plasmas **23**, 83103 (2016)

[17] S. Ruess, et al., 47th EuMC, 2017, 860-863 (2017)

[18] J. Gr. Pagonakis, J. L. Vomvoridis, IEEE Trans. on Plasma Science **32**, 890 (2004)

[19] F. Braunmueller, T.M. Tran, et al., Physics of Plasmas **22**, 063115 (2015).

[20] K. A. Avramides, et al., EPJ Web of Conference **32**, 04016 (2012)

Development of reflection gratings for advanced ECRH scenarios

Burkhard Plaum[1,*], *Martin* Schubert[2], *Achim* Zeitler[1], *Walter* Kasparek[1], *Carsten* Lechte[1], *Jörg* Stober[2] and ASDEX Upgrade Team[2]

[1]Institut für Grenzflächenverfahrenstechnik und Plasmatechnologie, Universität Stuttgart, Pfaffenwaldring 31, D-70569 Stuttgart, Germany
[2]Max-Planck-Institut für Plasmaphysik, Boltzmannstr.2, D-85748 Garching, Germany

Abstract. An existing framework for the design of reflection gratings was reworked. It takes the astigmatic complex beam parameters and the orientations of the beam axes of the incident and reflected beams as input and synthesizes a grating, which transforms the incident beam into the reflected beam. This is done by decomposing the 3D problem into a series of 2D reflections of plane waves. The 2D grating profiles are optimized in parallel on multiple computers. Finally, the 3D grating is derived using a simplified interpolation scheme.

1 Introduction

When heating fusion plasmas at higher harmonics of the electron cyclotron frequency, the absorption efficiency can drop. This leads to a significant transmitted beam power hitting the wall at the high field side (HFS). To protect the wall and to use the remaining power, one method is to place a specialized reflector at the HFS, which directs the beam back into the plasma for a second heating pass [1]. The reflector needs to conform to the tiles at the HFS, which are often non-planar. The direction of the reflected beam is chosen such that the absorption of the second heating pass is maximized and the wall area at the LFS, which is hit after the second pass, contains no sensitive components. Additional requirements are a refocusing of the beam and polarization independence of the reflection characteristics. All these conditions can only be fulfilled with a grating. The design process involves the decomposition of the 3D field problem into a series of 2D reflections of plane waves. After the 2D gratings are optimized, the final 3D grating is synthesized.

2 Design process

For the definition of the problem the astigmatic beam parameters of the incident and reflected Gaussian beams are required as well as the orientations of the beam axes. These are obtained by a series of beam tracing simulations using the TORBEAM code [2]. These runs are done for different typical heating scenarios and the most appropriate beam parameters are obtained by averaging [3]. Furthermore the basic shape of the reflector is needed. In the case of the high field side of a tokamak, it is a saddle-type shape defined by the curvature radii in the toroidal and poloidal directions. In local coordinates of the reflector, the shape is given in terms of the surface height $z(x,y)$ and the orientation of the normal vector $\mathbf{n}(x,y)$. Using these parameters, we can calculate the local wave vectors $\mathbf{k}_{in}(x,y)$ and $\mathbf{k}_{out}(x,y)$.

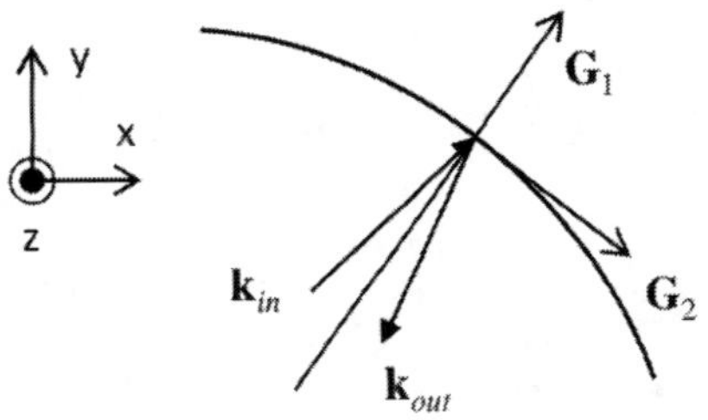

Fig. 1. Local k vectors and the resulting grating vectors (top view)

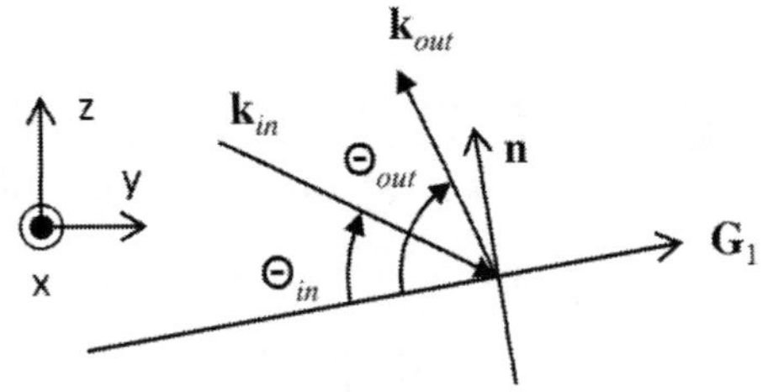

Fig. 2 Local k vectors (side view)

Figures 1 and 2 show the definition of the grating vectors. $\mathbf{G}_1$ is tangential to the surface and lies in the middle plane between $\mathbf{k}_{in}$ and $\mathbf{k}_{out}$. $\mathbf{G}_2$ is perpendicular to the middle plane and defines the direction of the groove. Parallel to the groove we have a reflection according to

* Corresponding author: plaum@igvp.uni-stuttgart.de

the reflection law (Fig. 1). Perpendicular to the groove we have the angles Θ_{in} and Θ_{out}, which, together with the diffraction order n, completely define one local 2D problem.

From an arbitrary start point and by using the known grating period, we can obtain a discretized form of the groove layout by integrating $\mathbf{G}_1$ with a standard Runge-Kutta method. An example is shown in Fig. 3.

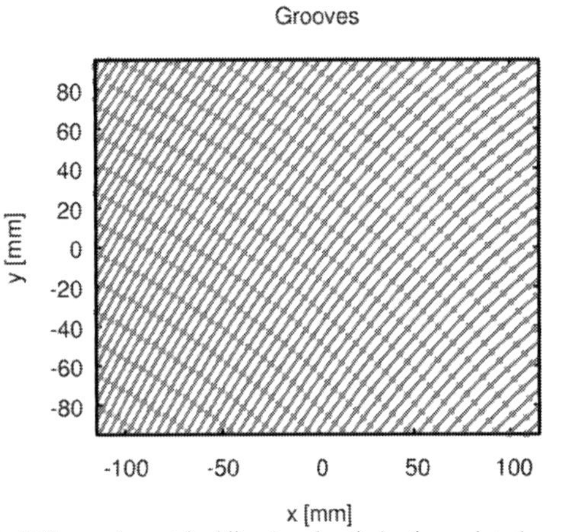

Fig. 3 Groove layout (red lines) and optimization points (green dots) for the 2D profiles

We now define the points, for which 2D gratings will be optimized. The distance of two neighboring points along the grooves is defined by the differences of the angles Θ_{in} and Θ_{out}:

$$|\Delta\theta_{in}| + |\Delta\theta_{out}| = 1° \tag{1}$$

Depending on the actual beam parameters, this can result in a large (≈1000) number of 2D profiles (green dots in Fig. 3). This number can be reduced by quantizing the values for Θ_{in} and Θ_{out}: If the angles are rounded e.g. to a resolution of 0.5°, a number of 2D gratings become identical and need to be optimized just once. Fig. 4 shows the number of distinct profiles as a function of the angular resolution. While the actual numbers can change strongly depending on the actual beam configuration, the reduction rate is always significant. Earlier works [4] showed that angular errors of up to 1° are acceptable. The angular resolution is a simple parameter for finding a compromise of accuracy and optimization time.

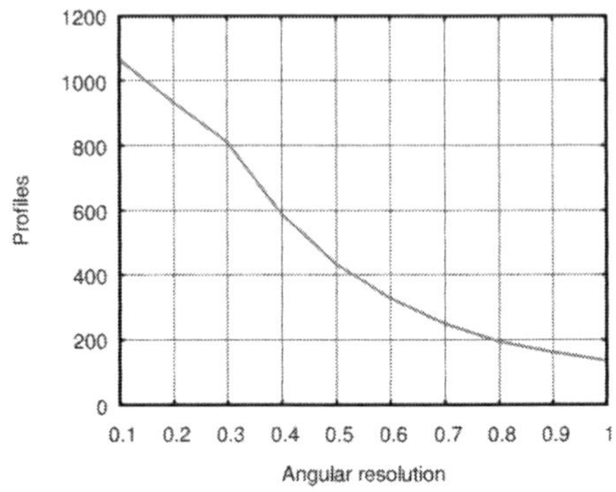

Fig. 4 Number of distinct profiles as a function of the angular resolution

The optimization goal for the 2D profiles is a maximum power in the desired diffraction order and minimum phase difference between the polarizations. The cost function is defined in terms of an efficiency η:

$$\eta = \frac{1}{4} I_n^{TE} + \frac{1}{4} I_n^{TM} + \frac{1}{2}\sqrt{I_n^{TE} I_n^{TM}}\cos(\Phi_n^{TE,TM}) \tag{2}$$

This results ideally in a focused reflected beam, which has the same elliptical polarization as the incident beam. Non-zero power in other diffraction orders will result in stray radiation. A non-zero phase difference will make the grating birefringent.

The profiles are modeled by Fourier series where the coefficients are the parameters for the optimizer. The DC-coefficient is not used for the optimization because it is not relevant for a plane wave reflection. Furthermore, a horizontal phase shift of the grating does not change the behavior because the BEM solver implicitly assumes an infinite periodicity of the 2D grating. Thus, we need only the cosine coefficient for the first order. The maximum order of coefficients is obtained from the wavelength and the sampling theorem. As a start value for the optimizer we use an Echelette grating.

The 2D profiles can be optimized on different CPUs on different computers simultaneously, because they are independent problems. For each 2D grating ten optimizations are run with different seeds for the random number generator. For the presented example, 5 Intel Xeon servers and one desktop PC with a total of 184 cores were running for 3 days.

Fig. 5 shows examples for optimized 2D profiles. After the optimization the profiles are shifted such that the maximum value (z = 0) is located at the boundaries. This ensures that no discontinuities can occur between adjacent grooves.

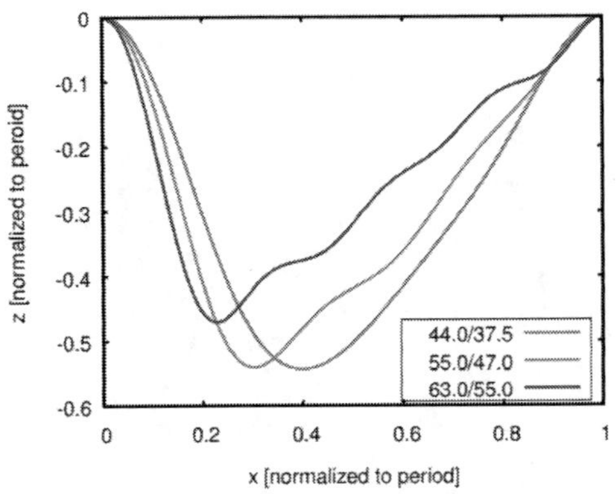

Fig. 5 Optimized 2D profiles for different combinations of Θ_{in} and Θ_{out}

The final grating is then calculated by interpolation of the 2D gratings. Fig. 6 shows the interpolation scheme. For a point (x,y) (cross in Fig. 6), we first identify the groove number.

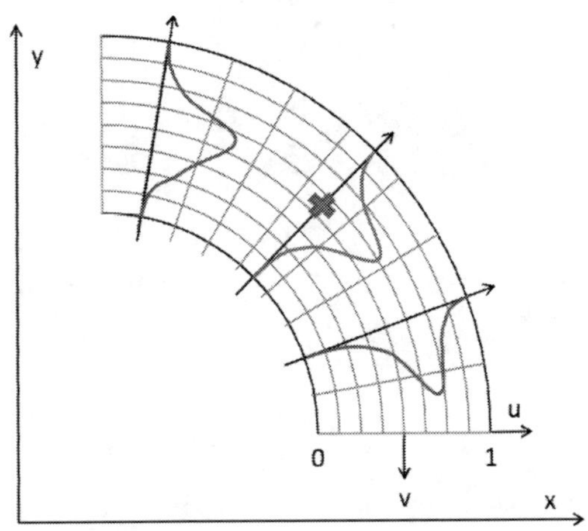

Fig. 6 Interpolation scheme for the final 3D grating. The cross is the point, for which the local *z*-coordinate is calculated. Note that in general the groove width is not constant.

Then, we transform the mirror coordinates x and y to groove-local coordinates u and v. The coordinate u corresponds to the grating vector $\mathbf{G}_1$ and is normalized to the grating period. The coordinate v corresponds to the vector $\mathbf{G}_2$ and is in metric units. We can now obtain a local groove profile by linear interpolation of the two neighboring profiles. This interpolation is done for the Fourier coefficients along the coordinate v. Finally, the actual local z-coordinate of the grating is obtained with the interpolated Fourier coefficients and the u coordinate.

This interpolation scheme ensures a smooth surface without any discontinuities. Since the actual interpolation is done in one dimension, the method is also quite simple.

In summary, the reworked optimization framework has the following new features compared to the earlier code:

- Generic coordinates for the beam parameters. The original code was tightly coupled to ASDEX-Upgrade for the mirror positions and TORBEAM for the beam parameters. It was not directly usable in other environments.
- Parallelized optimization of the 2D gratings
- Adjustable angular resolution to reduce the number of distinct gratings. The original code worked with fixed 1° steps.
- Simple 1D interpolation for the final grating. The original code used a generic triangulation and subsequent 2D interpolation.
- New analysis- and visualization tools

3 Results

Fig. 7 shows the interpolated grating, which was designed for the ECRH system of ASDEX-Upgrade [5]. One can see the saddle type basic shape. The orientation and period of the grooves correspond to Fig. 3.

Fig. 8 shows the efficiencies according to (2) of the 2D profiles. One can see that the efficiency increases monotonically from left to right. In two regions however, it drops to lower values. The increasing behavior of the efficiency corresponds to the angles Θ_{in} and Θ_{out} (See Figs 9 and 10) where a steeper beam incidence results in better efficiencies. The abrupt drop of the efficiency is in agreement with Fig. 11, where the number of existing diffraction orders is plotted. If the number of orders increases, the optimization problem becomes more difficult, because it includes the suppression of all diffraction orders except the wanted one. Furthermore, Fig. 8 indirectly confirms the convergence of the optimizer. A poor convergence (i.e. when the optimizer gets stuck at sub-optimal results) would result in more noise in the efficiency plot.

For cases, where several tiles are possible candidates for an ECRH reflector, a plot of the diffraction orders like in Fig. 11 can be helpful. It always corresponds to the final efficiency but can be calculated quickly before doing the actual optimization.

https://doi.org/10.1051/epjconf/201920304012

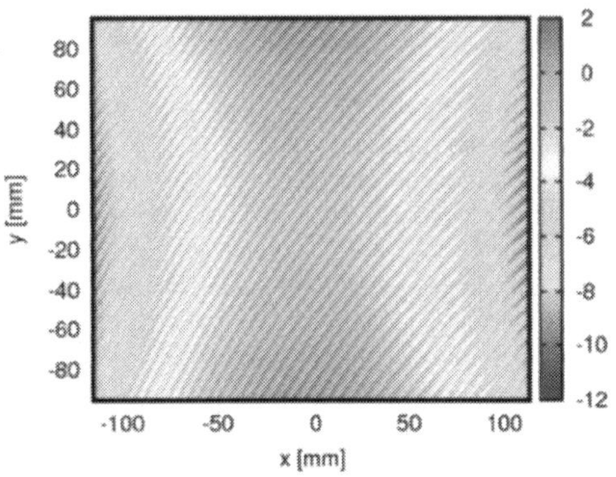

Fig. 7 Height profile (in mm) of the interpolated grating

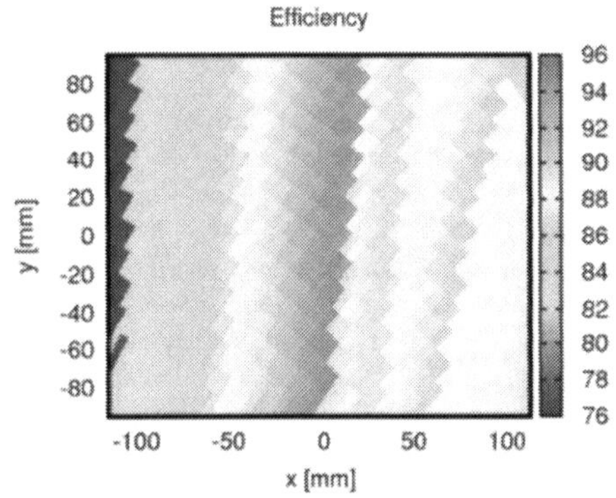

Fig. 8 Local (plane wave) efficiencies according to Eq. (2) of the grating

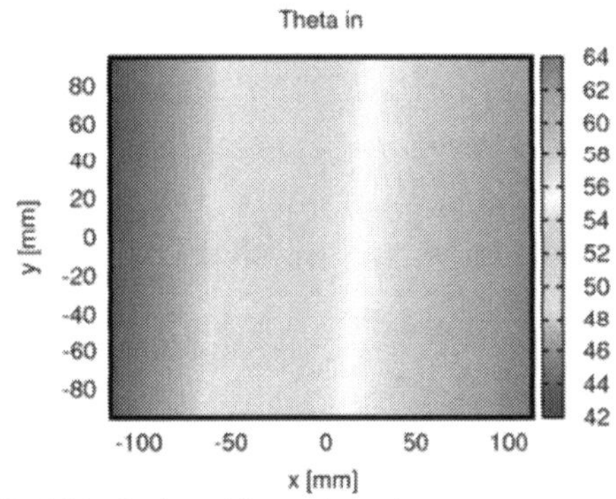

Fig. 9 Distribution of Θ_{in} on the grating

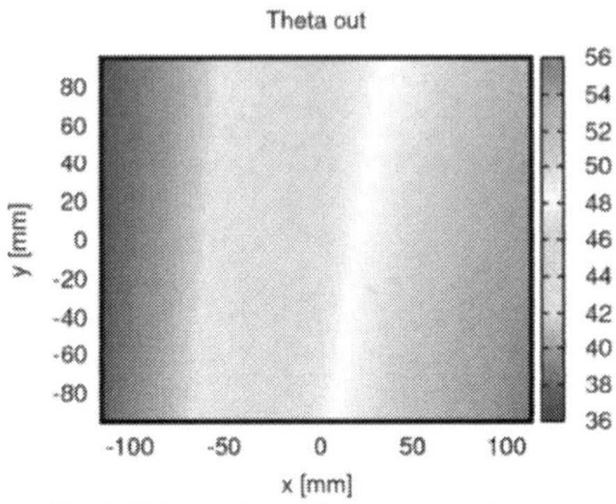

Fig. 10 Distribution of Θ_{out} on the grating

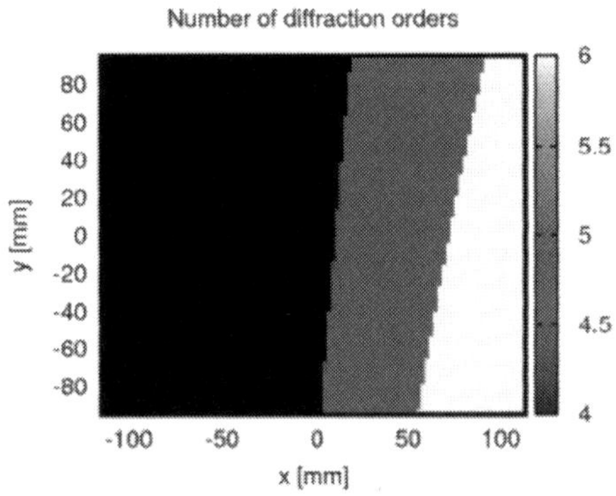

Fig. 11 Numbers of existing diffraction orders on the grating

4 Manufacturing

The reflector was manufactured from graphite with a numerical milling machine. An important issue is the selection of the milling cutter. The shape of the milling cutter limits the difference of the maximum and minimum slope angles on the grating. Typically, cylindrical cutters are preferred due to their robustness. However, they allow a maximum angle difference of 90°. The better choice is a dovetail cutter like in Fig. 12. It allows a maximum angle difference of e.g. 135°.

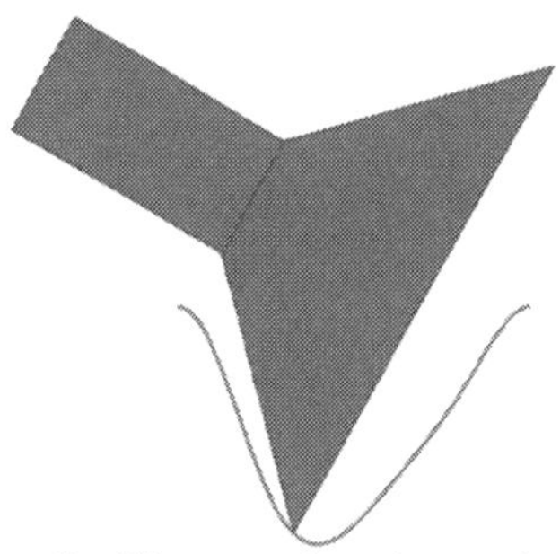

Fig. 12. Dovetail milling cutter for a slope angle difference of 135° in a typical groove profile.

The maximum allowed slope angle difference can be passed as a parameter to the optimizer. An example for the influence of this parameter on the achieved efficiency is shown in Fig. 13.

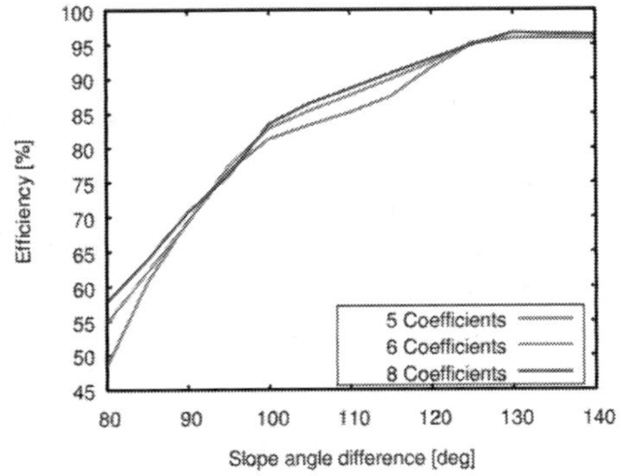

Fig. 13 Efficiency of an optimized 2D grating as a function of the allowed slope angle difference

One can see that for this example the limitation to 90° results in an efficiency below 70%, while for 135° it is above 95%. Also shown are the curves for different numbers of Fourier coefficients. According to the sampling theorem, 5 coefficients are sufficient for this example. We can see that increasing the number beyond 5 does not lead to significant improvements, especially if the 135° cutter is used. Fig. 14 shows the reflector in the milling machine.

Fig. 14. Reflector in the milling machine

After milling, the reflector was coated with a thin tungsten layer (13–15 μm) to become compatible with the ASDEX-Upgrade requirements.

5 Measurement

To characterize the reflector experimentally, a measurement was done with a vector network analyzer and an *x-y* scanner, Fig. 15 shows the setup. The reflector is mounted on a goniometer, which allows to arbitrary rotate the reflector (see Fig. 16). The incident beam was generated by a lens horn which produces a circular Gaussian beam. It is slightly different from the astigmatic beam used for the optimization. The important parameter, however, is the phase front curvature on the mirror, which is comparable in both directions.

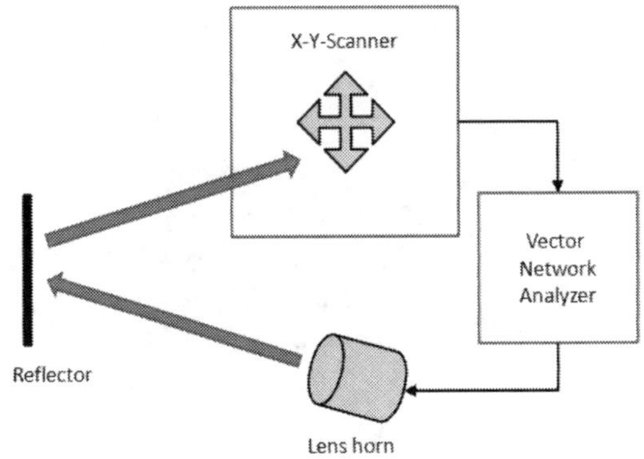

Fig. 15 Experimental setup

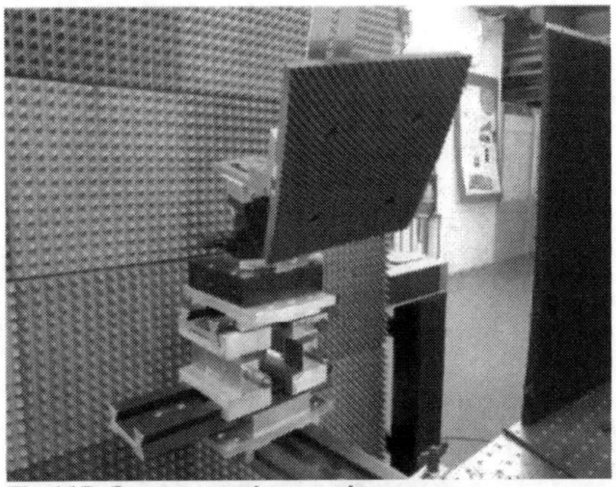

Fig. 16 Reflector mounted on a goniometer

The measured fields are shown in Figs 17 – 20. We can see a slightly elliptical beam with an astigmatic phase, which is similar to the design value. Also important is the similarity between the two polarizations, which was an optimization goal.

For a more detailed analysis it would be necessary to also measure the spurious fields due to unwanted diffraction orders. This would, however, require a field scan over a large solid angle, which is not possible with the current equipment.

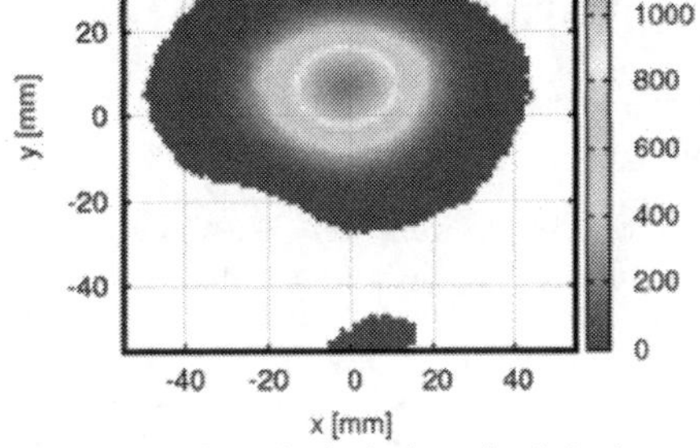

Fig. 17 Measured intensity for horizontal polarization

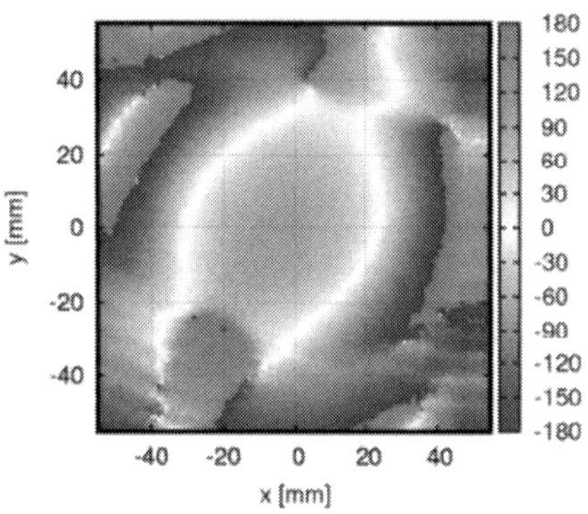

Fig. 18 Measured phase for horizontal polarization

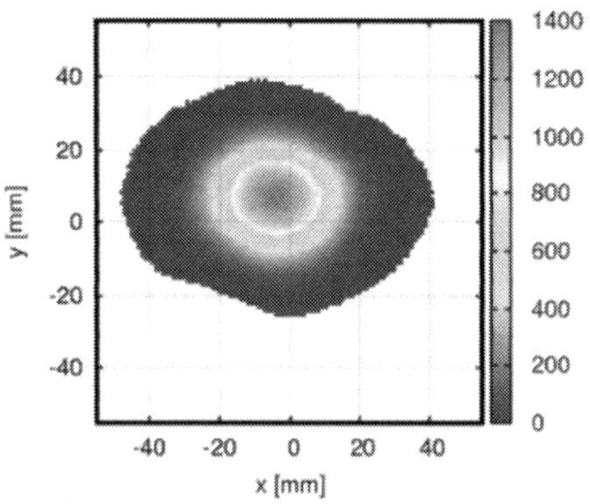

Fig. 19 Measured intensity for vertical polarization

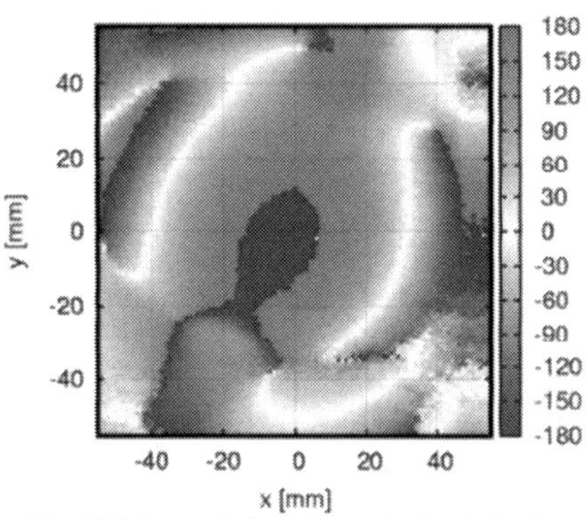

Fig. 20 Measured phase for vertical polarization

Conclusions and outlook

The optimization framework for specialized ECRH reflectors was reworked and improved. It is robust and can use an arbitrary number of computing nodes for the actual optimization with negligible overhead. The resulting surfaces have no discontinuities. The principal reflection characteristics as well as the polarization independence could be verified experimentally.

The parameters needed for the optimizer are in generic reflector-local coordinates, which makes it usable for other experiments as well. The base shape of the mirror can be flat or saddle-type. It is, however, trivial to add other shapes. For W7-X for example, a convex shape might be advantageous.

The described reflector is already coated with tungsten and built into ASDEX-Upgrade. It will be tested with high power during the next experimental campaign.

For aiding future developments, a series of 3D full wave simulations of the whole reflector was started with promising results [6].

References

1. H. Höhnle, J. Stober, A. Herrmann, W. Kasparek, F. Leuterer, F. Monaco, R. Neu, D. Schmid-Lorch, H. Schütz, J. Schweinzer, U. Stroth, D. Wagner, S. Vorbrugg, E. Wolfrum and the ASDEX Upgrade Team, *Nucl. Fusion*, **51**, 083013 (2011)
2. E. Poli, A. Bock, M. Lochbrunner, O. Maj, M. Reich, A. Snicker, A. Stegmeir, F. Volpe, N. Bertelli, R. Bilato, G.D. Conway, D. Farina, F. Felici, L. Figini, R. Fischer, C. Galperti, T. Happel, Y.R. Lin-Liu, N.B. Marushchenko, U. Mszanowski, F.M. Poli, J. Stober, E. Westerhof, R. Zille, A.G. Peeters, G.V. Pereverzev, Comp. Phys. Comm., **225**, 36-46, (2018)
3. M. Schubert et.al *This conference* (2018)
4. O. Mangold, *Effiziente holographische Gitterspiegel für die Elektronzyklotronresonanzheizung von Fusionsplasmen*, Doctoral Thesis, Universität Stuttgart, 2009
5. D. Wagner, J. Stober, F. Leuterer, F. Monaco, S. Müller, M. Münich, C. J. Rapson, M. Reich, M. Schubert, H. Schütz, W. Treutterer, H. Zohm, M. Thumm, T. Scherer, A. Meier, G. Gantenbein, J. Jelonnek, W. Kasparek, C. Lechte, B. Plaum, T. Goodman, A.G. Litvak, G.G. Denisov, A. Chirkov, V. Zapevalov, V. Malygin, L.G. Popov, V. O. Nichiporenko, V. E. Myasnikov, E. M. Tai, E.A. Solyanova, S.A. Malygin, ASDEX Upgrade Team, *Journal of Infrared, Millimeter, and Terahertz Waves*, **37**, 45-54 (2016)
6. C. Lechte et.al. *This conference* (2018)

Developments towards 1MW Gyrotron Test Facility at ITER-India

Vipal Rathod*, *E. Sharan* Dilip, *Ronak* Shah, *Deepak* Mandge, *Amit* Yadav, *Anjali* Sharma, *Rajvi* Parmar, *N.P.* Singh and *S.L.* Rao

ITER-India, Institute for Plasma Research, Bhat, Gandhinagar-382 428, Gujarat, India

Abstract. ITER-India, the Indian domestic agency for the ITER project, has the responsibility to supply a set of two high power Gyrotron sources (1 MW, 170 GHz, 3600 s) along with the auxiliary systems for Electron Cyclotron Heating & Current Drive applications. For such high power Gyrotron systems, one of the challenging areas is the system integration and establishment of reliable integrated system performance. ITER-India plans to establish the integrated Gyrotron system performance that essentially meets the ITER requirements in a Gyrotron Test Facility which is specifically being developed at ITER-India. This paper discusses about the recent updates towards the Test Facility development which includes, development of cost effective & modular Body Power Supply (BPS), Industrial grade prototype interlock & protection modules, a Gyrotron field simulator and cooling water distribution system.

1. Introduction

Gyrotron oscillators are capable of producing high CW RF power in the microwave and mm Wave frequency region and are widely used for ECRF applications on fusion devices [1]. ITER requires 20 MW of RF power at 170 GHz for which 24 sets of 1MW Gyrotron sources would be required [2]. ITER-India, the Indian domestic agency has the scope of providing two Gyrotron source sets [3]. The scope includes not only the Gyrotron tubes but also the associated auxiliary systems such as the control system, auxiliary power supplies, cooling manifold etc., along with the responsibility of demonstration of performance at site on a dummy load. The critical interfaces with High Voltage Power Supplies (HVPS), EC plant control system and waveguide transmission lines which are provided by other domestic agencies, are also to be taken into account while finalizing the system design. The integrated system performance with a reliability of >95%, as required by ITER is a challenge and requires considerable preparatory tests and debugging of the complete integrated system. Considering limited time window that would be available for each Gyrotron set to successfully complete the site installation and acceptance tests, it becomes essential that one has to recreate the closest possible test configuration and prior establish the performance in a test facility. With this objective in mind in Phase-1 of the project, **ITER-India** is developing a 1MW **G**yrotron **T**est **F**acility (IIGTF) with the help of an ITER like test Gyrotron along with the other subsystems that are either prototypes or close to actual ITER deliverables.

2. Gyrotron Test Facility

The ITER-India lab building which caters to several ITER related facilities also houses the IIGTF at level three of the building. The **M**ain **H**igh **V**oltage cathode **P**ower **S**upply (MHVPS) for the IIGTF is located on the ground floor, a configuration that is similar to ITER. The IIGTF layout (Fig. 1) is grouped into several areas such as the HV area, Gyrotron & waveguide area, Cooling Manifold area, Auxiliary & control cubicle area, RF diagnostics & low power mm Wave area and control room etc. The HV area consists of **B**ody High Voltage **P**ower **S**upply (BPS), MHVPS Interface unit, Heater Transformer, HV disconnector switch etc. The Gyrotron area consists of the Gyrotron tube, Super Conducting Magnet (SCM) with other auxiliary magnet coils, Matching Optic Unit (MOU) along with a set of waveguide components and a CW dummy load. The main control room is situated on the mezzanine floor from where the remote operation can be performed. The auxiliary power supplies and control cubicles are located under the control room. Cooling manifold area adjacent to the Gyrotron setup consists of the manifold along with a dedicated chiller for the SCM compressor.

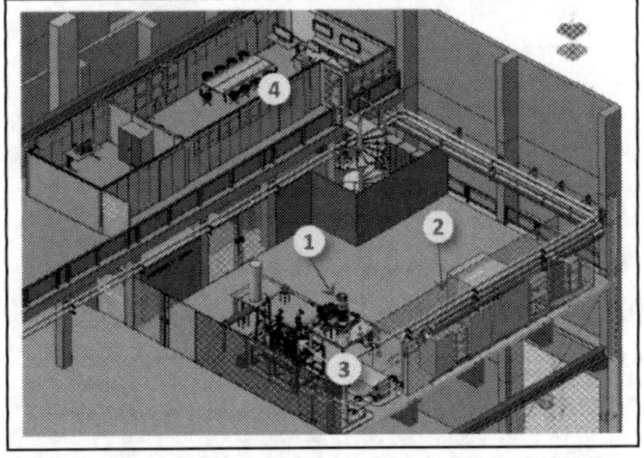

Fig. 1 : ITER-India Gyrotron Test Facility layout (1: Gyrotron set-up; 2: HV area; 3: Cooling Manifold; 4: Control room)

2.1. Test Gyrotron & Waveguide Set

To establish the integrated performance, a test Gyrotron close to ITER specifications, along with a set of waveguide components including a CW dummy load is under procurement (Fig. 2). The key specifications of the Gyrotron tube are shown in the Table 1.

Table 1 : Typical specifications of Test Gyrotron

Parameter	Value
Frequency	170 GHz
Output Power	0.96 MW (o/p of MOU)
Pulse	1000 s
Efficiency	≥ 50%

* Corresponding author: vipal.rathod@iter-india.org

A liquid He free superconducting magnet along with other auxiliary coils provide the necessary magnetic field for the Gyrotron. A short section (~3.5 m) of 63.5 mm diameter, evacuated corrugated waveguide line terminated into a water cooled CW dummy load is considered for testing the Gyrotron. To facilitate the measurements during initial short pulse operation, an atmospheric short pulse dummy load is also planned that can be selected through a manually operated SPDT waveguide switch. A power monitoring bi-directional miter bend coupler with a typical coupling coefficient of ~80 dB is also included in the line to monitor the forward and reflected power along with the oscillation frequency. A MOU with adjustable mirrors will be used to efficiently (96%) couple the Gyrotron output with the waveguide. Additional power monitoring provision in the MOU is also being considered. A TMP based vacuum line is prepared to evacuate the waveguide line up to 0.01 Pa.

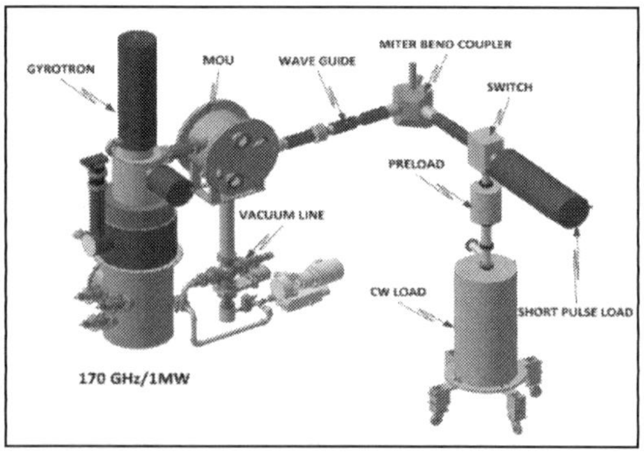

Fig. 2 : Schematic of Test Gyrotron & waveguide set test configuration

2.2. High Voltage & Auxiliary Power Supplies

For a diode type of Gyrotron, the common collector electrical configuration is considered for the IIGTF. For this configuration two HVDC power supplies namely MHVPS and BPS would be required. As shown in Fig. 3, the MHVPS drives the Gyrotron beam current and the BPS in addition to the MHVPS provides the necessary acceleration voltage.

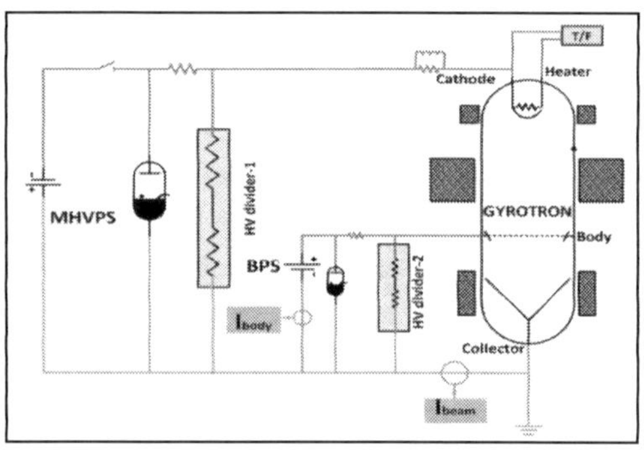

Fig. 3 : IIGTF HV Power Supply configuration

For IIGTF, an ITER like full scale prototype MHVPS with 55kV/110A in PSM topology is under development (Fig. 4). Long lead items such as the multi secondary transformers are already tested and installed at site. The manufacturing of Switched Power Supply (SPS) modules is currently ongoing.

Fig. 4 : Multi-secondary Transformer & 80kW SPS module mounted on test bench

For the BPS, where the required maximum ratings (35kV/100 mA) are modest in comparison to the MHVPS, it was felt that a PSM solution would be expensive and bulky. In view of this, an alternate cost effective modular solution using high voltage solid-state switch (push-pull type MOSFET) is being tested under a prototype development to meet the modulation requirement (DC-2 kHz). The concept of the prototype BPS is shown in the Fig. 5. The test setup consists of a conventional COTS power supply (100 kV/ 50 mA) with slow dynamics (~400 ms) terminated with a buffer capacitor which compensates the instantaneous peak loads during the switching transitions. The Gyrotron body circuit is represented with an equivalent RC load (350 kΩ or 850 kΩ ||1700 pF) which includes cable capacitance. A compact push-pull type solid-state switch (~ 300 x 350 x 150 mm^3) from M/s BEHLKE is used as the fast switching element that provides the necessary fast dynamics. Some of the results obtained with this prototype setup are presented in Fig. 6 to Fig. 8.

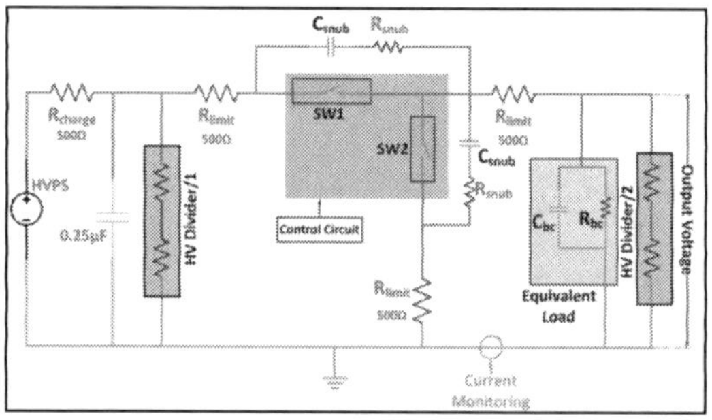

Fig. 5 : Schematic of the tests setup of prototype BPS using solid state switch

Fig. 6 shows the short pulse (50 μs) test results at 35 kV with an output voltage pulse rise and fall times about 5 μs. Fig. 7 shows the DC pulse at 35 kV of pulse length of 1 s. Fig. 8 shows the square modulation test results at 35 kV at 1 kHz for 5 ms duration. The voltage pulse characteristics across the load are quite satisfactory however, during the experiments it is also noted that the fast switching is resulting in excessive overshoots and undershoots across the switches which needs to be suppressed to safeguard the switch. To address the same, additional tests are ongoing with appropriate snubber elements. Further tests are to be carried out to assess the switch reliability. Also, the setup needs to be compactly organized to minimize the effect of stray reactive elements in the circuit.

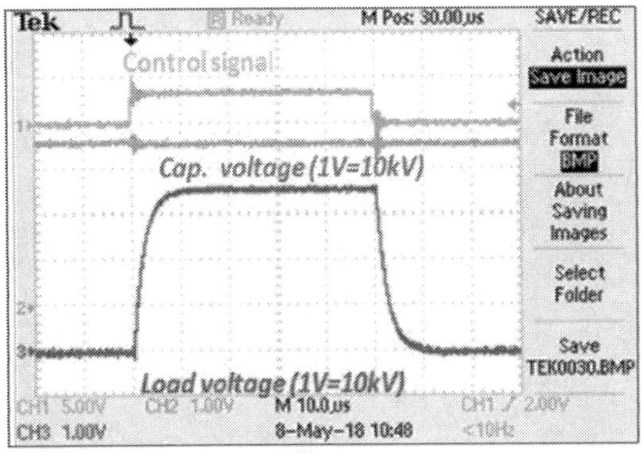

Fig. 6: Test result of 35 kV for 50μs (rise/fall time: ~5μs)

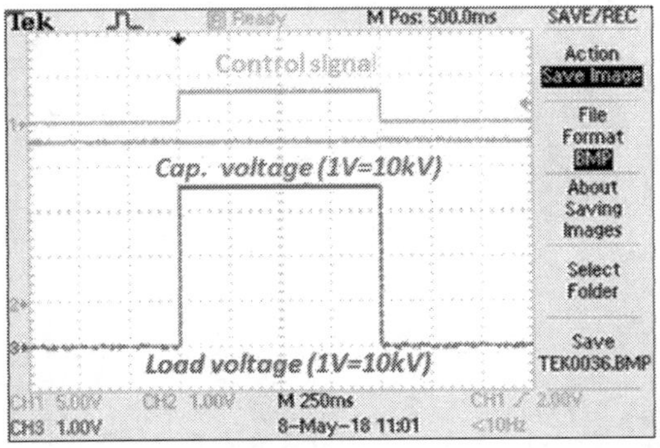

Fig. 7: Test result of 35 kV for 1 s

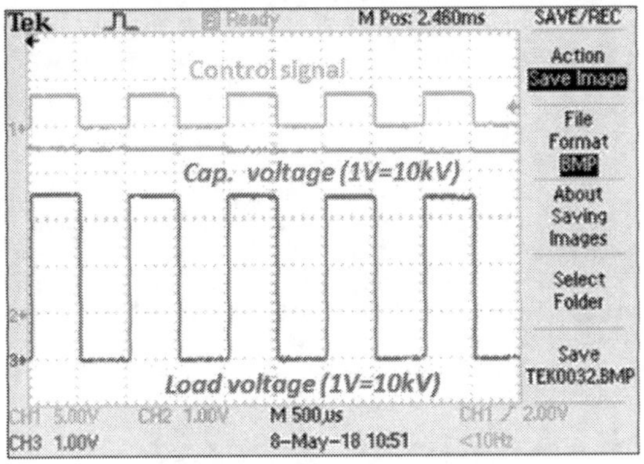

Fig. 8: Test result of 35 kV at 1 kHz Modulation

2.3. Control System

A Local Control Unit (LCU) is required for safe and reliable operation of Gyrotron system. A dedicated full scale ITER prototype LCU is planned for testing and commissioning of Test Gyrotron at IIGTF that will be developed in two phases using two different software platforms to minimize the operational risks. LabVIEW™ platform would be used in phase-1 and the ITER recommended CODAC platform & hardware would be used in phase-2 [4]. Fig. 9 shows the LCU architecture for Phase-1, having main functions of sequence control, monitoring, protection interlocks, data acquisition and Graphical User Interface (GUI). The sequence control and slow interlock functions of various sub-systems are implemented using S7-300 Siemens® Programmable Logic Controller (PLC). Real-time Control and Data Acquisition will be implemented using PCI eXtension for Instrumentation (PXIe) system. Critical Interlock & protections requiring a fast response time (< 10 μs) which are implemented using hardwired FPGA device. The auxiliary power supplies are controlled and monitored by slow controller using serial communication protocol. Slow controller communicates with fast controller using NI-OPC server. All cooling parameters are monitored using Remote expansion module (ET 200M) of PLC. The complete system operation and real time data acquisition will be performed using LabVIEW based GUI.

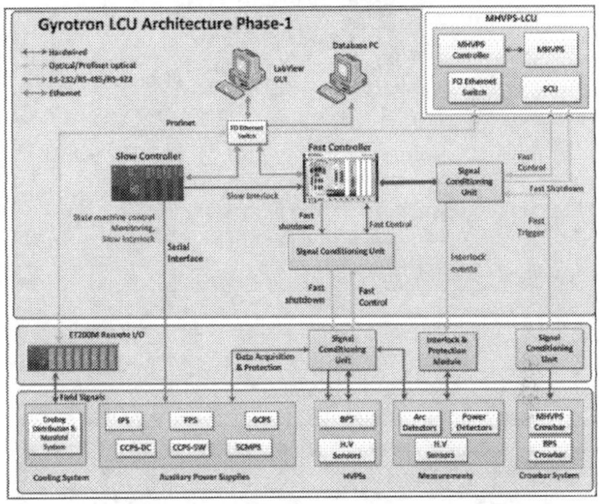

Fig. 9: LCU Architecture (Phase-1)

A Signal Conditioning Unit (SCU) is used for reliable transmission & interfacing of the field signals to control system. In this context, a prototype modular industrial grade fiber optic transmission link has been designed and developed based on the Voltage-to-Frequency (V/F) conversion technique (Fig. 10). Functional & EMC qualification tests have been carried out with satisfactory results [4] [5].

Fig. 10 : Prototype industrial Grade SCU (Fiber Optic link)

During the Gyrotron operation, critical faults such as internal arcs may arise which requires very fast interlock action to switch OFF the HV Power Supplies. Hence, a fast interlock system plays a very important role in ensuring the safe operation of Gyrotron system. Considering the criticality of fast interlock protections, two different types of prototype hardwired interlock modules are designed and developed, namely Distributed Interlock Module (DIM) and Centralized Interlock module (CIM). DIM is a single channel module distributed in the field where the analog signals are processed near to the sensors & the logic part is implemented in fast controller (Fig. 11). CIM is a standalone module placed in the field, where the analog & digital signals are processed in a centralized single unit. Because of the faults criticality, both the modules will be used with one as hardware redundancy.

Fig. 11: Prototype Distributed Interlock Module (DIM)

An Industrial grade prototype Centralized Interlock and Protection Module (CIM) based on ITER-India design has been developed successfully which is shown in Fig. 12. The main features are two tier configuration with an additional hardware redundancy; FPGA based hardwired modular design; Fail safe logic; Total module response time <1.5μs; Fiber optic interface; Fault Sequence & Spurious fault detection logic and EMC compliant as per IEC61000-4 (as a specific requirements). Functional & EMC qualification tests have been carried out with satisfactory results which are shown in Fig. 13 to Fig. 15. Full-scale fabrication of CIM system is currently ongoing.

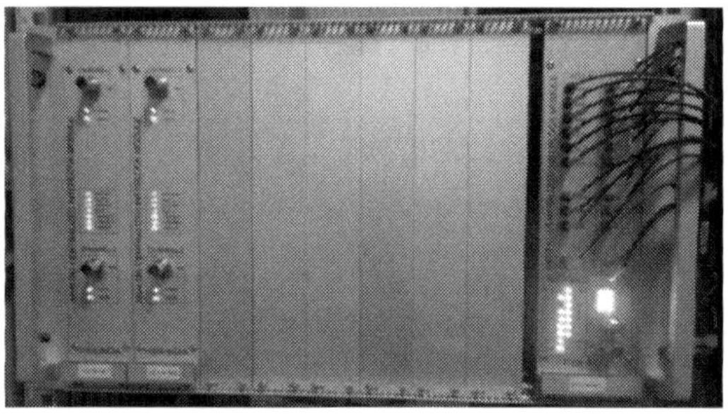

Fig. 12: Prototype Centralized Interlock and Protection Module (CIM)

Standard (Test Name)	Test Specifications / Levels	Results
IEC 61000-4-2:2008 (ESD)	Contact Discharge: ± 4 kV In-Direct Discharge : ± 8 kV	
IEC 61000-4-3:2006 (RS)	80 to 1000 MHz; 10 V/m	
IEC 61000-4-4:2012 (EFT)	± 1 kV	
IEC 61000-4-5:2014 (Surge)	± 1 kV	PASS Criterion A
IEC 61000-4-6:2013 (CS)	0.15 to 80 MHz; 10 Vrms	
IEC 61000-4-8:2009 (PFMF)	100A/m; 50 Hz; 3 axis	
IEC 61000-4-11:2004 (Volt. Dip)	70 % Res. Volt.; 25 cycle; 50Hz	
CISPR11:2015 (RE)	30MHz to 1GHz, 30-37 dBμV/m	

Fig. 13: Summary of EMC Test performance of CIM

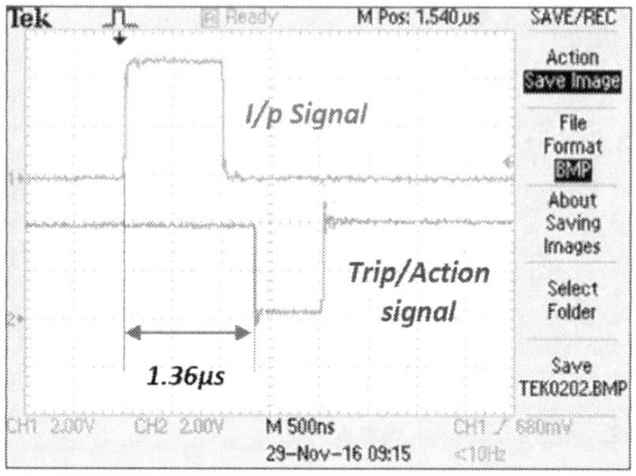

Fig. 14: Test Result of response time (~ 1.36 μs) of CIM

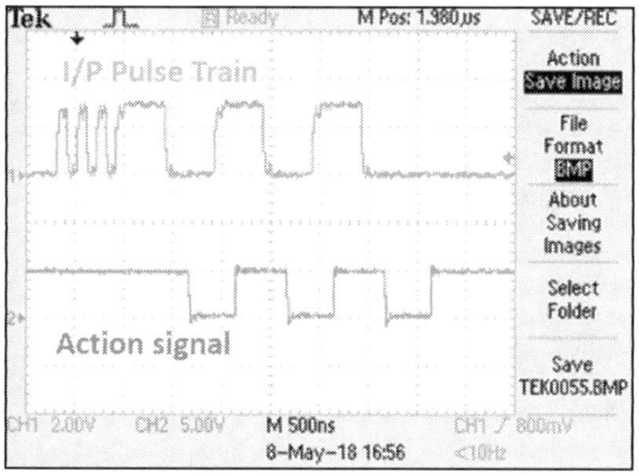

Fig. 15: Test Result of Spurious fault detection logic of CIM

Considering the LCU design and operational aspects, it is important to perform integrated testing & validation of the LCU functions, before using it for the actual Gyrotron operation. This will also improve the reliability and safety of overall system. As various Gyrotron sub-systems are under development & procurement phase for IIGTF, a Gyrotron field simulator is being developed that will emulate field sub-systems and will be used as a test bench to qualify the LCU functionality.

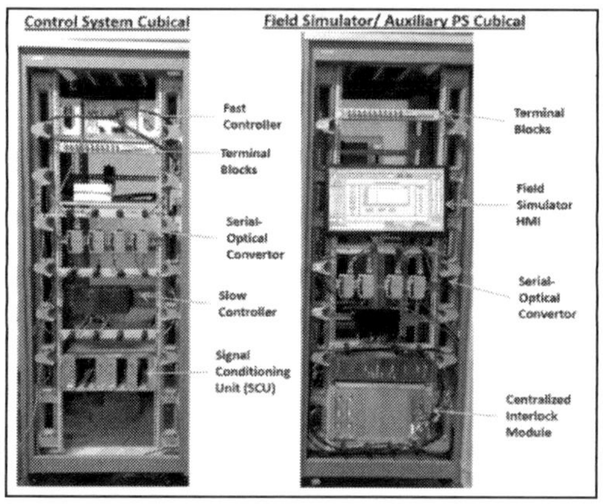

Fig. 16: Field Simulator & Control cubicle arrangement

Most of the auxiliary power supplies have serial interface for control, and hardwire interface for monitoring & protection, while the HVPS has a dedicated hardwire interface with LCU. LabVIEW GUI is used to simulate the behaviour of the auxiliary power supplies. The HVPS signals are simulated using Real time PXIe controller. A LabVIEW based GUI is developed to operate the field simulator. As shown in Fig. 16, the LCU components are arranged in a cubicle as per actual configuration, while the field simulator hardware is arranged in another cubicle that will be replaced with actual subsystems during the Gyrotron operation.

2.4. Cooling Distribution System

For 1MW Gyrotron test stand with a typical Gyrotron efficiency of 50%, a total of 2MW of thermal load would be generated. Half of the load would be received across various Gyrotron components (such as the collector) and the remaining half would be received in the output RF dummy load. In order to maintain the components well within their permissible operating temperature, the heat load must be effectively dissipated through active water cooling. For IIGTF cooling distribution system, a heat load of 2.5 MW at an effective efficiency of 40% is considered. Considering the typical flow rate requirements of 1MW class Gyrotron and the RF dummy load, a total of 2700 LPM is finalized for the IIGTF. An existing cooling water plant that can provide the required low conductivity water at room temperature with a pressure head of 6-7 bar is being utilized for the purpose. The main cooling parameters for the IIGTF cooling distribution system are listed in the Table 2.

Table 2 : Main Parameters of Cooling distribution system

Parameter	Value
Coolant	Water
Heat Load	2.5 MW
Flow rate	2700 LPM
Water Temperature	< 35 °C
Inlet Pressure Head	6-7 bar (typical)
Conductivity	< 1 μS/cm
Monitoring	Flow, Temperature, Pressure & Conductivity

Fig. 17 : Gyrotron Cooling Distribution Manifold

A cooling distribution system consisting of a main header & distribution manifold has been designed, manufactured and commissioned recently. A 6" stainless steel header line connects the plant and the Gyrotron distribution manifold. A cooling manifold with 22 inlet and outlets, which are grouped into 6 main branches depending upon the pressure and flow requirements, provides the dedicated input and output tapping points for each of the Gyrotron and waveguide cooling circuits. Each circuit is provided with necessary flow, temperature and pressure monitoring sensors. All necessary sensor parameters are locally displayed as well as remotely monitored and acquired through PLC system. Fig. 17 shows the actual image of the recently installed cooling manifold. Fig. 18 shows the general I&C configuration for the cooling distribution system. Fig. 19 shows the flow test results acquired through PLC in LabVIEW based GUI during the commissioning.

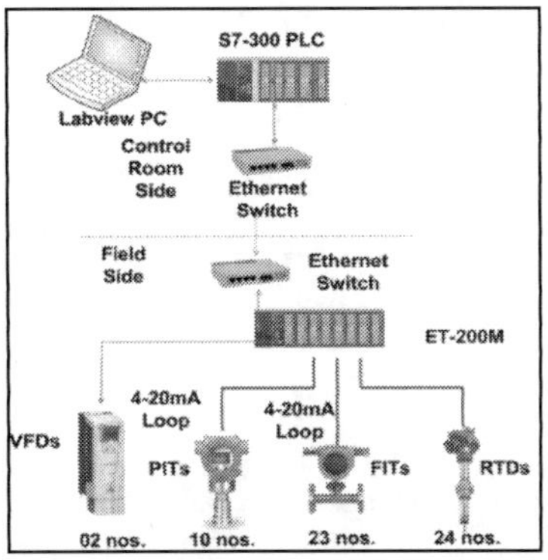

Fig. 18: General I&C configuration for the IIGTF Cooling distribution system

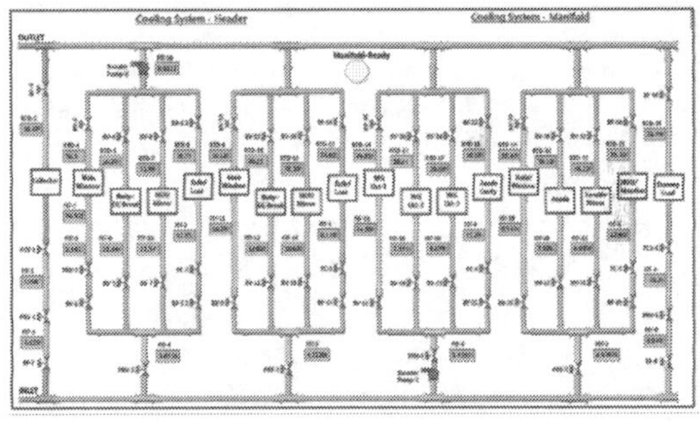

Fig. 19: Flow distribution & other parameters captured during commissioning

2.5. Diagnostics

For the IIGTF, different diagnostics are established or being planned to measure and characterize the Gyrotron output RF beam performance parameters such as the output RF power, RF frequency, RF beam mode purity etc. The generated output power would be measured and monitored using the reliable calorimetric measurements of the RF Dummy load water circuit. As the CW dummy loads typically have long time constants of the order of tens of seconds, a short pulse dummy load would be utilized in case of short pulse operation and the pulse integration techniques would be used to estimate the power. Also for instantaneous RF power monitoring, calibrated Schottky diode detectors mounted on the waveguide directional coupler would be used. For the frequency measurements, a spectrum analyzer suitably connected to the directional coupler coupling port would be used. As the spectrum analyzer has a minimum sweep time (~ 5 ms), a real-time frequency measurement setup is also being planned. With the real-time frequency measurements over an appropriate frequency band, it would be possible to detect the Gyrotron out of mode oscillation by detecting the frequency shift. There would be two band pass filters: one for wide band detection (166-172 GHz) and other is for narrow band detection (169.5-170.5 GHz). The detailed scheme using a heterodyne mixer and the output filter banks is shown in the Fig. 20.

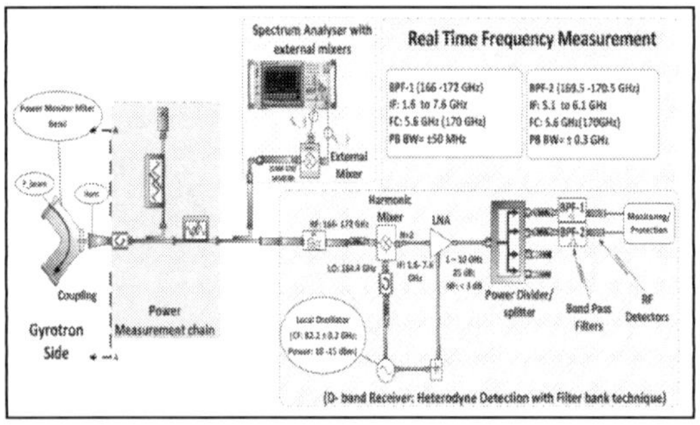

Fig. 20 : Gyrotron output RF beam Frequency measurement scheme

Good Gyrotron output beam mode purity of the order of 95% is required to minimize waveguide transmission line and coupling losses. Amplitude and phase profiles of the beam are needed to estimate the mode content. A non-contact measurement technique is typically employed to estimate the relative amplitude profile of a high power RF beam. The phase profile is determined using phase retrieval techniques in which radiated beam amplitude measurements taken at different distances along the direction of the beam propagation would be used to retrieve the phase [6]. Fig. 21 shows the schematic of the setup which includes an IR camera, a translating stage to mount a target paper along with an IR camera, a reflector and a beam dump.

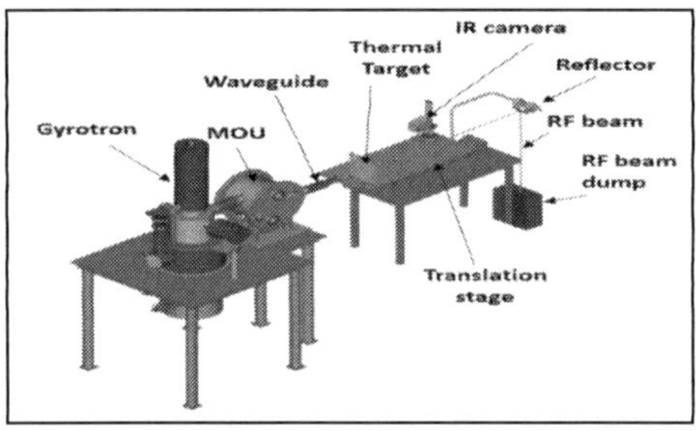

Fig. 21: Mode purity measurement scheme

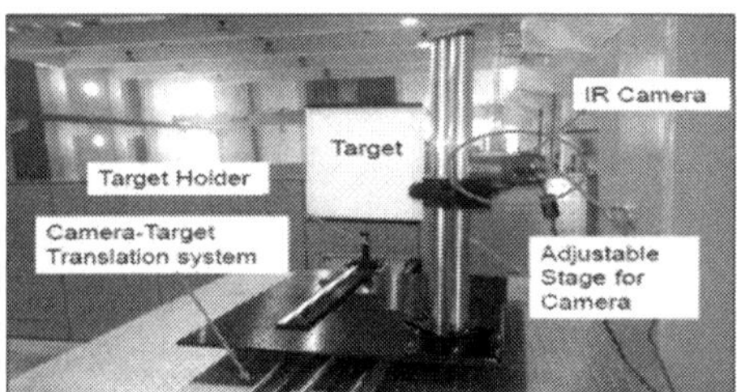

Fig. 22: Mode purity measurements Setup using IR camera

The temperature data of the target, after being irradiated by RF beam would be recorded by the IR camera at different locations. This data is used to retrieve the phase using a numerical program prepared in Matlab© based on irradiance moment theory [7]. The IR thermography setup as shown in Fig. 22 has been established at IIGTF for determination of the mode purity.

3. Summary

ITER-India, the Indian domestic agency for the ITER project is developing a 1MW Gyrotron test facility to establish the integrated Gyrotron system performance that meets the ITER requirements. The general arrangement of various subsystems of IIGTF and their layout has been finalized. To enable the integrated testing, a Test Gyrotron with close to ITER specifications (170 GHz, 1 MW, 1000 s) along with a corrugated waveguide set including RF load is finalized and the same is currently under procurement. Details of High Voltage power supplies and their development status presented. A PSM based MHVPS (55kV, 110 A) that also serve as a full scale ITER prototype is under development. For the Gyrotron BPS (35 kV, 100 mA), a cost effective and modular solution using high voltage solid-state switch is adopted for prototype development to meet the fast switching and high frequency modulation requirement. Promising initial test results on an equivalent RC load have been achieved and further testing is ongoing. Towards development of Local Control Unit (LCU), prototype design and development activities have been carried out. Under this, a prototype industrial grade Centralized Interlock Module (CIM) has been designed, developed and tested successfully against EMC compliance. The main features of CIM includes fail-safe logic, fault sequence & spurious fault detection logic using FPGA with a total response time ~ 1.5μs. Further a LabVIEW™ based Gyrotron field simulator is being developed to test the LCU's hardware & its software applications. In order to remove the significant thermal heat loads (~ 2.5 MW) across various components of Gyrotron system, the active cooling water distribution system (2700 LPM, 6 bar) has been designed, developed, and commissioned. Various diagnostics for measurement of output RF power, RF frequency, and RF beam mode purity are considered and some of the setups have been established. Additionally a real time frequency measurement system for spurious mode detection is also being taken up for development.

References

[1] M. V. KARTIKEYAN, E. BORIE, and M. K. A. THUMM, Gyrotrons: High Power Microwave and Millimeter Wave Technology, Springer-Verlag (2004).

[2] C. DARBOS et al., J. Infrared Millim. Terahertz Waves, 37, issue-1, 4-20 (2016)

[3] S.L. RAO et al., Fusion Science & Technology, 65, 129-144 (2014)

[4] V. RATHOD et al., Fusion Eng. Design, 112, 897-905 (2016)

[5] V. RATHOD et al., IEEE Xplore, pp. 687–691, ICIC (2015)

[6] J.P.Anderson et al., IEEE transactions on Microwave Theory and Techniques, vol. 50, Issue 6,pp. 1526-1535, (2002)

[7] P. C. KALARIA et al, IEEE Xplore, IRMMW-THz (2013)

Theoretical Study on the Operation of the EU/KIT TE34,19-Mode Coaxial-Cavity Gyrotron at 170/204/238 GHz

Tobias Ruess[1,*], *Konstantinos* Avramidis[1], *Gerd* Gantenbein[1], *Stefan* Illy[1], *Zisis* Ioannidis[1], *Jianbo* Jin[1], *Parth* Kalaria[1], *Ioannis Gr.* Pagonakis[1], *Sebastian* Ruess[1,2], *Tomasz* Rzesnicki[1], *Manfred* Thumm[1,2], and *John* Jelonnek[1,2]

[1]IHM, [2]IHE, Karlsruhe Institute of Technology (KIT), Kaiserstr. 12, 76131 Karlsruhe, Germany

Abstract. The 170 GHz 2 MW $TE_{34,19}$-mode coaxial-cavity modular short-pulse pre-prototype gyrotron at KIT was recently modified in order to verify the multi-megawatt coaxial-cavity technology at longer pulses. In parallel, theoretical investigations on a possibility to operate the 170 GHz $TE_{34,19}$-mode coaxial-cavity prototype at multiple frequencies up to 238 GHz have been started, with a goal to find a configuration at which the tube could operate in the KIT FULGOR gyrotron test facility using the new 10.5 T SC magnet. This paper indicates which adjustments have to be made and show the feasibility of the multi-frequency operation. Small modifications at the gyrotron cavity will support an RF output power of more than 2 MW at 170/204 GHz. Furthermore, a new gyrotron launcher has been designed capable of producing a Gaussian microwave beam with a Gaussian mode content of more than 96% at these frequencies.

1 Introduction

Gyrotrons are RF microwave sources used for fusion applications. These vacuum tubes are the sources for Electron Cyclotron Resonance Heating and Current Drive (ECRH&CD) for plasma heating and stabilization in fusion devices. For example, at the stellarator W7-X in Greifswald, Germany, ten gyrotrons operating at 140 GHz and at 1 MW (1800s) RF output power have been tested and demonstrated an excellent reliable performance [1]. The experimental reactor ITER, under construction in Caderache, France will use a 24 MW ECRH installation with 24 170 GHz, 1 MW gyrotrons [2]. In a next step, electricity will be produced by the subsequent project called DEMO (**Demo**nstration Power Plant) [3]. Fusion gyrotrons for DEMO and future FPP (**F**usion **P**ower **P**lants) will require multi-megawatt continuous-wave (CW) operation at multiple possible frequencies starting from 170 GHz up to significantly above 200 GHz. The EU DEMO, planned as next step after ITER, is under development within the Horizon 2020 EUROfusion programme [3].

The EU ITER gyrotron is designed with the conventional hollow-cavity technology [4]. This technology shows quite a good performance and a reliable operation whereby studies at frequencies above 200 GHz has been made [5]. However, there are two main issues regarding hollow-cavity gyrotrons: (i) increasing mode competition with raising mode order, which can be a problem at the operation above 200 GHz and (ii) high voltage depression caused by space charge effect of the electron beam. To overcoming these issues, a gyrotron using a coaxial insert (coaxial-cavity gyrotron) can be used. The coaxial cavity reduces mode competition and voltage depression [6,7]. The first KIT modular coaxial-cavity pre-prototype gyrotron reached an RF output power of more than 2 MW at 170 GHz in short-pulse operation [8,9]. The operating design parameters are summarized in Tab. 1. Pushing forward this technology, the pre-prototype has been upgraded towards longer-pulse operation of 100 ms, and towards 1 s in a second step [10]. For future DEMO gyrotrons, multi-frequency operation at e.g. 136 GHz, 170 GHz, 204 GHz, or 238 GHz would be highly beneficial. The aim of this work is to investigate if the operation at these frequencies is possible with the existing coaxial-cavity pre-prototype, using the KIT FULGOR gyrotron test facility [11] and a new 10.5 T Super-Conducting magnet, without a major effort on developing new components.

This paper is structured in the following way: Firstly, an overview about the mode selection for multi-frequency operation is presented in section 2. In the

Table 1. Design Parameters for the Coaxial-Cavity Gyrotron [8].

Operating mode	$TE_{34,19}$
Frequency, f	170 GHz
RF output power, P_{out}	2 MW
Beam current, I_B	75 A
Accelerating voltage, U_C	90 kV
Velocity ratio (pitch factor), α	~ 1.3
Magnetic field at cavity, B_{cav}	6.87 T
Outer cavity radius, R_{cav}	29.55 mm

* Corresponding author: tobias.ruess@kit.edu

following Section 3, initial studies regarding multi-frequency operation using the magnetic field profile of an existing magnet are reported. Further, a study about the geometrical issues regarding the new configuration of the new 10.5 T SC magnet from Tesla Engineering Ltd., Storrington, UK is started in section 4. Taken this into account, realistic start-up scenarios are simulated under the consideration of the proposal for the coil configuration from Tesla Engineering Ltd. and real electron gun parameters, shown in section 5. Comparisons are made between simulations addressing only the gyrotron cavity and simulations addressing both the cavity and the non-linear uptaper after the cavity. Additionally, a modification of the coaxial-cavity is proposed to increase the performance. In the last section the behaviour of the existing launcher is investigated and an advanced design is proposed showing promising results.

2 Mode Selection for Multi-Frequency Operation

The investigations are based on the existing coaxial-cavity of the KIT pre-prototype. This cavity has a total length of 68 mm and a midsection length of 16 mm. This corresponds to ~$9 \cdot \lambda_{170GHz}$ where λ_{170GHz} denotes the free-space wavelength at 170 GHz. The possible previously mentioned operating frequencies should correspond to the natural $\lambda/2$ resonances of the synthetic diamond output window using: $f_{op} = (N \cdot c_0) / (2 \cdot d_{window} \cdot \varepsilon_r^{1/2})$ with $N \in \mathbb{N}$, where c_0 is the speed of light in vacuum, d_{window} is the thickness of the window disk and ε_r is the relative dielectric permittivity of the window material ($\varepsilon_r = 5.67$ for the existing Chemical Vapor Deposition diamond window). Window matching is the first condition for multi-frequency operation. The second condition is given by the launcher performance. Possible modes for multi-frequency operation should exhibit a maximum deviation of the caustic radius R_C of <3-4 %, where the caustic radius is determined by $R_C = m \cdot R_o / \chi_{m,n}$. Here R_o is the cavity radius and $\chi_{m,n}$ the mode eigenvalue, which is the nth root of the characteristic equation in the coaxial cavity [6]. This condition enables the usage of the existing, or at least of a similar, quasi-optical system. The modes obtained, taking into account the mentioned considerations, are the $TE_{40,23}$ mode operating at 204 GHz and the modes $TE_{48,26}$ and $TE_{46,27}$ operating at 237 GHz and 238 GHz, respectively. Table 2 summarizes the properties in terms of window reflection and caustic radius of the chosen modes. The $TE_{40,23}$ mode ($\chi_{40,23} = 126.3$) has a frequency close to the natural resonance frequency of the diamond window, which results in a very low window reflection. But the difference in caustic radius with respect to the $TE_{34,19}$ mode is 2.01 % and it has to be checked if the operation of the existing launcher and mirror system is still acceptable for this mode. In future, a compromise of the window thickness can be considered to improve the reflection coefficient at 204 GHz. The operation at ~240 GHz can be realised with two possible modes. One of them ($TE_{48,26}$, $\chi = 146.8$) exhibits a relatively high window reflection, whereas the other ($TE_{46,27}$, $\chi = 147.5$) has a high difference of the caustic radius. The simulation results, in terms of beam-wave interaction, are quite similar for these two modes. This means that the mode selection for operation at ~240 GHz should be made after calculating the behaviour of the launcher and mirror system. In the following, results with respect to beam-wave interaction are reported only for the mode $TE_{48,26}$, for brevity.

Table 2. Mode Selection for Multi-Frequency Operation.

Mode	$TE_{34,19}$	$TE_{40,23}$	$TE_{48,26}$	$TE_{46,27}$
Frequency [GHz]	170	204.179	237.242	238.351
Window reflection [%]	0	0.026	0.472	0.101
Rel. caustic radius R_C/R_o	0.3232	0.3167	0.3270	0.3119
Diff. in caustic radius [%]	0	2.01	1.18	3.62

3 Simulations using the existing Oxford Instruments magnetic field profile

In initial studies, the KIT pre-prototype cavity without the non-linear uptaper was considered and already presented in [12]. However, the simulation results will be repeated within this paper for completeness. There, the magnetic field profile of the existing 7 T Oxford Instruments superconductive (SC) magnet was used as a typical profile. It should be mentioned, however, that the existing magnet cannot deliver the needed magnetic field strength for frequencies above 170 GHz. The appropriate super-conducting magnet is under procurement and is expected to be delivered by mid-2019 by Tesla Engineering Ltd. (Cf. Section 5).

The beam-wave interaction in the cavity for operation at 170/204/238 GHz was simulated using EURIDICE [13,14]. In order to be as realistic as possible at this initial stage of investigations, the simulations were done assuming a kinetic energy spread of $\Delta E = 0.2$ % rms and a relatively large spread of $\Delta\alpha = 20$ % rms for the electron velocity ratio α. The adiabatic start-up scenario was simulated for this setup at 204 GHz and 238 GHz by linearly raising the accelerating voltage. By adiabatic one understands, that the complete electron trajectory from the emitter to the cavity is assumed adiabatic. The most important competing modes at 204 GHz are shown in Fig. 1. The adiabatic start-up scenario simulation including 72 competing modes and a magnetic field of 8.15 T is shown in Fig. 2.

In the simulations, a lower velocity ratio $\alpha = 1.2$ instead of $\alpha = 1.3$ was used to reduce the risk of unwanted parasitic oscillations, especially at frequencies above 200 GHz. The simulation showed a good scenario starting with the $TE_{41,23}$ mode. At 75.5 keV the nominal $TE_{40,23}$ mode becomes dominant. In this case the operating point is defined by the Ohmic loading constraints. These constraints limit the insert wall loading to 0.2 kW/cm^2 and the cavity wall loading to 2 kW/cm^2. The maximum acceptable cavity loading defines the operating parameters of U_{beam} = 80.7 keV for the electron beam energy and I_{beam} = 70 A for the beam current. An RF output power of 1.8 MW radiated from the cavity at an interaction efficiency of 33 % is obtained at 204.14 GHz. The insert wall loading amounts to 0.065 kW/cm^2 and is far below the limit.

In order to improve the performance at the higher frequencies, the midsection length of the coaxial cavity was reduced to ~ $9\lambda_{204\ GHz}$ (13.6 mm) as a first proposed modification. Reducing the midsection length leads to lower quality factor and thus the wall loading is decreased as well. Hence, the operating parameters of the beam voltage, beam current, and magnetic field can be further increased without violating the maximum allowable wall loading. As a result, the RF output power was also increased. In this case, the operation parameters for the operation at 204 GHz are determined to be U_{beam} = 88 keV, I_{beam} = 75 A, and B = 8.23 T. The cavity RF output power was increased to 2.2 MW with an interaction efficiency of 33.4 %.

At the next higher frequency (~238 GHz) the number of possible competing modes is increased to 90. This effect comes from stronger mode competition due to a denser mode spectrum by raising the frequency. At ~238 GHz mode stability has been a problem, especially at high beam voltages and/or beam currents. The operation parameters have been reduced to U_{beam} = 60 keV and I_{beam} = 60 A. Therefore, the RF output power with the $TE_{48,26}$ mode drops to 1.04 MW at an efficiency of 30.2 %. One of the issues operating at higher frequencies with the same cavity is the comparatively long cavity, where electron overbunching can occur and so, the electron beam regains energy from the RF beam. To overcome this issue, the simulation was performed with the shorter midsection length of the coaxial cavity plus a reduced pitch factor (α = 1.1) which leads to stable operation at U_{beam} = 78 keV, I_{beam} = 70 keV, and B = 9.45 T. The cavity wall loading is 1.85 kW/cm^2 at that point. The output power reaches 1.6 MW with an efficiency of 30.4 %.

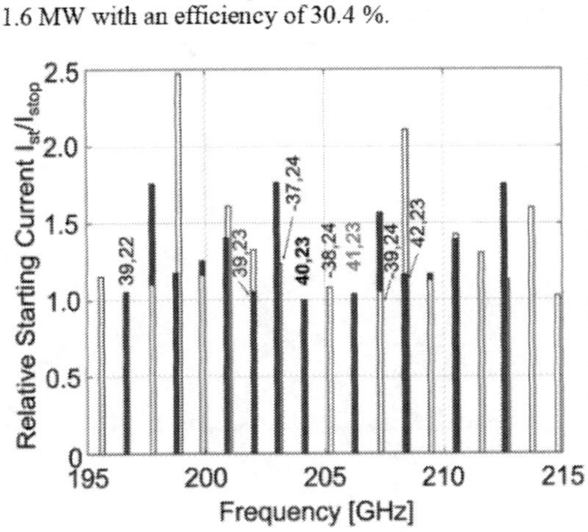

Fig. 1. Relative starting current of competing modes at 8.15 T and I_{beam} = 70 A (filled bars: co-rotating; empty bars: counter rotating modes) [12].

4 Feasibility study taking into account the field profile of the new 10.5 T magnet

The theoretical feasibility of operation at higher frequencies was shown in the previous section. Since the frequency of operation is proportional to the magnetic field, the magnetic field has to be increased for higher frequencies (e.g. 204 GHz leads to ~8.15 T and 238 GHz leads to ~9.45 T). The current test stand at KIT is limited to the maximum magnetic field strength of ~6.9 T of the existing Oxford Instruments magnet. Therefore, KIT ordered a new SC magnet including the following key features:

- Maximum field at cavity: 10.5 T,
- Warm bore hole diameter: 261 mm,
- Dipole coils for electron beam alignment,
- Cryogen-free cooling system.

The first question to be answered is to assess whether the currently operating coaxial gyrotron can also operate in the new magnet without any modifications. Therefore, the magnetic field configuration from Tesla has been used for interaction simulations. The simulated adiabatic start-up scenario, assuming a kinetic energy spread of ΔE = 0.2 % rms and a relatively large spread of $\Delta\alpha$ = 20 % rms for the electron velocity ratio α, shown in Fig. 3 has a smooth start up where only co-rotating modes have been excited.

In a second step, the new coaxial MIG with non-emissive coating of the emitter edges [15], delivered by Thales Electron Devices, is considered. The electron beam parameters have been obtained using ARIADNE [14,16]. The simulations show a huge number of trapped electrons and/or high alpha spread of around $\Delta\alpha$ = 40 % rms at the maximal velocity ratio of α = 1.12, due to the different length of the new magnet and, therefore, misplacement of the gun. For this reason, the distance between emitter and cavity has to be adjusted to the field profile of the new Tesla magnet. This issue can be solved by e.g. inserting a spacer into the beam tunnel. Further, the coaxial insert has to be redesigned, to keep

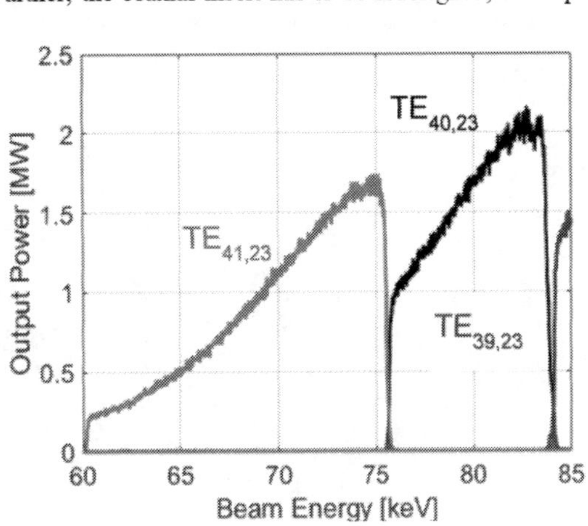

Fig. 2. Adiabatic start-up scenario with existing coaxial cavity at 8.15 T and I_{beam} = 70 A under the consideration of the magnetic field profile of the Oxford Instruments magnet.

the insert corrugations to the right place in the cavity.

This adjustment has been taken into account in the following simulations shown in this paper.

5 More Realistic Start-Up Scenario

After checking the existing setup of the pre-prototype, simulations using the longer beam tunnel have been performed. Realistic multi-mode start-up simulations have been made considering the coaxial cavity with the expected magnetic field profile of the new magnet and the corresponding beam parameters using the coaxial MIG with non-emissive coating of the emitter edges at 170 GHz ($TE_{34,19}$ mode) and 204 GHz ($TE_{40,23}$ mode). The more realistic scenario considers also the non-adiabaticity of electron trajectory close to the emitter Further, the simulation results with and without non-linear uptaper have been compared to each other. Additionally, a modification of the cavity has been taken into account for an improved performance.

5.1 Simulation results at 170 GHz using the existing coaxial-cavity design

Firstly, ARIADNE simulations using the new magnetic field profile have been performed and the obtained electron beam parameters have been used in EURIDICE. In this simulation, at 170 GHz the alpha spread is $\Delta\alpha = 6.08$ % for $\alpha = 1.3$. The energy spread amounts to 0.08 % and the kinetic energy spread to 0.12 %. The realistic start-up scenario at 6.86 T without non-linear uptaper is shown in Fig. 4. Taking the wall loading constraint into account, the operation parameters are $U_{beam} = 89$ keV and $I_{beam} = 75$ A, where the tube can deliver an RF output power of 2.3 MW at an interaction efficiency of 36.7 %. Initial studies including the non-linear uptaper after the cavity have been started. Figure 5 presents the start-up scenario using the same parameters as mentioned before for a better comparison. The shown start-up is not that smooth as before, but nevertheless the nominal $TE_{34,19}$ mode is finally excited. The wall loading constraint limits the operation to $U_{beam} = 90$ keV. The RF output power decreases to 2.25 MW at an interaction efficiency of 34 % and an insert loading of 0.12 kW/cm^2.

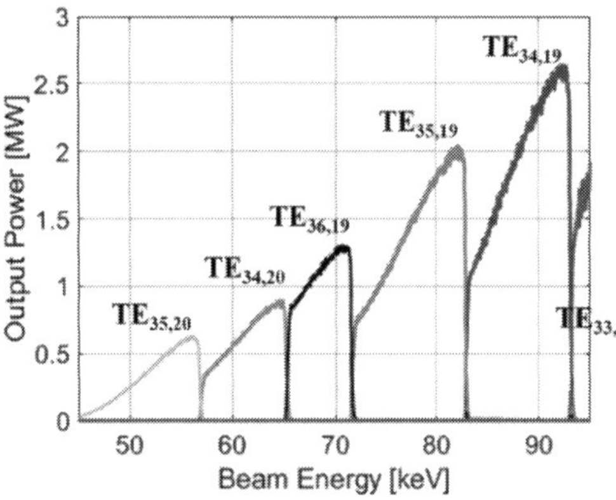

Fig. 3. Realistic start-up scenario with existing coaxial cavity at 6.86 T and $I_{beam} = 75$ A without non-linear uptaper using the magnetic field profile proposal from the new SC magnet and the gun parameters from the new coaxial MIG.

5.2 Simulation results at 204 GHz using the existing coaxial-cavity design

The coaxial MIG with non-emissive coating of the emitter edge have been simulated via ARIADNE for the operation at 204 GHz using the new magnet. The gun parameters have been selected to give a pitch factor of $\alpha = 1.2$ having an alpha spread of $\Delta\alpha = 4.3$ % and a kinetic energy spread of 0.1 %. The simulation shows a good scenario starting with the $TE_{-40,24}$ mode at ~48 keV in Fig. 6, where the minus refers to a counter-rotating mode. Starting from 60 keV the $TE_{41,23}$ mode is excited while the nominal $TE_{40,23}$ mode is getting dominant at ~75 keV. Starting from 83 keV the competing $TE_{-38,23}$ mode, which has nearly the same caustic radius, is getting stronger and so the nominal mode collapses. The operation is limited to 79.2 keV due to the wall loading constraint. The cavity delivers 1.7 MW RF output power without non-linear uptaper at an interaction efficiency of

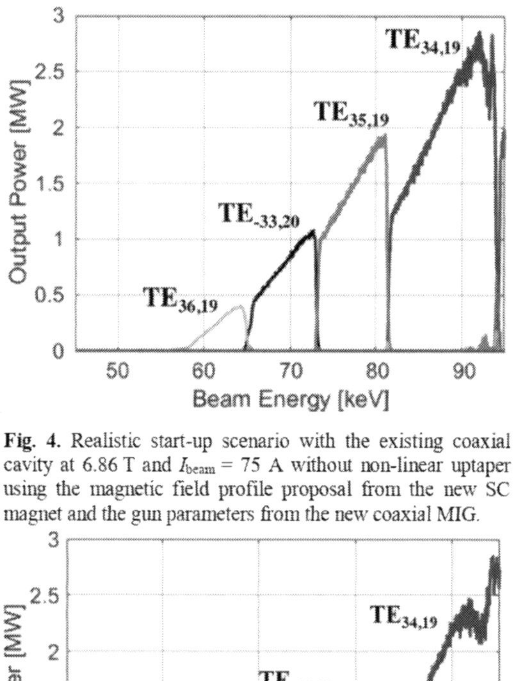

Fig. 4. Realistic start-up scenario with the existing coaxial cavity at 6.86 T and $I_{beam} = 75$ A without non-linear uptaper using the magnetic field profile proposal from the new SC magnet and the gun parameters from the new coaxial MIG.

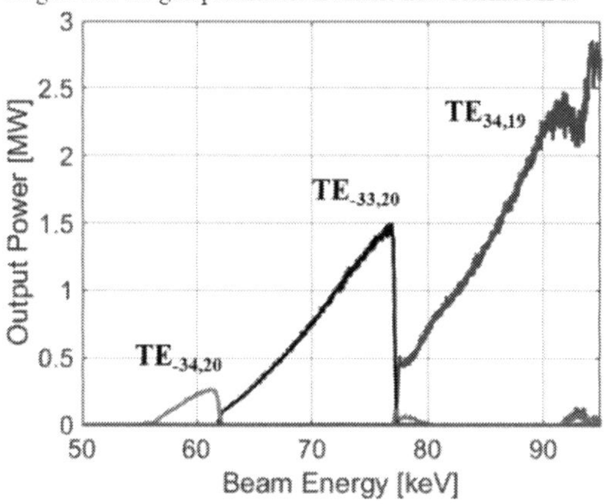

Fig. 5. Realistic start-up scenario with existing coaxial cavity at 6.86 T and $I_{beam} = 75$ A with non-linear uptaper using the magnetic field profile proposal from the new SC magnet and the gun parameters from the new coaxial MIG.

33 % and an insert loading of 0.06 kW/cm². The delivered RF output power is 0.1 MW lower than the results presented in section 3.

Initial studies considering also the influence of the non-linear uptaper at 204 GHz are being performed. Figure 7 presents the start-up scenario using the same parameters as mentioned before for comparing the results to each other. The scenario looks quite similar with and without the non-linear uptaper, as it should. The point at which the $TE_{40,23}$-mode collapses is nearly identical as before. Additionally, the $TE_{41,22}$ mode oscillates after the $TE_{40,22}$ mode. The delivered RF output power and the interaction efficiency are not influenced by the non-linear uptaper.

5.3 Further modifications

In order to get a dual-frequency gyrotron delivering 2 MW, or even more at both operation frequencies, modifications have to be done. For this reason, the cavity midsection length has been reduced from 16 mm to 13.6 mm, as mentioned in the previous section. The start-up simulation of the shortened cavity for the $TE_{34,19}$-mode case is shown in Fig. 8. Concerning the nominal operation parameters of $U_{beam} = 90$ keV and $I_{beam} = 75$ A and an increased magnetic field of 6.88 T, the cavity delivers 2.4 MW RF output power at an interaction efficiency of 37 % having a wall loading of 1.9 kW/cm². Compared with the longer cavity, the gyrotron can deliver 0.1 MW more RF output power with an increased interaction efficiency of ~1 percentage point at this operating point. However, concerning the Ohmic wall loading constraint the nominal operation point could be extended to $U_{beam} = 90.6$ keV, where the cavity delivers 2.5 MW at an efficiency of 38.2 %.

This modification has to be proven for the 204 GHz operation, as well. The simulation shows a good scenario starting with the $TE_{42,23}$ mode, over the $TE_{41,23}$ mode and finally the nominal $TE_{40,23}$ mode is excited until the competing $TE_{-38,23}$ mode raises, as shown in Fig. 9. The operation point is found by the wall loading constraint and is determined by U_{beam} = 86.5 keV, having an RF output power of 2.06 MW at an interaction efficiency of 33.5 %. Comparing these results with the existing cavity design, the RF output power can be increased by ~0.4 MW and the interaction efficiency by 0.5 percentage points.

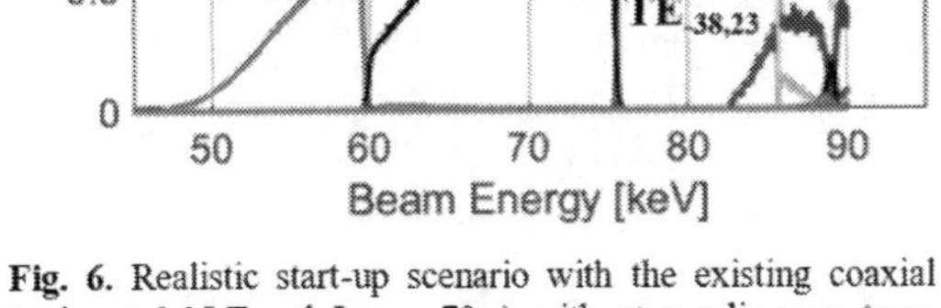

Fig. 6. Realistic start-up scenario with the existing coaxial cavity at 8.15 T and $I_{beam} = 70$ A without non-linear uptaper using the magnetic field profile proposal from the new SC magnet and the gun parameters from the new coaxial MIG.

6 Multi-Frequency Launcher Design

The operating modes excited in the cavity are converted into the fundamental Gaussian mode using an internal quasi-optical mode converter which contains a mirror-line launcher and three mirrors. The existing single frequency (170 GHz) launcher design is already presented in [10]. The simulation results show that in the case of fabrication error on the wall perturbation of ± 10 µm, the launcher will still provide a RF beam with high Gaussian mode content. The measurement results are in a quit good agreement with the simulations. The simulated fundamental Gaussian mode content is determined by 96.3 % at the launcher aperture at 170 GHz [17].

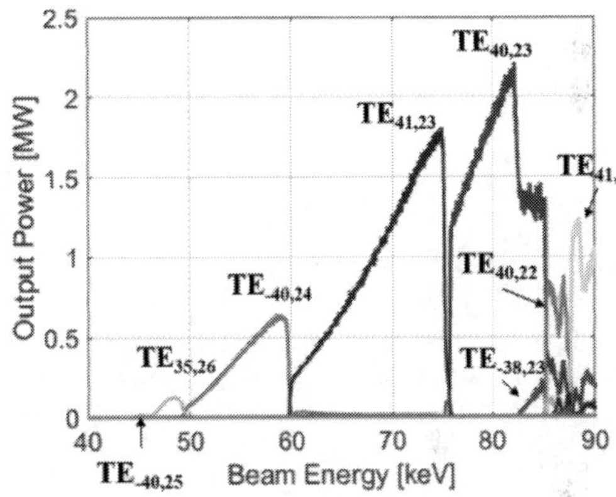

Fig. 7. Realistic start-up scenario with the existing coaxial cavity at 8.15 T and $I_{beam} = 70$ A including the non-linear uptaper using the magnetic field profile proposal from the new SC magnet and the gun parameters from the new coaxial MIG.

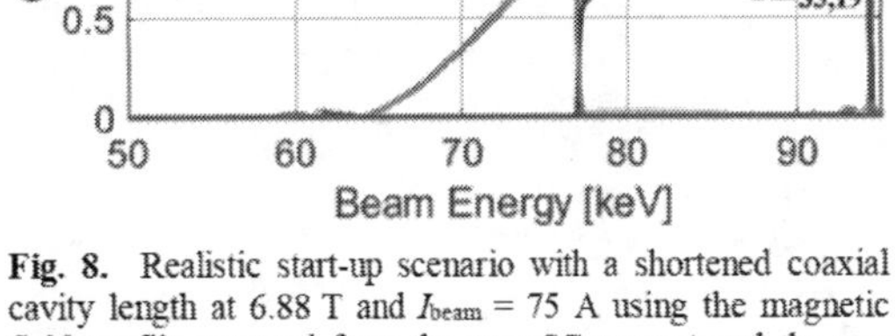

Fig. 8. Realistic start-up scenario with a shortened coaxial cavity length at 6.88 T and $I_{beam} = 75$ A using the magnetic field profile proposal from the new SC magnet and the gun parameters from the new coaxial MIG.

Table 2. Summary of the realistic multi-mode simulation results using EURIDICE using the new Tesla magnet configuration.

Mode	$TE_{34,19}$			$TE_{40,23}$		
Configuration	w.o. uptaper	w. uptaper	modification	w.o. uptaper	w. uptaper	modification
Frequency [GHz]	170.005	170.020	170.041	204.145	204.141	204.167
Beam voltage [keV]	90	90	90.6	79.2	79.2	86.5
Beam current [A]	75	75	75	70	70	75
Magnetic field [T]	6.86	6.86	6.88	8.15	8.15	8.23
Velocity ratio, α	1.3	1.3	1.3	1.2	1.2	1.2
Interaction length [mm]	16.0	16.0	13.6	16.0	16.0	13.6
Wall loading [kW/cm^2]	1.89	2.00	2.00	2.00	2.00	2.00
Insert loading [kW/cm^2]	0.11	0.11	0.13	0.06	0.06	0.07
RF output power [MW]	2.3	2.2	2.5	1.7	1.7	2.06
Interaction eff. [%]	36.7	34.0	38.2	33	33	33.5

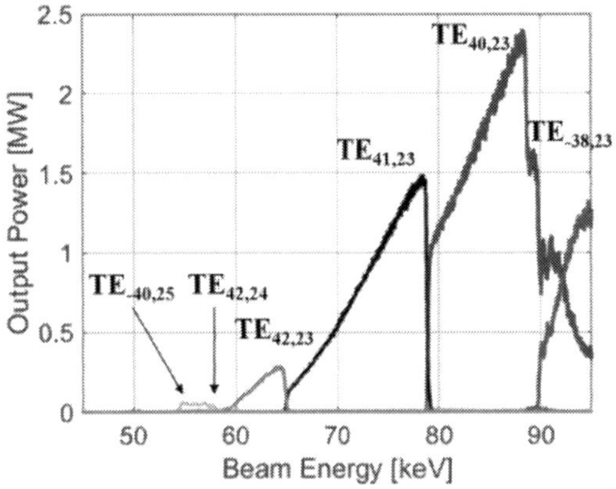

Fig. 9. Realistic start-up scenario with a shortened coaxial cavity midsection length at 8.23 T and I_{beam} = 75 A.

Having a dual frequency scenario, the existing mirror-line launcher with radius of 32.5 mm is tested to assess whether the performance is good enough at 204 GHz. First initial studies at 204 GHz for the $TE_{40,23}$ mode show a fundamental Gaussian mode content of only 91.6 % at the launcher aperture. Therefore, a new launcher was designed with good performance operating both at 170 GHz and 204 GHz. The new launcher is also a mirror-line launcher, and its radius is 32 mm. This dual-frequency launcher design shows promising results despite the difference of the modes' caustic radii of around 2 %. The Gaussian mode content was calculated to be 97.2 % at 170 GHz and 96.6 % at 204 GHz.

7 Summary and Outlook

The simulations of this paper show that the existing KIT $TE_{34,19}$-mode coaxial-cavity pre-prototype gyrotron could operate in the KIT FULGOR gyrotron test stand using the new 10.5 T magnet, under the assumption of enlarge the beam tunnel. Furthermore, the adapted gyrotron is capable of operation as a multi-purpose/multi-frequency device. Without any other modifications the simulations show that the gyrotron delivers at 204 GHz (170 GHz) 1.7 MW (2.3 MW) at an interaction efficiency of 33 % (36.7 %). A small modification of the cavity length would increase the RF output power by 0.4 MW and the interaction efficiency by 0.5 percentage points at 204 GHz. Operating a dual-frequency coaxial-cavity gyrotron in the new test stand, the following aspects have to be taken into account: (i) the gyrotron length has to be adjusted due to the height of the new magnet, (ii) as a consequence to this, the coaxial insert has to be modified by the same length, (iii) in principle the cavity can operate at both frequencies, but with a small reduction of the cavity length the results can be improved, and (iv) the launcher has to be improved for sufficient performance at 204 GHz.

This work has been carried out within the framework of the EUROfusion Consortium and has received funding from the Euratom research and training programme 2014-2018 under grant agreement No. 633053. The views and opinions expressed herein do not necessarily reflect those of the European Commission. Part of the simulations were performed on the EUROfusion High Performance Computer (Marconi-Fusion).

References

1. H. Braune, et al., *In Proc. IRMMW* (2016).
2. T. Omori, *et al.*, *Fusion Eng. Des.*, (2011).
3. G. Federici, *et al.*, in *Proc. 25th SOFE*, (2013).
4. Z. C. Ioannidis *et al.*, *IEEE Trans. on Electron Devices*, (2017).
5. P. C. Kalaria, *et al.*, *Physics of Plasma*, 23(9), 2016.
6. C. T. Iatrou, *et al.*, *IEEE Trans. Microwave Theory Tech.*, (1996).
7. K. A. Avramides, *et al.*, "*IEEE Trans. on Plasma Science*, (2004).
8. T. Rzesnicki, et al., "*IEEE Trans. on Plasma Science,* (2010).
9. S. Illy, *et al.*, EC-20 workshop, (2018).
10. S. Ruess, *et al.* , 47th EUMC, (2017).
11. M. Schmid et al., Fus. Eng. Des. **123**, 485, 2017.
12. T. Ruess, *et al.*, GeMiC (2018).
13. K. A. Avramides, *et al.*, EC-17 *Workshop (*2012).
14. K. A. Avramidis *et al.*, *IRMMW-THz*, Copenhagen, 2016, pp. 1-2.
15. I. Gr. Pagonakis, S. Illy, M. Thumm, *Physics of Plasmas*, 23, 83103 (2016).
16. I. Gr. Pagonakis and J. L. Vomvoridis. *IRMMW (*2004).
17. J. Jin, *et al.*, in *IEEE Trans. on Microwave Theory and Techniques* (2009).

Metrology techniques for the verification of the alignment of the EU gyrotron prototype for ITER

Francisco Sanchez[1,*], *Ferran* Albajar[1], *Alessandro* Lo Bue[1], *Stephano* Alberti[2], *Konstantinos* Avramidis[4], *Tullio* Bonicelli[1], *Alex* Bruschi[3], *Gerd* Gantenbein[4], *Jean-Philippe* Hogge[2], *Stefan* Illy[4], *Zisis* Ioannidis[4], *Alberto* Leggieri[6], *François* Legrand[6], *Ioannis* Pagonakis[4], *Etienne* Perial[6], *Tomasz* Rzesnicki[4], *Manfred* Thumm[4], and *Ioannis* Tigelis[5]

[1]European Joint Undertaking for ITER and the Development of Fusion Energy (F4E), Barcelona, E-08019, Spain
[2]Swiss Plasma Center (SPC), Ecole Polytechnique Fédérale de Lausanne, CH-1015 Lausanne, Switzerland
[3]Institute of Plasma Physics "P.Caldirola", National Research Council of Italy, Milan, Italy
[4]IHM, Karlsruhe Institute of Technology (KIT), Kaiserstr. 12, 76131 Karlsruhe, Germany
[5]Faculty of Physics, National and Kapodistrian University of Athens, Zografou, GR-157 84, Athens, Greece
[6]Thales Electron Devices, 2 rue Marcel Dassault, Vélizy-Villacoublay, F-78141, France

Abstract. The EU gyrotron for the ITER Electron Cyclotron (EC) heating system has been developed in coordinated efforts of the EGYC Consortium, Thales ED (TED) and Fusion for Energy (F4E) and under the supervision of ITER Organization Central Team. After the successful verification of the design of the 1MW, 170 GHz hollow cylindrical cavity gyrotron operating at the nominal $TE_{32,9}$ mode with a short pulse gyrotron prototype at KIT, an industrial CW gyrotron prototype was manufactured by TED and tested at ~0.8 MW output power and 180 s pulse duration, which is the limit of the HV power supply currently available at KIT. The experiments are being continued at SPC in 2018 to extend further the pulse duration, taking advantage of the existing CW full-power capabilities of the gyrotron test facility recently upgraded for the FALCON project. The gyrotron cavity interaction is very sensitive to the alignment of the internal mechanical parts of the gyrotron tube with the magnetic field generated by the superconducting magnet within a typical range of 0.2 – 0.5 mm. The control of the tolerances and deformations becomes therefore critical to achieving the target performances. With the EU gyrotron prototype it was possible to adjust the alignment of the gyrotron tube with respect to the magnetic field axis during the installation and commissioning phase. The actual shift and tilt movements were verified using advanced metrology methods such as photogrammetry. In this paper, the alignment control techniques and procedures will be discussed also in view of enhancing the reproducibility of gyrotron performance during series production.

1 Introduction

Europe contributes to the procurement of the high-frequency, high-power gyrotrons for the electron cyclotron resonance heating (ECRH) and current drive (ECCD) system of the ITER tokamak. In this framework, the European GYrotron Consortium (EGYC), in cooperation with Thales Electron Devices (TED) and under the coordination of Fusion for Energy (F4E) is developing the EU 1 MW-CW, 170 GHz ($TE_{32,9}$ hollow-cavity) gyrotron prototype.

According to the F4E project planning, the development of the European gyrotron for ITER was organized in two main steps. First of all, a modular short-pulse (SP) prototype was developed by the Karlsruhe Institute of Technology (KIT) in cooperation with TED in order to verify the scientific design of the gyrotron key components with pulses in the ms range. Then, a CW industrial prototype tube (with the same scientific design), TH1509, was developed in order to verify the ITER requirements, in terms of output RF power, efficiency, RF beam quality and pulse length.

The basic scientific design of the 170 GHz, 1 MW modular SP prototype was successfully verified in multiple experimental campaigns between 2015 and 2017. Based on the SP gyrotron design, the manufacturing of the first industrial CW prototype gyrotron (TH1509) was completed at TED in 2015. The testing of this CW industrial prototype was organized in two phases. For the first phase the tube was delivered to KIT in February 2016, where it was operated with the Oxford Instruments (OI) superconducting magnet (see Figure 1 below). Due to the limitation of the currently available HV power supply at KIT, the operation of the gyrotron at KIT is limited at high power to 180 s pulses. For this reason, after the successful completion of the experiments at KIT, the CW prototype was delivered to the Swiss Plasma Center (SPC), in Lausanne, where the tube is currently operated with a new cryogen-free superconducting magnet manufactured by Cryogenics (see Figure 2 below).

* Corresponding author: francisco.sanchez@f4e.europa.eu

One of the key goals at SPC is to extent the pulse length up to one hour. In this work we summarize the importance of the metrology for the correct alignment of the gyrotron with respect to the magnetic field axis (Section 2), we explain the available tools and techniques (Section 3), and we describe the results of the metrology campaigns carried out during the installation and RF high power tests on the TH1509 prototype at KIT and at SPC (Section 4).

Fig. 1. TH1509 gyrotron prototype (left) and the same gyrotron installed in the KIT test stand (right), in Karlsruhe (Germany).

Fig. 2. TH1509 gyrotron prototype installed in the SPC test stand, in Lausanne (Switzerland), with MOU and dummy load.

2 Importance of alignment sensitivity

The performance of a gyrotron (see Figure 3, [1]) can be significantly reduced by a misalignment of the electron beam generated in the electron gun and guided by the magnetic field lines of the superconducting (SC) magnet, with respect to the mechanical axis of the gyrotron cavity.

There are numerous sources of misalignment such as the tolerance chain during manufacturing and assembly of the different gyrotron components, stresses release during baking out of the gyrotron, imperfect sitting of the gyrotron on the SC magnet, mechanical vibrations during operation (due to, e.g., water cooling), thermal deformations during RF operation, SC magnet tolerances and deformations, as well as other external effects (e.g., tokamak stray fields and neighbouring SC magnets). As a consequence of these misalignment sources, either the electron beam or the gyrotron cavity can lose concentricity or get tilted with respect to the mechanical references, so that the gyrotron performance may drop. The effect of this misalignment has been theoretically predicted [2] and has recently been measured experimentally [3].

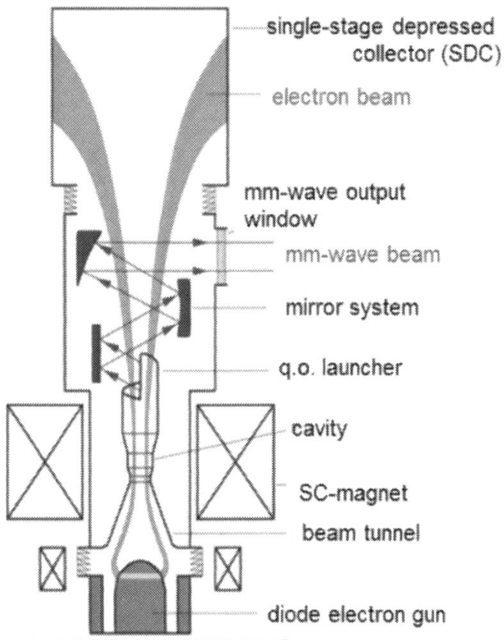

Fig. 3. Gyrotron schematics with electron beam correctly centred with respect to the superconducting (SC) magnet.

RF power measurements were done on the SP gyrotron at KIT when displacing the electron beam by means of additional magnetic fields in two (x, y) directions orthogonal to the gyrotron vertical axis. The magnetic fields were generated by two dipole coils integrated in the OI SC magnet. The results obtained are summarised on Figures 4 and 5 below: from the optimal position (nominal power extracted from the gyrotron), when the electron beam was deviated by a shift of ±0.25 mm, the output power was reduced by 10% from the nominal value. Similar results were obtained when measuring the gyrotron efficiency levels. Moreover, the maximum deviation of the tilt angle between the electron beam and the magnetic field angle was also measured experimentally, resulting in maximum acceptable tilt angles of 0.5 degrees (i.e., ~9 mrad).

By assuming a 10% maximum of allowable reduction of power and efficiency due to misalignments, the maximum deviation of the electron beam at the cavity would be of ±0.25mm. The budget of the previously mentioned misalignment sources results in higher overall displacements, therefore measurements and corrections

 https://doi.org/10.1051/epjconf/201920304015

are needed. In accordance to the international standards (see BS EN ISO 14253-1, 1999) and F4E procedures, the measurement uncertainty shall be equal or lower than 10% of the misalignment target, so that for a deviation of ±0.25mm, the measurement uncertainty shall be less than 25 μm.

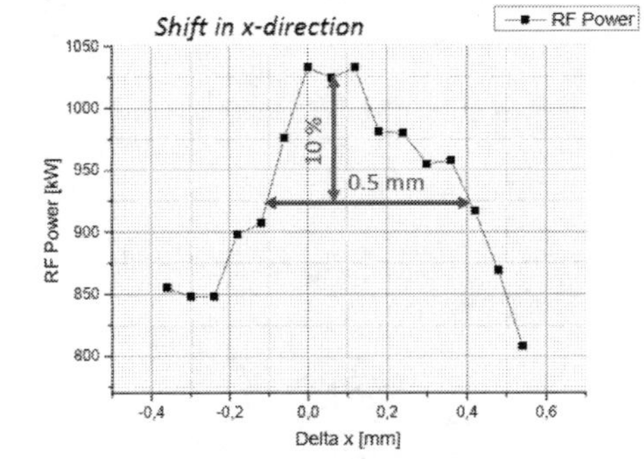

Fig. 4. Influence of the electron beam deviation (x-direction) on the RF power measurements carried out on the SP gyrotron.

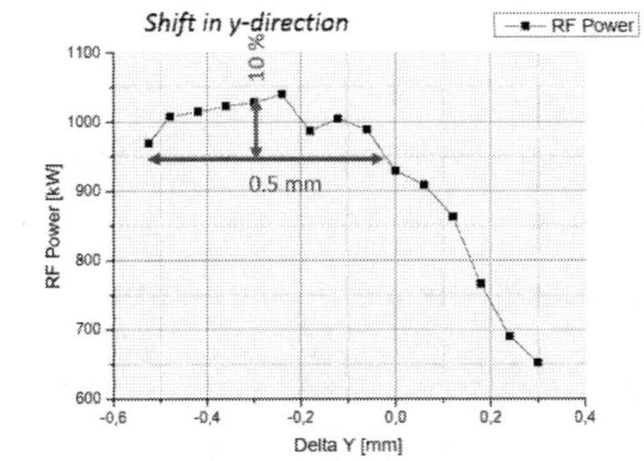

Fig. 5. Influence of the electron beam deviation (y-direction) on the RF power measurements carried out on the SP gyrotron.

Therefore, good alignment between the electron beam and the gyrotron cavity is crucial to obtain a good gyrotron performance, and the use of advanced metrology tools and techniques is highly advisable to comply with the required tolerances and measurement uncertainty. The following sections contain information about the available metrology tools and techniques used on the different surveys done for the EU gyrotron prototype for ITER (1MW-CW, TH1509).

3 Metrology tools and techniques

Advanced metrology tools and techniques are recommended to be applied for improving the gyrotron performance, in particular for the ITER gyrotrons, which require to operate in harsh conditions and with tight tolerances. Additionally, it is crucial to control and trace the precise positioning of the gyrotron tube into the magnet in view of ensuring the repeatability of gyrotron performance during the commissioning and maintenance phases. Thus, the following metrology techniques have been applied to the alignment of the EU gyrotron prototype with very good results in terms of repeatability and gyrotron performance.

The first tool used on the EU gyrotron was a laser tracker, whose principle is based on sending a laser ray to a target reference and comparing the forward to the reflected laser ray. This comparison is used to measure variations of the location of objects under analysis and to measure metrology features, such as holes, corners, edges and welding chamfers. The laser tracker is compatible with the required measurement uncertainty (25 μm) and offers, as main advantage, the possibility to do measurements of all kinds of metrology features in real time. However, the laser tracker targets are not magnetic proof and they cannot be used when the SC magnet is ON (so that it is impossible to use the laser tracker when in operation).

A second advanced tool used was photogrammetry, based on triangulation, getting pictures from at least two different locations (named as "lines of sight"), which are mathematically intersected to get the 3D coordinates of the points of interest. The process outcome are the XYZ coordinates of each one of the points, plus three aiming angles for each picture and camera calibration parameters (including uncertainty estimate). This methodology is even better than the laser tracker in terms of measurement repeatability and compatible with the required uncertainty of 25 μm. It is ideal to be used in the presence of high magnetic fields, as the target are magnetic and thermal (up to 200 °C) proven. The only disadvantage of the photogrammetry is that it cannot be used for measuring metrology features. As conclusion, a combination of a laser tracker metrology campaign with a photogrammetry metrology campaign would be ideal for the case of the ITER gyrotrons. Figure 6 shows the laser tracker and the photogrammetry camera used in the EU gyrotron prototype. Figure 7 shows the sort of images obtained with the photogrammetry.

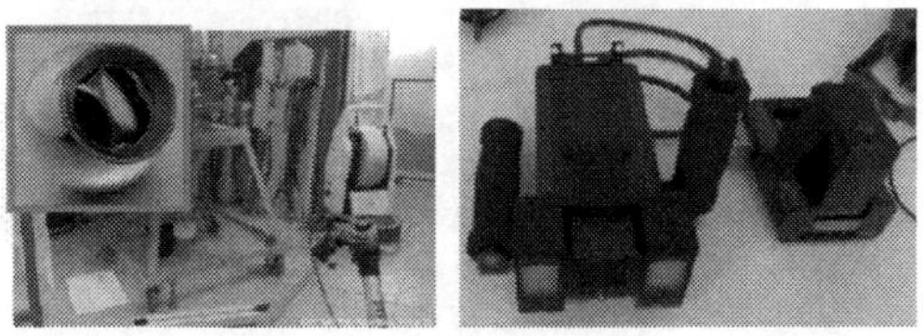

Fig. 6. Laser tracker (and its target), on the left. Camera used for the photogrammetry, on the right.

The metrology surveys shall have a good reference metrology network, based on a reference plane. Ideally, the network and reference plane would have to be established on the most stable component available in the facility. In the case of the EU gyrotron prototype, the SC magnet was chosen to establish a good reference plane and metrology network (see Figure 8).

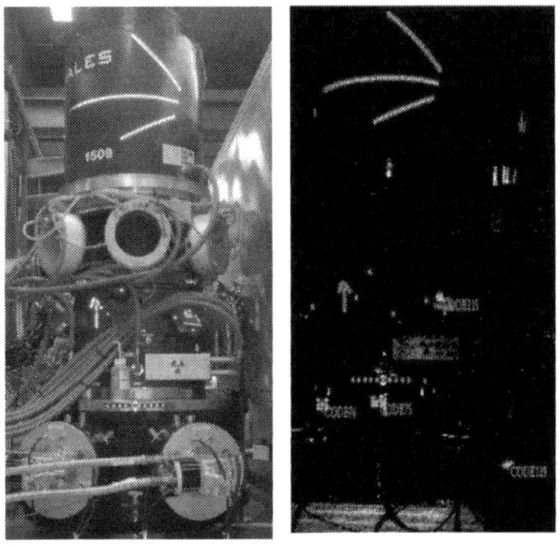

Fig. 7. Normal picture and photogrammetry image obtained in the different targets installed in the EU gyrotron prototype.

Fig. 8. Metrology network (laser tracker targets depicted in the red circles) as set on the SC magnet at the SPC tests stand (reference place set on the top plate of the magnet).

In order to ensure good reproducibility, it is recommended to plan regular metrology campaigns during the whole life-cycle of the gyrotrons: design, fabrication, assembly, installation, commissioning, operation and maintenance (replacement or refurbishment) for a relevant measurement of the gyrotron performance (in terms of voltage, current, calorimetry measurement of the power and efficiency). Aiming to fulfil the TH1509 operating parameters [4], correction techniques have been applied during design and manufacturing with the inclusion of a centering ring on the gyrotron side and an Adjustable Gyrotron Plate (AGP) on the SC magnet. In addition, mechanical compensations may be carried out to correct manufacturing tolerance chain errors, thermal stresses after baking out, potential misalignments during assembly and installation, influence of water cooling vibrations and thermal deformations after operation, unwanted displacements after possible earthquake events and/or seasonal effects. Finally, the metrology can also verify and ensure proper interface with other optical components (Matching Optical Unit, MOU) and dummy load mirror (used during the acceptance tests) [5]. A new application is under preparation to measure the internal irregularities of the transmission lines (circular corrugated waveguides, 50 mm diameter). For this application, F4E metrology group has designed a mole probe (see design in Figure 9) consisting of a main body provided with wheels to allow displacements inside the circular waveguides, a cable holder and sensors to measure the possible irregularities. This design is under fabrication and will be used soon for the ITER circular corrugated waveguides.

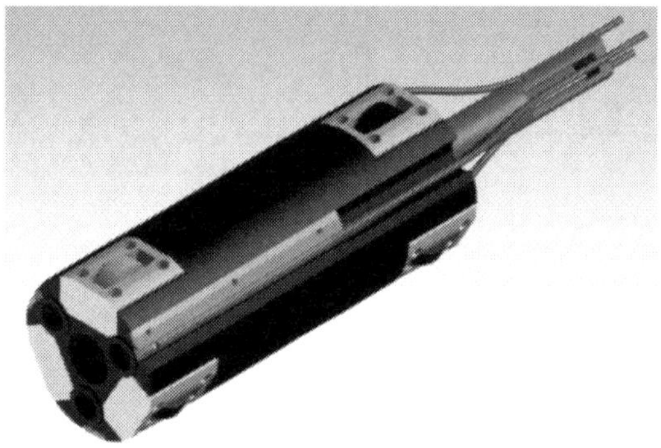

Fig. 9. Mole probe design by F4E to be used to measure irregularities in the ITER circular corrugated waveguides.

4 Results of the metrology campaigns

4.1 Metrology surveys at KIT

A laser tracker survey combined with photogrammetry was carried out on the TH1509 gyrotron prototype at KIT in June-July 2016. The objective being to measure the concentricity between the anode and the centering ring installed in the gyrotron (interface plane with the SC magnet). The concentricity was successfully verified. In addition, the metrology team measured and verified the alignment between the gyrotron and the SC magnet. In this case, there were some difficulties due to the crowded environment (many components close to the gyrotron) and the high magnetic field, which required to use photogrammetry once the reference plane was established at the bottom of the gyrotron centering ring (see Figure 10). As result of these measurements, a deformation of the centering ring was detected, caused by the weight of the gyrotron when sitting on three bars, which was the only way for the metrology targets to get access to the centring ring. As the survey was non-conclusive, it was recommended to modify the gyrotron sitting jig and to plan for additional metrology surveys. After completing the survey, the gyrotron was tested at KIT and the RF power obtained at the output of the RF window was around 800kW for 180 s [6]. In order to complete the extension of the pulse duration to comply with the ITER requirements [7], the gyrotron prototype was transferred to the SPC test facility in Lausanne [8].

Fig. 10. EU gyrotron prototype (TH1509) at the KIT test stand.

4.2 Metrology surveys at SPC

A laser tracker survey also combined with a photogrammetry survey was carried out on the TH1509 gyrotron prototype at SPC in 2017 (in two campaigns: March 2017 for the very short-pulse tests, and October 2017 for the long-pulse tests). The objectives being (similarly to KIT) to verify the concentricity between the anode and the centering ring, and the alignment between the gyrotron electron beam and the SC magnet axis. In this case, the reference plane was set on the top plate (AGP) of the SC magnet and the results showed a deformation of this plate provoked by shims installed during fabrication and installation of the magnet at the SPC facility (see Figure 11).

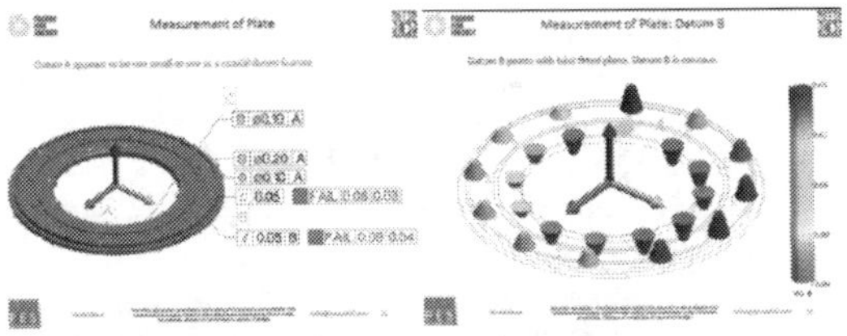

Fig. 11. Reference plane on the SC magnet top plate (AGP) and plate deformations measured by laser tracker.

SPC machined a new plate avoiding shims and ensuring a fat top plate to sit on the gyrotron. The metrology was used both to position correctly the new machined AGP and to align the gyrotron system. The RF tests performed after the metrology survey were successful and very short pulses were produced at the center of the output RF window (see Figure 12).

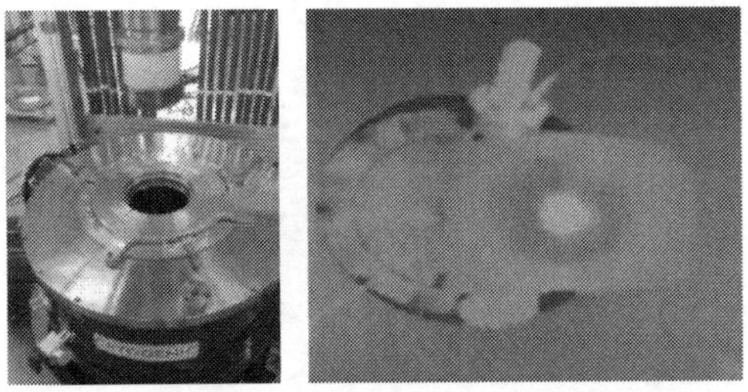

Fig. 12. On the left, SC magnet tope plate (reference plane), together with the XY table while installing the gyrotron. On the right, very short pulse centered at the output RF window.

The second metrology campaign was planned while the gyrotron prototype was in operation for finding the operational maps and RF stability regions which allowed to locate the gyrotron in the center of the region where the RF performance is optimal. This RF operation is usually done with repetitive horizontal displacements of the gyrotron by means of the so-called XY-table [9]. The operation is done with the SC magnet ON (i.e., in the presence of high magnetic fields) and the optimal metrology technique under this conditions was the photogrammetry (magnetic proof targets). Figure 13 shows the photogrammetry targets as seen by a normal picture and with the photogrammetry camera for data processing.

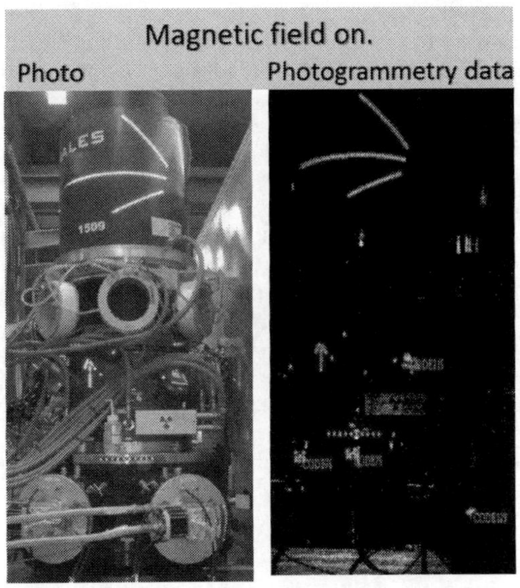

Fig. 13. Left: Photogrammetry targets on the EU gyrotron TH1509. Right: Targets as seen with the photogrammetry camera for data processing.

The metrology campaign results indicated good consistency between the horizontal displacements done on the gyrotron and the measurements taken by the photogrammetry camera. For example, when moving the gyrotron by 1 mm in x-direction, the metrology targets were measuring a displacement in the same direction in the range 0.936 to 1.015 mm. Additionally, a minor rotation of the gyrotron axis was detected with data extracted from the photogrammetry (maybe due to the different degrees of freedom allowed with the XY-table). Figure 14 shows further details on the data measured for this horizontal displacement. The possible tilting of the gyrotron when acting on the XY-table and when operating with the collector water cooling ON (gyrotron under the effect of vibrations) were also verified.

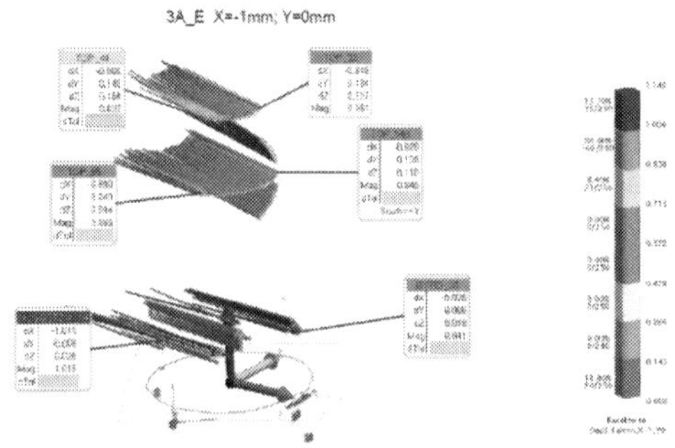

Fig. 14. Photogrammetry data obtained after a horizontal displacement of the gyrotron of 1mm in x-direction.

After concluding the metrology campaign and setting the RF stability regions, the gyrotron was moved to the center of the optimal operating region to perform the high-power long-pulse test campaign. The very good behaviour of the gyrotron in terms of vacuum stability and repeatability of the RF performances, allowed to get up to around 800 kW and pulses of up to 215 seconds. The limitation was found at the maximum energy dissipated in the RF Conditioning Unit (RFCU, aka MOU), whose internal parts are not correctly conditioned for CW operation. RF power coupled to some internal parts of the RFCU (mirror bellows and flanges) provoked unacceptable temperature increases and the stop of the RF operation when the RF power levels are not optimal yet. The RFCU is currently being modified to cope with CW operation. A next RF test campaign is planned for September 2018, where the power level and the pulse duration shall be consolidated for the ITER specifications (i.e., reach 1 MW-CW at 170 GHz, mode $TE_{32,9}$ and pulse duration of one hour).

5 Conclusions

The EU gyrotron prototype for ITER (1 MW-CW, 170 GHz cylindrical cavity, $TE_{32,9}$) under tests by the EGYC consortium at KIT and SPC has proven operation at power levels around 800 kW and 215 s pulse length. In order to get right RF performance, it is important to get a good alignment of the electron beam at the cavity center. In the example shown in the present paper, in order to avoid losing more than 10% of power/efficiency, the maximum misalignment has to be lower than ±0.25 mm (i.e., implying a metrology uncertainty of 25 μm). The metrology campaigns done on the TH1509 prototype both with laser tracker and photogrammetry at KIT and at SPC resulted in performing corrections with dipole coils and XY-table. In order to improve the RF performance and reproducibility of gyrotrons, it is recommended to plan for regular metrology campaigns during the different phases of its life-cycle: design, fabrication, assembly, installation, commissioning, operation and maintenance.

Acknowledgement

This work was done with the shared effort from F4E, the EGYC Consortium and the TED industrial partner.
This publication reflects the views only of the author, and Fusion for Energy cannot be held responsible for any use which may be made of the information contained therein.

References

1. Courtesy of S. Illy, IHM, Karlsruhe Institute of Technology (KIT).
2. K.A. Avramidis et al., Numerical investigations on the effects of electron beam misalignment on beam-wave interaction in a high-power coaxial gyrotron, 38th International Conference on Infrared, Millimeter, and Terahertz Waves (IRMMW-THz), Mainz, Germany, 14-19 September, 2013, Conference Digest.
3. I. Gr. Pagonakis et al., Status of the development of the 170 GHz, 1MW-CW gyrotron, Fusion Engineering and Design, 96-97 (2015), 149-154.
4. Private communication from A. Leggieri (TED) and information from the TH1509 Operation Manual.
5. W. Bin et al, Validation Experiments on the 2-MW CW 170-GHz Load for the European ITER Gyrotron, IEEE Trans. Plasma Sci. 45, Issue 3, 501-511 (2017).
6. Z. C. Ioannidis et al., CW experiments with the EU 1-MW, 170-GHz industrial prototype gyrotron for ITER at KIT, IEEE Trans. Electron Devices, vol. 64, no. 9 (2017).
7. C. Darbos et al., Status of the ITER Electron Cyclotron Heating and Current Drive System, Journal of Infrared, Millimeter, and Terahertz Waves, Volume 37 - Number 1 - January 2016, Special Issue: Recent Advances in RF Heating of Plasmas.
8. J-P. Hogge, Status and Experimental Results of the European 1 MW, 170 GHz Industrial CW Prototype Gyrotron for ITER, presentation at the 41st International Conference on Infrared, Millimeter, and Terahertz Waves (IRMMW-THz), Copenhagen, Denmark, 2016.
9. Z. C. Ioannidis et al., Report of recent experiments with the European 1 MW, 170 GHz CW and SP prototype gyrotrons for ITER, EC-20 proceedings, 2018.

Conceptual design of CFETR ECRH equatorial launcher and upper launcher

Yunying Tang[1,*], *Fukun* Liu[1], *Xiaojie* Wang[1], *Wei* Wei[2], *Liyuan* Zhang[1], *Handong* Xu[1], *Dajun* Wu[1], *Weiye* Xu[1],and *Jian* Wang[1]

[1]Institute of Plasma Physics, Chinese Academy of Sciences, Shushanhu Road 350, Hefei 230031, Anhui, China
[2] Hefei University of Technology, Tunxi Road 193, Hefei 230009, Anhui, China

Abstract. The Electron Cyclotron Resonance Heating (ECRH) system of China Fusion Engineering Test Reactor (CFETR) is designed to inject 20 MW RF power into the plasma for heating and current drive (H&CD) applications. The ECRH system consists of 20 gyrotrons, the associated power supplies, the transmission lines and one launcher. In order to compare the launcher performance from equatorial and upper ports, two types of launcher are designed in this paper. In equatorial launcher (EL), twenty in-vessel transmission lines are divided into four groups. Every five gaussian beams from in-vessel transmission lines face one fixed focusing mirror and one steering mirror. All gaussian beams are injected into the plasma with optimal toroidal and poloidal angles calculated by C3PO/LUKE code. In upper launcher (UL), twenty-one transmission lines are supposed and divides into three groups. Every seven gaussian beams are injected into one fixed focusing mirrors and all twenty-one beams reflected to one common steering mirror finally. The optical transmission characteristics and the convergence information of gaussian beams are checked and optimized. The EL or UL is installed on the equatorial or upper port with a port-plug modular structure.

1 Introduction

The objective for International Thermonuclear Experimental Reactor (ITER) [1] is to demonstrate the scientific and technological feasibility of applying fusion energy for peaceful purposes. Based on the existing physics and technology of ITER, CFETR works on the technology which cannot be completed by ITER. The mission and goal of CFETR are as follows: (1) Fusion power 200- 1500 MW; (2) Duty cycle time (or burning time) ≧50%; (3) Tritium should be self-sufficiency by blanket, TBR (Tritium Breeding Ratio) ≧1. It is challenged by integrated design considering all crucial elements [2,3], which include integrated plasma scenarios for steady-state operation and engineering optimization, particularly inboard radial space for tritium breeding modules and the tradeoff of plasma faced surface for tritium breeding and heating and current drive (H and CD), diagnostics, burning ratio of fuelled particles, plasma burning duty factor, RH, and so on.

To assist and sustain CFETR operation in various scenarios, basing on the design experience of EAST ECH system [4,5], an ECH&ECCD system [6,7] operating at 170 GHz, 20 MW power continuous wave (CW) will be installed for the CFETR with the power injected via two types of launchers, the equatorial launcher and the upper launcher.

In this paper, central heating and current drive are the main physical requirements for EC system. Because of the excellent local power deposition and simple coupling between wave and plasma, megawatt steady-state ECH system is in the first priority to be considered among all the auxiliary heating systems. ECCD maintains the non-inductive current drive and controls the current profile during CFETR steady-state operation.

The main purpose of this paper is to present the status of our current design work. The structure of ECH system is introduced in section 2. The designs of equatorial launcher and upper launcher designed basing on the initial small-size CFETR device [2,3] are described in section 3 and 4. Finally, the conclusion is given.

2 ECH system

ECH system is composed of microwave subsystem, high voltage power supply (HVPS) subsystem, interlock and control protection subsystem, vacuum pumping subsystem and water-cooling subsystem (Fig.1). The microwave subsystem consists of microwave source, transmission line and EC launcher, which show the generating, transferring and injecting process of microwave power respectively. HVPS which is mainly formed by cathode power supply and body power supply is in charge of power supply for microwave source. Each cathode power supply provides two gyrotrons [8-11]. PSM structure is chosen in the cathode power supply. It adopts all-controlling switching elements and the switching speed is very quick. Therefore the independent crow-bar protection for gyrotron do not need here. Independent body power supply, superconducting magnet power supply, filament power supply, gun coil power supply, ion pump power supply and collector coil power supply are configured to every gyrotron. Interlock and control protection subsystem, water-cooling subsystem and vacuum subsystem assure the safe and reliable operation of the overall ECH system. Interlock and control protection subsystem commands all the subsystems operating according to the setting procedures and parameters and monitors all kinds of operating parameters in real time. Once error occurs in the system, interlock and control protection subsystem will execute protective action immediately. Water-cooling subsystem cools all the microwave devices which are needed to be cooled in the ECH system. Vacuum subsystem ensures superconducting magnets, matching optics unit (MOU) and corrugated waveguides running in a good vacuum state.

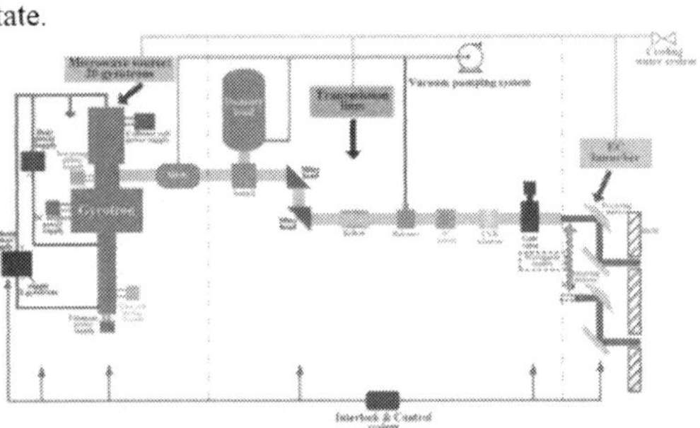

Fig. 1. The constitution of CFETR ECH system. It contains microwave sources, transmission lines, EC launcher, cooling water subsystem, vacuum pumping subsystem, interlock & control subsystem and HVPS.

In ECCD, the resonance condition for wave-particle in velocity space is given by the following equation [12]:

$$\omega = n\omega_{ce}/\gamma + k_{//}v_{//} \quad (1)$$

where ω is the wave frequency, ω_{ce} is the electron cyclotron frequency, n is the electron harmonic number, $\gamma = 1/\sqrt{1-(v^2/c^2)}$ is the relativistic form-factor, which is a consequence of the relativistic correction of the electron mass, c is the velocity of light in vacuum. The $k_{//}v_{//}$ term is the Doppler effect, which plays an important role in ECCD and causes both shift and broadening of the absorption line, where $v_{//}$, $k_{//}$ are the parallel components of the velocity of the resonant electrons and wave vector of the EC wave, respectively. In ECH system, two modes can be used to heat the plasma: O mode and X mode [13]. If the EC wave is injected into plasma from low field side, only O mode and X2 mode can be adopted. The probable resonance layers and heating methods in CFETR ECCD system are shown in Fig. 2. Taking the frequencies adopted in the present tokamaks into consideration, 110GHz, 140GHz, 170GHz and 230GHz are chosen in the analysis. In Fig. 2(a), when the center toroidal field is 5T, 140GHz, 170GHz and 230GHz satisfy the ECCD system demand. When the center toroidal field of CFETR is up to 7T in the future shown in Fig. 2(b), only 170GHz and 230GHz can be accepted. Combining with the mature manufacture technology of ITER 170GHz gyrotron, 170 GHz is chosen as working frequency of ECH system in CFETR according equation (1) and EC wave would be coupled to plasma with O1 mode.

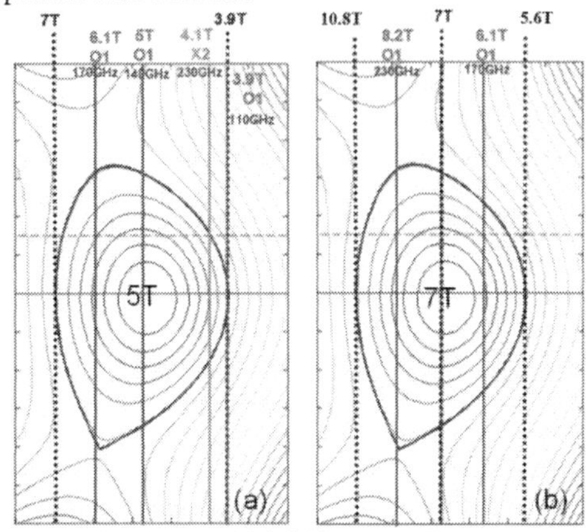

Fig. 2. (a) The center toroidal field B_T=5T; (b) B_T=7T.

The microwave subsystem is made up of RF source, transmissions line and launcher. The design of launcher is similarly ITER EL and UL[14-17], two types of launchers are designed. In Fig. 3, it shows ECH system for CFETR where the injected launcher is EL for instance. RF sources contain twenty 170GHz/1MW gyrotrons which take charge of generating microwave power. 20 MW microwave power is transferred by 20 transmission lines and injected into the plasma via EL. Transmission lines should satisfy a series of technical requirements, therefore there are different microwave components distributed along the line, such as MOU, corrugated waveguide, miter bends and some auxiliary equipments (Fig. 4). The diameter of corrugated waveguide is 63.5mm and the transferring mode in corrugated waveguide is HE_{11} mode. The miter bends in the transmission lines possess many functions, such as changing microwave power transferring direction, measuring the power [18], monitoring the arc and adjusting the polarization information [19]. The auxiliary equipments contain direct current (DC) break, vacuum pumping sections, chemical vapor deposition (CVD) window, gate valve and so on [20-22]. The DC break separates electrical potential along the transmission line to eliminate unnecessary electrical closed loops. The pumping section evacuates the transmission line through a gap of the waveguide and the whole transmission lines are evacuated to $\leq 2\times10^{-3}$ Pa in order to reliably transmit 1 MW RF wave. In each transmission line, there is one CVD window located close to the launcher in order to separate the high vacuum of CFETR vessel from vacuum transmission line. Gate valve is located at the end of transmission line and it is shut off immediately when one discharge is over.

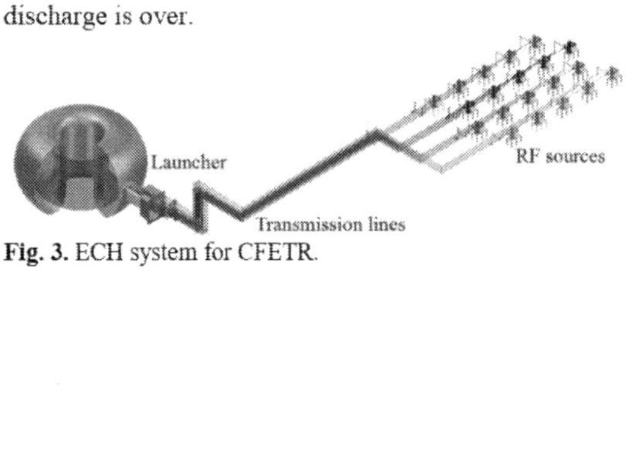

Fig. 3. ECH system for CFETR.

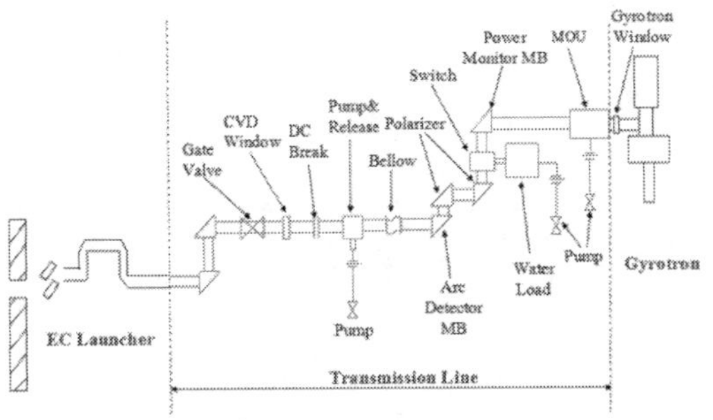

Fig. 4. Schematic of a generic TL.

3 Equatorial launcher

The wave with the specific mode, polarization information and incident angle is injected into CFETR plasma by the quasi-optical launcher which should be located on the special position for the optimal heating and driving current.

The EL is installed on the equatorial port with sectional area 2200mm × 1750mm. Total 20 MW microwave power from 20 gyrotrons should be injected into plasma via one EL. Before designing the EL, the injecting position of the microwave power is decided by C3PO/LUKE code [23] via calculating the driven current in different positions. In C3PO/LUKE code, the injection angles of the beams are defined as Fig. 5. θ is the poloidal injection angle and φ is the toroidal injection angle. The distance from the center of CFETR horizontal plane to the injection mirror in vertical plane is marked as Za in Fig. 6(a). The distance from the middle point between injection mirror and focusing mirror to the CFETR center in horizontal plane can be seen in Fig. 6(b) and is named Ra. The distance in horizontal plane between injection mirror and focusing mirror is 100cm.

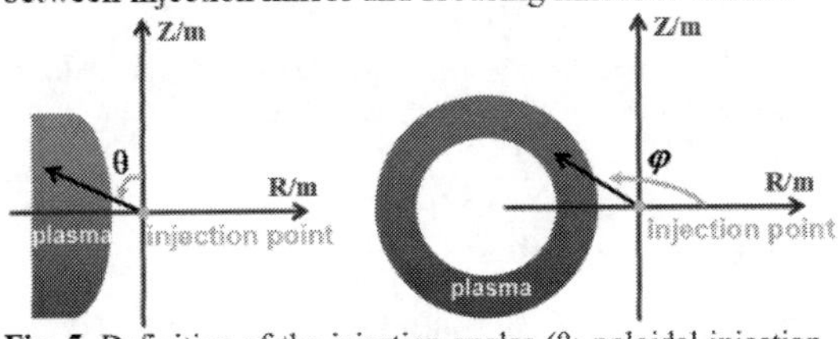

Fig. 5. Definition of the injection angles (θ: poloidal injection angle; φ: toroidal injection angle).

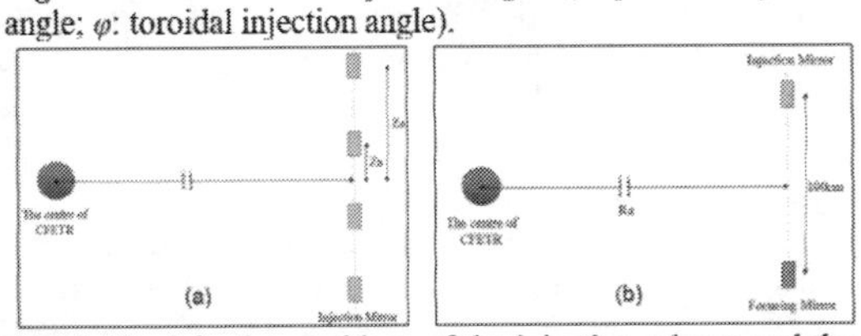

Fig. 6. The injection positions of the injection mirrors and the focusing mirrors.

With the profiles of T_e and n_e as shown in Fig. 7, the driven current is optimized with the different positions of injection mirrors and focusing mirrors. Combining the size and shape of the CFETR equatorial port, Ra is chosen as 1000cm and Za is ±22.5cm or ±67.5cm ("+" is higher than the center of CFETR horizontal plane and "-" is lower than the center of CFETR horizontal plane). The calculating configuration is ITER-like configuration which is not symmetrical about the center horizontal plane of CFETR, so the different driven current of four height positions of injection mirrors ((a)-67.5mm, (b)-22.5mm, (c)22.5mm, (d)67.5mm) are given in Fig. 8. The driven currents are gotten with toroidal angles and poloidal angles changing in a certain range. The maximum driven currents in four different positions are all about 29 kA when the injecting power is 1MW. The optimal toroidal injecting angles are all about 206° and the optimal poloidal injecting angles are about 83°, 88°, 94° and 98° respectively.

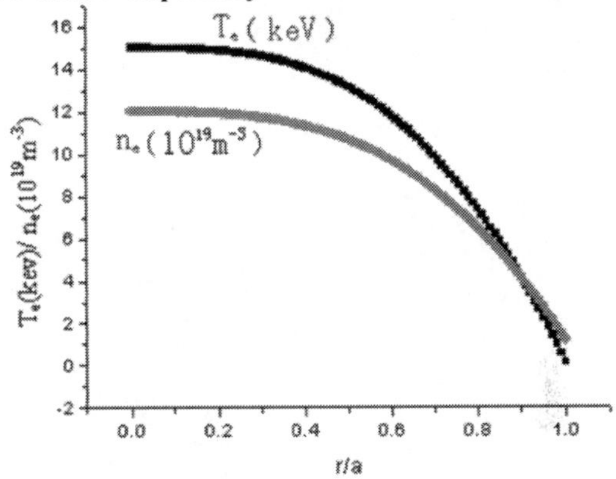

Fig. 7. The profiles of T_e and n_e.

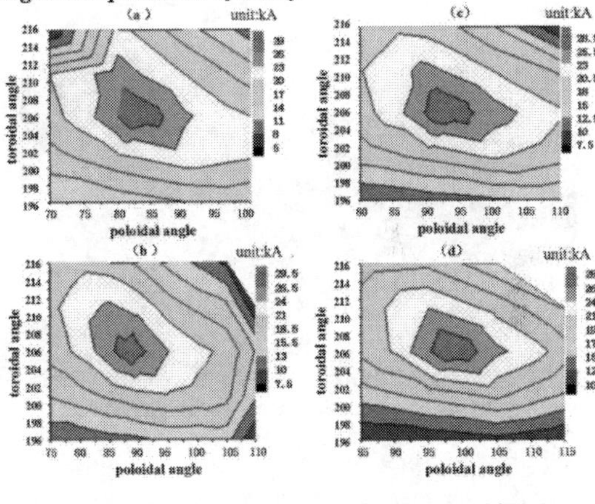

Fig. 8. In four heights ((a)-67.5mm, (b)-22.5mm, (c)22.5mm, (d)67.5mm) of the injection mirrors, the driven current vs different toroidal angles and poloidal angles.

EL is made up of in-vessel transmission lines (in-vessel transmission lines and ex-vessel transmission lines are separated by external surface of the flange), vacuum shielding, internal shield, focusing mirror, steering mirror, drive mechanisms of steering mirror, Blanket Shielding Module (BSM) [24], flange and so on (Fig. 9 and Fig. 10). The EL is installed on the CFETR equatorial port with port-plug modular structure. The in-vessel transmission lines are extended from transmission lines, restrain the microwave power and transfer the power with certain route. The optical mirrors contain focusing mirrors and steering mirrors which are in charge of changing the transferring route of gaussian beams. The vacuum shielding separates the high vacuum of CFETR chamber from external environment. One side of vacuum shielding connecting with tokamak has the same vacuum with the vacuum chamber. The other side is in the atmosphere facilitating the installing and repairing microwave devices, and controlling electric motors which are the driving mechanisms of steering mirrors. The internal shield and BSM will both shield the neutron radiation in the physics experiments in the future. The flange takes charge of the installation and jointing between EL and port.

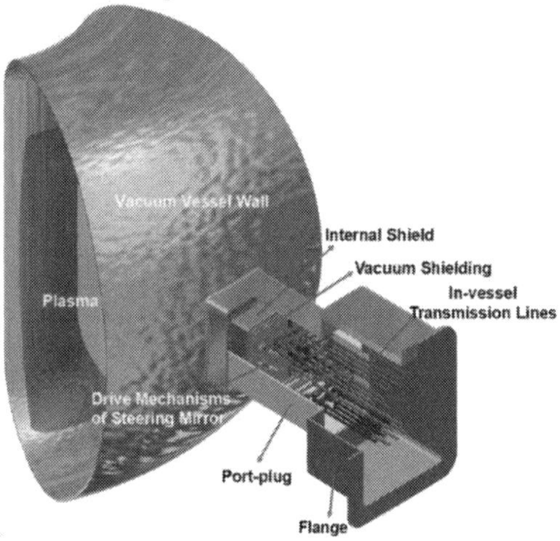

Fig. 9. The overall installation of the equatorial launcher.

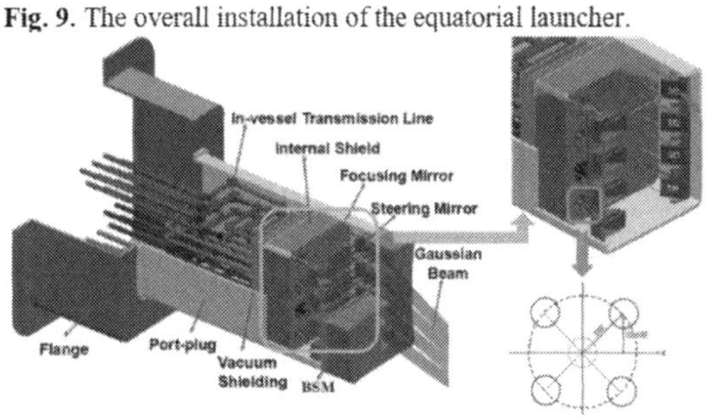

Fig. 10. The internal structure of equatorial launcher and the transferring route of gaussian beams.

Twenty in-vessel transmission lines are divided into four groups in EL and each group has five transmission lines. The top and the bottom of groups are arranged by four miter bends in every transmission line and the middle two groups are arranged by two miter bends. Miter bends increase the transmission loss, however, in order to converge the power from transmission lines into focusing mirrors, these are the integrant minimum quantities of bends. There are four focusing mirrors and four steering mirrors in ECH system. Five waveguide ends of the in-vessel transmission lines in each group face one fixed focusing mirror and five gaussian beams from the waveguide ends are injected into this focusing mirror, reflected to one steering mirror and then injected into the plasma of CFETR finally. The optical transmission routes are shown in the top right of Fig. 10. Four steering mirrors are fixed in poloidal direction and can steer in toroidal direction within a certain range of angles. The waveguide ends of each group are shown in the bottom right of Fig. 10. Four waveguide ends arrange around the middle end and they distribute symmetrically about the center. The distance between every ambient waveguide end and the center waveguide end is 102mm. The detailed designs of focusing and steering mirrors and the optical transmission characteristics of gaussian beams are optimized in Fig. 11. The distance from waveguide end to focusing mirror, from focusing mirror to steering mirror, and from steering mirror to plasma center are 500mm, 1000mm and 4300mm, respectively. When the focal length of focusing mirror is 7500mm, five gaussian beams are converged in plasma center shown in Fig. 11(a). The power density distributed in plasma center is shown in Fig. 11(b) with maximum power density 126.4MW/m^2.

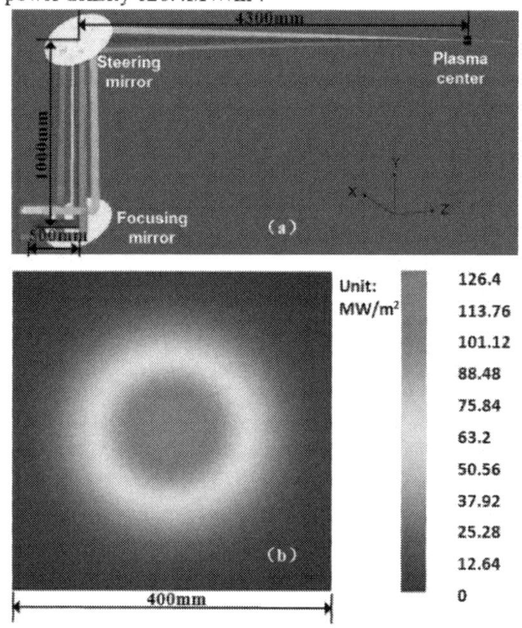

Fig. 11. The transferring process of five gaussian beams. (a)the transmission trace of gaussian beams; (b)the distribution of power density in plasma center.

4 Upper launcher

There are only eight equatorial ports in CFETR and maybe none of them will be allocated to the ECH system. In that case, whether the microwave power can be injected from upper port is very important and should be proved. The driven current is also calculated by C3PO/LUKE code.

The upper port is narrow and long compared with the equatorial port (Fig.12). The length of upper port in vertical direction is 13.29m and the horizontal cross-section of upper port is similar to a trapezoid whose wide length and narrow length are about 1772mm and 416mm respectively and whose height is about 3609mm. The narrow and long shape and irregular cross-section of the upper port bring great challenge to design the UL in the ECH system.

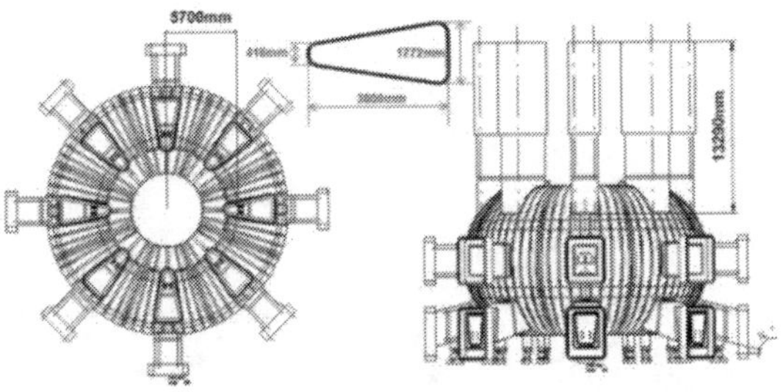

Fig. 12. The shape and size of the upper port.

On account of the special shape and size of the upper port, the injection position of the UL is pre-estimated before physics calculation. The horizontal distance Ra and the vertical distance Za (defined same as Fig. 6) are chosen 7.919m and 4.787m, respectively. The driven current is calculated and optimized in Fig. 13. The total driven current is 1.46MA and the coverage area of the current is about ρ~0.1-0.8 when 18MW microwave power is supposed to inject into the plasma. The adjustable ranges of the toroidal angles and poloidal angles are $\varphi \sim 210°$- $230°$ and $\theta \sim 135°$-$145°$.

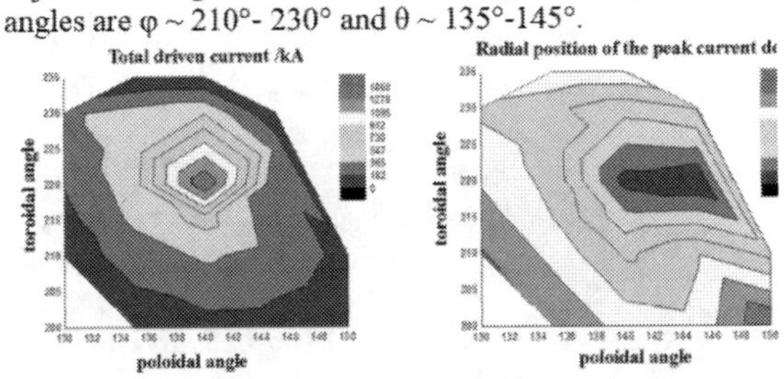

Fig. 13. Total driven current and radial position of the peak current density vs toroidal angles and poloidal angles.

During the UL design, twenty-one gyrotrons and transmission lines are supposed. The arrangement principle of the RF sources and transmissions lines is similar to the Fig. 3 and not given here. The UL forms one port-plug modular structure made up of multiple microwave transmission components. It contains twenty-one in-vessel corrugated waveguides, miter bends, optical mirrors and so on. Twenty-one gaussian beams from in-vessel transmission lines are divided into three groups, injected to three fixed focusing mirrors, reflected from one common steering mirror and injected into the plasma at last. The arrangement of twenty-one transmission lines located in the flange is shown in the top left of Fig.14. The distance between top and bottom groups is 400mm and the left group goes to the center of top and bottom groups is 350mm. In each group, ambient six transmission lines distribute surrounding one center transmission line equably and the distance between every two neighboring transmission lines is 100mm. From the right of Fig.14, the port-plug modular structure is composed of three sections which will bring the convenience of hoisting and installation in the future engineering construction. The upper section is mainly flange and twenty-one in-vessel transmission lines. The flange is in charge of the docking duty with the upper port. The middle section is constituted by the vacuum shielding, corrugated waveguides and miter bends. The bottom section primarily contains optical mirrors as well as their affiliated structure. Every seven gaussian beams from each group of transmission lines inject into the correspondent focusing mirror. The design of the optical mirrors and the transmission characteristics of gaussian beams are also optimized and the power distributions of all beams deposited in steering mirror and plasma center are shown in Fig.15. The focal length of focusing mirror is 10m.

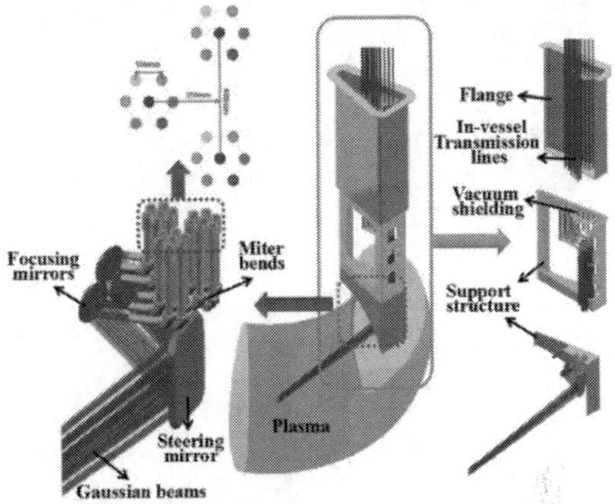

Fig. 14. The design structure of the upper launcher.

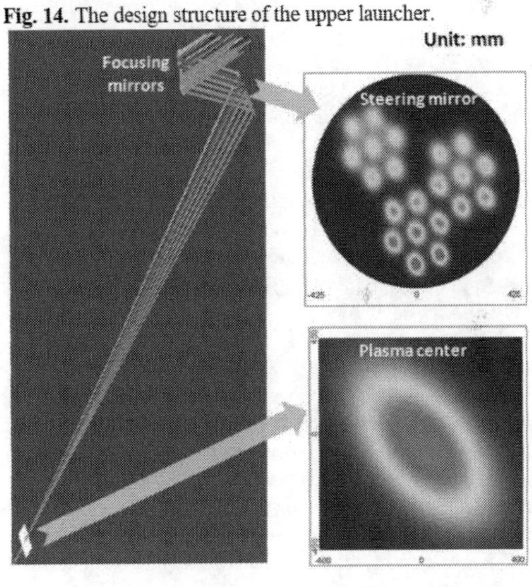

Fig. 15. The transmission characteristics of gaussian beams is analyzed. All guassian beams are reflecting to the steering mirror whose radius is 425mm and injected into plasma with spot length 800mm.

The UL is installed on the upper port as shown in Fig. 16. Total gaussian beams are reflected from the steering mirror, transfer through BSM and then are injected into plasma. From above physics calculation, the optimal toroidal angle is 220°, which means the total beams will steer 40° in toroidal direction. Oversize steering angle in toroidal direction is dangerous, because the powerful beams will be likely to burn the port edge. In order to prevent this case happening, the steering mirror is not symmetrical as three focusing mirrors and is moved 150mm towards the left in the opposite direction of the gaussian beams. The UL design is relatively simple so far. More detailed design and analysis will be carried on.

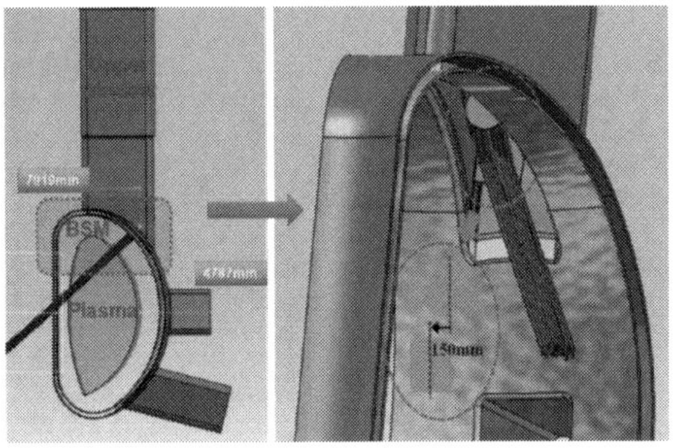

Fig. 16. The installation effect of the upper launcher.

5 Conclusions

According to the port position of CFETR tokamak, two types of injection launcher——EL and UL are designed with quasi-optics transferring pattern. The launcher is installed on equatorial port or upper port with port-plug modular structure. In the detailed research of the EL and UL, the injecting position and injecting angles of steering mirror are calculated by C3PO/LUKE code. Then according to the transmission characteristic of gaussian beams, the transmission routes and converging of all the wave beams are optimized by optical simulation software. After optimizing, the deploy of transmission lines in the launcher and the information of the optical mirrors are gotten. At last, the overall design of the EL and UL, including focusing mirror, steering mirror, driving structure of the steering mirror, in-vessel transmission lines, blanket shield module, internal shield module, vacuum shielding, supporting structure, flange and so on, are all explored.

Compared with two launchers, each launcher has its advantages and disadvantages. On the driven current, UL can get about 81kA driven current when 1MW microwave power is injected into the plasma, while EL can only get 29kA. The driven current of UL is better than the EL. On the engineering design, because of the major size and regular shape of the equatorial port, the layout of the microwave components in the EL is not as compact as the UL and more than one steering mirrors can be adopted in the EL. The difficulty on EL engineering design is lower than UL.

The present design of EL and UL is based on the initial small size CFETR, more research and design of ECRH launchers with respect to the present CFETR parameters are on-going.

Acknowledgement

This work has been supported by the National Key R&D Program of China under Grant No. 2017YFE0300200 and the Strategic Priority Research Program of Chinese Academy of Sciences under Grant No. XDA14040500.

References

1. R. Aymar, P. Barabaschi and Y. Shimomura, Plasma Phys. Control. Fusion **44**, 5, 519-565 (2002)
2. Y. T. Song et al., IEEE Trans. Plasma Sci. **42**, 3, 503-509 (2014)
3. B. N. Wan et al., IEEE Trans. Plasma Sci. **42**, 3, 495-502 (2014)
4. X. J. Wang et al., Fusion Eng. Des. **96-97**, 181-186 (2015)
5. H. D. XU et al., Plasma Sci. Technol. **18**, 4, 442-448 (2016)
6. Jiangang Li, et al., EPJ Web Conf. **149**, 01011 (2017)
7. Y. Y. Tang et al., Fusion Eng. Des. **94**, 48-53 (2015)
8. K. Sakamoto et al., Nucl. Fusion **43**, 729-737 (2003)
9. P. C. Kalaria, M. V. Kartikeyan and M. Thumm, IEEE Trans. Plasma Sci. **42**, 1522-1528 (2014)
10. R. K. Dhakad, G. S. Baghel, M. V. Kartikeyan and M. K. Thumm, IEEE Trans. Plasma Sci. **43**, 391-397 (2015)
11. B. Piosczyk et al., IEEE Trans. Plasma Sci. **32**, 413-417 (2003)
12. W. Wei et al., Chin. Phys. B **23**, 5 (2014)
13. W. Wei et al., Nuclear Fusion and Plasma Physics, **33**, 2, 147-154 (2013)
14. T. Omori, et al., Fusion Eng. Des. **86**, 951-954 (2011)
15. K. Takahashi, et al., AIP Conference Proceedings **1580**, 546-549 (2014)
16. K. Kajiwara, et al., Fusion Eng. Des. **89**, 6-10 (2014)
17. D. Strauss, et al., Fusion Eng. Des. **89**, 1669-1673 (2014)
18. D. J. Wu et al., IEEE Access, **5**, 6187-6191 (2017)
19. D. J. Wu et al., J. Fusion Energy, **33**, 634-639 (2014)
20. R. A. Olstad, J. L. Doane and C. P. Moeller, Fusion Eng. Des. **74**, 331-335 (2005)
21. R. W. Callis et al., Fusion Eng. Des. **84**, 526-529 (2009)
22. F. Gandini et al., Fusion Sci. Technol. **59**, 709-717 (2011)
23. W. Wei et al., Chin. Phys. B **25**, 15201 (2016)
24. P. Spaeh et al., Fusion Eng. Des. **89**, 960-964 (2014)

Plans for the electron cyclotron heating system on J-TEXT

D. H. Xia, C. H. Liu, Y. K. Jin, H. Y. Ma, Y. Z. Tian, Z. J. Wang

International Joint Research Laboratory of Magnetic Confinement Fusion and Plasma Physics, Huazhong University of Science and Technology, Wuhan, 430074, China

Abstract. A new 105 GHz/500 kW/1 s electron cyclotron resonance heating (ECRH) system has been designed and being constructed on J-TEXT Tokamak. This system mainly consists of a microwave source, a transmission line, a launcher and other auxiliary units. Based on corrugated waveguides, the wave from the gyrotron can be efficiently transmitted with HE_{11} mode to the steerable quasi-optical launcher for injection. The transmission efficiency is about 85%, and the injection angle of the wave can be adjusted by the flat mirror of the launcher. Commissioning of this electron cyclotron heating system is scheduled to be done at 2019.

1. Introduction

Electron cyclotron resonance heating (ECRH), recognized as a kind of efficiently auxiliary heating method, has been widely used on many tokamaks and stellarators [1-6]. Also, ECRH is proved to be helpful for current profile control, MHD stability, transport study, etc [7]. To fulfill the requirements of heating power on J-TEXT tokamak, an ECRH system with one 105 GHz/500 kW/1s gyrotron is being built.

The ECRH system under design mainly consists of a high power microwave source, a transmission line based on corrugated waveguide, and a launcher connected to J-TEXT tokamak. The tokamak has a major radius of 1.05m and a minor radius of 0.25~0.29m. The center-line toroidal field for normal discharge is about 2T. Considering the parameters above and requirements of our experiments, a 105GHz/500kW/1s gyrotron has been chosen as the microwave source. A transmission line is constructed based on circular corrugated waveguides to transmit electron cyclotron (EC) waves from gyrotron to the port of the tokamak with HE_{11} mode, which is proved to be an efficient way in reducing the loss of power. The launcher is designed according to the expected poloidal and toroidal injection angles and actual limitation of space.

This paper consists of following parts. The details of microwave source are given in Section 2. Section 3 introduces the design of the transmission line. Section 4 gives the parameters of launcher. Conclusions are shown in Section 5.

2. Microwave source

As the high power microwave source of ECRH system, the gyrotron delivers EC waves of megawatt level power at frequency varied from 28~170 GHz. For ECRH system, operating parameters of the whole system are mainly determined by parameters of the gyrotron. According to the theory of EC wave propagation and absorption in plasma, the frequency of gyrotron f is determined by the toroidal magnetic field B_T, as follows [8]:

$$f = nf_{ce} = \frac{eB_T}{2\pi m_e} \approx 28nB_T GH_Z \qquad (1)$$

Where f_{ce} is the electron cyclotron resonance frequency, e is the electron charge, m_e is the electron mass, and n is the number of the harmonic.

For the J-TEXT tokamak, the maximum toroidal magnetic field is 2.4 T at present [9]. Considering the needs of physics experiment and values, the 105 GHz/500 kW/1 s gyrotron, which is manufactured by GYCOM, has been chosen for J-TEXT tokamak. The parameters of this type gyrotron is shown in Table 1. When the EC wave is injected from the low field side of the J-TEXT port 4, the second harmonic of extraordinary mode (X2) will be absorbed at the toroidal magnetic field of about 1.88 T.

Table 1. Parameters of gyrotron.

Parameters	**Values**
Type	Depressed collector
Frequency	105 GHz
Output power	500 kW
Pulse length	1 s
Main output mode	TEM_{00} mode
Efficiency	45%
Output window	BN window

Table 2. Parameters of power supplies.

Parameters	**Values (max.)**
Cathode voltage	55 kV
Accuracy of cathode voltage	0.3 kV
Beam current	25A
Body voltage	27 kV
Accuracy of body voltage	0.5 kV
Body current	150mA
Voltage stability	1%
Filament voltage	30 V
Filament current	30 A
Ion pump power supply	3~4 kV

For this type 105 GHz/500 kW/1 s gyrotron, the power supplies consist of two high voltage power supplies (cathode and body), a filament power supply (PS), a super conducting magnet (SCM) PS and an ion pump PS. The main parameters and their maximum of these PS are shown in Table 2.

Table 3. Parameters of water cooling.

Cooling	**Pressure**	**Flow**	**Others**
Collector	< 0.5 Mpa	> 5 L/s	Temperature ≤ 25°C Resistance ≥ 1MΩ•cm
Cavity	< 0.5 Mpa	> 0.2 L/s	
Window	< 0.5 Mpa	> 0.3 L/s	
Body	< 0.5 Mpa	> 0.2 L/s	
MOU	< 0.5 Mpa	> 0.3 L/s	

Besides, the gyrotron needs water cooling for its collector, cavity, output window, body and matching optics unit (MOU). The main parameters of the water cooling system are shown in Table 3. The water resistance is not less 1 MΩ•cm, and the inlet temperature of the coolant is near or below 25°C. The maximum inlet temperature is 35°C.

3. Transmission line

A transmission line (TL) is responsible for transmitting the EC waves from gyrotron to launcher. The TL must meet the following requirements: high power capability, low attenuation, low probability of arcing, low difficulty in measuring the power of EC waves.

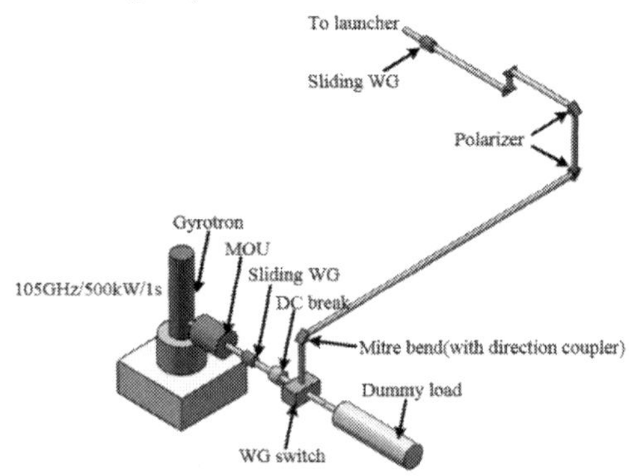

Fig. 1. Layout of a transmission line.

Layout of the TL is shown in Fig. 1 and the composition of the TL is given in Table 4. Since the power is not very high, the transmission line is under atmospheric conditions. The TL is equipped with several corrugated waveguides, two sliding waveguides, a waveguide switch, a DC break, several miter bends, two polarizers integrated into two bends separately, a direction coupler integrated into bend, and a dummy load. The inner diameter of the waveguides which are made of aluminum is 63.5mm, and the sinusoidal corrugations with the period of 1mm and the depth of 0.71mm are adopted. The sliding waveguide is used to compensate the thermal expansion. The DC break cuts off the DC connection between gyrotron and J-TEXT tokamak. With the waveguide switch, the EC waves can be switched to the dummy load for power measurement and calibration by calorimetry [10-11]. The miter bend changes the direction of propagation. The two polarizers work together to get the expected polarization of the EC waves. The direction coupler makes it possible to measure the microwave power online.

Table 4. Compositions of a transmission line.

Component	Quantity
MOU	1
Corrugated waveguides	~24m
Mitre bends	5
Polarizer	2
Direction coupler	1
DC break	1
Waveguide switch	1
Dummy load	1
Sliding waveguide	2

4. Launcher

As shown in Section 3, the EC wave is transmitted to the port 4 (233mm × 219mm) on J-TEXT tokamak. A quasi-optical launcher has been designed to inject the beam into plasma with the desired poloidal and toroidal direction. The structure of the designed 105GHz launcher is shown in Fig. 2.

In Fig. 2, the wave is firstly focused by the ellipsoidal mirror, and then it will be reflected to plasma by the steerable flat mirror. The poloidal and toroidal angles of the incident beam can be adjusted with the steerable flat mirror by two separate poloidal and toroidal drivers. The steering mechanism of the flat mirror is shown in Fig. 3. Considering the space limitation of port 4 and the steering mechanism, the poloidal and toroidal rotation angles are ±15°. More details about the design of ellipsoidal mirror and flat mirror can be found in Ref. [12].

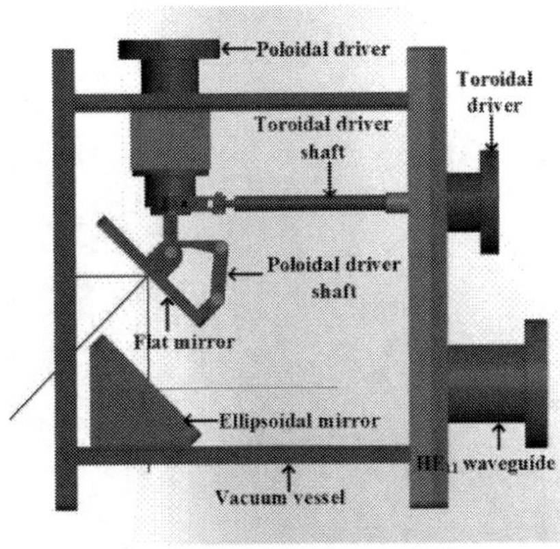

Fig. 2. Structure of the designed 105GHz launcher.

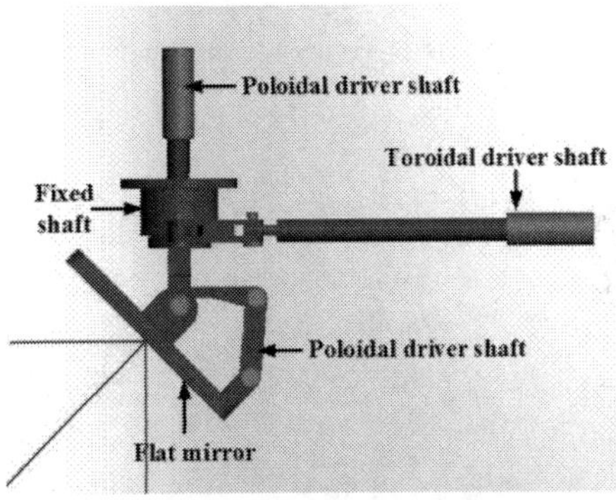

Fig. 3. Steering mechanism of the flat mirror.

Table 5. Parameters of ellipsoidal mirror and flat mirror.

Parameters		Values
Ellipsoidal mirror	F	400 mm
	OA	47 mm
	OB	33 mm
	OE	1.96 mm
Flat mirror	Length	100 mm
	Width	72 mm
	Thickness	10 mm

Considering beam width of the EC wave and space limitation, the calculated parameters of ellipsoidal mirror and flat mirror are given in Table 5, where F is the focal length; OA, OB and OE are the half-length, half-width and depth of the ellipsoidal mirror, respectively. Meanwhile, the length, width and thickness of the flat mirror are abbreviated as L, W and T.

Finally, the beam radius along the propagation path is plotted in Fig. 4, where the EC beam radius is about 14.09 mm at the center of plasma.

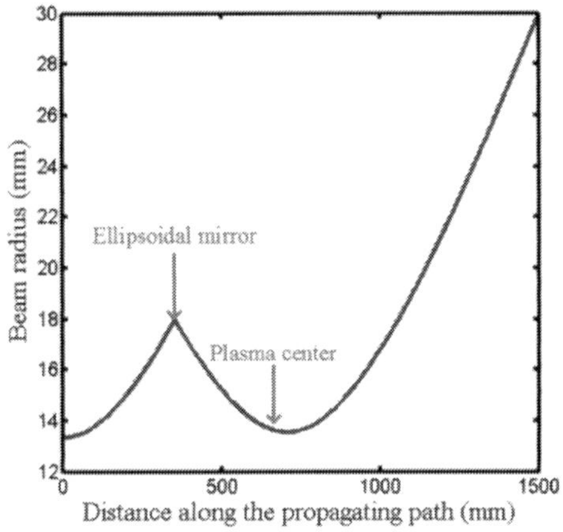

Fig. 4. Beam radius along the propagation path (plasma center, 589mm).

5. Conclusion

To expand the operation range of J-TEXT tokamak, a 105GHz/500kW/1s ECRH system is being developed. The system is based on the Gycom gyrotron, the transmission line with corrugated waveguides, and a quasi-optical launcher. Currently, the design of this system has been finished and all the subunits are on the way. The gyrotron will be delivered at the end of 2018, and the first commissioning of the ECRH system will be done at the middle of 2019. This ECRH system will be mainly used for plasma heating, transport, MHD stability, disruption avoidance, etc. We plan to carry out experiments with the 105 GHz ECRH system at the beginning of 2020.

Acknowledgments

The authors would like to thank the J-TEXT team with valuable discussions about the ECRH system design. This work is supported by the National Natural Science Foundation of China (Grant No. 11605066), and supported partly by National key research and development program (Grant Nos. 2016YFA0400600 and 2016YFA0400603).

References

[1] D. Wagner, J. Stober, F. Leuterer, F. Monaco, S. Müller, M. Münich, et al., J Infrared Millim Te, **37** (2016).

[2] Mirela Cengher, Lohr, Y.A. Gorelov, R. Ellis, E. Kolemen, D. Ponce, et al., IEEE Trans. Plasma Sci. **42** (2014): 1964–1970.

[3] M. Huang, J. Rao, B. Li, J. Zhou, Z.H. Kang, H. Wang, et al., *Proceedings of the 17th Joint Workshop on Electron Cyclotron Emission and Electron Cyclotron Resonance Heating* (2012), 04012.

[4] M. Thumm, P. Brand, H. Braune, et al., IEEE Trans. Plasma Sci., **36**(2008): 341-354.

[5] E. Westerhof, A. Hoekzema, D. Hogeweij, et al., Fusion Sci. Technol., **47**(2005): 278-283.

[6] T. Shimozuma et al., Fusion Sci. Technol., **58**(2010): 530–538

[7] G. Zhuang et al., Nucl Fusion., **53**(2013), 104014.

[8] V. A. Flyagin, A. V. Gaponov, M. I. Petelin and V. K. Yulpatov, IEEE T Microw Theory., **25**(1977): 514-521.

[9] G. Zhuang et al., Nucl Fusion, **57**(2017), 102003.

[10] Shidara H, Nagasaki K, Sakamoto K, et al. Fusion Sci Technol, **45**(2004): 41.

[11] Gorelov I A, Lohr J M, Baity F W, et al. *General Atomic Report*, GA-A24493.

[12] C. H. Liu, D. H. Xia, et al., Fusion Eng Des, **112** (2016): 404-409.

 https://doi.org/10.1051/epjconf/201920304018

In-situ real-time monitoring of spurious modes in HE_{11} transmission lines using multi-hole couplers in miter bends

Alexander Zach[1,*], *Walter* Kasparek[1], *Carsten* Lechte[1], *Burkhard* Plaum[1], *Francesco* Monaco[2], *Harald* Schütz[2], *Jörg* Stober[2], *Hiroshi* Idei[3], and *Thomas* Hirth[4]

[1]Institut für Grenzflächenverfahrenstechnik und Plasmatechnologie, Universität Stuttgart, Pfaffenwaldring 31, D-70569 Stuttgart, Germany
[2]Max-Planck-Institut für Plasmaphysik, Boltzmannstr. 2, D-85748 Garching, Germany
[3]Research Institute for Applied Mechanics, Kyushu University, Kasuga 816-8560, Japan
[4]Karlsruher Institut für Technologie, Kaiserstr. 12, D-76131 Karlsruhe, Germany

Abstract. Transmission of high-power millimeter waves for ECRH is often realised with oversized corrugated circular waveguides. Coupling from the gyrotron source to the waveguide is typically done via matching mirrors in free space. Small alignment errors of the system lead to the excitation of higher-order modes inside the waveguide beside the main transmission mode HE_{11}. Those modes have comparably higher losses and can in worst case result in local fields exceeding the breakdown limit of the medium inside the waveguide.
For alignment control over the whole pulse duration of the gyrotron, a set of hole-array couplers placed into a miter bend mirror probes the field inside the waveguide. The arrays are designed to detect the marker modes for beam offset and tilt ($LP_{11}^{(e/o)}$) as well as for beam waist mismatch (LP_{02}). In addition, a main mode coupler sensitive mostly for the HE_{11} content is used as a power monitor. By maximizing the signal of the power monitor and minimizing the content of marker modes, a first-order optimization of the coupling from free space to the waveguide can be achieved. Signal processing of the 140 GHz information is done at kHz range after downmixing, using a frequency shifted part of the power monitor signal.
As the measurement system is placed in a miter bend mirror, it can also be easily installed at various locations along the transmission line to check for possible misalignments of the waveguide connections between miter bends. Simulation and low power experimental results will be shown.

1 Introduction

An important problem in transmission of high-power millimeter waves for ECRH is to avoid spurious mode excitation in oversized corrugated circular waveguide (OCCWG) transmission lines. These modes have higher losses compared to the main transmission mode HE_{11} and can result in hot spots along the transmission line, which can also have influences on structure cooling demand. There are two main sources for spurious mode excitation that are related to misalignments. One is the coupling from the gyrotron millimeter wave source to the OCCWG, which is typically done with a matching optic consisting of two or more mirrors in free space. The other is mode conversion due to misalignments between OCCWG segments along the transmission line. Mode conversion at miter bends is another main source for spurious modes, but it can not be corrected by a better alignment.

The over 60 m long transmission lines at ASDEX Upgrade have a diameter of 87 mm each and are typically operated with up to 10 s pulses at 140 GHz and 1 MW in atmosphere. They are aligned with a laser system to an accuracy better than 1 mm. As the transmission line is not evacuated, miter bend mirrors can be easily exchanged, which gives the opportunity to place diagnostic systems at those positions. A spurious mode detector integrated in a miter bend mirror was developed to allow for faster, less complicated, and in-situ (re-)alignment. It can be used to confirm the alignment of OCCWG segments between two miter bend positions or to optimise the coupling from the gyrotron to the OCCWG.

Another use of the diagnostic system is as an in-situ real-time monitor of the mode purity in the transmission line, giving more information in comparison to using only power monitors. This covers also possible changes of beam parameters over the pulse duration, especially during the initial frequency chirp of the gyrotron due to thermal expansion of the cavity.

The following chapters describe the used method of mode detection and show results of the low power calibration for a prototype miter bend mirror diagnostic system.

2 Modes in the transmission line

In an OCCWG a couple of thousand modes can propagate. An estimated number in relation to the diameter D and the

*e-mail: zach@igvp.uni-stuttgart.de

free-space wavenumber k_0 is given by [1]:

$$(Dk_0)^2/8 \tag{1}$$

Linear polarized modes as described by Kowalski et al. [2] are used in this paper, as they directly relate to the most common misalignments.

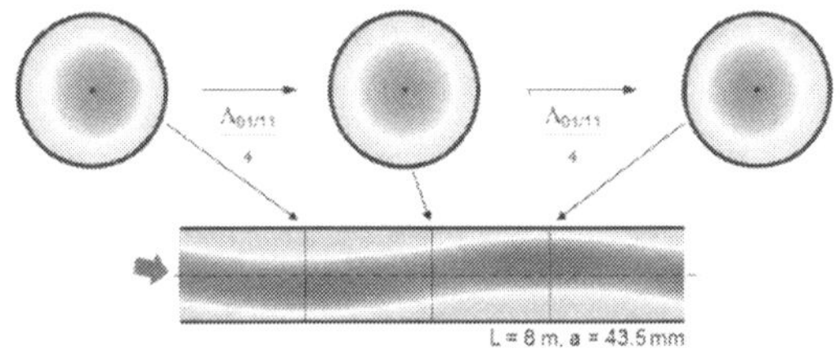

Figure 1. Resutling field in the OCCWG over a length of 8 m for a tilted input beam (0.15° tilted J_0 resulting in 98.8 % LP_{01} and 1.2 % LP_{11}).

The main transmission mode LP_{01} (or HE_{11}) propagates together with excited spurious modes forming an interference pattern inside the OCCWG. Fig. 1 shows such a pattern for an initially tilted beam composed of LP_{01} and LP_{11}. As the modes propagate with different wavenumbers

$$k_{mn} = k_0\sqrt{1 - 4X_{mn}^2/(k_0D)^2} \tag{2}$$

where X_{mn} is the nth root of the mth Besselfunction. The interference pattern after a quarter of the beat-wavelength corresponds to the maximum centroid offset.

Detection of the full spectrum of modes in an OCCWG would require information about amplitude and phase of the field over the whole cross section. But each kind of misalignment excites dominantly only a limited set of modes. In this paper they are hereinafter also called marker modes for that kind of misalignment. So a reduced mode spectrum for the detection of spurious modes due to misalignment can be assumed. The perturbations due to unconsidered modes scale with the residual power in those modes. The most important marker modes are shown in Fig. 2.

The LP_{11} modes are excited due to an offset or tilted input beam or such misalignments between OCCWG segments. For a gaussian beam input, also a minor amount of LP_{02} is excited. However, for small misalignments this effect is negligible and with corresponding small truncation losses the scaling of LP_{11} excitation is similar for offset and tilt as can be seen in Fig. 3 a). In the case of a mismatch between beam radius and waveguide diameter or a curved phase front at the OCCWG entrance the LP_{02} and LP_{03} modes are dominantly excited. For example, one can see the effect in low power 2D scans used for calibration such as in Fig. 3 b) and c). The field with 5 % LP_{02} content was created by moving the beam waist farther away from the input, thus creating a beam waist mismatch and a non-planar phasefront at the OCCWG input. In both settings some small amounts of other spurious modes are also present, as the generation of fields for the low power calibration was not done independently for each mode.

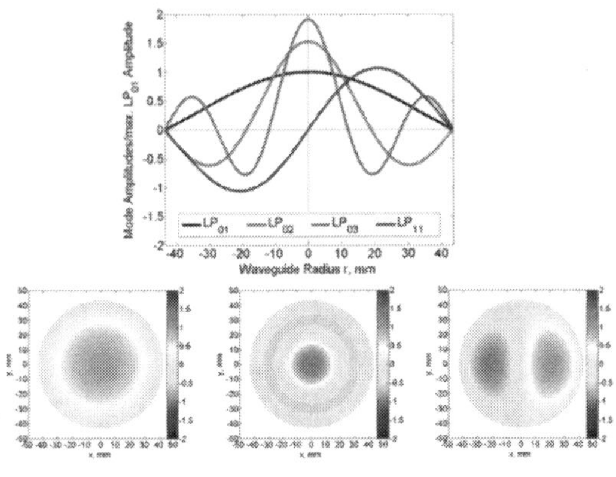

Figure 2. Mode amplitudes in the OCCWG cross section at y = 0 (top) and 2D amplitude distribution of main transmission mode LP_{01} (bottom left) and marker modes for LP_{02} (bottom mid, beam waist mismatch) and LP_{11}^{o} (bottom right, offset/tilt). Even first mode indices (m) relate to symmetric, odd first indices to asymmetric modes.

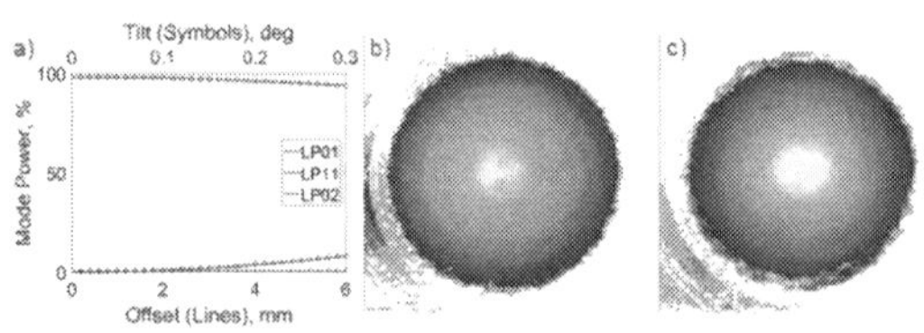

Figure 3. a) Mode content of LP_{01}, LP_{02}, and the dominantly excited spurious mode LP_{11} for an either tilted (symbols) or offset (lines) gaussian input beam. For negligible beam truncation, the values can also be related to the points of maximum tilt or offset in Fig. 1. b) Linear amplitude of 2D field scan with approx. 0.5 % LP_{02} content. c) Linear amplitude of 2D field scan with approx. 5 % LP_{02} content.

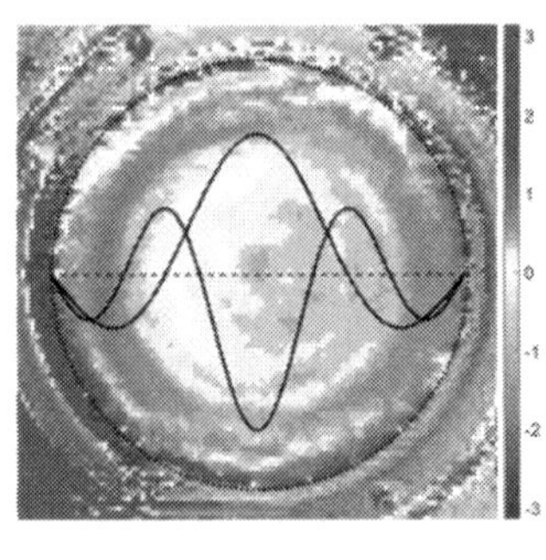

Figure 4. Phase of 2D field scan with 0.6 % LP_{02}, 0.4 % LP_{03}, and 2.7 % LP_{11}^{o} content (enhanced color scheme).

For strong misalignments, one can see the mode structure directly in the phase. In the phase plot in Fig. 4 one can see a ring of different phase that corresponds to the LP_{02} and LP_{03} modes, which have a phase offset to the main transmission mode LP_{01} of 36° and -66° respectively. A phase tilt between left and right which corresponds to

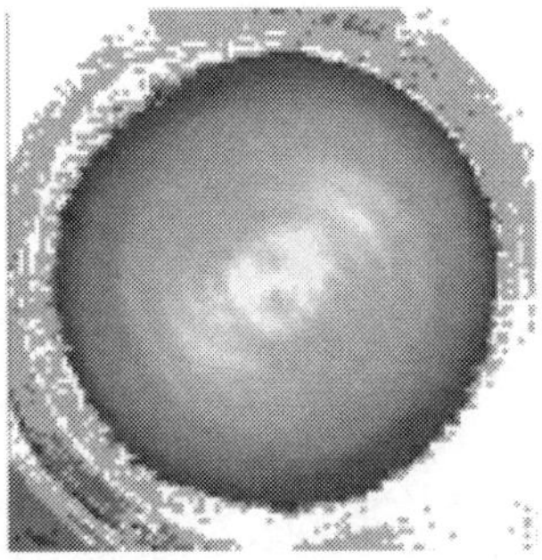

Figure 5. Linear amplitude of 2D field scan with approx. 1 % LP_{21}^{e} content (scaling factor $\sqrt{2}$ compared to Fig. 3).

the LP_{11}^{o} mode can also be seen. The field in Fig. 5 shows the result of an astigmatic input beam. The correspondent excited modes are the LP_{21} modes.

3 Detection of spurious modes in the transmission line

The field in the OCCWG cross section is sampled with hole couplers, which are integrated into a miter bend mirror. At 140 GHz the holes are in cut-off. This results in signal damping by about 60 dB. The signals are coupled to fundamental mode waveguides beneath the mirror surface as illustrated in Fig. 6. The detected signal of one channel is composed of coupling from several holes to the same waveguide, as

$$S = \sum_{H} A_H C_H e^{-ik_{mn}d_H} e^{-ik_g z_H} \tag{3}$$

where A_H is the field amplitude at the hole, C_H is the hole coupling factor, $e^{-ik_{mn}d_H}$ is the factor accounting for the propagation in front of the mirror surface and $e^{-ik_g z_H}$ for the propagation inside the fundamental mode waveguide from the hole to the detector. Directivity is achieved when hole signals coherently add up due to the phase velocity inside the fundamental mode waveguide being matched to the free space propagation in front of the mirror surface. This is done by matching the Brillouin angle of the fundamental mode waveguide (for reference see [1]) to the angle of incidence ($\theta = 45°$)

$$\theta = \cos^{-1}\left(\sqrt{1-\left(\frac{\lambda_0}{2a}\right)^2}\right). \tag{4}$$

Additionally, a correction of the waveguide depth a must be made for H-plane couplers (coupling through narrow waveguide wall) to account for a broadening effect in the fundamental mode waveguide due to the holes. This type of couplers has been used as power monitors for many years [3]. They are also compatible with water cooling channels integrated into the same mirror.

3.1 Interferometric detection

For a high power test, opposite couplers in a cross-like configuration were connected in a simple interferometric

Figure 6. Schematic of the coupling from OCCWG cross section to the fundamental mode waveguide behind the mirror surface. The waveguide depth is matched for the single hole contributions to coherently add up.

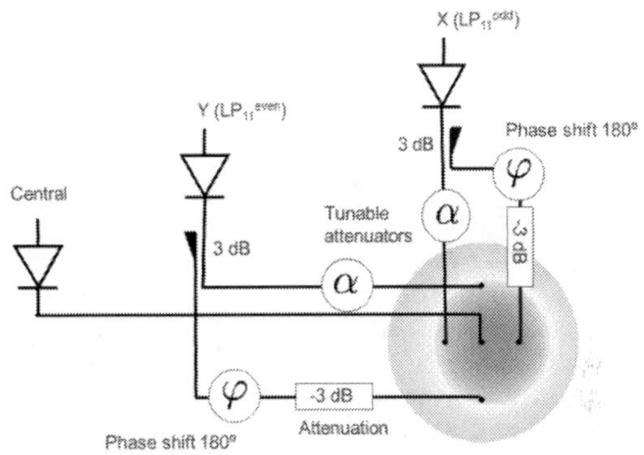

Figure 7. Schematic of an interferometric setup to measure direct error signals corresponding to the LP_{11} modes from coupled channels.

setup (see Fig. 7). If the connected channels are substracted, they detect the LP_{11} modes as the contributions of the symmetric modes (LP_{0n}, LP_{21}) cancel each other out.

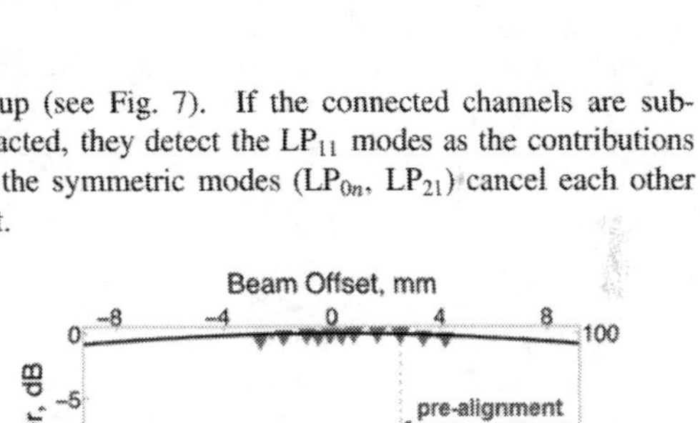

Figure 8. Measured signals and theoretical predictions for a simple interferometer setup. Beam offset and angle are coupled in this measurement.

The measurements were done in the ECRH-2 installation of ASDEX Upgrade. The miter bend with the couplers was installed instead of an already existing bend in the waveguide transmission line. The spurious modes were generated by tilting only one of the matching optic mirrors. That means the possible improvement of the

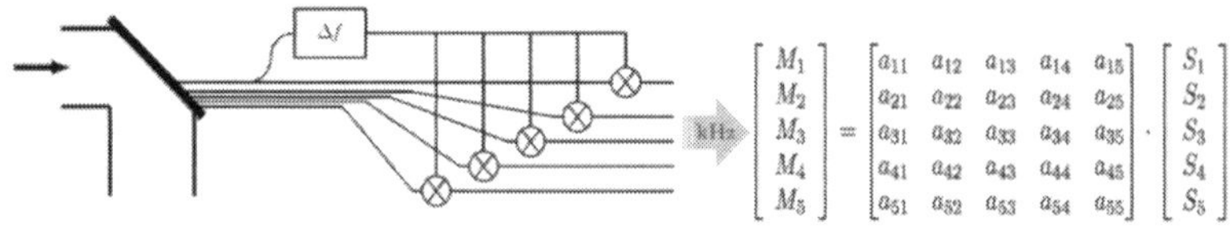

Figure 9. Schematic for the analysis of individual coupler signals

alignment was limited by the coupled tilt and offset of the input beam. The gyrotron power was 0.5 MW with pulse lengths up to 400 ms. Fig. 8 shows the measured signals along with theoretical predictions. A good agreement to the maximum signal from the center channel can be seen. Before the start of the measurements there was just a rough pre-alignment wih an LCD-target, so it does not indicate the accuracy of the mentioned laser adjustment at ASDEX Upgrade (see Sect. 1).

However, such an interferometric system gives no direct information wether the input beam is tilted or offset. Fine-alignment would have to be done interatively. For the detection of other modes (e.g., LP_{02}) an interferometric setup becomes very complicated and difficult to calibrate. Another way is to directly measure amplitude and phase of the single channels.

3.2 Analysis of Individual Coupler Signals

By implementing a heterodyne signal processing, the operating frequency of 140 GHz can be downmixed to a few kHz. The signals can then be directly measured with an oscilloscope and the mode powers can be calculated via a coefficient matrix obtained from the calibration. The schematic in Fig. 9 illustrates the realised method. One part of one channel signal is frequency shifted by reflection on a rotating grating and used as a local oscillator. The local oscillator signal is then split and coupled to the other signals and back to itself. After mixing at detection diodes, one obtains signals with the intermediate frequency in the kHz range. The phase of the downmixed signals is preserved.

4 Multiport Coupler Prototype

A prototype multiport coupler was designed and manufactured with some specification goals for spurious mode monitoring. It should directly detect the LP_{01} content on one of the channels by suppressing most of the marker modes for misalignment. The LP_{11} modes should be measured to determine offset and tilt of the input beam and the LP_{02} and LP_{21} modes should be measured. The system should also work with typical polarizations and have a bandwidth of at least 400 MHz.

The maximum hole diameter was set to $0.355\lambda_0$, because with larger hole diameters of $0.383\lambda_0$ arcing was observed at 0.6 MW during the high power test with the interferometric setup (see Sect. 3.1). Power monitors with $0.346\lambda_0$ hole diameters at the center of the OCCWG cross section are already successfully used at full power. The slightly larger maximum hole diameter is uncritical, as there are no such holes near the center, where the highest field strenghts are found. However the overall smaller holes for full power capability come at the cost of dynamic range for the low power calibration.

The manufactured prototye multiport coupler is shown in Fig. 10. It has three types of hole arrays. A main mode coupler accross the middle, two cross polarization couplers and four diagonal couplers. The symmetric layout was chosen to be able to use the multiport coupler in both directions in case of one direction giving better performance due to manufacturing tolerances.

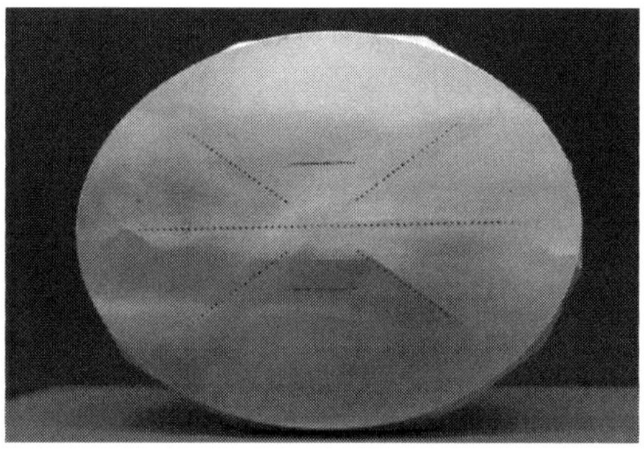

Figure 10. Coupling structures of a multiport coupler prototype.

4.1 Main mode coupler

The main mode coupler is designed to suppress all other LP_{0n} modes as well as all asymmetric modes (e.g., LP_{11}). Fig. 11 shows the suppression against the main mode for the coupler design. One has to keep in mind, that additionally the power in the spurious modes is typically at least 20 dB lower than in the main mode. Also one can see, that the suppression of the asymmetric modes is not as good, as for the LP_{0n} modes. This becomes clear, when looking at the coupling of the hole array in Fig. 12.

The coupling factors of the single holes are given by

$$C_H = r \cdot A(LP_{01}(r)) \tag{5}$$

where r is the distance from the cross section center and A is the amplitude of the LP_{01} mode at that position. The LP_{0n} mode coupling is suppressed along each side of the array, whereas the LP_{11}^{o} mode is strongly coupled on both the righthand and lefthand side and the suppresion

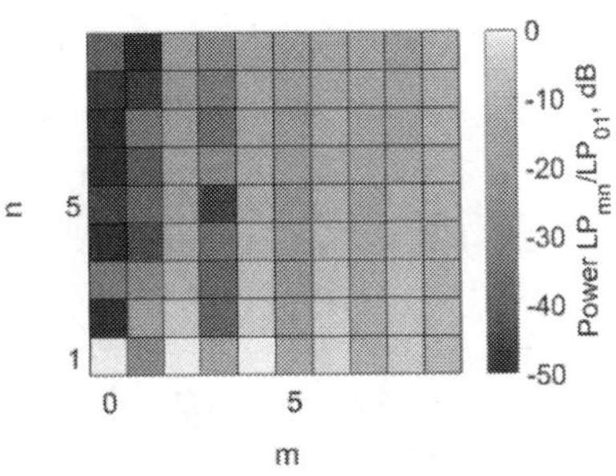

Figure 11. Calculated coupling suppression of spurious modes by the main mode coupler.

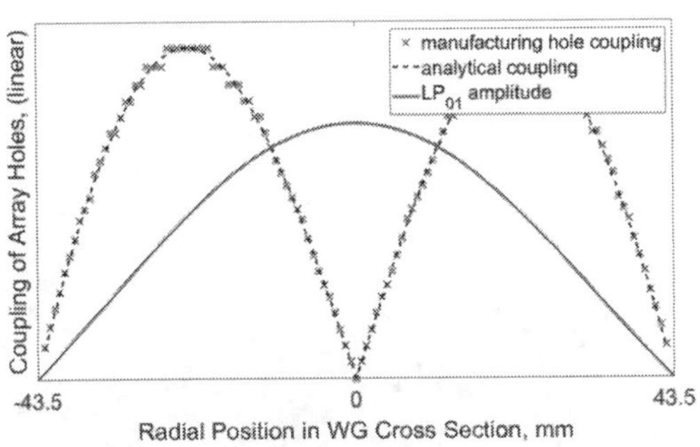

Figure 12. Coupling structure of main mode coupler.

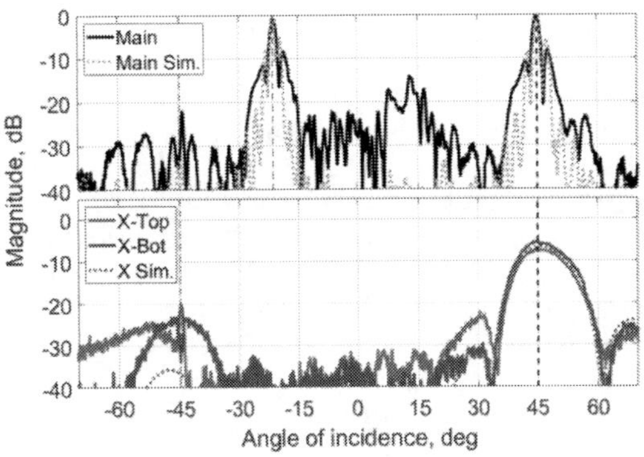

Figure 13. Radiation angle measurements and simulation for the main mode and the cross polarization couplers. At -45 degrees one can see the reflection of the main lobe due to non-perfect absorbtion at the other waveguide port. The second main lobe of the main mode coupler is related to the LP_{0n} suppression.

results from the 180° phase difference between the two sides. However, one of the both signal parts has to also propagate through the other half of the array, being attenuated more and partly reflected. A compensation for this effect was neglected in order to keep symmetry and due to the manufacturing constraint of discrete hole diameters. The symmetric LP_{21} mode is very strongly coupled by the main mode coupler and has to be accounted for.

4.2 Cross polarization couplers

The cross polarization couplers are tapered E-plane couplers (coupling through broad waveguide wall) set at the position of the LP_{02} phase jump. Because for E-plane couplers the position of the hole in the fundamental mode waveguide wall has to be more exact than for H-plane couplers, no special mode sensitive coupling was attempted.

Fig. 13 shows a nice agreement between radiation angle measurements for which the prototype was placed on a rotating table and simulations of the individual coupling arrays with CST Microwave Studio [4] for both the main mode coupler and the cross polarization couplers. The strong response of the main mode coupler between the main peaks is related to reflections in the measurement setup.

4.3 Diagonal Couplers

The diagonal couplers utilize both waveguide ports by coupling in the forward as well as in the backward direction. The coupling structure is designed to have a high mode selectivity for LP_{01} and LP_{02} towards opposite ports. Fig. 14 shows the absolute amplitudes of these modes. The waveguide depth is chosen that all hole signals coherently add up in the forward direction (see Sect. 3). By having an equal total coupling for the LP_{02} on both sides of its phase jump, this mode is suppressed in the forward direction. Backwards coupling for the LP_{02} is achieved by a distance of half the waveguide wavelength between the holes and an additional quarter wavelength at the phase jump position.

The LP_{01} mode coupling in the backwards direction is suppressed by having also equal total coupling on both sides of the phase jump position for this mode. Because such arrays have a high reflection for propagating waves in the fundamental mode waveguide, the coupling was also optimised to behave similar for both orientations of the array towards the field in the OCCWG (wether the first hole to couple is near the middle or on the outside of the cross section).

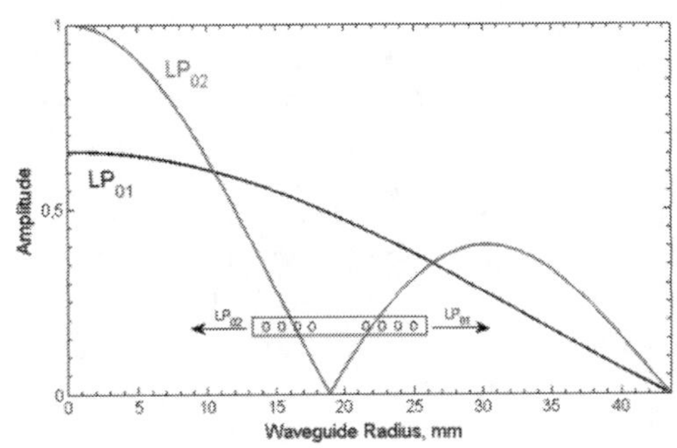

Figure 14. Absolute amplitude of the LP_{01} and LP_{02} mode showing the possibility of mode sensitive coupling in opposite directions inside the fundamental mode waveguide.

https://doi.org/10.1051/epjconf/201920304018

When looking at the transmission characteristics through the fundamental mode waveguides of the diagonal couplers, one can see in Fig. 15 a shift by approximately 2.2 GHz compared to the simulation. It is not yet fully understood, but the most probable error is the manufactured waveguide depth being 1.269 mm instead of 1.296 mm. It was maybe manually set in the machining program after being changed from 1.311 mm in the drawing. However, a remaining low mode selectivity still allows for full diagnostic capability.

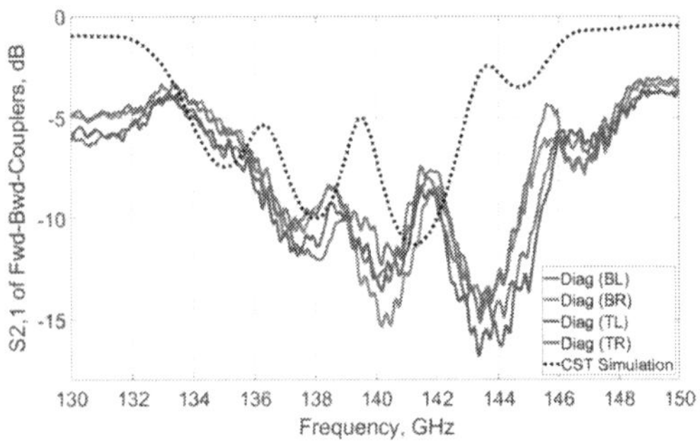

Figure 15. Transmission through the fundamental mode waveguides with the diagonal coupler structures. The simulation does not account for attenuation outside of the coupling structure region.

5 Low power results

The multiport coupler prototype diagnostic system was assembled and tested at low power (see Fig. 16). In preparation for high power tests, a calibration with 9 used ports and 9 different vertically polarized field configurations was performed. For additional tested field configurations a good agreement between the mode analysis of 2D scans and the results obtained from the multiport coupler is found (see table 1). An expected larger phase offset error at low mode power is observed. The minimum detectable LP_{11} content is limited to 0.2 % mostly by the accuracy of the 2D scans. This corresponds to a maximum offset of 0.9 mm in the beat pattern described in Sect. 2 and is in the same range as the laser adjustment.

As an example, one of the 2D scans is shown in Fig. 17. One can clearly see the offset and tilt corresponding to the large LP_{11} content in the amplitude and phase plot respectively.

6 Conclusions

A high power capable in-situ diagnostic tool for spurious modes due to misalignments in OCCWGs has been developed. Low power tests show promising results and high power tests are scheduled.

Table 1. Mode powers and phase differences from 2D scan mode analysis and multiport coupler measurements (MPC) for two tested field configurations. The LP_{21} modes are not shown, because their mode content is less than 0.2 %.

	2D (1)	MPC (1)	2D (2)	MPC (2)
LP_{01}	92.65 %	92.58 %	91.98 %	92.11 %
LP_{02}	0.43 % 1.5°	0.54 % -0.6°	0.16 % 12.2°	0.16 % 4.9°
LP_{03}	0.42 % -58.1°	0.35 % -68.3°	0.39 % -60.0°	0.29 % -60.4°
LP_{11}^{e}	4.59 % -15.5°	4.66 % -15.0°	0.18 % -26.1°	0.20 % -28.6°
LP_{11}^{o}	0.61 % -170.6°	0.69 % -181.2°	5.72 % -24.9°	5.60 % -24.7°

Figure 16. Assembled multiport coupler prototype diagnostic system consisting of the multiport coupler and the waveguide based signal processing unit. This unit is using a rotating grating and a coupling plate with integrated Talbot effect splitter as illustrated in Fig. 9.

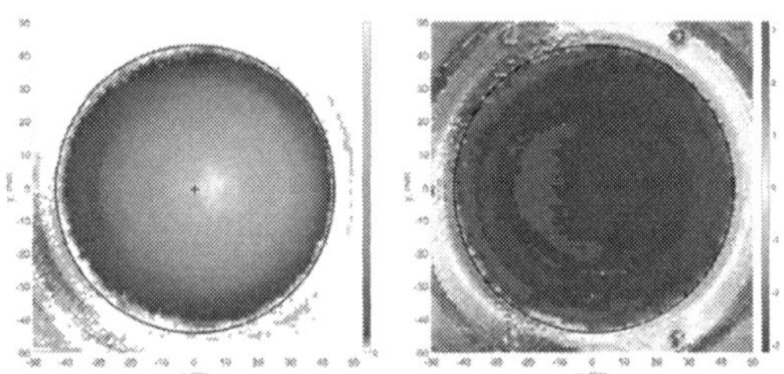

Figure 17. 2D scan with linear scaling of a field with a large LP_{11}^{o} content corresponding to (2) of table 1.

References

[1] H.-J. Hartfuss and T. Geist, *Fusion Plasma Diagnostics with mm-Waves - An Introduction* (Wiley-VCH, Weinheim (Germany), 2013) p177

[2] E. J. Kowalski et al., IEEE Trans. Microw. Theory Tech. **58**, 2772-2780 (2010)

[3] D. Wagner et al., J Infrared Millim Terahertz Waves **32**, 274-282 (2011)

[4] www.cst.com/products/cstmws

Study on the soft-start process of PSM high voltage power supply for ECRH

Jian Zhang[1], *Rui* Guan[1, 2,] *Fei* Guo[1,] *Yu* Zhou[1, 2], *Haozhang* Sun[1], *Junjiang* Wang[1, 2], *Yiyun* Huang[1,2]

[1]Hefei Institutes of Physical Science, Chinese Academy of Sciences, Hefei 230031, China
[2]University of Science and Technology of China, Hefei 230026, China

Abstract. The soft-start process of high voltage power supply (HVPS) based on pulse step modulation (PSM) for ECRH on EAST is introduced, which is the first procedure of system operation. The response process is detailed by proposing DC equivalent circuit model, process analysis and performance comparison is given under the conditions of different soft-start resistor parameters, and the theoretical analysis is proved by the simulation package ANSYS Simplorer simulations. The soft-start resistor is designed for the HVPS of 140GHz ECRH system for a smooth charging without overshoot of the capacitor [6], and the final experimental results show that it is in agreement with the theoretical analysis and is stable and reliable to the power devices.
Keywords. EAST, ECRH, pulse step modulation (PSM), soft-start

1 Introduction

The HVPS based on PSM[1] technology has the advantages of fast opening and closing, high reliability and modularization etc., which has been widely used in Tokamak auxiliary heating system. The structure of 140GHz ECRH HVPS system [2, 3] is shown in Fig.1, which is composed of 10kV breaker cabinets, 10kV contactor cabinets, multi-winding transformers and rectifying modules.

There are two steps to establish output DC voltage of power supply.

1) The first step is that the filter capacitor of rectifier module is charged by closing breaker and contactor.

2) The second step is that the output DC voltage is built by closing IGBT.

When the HVPS is powered up, to limit the inrush current of transformers and the charging current of capacitor, the series resistor needs to be connected in the 10kVAC side or in the module side. There are almost one hundred of rectifier modules in the HVPS system, so the series resistor[4,5,6] in each module will significantly increase the manufacturing difficulty, volume and cost of modules, therefore the series resistor in 10kVAC side is more economical and feasible[7]. The soft-start includes two processes: insert and removal of the resistor. In the following, the soft-start process will be analyzed theoretically and be verified by ANSYS Simplorer simulation and experimental results.

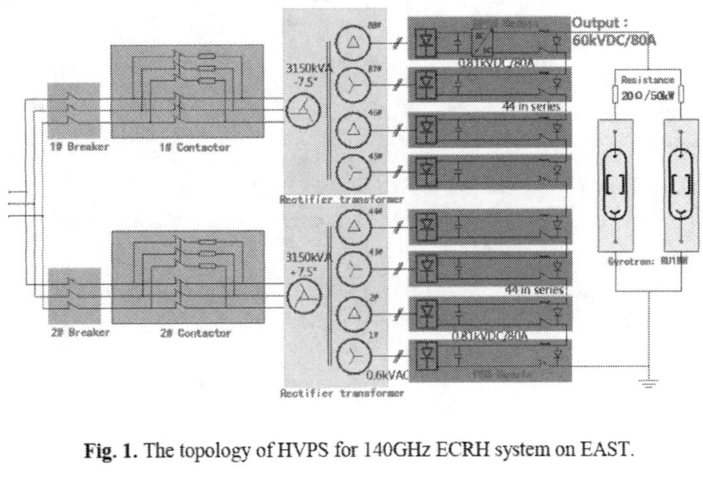

Fig. 1. The topology of HVPS for 140GHz ECRH system on EAST.

* Corresponding author: yyhuang@ipp.ac.cn

2 DC equivalent model of rectifier module

The equivalent circuit with the series resistor which is located in the 10kVAC side is shown in Fig.2.

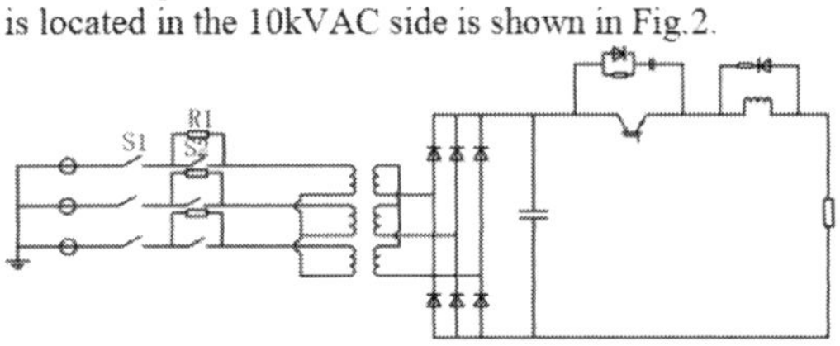

Fig. 2. The equivalent model of rectifying module.

The AC source charges the capacitor through the soft-start resistor, transformer, and three-phase uncontrolled rectifier, and then the DC output of the rectifying module is controlled by the IGBT switch. The process of charging the filter capacitor is called the soft start process. The parameters of each branch are equal, mutual influence can be ignored, and the current and voltage on each branch are approximately equal, so the model in Fig.2 can be simplified as DC equivalent circuit which is shown in Fig.3.

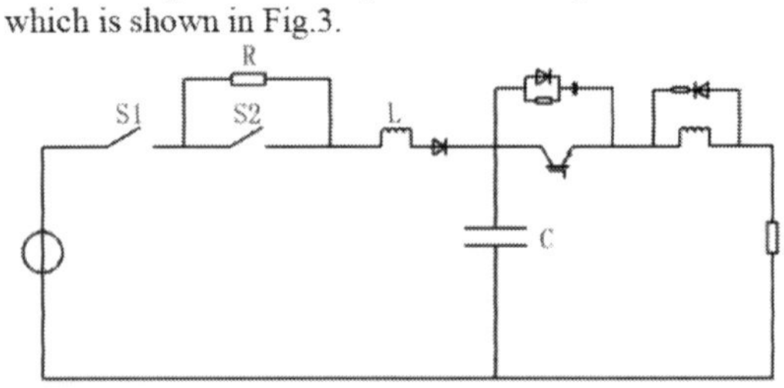

Fig. 3. The DC equivalent circuit of rectifying module.

The basic parameters of multi-winding transformer are shown in Table 1.

Table 1. The basic parameters of multi-winding transformer.

	Primary side	**Secondary side**
Rated power S_N	3150kVA	
Rated voltage U_{NL}	10kV	600V
Number of Winding	1	44
Short-circuit impedance U_K	4%	
Field current I_f	0.73A	

When the leakage inductance is converted to DC side, the leakage inductance can be calculated by the conventional method of transformer. The leakage inductance is expressed as follow equation:

$$L' = \frac{\left[\frac{\left(\frac{U_{2NL}}{\sqrt{3}}\right)^2}{\frac{S_N}{3N}}\right]U_K}{2\pi f} \quad (1)$$

U_{2NL}: line voltage of secondary side
N: number of secondary winding

Because the current always flows over two phases, when the filter capacitor is charged, the DC side resistor is two times of it in the AC side.
Equivalent inductance:

$$L = 2L' \quad (2)$$

Equivalent resistor:

$$R = \frac{2NR_{soft}}{K^2} \quad (3)$$

R_{soft}: soft-start resistor

3 The analysis of soft-start process

The soft-start process consists of two steps:

1) The first step is that the soft-start resistor is connected into the circuit, the filter capacitor is charged to the first steady state value through the soft-start resistor.

2) The second step is that the soft-start resistor is removed, so the filter capacitor is charged to the second steady state value.

3.1. The charging process with the soft-start resistor

The charging equivalent circuit with the soft-start resistor is shown in Fig.4.

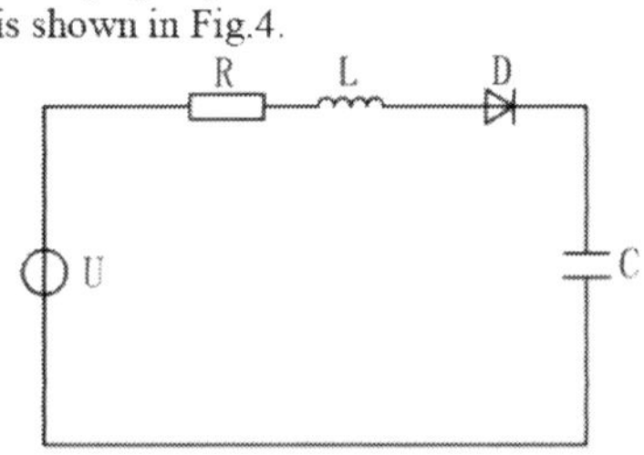

Fig. 4. The charging equivalent circuit with the soft-start resistor.

The transfer function of the filter capacitor voltage and step excitation is established as follow equation.

$$\frac{U_C(s)}{U_i(s)} = \frac{1/_{LC}}{s^2+s(R/_L)+(1/_{LC})} \quad (4)$$

And the characteristic equation of transfer function is $s^2 + s(R/_L) + (1/_{LC}) = 0$, so

$$\Delta = (R/_L)^2 - 4\frac{1}{LC}$$

When $\Delta < 0$, the equivalent resistor satisfies $R < 2\sqrt{\frac{L}{C}}$, the circuit is underdamped. It will cause overshoot of the output voltage. In order to avoid overshoot, Δ should be more than zero, the equivalent resistor should satisfy $R > 2\sqrt{\frac{L}{C}}$, so that the circuit is overdamped. Take note of (3), the soft-start resistor can be defined by:

$$R_{soft} > \frac{K^2}{N}\sqrt{\frac{L}{C}} \quad (5)$$

* Corresponding author: yyhuang@ipp.ac.cn

If the resistor is large enough to satisfy $R \gg 2\sqrt{\frac{L}{C}}$, the RLC circuit can be equivalent to a RC circuit,

so,
$$\begin{cases} U_C(t) = U(1 - e^{-\frac{1}{RC}t}) \\ i_C = \frac{U}{R} e^{-\frac{1}{RC}t} \end{cases} \quad (6)$$

U: Input voltage of DC equivalent circuit

The peak current of charging is express as follow equation:

$$i_{max1} = \frac{U}{\frac{2NR_{soft}}{K^2}} \quad (7)$$

In order to ensure that the rectifying bridge and capacitor can withstand maximum current, the charging current must be less than 300A, therefore the soft-start resistor should satisfy:

$$R_{soft} > \frac{K^2 U}{2N \times 300} \quad (8)$$

According to (5), when N=1, $\frac{K^2}{N}\sqrt{\frac{L}{C}}$ get the maximum $K^2\sqrt{\frac{L}{C}}$, so the soft-start resistor satisfy:

$$R_{soft} > K^2\sqrt{\frac{L}{C}}$$

The circuit is simulated by selecting different soft-start resistor, the results are shown in Figure 5. According to the design of the module, the simulation parameters are set as follows:

1) Input voltage of DC equivalent circuit U: 1000V.

2) Equivalent inductance of DC equivalent circuit L: 2mH.

3) Filter capacitor C: 10mF.

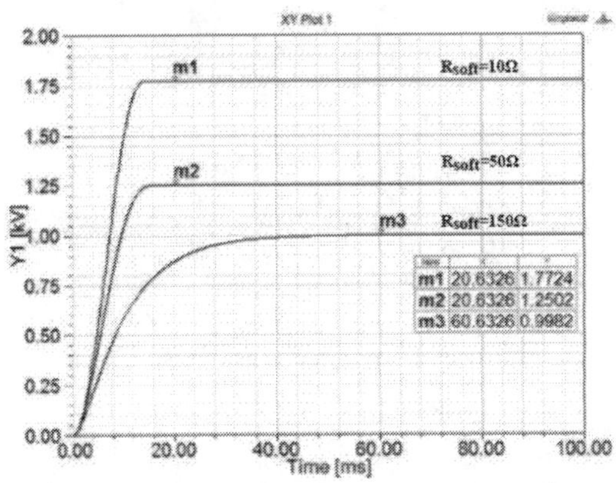

Fig. 5. The capacitor voltage of different soft-start resistor.

Table 2. The Influence of soft-start resistor on charge voltage.

R_{soft} (Ω)	Steady time T(ms)	Steady voltage U_c(V)
10	14	1772
50	16	1250
150	60	998

According to Fig.5 and Table 2, when the soft-start resistor is less than $K^2\sqrt{\frac{L}{C}}$ which is equal to 124Ω according to the simulation parameters, the circuit is underdamped, the steady voltage of filter capacitor is larger than 1000V, and furthermore, the overshoot voltage is smaller with the increase of the soft start resistor. When the soft-start resistor is larger than $K^2\sqrt{\frac{L}{C}}$ which is equal to 124Ω, the circuit is over damped, the steady voltage of filter capacitor is about 1000V, and the capacitor is charged without overshoot.

3.2 The charging process of removing the soft-start resistor

When the filter capacitor is charged to the first steady state value, the voltage is less than the rated value because of soft-start resistor voltage drop, so the soft-start resistor must be removed to get the rated value and reduce the power loss. The soft-start resistor is removed by closing S2 in Fig.3, then the capacitor will be charged again to the rated value. The equivalent circuit is shown in Fig.6.

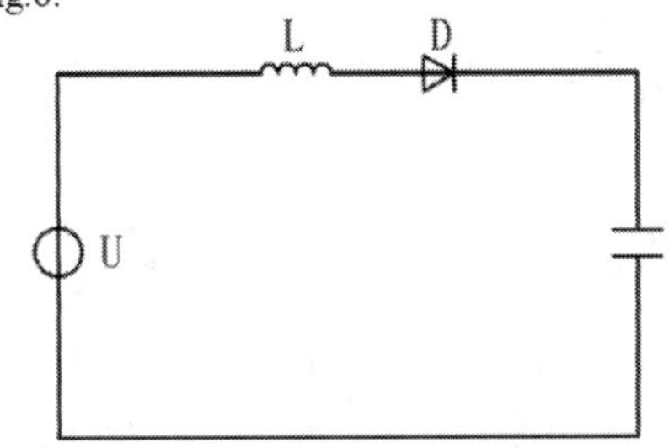

Fig. 6. The removing process of soft-start resistor.

The voltage drop of soft-start resistor which is produced by transformer magnetizing current is $\Delta U = I_f R_{soft}$, the excitation can be equivalent to:

$$u_i = \frac{\sqrt{2}\sqrt{3} I_f R_{soft}}{K} = \frac{\sqrt{6} I_f R_{soft}}{K}$$

The time domain expression is established as follow equation.

$$LC\frac{d^2 u_C}{dt^2} + u_C = u_i \quad (9)$$

According to the initial condition of filter capacitor: $u_C = U - \frac{\sqrt{6} I_f R_{soft}}{K}$, $C\frac{du_C}{dt} = 0$, the capacitance voltage and current are expressed as follow:

$$\begin{cases} u_C = U - \frac{\sqrt{6} I_f R_{soft}}{K}\cos\frac{t}{\sqrt{LC}} \\ i_C = \frac{\sqrt{6} I_f R_{soft}}{K}\sqrt{\frac{C}{L}}\sin\frac{t}{\sqrt{LC}} \end{cases} \quad (10)$$

According to (10), the voltage increment of filter capacitor is:

$$\Delta u_C = \frac{2\sqrt{6} I_f R_{soft}}{K} \quad (11)$$

The peak current is expressed as follow equation:

$$i_{max2} = \frac{\sqrt{6} I_f R_{soft}}{K}\sqrt{\frac{C}{L}} \quad (12)$$

In order to ensure that the charging current is less than 300A, the soft-start resistor should satisfy:

$$R_{soft} < \frac{300K}{\sqrt{6} I_f}\sqrt{\frac{L}{C}} \quad (13)$$

The circuit is simulated by selecting different soft-start resistor, the selection of resistor need to guarantee that the circuit is over damped. According to the simulation parameters, the soft-start resistor must be $R_{soft} > K^2\sqrt{\frac{L}{C}}$ and the simulation parameters are set as follows:

1) Input voltage of DC equivalent circuit U: 1000V.

2) Equivalent inductance of DC equivalent circuit L: 2mH.

3) Filter capacitor C: 10mF.

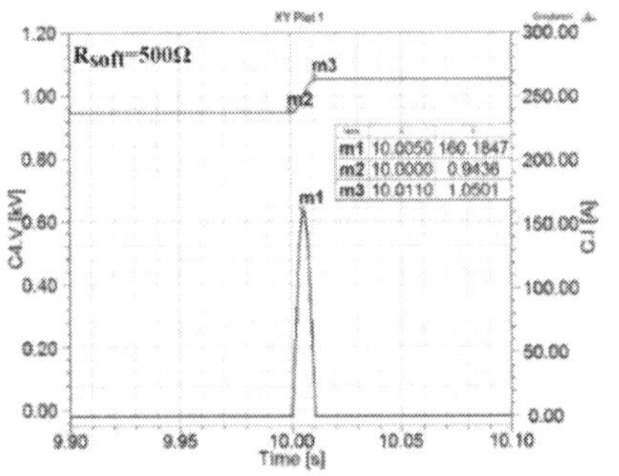

(a) R_{soft}=500Ω

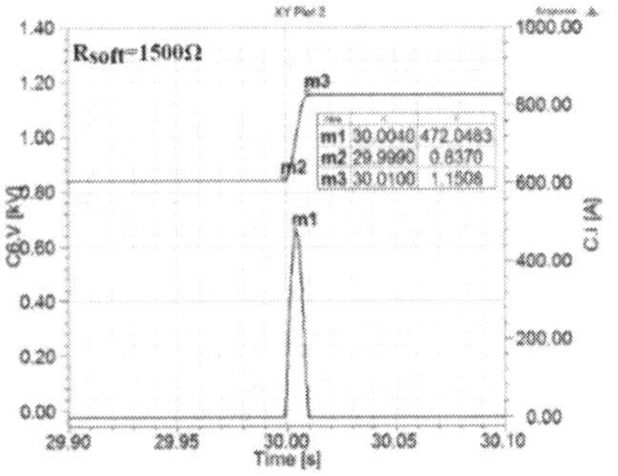

(b) R_{soft}=1500Ω

Fig. 7. The capacitor voltage of removing different soft-start resistor.

Table 3. The Influence of soft-start resistor on charge voltage and current.

R_{soft}(Ω)	I_{max2} (A)	U_1 (V)	U_2 (V)	Δ U
500	160	943	1050	107
1500	472	837	1150	313

I_{max2}: Maximum charging current of removing the soft-start resistor.

U_1: The initial voltage of the filter capacitor before removing the soft-start resistor.

U_2: The steady voltage of the filter capacitor after removing the soft-start resistor.

ΔU: The voltage change of the filter capacitor by removing the soft-start resistor.

According to (13), when the soft-start resistor is less than $\frac{300K}{\sqrt{6}I_f}\sqrt{\frac{L}{C}}$, which is equal to 968Ω, the rectifying bridge and capacitor can withstand maximum current 300A. Take note of Table 3, when R_{soft}=500＜968Ω, I_{max2}=160＜300A, and when R=1500＞968Ω, I_{max2}=472＞300A. According to (12) and Table 3, I_{max2} will enlarge with the increase of the soft-start resistor, and according to (11) and Table 3, △U will enlarge with the increase of the soft-start resistor. So in order to reduce voltage fluctuation of filter capacitor, the soft-start resistor should be reduced as far as possible.

According to the analysis, the design of the soft-start resistor must meet the following conditions.

(1)In order to avoid overshoot, the soft-start resistor must satisfy:

$$R_{soft} > K^2\sqrt{\frac{L}{C}}.$$

(2)In order to ensure that the rectifying bridge and capacitor can withstand the maximum charging current, the soft-start resistor must satisfy:

$$\frac{K^2U_i}{2N\times 300} < R_{soft} < \frac{300K}{\sqrt{6}I_f}\sqrt{\frac{L}{C}}.$$

4 The resistor design of HVPS for ECRH system

According to the multi-windings transformer parameters in the HVPS for 140GHz ECRH system, the leakage inductance of transformer can be calculated as follow:

$$L_1 = \frac{\left[\frac{\left(\frac{U_{2N}}{\sqrt{3}}\right)^2}{\frac{S_N}{3N}}\right] X_N\%}{2\pi f} = \frac{\frac{\left(\frac{600}{\sqrt{3}}\right)^2}{\left[\frac{3150000}{3\times 44}\right]}\times 4\%}{2\pi\times 50} = 0.6\text{mH}$$

So the equivalent inductance is:

$$L=2L_1=1.2\text{mH}$$

To ensure that the output voltage ripple is less than 1%, the filter capacitor is selected as follow:

$$C=10\text{mF}$$

According to condition (1), in order to avoid voltage overshoot the soft-start resistor must satisfy:

$$R_{soft} > K^2\sqrt{\frac{L}{C}} = 111\Omega$$

According to condition (2), in order to limit the maximum charging current the soft-start resistor must satisfy:

$$5\Omega = \frac{K^2U_i}{2N\times 300} < R_{soft} < \frac{300K}{\sqrt{6}I_f}\sqrt{\frac{L}{C}} = 1118\Omega$$

Actually the soft-start resistor is designed for 300Ω in the HVPS for ECRH system. Some experiment results are shown in Fig.8 and Fig.9.

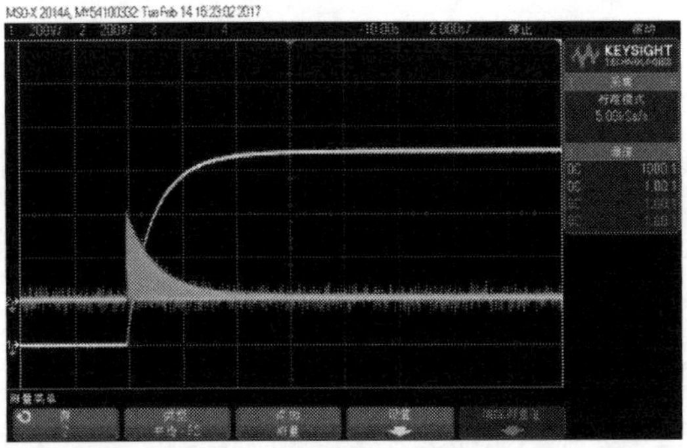

Fig. 8. The capacitor charging waveform with the soft-start resistor.

X-axis: 2s/div
Y-axis: 200V/div, 8A/div

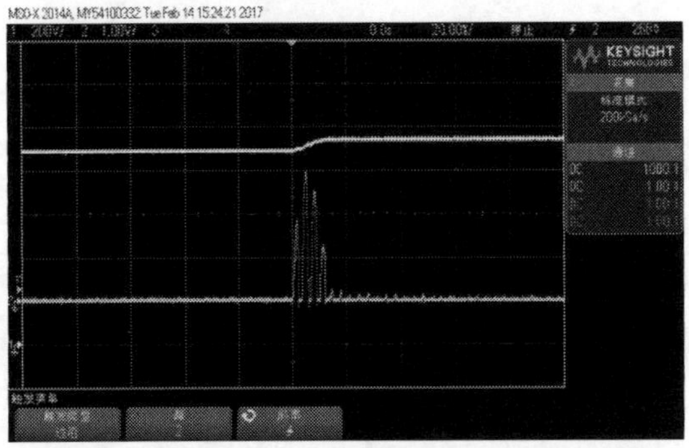

Fig. 9. The capacitor charging waveform of removing the soft-start resistor.

X-axis: 2s/div
Y-axis: 200V/div, 40A/div

In Fig.8, the maximum value of charging current with the resistor is 16A and the first steady state voltage is 870V. In Fig.9, the maximum current value of removing the resistor is 110A and the second steady state voltage is about 930V. The voltage change of the capacitor is about 60V.

According to (7) and (12), the calculation for the maximum value of charging current is 11A and 93A for different process. According to (11), the voltage change is 64V. The experimental results are in agreement with the theoretical analysis.

5 Conclusion

The soft-start process of HVPS based on PSM technology is studied in this paper. A simplified DC equivalent model is proposed, and the transient and steady state responses of the soft-start process are analysed in detail. The qualifications are obtained and proved by simulation results. The soft-start resistor has been designed for the HVPS of the 140GHz ECRH system, and the final experimental results show that it is coincident with the theoretical analysis.

This work was supported by the National Mangenetic Confinement Fusion Science Program of China under contract No. 2015GB103000, the National Key Research & Development Plan 2017YFE0300401, and CAS Key Technology Talent Program.

References

1. X. Hao, et al., Design and application of a new control system for Tokamak ECRH power supply [J]. *Radiation Effects & Defects in Solids*, **(3-4)**:1-10. (2016)
2. J. Zhang, et al., The Design of PSM-Based ECRH Power Supply Control System [J]. *Journal of Power & Energy Engineering*, **04**(4):91-102. (2016)
3. S. Ma, et al., Analysis and Design of the Module for PSM High-Voltage Power Supply [J]. *Journal of Fusion Energy*, **34**(2):261-266. (2015)
4. S.Ma, et al., High-voltage power supply for ECRH system on J-TEXT Tokamak[C]// *Fusion Engineering. IEEE*, 1-5. (2013)
5. L. Xia, et al., Analysis of the Soft-Start Circuit of the High Voltage Power Supply Based on PSM Technology [J]. *IEEE Transactions on Plasma Science*, **42**(4):1026-1031. (2014)
6. X. Hao. The Transient Analysis of PSM High Voltage DC Power Supply and the Research of Its Control Strategy [D]. University of Chinese Academy of Sciences. (2012)
7. L.Xia. The Study on Several Key Techniques of High Voltage Power Supply for Auxiliary Heating System in Tokamaks [D]. Huazhong University of Science and Technology (2015)

AUTHOR INDEX

AUTHOR INDEX

AUTHOR INDEX

AUTHOR INDEX

AUTHOR INDEX